ISBN 978-1-5277-7324-0
PIBN 10891046

1 MONTH OF
FREE
READING

at

www.ForgottenBooks.com

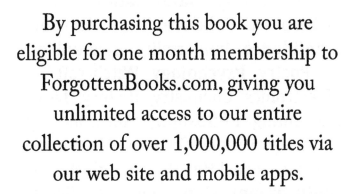

By purchasing this book you are eligible for one month membership to ForgottenBooks.com, giving you unlimited access to our entire collection of over 1,000,000 titles via our web site and mobile apps.

To claim your free month visit:

www.forgottenbooks.com/free891046

English
Français
Deutsche
Italiano
Español
Português

www.forgottenbooks.com

Mythology Photography **Fiction**
Fishing Christianity **Art** Cooking
Essays Buddhism Freemasonry
Medicine **Biology** Music **Ancient
Egypt** Evolution Carpentry Physics
Dance Geology **Mathematics** Fitness
Shakespeare **Folklore** Yoga Marketing
Confidence Immortality Biographies
Poetry **Psychology** Witchcraft
Electronics Chemistry History **Law**
Accounting **Philosophy** Anthropology
Alchemy Drama Quantum Mechanics
Atheism Sexual Health **Ancient History**
Entrepreneurship Languages Sport
Paleontology Needlework Islam
Metaphysics Investment Archaeology
Parenting Statistics Criminology
Motivational

RICHARDSON'S **and** WATTS'S

CHEMICAL TECHNOLOGY;

OR,

CHEMISTRY IN ITS APPLICATIONS

TO THE

Arts and Manufactures.

NEW EDITION.

VOL. I.—PART III.

CONTENTS OF THE VOLUMES

ALREADY PUBLISHED,

Chemical Technology; or, Chemistry in its Applications to the Arts and Manufactures.

Vol. I., Parts 1 and 2, 8vo. contain—FUEL and its APPLICATIONS. Profusely Illustrated with 433 Engravings and 4 Plates. 2nd Edition. £1 16s.

Vol. I., Part 3. 8vo. contains—The ACIDS, ALKALIES, and SALTS, their Manufacture and Applications. With 367 Illustrations on Wood. 2nd Edition. 1863. £1 13s.

Vol. I., Part 4., No. 1, 8vo. contains—Aluminium and Sodium, Tungstate of Soda, Silicate of Potash and Soda, Phosphorus, Lucifer Matches, Boracic Acid, Borax, Artificial Mineral Waters, &c. With Illustrations. 10s.

(The completion of this Volume is now printing.)

Vol. II., 8vo. contains—GLASS, ALUM, POTTERIES, CEMENTS, GYPSUM, &c. With numerous Illustrations. £1 1s.

Vol. III., 8vo. contains—FOOD GENERALLY, BREAD, CHEESE, TEA, COFFEE, TOBACCO, MILK, SUGAR. With numerous Illustrations and Coloured Plates. £1 2s.

Graham.—Elements of Chemistry; including the Application of the Science in the Arts. By T. GRAHAM. F.R.S., L. & E, Master of the Mint, late Professor of Chemistry at University College, London. 2nd edition, revised and enlarged, copiously illustrated with Woodcuts. 2 vols 8vo. 1850—1857. £2.

Ganot.—Elementary Treatise on Physics, Experimental and Applied. Translated from the Ninth Original Edition by Dr. ATKINSON, M.C.S., Lecturer on Chemistry and Physics, Royal Military College, Sandhurst. With 600 Illustrations. Post 8vo. 1863. 12s. 6d.

CHEMICAL TECHNOLOGY;

OR,

CHEMISTRY IN ITS APPLICATIONS

TO THE

Arts and Manufactures.

BY

THOMAS RICHARDSON, M.A., PH.D., F.R.S.E., M.R.I.A., &c.,
READER IN CHEMISTRY IN THE UNIVERSITY OF DURHAM;

AND

HENRY WATTS, B.A., F.C.S.,
EDITOR OF THE JOURNAL OF THE CHEMICAL SOCIETY.

Second Edition,
ILLUSTRATED WITH NUMEROUS WOOD ENGRAVINGS.

VOL. I.—PART III.

ACIDS, ALKALIES, AND SALTS,
THEIR MANUFACTURE AND APPLICATIONS.

LONDON:

H. BAILLIÈRE, PUBLISHER, 219, REGENT STREET.

NEW YORK:	MELBOURNE (AUSTRALIA):
BAILLIÈRE BROTHERS,	F. F. BAILLIÈRE,
440, BROADWAY.	COLLINS STREET EAST.
PARIS:	MADRID:
J. B. BAILLIÈRE & FILS,	BAILLY BAILLIÈRE,
RUE HAUTEFEUILLE.	PLAZA DEL PRINCIPE D'ALFONSO.

1863

WILLIAM HENRY COX,
5, GREAT QUEEN STREET, LINCOLN'S INN FIELDS, W. C.

PREFACE.

~~~~~~~~~

THE Preface of a book is generally that portion which the Reader first notices, but it is the last which the Author writes; and we confess to a feeling of satisfaction as we commence ours, for these volumes are a new work rather than a new edition, not one-tenth of the previous edition reappearing on the present occasion. No person who has not undertaken the labour of referring to the numerous Journals and Transactions of Societies now published in Europe and America, can form an idea of the amount of research involved in preparing a work on technical subjects, which aims at giving the results of the investigations and experiments of so many indefatigable labourers.

The principle of grouping kindred manufactures, first introduced by Dr. Knapp, has been retained as far as it was possible to do so, with the constant change which is taking place in the relations subsisting between the raw materials and manufactured articles now introduced into trade. These changes have induced us to include in the present edition, the following manufactures which found no place in the first, viz.: Bisulphide of Carbon; Potash, and some of its salts from sea-water, from the vinasse of the beet-root sugar manufacture, and from kelp; Iodine and Bromine; Aluminium and Sodium, which are now

manufactured on a large scale by Messrs. Bell, Brothers, of
Newcastle-on-Tyne; Aerated and Mineral Waters; Phosphorus
and Lucifer Matches; Phosphate, Hyposulphite, Stannate, and
Tungstate of Soda; Chlorate, Chromate, and Prussiate of Pot-
ash; Sulphurous, Oxalic, Tartaric, and Citric Acids; Grease;
Glycerin; Fireworks, &c.

We have also added an abstract of all the patents relating
to the most important manufactures described in these volumes,
which we hope will prove of some value to the parties engaged
in these trades.

The Appendix also contains several tables which will be
found of great utility to every one occupied with the daily practical operations of these manufactures.

In explaining, as far as we had the information, the processes
at present adopted in manufacturing the different products, we
have included an account of several plans which have been tried
on a large scale without success. In all such cases, we have
allowed the inventors to describe their plans, as much as possible, in their own words. These plans, for the present, have not
succeeded; but, in many instances, the idea involved in the
process is excellent, and the failure has often arisen more from
lack of means than from deficiency of theoretical knowledge.
The late celebrated Mr. Robert Stephenson once said, "The
plans which are feasible to-day, would have been considered
impracticable a few years ago, from the absence of skilled
workmen and the want of suitable appliances." Where we
have criticized any of these plans, it has been done in no un-
friendly spirit, as we have made our remarks with a feeling
of sympathy in the disappointments which inventors so often
encounter. These ideas may one day germinate, and we are
glad to give them currency.

This information, as well as the results of many costly experiments carried out for the mutual advantage of the public and the inventors, are stored up in the Blue Books for the benefit of our successors, and, but for the existence of the Patent Laws, would most probably have died with the patentees. This consideration leads us naturally to make a few remarks on the subject of patents, which has recently attracted some attention.

The first attack on the present Patent Law came from an unexpected quarter; and we are sorry to differ from one of the most generous and ingenious of modern inventors. The objections of Sir W. Armstrong amount to nothing short of the abrogation of all laws for the protection of inventions, mainly on the ground that they are obstructive in preventing all persons, except the patentee, from developing the idea which he is enabled to monopolize. If this objection of this celebrated inventor could be supported, it would be entitled to great weight in the decision of this important question. It is, however, generally recognized that an idea cannot be secured by patent. Before an idea can become the subject-matter of an exclusive right or privilege, it must assume the shape of a machine, process, or application. In every such case, the inventor must define exactly what he claims, as well as his *modus operandi*, and any other party who can carry out the same idea, or apply it in any superior or different method, is not debarred from securing protection under the present Patent Laws. Cases may, and do, sometimes arise, in which the second inventor cannot enjoy the benefit of his improvements without infringing the previous patent; but the restriction is mutual, as the first patentee cannot appropriate the fruits of the later improvement without compensating the author; and surely this is no sound objection to the establishment of a law for the protection of inventions.

· The objections of the late distinguished chairman of the Electric Telegraph Company to the existence of patent monopolies, are founded on a fallacy. He stated in 1851, that the company had given £140,000 for their original patent, and from £8,000 to £9,000 for the patent of Mr. Bain. Did these patentees not give the company a full equivalent for these sums? And, but for the existence of the Patent Laws, which recognized the natural right of these gentlemen, can any one believe that this company would have been in its present existing flourishing condition? As Mr. Ricardo insisted on forcing this view of the principles of free trade to such extreme lengths, he should have given us a proof of the sincerity of his objections to all monopolies, by proposing an agrarian law for the division of the land, and have admitted the public to the free use of the company's telegraphs. We would also refer him to Adam Smith and John Stuart Mill, the great advocates of free trade, who take a very different view of this question. The latter author, in his *Political Economy*, Vol. I., p. 497, says, " The condemnation of monopolies ought not to extend to patents, by which the originator of an improved process is permitted to enjoy, for a limited period, the exclusive privilege of using his own improvement. This is not making the commodity dear for his benefit, but merely postponing a part of the increased cheapness which the public owe to the inventor, in order to compensate and reward him for the service. That he ought to be both compensated and rewarded for it, will not be denied ; and also, that if all were at once allowed to avail themselves of his ingenuity, without having shared the labours or the expenses which he had to incur in bringing his ideas into a practical shape, either such expenses and labours would be undergone by nobody except by very opulent and very public-spirited persons, or the State must put a value on the service rendered by an inventor, and make him a pecuniary grant. This has been done in some instances, and may be done without inconvenience in cases of very conspicuous public benefit ;

but, in general, an exclusive privilege of temporary duration is preferable, because it leaves nothing to any one's discretion—because the reward conferred by it depends upon the invention being found useful, and the greater the usefulness the greater the reward—and because it is paid by the very persons to whom the service is rendered, the consumers of the commodity."

The late Mr. Brunel, Mr. William Hawes, and others, have raised the objection, that lessening the cost of patents would lead, and has led, to a multiplication of cheap and useless patents for the sake of advertising, and so rendered all patents comparatively worthless. As regards this objection, we have much pleasure in quoting the striking observations of the Editor of the *Record*, who says:—"We confess that the logic of this proposition appears to us somewhat obscure. If the patents are worthless, we should imagine they can injure none but the unfortunate patentees. If they are merely trivial, though, in consequence of their number, they may sensibly affect some important branches of the manufacturing interest, it would be a strange departure from the principles of free-trade, either directly or indirectly, to prohibit them on that account. But the dissatisfaction is not all on one side. If manufacturers complain, inventors complain too. They say that patents, though comparatively cheap, are yet far dearer than they ought to be; and that, although under the present system, a man has no need to expend £175 until he has every reasonable assurance that the proceeds of his invention will amply indemnify him for the outlay, yet that even then, such a sum may be very difficult to raise before any actual business has been done. And, apart from such considerations, it may be asked how far it is consistent with a sound political economy, to tax inventions of any kind; how far it is necessary that the Patent Office should require an overplus when all its natural expenses are fully met. If the invention be handsomely remunerative, the revenue will scarcely fail to reap its share of the benefit; but while the taxing of the fruit is all very well, let the tree be planted without needless impediment. Every year the Patent Office has a large surplus over and above the cost of maintaining it. Why should this be? If the United States can seal to an inventor his patent for £7; if France—not proverbially liberal in these matters—can do the same thing for £4, we may, one would think, do yet a little more in the way of facilitating the efforts of inventive genius."

The number of Patents secured in other countries also proves that no such excessive multiplication of patents has taken place in this country; in illustration, we copy an interesting table from the *Génie Industriel* of Armengaud.

Country.	Patents granted in 1858.	Population.
France	5820	35,781,628
United States	3710	23,191,948
Great Britain	1890	27,511,447
Belgium	1406	4,426,202
Austria	703	36,514,466
Sardinia	171	4,368,972
Saxony	107	1,828,732
Sweden	64	3,482,541
Victoria (Australia) . .	53	410,766
Prussia	49	16,923,721
Bavaria	41	4,519,546
Holland	39	3,203,232
Russia	26	69,660,146
Hanover	20	349,958

Another objection has been taken by some parties, that, under the present system, trifling improvements can be secured by patent. The history of nearly all our manufactures shows, however, that their progress has mainly consisted in the gradual introduction of small improvements; and fortunately so for all who have invested capital in their buildings and machinery, as little or no loss of plant and utensils is incurred in the adoption of such improvements. We may even go further and affirm, that it would not be difficult to collect some striking illustrations to show, that in many cases, trifling improvements have rendered some inventions available, which might otherwise have been postponed or overlooked for years. In fact, as Mr. Carpmael states, "The whole system of our manufactures is based upon patents," for the most part of comparatively little importance, but which, in the aggregate, have built up the imposing fabric of British industry. The present volumes are, in great part, a record of such improvements in the manufactures which are described.

These trifling improvements are generally introduced by practical men, whom it has been the fashion in some quarters to decry, but whose previous practical training of years had qualified them to turn to the commercial advantage of the country, the otherwise barren discoveries of mere principles or of theoretical knowledge. Are these men, then, to be deprived of the fruits of a life-long experience, because they ask for a fair remuneration of their exertions? Are the Patent Laws to be abolished, when, in the expressive language of the Editor of the *Economist*, the value of their improvements may be safely left to the "higgling of the market" to be ascertained? The simplicity of some of these improvements constitutes their value; and had they lain on the surface, we may fairly believe they would not have been left "to the promptitude of some stranger to appropriate."

We regret that the talented Professor of Metallurgy at the

School of Mines should also, to some extent, have joined in this outcry against the Patent Laws; but we think his remarks go to prove, that he has not had much acquaintance with the practical working of these laws. He remarks very justly, in another place, that the admirable facilities afforded by Mr. Woodcroft, leave those parties no excuse, who now re-patent existing improvements. The case, however, mentioned by Dr. Percy as illustrating the hardship inflicted by the present system, is only similar to many others which happen to every one who has had the misfortune to be involved in litigation under other laws; and the observations of the committee of the Society of Arts on this subject are as just as they are pertinent:—that it is a fallacy to deny rights lest the process established to redress wrongs should be inconvenient; while the thousand registrations effected under the Utility Designs Act have not encouraged litigation.

The observation of Mr. Revely, that Patent Laws are a product of the semi-barbarous age of Queen Elizabeth, and therefore absurd, is well met by Mr. Campin, who remarks, that it is equally valid against Magna Charta, the rights of the possessors of landed property, trial by jury, and many other things which date from times still further back, yet more barbarous.

The last attack has come from Professor Rogers, of the University of Oxford, which has called forth a most able reply from the Editor of the *Journal of Gas-lighting*; this, from its length, we have been obliged to print in our Appendix.

Patent-right may properly be called a "natural right, and ought to be recognized as peculiarly sacred." Inventors are surely as much entitled to a "positive recognition of their labours as property, as any other description of working people in the commonwealth." "To assert that a man is not entitled to reap any advantage from his labour and skill, developed in any way, not inconsistent with the good of the community, would be to advocate an anarchy that would sap the foundation of the rights of all property."

The rights of any man to remuneration for the inventive creations of his own mind, as Mr. Fairbairn, so truly said at Manchester, are obvious; and we can discover no distinction that can be drawn between the claim of patent-right and copyright, or in fact to the existence of a right in any other property. Mr. Sinnet strikingly remarks, "As an author sells a copy of his book, the embodiment of his thoughts—so an inventor sells a machine, the embodiment of *his* thoughts;"—or, as another able author writes, "By a somewhat singular inversion of the law of nature, a man's title to the property he possesses, has been held most sacred in proportion as he has had least to do with its production. We find the right to land gotten by inheritance more secure than that of personal property acquired by a man's industry, and this latter again, immeasurably stronger than that which a man has to the product of his own brains. Of all personal titles, that by invention,—or, to use popular language, copyright and patent-right,—was the last which the legislature recognized, and *it has always been the first which we have been willing to assail.*"

The claim of authors to copyright has never been contested; nay, it is conceded without exacting any premium on the concession; and it is now receiving a graceful acknowledgment from the French Government through M. Walewski, who has proposed to give to authors a greater interest in their productions than has ever yet been even suggested in any country.

In the recent discussion in the House of Lords on the Copyright (Works of Art) Bill, one of the greatest lawyers of the day, the Lord Chancellor, expressed himself surprised to find, that while a man was allowed a property in that which was, in the ordinary way, the work of his hands, it should be gravely contended that in those productions which were the creations of the mind, no such right should be admitted; and he agreed with the great Lord Mansfield, that in all works of the mind and of genius, the common law of this country ought to be held as giving an absolute property.

We trust that the rights of inventors may yet meet a similar reward in our own country, where, if it be not invidious to contrast the two, patent improvements are, in one sense, of more value to the community; and it is encouraging to notice, that in the discussion which took place on the 27th of May last, in the House of Commons on the motion of Sir H. Cairns, who introduced the question in a most able speech, the great majority of the speakers evidently sympathized with the just claims of inventors.

We would also refer our readers to the extract from the eloquent address of Sir D. Brewster, in the Appendix, in which this subject is examined by the learned Principal of the University of Edinburgh.

Any one who has had to negotiate the sale of improvements, constantly meets with high-minded men, who will listen to no communications unless the party has the protection of a patent, prudently refusing to perfect inventions for the benefit of unscrupulous competitors in trade. We believe, moreover, that the country would benefit to a still greater degree, if the facilities were still further increased and the expenses diminished, by which working men might reap the benefit of their suggestions, now too often appropriated by others.

It is surely a strange way of encouraging genius, to levy a heavy sum for securing to it the profits of its own labour, or, as Mr. Dickens's poor man of 1850 summed up his complaints, "Is it reasonable to make a man feel as if, in inventing an ingenious improvement meant to do good, he has done something wrong? How else can a man feel when he is met with such difficulties at every turn? All inventors taking out a patent *must feel* so. And look at the expense. How hard on me, and how hard on the country, if there's any merit in me (and my invention is took up now, I am thankful to say, and doing well), to put me to all that expense before I can move a finger."

It is often a dubious question for an inventor to decide, whether a certain discovery or application is worth the present expense of a patent ; sometimes those improvements which have cost him the most labour and the largest outlay to perfect, turn out of little or no value ; and as we know, in other cases, those which appear so simple, or not likely to meet with an extended application, are neglected at the time, but yet, in the long run, often prove of great importance.

The present system has some defects, but we apprehend that this is not peculiar to Patent Laws, and the great freedom in granting patents in this country (a very small number of applications being rejected here), contrasted with the difficulties elsewhere, doubtless attracts inventors from all parts of the world, while it encourages every one to give publicity to his ideas.

Our ideas regarding the improvement of the present Patent Laws embrace the following suggestions :—

1. That the expense of obtaining a patent should be reduced to the lowest sum necessary to cover office charges.

2. That in the first instance, as Lord Stanley stated in the House of Commons, the inquiry into the nature of the invention for which a patent is claimed, ought to be almost nominal and formal; for we agree with Sir H. Cairns, that the training of purely scientific or technical men, does not qualify such parties to be competent judges of the utility of a so-claimed improvement.

3. That in order to weed out useless patents, patentees should be compelled, at the expiration of, say three years from the date of filing the full specification, to make an affidavit that their invention has been in practical operation for at least twelve months previous; that this affidavit should be attested by two competent witnesses, and the whole published in the *Gazette*. Further, that, at the expiration of each succeeding three years, the same formality should be observed, but with this addition, that the process or machine should have been in use during the whole of the three preceding years.

4. That an analysis of all processes or machines relating to the various manufactures, should be published by the Commissioners of Patents, for the benefit of inventors and the purchasers of patents. If Mr. Woodcroft were to issue such a series of cheap blue books, he would add another benefit to those which he has already conferred on the manufacturing community. An intending purchaser could then examine, so to speak, the title of the patent for which he was negotiating.

5. That all litigation in patent property should be referred to a court specially erected to try such causes, before a jury of experts, each of whom would be liable to the usual challenge; while the judges should have the assistance of scientific and technical assessors, who ought to be subject to examination at the request of either party. The salaries and expenses of this

new court might be defrayed, partly out of the charges for granting patents, and partly by the litigants.

Two public associations have been formed for the defence and discussion of the laws relating to patent inventions, one in Lancashire, called "The Manchester Patent Law Amendment Association," and the other in London, "The Inventors' Institute." We print the Report of the latter in our Appendix, as well as the principal proposals of the former, in order that our readers may learn the opinions of these influential associations.

We must now draw the attention of our readers to the important information which Messrs. Stevenson & Williamson, the able managers of the Jarrow Chemical Works, have given us, to be found at page 318, Part III., and to express our thanks for their kindness in revising the proof sheets of the chapter devoted to the Soda Manufactures.

To Mr. E. A. Pontifex, we are under special obligations for having given us the full details of Mr. Ogsten's important researches in the manufacture of Tartaric Acid, in the works of his firm at Millwall, and to Mr. Ogsten himself, for much valuable information which has been incorporated in our chapter on this acid.

Mr. Blair, of Farnworth, has been equally liberal in placing at our disposal, for publication for the first time, the particulars of the great improvements he has introduced for burning sulphur, described at page 56, Part III.

Mr. Gossage, so well known for the valuable improvements he has introduced in the manufacture of alkali &c., has kindly revised our chapter on Soap, and introduced much important information. We are also indebted to Mr. E. R. Cooke, of the Soap Works, Bow, for much valuable information on this subject, and to Mr. A. Cruickshanks, of Gateshead, whose notes on this chapter are printed in the Appendix.

Among many friends who have aided us with useful information, we must enumerate the following gentlemen :—Messrs. Albright & Wilson, of Oldbury; Browell & Marreco, of Newcastle-on-Tyne; R. C. Clapham, of Walker; R. Irvine, of Musselburgh; P. Harris, of Birmingham; J. Bramwell, of the Heworth Chemical Works; E. C. C. Stanford, of London; Mons. Carré, of Paris; H. Darby, of London; A. W. Wills, of Wolverhampton; and T. Carr of Birkenhead.

We have also the pleasure to acknowledge the great assistance we have derived from the valuable works of Barreswil & Girard, Barruel, Bollaert, Dana, Dingler, Dorvault, Dufrenoy, Fresenius, Gerlach, Gmelin, Guillaumin, Harzer, Henry, Johnstone, Karsten, Liebig, Mohr, Morfit, Mulder, Muspratt, Parkes, Payen, Pelouze & Fremy, Pereira, Poggiale, Rammelsberg, Ure, Tomlinson, and Wagner; and among the many important Journals published in Europe, we would particularly mention that of the Society of Arts, and the "Jahresberichte" of Kopp and Will, and of Wagner, which we have found most useful.

In conclusion, we have to express our obligations to our enterprising Publisher for the liberality with which he has consented to illustrate the various manufactures described in these volumes.

THOMAS RICHARDSON,
HENRY WATTS.

September 1, 1863.
NEWCASTLE-UPON-TYNE, AND
LONDON.

CONTENTS.

CHEMICAL TECHNOLOGY;

OR,

CHEMISTRY APPLIED TO THE ARTS

AND

MANUFACTURES.

GROUP II.

PROCESSES OF MANUFACTURE EMPLOYED IN THE PRODUCTION AND APPLICATION OF THE ALKALIES.

THE more powerful salifiable bases—the hydrated oxides of the light metals, or the alkalies—are gifted with very energetic chemical forces, the action of which may be employed with equal facility, in the decomposition and transmutation of other substances, or in the production of salts.˙ Those salifiable bases which occur in large quantities, or which can be manufactured at a cheap rate, give rise, therefore, to a series of industrial employments, which are chiefly founded upon their powerful chemical action.

Among the different salifiable bases of this class, *soda* and *potash*—known in commerce in the state of carbonates—are the only two used in manufactures, and are consequently of pre-eminent importance. But, besides the direct applications of these bodies, certain other branches of manufacture, such as the making of sulphuric acid, which have not the production of potash or soda for their direct object, are, nevertheless, so intimately connected with this part of the subject, that a description of them belongs properly to this place.

SULPHUR.

Native sulphur.—In volcanic districts, as those of Toscana and Naples, in Italy, and those between Cattolica and Girgenti, on the south coast of Sicily, large masses of native sulphur are found deposited in layers of lime and clay marl. It is more than probable, that this sulphur was formed originally from the gaseous sulphuretted hydrogen, which, mixed with sulphurous acid, aqueous vapour, air, and sulphur, in a finely divided state, issues from all parts of the soil in the form of currents of gas, which are there called *fumaroles*. The glowing lava which simultaneously comes in contact with the ascending current of sulphuretted hydrogen from the interior, and with the atmospheric air, causes a partial combustion of the former, yielding sulphurous acid and water. The sulphurous acid thus formed, is decomposed in contact with the remaining portion of sulphuretted hydrogen into water and sulphur, and this is partly carried up with the current of gas, and partly sublimed in the cracks and crevices of the rocks.

Another explanation of the formation of these deposits of sulphur, has been suggested by Bischof, who has shown that by passing sulphuretted hydrogen through chalk, at a temperature of 212° F. in contact with atmospheric air, the chalk was converted into gypsum, and sulphur was deposited. Carbonate of lime and gypsum are generally found associated with these deposits of sulphur, and the production of sulphur in the springs near Aachen and Lubine, has evidently arisen in this way.

The composition of the so-called "sulphur-stone," which was recently imported into this country from the island of Milo, in the Grecian Archipelago, implies a similar origin. We are indebted to Mr. R. C. Clapham for the following analyses of two samples of this sulphur-stone:—

Sulphur	24·0	22·3
Gypsum	62·2	5·2
Carbonate of lime . .	...	63·4
Silica, &c.	6·0	2·3
Water	7·0	6·0
	99·2	99·2

The deposit of sulphur, noticed among the refuse heaps of tank waste from alkali works, evidently arises from a similar reaction.

By far the largest proportion of sulphur is produced in Sicily, and the imports into this country, almost exclusively from the Mediterranean, have risen from 333,731 cwts. in 1842 to 1,164,099 cwts. in 1859.

Payen gives the following as the average annual consumption in France:—

1820	kilog. 6,790,000
1825	„ 10,500,000
1830	„ 12,900,000
1835 to 1838 . . .	„ 19,000,000
1842 to 1846 . . .	„ 26,000,000
1852 to 1853 . . .	„ 29,360,000
1855 to 1856 . . .	„ 34,422,796

Small quantities are obtained in other countries, but the only statement we have met with respecting this point, refers to Prussia, where 10,592 centners were produced in 1858. For additional statistics, see Appendix to Part iii.

Extraction.—It has already been noticed that the deposits of sulphur are always associated with various mineral or earthy matters, and three processes are followed to separate the principal part of these impurities, which generally amount to more than one-half of the entire weight of the deposit.

When the deposit is rich in sulphur, it is melted in a cast-iron pot B (Fig. 1), heated by an open fire A. The melted mass is stirred with an iron rake to facilitate the separation of the earthy matters, which are allowed to fall to the bottom. The liquid sulphur is then removed by a ladle, C, thrown into an iron vessel, D, and allowed to solidify. The temperature ought to vary between 250° and 300° F., and never reach 480°, at which point the sulphur would take fire.

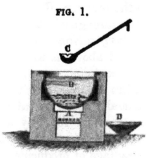

FIG. 1.

The residue which remains, and contains more or less sulphur, is removed, and may be treated by either of the following plans.

A small blast furnace, E (Fig. 2), constructed of fire brick or stone, is charged with the sulphur-stone at the bottom, which is ignited, and fresh charges of the sulphur-stone are thrown in, from time to time. The working holes, *f f f*, at the sides ad-

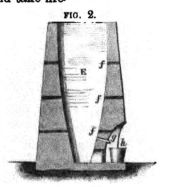

FIG. 2.

mit a small supply of air to support a combustion on the surface, by which means sufficient heat is generated to melt the sulphur, which runs off at the bottom through a pipe g into an iron pot h, where it solidifies.

The third plan is suitable for treating an impure sulphur-stone, containing from 8 to 12 per cent. of sulphur. It consists of a furnace (Fig. 3) sufficiently wide to receive two rows of earthen pots A A—the vessels for distillation—which are arranged in pairs, somewhat raised above the sole of the furnace, upon the sup-

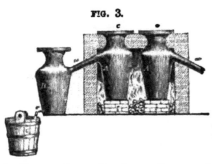

FIG. 3.

ports b, so that the necks of the pots are a little above the top of the furnace. Thus the mouths of the pots are free, and having been charged from without, they are closed by the lids c cemented on, and the distillation begins. The sulphur vapours pass over by the lateral tubes a, to the receivers B, where they condense to liquid sulphur, which flows through o into a vessel C filled with water, and there solidifies.

Purification.—The extraction effected in the manner described above, is so imperfect, that the product, *crude sulphur*, still contains a small percentage of earthy matter, and this makes a second purification, by distillation, necessary; this is generally performed at the sea-ports. A very suitable apparatus, invented in 1815 by a manufacturer named Michel, at Marseilles, and now everywhere used (Fig. 4), affords the double advantage of producing continuously *flowers of sulphur* (pulverulent sulphur), and *sticks of sulphur*.

The iron vessel a, which is large enough to contain a charge of from 10 to 12 cwts. of sulphur, is walled into a furnace, the interior of which has no other outlet than the arched channel x, leading to the chamber d d d, which has a capacity of from 6,000 to 7,000 cubic feet. The boiler and the chamber together form, therefore, a retort and receiver, upon a very large scale. The firing for the boiler f is maintained by a draught through the ash-pit c, by means of the chimney m. At p there is a door for charging and emptying the boiler; and in some works, an iron pot is built above the iron vessel a, in which the sulphur is melted by the waste heat from below, and allowed to flow through an open-

ing in the bottom, by
means of a valve opened
from above by an iron
rod attached to the
valve.

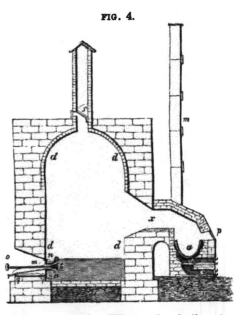

FIG. 4.

The chamber itself
has a separate outlet,
in which a valve, *s*, is
placed, opening out-
wards.

Soon after the heat
is applied, and the sul-
phur has become liquid
(at a temperature of
about 150° C. (302° F.),
it takes fire;* but this
extinguishes of itself,
as soon as the oxygen
of the air in the cham-
ber is converted into sulphurous acid. When the boiler has
attained a temperature above 316° C. (601° F.), the mass
begins to boil, and to give off vapours, which pass over to the
chamber through *x*, and condense against the cold walls. As
long as the temperature of the walls is below the melting point
of sulphur, 108° C. (226° F.), the sulphur can only condense in
the solid form. The sulphur, therefore, forms small microscopic
crystals, and these deposit in the form of a yellow dust (flowers
of sulphur). The constant condensation of the flowers of sulphur
heats, however, the walls of the chamber, and they acquire, at
length, a temperature of 108° C. (226° F.). When this period
arrives, on the third day, liquid sulphur appears in the chamber.
If the production of flowers of sulphur is to be continued, the
operation must be stopped during the night, that the chamber
may have time to cool. If not, the flowers of sulphur are
removed from the chamber by a door provided for the purpose,
and the heat being kept up, sulphur continues to distil over, and
collects at the bottom in the fluid state, as represented in the
figure. This second period has also its limit, like the first; for
as soon as the walls of the chamber have attained the temperature of 316° C. (601° F.), which occurs on the seventh day, the

---

* Violent explosions are not uncommon.

condensation diminishes so rapidly, that a stop must be put to
the process.

The sulphur can be drawn off, through an aperture n n, made
a few inches above the floor of the chamber. It is only necessary to push in the stopper h, which can be made to assume any
position by means of the spring m and its handle o. The sulphur flows through the gutter r, which is kept hot as far as q by
charcoal, into the wooden, somewhat conically bored moulds
(Fig. 5). The solidified sticks of sulphur are easily removed from
the moulds by the pestle n.

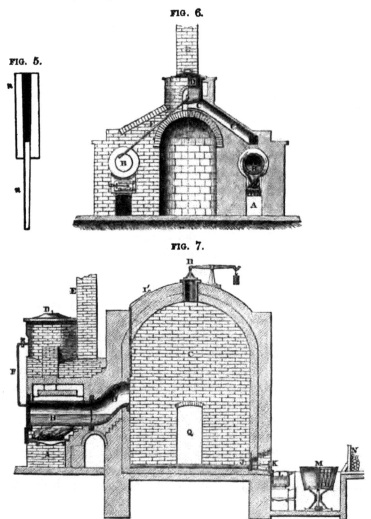

FIG. 6.

FIG. 5.

FIG. 7.

An improved arrangement for refining, has been described by
Payen, which is represented in Figs. 6 and 7. It consists of two
metal cylinders or retorts, B, closed at one end by a movable lid,
B'. One of the cylinders is straight, and the other curved like a
swan's neck. The fore part of the straight retort is heated by
the fire, A, and the flame afterwards passes through a flue, C,
previous to escaping by the chimney, E. At the top of the
flue, the waste heat is employed in melting the sulphur in a
metal pot, D, which is covered with a lid, and from which the
retorts are supplied. The other end of the retort opens into the
condensing chamber, G, possessing a capacity of about 3,000
cubic feet. This opening can be closed at pleasure from the outside by a damper I, by means of a connecting rod I'. At the end
an opening, Q, is made, formed partly of brickwork and partly
of iron. At the bottom of one of the sides of the chamber there
is an opening closed by a metal plate, in which a small hole can
be more or less closed by a conical damper, J K. The melted
sulphur flows through this hole into a pot, L,
heated by a fire below. Alongside of this pot is fixed
a vessel, M, divided into 6 or 8 partitions, shown
in plan in Figure 8, and on one side, a framework N
in wood is placed, for the reception of the rolls of
sulphur.

FIG. 8.

For making *lump, stick, or roll sulphur*, the apparatus
above described is worked in the following manner. The two
retorts are charged with about 6 cwts., when the doors are
carefully luted with clay. The fire under one of the retorts is
ignited, and when about half of the sulphur has been distilled
off, the other is set to work. The waste heat from the two fires
is sufficient to melt the 15 or 16 cwts. of sulphur with which
the vessel D has been charged. When the distillation of the
first retort is complete, it is charged again by means of the pipe
F', through which the flow of melted sulphur from the vessel D is
regulated by a cock.

During these operations, the air of the chamber G escapes
through the valve, which is regulated by a balance weight.

The distillation of a charge lasts 8 hours, and thus a retort is
filled every 4 hours. The heat produced, supports the temperature of the chamber at such a point, as to keep the sulphur
in the liquid state. When the melted sulphur has reached a

FIG. 9.

depth of about 4 to 6 inches, it is run off into the little pot L (Fig. 1), where the workman ladles it into the wooden moulds P P', Fig. 9. These moulds are kept cool in water, and as each is filled, it is placed in one of the partitions. When cold, the roll or stick of sulphur is expelled by a blow on the wooden plug O'.

Flowers of sulphur may be manufactured with the same apparatus; but in this case there are only two distillations in the 24 hours, and the charge is only 3 cwts. of sulphur. When the floor of the chamber is covered to a depth of 18 to 24 inches, the door G is opened, and the sulphur removed. We give below, the cost of the production of these two forms of sulphur, according to Payen, in French weights and values.

Working Expenses for one Day with this Apparatus.

Expenses.	Produce.	
	Roll Sulphur.	Flowers of Sulphur.
Rough sulphur, at 19 frs. per 100 k.	1,800 k. . frs. 342·0	300 k. . frs. 57·0
Fuel (wood or peat)	600 k. . ,, 18·0	201 k. . ,, 6·0
Wages . . .	. . . ,, 8·0	. . . ,, 3·0
General charges .	. . . ,, 10·0	. . . ,, 5·0
Packages & carriage	. . . ,, 9·0	. . . ,, 1·50
	frs. 387·0	frs. 72·50
	1,700 k. roll sulphur at 23·50fr.=399·50 80 k. residue at 7 fr.=5·60–405·10	270 k. flowers at 28 fr.=75·60 20 k. residue at 7 fr.= 1·40–77·00
	Profit...frs. 18·10	Profit...frs. 5·50

The demand for flowers of sulphur for protecting the vine from the *oïdium* disease, has induced some manufacturers to enlarge the chamber G to 11,000 cubic feet, which has increased the production threefold. See Appendix.

M. Bailly has introduced other changes, which have their peculiar advantages. The metal retort for distillation has a large belly, E, with two raised necks, one of which is closed by

a lid, and the other opens into the chamber. With this form
of retort the sulphur can never come in contact with the lid
or door. The position of the fire H I also prevents the direct
action of the hot fuel on the bottom of the retort, and the waste
heat passes under the melting pot A, which is furnished with a
kind of circular sieve or shelf to collect the heavy impurities, and
so secure a flow of pure melted sulphur, through the valve b, to
supply the retort below for distillation. The changing of the
supply pipe is also obviated by this arrangement. The position

FIG. 10.   FIG. 11.

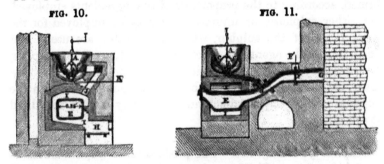

of the damper F F' presents obvious advantages over the former.
The retort can be charged with 20 cwt. at a time.

The great consumption of the flowers of sulphur, as a protection
against the vine disease, and the necessity of diffusing it over the
plant, has led to the use of a very effective utensil, which may
be termed a "sulphur-flower brush-box." Fig 12 gives different
views of this little instrument, which holds about 1 lb. of sulphur.

FIG. 12.

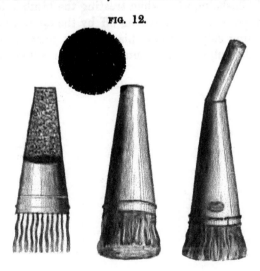

The top of the box is screwed on in the same way as a bayonet. This cover is furnished with five series of double folds of linen, kept in their places by thin iron wires, and the sulphur is thus scattered over the plant in a fine cloud.

Milk of sulphur is applied to medical purposes, and consists of sulphur in a most minute state of division. It is obtained by dissolving roll sulphur, or any other form of pure sulphur, in potash, soda, or lime, with a sufficient quantity of water. Sulphide of potassium, sulphite and hyposulphite of potash are formed, according to the proportions of the ingredients employed, and either sulphuric or muriatic acid may be employed for the precipitation of the sulphur, when the alkalies are used; but when lime is the agent for effecting the solution of the sulphur, it is necessary to use muriatic acid, as sulphuric acid would throw down the lime as gypsum, which would become mixed with the sulphur. The precipitated sulphur is carefully washed and dried at a low heat.

Sulphur from Pyrites.—The production of sulphur from the bisulphide of iron (FeS^2), a mineral disseminated throughout all formations, as *iron pyrites*, and *cockscomb pyrites*, depends upon the fact that this mineral is decomposed at a red heat into sulphur, and a compound of the same constitution as *magnetic pyrites*, $7FeS' = 6FeS$, $FeS' + 6S$. The mineral thus parts with 23 per cent, or $\frac{3}{7}$ of its sulphur.

The first attempt of this kind appears to have been made by Wyon and Chisholm, who, when treating the bisulphide, expelled as much sulphur as can be driven off by the application of heat. The remaining sulphur from this partially roasted ore, or any other metallic sulphide, was obtained in the following manner.

FIG. 13.

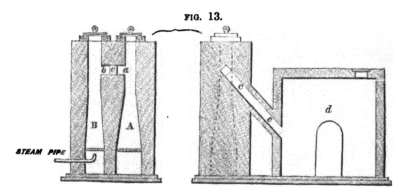

The furnace A (Fig. 13) was filled with the sulphur ore, and the furnace B with coke. The contents of both furnaces having been ignited, a current of steam was admitted into the latter, where it was converted into hydrogen and carbonic oxide. The two currents of gases from the furnaces meet at the junction of the two flues, where, at the high temperature prevailing, the hydrogen deprives the sulphurous acid of its oxygen, and the sulphur passes along a close flue c into a chamber d, where it is condensed.

Duclos followed, and proposed to pass two volumes of sulphuretted hydrogen and one volume of sulphurous acid, into leaden chambers, by which he expected to obtain all the sulphur, by the mutual decomposition of the two gases; but, as Muspratt has shown, only half of the sulphur is obtained, the other half forming pentathionic acid, S^5O^5. Duclos proposed to obtain his sulphuretted hydrogen by decomposing metallic sulphides by hydrochloric acid, or the alkaline sulphides by carbonic acid.

Spence has suggested two modifications, both of which are very simple, and adapted to some localities. He employs a kiln similar in form and dimensions to those in use where pyrites is burned for making sulphuric acid, and places in the centre of the kiln a stoneware pipe A, 12 or 15 inches in diameter, open at both ends, and in connection with an opening B; on the other side of the kiln there is an opening for conveying the gases and sulphur vapour through a flue to a chamber for condensation. The kiln is charged with pyrites through the doors F, and the pipes with coke through the opening E. The burning pyrites

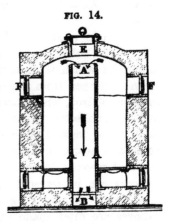

FIG. 14.

heats the coke in the pipes, and the sulphurous acid, following the course of the arrows, is decomposed by the hot coke into sulphur and carbonic acid. He, however, found that the same effect was produced by mixing the carbonaceous matters and pyrites in alternate layers in the kiln, igniting the whole, and allowing the sulphur to pass off into a chamber.

In Silesia, Bohemia, and Saxony, the decomposition of iron

pyrites is generally conducted in conical clay tubes *a*, which are placed in rows, from twelve to twenty-four, in a galley furnace (*Galeerenofen*); Fig. 15 is a side view. Both the wide and narrow ends of the tubes project beyond the sides of the furnace; the former for the purpose of charging the tubes with broken pieces of iron pyrites, about 70—100 lbs. at once; the other to open into an iron receiver *b*, filled with water; the narrow mouth is stopped by a perforated plate of clay, that the ore may not fall into the water from the inclined position of the tubes *c*. Not more than ¼ of the sulphur contained in the pyrites is actually obtained; for in attempting to expel all the sulphur present, the residue would melt, and the tubes would require to be broken up each time, in order to remove it for future operations. Green vitriol (sulphate of iron) is obtained from the residue by exposure to the oxidising action of the atmosphere.

FIG. 15.

The crude product, called drop-sulphur (*Tropffschwefel*), requires to be subjected to a second distillation, on account of the impurities which it contains, more particularly to free it from arsenic (derived from arsenical pyrites = sulphide of arsenic + sulphide of iron, which nearly always accompanies iron pyrites), and this is effected in cast-iron retorts, in the purifying furnace. Sometimes, instead of employing the furnace, the pyrites is roasted in heaps, or *meiler*, in a square brickwork space, over a layer of wood. By covering the heap with earth, the sulphur vapours can be forced to pass through a lateral tube into the condensing chamber. In uncovered heaps of this kind, deep grooves are often made in the upper colder parts, in which a portion of the sulphur collects.

The purified sulphur still contains selenium, arsenic, and bituminous particles.

Payen gives the following, as the cost of producing sulphur in this manner, in two different localities, one being favourably situated from the cheapness of the fuel, the other unfavourably, from the high price of coal.

Cost of Extraction of Sulphur in a Furnace containing 24 Tubes.

	Frs.	Frs.
Pyrites for 3 operations in 24 hours, 1,800 k.	2	2
Fuel, 1,000 k. coal	40	10
Wages	8	6
General charges	8	8
Produce, 252 k. sulphur . . .	58	26

The following plan for obtaining sulphur from iron pyrites, is pursued at Avoca in Wicklow. The ore, broken to pieces of the size of nuts, is thrown into a building of a cylindrical shape in quantities of 50 to 60 tons. The bottom is level, and in the walls there are four openings opposite to one another, which are connected by flues, crossing each other, in the floor of the building, and communicating with the interior by openings. The ore is then covered with a thick coating of peat and earth, and only one opening left for a draught pipe, which leads into a condensing flue 60 to 70 feet long by 6 feet high and 4 feet wide. The roaster is then set on fire at the four openings, and when the ore is fairly ignited, the openings are closed, and the mass allowed slowly to roast. This roasting lasts about six weeks, during which time the sulphur vapours condense in the large flues, and are afterwards collected, to be melted into blocks. When the roasting has been successful, each piece of ore is found to contain a nucleus of sulphide of copper, which is coated with a brownish red crust of oxides and sulphates. The roasted ore is heated with water, which dissolves the sulphates of iron and copper, and the latter is precipitated by iron in the usual way. The residue is then washed to separate the rich copper particles, which are sold as copper ore.

We are indebted to the kindness of Mr. R. C. Clapham, and of Messrs. Browell and Marreco, of Newcastle on Tyne, for the following valuable analyses of the different sulphur ores which are used in the manufacture of sulphur and sulphuric acid.

Tuscan Pyrites. (*Clapham.*)

Sulphur	48·4
Iron	44·5
Silica	3·6
Loss, &c.	3·5
	100·0

Spanish Pyrites. (Clapham.)

	1	2	3
Sulphur . . .	45·21	47·50	43·52
Iron	43·50	41·92	40·20
Lead	0·22	1·52	2·11
Zinc	...	0·22	0·32
Copper . . .	2·21	4·21	3·12
Arsenic . . .	0·41	0·33	1·10
Silica	7·10	3·40	8·00
Moisture . . .	1·21	·56	1·54
	99·86	99·66	99·91

Belgian Pyrites. (Clapham.)

	1	2	3
Sulphur . . .	42·80	35·50	46·20
Iron	36·70	38·60	40·50
Oxide of iron .	7·23	4·24	2·20
Lead	0·92	0·65	0·41
Zinc	0·40	5·26	0·22
Arsenic . . .	0·20	0·31	0·41
Alumina . . .	trace	...	...
Silica	8·86	14·90	9·10
Carbonate lime .	0·84	trace	...
Moisture . . .	1·46	0·56	0·42
	99·41	100·02	99·46

Cornish Pyrites. (Clapham.)

	1	2	3	4	5	6	7	8
Iron . . .	27·776	32·676	60·676	41·000	27·076	32·676	32·200	42·350
Lead . .	1·440	0·600	...	1·850	7·446	1·203	0·400	0·680
Copper . .	4·600	1·333	2·133	1·600	0·400	0·533	0·800	2·800
Zinc . . .	9·086	3·226	0·400	1·990	...	0·400	1·325	1·615
Sulphur .	25·272	24·013	34·880	33·725	26·009	30·803	34·345	30·325
Arsenic .	trace	...	...	...	0·500	0·403	0·910	1·160
Silica . .	31·676	38·676	2·000	20·000	38·676	33·333	29·000	17·000
Sulphate lime	...	...	...	...	...	...	...	0·596
Carbonate lime	...	...	...	...	...	...	...	3·579
	99·830	100·504	100·089	100·165	100·087	99·331	98·950	100·065

Irish Pyrites. (Clapham.)

	1	2	3	4
Iron . . .	32·222	35·000	34·650	42·400
Lead . .	2·963	1·600	1·080	1·593
Copper . .	4·133	2·400	2·400	1·333
Sulphur .	40·410	42·128	37·975	34·676
Arsenic .	...	0·602	0·400	0·183
Silica . .	17·676	18·676	22·500	20·000
	97·394	100·396	99·005	100·175

Browell and Marreco's Analyses of Sulphur Ore from

	Belgium.	Piedmont.	Pomeran.	Sweden.
Iron . .	45·50	40·26	42·93	42·80
Copper. .	trace	trace	2·87	1·50
Sulphur .	51·59	44·60	48·75	38·05
Arsenic .	trace	...	trace	...
Silica, &c. .	2·20	14·30	3·20	12·16
Moisture .	·28	...	·48	...
Oxygen and loss ...	...	...	...	5·49
	99·57	99·16	98·23	100·00

Dr. R. D. Thomson has also given analyses of the Wicklow pyrites, and of that found in coal beds known as "coal brasses."

	Wicklow.	Coal brasses.
Sulphur	47·41	53·35
Iron	41·78	45·07
Copper	1·93	...
Manganese	...	·70
Zinc	2·00	...
Arsenic	2·11	...
Silica	3·93	·80
Insoluble matter . . .	1·43	...
	100·59	99·92

In some places, copper pyrites (sulphide of copper) is made use of in the same manner; but the sulphur obtained from pyrites forms a very insignificant portion of that which is consumed in this manufacture.

Production of Sulphur from Alkali-waste.—Within the last few years, numerous patents have been enrolled for obtaining sulphur from what have hitherto been thrown away as waste products of manufacture. The first we shall notice, is the invention of the late Mr. Lee, who converted the sulphuretted hydrogen gas, obtained by an acid from alkali-makers' waste, or from the sulphides of potassium, sodium, or calcium, first into sulphurous acid by burning, and then into solid sulphur. The mode of converting the sulphurous acid into solid sulphur, is alike applicable, from whatever source the acid may have been obtained in the first instance. The following description of the process and apparatus, which has been partially carried out in Ireland, is taken from the *Repertory of Arts,*

Vol. v., p. 20, 1845, and from a memorandum drawn up by Mr.
Lee himself.

The sulphurous acid, in whatever manner it may be evolved,
is decomposed by passing through a fire of common coal or coke,
which fire maintains at a red heat, a flue or chimney filled with
bricks, in such a manner, that interstices are left, through which
the products of combustion are carried.

These products consist of some undecomposed sulphurous acid
gas, some sulphide of carbon, sulphuretted hydrogen, carbonic oxide
and carbonic acid gases, together with vapour of sulphur. In passing through the heated bricks in the flue or chimney, these
gaseous compounds react upon each other so as to produce solid
sulphur. The sulphides of carbon and hydrogen are decomposed
by the carbonic acid gas and undecomposed sulphurous acid gas,
producing sulphur and aqueous vapour, and the carbonic oxide
gas formed at the same time, as well as that produced by the
passage of the sulphurous acid gas through the fire, decomposes a
further portion of the sulphurous acid gas in its passage through
the heated bricks. Whenever a deficiency of sulphurous acid
gas occurs in this part of the apparatus, which is easily detected
by the escape of sulphide of hydrogen or carbon, it is supplied as
wanted. Thus the process is conducted quite as easily as the
manufacture of sulphuric acid. It is absolutely necessary that
the bricks should be at a full red heat. The sulphur vapour is
cooled down to a temperature of 350° F., and conducted through
an apparatus similar to the muriatic acid condensers (to be described hereafter), in which it is exposed to an extensive cooled
surface, on which it liquefies and filters down to a pan at the bottom,
from which it is run off. That part of the sulphur which escapes
liquefaction is collected in powder in an apparatus constructed on
the principle of the chamber for collecting lamp-black.

The following drawings represent that part of the apparatus
by which the sulphurous acid is decomposed.

Fig. 16, 1, is a longitudinal section.

Fig. 16, 2, is a transverse section of 1, through the dotted
lines 1 1.

Fig. 16, 3, is a sectional plan through the dotted lines 2 2,
Fig. 16, 1.

A is the flue through which the sulphurous gas passes, when
it is generated, to the fire-place B; C, the grate bars; D, the
ash-pit; E, the door closing the ash-pit; $F F F$ are holes closed

FIG. 16.

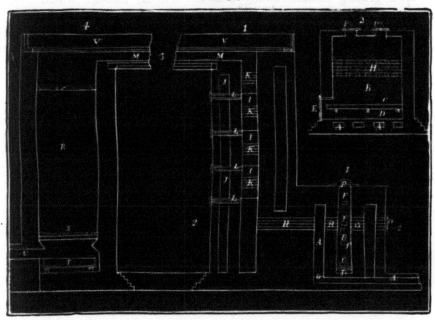

FIG. 17

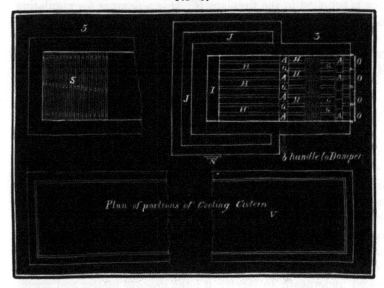

Plan of portions of Cooling Cistern

by fire-bricks, through which the fire is stirred; G is the fire-brick grating communicating with the flue A, through which the sulphurous acid gas enters the fire; $H\,H'$ are the flues carrying away the sulphur vapour and gaseous matter from the fire into the chimney or flue I, and constructed on the principle of, and similar to, the grating G. The chimney or flue I is constructed of common brick or stone, cased inside with fire-brick, in the interior of the walls, which are of great thickness. J is a space filled with ground coke or furnace ashes, or any other suitable material, that is a bad conductor of heat. KKK are brick arches or fire piles thrown across the chimney or flue I, and placed about two inches apart, upon which are piled fire-bricks or fragments of fire-brick, constructed in a way to form tortuous passages for the sulphur vapour and gaseous matter to traverse; they are placed in the chimney at intervals of 18 to 24 inches. $L\,L\,L$ are openings into the chimney, through which thermometers are inserted to ascertain its temperature; they are closed by an iron or brick stopper when not used; M is the flue carrying away the sulphur vapour and gaseous matter to the condenser hereafter described; N is the door at the bottom of the chimney for cleaning the chimney out; $o\,o\,o\,o$ are doors for closing the brick grating G, in which are registers for admitting atmospheric air when necessary; $P\,P$ are doors through which the coke is supplied to the fireplace B; Q is a damper for regulating the supply of sulphurous acid gas to the flue A'.

Figs. 16 and 17, 4 and 5, represent the apparatus in which the sulphur vapour, or part of it, is condensed and liquefied, and M is the flue in connection with the same flue M, Fig. 17, 1.

Fig. 17, 4, is a sectional elevation.

Fig. 17, 5, is a horizontal section through the dotted lines 3, 3, Fig. 17, 4.

The sulphur vapour and gaseous matter enter that part of the apparatus shown at Fig. 16, 4, which is called the condenser, by the flue M. R is the chamber or body of this condenser, filled with broken glass, bricks, stone, or coke, so as to expose extensive surfaces for condensing the sulphur vapour, on the principle of the condenser, so extensively used by the alkali makers, for condensing muriatic acid gas; S is an iron grating to support the materials with which the condenser is filled, and through which the liquefied sulphur runs into the iron pan T, placed underneath to receive it; the gaseous matter and uncondensed sulphur vapour also pass through this grating, and are carried out of the condenser by the flue U, into an apparatus constructed of brick or stone, and similar to the chambers used in the manufacture of lamp-black, in which the remainder of the sulphur is collected in the form of powder or flower brimstone; a draught is produced by connecting with the part of the apparatus last described, a chimney of sufficient capacity to command any draught that may be required.

The following is an estimate of net profit realized by this process, as estimated by the late Mr. Lee.

100 tons of pyrites at 13s., the price at the mine, averaging 38 to 40 per cent. of sulphur, and 2 per cent. of copper	£65	0	0
Expense of 60 tons of coal at 12s.	36	0	0
Carriage of sulphur to place of shipment, and freight to market.	36	0	0
Labour	20	0	0
	157	0	0

Produce, 30 tons of sulphur at £5 10s. . .	£165
2 tons of copper at £80	160
Gross value	325
Deduct expenses, &c. .	157
Net profit	£168

We have now to describe one of the most interesting and complete arrangements for producing sulphur, which has been patented by Mr. Gossage. We prefer to give the description of this process as much as possible in his own language.

He employs the hitherto worthless and most offensive material, known as "alkali or tank waste," and decomposes the sulphide of calcium it contains, by means of impure carbonic acid gas, so that the sulphuretted hydrogen produced, is mixed with other gases; or he effects the decomposition of "alkali-waste" by means of aqueous hydrochloric acid, thereby obtaining sulphuretted hydrogen mixed with carbonic acid gas. The mixed gases obtained by either of these processes, are not suitable for the manufacture of sulphuric acid by the combustion of the sulphuretted hydrogen which they contain. He therefore employs the

following processes, in order to obtain sulphur in a solid form.
He obtains his carbonic acid gas by the combustion of coke or
coal, using a furnace with a deep bed of fuel for effecting such
combustion, so as to cause the whole or nearly the whole of the
oxygen, contained in the atmospheric air, to combine with
the carbon or hydrogen of the coke or coal; or he obtains
carbonic acid gas by the calcination of carbonate of lime with
coke or coal in a lime kiln.   He conducts the gaseous products
of the combustion (these products containing carbonic acid,
carbonic oxide, and nitrogen gases) through the flue of a boiler or
evaporating pan, so as to utilize a portion of the heat, and afterwards through pipes, so arranged as to effect a.further cooling of
the gases, and he then causes these gaseous products to filter
through alkali-waste, (maintained in a state of dampness by
affusion of water and injection of steam) contained in suitable
vessels, to effect the absorption of the greater part of the carbonic
acid.   He uses for this purpose, the same vessels which are employed to effect the lixiviation of the compound called "black
ash" or "rough soda" (by which lixiviation "alkali-waste"
remains in the vessels as an undissolved residuum); and adapts
moveable covers and suitable pipes to these vessels for the purpose of confining therein, and conveying therefrom, the gases resulting from the action of carbonic acid on the "alkali-waste."
These gases contain sulphuretted hydrogen, also some carbonic
acid, and carbonic oxide, and the nitrogen previously existing in
the atmospheric air employed.

Aqueous hydrochloric acid may be employed instead of carbonic
acid gas, and the decomposition is easily effected, but Mr. Gossage
employs steam towards the end of the operation, to expel the last
portions of the sulphuretted hydrogen and carbonic acid.

The solid sulphur is obtained by decomposing the mixed sulphuretted hydrogen and other gases above named, by introducing
an additional quantity of carbonic acid gas, with some atmospheric
air or sulphurous acid gas, and heating the mixture of gases so
obtained to redness, when the hydrogen of the sulphuretted
hydrogen combines with oxygen (forming water), and the sulphur
is liberated in the form of uncombined sulphur vapour, which by
cooling, becomes condensed into liquid sulphur, and by further
cooling, the liquid sulphur becomes solidified.

The heating of the mixed gases, and the addition at the same
time of carbonic acid gas, is effected by causing the mixed gases

to become mingled with the highly heated gaseous products obtained by burning coke or coal with atmospheric air in a suitable furnace. This is accomplished, either by causing the mixed gases to filter through the burning coke or coal, together with a suitable quantity of atmospheric air, or introducing them into a flue, where they mix with the highly heated gases produced by the combustion of coke or coal.

It is noteworthy, that at this high temperature no excess of carbonic acid gas will cause oxidation of sulphur, which oxidation would be occasioned by an excess of atmospheric air. Mr. Gossage judges that the suitable quantity of atmospheric air or sulphurous acid gas has been employed, by testing the residual gases which pass off from the apparatus used for condensing the sulphur vapour, with moistened litmus paper for the detection of sulphurous acid, and paper impregnated with salt of lead for the detection of sulphuretted hydrogen.

Mr. Gossage's apparatus is represented in Figures 18, 19; Fig. 18 being a plan, and Fig. 19 a vertical section.

A A are vats used for lixiviating the compound called "black ash" or "rough soda," of the same construction as the vats which are used for what is called "continuous lixiviation," and having a perforated false bottom. These vats are provided with hydraulic lutes B B, into which covers, C C, can be introduced, so as to convert the open-topped vats into air-tight vessels.

D is a furnace for the combustion of coke or coal. The ash-pit of this furnace is closed by an iron door Y, and the feeding hole is closed by a door of fire-brick Z; these closings are made perfect by means of luting with fire-clay. The pipe S, which conveys gas from the fan F, is connected with the closed ash-pit of the furnace D. E is a flue connecting the furnace D with an apparatus of brickwork G. This is a square tower, constructed of double walls of brickwork, between which walls finely slaked lime is introduced for the purpose of rendering this apparatus impervious to air. A number of brick arches, H¹ H¹, are constructed, extending across the tower G, and closing the same except where perforations are left in them. These perforations are left near the outsides of one set of arches H¹, and in the middle of the other side of arches H². I is a cast iron pipe, connecting the tower G with another tower Q. This pipe is partly surrounded with a tank containing water, for the purpose of cooling the gases passing through such pipe. The

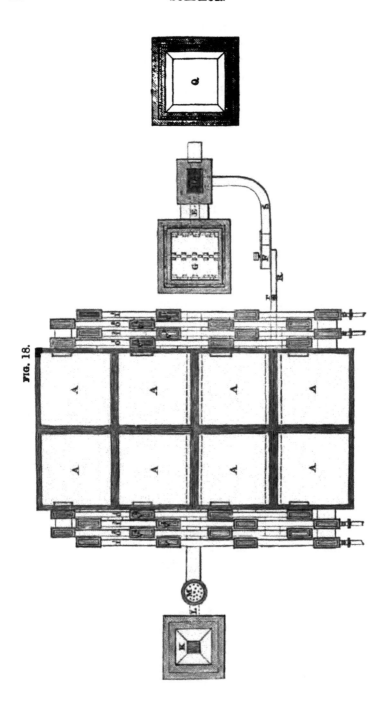

FIG. 18.

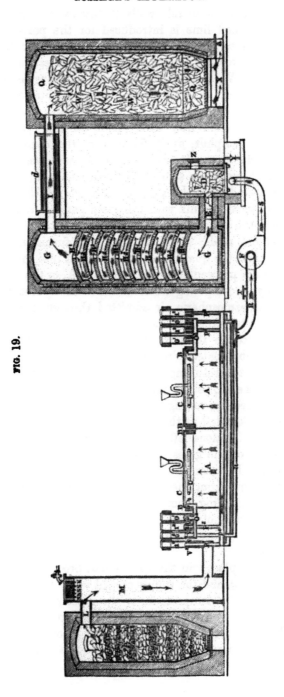

FIG. 19.

tower Q is constructed of double walls of brickwork, between which walls finely slaked lime is introduced for the purpose of rendering the apparatus impervious to air. The tower Q is nearly filled with broken bricks, clinkers, or other suitable materials, marked W W, capable of presenting extensive surfaces to gases passing through the tower, and thereby facilitating the condensation of the sulphur vapour. The condensing materials, W W, are supported upon iron bars. X is an iron pan for receiving the liquid sulphur, as it flows from the condensing tower Q. *a* is a flue conveying the non-condensed gases from the tower Q, to a chimney, or some other exit channel ; *r* is a branch pipe connected with the pipe R, through which atmospheric air is introduced to mix with the gases passing from the vats A A, or from the apparatus in which alkali-waste is decomposed by hydrochloric acid. This branch pipe is furnished with a regulator to control the quantity of air admitted into the pipe R. The mixed gases proceeding from the vats A A, or the mixed gases obtained by decomposing alkali-waste with hydrochloric acid, together with atmospheric air admitted through the pipe branch *r*, are conducted from the fan F, into the ash-pit of the furnace D, and percolate through the burning fuel contained therein ; being thereby heated to redness, and passing from thence through the flue E, the resulting gases are conducted into the tower G, in which the requisite mixing of the gases is completed. This mixing is effected by causing the gases to pass alternately through the perforations left at the middle of the arches H¹ H¹, and through the perforations left in the outsides of the arches H² H². The course of the gases through the tower G, is shown by the bent arrows. In order to condense the sulphur vapour, and to separate the sulphur from the gases with which its vapour is mixed, the gases proceeding from the tower G pass through the iron pipe I, and a supply of water is introduced into the tank *d*, through which this pipe passes, sufficient to effect the requisite cooling of the gases.

A thermometer is placed in the exit flue *a*, proceeding from the condensing tower Q, and the supply of cold water to the tank *d*, is so regulated that the temperature of the gases passing from the tower Q, is not less than 230° Fah. The gases passing from the tower G through the pipe I, become cooled below the temperature requisite for the existence of sulphur as a permanent vapour; and as these gases percolate through the materials contained in the tower Q, liquid sulphur becomes deposited on these materials, and flowing down the tower Q, collects in the iron pan X, whence it can be removed and allowed to solidify by cooling.

$I^1 I^1$, and $I^2 I^2$, are main pipes connected with the vats A A, by means of branch pipes $P^1 P^1$, and $P^2 P^2$, for the purpose of introducing gas into these vats. $O^1 O^1$, and $O^2 O^2$, are main pipes connected with the vats A A, by means of branch pipes $O^3 O^3$, for the purpose of conducting gas away from the vats. The main pipes $O^1 O^1$, are connected with the main pipes $I^2 I^2$. The main pipes $I I^1$, are connected with each other, and the main pipes $I^2 I^2$, are also connected with each other. $V^1 V^1$, and $V^2 V^2$, are a series of hydraulic valves, by means of which the communication between the main pipes $I^1 I^1$, or $I^2 I^2$, and the vats A A, may be opened and closed, as required. $V^3 V^3$, and $V^4 V^4$, are a similar series of hydraulic valves to be used for opening and closing the communication between the main pipes $O^1 O^1$, and $O^2 O^2$, and the vats A A. K is a lime-kiln, in which carbonate of lime is calcined in mixture with coke or coal. L is a pipe conducting gases from such lime-kiln into a cooling vessel M, into which streams of water are introduced by means of the pipes N N. The cooled gases pass from this vessel into the main pipes $I^1 I^1$, and from thence into such of the vats A A, as are in a suitable condition to receive them. After passing through these vats, the gases traverse the pipes $O^3 O^3$, into the main pipes $I^2 I^2$, and from thence into such of the vats A A, as are in a suitable condition for receiving them. After passing through these vats, the gases traverse the pipes $O^3 O^3$, and through the pipe R, into a fan F, by means of which the requisite current is obtained to draw gases through the vats, and transfer the gases proceeding from the vats through the pipe S, into the decomposing apparatus. T T are pipes for supplying steam to the mixed gases passing into the vats; and U U are pipes through which water can be dispersed over the surface of the alkali-waste contained in the vats. By these means, the waste can be kept suitably moistened.

By reference to the figures, and the foregoing description of the vats, with their main pipes and connections, it will be observed that each of the vats can be connected with one of the main pipes I^1_1 I^1, from which gas proceeding from the lime-kiln, is introduced under the false bottom of the vat, and after percolating through the mass of moistened alkali-waste, the gas is conducted through one of the pipes O^2, into the main pipes O^1, and from thence into the main pipes I^2 I^2, and from thence under the false bottom of a second vat, and is caused to percolate through the moistened alkali-waste contained therein, and after this second percolation the gases produced, are conducted through the pipes O^2, into the main pipes O^2, and from thence into the pipe R, to be transferred to the decomposing apparatus. Mr. Gossage employs a series of eight vats for the two purposes of lixiviating "rough soda" and decomposing the alkali-waste produced by the lixiviation, using four of these vats for the lixiviating process, and the other four for the gasing process, and conducting the two operations as part of one cycle, so that each of the vats is consecutively used for the gasing operation, which continually follows the lixiviating operation; and the alkali-waste, after complete lixiviation with water, is immediately and without any transference, subjected to the gasing operation, the vat containing the freshly lixiviated alkali-waste being immediately closed by its cover c. By these means the alkali-waste is not exposed to the injurious action of atmospheric air. The sulphide of calcium contained in alkali-waste which has been thus exposed to a current of mixed gases, proceeding from another vat during twenty-four hours, and subsequently exposed to a current of gas proceeding from a lime-kiln during a similar period, is converted almost entirely into carbonate of lime, and deprived of nearly all the sulphur it previously contained; but the period required, is influenced by the quantity of gas supplied for decomposing the sulphide of calcium. The solid residue being removed from each vat after the completion of the gasing operation, can be used for some of the purposes for which carbonate of lime is suitable, and the vat may then be employed for another lixiviating operation, and subsequently for another gasing operation.

Mr. Gossage, by a later patent, substitutes steam at a high heat in the place of muriatic or carbonic acid.

Instead of the alkali-waste, Thaulow, some years since, proposed to convert sulphate of lime into sulphide of calcium, by roasting it in close vessels with carbonaceous matters, and then decomposing the resulting compound with carbonic acid. The gaseous sulphuretted hydrogen might then be treated by any of the plans already described.

More recently Jacquemin, has revived this idea; but he passes a current of steam and carbonic oxide through the earthy or alkaline sulphuret.

Lastly, we must notice here, the proposal of Mc Dougal and Rawson, who pass sulphuretted hydrogen through a red hot tube, by which they state, the elements are separated, and the sulphur may be collected by any convenient plan.

SULPHUROUS ACID.

This acid is met with, as a natural product, among the gaseous exhalations of active volcanoes, and is produced whenever sulphur or metallic sulphides are ignited in contact with the air.

Sulphurous acid is gaseous at ordinary temperatures, but it assumes a liquid form when exposed to pressure or great cold. It is very soluble in water, which at 60° F. dissolves 37 times its own volume. It forms with water a hydrate, $SO^2.14HO$. It is not decomposed by a high temperature, except in contact with combustible matter, when sulphur is separated.

Sulphurous acid is employed in the treatment of cutaneous diseases, the patient being seated in a wooden box placed over a small fireplace (Fig. 20), in which the sulphur is burned, so that the vapour enters and fills the box. The top of the box is fitted with a hole through which the head of the patient passes, while the escape of the gas is prevented by folding a wet cloth round his neck.

Sulphurous acid also possesses bleaching properties and is employed for discharging colours from woollen and silk goods, which are exposed to the action of this gas, by being suspended in a damp state, from strings stretched across a chamber. It is also employed to blanch isinglass; to prevent too rapid a fermentation in wine vats; to impregnate casks in which wine, beer, and other fermented liquors are to be preserved, so as to stop the acetous fermentation; to bleach sponges, straw, the starch from wheat and potatoes, &c. &c.

The following plans are employed for the manufacture of this

FIG. 20.

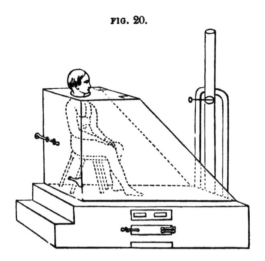

acid in solution. The sulphur is placed in an iron saucer in the furnace A (Fig. 21), and ignited. The flame ascends the sheet-iron chimney A B, which creates a draught, and the gas then passes through the pipe B B', made of sheet lead, kept cool by a current

FIG. 21.

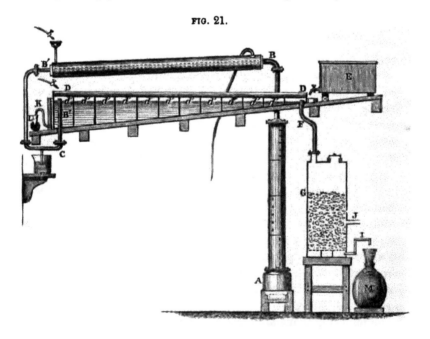

of water flowing in a contrary direction, between the outer and inner pipes. The pipe B B', making two bends, enters a leaden box B" (shown in section in Figure 21 and in plan in Figure 22), divided into several compartments d' d'', open alternately at the two sides. The sulphurous acid, in its passage, meets a stream of the liquid intended for its absorption, water or a solution of soda or potash flowing in an opposite direction and supplied from the tank E. The aqueous solution of the gas discharges itself through a water-lute K, into a funnel L.

The condensation is assisted by the action of a stream of cold water entering at D, and escaping at D'. Any sulphuric acid mixed with the sulphurous acid gas, condenses in the tube B B', and escapes at C, through a small pipe plunged in water.

FIG. 22.

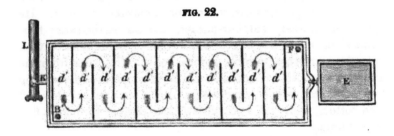

The uncondensed gas passes off through a pipe F, which conveys it into a cylinder filled with moistened soda-crystals, which are decomposed, and the waste gases escape, through the tube T, into a chimney. The sulphite of soda, more soluble than the carbonate, flows off through a water-lute I, into a carbon M.

In preparing sulphite of lime, the milk of lime employed for absorbing the gas, is kept in motion by an agitator passing through the compartments d' d''.

Another, but more expensive process, consists in heating concentrated sulphuric acid in contact with some organic substances, in a glass or earthenware vessel A (Fig. 23), and conducting the sulphurous acid gas through a series of Woulfe's bottles B, C, C', containing water or some alkaline solution. When the liquid is saturated, it is drawn through cocks at the bottom of the Woulfe's bottles. The change generally takes place as represented by the equation

$$2SO^3 + C = CO^2 + 2SO^2;$$

FIG. 23.

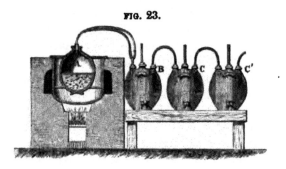

sometimes, however, a small quantity of carbonic oxide is produced.

Calvert has proposed to manufacture this liquid acid, when required in large quantities, in the following manner. He employs an ordinary sulphur-burner, and an earthenware tube immersed in water, leading to a tall wooden column filled with pumicestone or other suitable porous body. The gases from the burner pass up this condenser, where they meet a descending shower of water, which absorbs the sulphurous acid and flows off at the bottom, while the uncondensed gas escapes through a flue into a chimney.

Mr. Losh mixes 1 lb. of hyposulphite of lime or soda in 3 gallons of water and adds 6½ ounces of sulphuric acid of 1·848 sp. gr, by which simple method he obtains a solution of sulphurous acid. This process has just been patented by the inventor.

Estimation of Sulphurous Acid.

The quantity of sulphurous acid contained in a liquid may be determined by giving it a blue colour with a few drops of a solution of indigo, and then adding, by means of a burette, a test-solution of chloride of lime. The operation is conducted in the same way as when determining arsenious acid. The chlorine first destroys the sulphurous acid, and it is only when the transformation is complete, that the first drop of the chlorine causes the blue colour of the indigo to disappear. The number of degrees of the burette which have been employed are noted, and the volume of sulphurous acid is calculated : 1 volume of this acid corresponds to 1 volume of chlorine.

Dupasquier has proposed the employment of sulphurous acid

for the determination of iodine, and this method furnishes good results, if the sulphurous liquid is sufficiently dilute. The iodine may, therefore, be also employed for the determination of sulphurous acid. These two bodies, in presence of water, form hydriodic and sulphuric acid:

$$SO^2 + HO + I = SO^3 + HI.$$

On the other hand, these acids may react on each other, to produce sulphurous acid, water, and iodine. According to Bunsen, the latter reaction takes place more or less completely according to the greater or less concentration of the liquids. In this process, therefore, liquids sufficiently dilute must be employed; experience has shown that the sulphurous acid solution ought not to contain more than 0·03 to 0·04 in the 100 of anhydrous sulphurous acid.

To analyse sulphurous acid volumetrically, 3 or 4 cubic centimetres of a solution of starch are added, and then the normal iodine-liquor, until the liquid becomes blue. The volume of test-solution employed for the destruction of the sulphurous acid, gives the means of calculating the quantity of this body. The determination of sulphuretted hydrogen may be made exactly in the same manner.

Suppose that the number of burette-degrees of the iodine-solution required to decompose 100 degrees of the sulphurous acid solution is t, and the quantity of iodine in each degree of the normal solution is a, then the quantity of sulphurous acid SO^2 in 100 degrees of the acid solution is

$$x = \frac{SO^3}{I}\ at = \frac{32}{126\cdot4}\ at.$$

Bunsen's General Method of Volumetric Analysis.

Bunsen has established a general volumetric method upon the action of iodine on sulphurous acid. This method is applicable, for the most part, to those analyses in which the substances to be determined, are reduced or oxidised. A quantity of iodine is liberated corresponding to the substance to be analysed, and this iodine is determined by means of a weak test-solution of sulphurous acid.

When a large number of experiments are to be made, a considerable quantity of this liquor is previously prepared. For this purpose, a concentrated solution of sulphurous acid is added to 2 or 3 gallons of distilled water until 100 parts contain from 0·03 to 0·04 of this acid. The degree of concentration of the liquid is easily determined by means of starch and a test-solution of iodine.

To preserve a considerable quantity of the sulphurous acid solution, it is best to employ the apparatus recommended by Fresenius (Fig. 24). It consists of a large bottle, almost filled with water deprived of its air, and to which a sufficient quantity of a solution of concentrated sulphurous acid is added, so that 100 parts of the liquor shall contain the proportion of sulphurous acid already mentioned. This bottle is connected by means of a tube of india rubber and a glass tube bent at a right angle, with two other tubes of U shape, the first filled with fragments of phosphorus, and the other with hydrate of potash ; the oxygen of the air and the phosphoric acid are fixed by the phosphorus and potash.

To the lower part of the bottle a glass tube is adapted, communicating by means of an india-rubber tube, with another graduated tube, intended to measure the solution of sulphurous acid. The graduated tube contains 200 cubic centimetres and is divided into portions of 50 cubic centimetres each. A spring-clamp is used to close the two tubes of india-rubber at each end of the graduated tube.

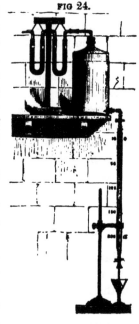

FIG 24.

Bunsen's method also requires test liquors of iodine and iodide of potassium.

To prepare the first, a known quantity of dry and pure iodine is added to a concentrated solution of iodide of potassium. The solution of iodide of potassium should not acquire a brown colour from hydrochloric acid. If the degrees of the burette correspond to ½ a cubic centimetre, the solution of the iodide is diluted, so that each degree shall contain 0·0025 gr. of the iodine employed.

As the iodine of commerce almost always contains traces of chlorine, it is necessary to determine the strength of the normal iodine-solution by direct experiment. This is most easily effected by means of bichromate of potash. The method is as follows :

When bichromate of potash is boiled with fuming hydrochloric acid, chlorine is evolved, and sesquichloride of chromium remains behind, together with chloride of potassium :

$$KO.2CrO^3 + 6HCl = Cr^2Cl^3 + KCl + 6HO + Cl^2.$$

For each equivalent of bichromate decomposed, 3 eq. of chlorine are evolved; and if this chlorine be passed into a solution of iodide of potassium, in such a manner that none shall escape, it will liberate 3 eq. of iodine. Hence, the relation between the weight, w, of bichromate decomposed, and that of iodine, i, liberated, is given by the equation

$$i = \frac{I^3}{KO.2CrO}\, w = \frac{381}{147\cdot4}\, w,$$

or $i = 2\cdot585\, w$;

and the problem to be solved is to find the relation between this quantity of iodine, and the quantity of the same element contained in a cubic centimetre of the normal iodine-solution. This is done as follows.

Take a small flask capable of holding, up to a certain mark on its neck, from 400 to 450 cubic centimetres of liquid; fill it up to this mark with the dilute solution of sulphurous acid already mentioned (p. 32); pour this quantity of liquid, which we will call a *flask-measure*, into the brown solution of iodide of potassium containing the iodine set free by the chlorine; and continue adding the sulphurous acid in these measured quantities till the brown colour is completely destroyed: let the number of flask-measures of sulphurous acid thus added be n.

The solution, which now contains an excess of sulphurous acid, is then mixed with gelatinous starch, and the normal iodine-solution added, from the burette, till a blue colour just begins to appear, showing that the excess of sulphurous acid is oxidised. Suppose the number of cubic centimetres of solution added for this purpose to be t': suppose also that each cubic cent. contains a quantity of iodine denoted by a: then the quantity of iodine by which the n flask-measures of sulphurous acid have been oxidised is

$$i + at'.$$

Next take one flask-measure of the same sulphurous acid; mix it with starch; and determine in the same manner the number of cubic cents. of normal iodine-solution required to oxidise it: if this number be t, the weight of iodine required to produce the effect will be at, and the quantity required to oxidise the n flask-

measure of sulphurous acid will be ant: but this same effect
was likewise produced by the quantity $i + at'$. Consequently,

$$i + at' = ant, \text{ and}$$
$$i = a(nt - t');$$

but we have already found $i = 2.585\ w$, where w denotes the
weight of pure bichromate of potash equivalent to the iodine:
therefore

$$2.585\ w = a(nt - t'), \text{ and}$$
$$a = 2.585\ \frac{w}{nt - t'}.$$

It will be seen at once that an essential condition of the
accuracy of this determination, is that none of the chlorine,
evolved in the distillation of the hydrochloric acid with bichromate of potash, should escape. To ensure this, the apparatus
represented in Figures 25 and 26, is used. On a glass tube

FIG. 25.   FIG. 26.

of about 4·5 millimetres internal diameter, a bulb of about 30 c. c.
capacity is blown, and a flask is thus obtained of the form shown
in Fig. 25. To a short piece of the same tubing a longer and narrower tube is fixed, drawn out at d, and bent at b, as shown in the
figure. If the neck of the flask and the adapter-tube be ground
flat and bound together at g with caoutchouc, so as to bring them
close together, an apparatus is obtained for the evolution of
chlorine, which is scarcely inferior to one consisting wholly of
glass. The caoutchouc tube must be freed before use, from adhering sulphur, by boiling with very dilute caustic soda and
thorough washing with water.

A retort of about 150 c. c. capacity (Fig. 26) holds the solution of iodide of potassium. The neck of the retort is widened
at a, to receive any solution driven back by the expelled air. In
order to make the determination, the bichromate of potash is
thrown into the flask, which is then filled to ⅔, with fuming hydro-

chloric acid ; the delivery-tube is attached, and placed so far in the retort (filled up to the commencement of the neck with iodide of potassium solution), that the chlorine which is not immediately absorbed, must collect at *b*. The flask is first gently heated till the decomposition is complete, then more strongly, in order to drive over every trace of chlorine into the iodide of potassium solution, by means of the gaseous water and hydrochloric acid. The retrogression of the iodide can scarcely take place, if care is taken, because it can only occur very slowly in consequence of the small volume of the apparatus and the fineness of the point *d*. After the hydrochloric acid vapours have been evolved for about five minutes, it may be assumed that all the chlorine is expelled. Without discontinuing the boiling, the delivery-tube is withdrawn from the retort, and the vapours are passed into some fresh solution of iodide of potassium, the boiling being continued for a few moments. If this solution remains uncoloured, the operation may be regarded as successful.

The method of volumetric estimation by means of iodine and sulphurous acid may be applied to a great variety of substances. In all the following examples, the quantity of substance weighed out is denoted by *w*.

1. To determine the percentage of pure *iodine* in a commercial sample :—Dissolve a weighed quantity in iodide of potassium ; decolorise the solution with sulphurous acid ; and oxidise the excess of that acid with standard iodine-solution as above : then the percentage of pure iodine is given by the formula

$$\frac{100}{w} a(nt - t'),$$

and if $\frac{100}{w} a$ be equal to unity, that is to say, if the quantity w weighed out be equal to $100a$ (e. g. 1 gramme, if $a = \frac{1}{100}$ grm.), the difference of the two measurements $nt - t'$ will at once give the percentage of iodine, without the trouble of calculation.

2. *Chlorine.*—The quantity of chlorine liberated in any reaction is determined, by passing it into a solution of iodide of potassium and proceeding as before : the amount is calculated by the formula

$$\frac{Cl}{1} a(nt - t), \text{ or } \frac{35\cdot5}{127} a(nt - t')$$

$$= 0\cdot286 \, a(nt - t').$$

3*

This method is directly applicable to the examination of *chloride of lime*, the value of which depends on the amount of chlorine liberated from it by acids. The percentage of available chlorine in a sample is given by the formula

$$\frac{100}{w} \cdot \frac{35\cdot5}{12\cdot7} \cdot a(nt - t') = 28\cdot6 \cdot \frac{a}{w} \cdot (nt - t'),$$

and if w be equal to 28·6 times a, the percentage is at once given by the difference of the measurements $nt - t'$.

3. *Chromic acid and Chromates.*—The percentage of pure bichromate of potash in a commercial sample is given by the formula

$$\frac{100}{w} \frac{KO.2CrO^3}{I} a(nt - t') = \frac{38\cdot69}{w} a(nt - t')$$

which may be simplified as before by taking $w = 33\cdot45\,a$.

The percentage of chromic acid in the same sample is

$$\frac{100}{w} \frac{2CrO^3}{I^3} a(nt - t') = \frac{25\cdot35}{w} a(nt - t').$$

4. *Peroxides.*—The quantity of oxygen in the peroxides of lead, manganese, &c., may be estimated by the same process: thus the percentage-amount of oxygen in peroxide of manganese is given by the formula

$$x = \frac{O^3}{I} a(nt - t') = 0\cdot1268\, a(nt - t'),$$

and if the weight of the sample be 100 times this, or 12·68 grm., the percentage will be determined at once by the difference of the measurements $nt - t'$.

Many other applications of the method will be found in Bunsen's Memoir (Ann. Ch. Pharm. lxxxvi. 265 ; Chem. Soc. Qu. J. viii. 281) ; see also Graham's *Elements of Chemistry*, second edition, vol. ii. p. 722.

Sulphite of Lime.

We have now to describe some processes for the manufacture of the sulphite and hyposulphite of lime, which it is also proposed to employ in the production of sulphurous acid. The cheapness of these salts and the facility with which the sulphurous acid can be obtained from them will probably give rise to a great extension of the use of this valuable bleaching and antiseptic acid.

Anthon believes that his plan for producing the sulphite of lime meets all the requirements of a good manufacturing operation. His salt consists of

1 eq. Lime	28 or 35·90 per cent.
1 „ Sulphurous acid . .	32 or 41·03	„
2 „ Water	18 or 23·07	„
	78	100·00

He can employ nearly the same arrangements as those hitherto used in the manufacture of bleaching powder. The gaseous sulphurous acid is conveyed into close cisterns, fitted with shelves, on which hydrate of lime in fine powder is spread to a depth of 1 or 2 inches, or into wooden cylinders revolving on their axes and containing the slaked lime. In the latter case, the absorption is quicker, in consequence of fresh surfaces being constantly presented to the gas.

The sulphurous acid may be produced by any of the well-known plans, but if obtained by the action of sulphuric or muriatic acid, it must be washed, previous to its entrance into the receiving vessels.

It is not absolutely necessary to employ fresh burnt lime, as the presence of a little carbonic acid is not injurious, otherwise than by lengthening the process. An indispensable condition, however, is the proper proportion of water; for CaO.HO absorbs only so much sulphurous acid as to produce a mixture of 1 eq. of sulphite of lime, and 1 eq. lime or of some basic sulphite of lime, the reason being, that 2 eq. of water are essential to the constitution of the sulphite of lime. It is necessary, therefore, that 18 parts of water be present with every 28 parts of lime. An excess of water must also be avoided, as its tendency is to cause the formation of a damp sulphite, which must be dried before packing and which injures the quality of this salt. The lime slaked so that this proportion of water is present, must also be passed through a fine sieve.

The saturation of the lime, of course, depends on the thickness of the layer and whether it is kept in motion as already mentioned, but in general the operation varies from 4 to 8 hours. The saturation is finished, when the evolution of heat ceases and when the pale yellow colour which the hydrate of lime acquires no longer increases in intensity. Another test consists in violently shaking a sample in a glass vessel; if it retains the

smell of sulphurous acid, the lime may be regarded as fully saturated. 100 parts of fresh burnt lime produce 275 parts of sulphite of lime.

Kuhlmann has proposed the same plan, and remarks that the
heat produced by the absorption of the gas, is sufficient to decompose the bisulphite when formed, and thus insures the absence of
this latter compound. He conveys the excess of sulphurous acid,
which escapes absorption when the gas is obtained by burning
sulphur or sulphur ores, to an ordinary chamber for conversion
into sulphuric acid. When a solution of sulphurous acid is wanted,
this salt is mixed with water in proper proportions and decomposed
by sulphuric or muriatic acid. In the one case, sulphate of lime
is formed, which remains behind, but in the other case, the
chloride of calcium dissolves, together with the sulphurous acid,
which must be taken into account in the application which is to
be made of the aqueous sulphurous acid. This salt is also suited
for the preparation of sulphurous acid, both in the solid and in
the gaseous form.

Hyposulphite of Lime.

A recent application of this salt has led M. Emile Kopp to propose the following plan for its preparation. He decomposes
chloride of antimony by hyposulphite of soda and obtains a sulphide of antimony of a most beautiful vermillion colour.

As the hyposulphite must be perfectly pure and in a state
of solution, M. Kopp's plan consists in diffusing slaked lime
in water, and boiling it with an excess of sulphur. When
the reaction is complete, the filtered solution consists chiefly
of polysulphide, mixed
with some oxysulphide
of calcium. This solu-
tion is run into large
cisterns C C', at dif-
ferent levels, which are
connected by pipes A'.
Wheels V V' are fixed
inside the cisterns, and
being kept in con-
stant motion, diffuse
the liquid in the form

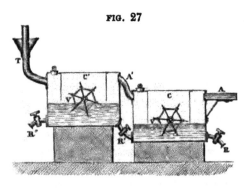

FIG. 27

of a fine rain, thus presenting a large surface to the sulphurous acid, which enters the apparatus through the pipe A. Vertical partitions interrupt the current of gas, so as to retard its progress. A reaction takes place, similar to that in the manufacture of hyposulphite soda; thus, $2CaS + 3SO^2 = S + 2 (NaO.S^2O^2)$. The liquor is occasionally tested by drawing off a sample from the lower cistern C by the cock R. As soon as a slight acid reaction is noticed, it is run off by the same cock, and the liquor from the upper cistern C' is admitted through the pipe R', while a portion of fresh solution is poured into this cistern by an opening in the upper side.

We think this is the proper place for giving an account of two recent patents for the manufacture of these and similar salts from the tank-waste of alkali-works, rather than when describing the production of soda.

The first is that of Messrs. Townsend and Walker. The patentees prefer to use soda-waste obtained in the ordinary process of lixiviating crude soda, as the black ash contains much sulphide of calcium insoluble in water; but if this waste is exposed to the atmosphere it absorbs oxygen, and a portion of it becomes soluble. The proportion of the soluble salts depends upon the extent to which the oxidation of the waste is allowed to proceed. If it has been sparingly oxidised, the solution obtained by lixiviating it, will chiefly contain soluble sulphide of calcium, and further oxidation will cause the sulphide of calcium to be converted into hyposulphite and sulphite of lime; but if the exposure to the air is continued for a sufficient length of time, all the sulphide of calcium becomes oxidised. Solutions are thus obtained at the various stages of the oxidation, containing soluble sulphide of calcium with varying quantities of hyposulphite and sulphite of lime, until the oxidation is complete, when the solution contains only the latter salts. To obtain the waste in the requisite state of oxidation for the following processes, it is exposed to the atmosphere, in a state of dampness, in heaps for some days, and the process of oxidation is promoted by occasionally turning over the heaps so as to expose fresh surfaces to the air, the waste being kept continually damp, by sprinkling it with water or with a solution of sulphide of calcium, until, by testing, it is found to contain little or no sulphide. The fully oxidised product, lixiviated with water, yields a solution of hyposulphite and sulphite of lime, the former salt predominating. If the process of oxidation is

arrested before its completion, and the product lixiviated with
water, a yellow solution is obtained, containing hyposulphite and
sulphite of lime and soluble sulphide of calcium, the presence of
the latter being indicated by the yellow colour of the solution, or
by the solution becoming black on the addition of a salt of copper or of lead. This dissolved sulphide of calcium is converted
into hyposulphite and sulphite of lime, by causing the solution to
absorb oxygen from atmospheric air, which is effected by allowing
it to percolate slowly through a tower containing pieces of coke
or other suitable material, similar to those used for condensing
muriatic acid, and admitting, at the same time, atmospheric air
through proper openings at the lower part of the tower. The
complete oxidation of the sulphide of calcium is effected by passing the solution repeatedly through such oxidising towers, but it
is better to complete the oxidation in one operation by regulating
the passage of the solution through the tower. The application
of heat to the solution, by injection of steam or by other means,
facilitates the process of oxidation. When the solution contains
much hyposulphite and sulphite of lime, and little sulphide of
calcium, the last portion of soluble sulphide of calcium is removed
by adding to it a small quantity of sulphuric, muriatic, or nitric
acid, or a solution of a soluble salt of iron, manganese, zinc, copper, or alumina. In either case, the decomposition of the soluble
sulphide is effected, and when the precipitate, which is formed, has
subsided, the supernatant liquor will be found free from sulphide.
The clear solution is run into a suitable vessel, and a solution of sulphate of soda of specific gravity about 1·18 is added, until the addition of more sulphate gives no further precipitate, when the solution
will contain hyposulphite and sulphite of soda, and a precipitate will
be formed, consisting of hydrated sulphate of lime. The precipitate
is generally found to contain more alkaline or earthy hyposulphite
and sulphite than is necessary, and is therefore washed with water
until not more than the requisite quantity remains. In ordinary
cases, it is found advisable to retain from 2 to 5 per cent. of these
salts mixed with the precipitate. This precipitate, when dried at
a temperature not exceeding 212° Fahrenheit, constitutes the
product which the patentees call "precipitated antichlore."
The product thus obtained, has been found highly valuable in the
manufacture of paper, as the hydrated sulphate of lime gives
weight and body, whilst the sulphurous compounds destroy
or neutralise any chlorine present in the paper pulp. From the

liquor and washings run off from the precipitated antichlore, hyposulphite and sulphite of soda are obtained in the form of crystals by concentrating them to a specific gravity of about 1·40; or the same salts are obtained in the form of lumps, by evaporating the solution to a specific gravity of about 1·65, when it solidifies on cooling. Sulphate of magnesia, sulphate of potash, or sulphate of ammonia, may be employed instead of sulphate of soda, in which case the products from the supernatant liquor and washings, will be respectively hyposulphite and sulphite of magnesia, hyposulphite and sulphite of potash, or hyposulphite and sulphite of ammonia.

The hyposulphite and sulphite of lime may be obtained in the solid form, by evaporating the liquor containing these salts to dryness. The residue, consisting of hyposulphite and sulphite of lime, may also be used as an antichlore, or the concentrated solution obtained before complete evaporation, will answer the same purpose.

From the insoluble residue, left after lixiviation of the oxidised soda-waste, which still contains much sulphite of lime, sulphite of soda or potash is obtained by boiling or steeping it with a solution of caustic soda or potash, or their carbonates, and then by the evaporation of the resulting solution, sulphite of soda or of potash is obtained in crystals or in lumps.

The uses to which any of the hyposulphites and sulphites mentioned, may be applied are, first, as antichlores, in the manufacture of paper, or in the treatment of cotton or other fabrics, after bleaching with chlorine or with compounds of that substance; second, as bleaching agents for woollen or other animal fabrics or straw, starch, oils, ivory, bone, and hair; and third, as purifiers and anti-ferments in the manufacture of sugar.

To obtain sulphur and certain useful sulphides from the solution of sulphide of calcium, it is mixed with chloride of manganese, or with the still liquor, resulting as a bye-product in the manufacture of chlorine, and these liquors may be treated in three ways.

Firstly.—The solutions may be mixed in such proportions, that the resulting precipitate will consist chiefly of sulphur. The proportions for this purpose will vary according to the greater or less amount of free acid in the still liquor; as a general rule, however, the proportions will be from 1 to 1½ gallons of unoxidised liquor, specific gravity 1·05, to one gallon of still liquor, specific gravity 1·20. It is known when enough of the unoxidised liquor has been added, by the precipitate beginning to assume a blackish colour. The precipitate is separated by filtering or draining, and when dry, may be used as sulphur in the manufacture of sulphuric acid, or for other purposes.

Secondly.—The solutions may be so proportioned that the resulting precipitate will consist chiefly of free sulphur and sulphide of iron. The proportions for this purpose, though liable to vary according to the amount of iron and free acid in the still liquor, will, as a general rule, be about three to four gallons of un-. oxidised liquor, specific gravity 1·05, to one gallon of still liquor, specific gravity 1·20. It is known when enough of the unoxidised liquor has been added, by the precipitate beginning to assume a buff or flesh colour ; or the same thing may be correctly ascertained by filtering a little of the mixture, and adding a few drops of unoxidised liquor to the filtrate, the separation of a flesh-coloured precipitate showing when enough of the unoxidised liquor has been added ; the formation of a black precipitate shows that more is required. The next operation consists in separating the precipitate by draining or filtering ; when dried, it may be used in the manufacture of sulphuric acid or for other purposes. By this process, the supernatant liquor is obtained free from iron and excess of acid, and consists chiefly of chloride of manganese, in a very suitable state for obtaining oxides of manganese free from iron.

Thirdly.—The solutions may be so proportioned that the precipitate will consist of free sulphur, sulphide of iron, and sulphide of manganese. The proportions will, in this case, be about nine gallons of unoxidised liquor of specific gravity 1·05, to one gallon of still liquor, specific gravity 1·20, but it is better to add the unoxidised liquor until no further precipitate is obtained. The precipitate is collected, drained upon filters, and cut into cubes of two or three inches, which are dried at a temperature not exceeding 250° Fah. The dried precipitate is then ready to be used in the manufacture of sulphuric acid, in an apparatus similar to that ordinarily employed in burning pyrites.

The processes which are next to be described, form the subject of a patent by M. Jullion, who employs two different methods of working, by which he can avail himself of the sulphur alone, which is contained in the waste, or of both the sulphur and a base, at the same time producing several secondary products which are marketable commodities. The products obtained by these means are, first, sulphide of magnesium, or hyposulphite of magnesia, from which magnesium-compounds may be prepared ; secondly, a salt of lime from which lime-compounds may be prepared ; thirdly, a salt which, on combustion, yields sulphurous acid; fourthly, precipitated sulphur, applicable to all purposes for which precipitated or sublimed sulphur is used.

Soda-waste or gas-lime is ground to fine powder, and, the percentage of sulphur having been determined, the powder is mixed with a considerable qantity of water, then an equivalent of carbonate of magnesia in the state of powdered magnesia limestone is added, and the whole is boiled for half an hour. By this process, an interchange of elements, or double decomposition, takes place, producing carbonate of lime and sulphide of magnesium. In order to separate these two compounds, the latter is rendered soluble in the water, with which it is mixed, by the addition of a second atom of sulphur, either in the free state or as a polysulphide of an earthy base, after which the lime may be separated by filtration. The liquid containing the sulphide of magnesium is saturated with sulphurous acid in order to form hyposulphite of magnesia from the carbonate or hydrate, which may be precipitated, or at once boiled down and evaporated to dryness, when it may be burnt in ordinary sulphur-kilns, so that the excess of sulphur may form sulphurous acid, and the residue may be employed to make Epsom salts, &c.

In the second process, the waste is oxidised to a state known to soda-makers as "yellow liquor." This is effected by spreading the waste on an inclined surface, and exposing it fully to the action of air and moisture by repeatedly turning it over. Both oxygen and moisture will be rapidly absorbed, and the greater part of the sulphur will run off as a yellow fluid, which is a solution of a complex "sulphuretted hyposulphite of lime." This is proved by the addition of a strong acid, such as hydrochloric acid, giving off sulphurous acid and sulphuretted hydrogen gases, at the same time that sulphur is precipitated in an oily or soft state.

In order to bring this compound into use, where a base is to be obtained, or free sulphur is required, it is in the first place saturated with sulphurous acid, which causes the precipitation of sulphur in an extremely minute state of division, well adapted to pharmaceutical purposes, or as a substitute for the sublimed sulphur of commerce. This sulphur is separated by filtration, and washed with clean water, and dried in a stove. The filtered liquid, which is a solution of hyposulphite of lime, is decomposed by boiling with powdered magnesian limestone; this produces carbonate of lime and hyposulphite of magnesia. The latter, by the ordinary process, yields both carbonate and calcined magnesia, and, by oxidation, Epsom salts or sulphate of magnesia.

When a lime-salt is wanted, the solution of hyposulphite of lime is filtered, and the lime precipitated as carbonate (such as is sold under the name of precipitated chalk) by a clear solution of carbonate of soda. Or the solution of hyposulphite of lime, being placed in a close vessel, is decomposed with an equivalent of hydrochloric acid, the sulphur being thereby precipitated and sulphurous acid given off, and the lime is precipitated with sulphuric acid in the form of sulphate, such as is sold to paper makers under the name of *pearl-hardening*. But when the sulphur only is to be recovered from the waste, it is better to convert it into "yellow liquid," which is boiled down in a common evaporating furnace, and when it is in a very concentrated state, about three times as much small dust-pyrites, as the salt from the yellow liquor would weigh if dried alone, is added, and when the whole has attained a pasty consistence, it is well stirred together, then raked out of the furnace and dried in a stove, after which the cake is ready for burning in kilns, as practised with ordinary pyrites.

SULPHURIC ACID.

This acid does not often occur in nature in the uncombined state. It is sometimes met with in the waters of coal mines, having been formed by the oxidation of the pyrites in the coal. We have known cases where it has destroyed the metal of the pumps, the leather of the clacks, and charred the wood of the open rods. It also exists in the waters of a torrent called the Rio Vinagre, which rises in the volcano of Purcé, in the Andes. Boussingault has calculated that this stream carries down, in the course of twelve months, about 15,000 tons of sulphuric acid, and 12,000 tons of muriatic acid. M. Degenhardt has found a source of these acids, even more abundant, in a spring in the Páramo of Ruiz, an active volcano of New Granada.

The water of these two streams has been analysed with the following results:

	Rio Vinagre. Boussingault.	Acid Spring, Lewy.
Sulphuric acid . . .	0·00110	0·005181
Hydrochloric acid . .	0·00091	0·000851
Alumina	0·00040	0·000500
Lime	0·00012	0·000140
Magnesia	traces	0·000320
Oxide of iron	traces	0·000365
Soda.	0·00023	0·000360
Silica	...	0·000183
	0·00276	0·007900

A stream containing this free acid flows from the crater of an extinct volcano, Mount Indian, in Java. Robert has also found free sulphuric acid in the hot springs which rise in the sulphur mines of Krisank, in Iceland. Caton states that this acid exists in considerable quantities in the free state in springs near Byron, in Tennessee.

Sulphuric acid was not known to the ancients, although they were well acquainted with sulphur, which was sometimes used in religious ceremonies. Vitriol, or sulphate of iron, is mentioned by Pliny, but he evidently knew nothing of its chemical nature.

The first mention of sulphuric acid occurs in the writings of Basil Valentine, a celebrated chemist of Erfurt. It was also known to Paracelsus, the Swiss alchemist, who died in 1541; but the first person who gave a tolerably correct account of it was Gerard Dorneas, in 1590.

When this acid was first prepared, green vitriol, or sulphate of iron, was the source whence it was procured, by distillation, and hence its name, *oil of vitriol*, by which it is even now called in trade.

The English makers were in the habit of placing 50 or more earthen vessels, called *long-necks*, in a reverberatory furnace, and distilling the oil of vitriol from the green copperas, collecting the acid in a glass receiver luted to each retort. The great cost of this plan, induced the manufacturers to substitute the use of sulphur, which was burnt along with nitre, in large glass globes, the product obtained, being afterwards concentrated by boiling it in retorts.

The plan of making this acid from sulphur is of English invention, and was first employed by Dr. Ward, who called it "oil of vitriol made by the bell ;" the acid thus prepared was sold at 1s. 6d. to 2s. 6d. per pound. Large glass vessels with wide necks, holding 40 to 50 gallons, were placed in two rows in beds of sand, and a few pounds of water poured into each vessel. A stoneware pot was introduced into the neck, and a red-hot iron ladle placed on it, into which a mixture of sulphur and nitre was thrown;—the neck being closed by a wooden stopper, the gases formed by the combustion of the sulphur and nitre were allowed to condense, and this operation was repeated from time to time.

The next great improvement was the introduction of leaden vessels, by Dr. Roebuck, of Birmingham, in 1786, who, with Mr. Garbett, established large works, in 1749, at Preston Pans, for making acid for bleaching linen. This was followed by the erection of similar works at Bridgenorth, and the third establishment was built at Dowles, in Worcestershire, by a Mr. Skey, who increased the size of the leaden chambers from 6 to 10 square feet, and whose works were in full operation as late as 1820.

The first manufactory for making this acid in London, was erected in 1772, by Kingscote and Walker, and consisted of 71 circular leaden vessels, 6 feet in diameter and 6 feet high. From that time, the production of this acid has gone on increasing in extent, and has been more widely spread over all parts of the country within the last few years, in consequence of its application for making manures.

Concentrated sulphuric acid, or the oil of vitriol of commerce, is composed of—

1 equivalent of anhydrous sulphuric
  acid (SO^3) . . . . 40, or 81·63 per cent.,
1 equivalent of water (HO) . . . 9, or 18·37 „

_____

100·00,

and has a specific gravity of 1·845 at 60° F. It is a heavy colourless oily liquid, boiling at 610° F., and may be distilled like water. It is one of the most powerful of the acids, and highly corrosive. It dissolves sulphur, acquiring a brown tinge, while charcoal communicates to it a reddish colour. The concentrated acid rapidly absorbs moisture from the air to the extent of 10 times its weight, increasing in bulk at the same time. When it is mixed with water, there is a great development of heat, the temperature of the mixture rising nearly 300°.

There are several hydrates of this acid, the lowest being that of the Nordhausen or fuming acid, which is composed of two equivalents of the anhydrous acid and one of water.

The next is the commercial oil of vitriol, containing one equivalent of each constituent.

The third, containing two equivalents of water to one of acid, has a specific gravity of 1·78, and is obtained by concentrating chamber acid at a temperature not exceeding 400° F.; when the heat rises above this point, the second equivalent of water is driven off. This hydrate freezes in the cold, forming large crystals, which remain solid until the temperature rises to 45° F.; this property of the hydrate often occasions accidents, by breaking the carboys in which weak acid is kept.

The greatest condensation takes place when the acid and water are mixed in the proportions necessary to form the fourth hydrate or acid having a specific gravity of 1·632, composed of one of acid to three of water. These hydrates are represented by the formulæ—

Nordhausen or Fuming acid. . . .	$2SO^3.HO$
Acid of 1·845 specific gravity . . .	$SO^3.HO$
,, 1·280 ,, ...	$SO^3.2HO$
,, 1·632 ,, ...	$SO .3HO$

Payen gives the following table of the temperatures at which sulphuric acid, in different states, either melts or remains liquid:

	Melting point.
Anhydrous, SO^3, first modification	+64·4° F.
,, second ,, ...	+212·0
With one equivalent of water, $SO^3.HO$	+50·9
The same with a very small quantity of anhydrous acid	−4·0
The same with a very small quantity of water . . .	−22·0
Sulphuric acid, $2 SO^3.HO$	+95·0
,, $SO^3.2HO$	+47·3
$SO^3 + 3, 4, 5,$ or $6HO$, remains liquid at −4·0	
and in vacuo at	−40·0

The two following tables will be found of some use to manufacturers.

TABLE I.

Variations in density of Concentrated Sulphuric acid, through change of temperature.

Temperature, Fahrenheit.	Specific gravity.	Temperature, Fahrenheit.	Specific gravity.
30°	1·8593	52°	1·8511
32	1·8563	56	1·8500
36	1·8546	60	1·8468
38	1·8532	63	1·8449
40	1·8527	68	1·8435
42	1·8520	70	1·8430
44	1·8522	74	1·8413
46	1·8519	80	1·8381
48	1·8517	84	1·8343

TABLE II.

Boiling points of Sulphuric acid at different strengths.

Specific gravity.	Dry acid, per cent.	Boiling point.	Specific gravity.	Dry acid, per cent.	Boiling point.
1·850	81	620°	1·769	67°	422
1·849	80	605	1·757	66	410
1·848	79	590	1·744	65	400
1·847	78	575	1·730	64	391
1·845	77	560	1·715	63	382
1·842	76	545	1·699	62	374
1·838	75	530	1·684	61	367
1·833	74	515	1·670	60	360
1·827	73	501	1·650	58—6	350
1·819	72	487	1·520	50	290
1·810	71	473	1·408	40	260
1·801	70	460	1·300	30	240
1·791	69	447	1·200	20	224
1·780	68	435	1·100	10	218

The theory of the process of making sulphuric acid by the English plan is explained in the following paragraph :—

1 part by weight of sulphur, when converted into sulphuric acid, takes up 1·5 parts of oxygen = 3 equivalents.*　When it is burnt in the air, which is easily effected at a temperature of 300° C. (572° F.), it combines with only 1 part or two equivalents of

* $O = 8 : S = 16.$

oxygen; the compound formed, sulphurous acid (SO^2), must therefore combine with another equivalent, $= 0.5$ parts, in order to become sulphuric acid. Sulphurous acid possesses the property, in contact with moist air, of gradually combining with this additional portion of oxygen, and the combination may be effected by the agency of platinum sponge, which then exerts the same power as in the platinum lamp; but this property of platinum sponge has not yet been turned to account on a large scale, partly because of the tediousness of the process, and partly from the cost of the platinum.

The oxidation may, however, be cheaply and rapidly effected by means of nitric acid, which, when brought in contact with sulphurous acid and vapour of water, converts the sulphurous acid into sulphuric acid, and is itself reduced to binoxide of nitrogen, which compound then takes oxygen from the air, and is converted into peroxide of nitrogen, and this last compound is decomposed by the water, reproducing nitric acid and binoxide of nitrogen, which then again takes oxygen from the air; and thus the process goes on continuously, with a comparatively small proportion of nitric acid, as long as sulphurous acid, vapour of water, and atmospheric air are supplied.

The reaction may be supposed to take place by the following stages:—

1. Sulphurous acid and nitric acid yield sulphuric acid and peroxide of nitrogen:

$$SO^2 + NO^4.HO = SO^3.HO + NO^4.$$

2. Peroxide of nitrogen and water yield nitric acid and binoxide of nitrogen:

$$3NO^4 + 2HO = 2(NO^5.HO) + NO^2,$$

the nitric acid being then ready to act on another portion of sulphurous acid, while the binoxide of nitrogen takes up two atoms of oxygen, and reproduces the peroxide:

$$NO^2 + O^2 = NO^4,$$

which is then decomposed by water as before.

Or, 3. The peroxide of nitrogen may be supposed to act on the sulphurous acid in presence of water, so as to form sulphuric and nitrous acids:

$$SO^2 + NO^4 + 2HO = SO^3.HO + NO^3.HO,$$

and the nitrous acid may then be resolved into nitric acid, water, and binoxide of nitrogen,

$$3 (NO^3.HO) = NO^4.HO + 2NO^2 + 2HO,$$

but the former view is perhaps the more probable (*vide* Péligot, Ann. Ch. Phys. [3], xii. 263).

In either case, it appears, that the nitric acid is ultimately reduced to binoxide of nitrogen, which acts as a carrier of oxygen from the air to the sulphurous acid.

There are, then, four agents engaged in the production of sulphuric acid; viz.,

Sulphur, yielding sulphurous acid.

A nitrate, yielding nitric acid and its derivatives.

The oxygen of the air.

Vapour of water.

The reactions above described require the use of chambers of very large dimensions:—1, because the gaseous sulphurous acid and the binoxide of nitrogen occupy a very large volume in proportion to their weight; 2, because the air which supplies the oxygen contains only 21 per cent. of its volume of that element; 3, because, in order that the reactions may be completed, the gases must remain in contact for a considerable time.

The most remarkable feature of the process is the small quantity of nitric acid and its derivatives required in proportion to the weight of sulphur burned. It is evident, indeed, from the preceding equations, that if no loss occurred, the quantity of nitric acid, once introduced into the circulation, ought to suffice for the oxidation of an unlimited quantity of sulphurous acid. This, however, is not actually the case, partly because, in operations on the large scale, it is impossible to avoid accidental loss; partly because it is necessary to provide for the continuous discharge of the nitrogen separated from the air, and this gas, in escaping, always carries away with it a certain quantity of the oxides of nitrogen.

A very large quantity of aqueous vapour is required in the process, partly to act on the oxide of nitrogen in the manner already described, partly to dissolve the sulphuric acid produced, and carry it down to the floor of the chamber.

If the quantity of water is deficient, another reaction takes place, attended with the formation of a white flaky crystalline body, consisting of the so-called *nitro-sulphuric acid*, S^2NO^9, or $NO^3.2SO^3$. Its formation is due to the action of peroxide of nitrogen on sulphurous acid, as shown by the equation,

$$2SO^2 + 2NO^4 = S^2NO^9 + NO^3.$$

It is instantly decomposed by contact with water, yielding sulphuric acid, nitric acid, and binoxide of nitrogen ; thus—

$$3S^2NO^5 + 7HO = 6(SO^3.HO) + NO^5.HO + 2NO^5;$$

that is to say, the same products as would have been formed without its intervention, if sufficient water had been originally present to give rise to the reactions previously explained.

It appears then most probable, that the formation of this crystalline compound is a mere accident of the process, arising from a deficiency in the supply of water ; indeed, when sufficient water is present, it cannot exist for an instant, being decomposed as soon as formed. This may be very well shown by connecting a large glass globe with three flasks, one evolving sulphurous acid, the second binoxide of nitrogen, and the third containing water. If the two gases be admitted into the flask without any steam, the crystalline compound forms all over the sides of the globe as an opaque crust ; but as soon as steam is admitted by boiling the water in the third flask, the whole of the compound disappears, and a solution of sulphuric acid, containing nitric acid, collects at the bottom of the globe, and so long as the supply of steam is kept up, the production of sulphuric acid goes on without a trace of the crystalline body being formed. Nevertheless, it has been maintained by several chemists, especially by Rose, Prevostaye, Gay-Lussac, and Gaulthier, that the formation of this substance is, if not an essential, at least an important condition in the production of sulphuric acid ; but as its formation may always be prevented by admitting sufficient water, and the ultimate products are the same, whether it is formed or not, it is difficult to conceive how this can be the case.

Manufacturing Operations.

We have already briefly noticed the first attempts to produce sulphuric acid on the manufacturing scale, and we may now allude to a succession of improvements : the continuous passage of the sulphurous and nitrous acids through the chamber ; the introduction of steam by Kestner ; the substitution of pyrites ores for sulphur by Hill ; the recovery of the nitrous acid by Gay-Lussac ; the concentration of the weak acid in platinum vessels ; the increasing size of the chambers ; and, we believe we may add, H. Blair's improved sulphur burner,—which have developed this manufacture to such an extent, that in the Lancashire district alone, about 700 tons of oil of vitriol are produced weekly, independent of the consumption of this acid in the soda trade.

The apparatus at present employed in the manufacture of sulphuric acid, will be most advantageously described in the following order—

1. The burners or furnaces for the combustion of the sulphur and sulphur ores.

2. The supply of the nitrous compounds.

3. The chambers and the mode of working them in this country and on the continent.

4. The recovery of the nitrous acid.

5. The concentration of the weak acid.

6. The manufacture of the fuming sulphuric acid.

1. *Production of the Sulphurous Acid from Sulphur or Sulphur Ores.*

The Sulphur Burner.—The oven or *burner* in which the combustion of the sulphur is effected, consists of a sole or bottom made of iron, termed the *burner-plate*, about 8 feet long by 4 feet wide. Three sides of this plate are turned up about one or two inches, and the front end has a slope forwards to facilitate the removal of the ashes left on the burning of the sulphur. The sides and arch of the burner are formed of brickwork. The door for charging the sulphur and withdrawing the ashes is an iron plate, which slides in a groove in a framework of iron, and is counterbalanced by a weight, which is suspended by a chain passing over a pulley in an upright pillar fixed on the top of the burner. An air channel or flue is built under the burner-plate to keep it cool, and prevent any sublimation of the sulphur. In the ordinary mode of working, this is a very important point, but, as will be shown, this very difficulty is one of the striking features of Mr. Blair's valuable improvement. The air is either admitted in a thin film by keeping the door slightly raised above the burner-plate, or through a slit in the door itself, which can be opened or shut at pleasure by a slide.

Figures 28 and 29, exhibit a range of these sulphur burners in section and front view. *a a* are the burners with their doors, *b b*. The front of the burners is formed of metal castings, *d d*, well bound and rivetted together, and through the openings, *c c*, the sulphur is charged; *e f g* represent the pillar, chain, and balance weight, for working the doors; *h* is the chimney or stalk, built of brickwork, carefully plastered to prevent any admission of air, and placed at the far end of the burners, facing the chambers; *i i* are the air flues for cooling the burner-plates, which are connected with a main flue, *k*, in connection with some chimney to create the necessary draught. The sulphurous acid from the burners passes into a main flue in the burner-stalk, and finally enters the chamber through iron pipes, *l l*, lined inside with fire-bricks.

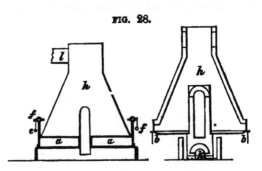

FIG. 28.

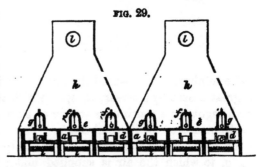

FIG. 29.

The size of the burner-plate of course depends on the quantity of sulphur to be burnt in a certain time, and four burners, with plates of the dimensions already given, will work off 20 cwts. of sulphur per day of 24 hours.

The size of the charges, and the mode of working, vary in almost every manufactory; the usual plan is to charge the burners in rotation, so that the quantity of gas entering the chambers shall always be the same, the weight of the charge varying from 56 to 112 lbs. One of the great drawbacks of this form of burner and mode of working, is the admission of large quantities of air at charging hours, which diminishes the productive power of the chambers, lessens the produce of acid from a given weight of sulphur, and necessitates a larger consumption of nitrous compounds than would otherwise be required.

Petrie has patented a plan for burning sulphur, by which the escape of sulphurous acid from the door of the burner is avoided, and the combustion of the sulphur rendered independent of the draught of the chamber. The burner, *a* (Fig. 30), is open at the

back, and fitted with an inclined grate, *b*, whose horizontal bars
are so arranged, that while drops of melted sulphur can fall into
the pan, no solid sulphur can pass through.

FIG. 30.

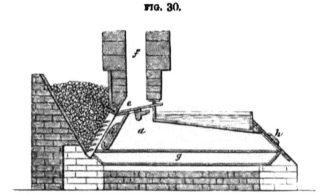

The sulphur, *c*, lies behind the grate, and it is protected from
too great a heat by a screen, *d*, upon which the slide, *e*, is placed,
which regulates a continuous supply of sulphur in proportion to
the air, measured for its combustion. The sulphurous acid passes
to the chamber by the stalk, *f*. The sulphur is burnt in the
pan, *g*, to which the air is supplied under the sliding door, *h*,
The burning sulphur is chiefly at the farther end, and the ashes
left are withdrawn through the door, *h*. An air flue under this
pan keeps the bottom plate cool.

Petrie has also devised a kiln for burning materials containing

FIG. 31.

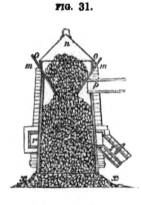

only a small proportion of sulphur,
such as sulphur ashes, &c. It is made
of metal plates surrounded by brickwork, or it may be built entirely of fire-
bricks. It is filled by the funnel, *m*,
which is closed with a cover, *n*, and the
edges are filled up with sand, *o*. It is
fed with heated air at *s*, which passes
round the kiln in the flue, *r*, and
enters by an opening at *q*. The
exhausted ore forms the sloping mass
at *x x*, which is removed from time to
time, and admits a small supply of
air.

By the kindness of Mr. Harrison Blair, we are fortunate in being able to give a description of his valuable sulphur burner, of which Fig. 32 gives a perspective view, plan, and section.

The difficulties attending the ordinary methods of burning brimstone in the manufacture of sulphuric acid, either in brick or in cast-iron furnaces, consist chiefly in obtaining the proper admission of atmospheric air to provide for the combustion of the sulphur, and its subsequent conversion into sulphuric acid, without, on the one hand, producing so high a temperature as to cause sublimation, and, on the other, not having the heat requisite to burn the whole of the sulphur from its impurities; whilst in both instances, in consequence of the frequent interruptions in the operation caused by the charges being necessarily small and quickly burned off, a larger quantity of atmospheric air is admitted into the chambers than is requisite, and by diluting the gases, retards the formation of sulphuric acid.

To remedy these defects, a furnace has been planned and erected by Harrison Blair, of Kearsley, in Lancashire, in which the sulphur is partly burned, and partly sublimed, in one compartment, and the combustion completed in another, by the addition of a further supply of atmospheric air. By this arrangement, although both compartments are at a red heat when the furnace is in full action, sublimation is next to impossible, and at the same time, the operation of burning is so nearly continuous, that it is suspended only once in twenty-four hours, for the purpose of withdrawing the residue.

In the accompanying ground plan and section, A is the bell of the brimstone furnace, which is slightly dished and sloping to within two feet of the door, where there is a raised floor for the purpose of permitting the residue, which is drawn from the floor, to remain until the sulphur is completely burned out, when it is pulled out and replaced by that which, at the same time, is drawn from the floor, this being done once every twenty-four hours. The door, B, is a loose iron plate placed in an iron frame, a little out of the vertical, so as to make an almost airtight arrangement, and easily removable when required. This door is perforated with a number of holes, which can be closed by a slide, so as to regulate the admission of air into the furnace. The brimstone is fed through a hopper, C, from which an iron pipe, 7 inches in diameter, is continued to within six inches of the floor of the furnace, this pipe being protected by one of larger diameter.

A fire-clay damper, D, regulates the passage from the brimstone furnace to the combustion oven, by opening or closing which, a larger or smaller quantity of brimstone may be burned in a given time.

FIG. 32.

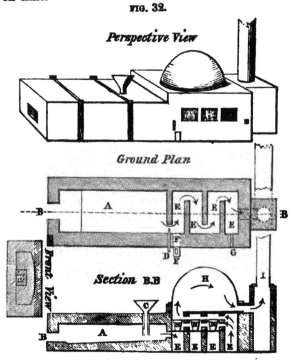

Perspective View

Ground Plan

Section B.B

The sulphurous acid mixed with the vapour of sulphur, passing from the brimstone furnace to the combustion oven, E E E E, is met by a current of atmospheric air, admitted by the aperture, F, which is furnished with a damper of 3 inches by 8, to regulate the quantity required for perfect combustion of the vapours of sulphur; this is ascertained by taking out the small stopper at G, when no flame should appear by the admission of air through the aperture.

The roof of the combustion oven is of tiles, which also form the floor of the nitre oven, supported on dwarf walls, between which the gases traverse, in the direction of the arrows, into the nitre oven, where reticulated brick walls support the tiles forming its roof, and permit the passage of the heated gases over the nitre dishes, N N N. These dishes are renewed every two hours in alternate sets, iron doors corresponding with each compartment providing for their removal, so that a set remains six hours in the oven.

The gases now mixed with those from the decomposition of the nitre, pass under a cast-iron dome, H, for the purpose of being deprived of a portion of their heat, and thence by a cast-iron chimney, I, 24 feet in height, to a small cooling chamber 6 feet wide, 18 inches high, and 18 feet long, the roof and floor of which are covered with water, which communicates with the sulphuric acid chamber.

Steam has recently been admitted along with the air into the combustion oven, and is supposed to cause a more rapid formation of sulphuric acid.

In a furnace of the dimensions of 2 feet by $4\frac{1}{2}$ feet, 26 tons of brimstone have been satisfactorily burned in a week, and the same furnace, by admitting less atmospheric air, will burn only 5 or 6 tons in the same space of time; it has also been found, that a much larger quantity of brimstone may be safely burned in the same space of chambers, than before this plan of furnace was adopted.

Burners and Furnaces for Sulphur Ores.

These ores, of which we have given the analysis at page 14, are all compounds of two equivalents of sulphur with one equivalent of iron, $Fe S^2$, and the first effect of the heat, with free access of air, is to convert one equivalent of the sulphur into sulphurous acid, leaving the protosulphide of iron, FeS, which, however, by longer exposure to the same conditions, parts with its remaining sulphur, leaving an impure oxide of iron, containing more or less sulphur, and some basic sulphate of iron, $Fe^2O^3.SO^3$.

Dr. Manson gets the credit of having first suggested the use of pyrites, but it is said that Hill, of Deptford, used this ore as long ago as 1818.

FIG. 33.

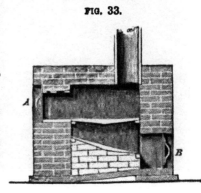

FIG. 34.

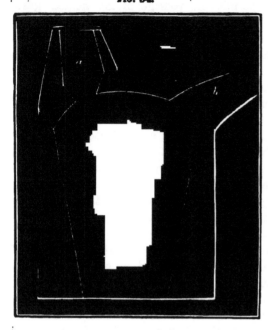

FIG. 35.

We give a form of pyrites burner first introduced by Mr. Farmer, which was very extensively used at one time. A (Fig. 33) is the door for introducing the ore, a the principal grate; whatever falls unburnt through this, collects in the second grate, b, and is removed afterwards through B. The draught is regulated by B and A, whilst the sulphurous acid is conducted by x to the chamber.

Another plan of furnace for using the pyrites, consists of a square furnace, strongly bound round with iron, somewhat similar to Figures 34 and 35, about 10 feet high, and 4 feet wide inside at the top, and 3 feet at the bottom. The opposite sides of the furnace are similarly constructed. Thus, in Fig. 34 the opening a is only temporary, to allow the gases of the coke or fuel, which is burnt in the furnace at first to bring it up to a sufficient heat, to pass off into the atmosphere; the other opening, *b*, is a flue, about 12 inches square, for the passage of the sulphurous acid to the chamber. In this flue the ordinary nitre cups are placed. Figure 35 represents the two other sides, where *a a a* show the iron straps for strength-

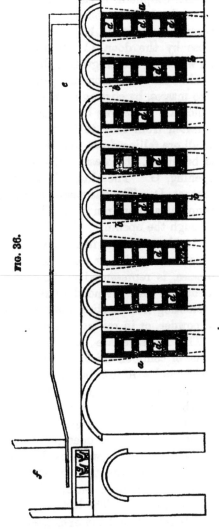

FIG. 36.

ening the furnaces; *b b* are doors for charging the furnace with the pyrites, broken into small pieces; the doors lower down, are intended to allow the workmen to keep the materials free and as open as possible, for the passage of the air and gas; *cc* are the openings through which the residue is drawn out.

A modification of this plan is shown in Figure 36, where a range, *a a*, of burners is represented. The dotted lines show the shape of the interior, *b b*, of the kiln, while an iron casting, *c c*, is built into the front face. There are several openings, *d d d*, left in this casting, to en-able the workman to stir the burning ore from time to time. The gases, from the different kilns, all pass into the horizontal flue, *e e*, and finally into the upright stalk, *f f*, where the position of the nitre cups, *g g*, is indicated.

The great variety of sulphur ores, now furnished by commerce, from Ireland, Cornwall, Belgium, Spain, Portugal, Italy, and

Norway, differing far more from each other, by peculiarities as
regards their fusing point, the rate at which they burn, and the
small they make, than by the difference in their percentage of
sulphur, has led to corresponding attempts to construct kilns,
suited to meet these peculiarities.

One of the most successful of these new burners, as improved
by Mr. Clapham, is shown in Figure 37, where AA represents
a single kiln or burner. 3½ to 4 cwts. of sulphur ore are
thrown in, every 12 hours through the opening, B, which is made
close by the door, C. In this door there is a small slit, D,
for watching the progress of the combustion, and for the admission
of air. As the combustion progresses, the ore is kept from fusing
into masses, by stirring with a poker through the openings, EE,
which are kept close by the slides, FF. The loose grates, H, are
agitated from time to time through the doors, KK, the dust or
small ore falling down into the ashpit, I ; and when the sulphur
is burnt off, the waste exhausted ore is removed through the
opening, G. The slits, LL, are used for the same purpose as
that at D. The gas passes along the horizontal flue, O, which
terminates in an iron pipe. The nitre pots are placed in this flue
through the openings, M, which are kept closed by the slides, N.

Another form of furnace, intended especially for burning ores
which melt easily, or which contain much organic matter, as coal
brasses, is represented in the next woodcut. The same arrangements are also intended to render the waste heat available, in
promoting and assisting the combustion of sulphur ores in a state
of powder, or in raising steam, &c.

The furnaces are built with a fire-brick lining inside, with
stone or common brick on the outside, cased with metal plates,
and bound with iron binders.

The furnace A (Figs. 38, 39) is charged with the lump sulphur-
ore through the door B, and as the sulphur is burnt off, the ore is
pushed forward on the grates, a fresh charge being thrown in at
the same time. This operation is repeated from time to time, the
ore being moved forward by the workmen through the side doors,
CCC, until, by the time it reaches the other end, all or nearly
all of the sulphur is burnt off, when it is pushed over the grates
into the kiln, D. The last portions of the available sulphur are
here burnt off, and the exhausted ore is withdrawn through the
door E (Fig. 38). The small sulphur-ore is burnt in the furnace F, which is similarly constructed, but with a solid bottom or

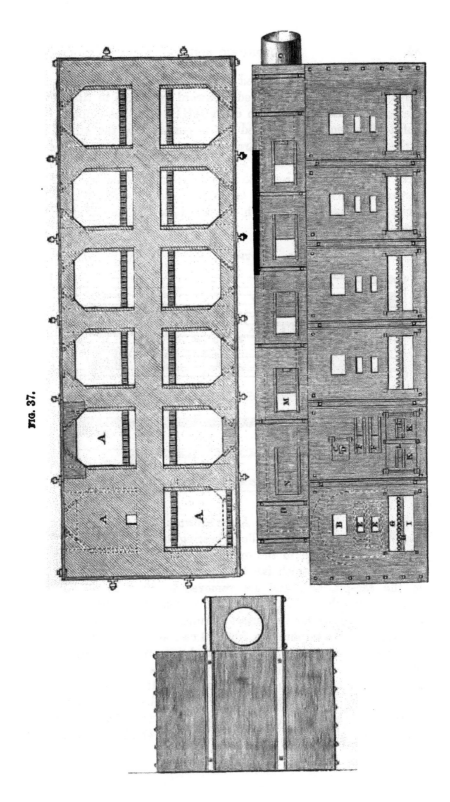

FIG. 37.

sole. The small ore is charged at the farther end of this furnace
through the door G, and moved backward from time to time, by
means of rakes through the doors HHH, until the sulphur is
burnt off, when the exhausted ore is pushed into the kiln, D,
through an opening in its arch, which is closed by a damper
during the working process. The heated sulphurous acid passes
from the furnace A, through the flue K, over the small ore in the
furnace F, whose combustion is thereby assisted, and the united
gases then pass away through the flue L, to the sulphuric acid
chamber.

FIG. 38.

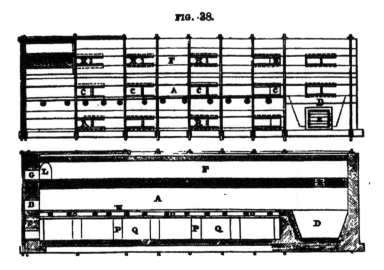

The grates used in furnace A are shown at N; they lie across
the furnace, resting on hollow bearing bars or pipes, OOO,
which are supported in the middle by a brick wall, PP. The
dust and small ore which fall through these grates, are collected
in the ash-pit, QQ, and are removed as often as necessary
through the ash-pit doors, RR. The weight of the charge, and
the time required for its perfect combustion, vary according to
the nature of the sulphur ore, from 10 cwt. every 4 hours.

The arrangement for raising steam by means of the hitherto
waste heat evolved in burning sulphur ores is shown in Figure
39. AA are ordinary tubular boilers, about 12 feet long, fitted
with grates through the whole length of the tube. The charging
doors are marked BB, and the ash-pit door C. At the other
end of the boiler, a kiln, D (Figures 40 and 41), is constructed,

similar to that already described ; a flue, E, connects the various boilers, and conveys the sulphurous acid to the chamber ; it may also be employed for placing the nitre cups through the doors FF. That portion of the boiler tube, GG, exposed to contact with the burning sulphur ore, is lined with half-thicks of fire-clay. The sulphur-ore is thrown on the grates, through the charging doors, BB, and is pushed forward, as the combustion of the sulphur progresses to the other end, where it is drawn into the kiln D, through the door I, and at last removed at K.

FIG. 39.   FIG. 40.

FIG. 41.

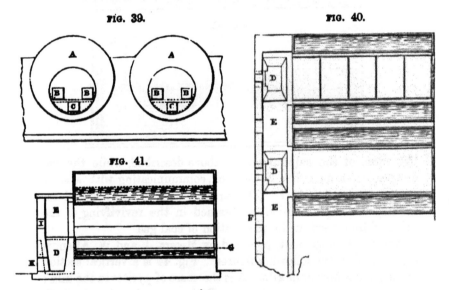

Some of the sulphur-ores are very soft and easily disintegrate by exposure to the air. The small ore, in powder more or less fine, is of course not adapted for combustion in any of these kilns or furnaces, except in the upper flat of the furnaces just described. Some manufacturers make this small ore into balls or bricks with clay, dry, and then burn them in the usual way. This plan is attended with great expense and a loss of sulphur.

Mr. Imeary has employed, with success, a furnace similar to the one last described, but with a bottom made of fire-brick flags, and heated from below by means of flues running from end to end, the fires being placed in the front end of the furnace.

Mr. Cowen has also patented a plan on a similar principle, for burning off the sulphur from valuable ores which have been crushed for washing. His arrangement consists in fixing a bench of clay retorts, *aaa* (fig. 42), on flues, *b*, heated by one or more fires, *c*, the one end of the retort being closed by doors, *ddd*,

FIG. 42.

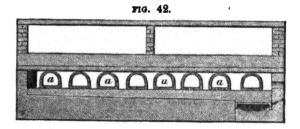

FIG. 43.

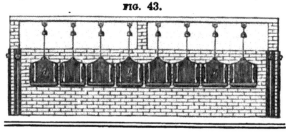

like those of the sulphur burner above described, while the gas escapes into a main flue at the back, communicating with a stalk leading to the chambers. This arrangement is well adapted for burning the sulphur which is obtained in the revivifying of the oxide of iron, used in the purification of coal gas.

Another arrangement, patented by Messrs. J. and W. Allen, is represented in the following figures. Fig. 44 is a vertical section of the pyrites burner or kiln. A is the mass of pyrites in the kiln;

FIG. 44.

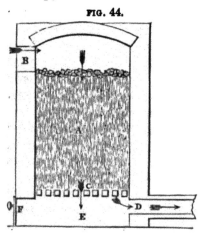

B, the open charging door; C, the grate bars on which the mass of pyrites rests; E, the ash-pit; F, a door for taking out the burnt residuum from the ash-pit; D is a flue lead-ing from the part below the bars or ash-pit, to the ordinary leaden chamber.

When first set in operation, the kiln is disconnected from the chamber and then lighted and filled in the ordinary way, being open to the natural upward draught of air; but when properly filled with burning pyrites, the flue D is connected with the chamber, and the draught is reversed by closing the door F, and opening the door B; the current of atmospheric air now enters the upper part of the kiln through the open charging door B, and, with the sulphurous acid formed at the top of the kiln, passes down through the mass of burning pyrites and red-hot burnt pyrites or peroxide of iron, by which the sulphurous acid is converted into sulphuric acid, and passes through the grate bars C, and away to the leaden chamber by the flue D; the door F is kept shut except when withdrawing the residuum. The grate bars C may be dispensed with by allowing the mass to rest on the floor of the ash-pit, but the patentees prefer using grate bars, as they secure a more equally diffused draught. Instead of being conveyed to a leaden chamber, the sulphuric acid may be pumped, or drawn by a chimney from the flue D to any suitable condensing apparatus.

French Furnaces for Sulphur Ores.

We add (Fig. 45) a wood-cut of the improvements introduced at Chessy, in France, for burning these sulphur ores. There are 16 furnaces arranged around a chimney, H, which terminates in a metal pipe, I I, to which an iron elbow-pipe is added, which is connected with the pipe K of the chambers. The furnaces are built in groups of four, all connected by the transverse flue C, which leads to the upright chimney H, through the ascending flue D.

Each furnace is furnished with a damper, a, for regulating the draught.

The central chimney is constructed with a circular flue, into which the gas from all the furnaces enters, so as to pass through a fire-brick flag pierced with holes and covered with coke, with the object of arresting all the particles of dust, which are carried along with the gases.

The ore is thrown into the furnaces through an opening, near the top, in the front of the furnaces, and the burnt ore is removed, once or twice in the twelve hours, by the openings b at the bottom.

In other parts of France, the process of burning the sulphur ores depends upon the mechanical condition of the mineral, and three plans are adopted for the purpose: 1, for small ores; 2, for lump ores; and 3, for the two descriptions combined.

FIG. 45.

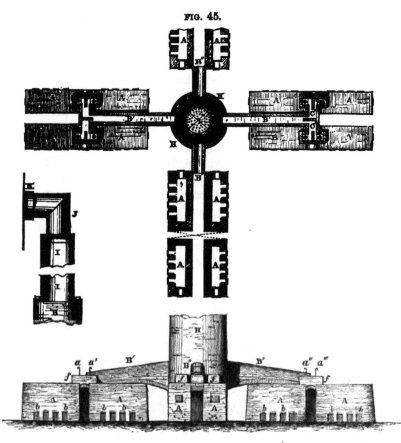

1. *Ores in powder* are burnt in ovens very similar to those where sulphur is used. The bottom plate of the burner is heated by a fire underneath, and the ore is spread over the surface of the plate to a depth of from 2 to 4 inches, and occasionally stirred to present a fresh surface to the current of air passing over it. After six hours, it is raked out into a recess below, where the remaining sulphur is burnt off.

2. *Ores in Lump.*—The ore is broken down to the size of nuts and washed to remove any adhering gangue. The ovens (Fig. 46) are built of brick and arched, about 4 feet 8 inches high by 4 feet 8 inches × 2 feet 4 inches, with a grate about one-third from the bottom. The sides of the oven are first raised to a red heat by burning coal or coke in the usual way, and then the ore is thrown in at O, which is closed by a cover, M. The sulphurous acid

passes to the chambers through a pipe fitted in at D.   When the
charge is nearly burnt off, it is allowed to
fall through the grates, with the aid of a
poker, into B, where it meets the supply of
fresh air, and where the combustion is
completed.

FIG. 46.

At Marseilles, this plan is modified by
the following arrangements : The furnace
is composed of two ovens, A (Fig. 47), provided with iron grates.' Between the ovens,
a basin of stone, B, is built, in which the
nitrate of soda is decomposed.   The air
is supplied below the grates, and the gases
from the pyrites and nitre, mix in a flue at the back of the furnace, whence they pass on to the chambers.

FIG. 47.

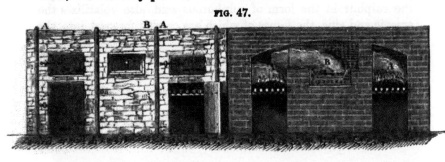

The pyrites in pieces of the size of an apple, is thrown on the
grates in charges of about 3 cwt., which are burnt off in 3 to 4
hours, when the residual roasted ore is withdrawn, though it still
contains from 10 to 15 per cent. of sulphur.   The workman has
to judge, from the appearance of the burnt pieces, which of them
contain too much sulphur to permit of their being thrown away,
and such pieces are mixed with the next charge to be re-burnt.
Each oven burns about 20 cwt. per 24 hours.   Any sulphur ore
which melts during the operation, is forced between the bars and
falls down into the ash-pit.

3. *Lump and small Ores.*—The arrangement for burning
these two kinds of ore has been devised by MM. Usiglio and
Dony.   The ovens just described, are covered with a metal pan,
arched over in brick, and heated from below by the burning of
the lump pyrites.   The pyrites in powder is thrown into this
metal pan, and the operation is conducted in the usual way.

5*

Burners for Impure Sulphur Ores.

There are several sulphur ores too impure to employ with the plans in ordinary use, and Mr. Oliver has attempted to get over the difficulty by the following arrangements. The class of ores which Mr. Oliver employs, are those pyrites which contain arsenic, zinc, and other impurities; and before the sulphur present in this class of ores, can be rendered available for the manufacture of sulphuric acid, the practical inconveniences arising from these impurities must be so far overcome, as to render the manufacture of a merchantable article, compatible with the physical character of these ores.

Now, the principal difficulty experienced in manufacturing pure sulphuric acid, when these ores are used as a source of sulphur, arises from the fact, that the temperature necessary to expel the sulphur in the form of sulphurous acid, also volatilizes the arsenic and zinc, the fumes of which are carried forward into the chambers, and contaminate the acid.

The patentee proposes to overcome this difficulty by the following arrangements between the burners and chamber.

Figure 49 represents a longitudinal vertical section of a set of pyrites burners, *h h*, with evaporating pans, L L L; precipitating chamber, F F; cooling flue, Q Q; and sulphuric acid chamber No. 1, upon the line A A, Figure 48, and also a vertical section

FIG. 48.

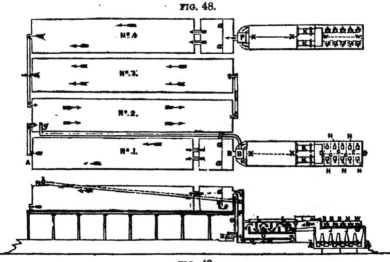

FIG. 49.

of the ascending flue P upon the line B B, Fig. 48. The burners are constructed of the ordinary materials, and the ground plan forms a parallelogram about 30 feet by 15, but these dimensions may be modified as circumstances may require. The building is divided into two ranges of burners, by a centre wall, D D (Figure 48), extending the entire length, and of the same level as the front walls, and upon which the crown arch of each duction flue, J J, rests, as represented by the line D D (Figure 48). Each range of burners is divided into six compartments by five blocks of brick-work extending from the centre to the front walls, as shown, C C C C C (Figures 48 and 49), and terminating about 2 feet 6 inches below the crown arch, E, forming the flue communicating with the precipitating chamber, F F. About 1 foot 6 inches from the bottom of each burner, a fire grate and slab with luted door is introduced, the area of the burner at that level being about 2 feet by 1 foot 6 inches, and scaling outwards to the top of the brickwork represented by c c c c c, where the area is about 4 feet by 3 feet. The burners are charged through the crown arch, by withdrawing the sliding bottom attached to each of the hoppers, H H, &c., which contains the charge of pyrites in a dry state, and to facilitate the descent of the pyrites, as the process of calcination advances, each burner is provided with luted openings in the front. In front of the recess forming the ash-pit of each burner, I I I I I I (Figure 49), an air-register is placed, by which the quantity of air required to permeate the superincumbent mass of incandescent ores, in the respective compartments, may be regulated. The gases and metallic fumes generated, are carried along the flues, J J (Figure 49), through the tunnels, K K, which are immersed in water to cool the gases, and facilitate the precipitation of the metallic fumes as they pass through the precipitating chamber, F F, which is of much greater sectional area than any of the flues. The gases pass along with an undulating motion in the precipitating chamber, by the intervention of the cross walls, m m, &c., (Figure 49). The floor of the precipitating chamber, n n, forms the roof of the subjacent flue, Q Q, which connects the precipitating chamber with the sulphuric acid chamber No. 1 through the ascending flue, P. A stream of highly compressed air, which has parted with a considerable amount of its latent heat, is introduced into the cooling flue, Q Q. The introduction of this air into the cooling flue, say, at a pressure of two atmospheres, at a point near to the exit flue from the precipitating chamber, enables it to abstract a large amount of heat from the heated gases in the flue, and on assuming its original volume, it causes a precipitation of the metallic oxides.

The precipitating chambers and subjacent flues are divided by a centre wall, as shown by the dotted lines X X, so that the communications between the pyrites burners, and the precipitating chamber and cooling flue on either side of the centre wall, may be intercepted by a damper situated at the end of the tunnels, as shown by the dotted line Y, by which arrangement the deposited metallic oxides can be removed through lateral openings in the front walls, without interrupting the working of either the pyrites burners or the acid chambers.

The nitrate of soda or potash is decomposed in a different apparatus from the burners, where the sulphurous acid is generated, as shown at Z (Figure 49). The nitrous gas generated, is conducted into the small mixing chambers attached to Nos. 1 and 4 acid chambers, where it meets the sulphurous acid ascending the upright flues, as represented by P P P (Figures 48 and 49). The gases pass underneath the leaden curtain represented by the letters a a, and Nos. 1 and 4 mixing chambers, and proceed forward in the direction of the arrows; and the gases passing through the working chambers Nos. 1 and 4 are made to circulate through No. 2 and No. 3 in succession.

Whatever the arrangement or kind of kiln employed, it is necessary that the gases should be cooled, by passing through an iron pipe from 20 to 30 feet long, according to the quantity of ore burnt in a given time. This has also another advantage of allowing the solid particles or dust to deposit before entering the chamber.

This dust has been examined by Mr. R. C. Clapham, to whom we are indebted for the following analysis :

Sand, &c.	. .	2·333	Arsenious acid .	58·777
Oxide lead	. .	1·683	Sulphuric acid .	25·266
Peroxide iron	.	3·700	Nitric acid . .	trace
Oxide copper	.	trace	Water . . .	8·000
Oxide zinc	. .	trace	.	99·759

The following details of the cost of materials and labour to make the same quantity of acid from sulphur and sulphur ores, possess some value, as the result of the practical working of chambers, by a manufacturer of great experience. The prices of the materials are assumed, to show the relative value of the ores as compared with sulphur.

Sulphur.

Tons.	cwt.	qrs.	lbs.				
5	10	0	0	Sulphur, at £7	£38	10	0
—	5	2	0	Nitrate of soda, at 18s. .	4	19	0
				2 Men at 21s. per week .	2	2	0
					£45	11	0

Belgian Pyrites.

13	15	0	0	Pyrites, at 45s.	£30	18	9
—	6	2	11	Nitrate of soda, at 18s. .	5	18	9
				8 Men at 21s. per week .	8	8	0
					£45	5	6

Irish Pyrites.

18	0	0	0	Pyrites, at 33s.	£29	14	0
—	6	2	11	Nitrate of soda, at 18s. .	5	18	9
				9 Men at 21s. per week .	9	9	0
					£45	1	9

Other sources of Sulphurous Acid.

M. Gossage was the first to propose *sulphuretted hydrogen* as a source of the sulphurous acid. He obtained his supply of this gas, by decomposing tank or alkali waste by carbonic or muriatic acid. The presence of large quantities of carbonic acid and the difficulties of managing so dangerous a gas, induced parties to abandon the plan; but we know of one acid manufacturer who makes sulphate of iron by decomposing refuse protosulphide of iron by sulphuric acid, and burns the sulphuretted hydrogen which is evolved, in the same manner as coal gas, conveying the sulphurous acid and aqueous vapour to his chambers.

Mr. Cookson has proposed to mix lead ores with metallic iron and a flux, and reduce the lead by fusion in a crucible or furnace. A mat is formed, consisting chiefly of sulphide of iron, which falls to powder on exposure to the air. This powder he makes into a paste with water, moulds it into bricks, and burns them as pyrites in an ordinary kiln for the manufacture of sulphuric acid.

Previous to Mr. Cookson's patent, a very feasible plan was tried by Mr. J. L. Bell, who smelted the exhausted pyrites with the tank or alkali waste and some clay or burnt bricks, in an ordinary blast furnace. The furnace being tapped in the usual

manner, the mat or "factitious pyrites," as it was termed by Mr.
Bell, flowed out, and when the hearth was emptied, the slag ran
out afterwards, as in the smelting of iron. This factitious pyrites
was allowed either to weather in the air, or to flow into water,
kept below the boiling point. The fine mud or slime obtained,
was dried into cakes at a temperature below 212° F. on an
ordinary brick flat, and was afterwards burnt in kilns in the
same way as pyrites. Various causes have induced the patentee
to abandon this ingenious process.

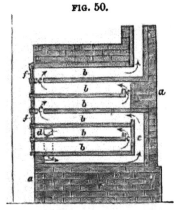

FIG. 50.

Blair has proposed to burn the
sulphuretted hydrogen from tank
waste in ordinary burners, but only
to admit as much air as would consume the hydrogen, by which he
expects to obtain the sulphur, which
he collects in a chamber, while
Favres suggests passing this gas,
obtained from the same source in
the usual way, over burning sulphur
ores in any of the kilns in use.

As we formerly mentioned, the
renovated oxide of iron from gas
purifiers is now being used as a
source of sulphur, and Mr. Hills employs a burner of peculiar construction for the combustion of this material. It consists of fire-brick
furnaces, $a\ a$, having a series of shelves, $b\ b$, composed of fire tiles,
leaving a space of 5 or 6 inches between the shelves. The sulphur
oxide of iron is placed on these shelves in thin layers. The shelves
are connected alternately, at back and front, by the apertures c, and
the gas from the lower shelves passes, by a side descending flue, d,
to the back of the furnace, where it rises by another side flue, e,
to enter the third shelf. Air is admitted, as may be found necessary, at the ends of the shelves by sliding doors, $f\ f$.

Treatment of Exhausted Poor Cupreous Ores.

Some of the sulphur ores, as the analyses (p. 14) indicate, contain sufficient copper to pay for smelting the ores, when the
sulphur has been burnt off. The process pursued is similar to
that followed in copper works, but we are induced to give a
description of the following plan on account of its simplicity.

Fig. 51 is a vertical section of a cupola for smelting the cupreous residue of sulphur ores. A is the mass of materials

FIG. 51.

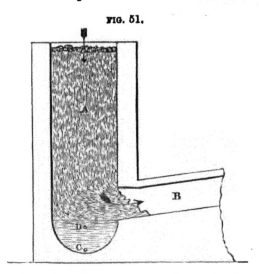

under treatment and the fuel; B, the flue leading from the lower part of the cupola to the chimney; C is the tap hole for the regulus; and D, the tap hole for the slag. When first set in operation, the cupola is filled with coke or coal, and lighted at the top, when the downward draught soon causes the whole mass to ignite, and the material intended to be smelted may then be added with fresh fuel.

The atmospheric air passes down through the mass A, and on to the chimney, by the flue B, causing intense combustion in the body of the cupola. The slag and regulus flow together to the bottom of the cupola, and may be tapped off, from time to time, at their respective tap holes. Or, a horizontal furnace, as A in the vertical section (Fig. 52) may be placed between the cupola B and the chimney, in which case the bottom of the cupola will incline into the furnace, and the slag and regulus will flow into the same, whence the regulus may be tapped off at the tap hole E, and the slag raked out at the furnace door D; the heat and flame passing down through the hole F at the lower part of the cupola, over the bed of the furnace, to the flue C, leading to the chimney. A saving of fuel in this operation, may be effected by covering the top of the cupola, except when putting in materials, and heating

FIG. 52.

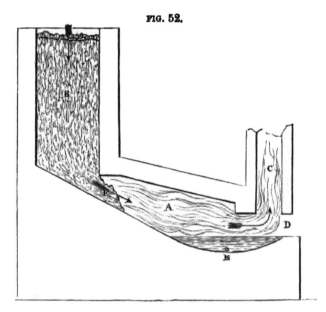

the atmospheric air supplied to the upper part of the cupola, by causing it to pass through iron pipes heated by the waste gases which escape to the chimney by the flue C.

2. SUPPLY OF NITROUS COMPOUNDS.

Various plans have been followed at different times to obtain a supply of the nitric acid in making sulphuric acid. The first process was to ignite a mixture of 8 parts of sulphur and 1 of saltpetre, whereby sulphurous acid and nitric oxide were evolved, and sulphate potash remained behind—

$$KO \cdot NO^5 + S = KO.SO^3 + NO^3,$$

the excess of sulphur yielding the sulphurous acid.

Another plan was tried in France, to combine the manufacture of oxalic acid with the production of nitric oxide. For this purpose, molasses were heated with nitric acid, and the vapours, consisting of nitric oxide and peroxide of nitrogen, were conveyed to the chamber. This plan has been abandoned on account of the low price of oxalic acid.

When a demand for caustic baryta existed, some time since, nitrate of baryta was heated in a close vessel, and the gases evolved, were made use of in the working of the chambers, but this was of course an exceptional state of affairs.

An aqueous solution of the alkaline nitrates, run into the chambers in a small stream, has also been used with but partial success, and we believe but two plans now prevail. One consists in employing nitric acid in a liquid form, in saucers; the other, adopted in this country, where a given weight of nitrate of soda is placed in metal cups, technically called *nitre pots*, of various shapes, cast with short legs to raise them an inch or two from the ground, to which the equivalent of sulphuric acid is added, and the workman places them in the sulphur burner or kiln by means of an iron hook or bar. The heat of the burning sulphur or sulphur ore is sufficient to effect the liberation of the nitrogen-compounds, by the action of the sulphuric acid on the nitre. In the making of this acid, it is an important point to send these gases into the chambers in the highest stages of oxidation; and too high a heat in decomposing the nitre, has a tendency to produce the lower oxygen-compounds of nitrogen. Hence, in some manufactories, to avoid the irregularities of temperature, incidental to the burners and kilns, this decomposition is made in separate vessels, heated by a special fire, as being more under control.

The great point, however, in saving nitre has been found to consist in employing chambers with a large capacity in proportion to the quantity of sulphur burnt. It is, however, difficult to state what the most economical relation really is, for almost every manufacturer has a different standard. Much more space is required when burning pyrites than with sulphur, for the obvious reason, that the residual nitrogen in the former case, is much greater than in the latter, the iron of the pyrites requiring nearly half as much air for oxidation as the sulphur with which it is combined.

3. THE CHAMBERS.

The chamber itself is simply a large room lined with lead, the top and sides being made with sheet lead of 5 lbs. to the square foot, while the bottom lead should weigh 8 to 10 lbs. The floor of the chamber sometimes rests upon the ground, but is more generally supported on a series of stone or brick pillars, the height of which varies from a few inches to several feet, the space underneath in the latter case being employed as a warehouse or serving as the locale for the burners, &c.

The sides of the chamber consist of a wooden framework, which also carries the timber supporting the roof or leaden covering. The sheets of lead are fastened to the timber framing by means of small pieces of sheet lead, called *straps*, one end of which is soldered or burnt in to the sheet lead, and the other nailed to the wooden rail or upright.

The sides of the chamber also hang from the timber framing, running along the top of the uprights; the roll of sheet lead being raised by suitable tackle, one end is nailed down to the wood, while the roll is allowed to uncoil itself, until the lower end nearly reaches the bottom of the chamber.

The bottom of the chamber is turned up at the sides from 12 to 18 inches, forming as it were an immense leaden cistern, while the sides above described, hang with the lower end pressing closely against the turned up edge of the bottom. The liquid in the chamber rising above the lower end of the side sheet, the whole is *luted*, and then all connection with the outer air is cut off.

The front side of the chamber is provided with an opening about two-thirds up, into which an iron pipe is fixed and properly supported. This pipe is lined with fire bricks to protect the metal from the action of the acid gases.

Another opening at the opposite end, and much lower down, made of lead, carries away the escaping gases to another chamber or some condensing apparatus. The size of these depends on the volume of the gases to be admitted during a given time.

The sides are also furnished with a man-hole made of lead and luted, which is useful for examining and repairing the interior of the chamber from time to time.

The following arrangement is employed for testing and drawing off the acid when required.

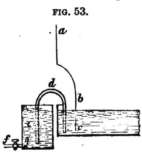

FIG. 53.

The side of the chamber, *a*, near the lower end, is fitted with a semi-bell shaped lead recess, dipping below the acid, *c*. A syphon, *d*, always ready for action, is placed with the shorter arm in the chamber, and the longer one in the leaden vessel, *e*, which is thus always kept full of acid. A lead pipe, *f*, is fitted to this vessel, its end turned up inside, to keep clear of sediment, and furnished with a cock, so that the acid can be run away at any moment, without loss of time.

The sides of the chambers are also fitted with another very simple but important contrivance for indicating the progress of the condensation at all times. This is termed the *drip pipe*, which is shown in the wood-cut.

A leaden pipe, enlarged at the end *a*, is fixed in the chamber side, *b*, while the other end, *c*, forms a syphon, the liquid from which drops into a leaden vessel, *d*, standing on a leaden shelf, *e*, supported against the chamber. The acid formed and condensing inside the chamber, rains, so to speak, into the rain-gauge, *a*, and, filling the pipe, overflows into the vessel *d*, where it can be tested and measured, thus at one and the same time, indicating the strength and rate of acid made in a given period. It is needless to draw attention to the value of this gauge or *drip pipe* in the management of a chamber.

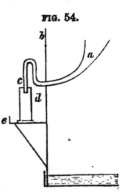

FIG. 54.

This place is, perhaps, the most suitable for describing the useful apparatus, devised by M. Richemond, by which a jet of mixed hydrogen gas and atmospheric air is ignited, producing great heat, and which is now universally employed in fusing the two edges of the sheet lead, in building these acid chambers. The two edges of the sheets of lead, perfectly clean and free from dust, are laid side by side, and the jet of flame applied until the lead melts, uniting the sheets ; sometimes a thin strip or bar of lead is held in contact with the sheet lead, in front of the flame, and as it melts with the edges of the sheets, it is moved forward with one hand, while the jet is directed against it with the other hand. The apparatus consists of two pieces, a portable bellows, 1 and 2, and a generator of hydrogen gas, 3 and 4. The latter is composed of two cisterns, lined with lead, *a* and *b*, connected by a pipe, *c*, which descends from the lower side of *a*, to the bottom of *b*, an inch or two above which, there is a shelf, *d*, on which the zinc is placed. When the cock, *e*, is opened, the sulphuric acid, with which the cistern, *a*, is filled, flows down, dissolving the zinc and generating hydrogen gas, which escapes by the pipe, *f*, passing into a small vessel, *o*, partially filled with water to wash the gas, which finally issues through the pipe *h* for use. The gas pipe and the bellows pipe are connected with the little piece of apparatus 5, each arm of which is fitted with a cock; and

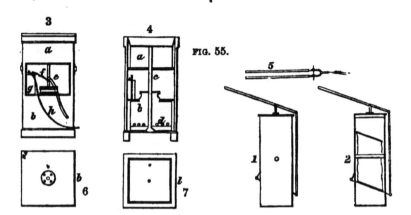

FIG. 55.

when at work, the plumber regulates the supply of gas and air, to suit the force and kind of jet which issues from the orifice at *i*. 6 and 7 show the top fittings of the gas apparatus.

The size of the chambers employed, varies in every manufactory, as well as the mode of arranging them. Some are worked in sets of three or of five, thus:

FIG. 56.

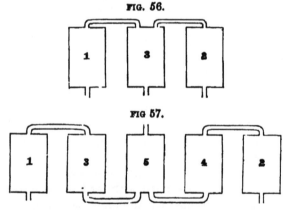

FIG 57.

Nos. 1 and 2 being in connection with the burners are called working chambers, and discharge their uncondensed gases into No. 3, or receiving chamber, which communicates with a condenser or chimney.

Nos. 1 and 2 are the working chambers, Nos. 3 and 4 the receiving ones, and No. 5 acts as a condenser.

Instead of using several chambers, connected with each other in the manner described, one roomy and very long chamber was

FIG. 58.

divided by partitions into a number of compartments, and.in such a manner that the gases, by ascending and descending, were forced to mix intimately with each other. Although a saving of lead for the sides is effected by this plan, the rapid corrosion of the partitions has led to its disuse.

The draught of the chambers and burners is obtained by connecting them with an ordinary chimney; but Mr. Bell has patented a plan, in which he uses a steam jet, which gives him more power and control. He avails himself of this increased power of draught to employ a number of coke columns, through which the residual gases pass, and can be condensed. In Fig. 59, showing this plan, A is a section, and B a ground plan, C is the last chamber, and DDDD are the coke columns. At the end

FIG. 59.

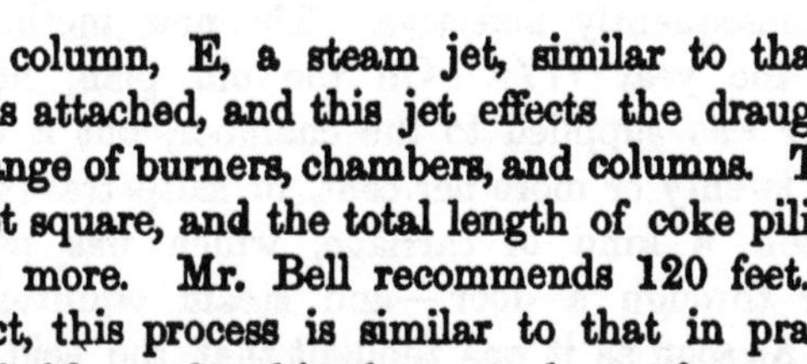

of the last column, E, a steam jet, similar to that described further on, is attached, and this jet effects the draught through the whole range of burners, chambers, and columns. The columns are four feet square, and the total length of coke piling, may be 200 feet or more. Mr. Bell recommends 120 feet. In every other respect, this process is similar to that in practice. The advantages held out by this plan, are the perfect control of the draught under all circumstances, the increased quantity of sulphur which may be profitably burnt in the same time in the same chamber room, and the saving of nitrate. The condensing power of the coke column is very great, and enables the manufacturer to dispense with chamber room, or increase his make, while the stream of acid which runs from the condensers, carries back to the chambers all nitrous gases which had escaped.

We quote from Muspratt the dimensions of some chambers at work in this country, one at Bradford, belonging to Mr. Schole-field, 75 feet long, 35 feet wide, and 35 feet high; one at Newton Heath, Spence's, 75 feet long, 40 feet wide, and 40 feet high; one at Flint, belonging to Muspratt Brothers, 140 feet long, $24\frac{1}{4}$ feet wide, and $19\frac{1}{2}$ feet high. Smaller chambers are believed, however, when worked in sets, to be practically more useful than such large ones as those mentioned. A very good working chamber is 100 feet long, 20 feet wide, and 18 feet high, and the quantity of sulphur which may be burnt to the best advantage, may be taken at 1 cwt. of sulphur per 24 hours for each 3000 cubic feet of chamber space.

We may add that it has been proposed by Mr. Wilson, to construct chambers of glass by means described in his patent for this purpose, and it has also been suggested to build them of fire-bricks, laid in sulphur and clay as the cement, but we do not know whether either idea has been carried into practice.

Mr. Gossage has also patented a plan of building chambers, distinguished by being high, rather than long, up which the gases would ascend, and to build such chambers of large slates attached to the ordinary frame-work of wood, and covering them with a cement of sulphur and fine sand.

The Working of the Chambers.

The production of sulphuric acid in these chambers is a continuous process. This was not the case upon the old method, which was consequently defective. The new method was introduced in the year 1774. On the old plan, no constant current of air was supplied to the chamber, but a mixture of sulphur with twenty or more per cent. of saltpetre was burnt,— originally upon a kind of carriage, which was pushed into the chamber through a door,—and steam admitted at the same time. As soon as it was thought that the acid, originally on the floor of the chamber, had absorbed the newly-formed portion, the residual gas was allowed to escape by the chimney, and the chamber again filled with air, for a fresh operation. In this manner, only so much sulphuric acid could be formed as would correspond with the quantity of air (oxygen) in the chamber, and this would naturally depend upon the capacity

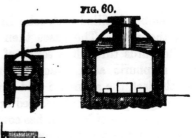

FIG. 60.

of the latter. All the sulphurous acid produced in excess, as well as, in the opposite case, a quantity of nitric oxide, which had as yet taken no part in the production of acid, was therefore lost each time that fresh air was admitted into the chamber.

We will now describe the working of the chambers as pursued in Germany, France, and this country.

The chamber arrangements at Kunheim represented in figures 60 and 61 may serve as an illustration of the process as carried on in Germany.

120 lbs. of sulphur are burnt per hour on the ordinary iron plate, over which a boiler is built, in which the water is heated for the steam boiler (Fig. 61). A cast-iron pipe passes through this heater, and leads the gases to the first chamber. A steam jet assists the draught. The sulphurous acid passes into the second chamber, where the nitric acid falls over a terrace of earthenware saucers, to diffuse it as much as possible. The gases and vapours then pass into a large chamber, 100 feet long by 20 feet high and 30 feet wide. Steam is admitted into the chamber from different sides, in order to mix the gases and supply the necessary water. The

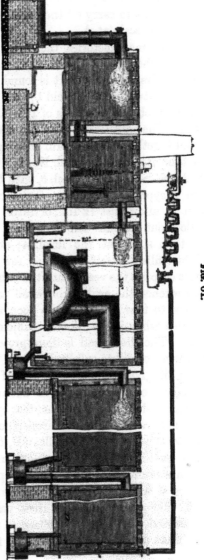

FIG. 61.

next chamber is filled with coke in order still more effectually to mix the gases. The gases then pass into the fifth chamber, and afterwards through a series of earthenware jars filled with strong sulphuric acid, which absorbs the nitrous compounds. From the last, the saturated acid flows into the second chamber. The strong sulphuric acid is forced up into a vessel to supply these jars, which consist of large dishes covered with bell-shaped domes, connected at the top by pipes, which convey the gases, and an overflow pipe to carry off the acid. One of these jars is shown on an enlarged scale at A.

In some manufactories, the nitric acid is exposed in flat glass saucers in the second chamber, where it is decomposed by the gases from the first chamber. It is found beneficial in practice to allow the concentrated nitric acid to flow in at intervals of about half an hour each, in quantities of 12 lbs., or about 600 lbs. in the 24 hours. The nitric acid flows from a series of glass carboys, connected together so as to form one large reservoir, through the narrow tube, a, into the vessel, N, Fig. 62 (outside the chamber), in which the bellshaped vessel, b, with the wide exit tube, c, together form a syphon. The size of a is calcu-

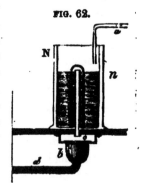

FIG. 62.

lated to allow just 12 lbs. of nitric acid to flow out in half an hour, and this will then stand at the level, n; at this height c fills of itself, and in a few moments empties the vessel into the tube, d, from whence the acid is conveyed to the saucers, to flow off on to the floor of the chamber when it has become sufficiently weak. The sheet lead is there protected by a layer of dilute sulphuric acid.

Fig. 63 gives an excellent idea of this manufacture as conducted in France. A is the usual sulphur burner; a showing the sliding door by which the charge is thrown into the burner. The sheet-iron pipes, BB, convey the sulphurous acid to the chimney or stalk, CD, from which they pass into a small lead chamber, EE', called in France *premier tambour*. Steam is admitted from a straight pipe, in the direction DE, which creates a draught and facilitates the decomposition of the nitric acid. The gaseous mixture then passes through a pipe, E", into the second chamber, E''', where it meets the current of nitric acid from the earthenware vessels, flowing in a stream on the

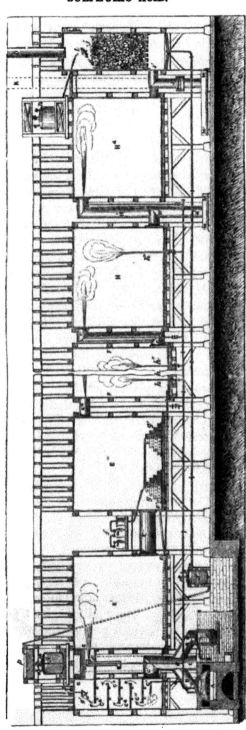

FIG. 63.

earthenware shelves, *gg'*. The sulphurous acid is converted into sulphuric acid, which flows back into the first chamber through a small pipe *ee'*, where the nitric acid and peroxide of nitrogen, absorbed by the liquid acid, are decomposed by the joint action of the sulphurous acid and steam, into deutoxide of nitrogen, which is evolved and reconverted into peroxide of nitrogen, and joins the gaseous current passing into the chamber E'''. The gases then pass on through the pipes E⁴ into the principal chamber, FF. A jet of steam from the centre of the pipe E⁴ agitates and mixes the gases, which effect is also promoted by the action of three more jets, *hh'h''*, directed from below. The gases pass from the lower part of this chamber through pipes GG' into the higher sides of the next chamber, H, where the steam jets G' and *h'''* are intended to act in the same way. The gases and vapours descend from the chamber H by a pipe, H'H'', into a close vessel, MM, fitted with compartments to facilitate the condensation. Near the bottom of this vessel a pipe, H''', is attached to carry the uncondensed gases into a last chamber, H', from whence they circulate through another condensing vessel, MM', from which they escape into a chimney, JK. A damper is fixed at J for regulating the draught.

In working chambers on the English system, much of the complication of the preceding plan is avoided. The chambers are larger, arranged side by side, and the multiplication of the steam-jets is avoided. The sulphuric acid is run from the working chambers to the receiving chamber to save the nitric acid. There are no condensing vessels, and the wear and tear attending so many pieces of apparatus is saved.

Fig. 64 will give an idea of the arrangements adopted in this country. The receiving and working chambers, AB, are built on a flooring supported on brick or metal pillars, at varying heights above the ground, and the space below is used for a warehouse for the raw materials, or as a locale for the furnaces, &c. The sulphur burners are shown at C, the details of which have been given at page 52. DD represents the absorbing column of Gay-Lussac's process, more fully explained in the following pages. E, steam boiler; F, the lead pan for concentrating the sulphuric acid, which we have described at page 102. GG represents the timber framing, which is covered and boarded in, to protect the chambers and internal apparatus from the weather.

FIG. 64.

4 THE RECOVERY OF THE NITROUS ACID.

The great cost in this manufacture is the nitric acid, and no plan for saving this material has been so effective as that of Gay-Lussac, which, as explained by himself, consists in—

1st, Absorbing the nitrous vapour that escapes from the chambers by means of concentrated sulphuric acid, 1·760 specific gravity : the product is called "nitrous sulphuric acid."

2nd, Causing this nitrous sulphuric acid, by the aid of a jet of steam, to yield its nitrous gas to the sulphurous vapour proceeding from the sulphur-burner.

The steam dilutes the acid, destroys its absorbing power, and the nitrous gas becoming free, mixes with the sulphurous gas, and passes with it into the chamber, where it performs the same part as an equivalent quantity of nitrate, decomposed in the usual way, would do.

Figs. 65 and 66 are a side view and a plan, of a chamber with the apparatus complete, as proposed by Gay-Lussac, and which consists of two principal parts :

1st, A leaden column, H, filled with coke, placed near the escape pipe of the chamber, for the absorption of the nitrous gas ; this is called the "absorbing column."

2nd, A leaden column, I, also filled with coke, placed between he sulphur burner and the chamber, where the nitrous sulphuric acid yields its nitrous gas to the sulphurous gas and steam ; this is called the "denitrating column."

Absorbing column.—The vapours from the chamber enter the bottom of the column, while sulphuric acid of 1·706 specific gravity is poured in at the top. The vapours and the acid are thus kept for some time in contact with each other, and the acid absorbs the nitrous gas contained in the vapours. The vapours not absorbed, are allowed to escape from the top of this column into a chimney, while the nitrous sulphuric acid runs from the bottom and is collected in a cistern,' R.

A chamber for which one ton of sulphur is consumed per day, will require a column about 30 feet high by 2 feet 10 inches diameter ; a larger consumption will require a larger column ; but as the absorption is most completely effected in a narrow and high column, it is preferable to increase the height rather than the diameter of the column ; and in no case should the diameter exceed 3 feet, because it would be very difficult to distribute the acid uniformly over the whole of a column of larger diameter. When the consumption of sulphur is so large, that an elevation sufficient for the absorption, cannot conveniently be given to the column, it is much better to use two columns of a small diameter than a single large one, and two escape pipes instead of one, to the chamber, each pipe being connected with an absorbing column, and the gases, as they issue from these columns, should be conducted into the flue communicating with a chimney.

FIG. 65.

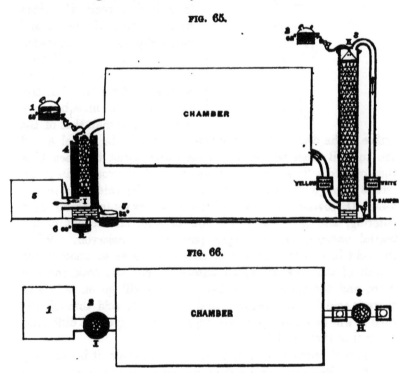

FIG. 66.

As the escape of vapours from a vitriol chamber should always be the same, it becomes of essential importance that perfect regularity should be obtained in the supply of acid for absorbing the nitrous gas. This is effected by the following contrivance, which is represented by Fig. 67.

It is evident that if the reservoir which holds the acid be open, it will be impossible to obtain a regular flow of acid, as the level will be continually varying; thus, when the reservoir is full, the weight of acid would cause it to run out very quickly,

FIG. 67.

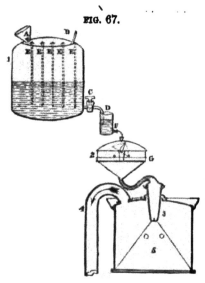

but as the level descends, the acid would run more slowly; it is therefore necessary to make use of a closed reservoir.

This reservoir is filled with sulphuric acid 1·760 specific gravity, by means of a funnel, A, the air escaping through the small tube, B; near the bottom is a leaden cock, C, which must be perfectly tight, and which dips into a small leaden vessel, D, with a small glass tube, F, at the bottom for the escape of the acid. The section of this tube must be somewhat smaller than that of the orifice of the cock. When the reservoir is full, the funnel A and the tube B must be carefully closed with corks, the cock C is then opened. The acid flows faster through the cock than it can escape through the glass tube F, so that the acid rises in the vessel D, covers the orifice of the cock, prevents the air from entering, and soon stops the flow of acid, on account of the partial vacuum in the upper part of the reservoir. When the acid has so far run out of the vessel D as to uncover the mouth of the cock, air again enters through the cock, the acid flows, and the vessel fills as before. This will go on as long as there is acid in the reservoir, and the flow of acid through the tube F will be almost perfectly regular, as the greatest difference of level in the vessel D is so very small.

A partial vacuum being formed in the reservoir, it is necessary to support its cover; this is done by means of five or six bars of iron, E, covered with lead, resting on the bottom, and soldered to the cover.

It is equally important that the acid should be distributed uniformly over the whole surface of the coke. This is effected by means of cones of varying sizes, one placed above another, in the top part of the column. These cones being all, except the middle one, truncated, receive the acid at the top, which flows over the surface, and falls in a number of circular streams, which spread over the whole surface of the coke. If the acid fell on the cones in a regular stream, there would not be a sufficient quantity to cover all their surfaces; it is, therefore, necessary that a certain volume of acid be supplied suddenly at regular intervals. This object is accomplished by means of the oscillator, G.

This oscillator consists of a sort of leaden box, with a partition in the middle, which divides it into two equal parts; this is supported on an axis on which it oscillates, the contrivance being so arranged that while one side pours out the acid it has received, the other is being filled, and only falls to pour it away, when the calculated quantity has been supplied. Every time the oscillator falls, its contents run into an S pipe, and from thence into a short pipe placed just above the tops of the cones, and fitting closely on the upper one; this pipe being suddenly filled with the acid, the acid escapes and flows over all the cones. The oscillator should deliver the acid, at least, twenty times per hour.

For a daily consumption of 1 ton of sulphur, 1200 pints of acid, 1·760 specific gravity, will be required, and proportionately for a larger or smaller consumption.

Denitrating column.—The nitrous sulphuric acid that flows out of the absorbing column, is conveyed into a close reservoir, similar to the one above described, which is placed, with its small leaden vessel, above the denitrating column. The acid flows from this into the column, and after having parted with the whole of the nitrous gas that it contained, it runs into a cistern at the bottom, while the sulphurous gas, together with a certain quantity of steam supplied by a pipe placed at the bottom of the column, carries away the nitrous gas into the chamber.

A daily consumption of 1 ton of sulphur requires a denitrating column about 10 feet high by 2 feet 6 inches diameter. This, of course, varies according to the consumption, but for this, as well as for the absorbing column, an increase should rather be effected in height than in diameter.

As it is important to ascertain the nature of the vapours before they enter into the absorbing column, and when they issue from it, a glass case forms part of the pipes which conduct them to and from the absorbing column.

The vapours should always appear yellow in the former, and white in the latter.

A damper is placed under the latter case to regulate the draft;

as too great a quantity of air in the chamber would dilute the nitrous gas and render its absorption more difficult.

The acid used, should be quite clean.

The coke used, should be well burnt and of good quality. In the absorbing column, the pieces ought to be of about the size of an egg, while in the denitrating column, they ought to be about the size of the fist. The coke in both columns, is supported on a kind of grate formed of iron bars covered with lead, and placed just above the entrance of the gas. Dust should be avoided, and the pieces should be placed carefully, so as not to obstruct the passage through the grate.

Directions for the working of the process.—It is well known, that in order to convert the whole of the sulphurous gas into sulphuric acid, it is necessary to have an excess of nitrous gas in the chamber, so that the vapours which escape, may be quite yellow, the white colour of these vapours being caused by an excess of sulphurous gas.

But in consequence of the comparatively greater value of nitre, manufacturers have hitherto preferred losing sulphurous, rather than nitrous gas.

The advantage of this invention being to save the whole of the nitrous gas contained in the vapours that pass from the chambers, it is evident that, however great may be the quantity of nitrous gas employed, it will not be lost, and, consequently, manufacturers can use a sufficient quantity of nitrate to convert the whole of the sulphurous gas into sulphuric acid.

This is more fully effected by introducing a jet of steam near the escape from the chamber, in order to condense the last traces of sulphurous gas which might have been carried away from the chamber.

The vapours should always appear perfectly yellow in the first glass case, as the nitrous gas is more readily absorbed, when it is quite free from sulphurous gas.

The first thing, therefore, to be done before applying the apparatus, is to render the vapours quite yellow, using as much nitrate as will be necessary to produce that effect.

The concentrated sulphuric acid, 1·760 specific gravity, should be poured into the absorbing column as soon as the vapours have become yellow.

The acid that runs out of the absorbing column is poured into the denitrating column, and a quantity of nitrate used, merely sufficient to maintain the yellow colour of the escaping vapours. In the course of a week, the proportion of nitrate used, should be reduced from 10 per cent. to 3 or 4 per cent.

Should the gas in the second glass appear white, it will be a proof of the entire absorption of the nitrous gas.

Should it appear yellow, it will result from some defect in the working, probably from the acid used, not being sufficiently concentrated, or from the draft being too great, in which latter case, the damper may be partially closed.

It is necessary that the acid should be at least 1·760 specific gravity. Weaker acid would not absorb the nitrous gas so readily, nor so well.

When it has passed through the absorbing column the acid should be about 1·717 specific gravity.

If it is stronger, the quantity of steam in the chamber is not sufficient, and must be increased, as the use of steam is necessary for the total absorption of the sulphurous vapours.

If the acid is weaker, it proves that there is too great a quantity of steam, which might prevent the acid retaining the nitrous gas, after having left the absorbing column. To obviate this, the supply of steam should be decreased.

The nitrous sulphuric acid generally loses 8° Twaddell in becoming properly denitrated; thus the acid, passing from the denitrating column, should be 1·615 specific gravity.

Should the vapours become pale in the first glass case, the best thing to be done is to close, during an hour or so, the doors of the sulphur burner; the air being then no longer admitted into the chamber, the vapours soon become yellow again. If this fails, an increased quantity of nitrate must be used.

10 pints of good nitrous sulphuric acid will yield sufficient nitrous gas to save 1 pound of nitrate.

Arrangements for pumping up the Acid.

The following arrangements (Fig. 68), are employed for raising the acid to the cisterns, which are placed above the columns.

A is a strong leaden cistern, which has lately been replaced by cast iron, B is the shaft of an air-pump, C is a safety valve, D is a pipe reaching from the bottom of the cistern A to the cistern E at the top of the coke column.

The valve C being opened, and the cistern A filled with the

FIG. 68.

concentrated sulphuric acid, the valve is closed, and the air-pump set to work, which forces the air into the cistern at the top, and causes the acid to ascend the pipe D into the cistern E. When the cistern A is emptied, the operation can be repeated as often as necessary.

Testing process.

In order to ascertain the degree of absorption, the nitrous sulphuric acid should be tested in the following way:

A small tube, measuring 20 cubic centimetres, is filled with chloride of soda, 100° Gay-Lussac; its contents are poured into a phial, which contains about two-thirds of a pint of distilled water.

FIG. 69.

Another glass instrument, measuring 20 cubic centimetres, divided into 200 parts, is filled with the nitrous sulphuric acid to be tested, and the acid is gradually poured from it into the phial, which contains the solution of chloride of soda, until, on adding a few drops of a solution of indigo in sulphuric acid, the blue colour is no longer destroyed.

FIG. 70.

If 17 or 20 parts only have been used, 10 pints of the acid tested will cause a saving of 1 pound of nitrate.

The more acid required to destroy the colour of the indigo, the less nitrous gas it contains.

In order to ascertain whether the sulphuric acid still contains nitrous gas, after it has run out of the denitrating column, a drop of indigo solution may be poured into the acid, which contains no nitrous gas, if the indigo preserves its colour.

We are indebted to Muspratt's Chemistry for an account of a very elegant plan of estimating the value of this nitrous acid, which is founded on the reaction which ensues on mixing urea with nitrous acid, as explained by the following equation:

Urea.   Nitrous acid.   Carbonic acid.   Nitrogen.   Water.

$$C^2N^2H^4O^2 + 2NO^3 = 2CO^2 + N^4 + 4HO.$$

Mr. Hart employs nitrate of urea, as it is easily procured in a pure state. He dissolves 20 grains in 2¼ ounces of water, and raises the solution to a boiling temperature. He covers a white plate with drops of a test solution, formed of thin starch water, holding a little iodide of potassium in solution. He then fills an alkalimeter with the nitrous sulphuric acid, which he adds to the urea solution, stirring all the time, until a drop mixed with the starch liquid gives a blue colour, indicating the presence of free undecomposed nitrous acid, liberating the iodine which gives the characteristic reaction with the starch. The calculation is based on the fact that 20 grains of nitrate of urea require 12·35 grains of nitrous acid for decomposition, which again corresponds with 27·64 grains of nitrate of soda.

This process of Gay-Lussac has been very generally introduced, but the manner in which it is carried into practice has been greatly modified. It was soon found that no material or combination of materials could be used in the construction of the denitrating columns, which could resist the great heat and intense chemical action that ensued, when sulphurous acid was passed through the nitrous sulphuric acid. The denitration was then attempted by passing steam up through the nitrous sulphuric acid, the nitrous gases escaping into the chamber, and the diluted sulphuric acid, which was only 110° to 120° T., running off at the bottom. This plan is now in common use, and the columns are either constructed in segments of cast-iron bolted together, or of lead lined with fire-brick, or of fire-bricks carefully laid in fire-clay mixed with coal tar. It is, of course, attended with this drawback, that all the sulphuric requires to be concentrated again and again, entailing both the expense of the process, and the waste of acid.

In some large manufactories, the original proposal of Gay-Lussac is carried out by the following arrangements. Instead of several small absorbing or nitrating columns, one large tower of lead, 14 lbs. to the foot, is erected and lined with half-thicks, made to suit the curve. The tower is 7 to 8 feet in diameter, and 50 to 60 feet high, and filled with hard burnt coke. The denitrating column is also increased in dimensions, being made 6 feet in diameter, and 20 feet high. It is built of fire-bricks, laid in china-clay, and the whole is enclosed in a casing of 10 lb. lead. It is fitted with fire-bricks placed crosswise, leaving open spaces between the bricks. A stream of nitrous

sulphuric acid and of chamber acid is allowed to run over
the bricks, and a limited quantity of sulphurous acid passed
upwards to denitrate the acid, which does not fall below
140° to 145° Twaddell. The expense of concentrating the sulphuric acid is thus almost entirely saved, while the chamber acid
itself is denitrated previous to use in the works. A stream of
acid flowing through a ¾ inch pipe is denitrated by a current of
gas from ten ordinary pyrites burners.

In other manufactories, all the chamber acid is denitrated
in the ordinary columns, but the adoption of one large tower is
now gaining the preference.

With these modifications the consumption of nitrate of soda
has fallen to 2 or 3 per cent. when sulphur is burnt, and to 3 or 4
per cent. when pyrites is used as the sources of the sulphurous acid.

The following modifications of Gay-Lussac's apparatus are
described in Muspratt's Chemistry.

Instead of the closed cisterns, the following simple plan is

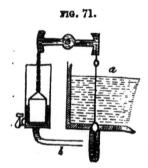

FIG. 71.

substituted : a represents the cistern containing the supply of acid, adjoining
which is a circular vessel of lead of the
same height as the cistern, 12 inches
diameter, and connected with it by the
pipe, b. The upper part of this tube is
conical, and a conical lead plug is accurately fitted into it. This plug is cast on a
long iron rod, which projects 8 or 10
inches below the plug, all the iron being
carefully coated with lead, to protect it
from the acid. This extension of the rod serves to keep the plug
in its seat. A leaden bucket hangs in the circular vessel, suspended by a chain from one arm of a lever, freely moving on its
axis. The other end of this lever is connected by a chain with a
hook at the upper surface of the plug. The lever is fitted with
semicircular pieces of iron, over which the chains work, so as to
move in a vertical direction. The cylindrical vessel is fitted with
a tap in connection with the delivery-tube. It is easy to see
that the upward movement of the bucket in the cylinder, will
send the plug into its seat, and by setting the apparatus to work
at any given pressure, the flow of acid will be uniform.

Another modification relating to the mode of using this
nitrous sulphuric acid, is thus described by Muspratt. A round

leaden vessel, 12 inches high and 18 inches diameter, is placed within the chamber, as near the burner-stalk as possible. Three pipes pass through the sides of the chamber into this vessel, conveying steam, water, and the nitrous sulphuric acid. The round vessel is filled with water, which is raised to the boiling point by steam, and then a stream of nitrous sulphuric acid admitted, when the evolution of the nitrous acid follows. This appears to be a very simple and effective arrangement.

Another arrangement for denitrating the nitrous sulphuric acid has been proposed, and consists in placing a small vessel of earthenware or hard stone, so as to form a part of one of the sides of the chamber. The nitrous sulphuric acid flows from a cistern, P, through the pipe R, the supply being regulated by a cock, Q. A jet of steam is admitted from the steam pipe, CC, through a rose at the end of the pipe inside the small vessel, which decomposes the nitrous sulphuric acid; the nitrous products escape into the chamber through the pipe T, while the sulphuric acid runs into the chamber by the pipe V.

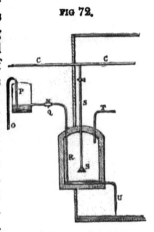

FIG 72.

The last plan we shall notice for saving waste nitrous gases has been described by Payen. It consists in passing them through a series of Woulfe's vessels, as shown in Fig. 73.

The gas, as it leaves the chamber, H', passes through a cooler or condensing vessel, JJ', following the course indicated by the arrow, and then passes through one bottle after another, ranged in ranks, ll'l"; the residual gas escapes into a chimney, M, through a kind of steam-chest, N, when the draught is regulated by a damper, which slides over a diaphragm pierced with holes. As the nitrous sulphuric acid is formed, the vessels are emptied from time to time, those of the first range being filled from the second, the second from the third, and these last being supplied with sulphuric acid of 1·75 specific gravity. The nitrous sulphuric acid is conveyed to the chamber in the manner already explained.

Payen describes the following mode of working the plan of Gay-Lussac.

The gases having passed through the condenser, M'M', Fig. 63,

FIG. 73.

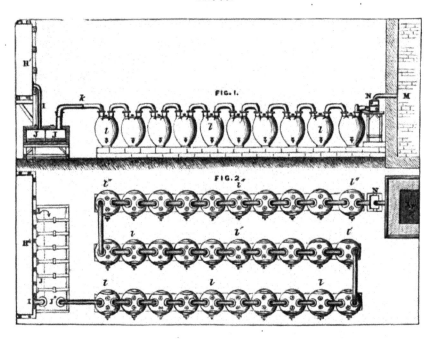

p. 83, escape through a pipe, into the lead columns, $J''j$, filled with coke; near the upper surface of which is a small triangular box, l, supported by an iron axis covered with lead. This box or hopper is divided into two parts, which are filled alternately with concentrated acid from a vessel, n, always full. As soon as one compartment is filled with acid, the box, losing its centre of gravity, falls to one side, and pours out its contents, during which the other compartment is being filled, which, in its turn, is emptied towards the opposite side. The chamber gases, meeting the descending current of strong acid, are deprived of their nitrous compounds, and the residual gas escapes through the chimney, O, into the air. The nitrous sulphuric acid flows off below, and is conveyed into a vessel, J'', from which it is forced up into the cistern, p. It is allowed to flow from this cistern into a little box, q', similar to that just described, from which it pours in cascades over sheets of lead fixed in a small chamber, $c''c^9$. The gas from the sulphur burners, enters this chamber through the pipe, CC', and following the course of the arrows, along with a jet of steam admitted at C'', denitrates the nitrous sulphuric acid, in the manner already explained.

Another plan of saving the waste nitrous compounds has been proposed by Kuhlman, which consists in absorbing them by means of carbonate of baryta, crystallizing out the nitrate of barytes, and then heating this latter, as has been mentioned, in a close vessel, passing the nitric acid into the chamber, and obtaining caustic baryta.

A modification of this idea has been patented in this country, in substituting lime for the baryta, and decomposing the nitrate of lime by sulphuric acid, in the same way as when nitrate of soda is employed. In this case the sulphuric acid is lost.

Another plan has been in operation at more than one establishment with fair success. It consists in allowing a stream of ammoniacal water from gas works, to flow down a coke column, up which the waste chamber gases are ascending. The ammonia is converted into sulphate of ammonia, and where there has been an excess of sulphurous acid, it has been found, that on heating the liquor in contact with the air, the sulphite and hyposulphite of ammonia have rapidly passed into sulphate.

Mr. Oliver, with a view of preventing any escape of nitrous compounds, has patented the idea of throwing in a current of sulphurous acid in different chambers, but we cannot see the advantage of this proposal.

5. CONCENTRATION OF THE WEAK ACID.

After the full details we have already given of this manufacture, it is scarcely necessary to remark, that the strength of the acid in the chambers, has an important influence on the consumption of the nitre. Strong sulphuric acid absorbs sulphurous as well as nitrous acid, and hence it has been found, that the most economical strength at which to keep the acid in the chamber, varies from 1·600 to 1·650. As acid of this strength contains a large percentage of water, the separation of a portion is effected by various plans for concentration, which we will now describe.

The concentrating apparatus consists either of leaden pans and a platinum retort, or of glass bottles. The former, Fig. 74, are erected over a fire, A, and supported by iron plates, aa, against which the flame beats; dd are incisions through which the lower pans are filled. The evaporation precipitates a little sulphate of lead and oxide of iron (anhydrous), whilst the boiling point of the liquid, at about 65 per cent. of hydrate, rises so high as to endanger the pans from the great heat

FIG. 74.

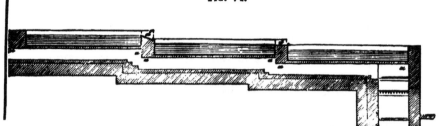

and to cause a useless loss of acid by evaporation. When, therefore, about 11 per cent. of water has been evaporated, with which nitric oxide, nitric and sulphurous acids, pass off, the acid, at about 1·7 specific gravity, is conveyed through the syphon, x, into the platinum retort, Fig. 75, the fire under which assists the working of the pans. It is necessary to avoid the presence of the nitrous compounds in the acid going to the platinum vessel, the metal of which would be attacked. This is accomplished, when necessary, by adding a little sulphate of ammonia to the acid in the leaden pan. The syphon x, which is here represented as closed, is

FIG. 75.

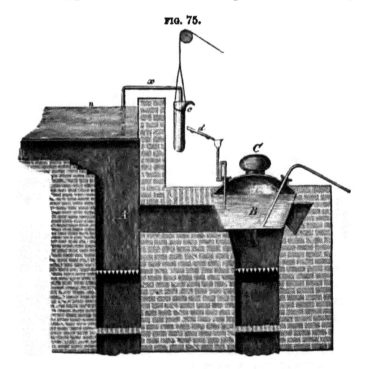

worked without a stopcock by the vessel c, on filling the retort, in a remarkably simple and ingenious manner. When this vessel is lowered with its spout to the spout d, the outer limb of the syphon, which is constantly full, becomes lengthened below n, and acid consequently flows out.

These platinum vessels are made in London, and vary in size. They are charged according to weight, and we give here the size and weight of some actually at work, viz., to hold

10 gallons, about 340 ounces, at £1. 6s. 6d. per oz.
15   „   „   440   „   „   „
20   „   „   520   „   „   „
30   „   „   650   „   „   „

The syphon, filling cup, gauge pipe, floats, &c., weigh from 100 to 120 ounces, which are charged at £1. 8s. per oz. A still, holding 50 gallons, produces a carboy of acid per hour.

During the boiling, the contents of the retort separate into commercial acid and some acid water, which is conducted by the tube in the capital, and its spiral leaden continuation, to one of the pans, where it is used to concentrate a quantity of weak acid from the chamber. If the heat were further increased, the boiling point would suddenly rise to 619° F., and hydrated sulphuric acid would distil over, which of course is not desirable.

FIG. 76.

The preceding wood-cut (Fig. 76) shows the manner in which

7*

the platinum still is erected in the concentrating house of the
vitriol works.

The high price of platinum vessels renders it very much to the
interest of the manufacturer that they should be in constant use ;
yet it is impossible to draw off so powerful an acid at a high
temperature into the glass carboys, in which it is sent out, and
leaden coolers cannot be used. Hence arises the necessity for the
platinum syphons, Figs. 77 and 78, which at the same time answer
the purpose of coolers. The syphon *abc* is let into a wide tube,
dd, which is supplied with a current of cold water through *e*.
The water, after becoming warm, flows off at *g*. The cooling is,
therefore, effected by surrounding the hot acid in *bc*, with a current of cold water passing in an opposite direction, the effect of

FIG. 77.

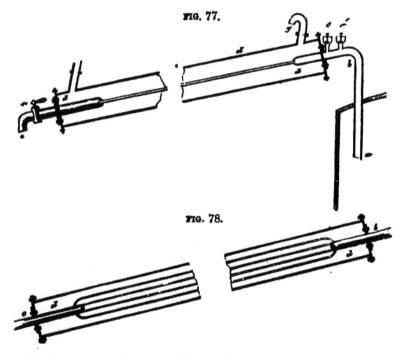

FIG. 78.

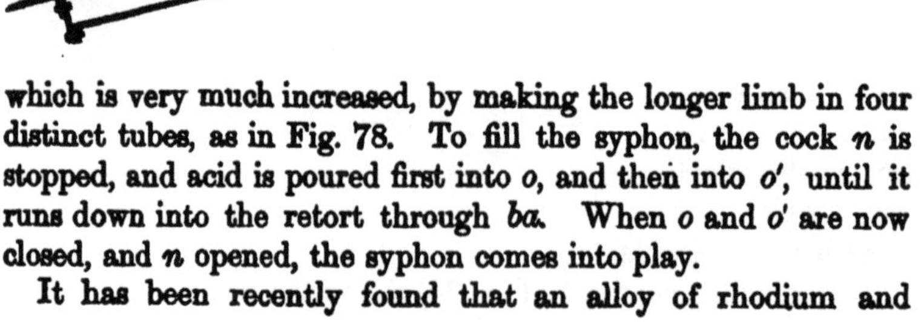

which is very much increased, by making the longer limb in four
distinct tubes, as in Fig. 78. To fill the syphon, the cock *n* is
stopped, and acid is poured first into *o*, and then into *o'*, until it
runs down into the retort through *ba*. When *o* and *o'* are now
closed, and *n* opened, the syphon comes into play.

It has been recently found that an alloy of rhodium and

iridium, to the extent of 10 to 15 hundredths, with the platinum employed in making the tubes, syphons, cocks, &c., answers better than the latter metal alone, in resisting the friction and the action of the sulphuric acid. This alloy is less ductile than platinum, and cannot be employed in the manufacture of the large stills, or in the long tubes made without soldering.

Clough has proposed to concentrate sulphuric acid in the following apparatus, by which he dispenses with the use of the platinum still. It is well known that concentrated sulphuric acid, when hot, acts rapidly on lead, while it has little or no action in the cold. He employs a large lead pan cased in an iron pan, the intermediate space being kept full of cold water. The lead pan is lined inside with fire-bricks, and the whole covered with a fire-brick arch. A fire-place is built at one end, and the flame carried over the surface of the acid in the pan. The heat of the acid only slowly penetrates through the brickwork to the lead pan, which is moreover kept cool by a supply of cold water to the space between it and the iron casing.

The lead pan is doubtless protected by this plan of concentrating, but as Kopp remarks, and as every manufacturer has found, when employing surface heat in concentrating this acid, there is a most serious loss of acid which often escapes in dense vapours.

Kuhlmann, in France, and Pontifex, in this country, have proposed to conduct the concentration of sulphuric acid in vacuo.

Anthon has lately made some experiments on the quantity of lead dissolved by sulphuric acid, in different stages of concentration, and the results show how very objectionable this plan is, when carried beyond certain limits. Anthon found in the concentrated acid, after being cooled down to 68° F., the following quantities of sulphate of lead.

Acid of 1·724 sp. gr. contained $\frac{1}{1000}$th of sulphate of lead.
    „   1·791 „   „ $\frac{1}{50}$th   „
    „   1·805 „   „ $\frac{4}{100}$ndths   „

There is, however, a large consumption of acid of 1·700 sp. gr., which can be safely made in leaden vessels, and this is manufactured in some places, by applying the heat to the surface of the acid, as shown in the figure, where the leaden pan *a* is built into the brick-work, and the sides protected from the action of the flame, in addition to which some manufacturers attach a large pipe to the sides of the pan, through which a current of cold

FIG. 79.

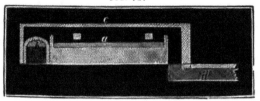

water is allowed to flow. The heat from the coke fire *b* passes over the acid, and all the gaseous products escape through the flue *d*.

' A modification of this plan is to pass the gas from the pyrites kiln over the surface of the acid in the furnace *c* in the above figure ; but, instead of the underground flue, a pipe at the upper side conveys the gas and vapour to the chambers. This, of course, saves the loss of the nitrous compounds which exist in the chamber acid, with which the lead pan is filled for concentration.

Fig. 80 represents another plan for concentrating chamber acid in close pans, in which a similar leaden vessel is employed, but it is covered with lead supported in the same way as the chamber sheets.

For a long time platinum stills only were employed in the

FIG. 80.

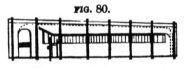

FIG. 81.

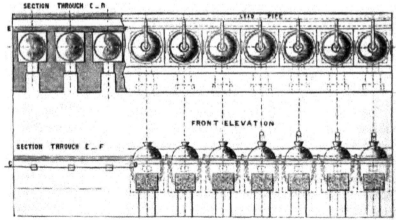

production of acid of 1·845, but within the last few years glass vessels have come into very extensive use. We annex detailed drawings of the arrangements now in use, where glass vessels (Fig. 81) are employed for this purpose. The glass retort or balloon is placed in a metal pot, and a thin layer of sand on the bottom, keeps the two from coming into immediate contact. Each retort is heated by its own fire, and the water which distils off, containing a little acid, is carried by an earthenware pipe to a leaden tube at the side, where a portion condenses, and the remainder escapes into the chimney. We have shown one range of 10 glass retorts, but in large establishments the number is doubled and they are placed back to back. It is necessary to conduct the operation in brick buildings with a ceiled roof, so as to prevent all currents of cold air from entering, which have a tendency to crack the glass vessels. In some places, that portion of the retort which rises above the rim of the metal pot, is protected by an iron hood made with a flange to rest on this rim, and in two halves, so as to be easily removed when necessary.

After some hours, the brown colour of the chamber acid disappears, which occurs when the acid has reached the highest point of concentration. The contents of the retort are allowed to cool, when they are drawn off into the carboy by means of a syphon, similar to that known as Mitscherlich's. The only precaution necessary, is to allow the water with which the syphon is filled, to escape before running the contents of the retort into the carboy. The warm retorts are filled with partially concentrated warm brown acid, by which a saving is effected, and more oil of vitriol obtained in a given time from the retort. Mr. Hills proposes, with the same object, to draw off only a part of the concentrated acid, and to fill up the retort with hot partially concentrated chamber acid.

It sometimes happens that the ebullition in these retorts is very violent, and it has been proposed to obviate the danger of breaking the glass from this cause, by putting a little sand or sulphate of lime in the retort, which is said to moderate the boiling.

In some localities in France, where glass is very cheap, the concentration of this acid is performed in glass retorts, which hold about 1 cwt. of acid of 1·66 sp. gr. The retorts are arranged in rows, as shown in Fig. 82. They are placed in metal pots, and the space between the glass and metal is filled with sand or fine

clay.   The neck of the retort passes into a glass tube, which conveys the weak acid water into a carboy.

FIG. 82.

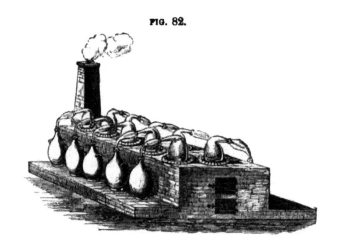

The range of retorts is heated by one fire, which requires great
care in its management to prevent breakage.   When the concentration is finished, the retorts are allowed to cool, and the oil of
vitriol is syphoned off into carboys.

We conclude this account of the concentration of acid, with a
description of a very ingenious plan by Mr. Gossage.

Fig. 83 is a longitudinal vertical section of the apparatus for the
concentration of sulphuric acid, consisting of a tower, T (containing a bed of siliceous pebbles or other suitable material), and
an apparatus, H, for the production of heated air.   Fig. 84 is a
transverse vertical section of the tower.   The tower, T, is built of
siliceous flagstones from Yorkshire ; A is the casing of flagstones
joined together ; this casing of flagstones is lined with an internal
casing, B, of well-burned semi-vitrified bricks ; and the casing of
flagstones is coated externally with a covering of planks, C.   The
whole is held together by bars of iron, DD.   The bottom of the
tower is formed of a thick siliceous flagstone, K, in which channels, aa, are cut contiguous to the lower part of the stone casing,
AA ; also another channel, b, extending from the channel a' to
the outside of the bottom, K, these channels being formed for the
purpose of conducting the concentrated acid from the tower
and delivering it into a suitable receiver, I, where it is allowed to
cool.   E is a perforated channel or flue communicating with the

FIG. 83.

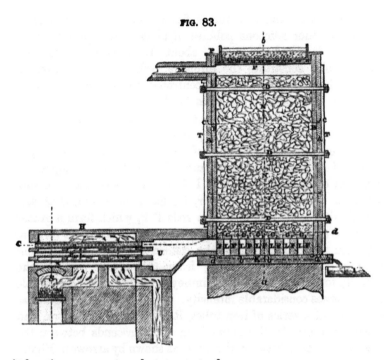

air-heating apparatus by means of a flue, U, and extending from that side of the tower to the opposite side. FFF are walls extending from each side of such channel or flue E to the adjacent side of the tower; these walls, FFF, being so placed as to leave spaces or air channels, LL, between them; these spaces or channels communicating with the channel or flue E, by means of the perforations in the walls. The spaces or channels LLL are covered by means of bricks or stones, GGG, placed at some distance from each other; thus a complete grating of stonework or brickwork is constructed, extending over the bottom of the tower, T, through which the heated air from the apparatus H can pass, and through which the acid can flow as it descends from the bed of materials in the tower, T. Upon this grating a covering of siliceous

FIG. 84.

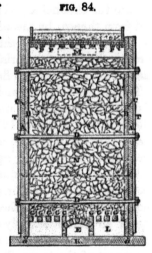

pebbles about the size of a man's fist is spread, and upon these are placed other siliceous pebbles of about half this size, until the tower is filled to within about 18 inches of the top. These pebbles constitute the diffusing bed, N, of the tower, T, through which the acid, to be concentrated, percolates. A leaden cistern, O, perforated with holes in the bottom is fixed at the upper part of the tower, T. The bottom of this cistern is covered with a layer of small siliceous pebbles of such a size as not to pass through the holes, a layer of coarse sand follows, and upon this a bed of fine clean white sand. The cistern O is supported by means of bent rods of iron, P P, covered with lead and supported on the sides of the tower, T; flat rods of iron, Q Q, also covered with lead, rest upon the rods P P, which form a grated flooring to sustain the bottom of the cistern. The acid for concentration is introduced into the cistern O from any convenient reservoir. M is a flue leading from the upper part of the tower, T, to some convenient chimney draft, which it is desirable should possess considerable intensity. The air-heating apparatus, H, consists of a series of iron tubes, R R R, set in brickwork, and heated by a fire, S, the flame from which ascends between the tubes and descends between the same, as shown by arrows in Figure 81; air enters at the open end of the tubes and passes as heated air into the channel U, which communicates with the channel E, and from thence, by means of its own ascending power, assisted by the chimney-draft, this heated air is caused to percolate through the bed of materials, N, contained in the tower, T, and to pass off through the exit flue, M.

The fire, S, is lighted, and the heated air passes through the bed of diffusing materials, N, until the air, escaping by the flue M, has a temperature of about 212° F. The weak acid from the cistern O passes through the perforations in its bottom, and is distributed over the surfaces of the upper layer of pebbles in the tower, T, and from thence percolates through the bed of pebbles, thus becoming diffused in thin films on the surfaces of the pebbles, and thereby exposed to the evaporative action of the currents of heated air simultaneously traversing this bed of materials, and thus becomes concentrated. The concentrated liquid arrives at the lower part of the tower, T, and being connected by the channels a, passes through the channel b into the receiver, I. The supply of acid to the cistern O is regulated according as it passes from the tower, T, sufficiently or insufficiently concentrated.

6. MANUFACTURE OF SULPHURIC ACID WITHOUT CHAMBERS.

It has been proposed to manufacture sulphuric acid without the leaden chambers, and Döbereiner was the first to suggest the use of spongy platinum, which Phillips proposed to employ in the form of balls heated to redness, while a mixture of air and sulphurous acid was passed over them. We are not aware that any experiments have been made on a scale sufficiently large to test the commercial value of this suggestion, but we have seen the process at work in a laboratory : the operation was as perfect as it is possible to conceive.

Schneider, of Charleroi, and Laming, of London, have proposed to substitute pumicestone for spongy platinum ; we will describe the apparatus of the former, who exhibited a model at the Belgian Industrial Exhibition in 1847.

A is the sulphur-burner, enclosed in brickwork, B, to retain the heat ; C is a fire-place to heat the burner, when starting to work ; D D are doors for introducing the sulphur and withdrawing the ashes ; E is an iron pipe for carrying the gas from the

FIG. 85.

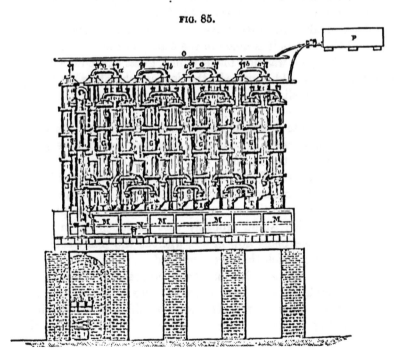

FIG. 86.

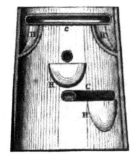

FIG. 87.

burner; F another pipe of lead, and G G G are upright pipes of earthenware.

H H are saucers filled with pumicestone, and I I earthenware plates for carrying the acid from one pipe to another.

L, lead cock for the flow of acid into the cistern M, from which the acid is run off to the concentrating vessels through the cock N.

The upright pipes are connected by smaller ones marked O O. The water is supplied from the cistern P, through the pipes a a, and the steam regulated by the cocks b b.

c c are small pipes with three openings on the upper side to carry off the acid as it is made to the centre of the upright pipes.

When this plan was first made public, great expectations were raised that it would supersede the leaden chambers, but they have not been realized.

Margueritte has patented the following processes. He heats chloride of magnesium mixed with clay, and decomposes phosphate of lime by the muriatic acid which he obtains. He then decomposes sulphate of lime by the phosphoric acid, reproducing his phosphate of lime, and obtaining sulphuric acid. It is evident that a large consumption of muriatic acid would take place in this process, which would probably cost as much as sulphuric acid made on the plans now in use, without the many collateral advantages with which they are combined.

A somewhat similar plan has been proposed by Shanks, for obtaining sulphuric acid from gypsum. He mixes 86 parts of gypsum, finely ground, with 140 parts of chloride of lead in a large quantity of water heated to 140° F. On agitating the whole, a change of elements ensues, $CaO.SO + PbCl = CaCl$

+ PbO.SO^3; the latter falls to the bottom as an insoluble powder. The solution of chloride of calcium is run off, and muriatic acid added to the sulphate of lead, PbO.SO^3 + HCl = PbCl + HOSO^3; the latter is removed and boiled up to strength, while the chloride of lead serves for the decomposition of a fresh portion of gypsum.

We must also notice the plan patented by McDougall and Rawson, which consists in passing a current of sulphurous acid through nitric acid; sulphuric acid and binoxide of nitrogen are formed, but the latter is converted into peroxide of nitrogen by the oxygen of the air which accompanies the sulphurous acid; and as it passes into water at 100° C., it is converted into nitric acid and binoxide of nitrogen. The former remains in solution, while the latter, as it escapes in a gaseous form, again combines with atmospheric oxygen, and thus the process goes on without loss of the nitrogen compounds. The great difficulty of this process would consist in managing the apparatus, and preventing the great wear and tear of plant.

Persoz's process likewise depends upon the oxidation of sulphurous acid by the action of nitric acid and water,

$$SO^3 + NO^4.HO = SO^3.HO + NO^4;$$

the sulphurous acid being made to pass, either through weak nitric acid heated to 100° C., or through a mixture of nitric acid or a nitrate with muriatic acid, whereby hypochloronitric acid, NO^{3}Cl^2, is formed; and on the fact that peroxide of nitrogen or hypochloronitric acid, subjected to the action of air and steam, yields nitric acid, which again oxidises the sulphurous acid.

His apparatus consists of three principal parts, A, B, and C. The form and dimensions of A depend upon the nature of the material from which the sulphurous acid is obtained. The sulphurous acid meets the heated nitric acid in B, and the part C is intended for the recovery of the nitric acid.

Persoz employs a peculiar arrangement for burning his sulphur. It consists of a metal or clay retort, d, similar to a gas retort, and heated by a fire, g, underneath. Air is forced into the retort by a cylinder or fan by the pipe e, and the sulphurous acid escapes through the pipe f. To prevent the sublimation or loss of sulphur by the force of the current, Persoz fits a series of tubes of porous clay into the upper side of the retort, the lower end being closed. These tubes are filled with sulphur from the outside, and as the sulphur melts, it filters through the clay and takes fire as it makes its appearance in the retort. The lower end is, therefore, surrounded with flame, which is supported by the trickling of the sulphur through the clay and the supply of air from the blast.

The pipe f passes into a large Woulfe's vessel, h, which is sup-

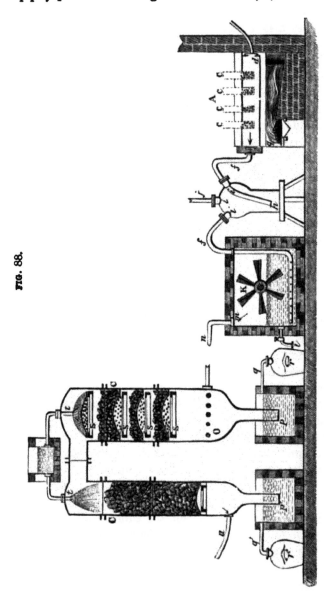

FIG. 88.

plied with air through the cock *j.* All the gases and air then pass into the vessel B, in which a fan, K, is fixed to keep the nitric acid inside in motion. The vessel B is composed of two parts, so as to be easily taken to pieces, and it can be emptied at pleasure by the cock *l.* There is an opening at *m*, which connects this vessel with the chimney, *c.* This chimney is fitted with several openings, *o*, for as many pipes from B. The chimney tapers towards the lower end, which dips into the liquid nitric acid, *p*, which overflows through *q* into the carboy, *r.* The chimney is divided into three or four compartments by the perforated shelves, S S S S, through which the acid vapours ascend. A jet of water from the pipe *t* is spread by a rose over all the surface, to absorb the nitrous gases. There is a second chimney, C′, similar ·to the first, to complete the absorption, from whence the permanent waste gases escape into the air through the pipe *u.* The condensed sulphuric acid overflows through *p′* into the carboy, *r′.*

Persoz employs the same apparatus for the decomposition of sulphurous acid by sulphuretted hydrogen in contact with water or steam.

Keller obtains sulphuric acid from sulphate of lead in the following way. He obtains sulphuretted hydrogen from any well-known source, proposing, among others, to expose sulphate of lime or baryta with coal and sand in a heated state, to steam, and conveys the sulphuretted hydrogen into vessels filled with water, in which sulphate of lead is kept suspended. The sulphuric acid dissolves in the water, while the sulphide of lead precipitates, and, by roasting, furnishes a fresh supply of the metallic sulphate.

This weak sulphuric acid he proposes to concentrate in vacuo. The apparatus may be made of platinum or platinized metal, or even of earthenware. The two halves of the metal still are lined with lead, cast on them, up to the neck. The two halves are screwed together at the rims with the lead between, which makes a better joint than soldering. The still is connected with a reservoir to receive the vapour where it is condensed, thus forming a vacuum, and thus the process is continued.

In addition to the plans and processes we have either noticed or described, there are several others, the outlines of which will be found in our appendix, in the abstract of all the English patents relating to sulphur and its oxygen-compounds.

7. IMPURITIES IN SULPHURIC ACID.

English commercial oil of vitriol is not sufficiently pure for all purposes: it contains, besides hydrochloric acid (from the common salt of the impure nitrate), traces of nitrous acid, nitric acid, and particularly nitric oxide, which are detected by the purple colour which it assumes with green vitriol. On distilling the acid, sulphate of nitric oxide, being less volatile, is contained in the residue. The presence of the oxides of nitrogen is objectionable to the manufacturer, on account of their communicating to the acid, the property of attacking platinum. To get rid of that portion which would otherwise reach the platinum retort, $\frac{1}{10}$ to $\frac{1}{4}$ per cent. of sulphate of ammonia may be added, before the heat is applied; the ammonia, and the oxides of nitrogen, are then converted into water and nitrogen gas. (Pelouze.) Lowe has also proposed to expel the nitrous compound by oxalic acid, which is decomposed by the strong acid at 230° F., into carbonic acid and carbonic oxide. The latter in its nascent state deoxidises the higher nitrous compounds, which escape either as nitrogen or nitrous oxide together with the carbonic acid. Besides these impurities, there are also sulphate of lead, which is separated by dilution, anhydrous persulphate of iron (as a white deposit), which vanishes on dilution, selenium, and arsenic, from the sulphur or pyrites. This latter substance, Lowe says, can be driven off by mixing some common salt with the hot acid, when the chlorine combines with the arsenic and is evolved as chloride of arsenic.

From recent experiments by Professor Bloxam[*] it appears that all the sulphuric acid of commerce—even that which is reputed to be quite free from arsenic—gives indications of that metal when examined by delicate tests, viz., by Marsh's test, or by decomposing the acid diluted with water by the electric current, and passing the evolved hydrogen through a glass tube heated by a lamp-flame, as in Marsh's apparatus: every sample thus examined gave rings of metallic arsenic in the reduction tube. Moreover, it was found that the arsenic could not be completely eliminated either by fractional distillation of the acid, as proposed by some chemists, or even by distilling it with common salt, the residual acid still exhibiting traces of arsenic in every instance. Now it is of great importance to the analytical chemist, especially in ex-

[*] Journal of the Chemical Society, xv. 52.

aminations for arsenic in cases of suspected poisoning, that the sulphuric acid which he uses, whether to evolve hydrogen gas with zinc, as in Marsh's apparatus, or for acidulating the liquid to be subjected to electrolysis, or in any other part of the process, should be quite free from arsenic. For, although the quantity of that metal present, in so-called pure sulphuric acid, may be so small as not really to deceive an experienced analyst, still as the evidence has to be judged of by persons who are not chemists, it is much more satisfactory, if he can say positively, that his reagents contained no arsenic whatever.

With this view, as the entire purification of the commercial acid appeared hopeless, Professor Bloxam endeavoured to prepare the acid in a state of purity in the first instance. For this purpose, sulphurous acid evolved in the ordinary way from sulphuric acid and copper, and nitric oxide evolved from nitric acid and copper, were made to pass into a large glass globe, into which air and vapour of water were also introduced at intervals, as described at page 51 ; but the sulphuric acid thus obtained was still contaminated with arsenic, proceeding partly from the copper, and partly from the oil of vitriol used to evolve the sulphurous acid : the substitution of mercury for copper diminished the quantity of arsenic, but did not remove it altogether, and the acid thus prepared was found to contain small quantities of mercury. But the difficulty was at length overcome, by evolving both the sulphurous acid and the nitric oxide, at comparatively low temperatures, viz., the sulphurous acid by decomposing hyposulphite of soda with sulphuric acid, the nitric oxide by heating nitric acid with protosulphate of iron. Both gases were then evolved at temperatures not much above the boiling point of water, and the sulphuric acid produced by their mutual action, was found to be *absolutely free from arsenic.* This process could not be economically carried out on a manufacturing scale ; indeed, for all ordinary purposes, there is no demand for chemically pure sulphuric acid ; but it might be worth the attention of the pharmaceutical chemist, to prepare pure sulphuric acid in this way for the use of the analyst, who would not object to pay a higher price for so indispensable a material.

Hayes makes use of the property which the acid $SO^3.2HO$ possesses of crystallizing, to free it from impurities. He first destroys the colour in the usual way by adding nitre to the hot acid, and afterwards gets rid of the nitrous compounds by sulphate of ammonia. The acid is now concentrated to 1·780, allowed to stand till quite clear, and then run off into shallow vessels, where it is cooled down to 0° C. When half of the acid has crystallized, the liquid portion is run off for ordinary use, the crystals are washed with pure acid and then melted.

7. *Manufacture of Fuming Sulphuric Acid.*

Of those sulphates which part with their acid at a red heat, without decomposition of the latter into sulphurous acid and oxygen, the persulphate of iron is the only one that can be used in the manufacture of the fuming acid. This substance is always evolved when green vitriol, a cheap, easily obtainable salt, is heated to redness. The starting point of the manufacture is, therefore, green vitriol (hence the name, *oil of vitriol*), or crystallized proto-sulphate of iron ($FeO.SO^3 + 7HO$), 6 equivalents of its water being driven off, before decomposition ensues ; the 7th equivalent, which is more intimately combined, is only expelled when that process begins. In the vitriol manufactories, the impure vitriol obtained by evaporation of the mother-liquors, which has no commercial value, is subjected to two consecutive operations, the one of which removes its water, the other its acid.

The furnaces.—Both these operations are effected simultaneously in the same furnace, the arrangement of which, as constructed at Hermsdorf, is seen in profile at Fig. 89. The furnace used at Radnitz, in Bohemia, is represented at Fig. 90. In the galley furnaces (*Galeerenöfen*) in general, as represented in Figures 89, 90, 91, the same fire heats two rows of vessels, $a\,a$, in which the decomposition is effected ; these pots are walled in at the necks, $c\,c$, and in such a manner, that the moveable receivers, $b\,b$, can easily be inserted, and made tight by cement. If, on the contrary, the necks of the retorts, a, were inserted into the receivers, b, the cement would be liable to fall into the acid. The grate, d, is carried throughout the whole length of the furnace to the chimney, l ; e is the ash-pit, and m the drying chamber for separating the water from the vitriol, which is heaped up upon the projecting plate, n. The walls, $c\,c$, are overlaid at the top with plates of clay, h. The Bohemian furnace is not essentially different from the one described, except that it is adapted for a double range of retorts. The retorts are 1¼ feet long, and 4 inches wide at the neck ; the receivers are of the same length, and are 1¼ inch wide at the

FIG. 89.

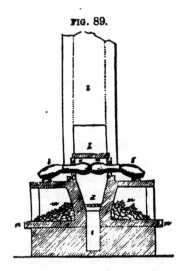

FIG. 90.

mouth; both retorts and receivers are from 5 to 6 lines thick
and made of earthenware. The furnaces contain from twelve
(as in Fig. 89) to thirty retorts on each side, which are placed
3 inches apart.

The present arrangement adopted abroad for making this
fuming acid is shown in Fig. 91. The earthenware tubes or retorts are arranged
in two rows to the
number of 200,
each containing 2
cwt. of dry sul-
phate of iron.
The furnace con-
sumes about 30
cwt. of lignite in
working off a
charge, and the
yield of acid is
about 45 per cent.
of the sulphate
employed. The
process of distil-

FIG. 91.

lation lasts 24 hours, and another day is occupied in cooling the
furnace, removing the apparatus, and preparing for the next
charge, so that three distillations are managed in a week.

8*

Process of distillation.—When the retorts are charged with
the calcined vitriol, a fire is made with dry pine wood, which
produces a long flame. The first portions that pass off, consisting
of very weak sulphuric acid and much sulphurous acid, are not
collected. As soon, however, as the grey, fog-like vapours of
anhydrous sulphuric acid appear, the receivers are connected, each
containing about an ounce of rain-water, and the distillation
begins. The heat is kept up until the retorts have been exposed for some time to a white heat, in all about thirty-six
hours; as long, indeed, as acid passes over. When this ceases,
and the receivers have become cool, they are at once removed,
the retorts emptied, and charged anew, which is done by means
of iron shovels; the same receivers are again adapted (without
having been emptied), and the process is then repeated. The
acid in the receivers is not so saturated with anhydrous sulphuric
acid, as to attain the strength of commercial oil of vitriol, until
this process has been repeated four times. In this manner,
45 to 50 per cent. of the dry vitriol is obtained as acid, which is
sent out in stoneware jars, with screw stoppers, covered with
cement. The residue is a reddish-brown earthy mass, called
Colcothar, or *Caput mortuum vitrioli*, which is used as a paint.

When anhydrous vitriol ($FeO.SO^3$) is heated to redness, the
protoxide is converted into peroxide, by taking from a portion of
the sulphuric acid, the requisite quantity of oxygen, the former
becoming reduced to sulphurous acid;

$$2 (FeO.SO^3) = Fe^2O^3.SO^3 + SO^2.$$

Basic persulphate of iron is, therefore, formed, which at a still
higher temperature, parts directly with its acid. From the great
tendency which protoxide of iron has to combine with oxygen, or to
become converted into peroxide, simple exposure to the air, or slight
roasting, is sufficient to effect this conversion when $2FeO.SO^3$ or
$2Fe + 2O + 2SO^3 + O^3$ (from the air) become $Fe^2O^3 + 2SO^3$ or
½ basic persulphate of iron, which, at a high temperature, parts
with the whole of its sulphuric acid, without any formation of sulphurous acid, and, therefore, without loss. This is the case in those
vitriol works, where vitriol is employed that has become almost
entirely peroxidised by exposure to the air, when for every part of
oxygen absorbed from the air, 5 parts more of anhydrous acid are
obtained. The chambers, M, therefore, besides separating the water
from the vitriol, also roast it to a certain extent. The amount of
produce proves what is here stated: for whilst protosulphate of

iron ought to produce 30·3 per cent. and the persalt 53·3 per cent. of dry acid, 50 per cent. of oil of vitriol is obtained, which is equivalent to much more than 30·6 per cent. of anhydrous acid.

A fuming acid, similar to that described, is now made by heating dry bisulphate of soda in vessels like those used in the manufacture of nitric acid. The products of the distillation are collected in carboys, and the cost of preparing it in this way, is much less than that of the German acid. Since ordinary sulphuric acid prepared in the leaden chambers has become so cheap, manufacturers have begun to use it in the vitriol works, either for dissolving colcothar (to form persulphate of iron) which is then decomposed in the manner described, or for replacing the water in the receivers. In the latter case, the product is of course contaminated with all the impurities of the ordinary acid.

Waltl has recently proposed to employ dry sulphate of alumina for the manufacture of fuming sulphuric acid.

Fuming oil of vitriol is an oily brownish liquid of 1·9 specific gravity. Its chief constituent is the hydrate $HO.2SO^3$; in the weaker acid, this is mixed with the simple hydrate $HO.SO^3$; and in the stronger, it contains anhydrous sulphuric acid SO^3 in solution. The latter substance is so volatile that it escapes, even at ordinary temperatures, from the oil of vitriol, and uniting with the moisture of the atmosphere, is condensed in the form of a visible cloud as $HO.SO^3$: hence the fuming. Together with selenium and earthy particles, sulphurous acid is present among the impurities, and is only gradually given off on dilution; it has been found to destroy the spongy platinum in the platinum lamp, when fuming acid is used to supply the hydrogen.

ACIDIMETRY.

This name is given to the volumetric processes employed for the testing of acids. The strength of acids is generally determined by means of a normal alkaline solution, prepared with caustic potash, caustic soda, saccharate of lime, or dried carbonate of soda. Potash or soda is generally preferred, and an alkaline solution is considered normal, of which a determined volume will saturate a known quantity of monohydrated sulphuric or other acid of known strength.

To make an acidimetrical assay with alkaline liquors, a few

drops of tincture of litmus are added, and then the test-liquor by means of a burette ; neutral alkaline salts are first formed, which do not act upon the litmus reddened by the acid ; but when the point of saturation is passed, the litmus regains its original colour. It is advisable to make at the same time, a check assay with water coloured with the same quantity of litmus.

To prepare the tincture used in acidimetry, a quantity of powdered litmus is digested in 7 or 8 parts of water, filtered, and an equal volume of alcohol added. The tincture of litmus loses its colour in well-stoppered bottles ; it ought, therefore, to be kept in jars only partly full and open. The tincture intended for acidimetrical testing, should be very sensitive ; for this purpose, it is divided into two equal parts, one of which is reddened by a drop or two of nitric acid, and the two liquors are then mixed. The mixture retains the blue colour of the litmus.

Acidimetrical assays may be made by other processes ; thus. sulphuric and hydrochloric acids may be estimated by means of test-liquors of chloride of barium or nitrate of silver ; nitric acid, by protosalts of iron and permanganate of potash, &c.

Testing for Acids found in commerce.

M. Violette has proposed to employ the alkalimeter of Gay-Lussac for testing the acids of commerce. This process is founded on the following fact : If 1 gramme* of acid whose equivalent is known, is taken for the purpose of testing, and 1 gramme of monohydrated sulphuric acid, whose equivalent is 49, and if the quantity of real acid contained in 100 parts, is represented by T, the general formula will be

$$T = 100 \times \frac{e}{49} \times \frac{p}{p'} ;$$

p and p' represent the quantities of potash or saccharate of lime, which are required to saturate the two acids, and which are determined by experiment ; e expresses the equivalent of the acid tested. Thus, it will be 49 for monohydrated sulphuric acid, $SO^3.HO$; 63 for hydrated nitric acid, $NO^5.HO$; 36·5 for hydrochloric acid, HCl ; 60 for hydrated acetic acid, C^4H^4O ; 22 for anhydrous carbonic acid, CO^3 ; 71 for anhydrous phosphoric acid, PO^5, &c.

M. Violette employs saccharate of lime to saturate the acids.

---

* If English weights and measures be preferred, it is merely necessary to substitute grains for grammes and grain-measures for cubic centimetres.

To prepare it, 1 litre of water is placed in a flask or decanter, and 100 grammes of sugar are added, and when it is dissolved, 50 grammes of slacked caustic lime in powder.

The mixture is agitated frequently, for some hours, and then filtered. This liquor, which ought to be kept in well-stoppered vessels, is prepared, so that about 50 divisions of the alkalimetrical burette, will saturate 10 cubic centimetres of normal sulphuric acid. The solution of saccharate of lime has the disadvantage of changing rapidly during the heat of summer. Whatever the liquor employed, it is necessary occasionally to verify the strength.

The instruments required for acidimetric operations are five in number; viz.,

1st. The burette (Fig. 92), divided into 100 divisions, each of

FIG. 92.   FIG. 93.   FIG. 94.   FIG. 95.

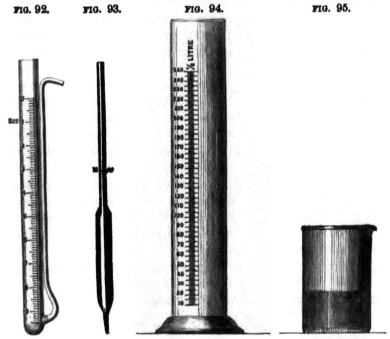

1 cubic demi-centimetre, for the purpose of measuring the test-solution of saccharate of lime;

2nd. The pipette (Fig. 93) of the capacity of 10 cubic centimetres up to the mouth, for measuring the acids. It is filled by immersing it in the liquid above the mouth. The forefinger is then placed on the upper orifice, and the liquid allowed to run out until it reaches the level of the mark ;

3rd. A test-glass containing 500 cubic centimetres (Fig. 94) ;

4th. A precipitating jar (Fig. 95) with flat bottom, for testing the acids ;

4th. A glass rod for agitating.

Estimation of Commercial Sulphuric Acid.

10 cubic centimetres of normal sulphuric acid are taken in the pipette, at a temperature of 15° centigrade, and poured into the jar (Fig. 95), and a few drops of tincture of litmus are added, which assumes a fine red colour. The graduated burette is then filled with a solution of potash or saccharate of lime, and the liquor added drop by drop to the sulphuric acid, while constantly agitating the jar ; by degrees the liquid becomes of a rose colour and at last blue, as soon as the saturation is complete. The number of degrees employed are then read off from the burette. Suppose them to be 50 ; then 10 cubic centimetres of normal sulphuric acid have been saturated by 50 cubic centimetres of the alkaline solution.

In a second operation, 50 grammes of the sulphuric acid to be tested, are weighed out and poured into the test-glass containing 500 cubic centimetres, and water added to about 1 centimetre below the mark. It is now agitated, and when the temperature of the mixture, which is at first raised, falls to 60° F., sufficient water is added to complete the 500 cubic centimetres or ½ a litre of liquid. 10 cubic centimetres of the acid solution containing 1 gramme of the acid to be determined, are taken up by the pipette and poured into the jar (Fig. 95), a few drops of the tincture of litmus are added, and the acid is saturated with the test alkaline liquor, as in the preceding operation.

The number of divisions employed are noted. Suppose them to be 46. The following formula will give the quantity of real monohydrated acid contained in 100 parts.

$$T = 100 \times \frac{46}{50} = 92.$$

Thus the acid tested contained 92 per cent. of mono-hydrated sulphuric acid and 8 per cent. water, and by multiplying 92 by ·8163, the quantity of anhydrous sulphuric acid is obtained.

M. Levol has proposed to determine the sulphuric acid contained in the sulphates, by means of nitrate of lead. If the liquor contains free acid, it is saturated with magnesia, and the excess separated by filtration, then a solution of iodide of potassium containing 10 per cent. of iodine is added. A solution of nitrate of lead is added, and a yellow precipitate appears, which dissolves on agitation. The operation is not complete until the mass takes a permanent yellow colour.

M. Levol employs for the determination of sulphuric acid a liquor containing 0·0413 gr. of nitrate of lead in the cubic centimetre, representing 0·010 gr. of sulphuric acid, or 1 gramme for the 100 cubic centimetres. This gives the result at once in 100 parts.

The acidimetric method of Fresenius and Will depends upon the weight of carbonic acid which is evolved by a given measure of acid. Bicarbonate of soda, dried in the air, is used, because it contains a larger percentage of carbonic acid than any other similar salt. It is not necessary that it should be perfectly dry or free from other salts, provided no neutral carbonate of soda is present, which can be easily removed by washing with cold water. When the acid to be tested, has been weighed, diluted with water, and introduced into the flask A (Fig. 96), a short

glass tube in the shape of a thimble, and closed at the end, is filled with bicarbonate of soda, and this is hung by a silk thread, between the cork and neck of the flask, as shown in Fig. 97, so that the salt and acid do not come into contact. Ordinary concentrated sulphuric acid is placed in B. The apparatus is now placed in the balance with a counterpoise, when the little tube is dropped into the acid, the cork being quickly removed and re-inserted.

FIG. 96.

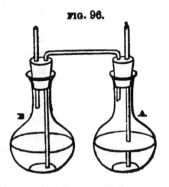

The apparatus is shaken, and when the operation is finished, the carbonic acid still remaining in the flask A is expelled, by placing it in hot water for a few minutes, and drawing air through the apparatus. When cool it is weighed, and every equivalent of carbonic acid expelled, indicates an equivalent of sulphuric acid in the specimen.

A simpler apparatus for these determinations, which gives equally accurate results, is a small light glass flask of 3 or 4 oz. capacity, having a lipped edge, and fitted with a cork perforated with two holes. Into one of these apertures is fitted a bent tube, *a*, carrying a drying tube, *b*, filled with chloride of calcium, and into the other, a narrow tube, *c*, reaching nearly to the surface of the liquid, and bent at an obtuse angle above the cork. A convenient quantity of the acid whose strength is to be determined, having been weighed out in the flask, a quantity of acid carbonate of soda or potash more than sufficient to neutralise the acid, is placed in a small test-tube about an inch long, and having its lip slightly turned over, so that it may be suspended by a thread. This tube is then let down into the flask by the thread, but not low enough to come in contact with the acid; the thread is fixed in its place by inserting the cork into the neck of the flask, and the whole apparatus is weighed. The orifice of the bent tube *c*, is then closed with a plug of cork or wax, the cork of the flask loosened sufficiently to allow the short tube *t*, containing

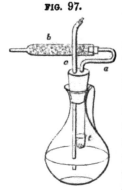

FIG. 97.

the alkaline carbonate to drop into the acid, and the cork immediately tightened. The carbonate is now decomposed by the acid, and carbonic acid escapes through the drying tube, the chloride of calcium retaining any moisture that may be carried along with it. When the effervescence ceases, the flask must be warmed to ensure the complete removal of the carbonic acid from the liquid, and after it has cooled, the plug must be removed from the bent tube *c*, and air drawn through the apparatus by applying the mouth to the extremity of the chloride of calcium tube, in order to remove all the carbonic acid remaining in the flask, and replace it by air. The whole is then again weighed, and the loss of weight gives the quantity of carbonic acid which has escaped. At the completion of the experiment, a piece of blue litmus paper must be thrown into the liquid in the flask; if it remains blue, the determination may be considered exact; but if it is reddened, there is still free acid in the flask, showing that the quantity of carbonate introduced was not sufficient to decompose it. In that case, a second small tube containing alkaline carbonate must be introduced as before, the apparatus again weighed, and the whole process repeated. The second loss of weight added to the first, gives the total quantity of carbonic acid evolved.

BISULPHIDE OF CARBON.

This substance, composed of 2 equivalents of sulphur and 1 eq. of carbon, is formed whenever sulphur is brought in contact with carbon heated to redness in a close vessel.

At ordinary temperatures it is a colourless liquid of 1·293 specific gravity, volatile, with a strong unpleasant odour, which, however, is removed by agitation with acetate of lead, or with strong sulphuric acid and bichromate of potash, and redistillation. The separation of sulphide of lead, when the acetate is employed, seems to indicate that the offensive odour is due to sulphuretted hydrogen. When thus rectified, it has an ethereal order, resembling a mixture of garlic and horse-radish. It boils at 110° F., and when ignited in the air, burns with a blue flame. Its volatility is so great that at a temperature of 65° to 70° Fah. it is easily ignited, even when the flame is at some distance, and when its vapour is diffused through the air, the mixture is explosive, which necessitates great care in its manufacture. It is nearly insoluble in water, but soluble in alcohol, ether, benzene, and turpentine, in all proportions. It dissolves sulphur, phosphorus, and iodine, very readily, and it has been proposed to apply this property to remove the bisulphide of carbon which exists in coal gas, by passing this gas through a layer of pieces of sulphur. It combines with the protosulphides of the metals, and has in consequence received the name of sulpho-carbonic acid.

Seyferth gives the following table of the tension of the vapour of bisulphide of carbon at different temperatures.

110° F. or 45·5° C.	1	atmosphere.
57·7	1½	do.
66·9	2	do.
74·3	2½	do.
80·6	3	do.
86·1	3½	do.
90·9	4	do.
94·5	4½	do.
99	5	do.

The Manufacture of Bisulphide of Carbon.

The application of this substance for dissolving and softening
caoutchouc in the process of its vulcanisation, in the gutta-percha
manufacture, and to other purposes, has induced various parties to
produce it on a manufacturing scale, and we will now describe
some of the processes employed for this purpose.

Process of Peroncel.

Mons. Peroncel was one of the first to commence this manufacture in France, and he employed a metal cylinder, A, carefully
coated on both sides with fire-clay. It was fixed on a solid metal
support, B, and the whole was surrounded with brickwork, C C.

FIG. 98.

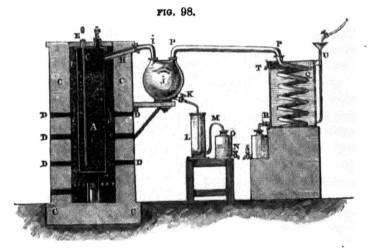

The lid was cast with two openings, E E, through one of which
an earthenware pipe, c, descended nearly to the bottom of the
cylinder, the lower end resting on hard burnt coke, with which
the cylinder was filled. As the carbon passed off, during the
operation, in combination with the sulphur, its place was supplied
by fresh charges of coke, which were filled in through the other
opening, E, in the lid.

A metal arm, H, cast with the cylinder, was connected with
a pipe, I, which was ground in at one end, the other passing
into an earthenware carboy, J'. A pipe with a cock, K, was
attached to the bottom of the carboy, which carried the liquid
into a florentine receiver partially filled with water. This re-

ceiver discharged the sulphide of carbon into a close vessel, O, by a pipe, M, from which it could be drawn off by a cock, N.

The uncondensed vapour escaping from the carboy, passed through another earthenware pipe, P P, to a worm, Q, made of zinc or earthenware, which was kept cool by a current of water entering by the funnel pipe, U, while the overflow was carried off at T.

The sulphide of carbon from the worm, ran off at R into another close vessel, S, furnished with a cock.

The space between the metal cylinder, A, and the brickwork, C C, is partially filled with fuel, which, when ignited, is supplied with air through the openings D D, and the products of the combustion escape at *i*.

Instead of the iron pipe, I, an improved connecting arrangement is shown in Fig. 99, where the beak of an alembic, O, made of glass, passes into the carboy,—the advantage of which is, that the vapours of sulphur are seen, when the sulphur is thrown into the cylinder too rapidly for combination with the carbon.

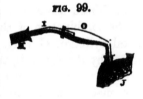

FIG. 99.

The fuel is now ignited, and the cylinder gradually heated until it becomes bright red, when pieces of sulphur are thrown down the pipe c, at intervals of a few minutes. The sulphur is converted into vapour, which, coming in contact with the red hot coke, combines with the carbon, producing the sulphide of carbon.

The cylinder lasts only about a week; when made about 6 feet long by 1 foot diameter, it will produce about 2 cwt. of sulphide of carbon per day.

M. Deiss has introduced a great improvement in this arrangement, by substituting cylinders of fire-clay, about 5 feet high and nearly 20 inches diameter. He places four of these cylinders in one furnace, and heats them by a fire placed at one side. Each cylinder is fitted with three openings in the lid instead of two, one of them being much larger than the other, for feeding in the coke during the operation, so as not to interrupt the working.

M. Deiss gives the following statement of the cost of the bisulphide, when made by his plan :—

Sulphur in rolls, 733 kil. at 24 fr. . . . . 175·92
Charcoal, 160 kil. at 4 fr. . . . . 16·00
Labour, 2 men, per shift . . . . 18·00
Fuel, coke, 28 hect. at 1 fr. 24 c. . . . . . 31·25
Rectification, general charges . . . . . . 13·00
Cleaning the apparatus . . . . 20·00

                                        274·17
Sulphur recovered, 300 kil. at 24 fr. . . . . 72·00

                                        202·17

Produce, 433 kil. of bisulphide, or the kil. costs
    nearly 47 c.

Apparatus of Gérard.

M. Gérard has introduced the following improvements.

The combination of the sulphur and carbon takes place in the metal cylinder, C (Fig. 100), which is elliptical in form, and is heated by the fire, AB, the flame from which passes all round, so as to heat the whole of the exterior surface.

A metal arm at the lower end, is connected by a flange to an outer pipe, which is fitted with a metal lid, E, with a hinge-joint. The sulphur is thrown in through this pipe.

The top of the cylinder is cast with a wide pipe, which can be closed at pleasure, by a thick metal damper, E'. Another pipe, F G, cast in one piece with the upper one, slopes forward, the flange on the lower end being screwed tight up to the flange of the upper portion, which is made of sheet iron. At the lower end of this latter pipe there is an opening closed by a lid, through which a plug or stopper can be inserted, to cut off the connection with the vessel H, when necessary.

The condenser is formed of three cylindrical vessels, JKL, which are connected with one another by three vertical pipes, the upper one having a pipe, M, which passes through the roof to allow the uncondensed gas and vapour to escape into the air These vessels act both as condensers and as receptacles for the liquid sulphide of carbon, which is drawn off by the cock N. These vessels are kept cool by the water which surrounds them.

The operation is conducted as we have already explained, and the sulphur, which escapes in a free state, is retained in the ves-

FIG. 100.

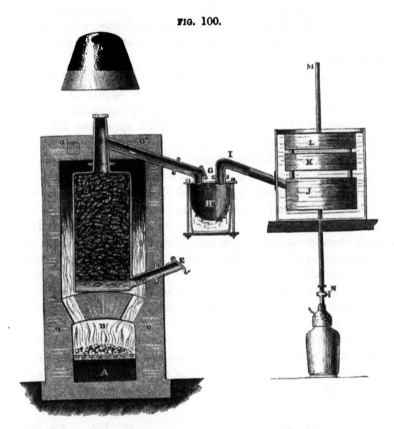

sel H, along with a portion of the sulphide of carbon in which it is dissolved.

This apparatus produces daily about 5 cwt. of this substance.

The rectification is easily performed in an alembic made of zinc, the cucurbite being heated in a water-bath, and the vapour condensed in a worm in the usual way.

Seyferth's Process.

The apparatus used in this process by which the inventor, Seyferth, states he can produce bisulphide of carbon at from 2d. to 3d. per lb., consists of an iron cylinder, cased in masonry, and built upon the open arch of the fire-place, from which the flames pass in a circular flue round the cylinder, and escape to the chimney from the upper part. The cylinder is filled with charcoal through an opening in the top. The sulphur is thrown in by a pipe passing nearly to the bottom of the cylinder. The bisulphide of carbon passes into the first condenser, where some is deposited along with uncombined sulphur, while the bulk escapes into the main condenser.

Seyferth proposes to employ the vapour of bisulphide of carbon to obtain a motive power, and describes an apparatus for the condensation of the vapour, which is equally applicable in the manufacture of the substance itself. This apparatus for condensation is constructed as follows: the vapour of the bisulphide of carbon enters a small square vessel, D (Figures 101, 102, 103), by the pipe, A. This vessel is placed in a water cooler, NN, and is divided into compartments by the metal plates, hh, which are pierced with small holes, and are open at the alternate ends, so that the vapour is made to traverse as great a distance as possible. The holes in these sieve-like plates, are filled with cotton threads, in order to present a large cooling surface. The water for condensation is admitted through the cock, B, and, falling on the first plate, drops through the holes on the surface of the cotton-threads, and so on through the whole series, until it reaches the lower open space of the vessel, where it rises to the level, U. The cotton-threads act beneficially, not only by spreading the water, but by cooling the inner portions of the steam and vapour as it passes across the falling shower of water, from side to side of the vessel, D.

FIG. 101.   FIG. 102.   FIG. 103.

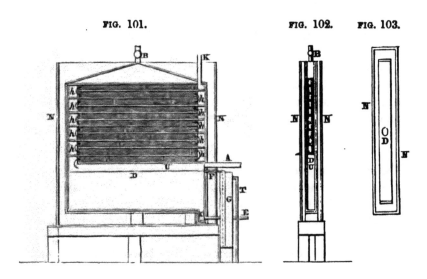

The condensed bisulphide of carbon separates itself from the water, and is collected in a recess, C, in the bottom of D, whence it flows through E into another vessel.

When the water has risen to the level of U, in the reservoir, D, it flows off through a pipe, F, to the bottom of a narrow deep vessel, G, to be finally drawn off by the pipe T. Any bisulphide of carbon which has been carried over mechanically, mixed with the water, collects in the bottom of G, whence it is removed from time to time.

The pipe K, in the top of the condenser, serves to open a communication with the external air, or for connection with a second condenser in case of need.

Salomon of Baltimore first suggested the use of bisulphide of carbon as a motive power, and Hand of Paris has had an engine of 3-horse power at work, constructed at the expense of the Prince of Hohenlohe.

Process of Messrs. Galy-Cazalat and Huillard.

Fig. 104 is a vertical section of an apparatus, patented in this country by Messrs. Galy-Cazalat and Huillard. It consists of a furnace of cylindrical form, the sides being formed of fire-bricks, which are 4 inches thicker round the fire-place, F, than round the upper chamber or oven, A ; *ab* and *cd* are grates of fire-clay, their bars and openings being placed alternately. This part of the apparatus is called the " caloriphore," and forms the principal reservoir of heat. The top of the upper chamber, U, is covered by an arch, L, and the whole is enclosed in a sheet iron case, B. C is a door, for admitting air to the fire-place; F, which does not require any grate or ash-pit; H is the fire-door for introducing the fuel; D is a door for introducing the bars of the caloriphore; P is a

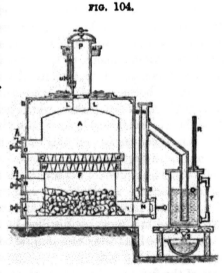

FIG. 104.

chimney, which can be closed by a door or lid on the top; M is a cistern or vessel containing melted sulphur, which is kept fluid by the heat of the chimney, or otherwise; O is a cock in a pipe leading from the vessel M into the chimney; F is a pipe leading from the lower part of the fire-place into a vessel, Q, which is partly filled with water; R is a pipe for the escape of gas; S is a cock for drawing off the liquid from the vessel Q; T is a glass tube, or water-gauge, for showing the height of the liquid in Q. The case B is air-tight, and all the doors are made to close air-tight. A fire is lit in the fire-place, F, which is filled with coke through the door H, which is then closed, while the door C and the lid on the top of the chimney, P, are left open. The combustion goes on, and the interior of the oven, A, and all the grates, a, b, c, d, become heated. Small cocks, hh, are placed on the doors H and D, and by opening these cocks, the interior of the furnace may be inspected from time to time. When the upper grate has acquired a cherry-red heat, the temperature of the furnace is sufficient to volatilize a large quantity of sulphur. The door C, and the top of the chimney P, are then closed by their doors or lids, which may be constructed and made air-tight, like the lids of gas retorts, or they may have a projecting bead or rib entering a groove which contains asbestos. The hot furnace being thus closed, the melted sulphur is introduced by opening the cock O, and falls upon the grates, which immediately convert it into vapour. The superheated vapour of the sulphur, in escaping from the furnace, is compelled to traverse the mass of incandescent coke from top to bottom, and in doing so, combines with the carbon to form bisulphide of carbon, which passes off in vapour through the tube N, which conducts it into the cold water in the vessel Q, where it condenses and falls to the bottom. Some sulphuretted hydrogen gas escapes at the pipe R. This operation does not yield the bisulphide of carbon in a state of purity, as it still contains some free sulphur, which gives it a yellow colour. It is separated from the water, by allowing it to run off through the cock S, and is afterwards distilled by means of a water bath, at a temperature but little exceeding its boiling point, 111° F. After the sulphur has been allowed to flow in for a certain time, the heat of the furnace falls; the cock O is then closed, and the door C and the damper on the top of the chimney, are opened. This sets the coke again in combustion, which increases its tempera-ture, and that of the grates and oven A. The operation is repeated and continued as long as required, fresh quantities of coke being introduced at H, and the vessel M supplied with sulphur when necessary.

Wagner has suggested that bisulphide of carbon might be produced by distilling some of the metallic sulphides with carbon in close vessels, thus—

$$\left.\begin{array}{l} 2\ PbS \\ C \end{array}\right\} \text{ would give } \left\{\begin{array}{l} 2\ Pb \\ CS^2. \end{array}\right.$$

This substance is already manufactured on the Continent to a large extent, Marguart of Bonn having made it since 1850, in which year he was producing 12 cwt. daily.

The Storing of Bisulphide of Carbon.

It is impossible to use too many precautions in manufacturing this substance, its volatility at ordinary temperatures being so great, as to render the atmosphere of a workshop unfit for respiration, while the introduction of a light is not merely followed by an explosion, but by the production of gases which cannot be respired, thus—

$$CS^2 + air \text{ (or } O^6 + N \text{ in excess)} = 2SO^2 + CO^2 + N.$$

The different operations ought to be carried on in distinct buildings, which must be thoroughly ventilated. For similar reasons, this substance must be stowed away in a suitable vessel, such as M. Gerard has proposed. This vessel is a zinc cylinder (Fig. 105), capable of containing from 20 to 30 gallons. A large pipe, BB, fits into a recess in the bottom, and rests by its rim on a short upright pipe of the vessel itself, which serves as an opening at the same time. A fold of paper round this tube, in which the rim of the tube BB rests, makes it perfectly air-tight. The vessel is filled by a funnel, during which operation, the air is allowed to escape through the cock, C. A little water is poured into the tube BB, to cover the surface of the bisulphide of carbon, and it is also

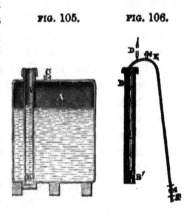

FIG. 105.   FIG. 106.

fitted with a lid. The liquid is drawn off by means of a syphon, constructed as shown in Figure 106.

9*

The continual breathing of an atmosphere, more or less impregnated with sulphide of carbon, exerts a slow but very pernicious action on the workmen. An antidote has been found in giving the men a solution of carbonate of protoxide of iron in aqueous carbonic acid to drink, which either prevents or removes the symptoms of weakness, pale complexion, and loss of memory, with which the workmen are attacked.

Rectification.

Bonière has found that the solvent power of bisulphide of carbon is materially increased by its perfect rectification, whereby it acquires an odour similar to that of chloroform. He employs for this purpose the apparatus shown in Figure 107. The cistern, H, for holding the raw bisulphide of carbon, is fitted with a glass tube, J, to show the level of the liquid ; it has likewise an opening, L, which can be made tight by a stopper screwed into it, and an air-cock, K.

The bisulphide is run into a water-bath still, A 'B', through the pipe, P, which is connected with a glass pipe and cock, P and M. This still is connected with a range of similar stills and water-baths, AB. They are connected with one another, and with the condenser, S, by means of the bent pipes, F. The water-baths are heated by steam, which is supplied by the pipe, C, and is admitted to the steam casings, BB', by the branch pipes, C', which are fitted with cocks to regulate the admission of the steam and the rate of distillation.

The water-baths, A, are placed in casings, and are covered with domes, *b*, which, with the exception of the first, have each fittings for two pipes. The first of each of these connecting pipes, is attached to another pipe, I, inside of each water-bath, and is pierced with a number of small holes, *o*, at the lower end. The last in the series is connected with the condenser. The dome covers are screwed closely to the steam casings, to which thermometers are attached, for noting the temperature of the interior. The condensed water of the casings, is conveyed away by the pipes DD'.

The raw bisulphide is allowed to run into the still A', which contains a potash-lye warmed up to 140° to 150° F., and in consequence, distils over from one vessel to another, through the whole range. The other stills are filled with different solutions of alkaline salts, salts of lead, copper, iron, &c., and the bisulphide

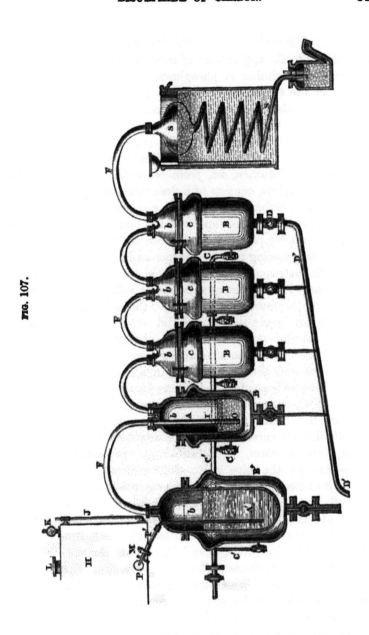

FIG. 107.

at length reaches the condenser, and flows through the pipe S
into a florentine vessel, containing some distilled water.

Applications of Bisulphide of Carbon.

In addition to the applications of bisulphide of carbon which we have named, its solution of phosphorus is employed to deposit a thin film of the substance on delicate articles which are to be coated with metals; but perhaps the most unexpected use of the bisulphide, is that of M. Moussu, for extracting bitumen from mineral bodies which contain so small a percentage, as not to pay when subjected to distillation in the ordinary way.

Extraction of Bitumen, &c., from Minerals, Bones, &c.—The apparatus he employs, consists of a close reservoir, 1, Fig. 108, for the bisulphide of carbon, mounted with a cooler, which is intended for the condensation of the vapour of the bisulphide, after it has extracted the soluble part of the substance operated on. It is kept cool by a worm passing round its upper and lower ends, which is supplied with a stream of water from a higher level. The filters 4 4', are charged with minerals, bones, or substances containing the bitumen, sulphur, fats, or oils, through the openings, 5 5, which are made tight by a lid screwed down upon a flange, and they lie on a false bottom pierced with holes. The liquid bisulphide of carbon is admitted from below, by the pipes, BB, and flows from the reservoir, 1, through the cocks, B'B'. It dissolves the soluble portion, as it filters up through the mineral, and flows off at the top through the pipes, C C and D D, to the bottom of the boiler, 3, with the bitumen, &c., in solution. When it is filled, a current of steam is admitted from the steam pipe, F, into a worm, T, volatilising the bisulphide, which ascends through the pipe A to the condenser, 2, whence it flows to the reservoir, 1. The fatty or bituminous matters are drawn off by the cock R, and the bisulphide of carbon which remains absorbed by the minerals, &c., in the filters 4 4', is expelled by a current of steam admitted into the steam-jacket of the filters, by the pipe-valves, G and H. The bisulphide of carbon is conveyed to the condenser, 2, by the pipes, EE. This admirable apparatus, therefore, enables the manufacturer to carry on his operation as often as he pleases, with very little loss of the bisulphide of carbon. We add one or two explanatory remarks as to the other details of the apparatus.

I I. Pipes for carrying off the water from the condensed steam.

J J and K. Pipes connected with an air-pump for emptying

FIG. 108.

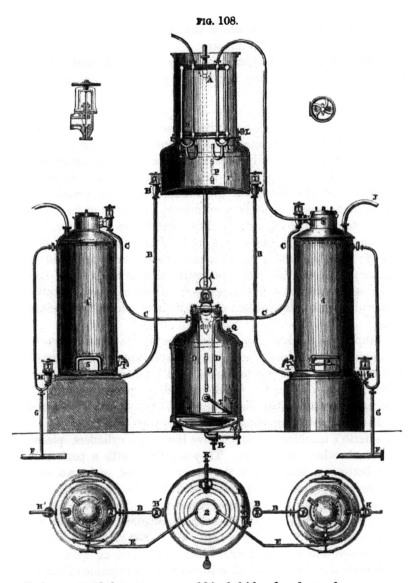

all the parts of the apparatus of bisulphide of carbon, when necessary, by creating a vacuum.

L. Cock for supplying water to the condenser.

M. Supply-pipe for water.

N. Exit-pipe for water.

O P. Gauges for showing the levels of the liquids.

Q. Gauge-valve of boiler.

U U. Luting tubes, by which the condensing bisulphide of carbon overflows into the reservoir.

M. Moussu is able to extract 12 per cent. of bitumen from minerals by this apparatus, which only yield 7 to 8 per cent. by distillation.

Extraction of Oil from Seeds.—The apparatus above described can, of course, be employed for the extraction of oil from seeds, but Seyferth has constructed an apparatus specially for the purpose. As Wagner properly remarks, this or any other apparatus, to be of manufacturing utility, must protect the workmen from the action of the bisulphide of carbon, while the quantity of bisulphide necessary, must be a minimum, with little or no loss of this substance. The advantages of this system are great; for, 1st, from 25 to 40 per cent. more oil is obtained from the same quantity of seed; 2nd, the cost of plant is much less; 3rd, the labour is cheaper; and, 4th, the cake is of equal value as cattle food; and lastly, seeds with only 6 to 8 per cent. of oil, can be profitably treated by this system, which cannot be used under the ordinary plan.

Sheep's wool and woollen goods, either in the natural state or saturated with oil, during any of the manufacturing operations,—and woollen lumps or cotton waste, used in cleaning machinery, which are now lost when filled with oil or grease,—can all be profitably treated with this new agent, the worn-out material rendered as good as new, and the oil recovered.

Seyferth's machine consists of five large iron cylinders, placed near each other, in a circle. They are fitted with a perforated false bottom, a little above the lower end, and with two wide man-hole doors, one corresponding with this false bottom, through which to withdraw the exhausted material, and the other in the top, through which to fill the cylinder with the seed, wool, &c. Each cylinder is fitted with a tube which rises from the bottom up through the top, and with which the adjoining cylinder is connected, the object of which is to cause the bisulphide of carbon to pass through the oily material in one cylinder after another, until it is fully saturated with oil. The top cover of each cylinder is also fitted with a kind of capital or still-head, through which to draw off any bisulphide of carbon that may remain behind, and convey it to the condenser.

A reservoir for the bisulphide of carbon, equal in capacity to two of the cylinders, is fixed near the five cylinders, and alongside a condenser, similar to that previously described. There are also two large copper stills heated externally by steam.

_ The cylinders are filled with the oily materials, either in their natural condition, or when necessary, after being crushed through rollers. The textile materials are pressed tight, but the oil seeds are thrown in, so as to be open and loose.

When three of the cylinders are filled, the materials are covered with a perforated plate, which is fixed in its position, and the covers and man-hole doors are now made fast. The bisulphide of carbon is then allowed to flow into the first cylinder, and to remain about 15 minutes in contact with the materials, to afford time for the solution of the oil.

The pipe which connects the second cylinder with the first, is opened, and the bisulphide of carbon allowed again to flow into the first, when the pressure it exerts, drives out the partially saturated bisulphide of carbon into the second cylinder. Another 15 minutes are allowed to elapse, and the operation repeated, till the third cylinder is filled. In another quarter of an hour, the third cylinder is connected with the copper still, and a supply of fresh bisulphide of carbon run into the *second* cylinder, which drives out the fully saturated fluid of the third cylinder into the still, while the fluid in the *first* cylinder is drawn off into a lower reservoir. The whole range of five cylinders is gradually brought into operation, which is conducted in the manner already explained.

The first cylinder is now filled with water to wash out the bisulphide of carbon, which afterwards separates from the water, and is run into the lower reservoir. This cylinder is then filled with hot steam, and the vapour of the bisulphide of carbon conveyed to the condenser.

The water from which the bisulphide of carbon has separated, is pumped back to the upper reservoir, and the bisulphide of carbon from this source, as well as from the condenser, is also pumped back to the higher cistern. In this way, no loss can take place when all the vessels are tight, and the same bisulphide of carbon and water are employed continuously for any length of time.

The oil remains behind in the copper still, and is found to retain a slight taste of the bisulphide of carbon, which is completely removed by agitating the oil with 10 per cent. of alcohol. The mixture, by standing, separates into two layers, and the alcohol is drawn off; the oil can then be again treated with fresh

alcohol if necessary.   The impure alcohol is rectified by distillation with lime.

This purification of the oil is not required when it is to be used
for lighting purposes, nor in the case of linseed and similar oils,
nor where the oils are to be refined with sulphuric acid, bleaching
powder, or chromate of potash.

The oil-cake which is left, is of superior quality, as it contains
by this system, all the nitrogenous constituents of the seed.

The bisulphide of carbon may be rendered perfectly pure and
free from smell, by redistillation, and in this state, if mixed with
a little ethereal oil to give a pleasant aroma, it is an excellent
material for domestic use, to remove all stains of grease, oil, &c.

Seyferth has applied his plan to the manufacture of rape oil,
and has found that when prepared by means of the bisulphide, it
contains a much larger quantity of solid fats than when manufactured on the old plan, in some degree analogous to the stearine
manufacture, by pressure and heat.   His oil is remarkably
adapted for use as a friction oil, and is much cheaper than the
oils generally employed for this purpose.

This idea of Seyferth's has been applied by Berjot, in the construction of an apparatus, which he calls an *Elæometer*, for rapidly
and easily determining the quantity of oil in seeds, &c.

It consists of a glass vessel, A (Fig. 109), into which a glass
cylinder, B, is accurately fitted.   A small air-pump, P, is inserted
in the arm, t.

FIG. 109.   FIG. 110.

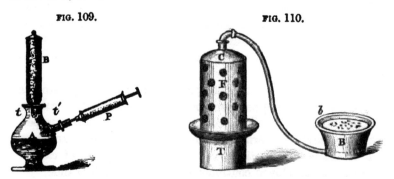

He employs along with it a small copper vessel, C (Fig. 110),
and a boiling apparatus, FT, a tinned copper dish, b, with a water
bath, B, which may be connected with the former by a caoutchouc
tube, as shown in the woodcut.

50 grains of the finely ground seed are placed in the cylinder B,

covered with a diaphragm, upon which other 50 grains are placed, and covered with another diaphragm. Bisulphide of carbon is poured in until the whole of the seed is fully moistened. After the lapse of a few minutes, the air-pump is worked, and a vacuum made in the glass vessel A, when the atmospheric pressure drives out the bisulphide with the oil in solution. This is repeated until the bisulphide appears colorless, and leaves no oily stain on bibulous paper. About 400 to 450 grains of bisulphide are sufficient for the exhaustion of the seed.

The bisulphide solution is placed in the saucer *b*, and heated by the steam from the apparatus FT. As soon as all the bisulphide is expelled, the saucer is placed over the lamp, heated to boiling, and allowed to cool, when it is weighed. The following table of some results obtained by means of the bisulphide, is added to give the reader an idea of its value.

Per cent. oil.

Rape seed	40 to 45
Linseed	34
Hemp seed	28
White mustard seed	30
Black „	29
Seed of the water melon	36
„ *Arachis hypogœa*	38
„ *Sinapis arvensis*	42

100 parts of the following seeds dried at 212° F., gave

Per cent. oil.

Seed of the *Fagus sylvatica*, harvest 1857	23·2
„ „ „ 1858	25·4
„ „ 1859	18·9 to 22·6
„ *Silia parvifolia*	39·2 to 41·8
„ *Pinus sylvestris*	20·3 to 23·4
„ *Pinus picea*	· 17·8
„ *Pinus strobus*	29·8
„ *Pinus canadensis*	11·4 to 12·9
„ Hazel nut (*Corylus avellana*)	55·8
„ *Pinus Cembra*, in natural state	29·2
„ „ shelled	36·5

Extraction of Spices, &c.—In addition to the above industrial application of bisulphide of carbon, Bonière has devised an apparatus for extracting the active aromatic principle from various spices, and other substances used for seasoning food, such as garlic, onions, &c., also by the aid of this powerful solvent. These *épices solubles*, or soluble spices, as Bonière calls them, consist of common salt, sugar, gum, or lactin, &c., which are impregnated with the peculiar aromatic principles of the spice, &c.

The apparatus he employs for treating pepper, for example, is shown in figure 111.

The pepper is first ground, and then placed in sieves, P, made of iron wire. These sieves are made of two hoops connected by four upright pieces, which form the framework, the bottom and sides of which are covered with wire gauze. A hook is attached to the upper end of these uprights, to facilitate the removal of the sieves from the tank or vessel, B. The lowest sieve rests on a ledge on the inside of the tank, and thus supports all those placed

FIG. 111.

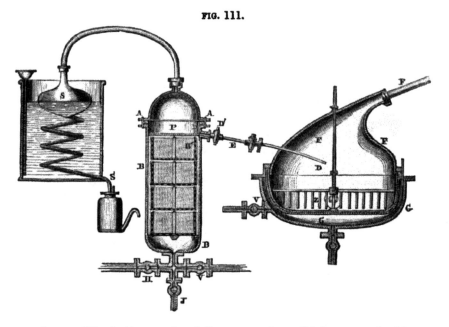

above. The bottom only of the upper sieve, B', is covered with wire gauze, the sides being made of tin plate, in which there is but one opening to allow the liquid to pass through the pipe D. This sieve is prevented from being forced upwards by the pressure of the liquid, by small knobs tapped in, through the sides of the tank B near the rim, P'. When the sieves are filled with the ground pepper, the bisulphide of carbon is run into the exhausting vessel or tank B from a reservoir, H, similar in construction to that already described; the cocks, J and V, being closed, and H opened. The bisulphide rises, passes through the sieves, and dissolving out the active principles of the pepper, runs into the boiler, F, through the pipe, D, into which a glass tube, E, is fitted. The boiler, F, contains either salt, sugar, lactin, dextrin, saltpetre, or any other material, according to the special object of the process. The boiler is made of cast-iron, glazed with earthenware, or of copper, zinked inside, and fitted with a steam casing. It is also connected with a condenser by a pipe, F', and has inside, a stirring or grating apparatus, Z. The liquid which flows into the still or boiler F, is heated up to 140° to 150° F., by the admission of steam into the casing, G, through the steam-pipe, V', when the bisulphide distils over, and is condensed in the apparatus already described, while the salt or sugar remains behind, retaining all the aromatic or acrid principles of the pepper.

When the liquid passing through the glass tube E, indicates by its colour, that the pepper is exhausted, the cocks D' and H are shut. The bisulphide left in the tank B, is run off through the pipe J, after which the cock J is shut, and steam admitted by opening the cock V. The bisulphide absorbed by the exhausted pepper, then distils over, and is collected in the condenser, S.

It is found that the pepper is not only fully exhausted by this plan, but that the salt or sugar retains all the aromatic principles, perfectly free from smell or taste of the bisulphide, which, on the other hand, remains perfectly pure, and fit for further treatment.

When garlic, onions, or similar materials are to be operated on, they are first pressed, and the juice heated with the bisulphide, in which the active principle dissolves. The two liquids are separated by decantation, and the cake placed on the sieves, when it is treated in the same way as the pepper.

COMMON SALT, SEA SALT, CHLORIDE OF SODIUM.

Great interest attaches to this substance, which is so widely diffused throughout nature, which plays so important a part in the economy of life, and which now forms the basis of so many manufactures. Salt is repeatedly mentioned in the Scriptures; and from the most ancient of all, the Book of Job, it is clear that salt was then used with food. Its use was enforced by the Levitical law to indicate the purity and sincerity of the worshipper, as well as the fidelity and incorruption of the covenant made with the chosen people. Rock-salt was used in the Temple, and in a chamber appropriated for the purpose, it was stored up in considerable quantities for use. It was mixed with clay or sand, and when it wasted away, from exposure to the air and absorption of moisture, the residue was thrown out to mend the roads, a practice to which the Saviour alludes, speaking to his disciples, "Ye are the salt of the earth; but if the salt have lost its savour, wherewith shall it be salted? it is thenceforth good for nothing but to be cast out and trodden under foot of men." And in a sense similar to that of the former part of this passage, Livy calls Greece, *Sal Gentium*, or the salt of all the nations, referring to the intellectual character of that people.

Salt among Eastern nations is an emblem of hospitality, and their federal engagements are generally ratified by salt, to which custom there are many references in the Scriptures. It is also a type of allegiance, and a memorable reference to this ancient symbolic meaning, was once made by the late Duke of Wellington, when alluding to his duty to the Crown. It is well known that salt in moderate quantities acts beneficially as a fertilizer, but that when present in excess, it produces sterility and barrenness; to this circumstance allusion is made in the book of Zephaniah, and it is abundantly exemplified in the salt deserts of South America.

There are also many curious customs connected with salt, in different parts of this country; for example, salting a house if new, or on being entered by a new tenant, for *luck;* placing salt on the corpse after being laid out; and presenting an infant on its first visit, with three articles, one of which must be salt.

The necessity which exists among human beings and the lower

animals for a supply of salt, is well known. It is certainly everywhere present in the solids and liquids of our bodies, and supplies the soda and muriatic acid which, in one form or another, perform such important duties in digestion and respiration. The consumption in this country is estimated to be 20 lbs. per head per annum, in America the same, but in France it is only 15 lbs. The love of salt in some countries is remarkable : in Africa the children look as anxiously for a piece of salt, as more favoured little ones here, long for sugar ; and in Abyssinia, where the inhabitants suffer much from want of salt, every man carries a bit of salt, which is always offered to a friend to lick when they meet. Its antiseptic property, its value as a manure, and the necessity of its presence to render food palatable, have doubtlessly given origin to these figurative applications and customs.

Chloride of sodium consists of one equivalent of each of its constituents, or of 59·67 chlorine and 40·33 sodium in 100 parts. It crystallizes in cubes, decrepitates when heated, melts at a red heat, and may be volatilized by increasing the temperature. Its solubility in water is scarcely affected by the temperature. Fuchs states that 100 parts of water dissolve 37 parts of the salt at all temperatures ; but, according to Gay-Lussac and Poggiale,

				Poggiale.	Gay-Lussac.
100 parts of water at 32° F. dissolve of salt				35·52	— parts
„	57·2	„		35·87	— „
	58			—	36
	140			—	37
	212			36·61	— „
„	225	„		—	40·38 „
„	229·46	„		40·35	— „

Or, stating the same point in another form, 100 parts of a saturated solution contain

At	1° C.,	26·53	chloride sodium, according to	Unger and Gay-Lussac.
„	17	26·40	„ „	Gay-Lussac.
„	18·8	26·75	„	Karsten.
„	25	26·30	„ „	Kopp.
„	100	28·22	„ „	Unger and Gay-Lussac.

Or, 27·00 chloride sodium, at every temperature, according to Fuchs.

Chloride of magnesium exerts great influence on the solubility of this salt ; for example, water containing 27·4 per cent. of the former in solution, can only dissolve 2 per cent. of the chloride of sodium, so that the more sea-salt or brine is concentrated, the greater is the separation of the latter ; and again, water, containing 22·1 to 23·8 per cent. of the chlorides of sodium and magnesium in solution, cannot retain any sulphate of lime in solution, which, on the other hand, is rendered more soluble by the presence of the chlorides of potassium and calcium.

Chloride of sodium has a pleasant saline taste, deliquesces slightly in the air, and has a specific gravity of 2·011.

The salt of commerce is obtained from two sources, from mines and by the evaporation of sea or brine waters.

Rock Salt.

As a rock, called *rock salt* by mineralogists, chloride of sodium forms a distinct member in the series of stratified rocks, occurring with limestone, clay, chalk, gypsum, marl, stinkstone, slate, and not unfrequently in bituminous formations. Sedgewick and Murchison, writing of the deposit of salt among rocks of almost all ages, as an interesting and important fact, say—"Salt is daily accumulating in certain inland lakes and marshes. In Poland it probably exists principally, if not entirely, among tertiary rocks; in the Austrian Alps it is placed in the oolitic system; in Switzerland it is referred to the lias; in Wirtemberg it is in the Muschelkalk; in England, the greatest salt-mines are in the new red sandstone, but there are two or three copious salt-springs in the coal formation, from which salt has been largely extracted. In certain parts of the United States, salt-springs issue from the old transition slate rocks. And lastly, a spring containing a great proportion of salt, rises near Keswick, from the lowest division of the slate rocks of Cumberland. The layers of salt vary in thickness, from a few inches to several hundred feet.

In this country, extensive beds of salt are found, as well as brine-springs. These latter are chiefly found in Cheshire; in Wevenham, brine has been worked from the time of William the Conqueror. These springs occur at various depths; at Nantwich, being met with about 10 or 22 yards from the surface, while at Winsford, it is necessary to sink 55 to 60 yards to tap the spring.

Brine pits were worked at all the wiches in Cheshire in the time of Edward the Confessor, but it was not until 1670 that the beds of rock-salt were discovered near Northwich, and since then, it has been found extending over the whole of the broad valley from near Malpas to Congleton, sweeping in the form of a crescent between the red sandstone hills of Shropshire and the

Staffordshire coalfield on the one side, and the high ground of the Peckfortem hills and Delamere range on the other side. We give here a section of one of these salt mines—Wharton, on the river Weaver, in Cheshire, copied from Mr. Tomlinson's interesting little volume on Salt.

FIG. 112.

An extensive bed of salt has also been discovered, and is now worked, near Belfast, in Ireland.

FIG. 113.

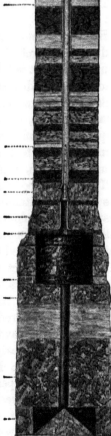

The most extensive deposit of salt on the continent of Europe, extends from Moldavia to Suabia, and includes the salt mines of Wallachia, Transylvania, Gallicia, Upper Hungary, Upper Austria, Styria, Salzberg, and the Tyrol.

From the section, Fig. 113, of the deposit at Wimpfen, at Wirtemberg, it will be seen how the gypsum is enclosed by a deep layer of shell-limestone, containing the rock-salt as a separate mass.

The most celebrated of these mines, is that of Vieliczka, which was discovered in the year 1251, and has been worked for some centuries. The mines of Bochnia and Vieliczka furnish about 45,000 tons per annum. They extend over a space of 35,000 square fathoms, and the galleries are 30 miles long. They belong to the Austrian government, as do those of Ischl, of which we give a section, showing the position of the salt mass among beds of limestone rock. It is worked at different levels which are approached by twelve horizontal galleries. The salt mass, S (Fig. 114), as at Hallein, is separated from

FIG. 114.

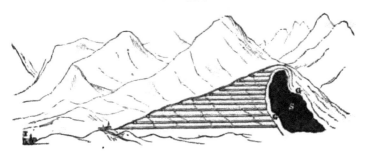

the surrounding limestone by bands of dark coloured gypseous marls, G, not saliferous. The position of the town of Ischl is shown at I.

The Austrian Government also work the salt-mines at Halle, in the Tyrol.

Rock-salt occurs in some parts of the vast empire of Russia; and in the parched and undulating steppes of the Virghis, in South Russia, it is found in abundance. These steppes are chiefly made up of reddish sandy marl and whitish gypsum, in which the rock salt appears as a vast, irregularly formed mass. Where the locality answers, the miners expose a broad surface of salt, and it is quarried at different elevations, the height of the bed of salt being about 70 feet. The salt is so white and pure that it is pounded for use, without further preparation.

The next important deposit of salt is at Cordona, in Spain, and the mines are situated at the head of the valley, the bold cliffs of greyish white colour, consisting of one mass of salt, or, as Ford calls it, "one mountain of salt." The sides and bottom of the valley are composed of reddish brown clay, similar to that found in the salt district of Cheshire. The summits of the ridges which bound the valley, are formed of a yellowish grey sandstone. The great body of the salt forms a rugged precipice, 400 to 500 feet high, and covered by the thick bed of the clay, just mentioned. There is no excavation, but the salt is procured by working down perpendicularly, as in an open quarry. The depth of the salt has never been determined.

The general appearance of the salt is greyish white, owing to the action of the weather and rain, but the body of the salt itself is so white that it requires no preparation except grinding, to convert it into snow-white culinary salt.

There is a brine spring at the foot of the great precipice, so saturated, that when the hand is dipped into it, and exposed to the air, it is instantly covered with a thin film of salt.

Of all the countries of Asia, Persia is the most abundantly supplied with salt, and mines of this mineral are found in different parts. The salt found at Nishapore is so beautiful and clear, that it is said, objects are visible through it, even when two inches thick.

The salt deserts which occur in various parts of Persia, form one of the most striking objects of its scenery, and are distinguished by the saline efflorescence with which they are covered.

There is also said to be an immense bed of rock salt in Scinde, 20 miles in length by 15 miles in breadth, and averaging 3 feet in thickness.

Beds of common salt of great extent and thickness, are found in many places, in a group of mountains called the Salt Range, stretching from the base of the Suliman Mountains in Afghanistan, in an easterly direction to the river Jailum in the Punjaub. One of these deposits, now worked, is 200 feet thick, and varies in colour, from white to flesh colour and brick red. When Burnes visited it in 1822, the quantity raised in a year amounted to 80,000,000 pounds.

In South America, the Andes are said to contain deposits of rock-salt equal to any in the world; and north of Copiapo, in Bolivia, the desert of Atacama exists, in which abundance of salt is found. Mines of salt are worked at Tocallo; and in Peru there are extensive districts, lying between the Andes and the coast, of absolute sterility, in which large deposits of salt exist. Mr. Smith describes it as continuing for leagues on level plains, on the sides of mountains, and at the bottom of deep hollows, like dry lakes or ponds.

Brine or Salt Springs.

We have already alluded to the brine-springs in this country; and some of the more celebrated in other countries, require a brief notice.

One of the most remarkable, is that of Moutiers, in Sardinia, where nearly 3,000,000 lbs. of salt are obtained from water, not nearly so strong as sea-water, which is managed by a process to be more fully detailed. The springs rise at the foot of a rock of limestone, accompanied by a discharge of carbonic acid and sulphuretted hydrogen gases : they contain from 1·50 to 1·83 per cent. of salt.

Salt is also obtained from numerous salt-springs in Germany, at Halle, Schönebeck, Manheim, &c., by the same process, and the production of some of these localities is very great, that at Halle alone amounting to 300,000 tons per annum.

Tuscany has a similar establishment at Volterra, and up to 1834, the production supplied an annual demand of above 17,000,000 lbs.

Salt-springs are very common in the Central States of the American Union, and works exist near the town of Onandago, for the manufacture of salt from the springs in the neighbourhood. The most remarkable source of salt in the States is found in the Great Salt Lake, which is probably derived from some bed of rock-salt in or near the Wha-satch Mountains.

In China, in the province of Sza-Tchhouan, there are a number of salt-wells, which evolve so large a quantity of inflammable gas, that it is employed in the evaporation of the brine-water. M. Imbert reports that there are several thousands of these wells in the vicinity of the town of Ou-Thouang-Khiao.

It is highly probable that the deposits of rock-salt above described, have all been formed from saline lakes, and hence the salt occurs in more defined and rounded masses, as compared with the enormous extent of the other members of the same formation. In the high lands of Asia and Africa, there are often extensive wastes, the soil of which is covered and impregnated with salt, without having ever been covered by other deposits. Salt lakes themselves are not uncommon, and occur on the banks of the Wolga, in South Africa, in Germany, and in the neighbourhood of the Caspian Sea.

The tables which follow, give the analyses of these lakes and salt seas. The simultaneous occurrence of soluble sulphates, or other chlorides, as well as the corresponding compounds of bromine and iodine, which has everywhere been observed, is of equal importance as regards the history of its origin, and the mode of obtaining sea salt.

Analyses of the Saline Matter contained in the Water from different Lakes.

Locality.........	Lake Van.		Lake near Ararat.	Leacher See.	Lake of Ardchek.
Analysts	Chancourtois.	Richardson and Browell.	Abich.	Bischof.	Richardson and Browell.
CONSTITUENTS.		grs.			grs.
Carbonate of Soda ...	86·13	451·95	37·0	1·1259	616·77
Sulphate of Soda	33·30	270·25	55·7	·0959	320·13
Chloride of Sodium ...	93·80	643·81	213·6	·1791	627·63
Sulphate of Potash ...	5·50	—	—	—	—
Carbonate of Lime ...	—	—	—	·5898	traces
Carbonate of Magnesia	5·50	10·68	—	·2112	12·17
Oxide of Iron	traces	} ·36	—	—	·55
Alumina	—		—	—	—
Silica	1·80	—	—	·0295	—
Organic matter	—	traces	—	—	traces
	226·90 per 10,000 parts water.	1377·05 per gallon of water.	306·3 per 10,000 parts water.	2·1814 per 10,000 parts water.	1577·85 per gallon of water.

Analyses of the Water of Salt Lakes.

Locality	Elton Lake.			Bitter Salt Lake.	Bogdo Lake.	Stepanowa Lake.	Putrid Lake.
Analysts.........	Göbel, in April.	Erdmann, in August.	H. Rose, in October.	Göbel.	Göbel.	Göbel.	Göbel.
CONSTITUENTS.							
Chloride of Sodium ...	13·124	7·451	3·83	10·54	19·000	22·43	14·20
Chloride of Potassium	0·222	—	0·23	—	0·199	—	—
Chloride of Calcium..	—	—	—	—	0·989	—	—
Chloride of Magnesium	10·542	16·280	19·75	9·91	5·435	0·91	1·93
Bromide of Magnesium	0·007	—	—	—	0·006	—	—
Sulphate of Magnesia.	1·665	2·185	5·32	8·22	—	0·69	1·21
Sulphate of Lime ...	—	0·036	—	—	0·028	0·05	—
Carbonate of Magnesia	—	0·038	—	—	—	—	—
Organic Matter	traces	0·505	traces	—	—	—	—
Sulphide of Calcium...	—	—	—	—	—	—	0·04
Water	74·440	73·505	70·87	71·33	74·343	75·92	83·62
	100·000	100·000	100·00	100·00	100·000	100·00	100·00

Lastly, the water of the ocean,—which from the geological processes concurring in its formation is necessarily a solution of salt,—contains chloride of sodium as its chief saline ingredient. On account of the unequal amount of evaporation, however, the water of the sea has not always the same composition. The following tables of analyses, give an excellent idea of the variation in the composition of sea-water.

	North Sea					Atlantic Ocean	Harbour of Callao	Harbour of Tocopilla	Cape Horn	Mediterranean Sea				English Channel
Locality / Analyst	Marcet	Clemm.	Bacha	Figuier and Mialhe	Bischof	Bibra	Bibra	Bibra	Bibra	Vogel	Calamai	Calamai	Usiglio	Schweitzer
CONSTITUENTS.														
Chloride of Sodium	80·90	77·94	77·41	78·71	79·14	74·20	75·80	77·20	75·75	68·02	76·73	76·33	78·14	27·05948
Chloride of Potassium	—	4·24	3·31	0·25	2·06	3·80	3·68	3·72	3·26	—	2·86	3·24	1·84	0·76552
Chloride of Magnesium	8·56	7·59	9·70	8·74	8·54	11·04	8·87	8·09	8·84	14·23	8·90	8·82	8·55	3·66658
Bromide of Magnesium	—	—	—	0·09	—	—	—	—	—	—	—	—	1·48	0·02929
Bromide of Sodium	—	—	—	0·81	—	1·09	1·23	1·20	1·21	—	—	—	—	—
Sulphate of Lime	4·01	3·77	3·65	3·71	3·89	4·72	4·54	3·94	5·18	0·41	2·07	2·60	3·60	1·40662
Sulphate of Magnesia	6·53	6·46	6·53	7·74	6·37	5·15	5·88	5·85	5·76	16·28	9·44	9·01	6·51	2·29578
Carbonate of Lime	—	—	—	0·40	—	—	—	—	—	{0·41}	{—}	—	0·30	0·08801
Carbonate of Magnesia	—	—	—	—	—	—	—	—	—			—	—	—
Oxide of Iron	—	—	—	—	—	—	—	—	—	—	—	—	0·01	—
Silicate Soda	—	—	—	0·05	—	—	—	—	—	—	—	—	—	—
Water														964·74872
	100·00	100·00	100·00	100·00	100·00	100·00	100·00	100·00	100·00	100·00	100·00	100·00	100·00	1000·00000
Percentage of saline matter in the water	3·76	3·19	3·05	3·27	3·28	3·44	3·28	3·68	3·48	3·62	2·91	3·43	3·77	
	1	2	3	4	5	6	7	8	9	10	11	12	13	14

In the analyses 1 to 13, the composition of the saline matter is given in 100 parts, but in No. 14, as in the previous tables, the analysis is of the water.

No. 2. Water collected on the English Coast.　　　　No. 11. Water collected from the Lagunes of Venice.
3. „ 　 on the Coast of Heligoland.　　　　12. „ 　 from the Harbour of Leghorn.
4. „ 　 some leagues from the Coast off Havre.　　　　13. „ 　 near Cette.
5. „ 　 between Belgium and England.　　　　14. „ 　 off the Coast near Brighton.
10. „ 　 some miles from Marseilles.

Locality.	Caspian Sea.		Black Sea.	Sea of Azof.	Dead Sea.				
Analysis.	Göbel.	H. Rose.	Göbel.	Göbel.	Gmelin.	Marchand.	Hempath.	Booth and Muckle.	Bourton and Henry.
Chloride of Sodium	0·3673	0·0754	1·4320	0·9658	7·078	6·578	12·110	7·855	7·003
Chloride of Potassium	—	—	0·0189	0·0128	1·674	1·398	1·217	0·659	3·166
Chloride of Magnesium	0·0632	—	0·1304	0·0887	11·773	10·543	7·822	14·590	5·696
Chloride of Calcium	—	—	—	—	3·214	2·894	2·456	3·108	0·680
Chloride of Manganese	—	—	—	—	0·212	—	0·006	—	—
Chloride of Iron	—	—	—	—	—	—	0·003	—	—
Chloride of Aluminium	—	—	—	—	0·090	0·018	0·056	—	—
Chloride of Anemonium	—	—	—	—	0·008	—	0·006	0·087	trace
Bromide of Potassium	trace	—	—	—	—	—	—	—	—
Bromide of Magnesium	—	0·0036	0·0005	0·0004	0·439	0·251	0·251	—	—
Sulphate of Soda	0·0490	0·0406	—	0·0288	—	—	—	—	0·238
Sulphate of Lime	0·1239	—	0·0105	0·0764	0·053	0·088	0·068	0·070	—
Sulphate of Magnesia	—	0·0018	0·1470	0·0022	—	—	—	—	—
Bicarbonate of Lime	0·0171	—	0·0859	0·0129	—	—	—	—	—
Bicarbonate of Magnesia	0·0013	0·0440	0·0209	—	—	—	—	—	—
Silicate of Soda	—	—	—	—	—	—	—	—	—
Silica	—	—	—	—	—	0·008	trace	—	0·200
Nitrogenous Organic matter	—	—	—	—	—	trace	0·062	—	—
Bituminous matter	—	—	—	—	—	—	—	—	0·958
Water	99·3706	99·8846	98·2339	98·8120	75·459	78·227	75·944	78·681	85·069
	100·0000	100·0000	100·0000	100·0000	100·0000	100·0000	100·0000	100·0000	100·000
	1	2	3	4	5	6	7	8	9

1. Collected about 1 mile S.W. of the Island of Pischmoi.
2. „ about 50 miles from the north of the Volga.
3. „ about the middle of the S. coast of the Crimea.
4. „ between Kertsch and Mariapol.
5. „ Transactions of the Wirtemberg Nat. Hist. Society.
6. Collected from the Northern point of the sea.
7. Journal of Chemical Society, II, 336.
8. Collected by the United States Expedition.
9. Collected near Western side, two hours' distance from the Jordan.

Production of Salt from Sea-water.

It is seldom that artificial evaporation is employed for separating the salt from sea-water; when it is practised, the same mode is adopted as will be described hereafter with reference to brine-springs; sometimes, as in Siberia, frost is made subservient to this object, for salt water separates on freezing into ice (free from salt), and a strong saline lye.

Dr. Kane states that, if the cold is sufficiently intense, the water which is obtained on melting the ice, is fit for drinking. This plan is pursued in Russia, Sweden, and other northern countries, where the sea water is frozen in reservoirs, and the concentrated brine-liquor boiled down to salt. The salt obtained by this plan is not pure, as the following analyses by Hess will prove :—

Constituents.	Irkontsk.	Selenguisk.	Oustkoustk.	Okhotzk. 1st soccage.	Okhotzk. 2nd soccage.	Okhotzk. 3rd soccage.
Chloride of Sodium .	91·5	74·7	76·3	86·0	77·6	79·1
„ Aluminium	2·6	6·5	1·2	3·6	6·2	7·8
„ Calcium ·	1·1	1·4	3·8	0·9	0·9	0·7
„ Magnesium	2·0	3·6	3·6	2·0	1·7	0·8
Sulphate of Soda . .	2·8	13·8	12·6	7·5	13·6	11·6
„ Lime · ·	—	—	2·5	—	—	—
	100·0	100·0	100·0	100·0	100·0	100·0

The evaporation is generally effected by the influence of the air and sun in what are termed, in some localities, *Salterns* or *Brine-pans*. Hayling Island, near Portsmouth, has been celebrated for many centuries for its manufacture of salt carried on in this way. It was made here long before the Conquest, and is praised by St. Augustin for its excellence. It is manufactured only during the four summer months. The brine-pans vary in size, up to a quarter of an acre, in which, in favourable weather, the sea-water becomes brine in about seven days. It is then pumped by windmills into reservoirs or pits, whence it is run into iron pans, and boiled for 12 hours, during which it is skimmed to remove impurities. The crystals of salt begin to appear towards the end of the day, and are shovelled out, hot and wet, into wooden troughs, with holes in the bottom to allow the *bitters* to run off.

A somewhat similar plan is pursued at Lymington, but the

brine is there run through a series of salterns, before it is pumped into the boiling pans. The drainings from the troughs drop on upright stakes of wood, on which stalactitic masses accumulate, and in ten or twelve days weigh from 60 to 80 lbs.; these masses are called *salt-cats*. The residual liquor, called *bitters*, *bittern*, or *bittern-liquor*, is employed in the manufacture of carbonate or sulphate of magnesia.

For many centuries, salt was also obtained from sea-water in the Firth of Forth, and other places on the east coast of this country, by evaporating it in large shallow iron pans, saltwork of this kind being given, in former days, as donations to abbeys.

The above plans are a combination of the natural and artificial systém of evaporation, dependent on a supply of cheap fuel. In most southern latitudes, however, the sun takes the place of the fuel, and the same process apparently is followed, whether it be in Java or in Spain; this process will be presently described. The necessities arising from the peculiarities of the locality, have led, however, to a modification of the usual process, which is practised in three countries wide apart, viz., on the south coast of Java, in Lower Normandy, and in some parts of Brazil.

On the boisterous south coast of Java, the sand on the beach is made into ridges and furrows, and the natives, having filled a pair of watering cans from the surge, run along the furrow, pouring a shower of sea-water on the ridges. The heat of the sun rapidly dries the outer layer of sand, which is scraped off by a kind of hoe, placed in rude funnels, and sea-water is run through it, by which means a strong brine is obtained, which is boiled down to salt, over the ordinary fires of their huts.

In Normandy, the sand on the sea-shore, left dry by the tide, is collected by a long broad scoop, drawn by a horse; this is formed into a kind of filter, through which the sea-water is allowed to percolate, thereby adding to its strength, and is then evaporated to salt in leaden boilers.

In Brazil, an artificial hollow extends along the river Rio de Salitre, about six leagues from Ioazeiro, the bottom of which is formed of a fine soft ochre-coloured earth. The annual floods melt the saline particles which this earth contains, and a salt-pool is formed when the river falls. The salt is left, when the water evaporates by the heat of the sun. The soil is of similar formation along the bed of the San Fransisco for nearly two degrees of longitude, and nearly thirty leagues in breadth,

and the hollows, such as above named, are scattered over the
whole district, forming the salt-mines of the country.

The evaporation by the sun and air in southern latitudes is
carried on in what are termed *salt-gardens* (Fig. 115), which are

FIG. 115.

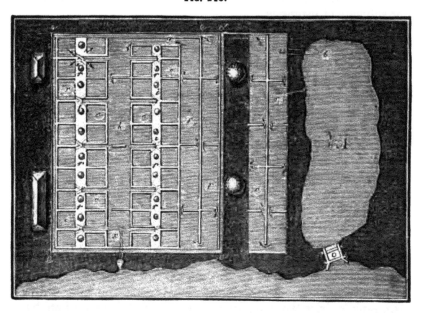

laid out upon a clay soil on the sea-coasts, and, being protected
from the influence of the tides, are cultivated from March to
September. These salt-gardens are nothing more than a series
of shallow ponds, intended to spread the water over a very large
surface with hardly any depth of liquid, so that, by increasing the
evaporating surface as much as possible, the drying action of the
air is rendered more available, and salt is deposited in the furthest
pools, while the nearer ones are constantly receiving fresh supplies
of salt water. The collecting pond, A, is filled, on the flow of the
tide, through the flood-gate, *a*, to a height of from 2 to 6 feet, in
which the evaporation begins; but the principal object in this
pond is to allow the mud to deposit. The pipe *b* then carries
the clear water from the collecting pond to the perfectly horizontal but very shallow front pool, *ccc*, whence, by means of a
second pipe, *d*, it is circulated through a channel, *ee*, 16,000 feet
long; it then enters the ponds, *fff*, and lastly runs through the

open channel, *g*, to a third series of ponds, *hhh*. At this point, the evaporation has proceeded so far that the salt begins to crystallize in the further reservoirs, *ooo*, of which there are four rows. From the reservoirs *hh*, numerous channels, *ii*, branch out, which supply the crystallizing ponds *oo*. The manner in which the water reaches these through the gutters in the sides, is shown in the drawing. The saline incrustation with which the surface of *ooo* becomes gradually covered, is broken up and collected, with rakes, into small heaps, *rrr*, on the sides, and from these the mother-liquor runs off into the ponds *oo* and *hh*. When no more salt separates by crystallization, the lye is allowed to run off through *x* into the sea. The salt as at first collected, would contain too much impurity, chiefly·consisting of chloride of magnesium; the smaller heaps *rr* are therefore made up into larger square heaps, *mm*, or round heaps, *nn*, which are allowed to remain for a long time covered with straw. The rain is thus kept off, and the moisture of the atmosphere is sufficient to dissolve the chloride of magnesium, which is thus gradually separated from the mass.

Although the entire surface of the ponds amounts to many hundred acres, the process so entirely depends upon the sun and wind, that in wet weather the evaporation sometimes ceases altogether.

The following analyses of sea-salt show the nature and amount of its impurities:

Locality.	Chloride of Sodium.	Sulphate of Magnesia.	Chloride of Magnesium.	Sulphate of Lime.	Water.	Insoluble Matter.	Analyst.
St. Ubes, Portugal	96·50	0·25	0·32	0·88	1·95	0·10	Karsten.
„ 1st sort	95·19	1·69	—	0·56	2·45	—	Berthier.
„ 2nd sort	89·19	6·20	—	0·81	3·60	0·20	„
„ 3rd sort	80·09	7·27	—	3·57	8·36	0·20	„
Figueras	91·14	3·54	0·70	0·33	4·20	—	„
Lymington	93·70	3·50	1·10	1·50·	—	0·20	Henry.
„ cat salt	98·80	0·50	0·50	0·10	—	0·10	„
Marennes ......⎫ France, white ⎬	97·20	0·50	0·40	1·20	—	0·70	Berthier.
„ yellow	96·70	0·66	0·23	1·21	—	1·20	„
„ red	96·78	0·60	0·68	1·09	—	0·85	„
„ green ...	96·27	0·80	0·27	1·09	—	1·57	„

Great improvements have been introduced in the management of these salt-gardens by M. Balard, and we are glad to be able to give the following detailed account of his processes for extracting common salt and other salts from sea-water, which is preceded by some notices of the salt-works in the basin of the Mediterranean.

1st. *Salt-works on the coast of Spain.*—The most important are situated in the Bay of Cadiz, and produce about 200,000 tons yearly, of which quantity about 160,000 tons are exported. America generally takes for New Orleans 50,000 tons, and Brazil about as much ; the remainder goes to Norway, Sweden, and Calcutta.

Torrevieja, and the island of Iviza, produce much less, and the export is limited, these places not enjoying the commercial importance of Cadiz. The quantity exported is probably not more than 20,000 or 30,000 tons yearly.

2nd. *Salt-works on the French coast of the Mediterranean.*— They may be divided into several groups, according to the departments to which they belong. These are the groups of the Eastern Pyrenees, of Ande, Hérault, Gard, Bouches du Rhone, and Var. The last three groups are the most important. The whole united produce is about 200,000 tons, and the export from 70,000 to 80,000 tons to various parts, including the western coast of France.

3rd. *Salt-works of Italy.*—There is a small one near Ostia, at the mouth of the Tiber, and another at Comachio in the lagunes at the mouth of the Po. These two salt-works are scarcely sufficient for the supply of the Roman States ; none therefore is exported.

4th. *Salt-works of Sardinia.*—These salt-works are situated at Cagliari, in one of the most beautiful harbours in the world, manufacturing and exporting abroad, or to the Piedmontese states, about 80,000 tons. It is one of the best localities for loading foreign vessels, and one where they meet with the quickest despatch. The principal exports are made to Norway, Sweden, the countries of the Levant and America, but the latter country takes much less here than at Cadiz.

5th. *Salt-works of the Island of Sicily.*—The principal are situated at Trappani, and produce annually from 125,000 to 150,000 tons, which is disposed of in the Kingdom of Naples, in

Lombardy by way of the Adriatic, and in Venice, Greece, Turkey, Sweden, Norway, and Prussia.

These are the principal salt-works in the Mediterranean, and the only ones which have any importance as places of export. There are some small establishments on the coasts of Illyria and Dalmatia, at Tunis, and on the coast of Egypt; but they are, at most, sufficient only to supply the wants of the countries in which they are situated.

The sea of Azof has large natural salt-works near Kertch, whence Russia draws her supply; they are called *Linians*.

In the Atlantic, on the Tagus, the famous salt-works of Setubal are situated, which produce at least 150,000 tons yearly.

Manufacture of Sea-salt and other products from Sea-water.

A salt-work is a surface of which the greater part is generally about one foot below the main level of the sea, so that in the Mediterranean, where the level is constant, a small flood-gate is opened to allow of the entrance of the salt water. These surfaces are divided by small banks of earth, 2 feet high, and from 1 to 2 feet thick. The water circulates by means of openings in these divisions. It enters with a density of 1·025, and continues to increase in density by evaporation, under the action of the dry winds, until it attains a sp. gr. of 1·143. Its volume is then reduced to one-sixth of its first proportions. It now begins to deposit sulphate of lime, till the density reaches 1·210. It is then conveyed by means of pumps worked by horses or steam power, to reservoirs slightly more elevated, the bottoms of which are generally composed of hard clay, well beaten in, where the precipitation of the sea salt, NaCl, takes place. The volume is by this time reduced to $\frac{1}{10}$th of its original bulk. From the density of 1·210, representing 25° on Beaumé's hydrometer, up to 1·286, equal to 32° on the same instrument, the evaporation takes place in the open air, the liquid constantly depositing chloride of sodium almost pure, which goes into the market without any further preparation. In order to obtain it, care is taken to run off the liquor marking 32° Beaumé, which covers the bed of salt about 2 inches thick. It is then collected by means of shovels, and thrown into heaps called in the south of France, *Caniells*, and in Italy, *Cumuli*. It is then allowed to drain, after which it is ready for sale.

The liquors marking 32°, which were run off, are called bitterns; the evaporation is continued by exposing them afresh on surfaces

of clay, of about an acre in extent. They there attain by evaporation, a density of 1·321, marking 35° Beaumé.

From 32° to 35° they deposit a mixture, in about equal proportions, of sea salt and sulphate of magnesia with 7 equivalents of
water.

When the liquors reach 35° Beaumé, they still contain a considerable quantity of sulphate of magnesia, also of chloride of
potassium and an abundance of chloride of magnesium. Two
methods of treatment may then be followed.

1st. The evaporation is continued, and up to 37·5° Beaumé,
they deposit a mixture consisting of about 10 per cent. of sulphate
of magnesia, 25 per cent. of a double sulphate of potash and magnesia, a small quantity of a double chloride of potassium and
magnesium, and about 30 per cent. of chloride of sodium.

Or, 2nd. The liquors of the strength of 35° Beaumé are passed
into covered or uncovered reservoirs, but in the latter case of
such a depth (12 feet) that the fall of rain may not sensibly
diminish their strength. These operations naturally take place
in summer. Nothing is now done until the months of November
and December, in order to take advantage of the periods when
the temperature falls to + 5° C. The water contained in the
reservoirs, is then exposed on the evaporating tables with a depth
of 6 inches. On cooling, the whole or the greater part of the
sulphate of magnesia is deposited perfectly pure. The liquors are
then pumped back into the reservoirs and preserved until the
following summer. In the summer they are again exposed afresh
to evaporation, and a deposit obtained, which contains only a
double chloride of potassium and magnesium, and a little sea
salt. Two successive refinings give pure chloride of potassium,
containing 99 per cent.

The chloride is treated with acid, &c., and carbonate of potash,
$KO.CO^2$, is obtained.

By the first method, by continuing the evaporation the same
year, a mixture is obtained, composed of sulphate of magnesia
and potash, $KO.MgO.2SO^3$, chloride of sodium, NaCl, chloride of
potassium, and basic chloride of magnesium. These products are
refined, and one-fourth of the weight is obtained under the form
of $KO.MgO.2SO^3$ pure. The salt originally obtained gives its own
weight of alum, but the pure double salt, obtained by refining,
gives 2 or 3 times its weight. About one half of the potash is
thus secured. The liquors containing the remainder are suitable

for making alum ; or rather, they are again evaporated on the surface of the soil, whereby sulphate of magnesia is obtained, together with a double sulphate of potash and magnesia.

This salt, $KO.MgO.2SO^3$, is then treated in a furnace with chalk and charcoal to transform it into crude potash, and by lixiviation into refined pearlash, $KO.CO^3$. It is exceedingly difficult to fuse, requiring a very strong heat in the furnace, and the loss of potash by volatilization is sometimes as much as 20 per cent. The infusibility is caused by the presence of the magnesia.

Beyond 37° Beaumé the liquors do not retain any potash, but they still contain a considerable quantity of hydrochlorate and sulphate of magnesia, and it is difficult to obtain these salts otherwise than by furnace evaporation.

The Bitterns from 32° to 35° deposit a mixture of sea salt, NaCl, and sulphate of magnesia, $MgO.SO^3 + 7HO$: this salt is set aside in summer, kept till winter, and operated upon when the temperature approaches the freezing point. The strength of the liquor need not exceed 32° Beaumé, which corresponds to a density of 1·286. This solution is exposed to the open air, and an abundant deposit of hydrated sulphate of soda obtained. The liquors are then run off, and the sulphate is collected, and heated in a furnace to expel its 10 atoms of water.

Some experiments have been made with the view of transforming the sulphate of magnesia into sulphate of soda, by heating it with excess of chloride of sodium. These experiments succeed well in a muffle, but very imperfectly when tried on the large scale in a furnace.

Mean produce of the Salt-works.

100 hectares of surface, or 250 English acres, will produce—
10,000 tons of sea salt, NaCl.

800   ,,   salts from 32° to 35°, corresponding to 160 tons, $NaO.SO^3$.

600   ,,   ,,   from 35° to 37½°, corresponding to 40 tons, $KO.CO^3$.

There are salt-works where, with scarcely any expense, 10,000 acres could be applied to the purposes of evaporation.

The price of the sulphate of soda by the above process is 40f. or 32s. per ton.

The ton of carbonate of potash costs 400f. or £16.

Rock-salt.

The mode in which salt is obtained from the deposits of rock-salt, depends very much upon the locality, upon the depth of the deposit, the price of fuel, the rate of wages, &c. &c. In some places it is a mining operation, and is carried on by means of shafts and horizontal galleries, as at Wieliczka, in Gallicia, (where the layer is 500 miles long, 20 miles broad, and 1200 feet deep,) Cheshire, and in other places. It depends upon the degree of purity of the rock-salt, whether it can at once be brought into the market, or must first be purified by solution and recrystallization. This is the case with some of the rock-salt in Cheshire, at Cordona, in Spain, and in Russia, as we have already noticed ; in general, however, it varies throughout the mass, is often coloured red, either from clay or bitumen, and particularly from the same kind of infusoria that are still found inhabiting salt-lakes. Rose has also observed, in the rock-salt of Wieliczka, a particular kind of hydrocarbon, C^9H^8, which is enclosed in a high state of condensation, and is evolved on dissolving the salt, with a peculiar crackling sound. In some places, as at Wimpfen, fresh water is let down through a bore extending to the middle of the salt bed, and pumped up again as a saturated solution. For this purpose, pumps and pipes descend the whole depth of the bore. The solution of salt is raised in the pipes, whilst between these and the sides of the bore, fresh water flows down. Thus a wide cavity is gradually formed in the bed of salt, in which fresh water and a solution of salt are contained. Now, as 1 cubic foot of the saturated solution weighs 6·4 lbs. more than the same volume of fresh water, the latter will chiefly occupy the surface of the liquid in the cavity. The pump must, therefore, work from the bottom of the cavity, and the suction-pipe be sufficiently deep for that purpose. The valves, however, may be placed much higher up, because the liquids on the inside and outside will balance each other, and the solution of salt will be as much below the level of the water, as its specific gravity is greater. At 1200 feet depth of bore, for instance, the saturated solution of salt will of itself stand at 1000 feet, and the pump will only have to raise it 200 feet. The following table gives the composition of rock-salt from several localities :

Analyses of Rock Salt.

	Wieliczka	Berchtesgaden	Hall, Tyrol	Hallstadt, Austria	Schwäbisch Hall	Wilhelmsglück	Vic.	Djebel Meiah.	Algeria.		Holston, U. States.	Boring, at Stassfurth, Thuringia.		Cheshire.	Cordova.
									Ouled Fibrous.	Kebbah Impure.					
Chloride of Sodium ...	100·00	99·85	99·43	98·14	99·63	98·36	99·30	97·0	98·89	72·16	99·55	94·57	25·09	98·30	98·55
Chloride of Potassium	—	trace	trace	trace	—	—	—	—	—	—	trace	—	—	—	0·99
Chloride of Calcium ...	—	trace	0·25	—	0·09	—	—	—	1·11	1·65	—	0·97	—	0·05	0·02
Chloride of Magnesium	trace	0·15	0·12	—	0·28	—	—	—	—	5·57	—	—	—	—	—
Sulphate of Soda	—	—	—	—	—	0·08	—	—	—	—	—	—	1·57	—	—
Sulphate of Lime	—	—	0·20	1·86	—	0·55	0·50	3·0	—	10·72	—	0·89	1·23	1·65	0·44
Sulphate of Magnesia...	—	—	—	—	—	—	—	—	—	2·06	—	—	42·07	—	—
Carbonate of Lime......	—	—	—	—	—	0·52	—	—	—	3·71	—	—	—	—	—
Carbonate of Magnesia	—	—	—	—	—	0·18	—	—	—	2·89	0·45	—	—	—	—
Clay and Oxide of Iron	—	—	—	—	—	0·53	0·20	—	—	1·24	—	1·12	0·46	—	—
Insoluble matter.........	—	—	—	—	—	—	—	—	—	—	—	2·23	1·37	—	—
Water	—	—	—	—	—	—	—	—	—	—	—	0·22	28·21	—	—
	100·00	100·00	100·00	100·00	100·00	100·00	100·00	100·00	100·00	100·00	100·00	100·00	100·00	100·00	100·00

Among the products thrown out of volcanoes, a substance is found which is allied to rock salt by its composition. Thus, in 1822, an immense quantity of saline matter was (1) thrown up from Vesuvius, and similar matter accompanied the eruption in 1856 (2 and 3), but that in 1855 was nearly pure common salt (4).

	1	2	3		4
Chloride of sodium	62·9	46·16	62·45		94·30
„ potassium	10·5	53·84	37·55		—
Sulphate of lime	0·5	—	—		0·70
„ soda	1·2	trace	—		0·20
Silica	11·5	—	—	SO³.MgO	0·40
Oxide iron	4·3	—	—	SO³.KO	1·00
Alumina	3·5	—	—	ClMn	0·60
Lime	1·3	—	—	—	—
Water	3·7	—	—	—	—
	99·4	100·00	100·00		97·20
	Laugier.	Bischof.	Scacchi.		Deville.

Natural brine wells.—Salt wells, which may be artificially constructed in the manner above described, are also frequently found ready formed in nature, wherever a spring, during its course, has come in contact with a bed of rock-salt (*Soolen*). It is rare, however, that these are so highly saturated as the artificial springs, although this is actually the case with that of Lüneburg, which contains 25 per cent.; but they are generally very slightly impregnated, or have become weakened by an after addition of fresh water. This difference in the strength, and many other circumstances attending the occurrence of salt, will be seen from the following tabular view.

Analyses of the Salts from Saline and Brine Springs.

Locality	Halle	Schöne-beck	Artern	Dürren-berg	Kosen	New Saltwork	Tuism	Nauheim	Soden	Clemens-hall	Reichen-hall	Kissin-gen	Cheltenham			Caledonia, Canada	Tarentum, U. States
Per cent. of Salt in water	12·98	11·10	26·50	3·09	4·96	4·35	4·39	2·87	1·27	26·39	23·30	1·04	0·92	0·82	1·06	7·64	4·09
Chloride of sodium	94·43	99·72	96·35	96·68	97·60	80·90	91·24	89·23	86·01	98·17	96·99	75·30	74·53	53·06	80·90	84·31	78·24
" potassium	0·21		0·45				0·73	1·83	1·81		0·20	0·62				0·39	13·96
" calcium	1·03		1·59	1·49				6·74	9·24	trace		6·46					7·17
" magnesium	1·59	0·67	1·99			0·65		1·18			0·20				0·41	{B+K 0·25}	0·93
" barium											0·01						trace
" ammonium						0·006		0·04				0·03	0·61	0·35		0·02	
Bromide of sodium															0·06	0·06	
" calcium																	
" magnesium	2·23	1·34	1·10	0·99	0·62	0·11	2·53	0·18	0·65	0·08	0·26		0·46	25·19	11·19	2·31	
Iodide of sodium		2·55	1·31	0·04	0·47	7·04	3·38			1·68	1·79	3·18	17·48	1·99	1·97	1·54	
" magnesium	0·37	1·18		0·53	0·38	6·93	0·78				0·73	1·50		11·32		6·77	
Sulphate of potash				6·77	9·08	0·002											
" soda								7·43	7·73		0·004	9·37	8·18	3·96	2·30		
" lime	0·44		0·15	0·08	0·28	2·53	0·40		0·77	0·07	trace	1·88	1·19	0·34	0·92		
" magnesia	0·05		0·02	trace	1·67		0·23	0·29		0·003	0·62	1·76					
" alumina	0·03				0·16		0·07	0·50									
Carbonate of soda						0·002											
" lime																	
" magnesia																	
" iron																	
" manganese																	
Silicate of soda																	
Phosphate of iron																	
Oxide of iron	0·02	0·02		0·01	0·02	0·11		0·07	0·50		0·006		0·48	0·18	0·27	trace	
" manganese																0·56	
Alumina																	
Silica	0·02	0·02		0·02				0·07					0·61	0·94	2·48	3·02	
Organic matter																	
Oxalic acid																	
Carbonic acid																	
	100·00	100·00	100·00	100·00	100·00	100·00	100·00	100·00	100·00	100·00	100·00	100·00	100·00	100·00	100·00	100·00	100·00

Constituents in 100 parts of brine.	Cheshire.		Worcestershire.	
	Marston.	Wheel-och.	Droit-wich.	Stoke.
Chloride of sodium	25·222	25·333	22·452	25·492
„ potassium	—	—	—	—
Bromide of sodium	0·011	0·020	trace	trace
Iodide of sodium	trace	trace	trace	trace
Chloride of magnesium	—	0·171	—	—
Sulphate of potash	trace	trace	trace	trace
„ soda	0·146	—	0·390	0·594
„ magnesia	—	—	—	—
„ lime	0·391	0·418	0·387	0·261
Carbonate of soda	0·036	—	0·115	0·016
„ magnesia	0·107	0·107	0·034	0·034
„ manganese	trace	trace	—	—
„ lime	—	0·052	—	—
Phosphate of lime	trace	trace	trace	trace
„ ferric-oxide	trace	trace	trace	trace
Alumina	trace	trace	—	—
Silica	—	—	trace	trace
	25·913	26·101	23·378	26·397

Constituents of the brine.—The impossibility of conveying salt to a distance, on account of the lowness of its price, the cost of carriage, and the different peculiarities of locality, sufficiently explain why, in some places, very weak brine, and in others, brine containing ten times as much salt, may be worked with advantage. Thus, for instance, the salt-works at Salzhausen can only be carried on in consequence of fuel being obtained without cost, the refuse of the brown-coal works upon the same spot, supplying a source of heat.

Borings.—It is sometimes possible, by a suitable arrangement of the borings, to bring the brine to the surface by means of natural hydrostatic pressure (Artesian). The raising is then effected without the use of machines, and without diluting the brine with fresh water. Borings are also frequently made for the purpose of obtaining the brine nearer to its source, as at Rodenberg, in the principality of Schaumberg, where a very weak spring (0·6 per cent.) was obtained at an 8·5 times greater state of saturation (5·1 per cent.) by means of a boring. The new salt-work at Prussian Minden, is peculiarly adapted to give an idea of the importance of such undertakings. The new boring there is 4½ inches in diameter, and was begun in lias, and had attained in May, 1843, a depth of 2515 feet under the surface, and 2105

feet under, the level of the sea, having passed through the new
red sandstone formation, and arrived, still unfinished, at the Muschelkalk. 84 cubic feet of brine, containing 4 per cent., issue
from it per minute; therefore, 567,670 cwts. of salt per annum.
The shaft at Schönebeck produces per minute 20 to 25, and the
spring at Artern 211 cubic feet of brine.

The effect of increasing the depth of the boring on the strength
of the brine, is well illustrated in the boring at the Dürrenberg
salt-works, where the strength of the brine varied with the depth
as follows :

At	37·6 feet the brine contained . .	4·38 per cent.
	55	do. . .	3·79
	74·5	do. . .	2·95
	91	do. . .	3·21
	280	do. . .	4·64
	285	do. . .	3·82
	305	do. . .	4·94
	463	do. . .	2·50
	502	do. . .	2·90
	588	do. . .	4·25
	706	do. . .	3·30
	815	do. . .	5·25
	965	do. . .	6·82
	986	do. . .	9·26
	994	do. . .	12·14
	1024	do. . .	17·00
	1026·6	do. . .	19·16
	1028·5	do. . .	17·06
	1055	do. . .	16·86
	1064	do. . .	16·57
	1133	do. . .	15·15
	1138·5	do. . .	14·08
	1158	do. . .	14·66

Observations taken in sinking the Bore-hole for Brine at Neusalzwerk.

Depth of the bore-hole.	Quantity of brine per minute.	Percentage of salt in brine.	Tempera-ture of the brine.	Tempera-ture calcu-lated.	Mean tempera-ture of soil.
Feet.	Cubic feet.		R.	R.	R.
600	0·67	1⅝	12·5°	13·05°	8°
654	1·00	2½	13	—	—
692	1·18	—	14	—	—
775	1·50	2½	15·5	—	—
793	1·36	—	15·75	—	—
820	1·25	2½	14·25	14·95	—
848	0·95	—	14·5	—	—
923	0·78	—	15	—	—
960	1·14	1½	15	—	—
975	1·05	—	14·75	—	—
1004	0·90	1¾	15	16·46	—
1003	2·14	1	16·5	—	—
1039	4·60	⅞	17	—	—
1045	4·60	—	18	—	—
1100	4·60	—	18	17·26	—
1111	6·00	—	17·75	—	—
1178	5·00	—	17·75	—	—
1182	4·60	—	18	—	—
1225	5·00	⅞	18	—	—
1298	7·50	—	18	—	—
1343	7·50	—	18	—	—
1382	6·70	—	18	—	—
1418	6·00	2¼	18	19·94	—
1464	5·50	—	18	—	—
1478	4·70	—	18	—	—
1494	5·75	—	18	—	—
1525	6·00	—	18	—	—
1575	6·50	—	18	21·27	—
1586	6·00	3¾	19·5	—	—
1595	8·60	4½	20	—	—
1615	10·00	—	20·5	—	—
1633	10·00	5	21	—	—
1640	10·00	—	21	21·82	—
1664	15·00	5	22	—	—
1690	18·00	—	22	—	—
1713	20·00	—	22	—	—
1763	20·00	5	22·5	22·85	—
1783	20·00	5	23	—	—
1897	25·875	5	23·5	—	—
1951	45·00	4	25	24·44	—
1981	45·00	4	25	—	—
2020	45·00	4	25	—	—
2160	60·00	4½	26·2	26·20	—

In this country, all the salt is obtained by the evaporation of strong brines by artificial heat. The brines used are nearly saturated solutions, those of Cheshire containing 24 per cent., and those of Worcestershire, 26 per cent. of common salt; when a brine is weak, it is customary to add rock salt, so as to bring it up to saturation.

The brine springs of Cheshire do not rise to the surface, and an iron pipe is placed in the boring, up which the brine will rise to a certain height, whence it is pumped into reservoirs at a sufficient elevation to feed the pans. The brine is clear, of a sea-green colour, and free from suspended matter. The evaporating pans are made of boiler plate, and vary in size according to the kind of salt to be made. Those in which the salt is evaporated at a boiling heat, are 20 feet square and 18 inches deep, and for lower heats, they are made 40 feet by 20 feet broad, and 2 feet deep. They are set on the brickwork with a slight inclination from the end above the fire, and each working pan is heated by three or four fires. In many works, three pans are set in a series, the fires being placed under the two farthest removed from each other, and the flues unite and pass under the third pan. The flue in passing to the chimney, heats a chamber built over it, which is used as a drying stove for the squares of salt. Along each side of the pan house, there is a walk; and next the walls, long benches, 4 or 5 feet wide, are fixed, on which the salt is placed in conical baskets, to drain, after it has been taken from the pan.

The pan is filled to three-quarters of its contents, and a little glue or waste grease is thrown in, to assist in forming a scum, while the liquor is rapidly raised to the boiling point, and the scum removed by lightly skimming it with a piece of wood. When half the water has been evaporated, fresh brine is run in, and the salt is removed every 12 hours, by means of a scoop. The salt is

FIG. 116.

placed in wooden boxes, 18 inches high, 9 inches broad at the top, and 7 inches at the bottom, and slightly pressed down. The boxes or tubs are then placed on a perforated floor, to allow the mother-liquor to drain away through the slits in the bottom of the box. After 12 hours' draining, the box is turned up, leaving the squares of salt, which are then taken to the drying chamber.

Instead of drying the salt in large blocks or baskets, a very excellent plan is adopted at Allendorf, somewhat similar to the arrangements in this country for drying malt. The air is heated by the furnaces, and is conveyed in iron pipes to the drying room, which it enters through the tube *a.* This room is made of wood and perfectly air-tight, fitted with a set of wooden gratings, *c*, on which the salt is placed to a depth of about 4 inches. Below the gratings there are some boards, *g*, to catch any salt that may fall through *c*. The salt is filled in and removed by the opening *f*, made air-tight by a closely fitting cover. The hot air emerging into the open space *d*, passes up through the grating and salt, into the upper space, whence it is drawn off by the pipe *b*, in connexion with a chimney.

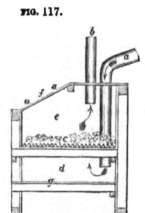

FIG. 117.

Varieties of Salt.

There are various kinds of salt made, called *stoved* or *lump salt*, *common salt*, *large-grained flaky*, and *large-grained* or *fishery salt*.

The first or *stoved salt* we have already described, but, instead of placing it in the wooden boxes or moulds, it is often placed in a conical wicker-basket, or *barrow* as it is called, which is allowed to drain in one of the side benches, and then taken to the chamber for drying (Fig. 118).

In making *common salt*, the pan is filled only once in 24 hours, and the brine, as in the former case, is first brought to a boiling heat, after which the fires are damped, and the crystallization carried on with the brine heated to 160° or 170° F. The salt is drained in baskets, but not dried artificially.

The *large-grained flaky salt* is made at a temperature of 130° to 160° F., and the pan is filled once in 48 hours. As this salt is often made between Saturday and Monday, during the damping of the fires, it has been called *Sunday salt.*

FIG. 118.

The *large-grained* or *fishery salt* is obtained from brine heated only to 100° or 110° F., at which temperature there is no agitation produced in the liquid, while the evaporation progresses very slowly, and thus the crystallization of the salt in large cubical crystals is promoted. It occupies from five to six days to complete the evaporation.

Improved processes for Manufacturing Salt.

Forty-five patents for this purpose have been taken out in this country, since the year 1801; but of all this number, only four or five embrace, what may be termed, a chemical improvement. That of Mons. L. J. F. Margueritte will be explained more at length when we describe the extraction of the salts of potash from kelp, &c. Manley proposed to precipitate the earthy impurities by adding barilla, or a mineral alkali, to the brine, and the Messrs. Loudon added the same substances to the rock salt, when in a state of fusion, and drew off the pure salt into receiving vessels. Mr. Arrott adopts a totally different principle.

He takes advantage of the greater solubility of common salt in hot solutions of the chlorides of calcium and magnesium than in cold solutions. He mixes these salts with the brine and concentrates it in closed boilers, by which he can safely avail himself of the steam which is generated, without any risk of the deposition of the salt. The concentrated liquor from the boiler, is allowed to flow into an open pan, where the salt deposits on the cooling of the liquor.

Mr. Arrott employs the same principle when treating rock-salt. He acts on the rock-salt with a solution of one of the earthy solutions heated to 224° F., and allows the liquor to cool, when the salt deposits. The same solution is again heated, and takes up a fresh portion of salt, which separates on cooling.

The mechanical arrangements are ingeniously contrived in each case, that the operation may go on continuously; but it is

obvious, that the salt so obtained, must always be mixed with the
adhering solution of the earthy chlorides, from which it must be
freed at some cost, both of labour and produce.

All the other processes may be said to aim at—
1. Economy in the fuel, or
2. Preservation of the pans.

In the first case, some of the patentees modify the number,
shape, and setting of the pans; or convey the fresh brine to the
pans through pipes warmed by the waste heat of the furnaces;
or employ hot air or steam, in pipes immersed in the brine; or
force hot air through or over the brine; or expose the brine, in
thin films or spray, to the action of the heated gases from furnaces;
or subject the brine, in fibrous or other materials, to the action of
currents of heated air; or conduct the evaporation in closed pans,
and employ the steam from the brine, for heating a second set of
pans filled with brine of a lower specific gravity; others create
an artificial draught, to carry off the steam more rapidly from
the surface of the brine as it boils.

There is one plan in operation which consists in dissolving rock
salt in sea-water, and either evaporating the liquor by the waste
heat of a lime kiln, or in the ordinary manner by fires underneath
the pans, and making use of the waste heat to dry levigated chalk
which is used for whitening.

As regards saving the wear and tear of the pans, some of the
proposals consist in arching the bottom of the pans over the furnace; others employ revolving pans, which never remain long
exposed to the direct action of the fire; others deepen the farther
end of the pan to receive the salt; others employ coolers, or construct the pans with sides, extending beyond the furnace, where
the salt may deposit; and others substitute inverted cones into
which the salt may fall; the object in each case being to prevent
the deposit of salt on that part of the pan bottom directly over
the fire; in other patents, the salt is constantly removed by a
mechanical arrangement of scrapers or rakes, into pockets at the
sides of the pans; or the pans are constructed with double bottoms
and sides, the intermediate space being filled with water heated
by the ordinary furnaces; and in one case the exposed part of
the pan is covered with a fusible alloy, to which the *pan scratch*
attaches itself, and from which it is easily removed.

In some patents, both objects are attempted by a combination
of one or more of the plans we have thus briefly indicated, but

we are not aware that any of the numerous suggestions have been carried into constant and daily practice.

We give the details of one plan, patented by Mr. Wilkinson, for employing the waste heat of coke ovens, which has been very successful, and is now in operation at Newcastle-on-Tyne.

This process is illustrated in the accompanying wood-cuts, in which Fig. 119 represents in plan view, a range of twelve coke ovens, with an evaporating and crystallizing pan (used for the manufacture of common salt) attached, and Fig. 120 is an elevation of the same range. AA are the flues leading from the ovens BB, and connected with the flue C (shown by dots in Fig. 119), which run underneath the evaporating pan D; EE are two auxiliary furnaces, for keeping up the temperature of the flue C, and burning the combustible gases which are given off from the coke ovens; F is a crystallizing pan, and G, a receptacle or well into which the salt is collected; HHHH are furnaces attached to the crystallizing pan, and intended to regulate the concentration of the liquor, as circumstances may require. Fig. 121 represents in end-elevation, the range of ovens and pans shown at Figures 119 and 120; and Fig. 122 is a longitudinal section of one of these ovens.

The mode of operation pursued in the manufacture of the common salt is as follows:—The evaporating pan D is supplied with salt water from a reservoir or other convenient source, and in order to apply the heat equally to the pan D, each alternate oven is charged at proper intervals of time, dividing the range into three sections of four ovens each, and charging one section per day. By this means, the combustible gases liberated from each newly charged oven, are brought into contact with the heated flue of the next oven, which is in a more advanced stage of the process, and by introducing a proper quantity of air to support combustion in the circulating flue, nearly all the combustible gases are consumed; QQ are apertures to introduce air to support combustion at any point, which may be found necessary in the circulating flue. Each of the ovens is provided with a damper, R, by which the communication with the circulating flue C, may be stopped at pleasure. The two auxiliary furnaces, EE, are used to keep up the temperature of the circulating flue, and at the same time to control the evaporating pan.

When the liquor or brine is sufficiently concentrated in the evaporating pan, a portion is run into the crystallizing pan by

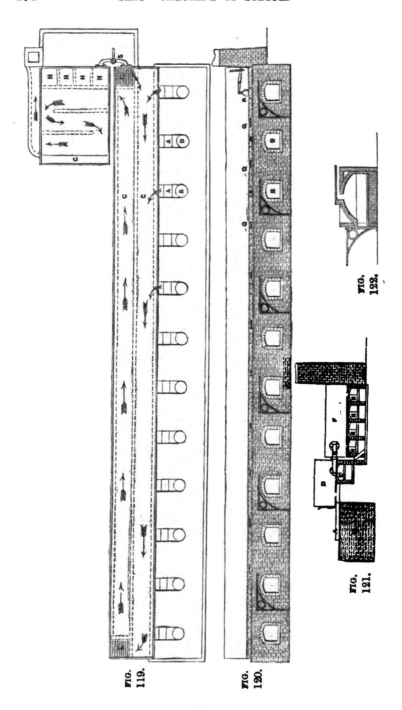

FIG. 119.

FIG. 120.

FIG. 121.

FIG. 122.

means of the connecting pipes, S (Figs. 119 and 121), and the liquor is purified before precipitating or crystallizing the salt, by a further concentration of the liquor by the furnaces HH.

Manufacture of Salt in Germany.

The process adopted in Germany, where fuel is dear, differs from that pursued in England, and consists of—

First—*The graduation.*—The greater number of brine springs are far too dilute, to repay with the present price of salt, the cost of evaporation by means of fuel. At Salzhausen, for instance, the production of 1 cwt. of salt presupposes the evaporation of 339 cubic feet of brine ; at Schönebeck the annual produce of 575,000 cwt. of salt, is obtained by the evaporation of 19 million cubic feet

FIG. 123.

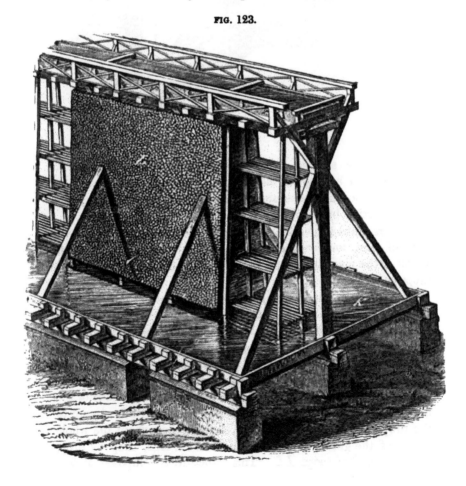

FIG. 124.

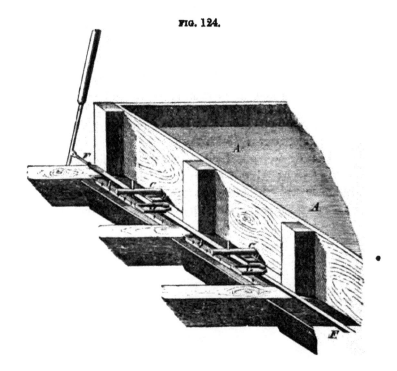

of water.   In all the brine springs, therefore, which are far
removed from a state of saturation, the greater portion of the
water is removed by evaporation in the air, "graduation;" the
smaller portion by "boiling."

The graduation house is intended to distribute the brine in the
form of rain, and expose it to the air in this state, while the
action of the air is increased by stopping and retarding the single
drops as they fall.

The brine is caused to fall from the trough or cistern *A* (Fig.
124), into the tank *K*, the retardation being effected by means of
a wall of twigs or thorns, *LL*, and its distribution in the form of
rain, by means of a series of perforated tubes and plugs (*Gesch-
windstellung*) (Fig. 123).   The motive power raises the brine
into a large reservoir, generally placed in a tower, whence it must
flow freely into the troughs *AA*, as it is wanted.   By means of the
horizontal pipes, *CCC*, the brine is conducted in a thin stream to
the dropping channel, *BB*, which extends throughout the whole
length of the graduation, and from thence it falls, drop by drop,
upon the wall of twigs, *LL*.   This structure is composed of fag-

gots of black-thorn, placed between the lathwork, *ll*, in a horizontal, even manner. The protecting board, *II*, prevents the wind, which must pass through the thorns, from giving a wrong direction to the drops which are constantly falling on the outer side. That the air may exert its full influence, the whole structure for graduation is erected in an airy situation, and in a direction at right angles to that of the prevailing wind. It is obvious, that this arrangement must expose the *extended surface* of the brine for a *longer time* to a constant current of air. If the wind changes, and threatens to carry the brine away from the wall and over the structure, the graduation must be reversed to the opposite surface of the wall of thorns, and this is done by a simple movement of the lever, *E* : for which purpose *E* is attached to the wooden rod, *FF*, supporting the boxes, *GG*. The lever brings the wooden rod forward, and with it the boxes *GG* are moved into a position just under the horizontal pipes, so that their narrow lips at the back, project over the cross channels *HH*. Thus the brine is intercepted above the channels *B*, and carried to the other side, and opposite surface of the thorns, by a channel precisely similar to *B*. That the whole arrangement of spigots, channels, &c., may be easily managed, planks for walking are laid on both sides of *A*, and these are furnished with a railing. The erection for graduation here described, as it is practised in Salzhausen, is known as the " *one-walled* " graduation house, and is used in small works, where building material is scarce. The walls of thorns are, however, frequently made in pairs (Fig. 125), and sometimes the outer surfaces, *m m*, only are used—" *surface graduation* ; " at

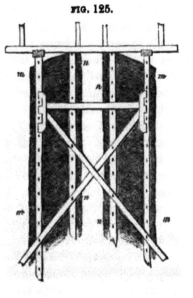

FIG. 125.

others, the inner surfaces, *nn*, are employed at the same time—" *cubic graduation.*" This latter practice does not quite double the effect, but (from observations made at Dürrenberg) it increases it in the ratio of 5 to 8 or 9. In each of these operations, the brine must be allowed to fall 3, 4, 6, or even 8 times through the thorns. For this reason, the graduation houses are partitioned into several compartments, the foremost of which serves for the first, the second for the next fall, and so on. At Schönebeck, the effective thorn surface comprises 390,000 square feet, and evaporates, on an average, during the day, 3·7 cubic feet of water from each square foot; or during the year (= 258 working days) above 12 millions of hogsheads.

According to an old-fashioned plan, the graduation was effected by distributing the brine over flat, inclined wooden surfaces ("*Pritschen*"), or over ropes stretched backwards and forwards for a length of several thousand feet. Thus, for instance, at Moutier, in France, where salt is crystallized during the whole summer, without any evaporation by fire, solely by graduation, the brine is caused to pass ten or more times over these ropes. The thorn walls, introduced into Saxony from Lombardy, in the year 1559, have superseded every other plan in Germany. It is obvious, that graduation proceeds best with a moderately warm wind and sunshine; that a moist, calm atmosphere is less favourable to it; and that in rainy weather it is altogether stopped, whilst the wind, when it acquires a certain force, is liable to carry the brine entirely away from the brine cistern. Frost is also prejudicial; for Berzelius observed, that below —3° (27° F.) sulphate of magnesia, with a portion of chloride of sodium, became converted into chloride of magnesium, and Glauber's salt—$MgO.SO^3 + Na.Cl = Mg.Cl + NaO.SO^3$— and that this decomposition is not reversed when the weather becomes warmer. Salt is, therefore, not only lost in this manner, but the quantity of chloride of magnesium is increased, which is detrimental in the boiling process. Graduation is, consequently, limited to the more favourable time of the year, and can only be practised during 200 or 260 days. The quantity of brine allowed to flow over the thorns, must be proportioned to the power of the wind. Nevertheless, a considerable loss is unavoidable during the graduation (12·4 per cent. at Schönebeck), which is partly occasioned by small drops being blown away, and partly from salt evaporating with the water, which is probable, judging from similar observations made with boracic acid. At Nauheim, a glass plate, removed to a distance of 600 feet from the building, and placed upon a high pole, was found covered, after some time, with a thin incrustation of salt.

The changes which the brine undergoes in passing through the thorns, are various. The carbonates of the earths are dissolved in the brine as bicarbonates; all the free carbonic acid, and the half of that combined with the earths, escapes, partly in passing through the pumps, and still more during graduation; and the earths are deposited as insoluble simple carbonates, while the greater portion of the gypsum crystallizes, in consequence of the diminished amount of water. Gypsum, according to Berthier, is most soluble in brine of specific gravity 1·033, and is, therefore, not deposited at first from very weak brine. In consequence of these depositions, the thorns become gradually covered with a thick coating (*thornstone*), which, inasmuch as it at last fills up the interstices, and stops the draught of air, renders it necessary to renew the thorn wall every 5, 6, or 8 years. This coating has been analysed, and the following table gives the composition from the different falls.

Analyses of the deposit, called Thornstone, on the thorns at Kösen, from the different compartments.

Constituents.	1	2	3	4
Silica	0·268	0·230	0·152	0·234
Oxide of iron	0·147	0·113	0·074	0·057
Carbonate of lime	3·414	0·600	0 536	0·470
Sulphate of lime	75·027	77·566	76·725	75 319
,, potash	0.072	0·116	0·239	0·161
,, magnesia	—	trace	—	—
Chloride of potassium	0·478	0·217	0·429	0·854
,, sodium	0·202	0·240	0·831	2·115
,, magnesium	—	—	0·102	0·117
Organic matter and water	20·392	21·918	20·912	20·673
	100·000	100·000	100·000.	100·000

In the brine cisterns, precipitates of a similar composition, fall as a fine mud, sometimes accompanied by a greyish, thick, scum-like mass, filled with bubbles, which is almost entirely composed of living infusoria, evolving large quantities of pure oxygen. The principal change which takes place in the brine, is naturally the progressive evaporation of the water; and the manner in which this progresses, although variable on account of locality and the weather, may be seen from the following general view of the graduation at Dürrenberg and Kösen.

In the beginning 1 cubic foot contains 2·5 lbs. of salt; 100 lbs.

of salt are therefore dissolved in 38·3 cubic feet of water. After
the first graduation the contents of 1 cubic foot are 3·9 lbs. ; 100
lbs. being dissolved in 24·7 c. f. of water, the evaporation therefore
being 13·6 c. f. After the second graduation the contents of 1 cubic
foot are 5·6 lbs., 100 lbs. being dissolved in 16·6 c. f. of water,
the evaporation therefore being 8·1 c. f. After the third graduation
the contents of 1 cubic foot are 8·0 lbs., 100 lbs. being dissolved
in 11·3 c. f. of water, the evaporation therefore being 5·3 c. f.

Analyses of the original and graduated Brines at Kösen.

Constituents.	Original Brine.		First fall.	Second fall.	Third fall.	Brine for boiling.
	Higher shaft.	Lower shaft.				
Water	95·042	96·656	90·598	85·817	79·008	74·017
Chloride of sodium ...	4·343	2·741	8·523	13·169	19·886	25·811
Sulphate of soda	0·028	0·141	0·054	0·069	0·153	0·139
„ potash ...	0·031	0·030	0·061	0·074	0·102	0·163
„ magnesia	0·103	0·076	0·218	0·317	0·465	0·632
„ lime	0·438	0·334	0·538	0·551	0·385	0·238
Carbonate of lime	0·014	0·021	0·008	0·003	0·001	—
„ magnesia	trace	trace	—	—	—	—
Oxide of iron	0·001	0·001	—	—	—	—
	100·000	100·000	100·000	100·000	100·000	100·000

Whilst, therefore, the evaporation diminishes, the loss in graduation increases with the strength of the brine, so that, at last, a
period arrives, when the loss of salt by the wind, and the advantage
of further removal of water, compensate each other. Upon this,
and upon the price of fuel, must depend the extent to which graduation is carried. In general, the *brine fit for boiling* should
not contain more than 23 per cent. ; at Nauheim it contains
16·47, at Rodenberg 16·38, at Dürrenberg 22, at Schönebeck
17·5, at Artern 21 per cent. of salt ; all natural or artificial
brine of this or still greater strength, as, for instance, that of
Lüneburg (25 per cent.), Reichenhall (23 per cent.), or Wimpfen
(25 per cent.) can of course be boiled down at once.

2. *The boiling.*—As graduation goes on only during the
fine season of the year, and the boiling during the winter, the
concentrated brine is collected in large reservoirs, protected from
the frost, and is supplied from thence to the *pans* in the boiling
houses (*Salzkothen*), which are thus rendered independent of
the irregular progress of graduation. The mode of construction

will be seen in Fig. 126, which is a perpendicular section, and
Fig. 127, which is the horizontal cross-section at the height of the

FIG. 126.

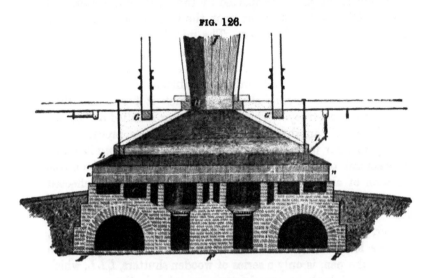

FIG. 127.

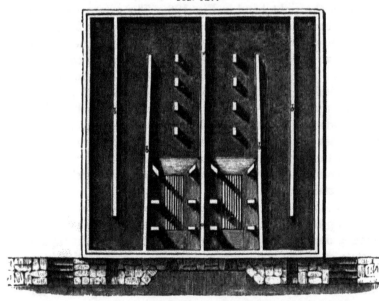

bottom of the pan. The pan A is a flat, four-sided vessel, of
sheet-iron, with a flat bottom, somewhat deepened towards the

12*

middle, and several yards in length and breadth. Some are above 60 feet in length and half that width, others 20 feet square. The bottom of the pan is supported by the small pillars *aaa*, and the pieces *CCC*, which are built upon the foundation *F*, and form at the same time the flues *ccc*. The middle wall *d* divides the pan into two halves, each of which has its own grate, so that the pan is heated by two separate fires. The flues are arranged to disseminate the flame as uniformly as possible, so that it passes out behind at *e* to the drying chambers, which it heats, and then escapes by the chimney. This kind of firing is called a "*circulating hearth;*" another kind in which the flues proceed in a radiating manner from the grate is called a "*radiating hearth.*" It is essential to rapid evaporation, that there should be a free circulation of air above the surface of the fluid. This is effected by a roof-shaped hood of boards, *GG, the steam or vapour trunk*, which, hanging from above, opens into the roof of the chamber, where (at *H*) it is connected with the vapour chimney, *II*. The lower margin with which the steam trunk appears to rest upon the pan, is only a series of wooden shutters, *LLL*, which can be turned back as occasion may require, in the manner represented on the right hand in Fig. 126. The external air thus passes in a constant current over the surface of the fluid, and becomes saturated with aqueous vapour, which is carried off in the form of a visible cloud through *I*, where it partly condenses and collects. The liquid which collects, has been found to contain, particularly during the boiling of the brine, about 1 per cent. of salt. To preserve this, and to prevent its running back into the pan, conducting tubes leading to a tank, are fixed at *H*.

The process of boiling consists of two distinct operations; it begins with the further purification and evaporation of the brine up to the point of saturation—the *schlotage;* and finishes with the crystallization of the salt—the *soccage.*

The Schlotage.—When the pans have been rather more than half filled (up to *nn*, Fig. 126) with clear brine from the reservoir, in which a deposit is also formed, the brine is brought rapidly into a state of violent ebullition, and the evaporated portion is replaced, from time to time, by fresh brine. The surface soon becomes covered with a dirty brown scum, consisting of decomposed crenic and apocrenic acids, and a bituminous matter. This scum has been examined, and its composition is given in the following analyses.

Analyses of Scum during the Evaporation.

	Stassfurth.	Dürrenberg.
Oxide of iron	1·158	0·285
Alumina	0·362	0·123
Silica	0·431	0·149
Carbonate of lime	4·387	1·542
Carbonate of magnesia	0·971	0·211
Magnesia with traces of chlorine	0·221	—
Sulphate of lime	54·908	62·436
Sulphate of potash	1·694	1·687
Sulphate of magnesia	0·762	1·440
Sulphate of soda	2·964	6·856
Chloride of sodium	26·090	14·909
Organic matter	0·796	0·148
Water	5·256	10·214
	100·000	100·000

This scum, and the salts precipitated at the same time, collect into a thick mud, which is partially removed by means of rakes; but some attaches itself to the bottom of the pans (the *scale*). After 12 or 15 boilings, it increases, often to the thickness of an inch, and must then be broken up with chisels. The salts are chiefly gypsum and sulphate of soda, probably in combination, forming an insoluble double salt, which encloses a considerable quantity of chloride of sodium, and small portions of other metallic chlorides, as will be seen by the following analyses :

Analyses of the Mud from the Brine.

Constituents.	Schöne-beck.	Stass-furth.	Halle. Large pan.	Halle. Small pan.	Dürren-berg.	Kösen.	Moutier. At the first.	Moutier. In the middle.	Moutier. At the end.
Silica	0·020	0·328	0·193	0·103	0·135	0·083	—	—	—
Oxide of iron	0·060	0·146	0·116	0·077	0·083	0·086	—	—	—
Alumina		0·068			—	—	—	—	—
Carbonate of lime	—	1·834	3·600	1·931	0·448	0 077	—	—	—
„ magnesia	0·567	0·150	0·720	0·897	0·385	—	—	—	—
Sulphate of magnesia	0·409	1·670	0·320	0·233	0·226	0·094	—	—	—
„ lime	22·649	65·956	64·027	51·315	50·773	29·904	28·0	41·1	10·1
„ soda	13·499	2 531	3·405	1·677	7·452	28·242	24·5	52·6	25 7
„ potash	1·839	0·742	1·479	1·664	0·991	0·712	—	—	—
Chloride of sodium	57·102	23·253	26·140	42·104	32·951	33·060	47·5	6·2	64·2
„ magnesium	—	—	These samples		—	0·058	—	—	—
Water	2·855	3·322	were dried.		6·556	7·084	—	—	—
	100·000	100·000	100·000	100·000	100·000	100·000	100·0	100·0	100 0

Analyses of the Panstone or Scale.

Constituents.	Schöne-beck.	Stassfurth.			Halle.	Dürren-berg.	Kösen.	Moutier.
		Schlotage pan.	Soccage pan.					
Silica	0·023	0·061	—	0·533	0·150	0·065	—
Oxide iron and alumina......	0·030	0·053	0·022	0·304	0·133	0·118	—
Carbonate of lime	—	0·123	0·060	1·265	0·095	0·467	—
,, magnesia	0·408	0·140	trace	1·905	0·210	—	—
Sulphate of potash	2·125	0·435	0·791	—	1·017	7·339	—
,, soda	20·667	4·000	40·407	—	9·055	20·925	20·6
,, lime	27·381	76·849	45·779	62·981	71·941	20·198	11·8
,, magnesia........	1·638	1·114	1·130	—	0·433	1·438	3·3
Chloride of sodium	44·284	15·065	10·566	29·028	10·771	53·426	63·4
,, potassium	—	—	—	1·310	—	—	—
,, calcium..........	—	—	—	2·431	—	—	—
,, magnesium	—	—	—	0·243	—	0·318	0·8
Water.....................	3·444	1·540	1·255	—	6·195	1·686	—
	100·000	100·000	100·000	100·000	100·000	100·000	100·0

Both these depositions are, therefore, a new and increasing source of loss during the boiling. In the meantime, the solution of salt becomes more concentrated, by the constant evaporation and renewal of the brine, until at last it crystallizes. Imagine a pan containing 1600 cubic feet of brine (therefore, 176 cwts. of salt), to be refilled as often as ¼th is evaporated, after the first addition there will be $176 + \dfrac{176}{4} = 221$ lbs. of salt in the pan, after the second addition $176 + 2\dfrac{176}{4} = 286$ lbs., and so on.

When, therefore, at the expiration of 20 or 24 hours, a crust of crystals begins to form over the surface, the fire is slackened, until the temperature of the brine falls to 90° (194° F.), or 75° (167° F.), when with slow evaporation, the *soccage* begins and lasts for several days.

The Soccage.—During this time the surface is covered here and there with small floating crystals, which gradually grow into the well-known 4-sided funnels, and soon sink to the bottom, when agitated by the vapour in making its escape. When the pan is kept at a high temperature, the crystals have no time for growing, and salt of a finer grain falls to the bottom; at the lowest possible temperature, they remain floating a longer time, and produce salt of coarse grain. In the former case the process is rapid, and slower in the latter. The process, however, or the temperature of the soccage, is not entirely at the command of the workman, because the chloride of magnesium is always a source of difficulty when little or no sulphate of soda is present. These two salts decompose each other in the pan, giving rise to chloride of sodium and sulphate of magnesia ($MgCl + NaO.SO^3 = NaCl + MgO.SO^3$).

This has been observed in an interesting manner with the brine from Rodenberg. The more concentrated brine, which contained chloride of magnesium, but no Glauber's salt, became constantly covered all over its surface, at the ordinary temperature of soccage, with a continuous scum of salt, which could not be permeated or broken up by the vapours, and when removed, was immediately reproduced and entirely prevented evaporation. Thus no coarse-grained salt could be produced, as it is wanted in commerce : the evil could only be remedied by reducing the temperature, which was of course attended with loss of time. An effectual remedy was accidentally found for this evil, by mixing the weaker brine (in which there is no chloride of magnesium, but only Glauber's salt) with the former, the effect being doubtless due to the mutual decomposition of chloride of magnesium and sulphate of soda in the mixture, into chloride of sodium and sulphate of magnesia. The result was the same, when sulphate of soda was added at once without diluting the brine by the addition of so much unnecessary water. During Sunday, when all work is stopped, the large crystals may be seen growing at the bottom (*Sunday salt*), for the salt not being quite so soluble in the cold, a portion is forced to crystallize, as soon as the temperature is lowered, and this attaches itself to the other crystals already in the pan. It is evident that the purity of the salt must diminish gradually towards the end of the process of soccage; thus Berthier found in the salt of Moutier—

	Salt.	Chloride of magnesium.	Gypsum.	Sulphate of magnesia.	Sulphate of soda.
At the beginning . .	94·64	—	1·56	—	3·80
In the middle . . .	93·59	0·61	—	0·25	5·55
Towards the end . .	85·5	2·0	—	12·5	—

For this reason the soccage must be stopped before all the salt is deposited. During the whole process, the salt is drawn up from the bottom with long rakes to the edge of the pan, and placed either in wicker baskets of peeled willow, or heaped upon the boards which are thrown back for the purpose, when in both cases the brine runs back to the pan. The moist salt, either in the same baskets or spread out upon hurdles, is then placed in the drying chamber as long as it loses moisture, when it is packed up for sale.

A general view of the nature of the brine, and the mode of procedure in the salt-works, does not lead us to expect pure chloride of sodium, and this is confirmed by the analyses. There has been found, for example, in salt from—

Analyses of Salt.

Constituents.	Schöne-beck.	Stass-furth.	Dürren-berg.	Ludwigs-hall.	Cheshire, stoved.	Lyming-ton cat-salt.	Scotch, Common.	St. Malo, Sea-salt.	Moutier, Summer salt. Brine cistern.	The ropes.	St. Ursa.	Cadiz.	Onondaga.	Nancy.	Florida.	Spencer, New York.	Pomeroy, Ohio.
Chloride of sodium ..	96·402	97·094	96·061	99·45	96·250	98·90	98·55	96·00	98·67	97·17	89·04	91·37	92·98	96·03	99·35	96·29	93·83
" potassium....	—	—	—	—	—	—	—	—	—	—	—	—	—	—	—	trace	trace
" calcium ..	—	—	—	—	0·025	—	—	—	0·18	0·25	3·38	1·06	1·10	0·86	0·01	trace	0·69
" magnesium	0·080	0·246	0·218	0·86	0·075	0·50	2·60	0·30	0·75	3·00	1·44	0·99	0·73	0·95	0·09	0·27	0·31
Sulphate of soda......	—	—	—	—	—	—	—	—	—	—	trace	trace	trace	trace	—	trace	—
" potash....	0·414	0·390	—	—	—	0·10	1·90	3·35	—	—	0·98	0·33	0·23	0·48	0·05	1·39	—
" lime......	0·732	0·737	1·277	0·28	1·350	0·50	1·75	0·45	0·4	0·58	—	—	—	—	—	—	—
" magnesia	0·471	0·180	0·279	—	—	—	—	—	—	—	—	—	—	—	—	—	—
Silicate of soda	—	—	—	—	—	—	—	—	—	—	0·03	0·03	trace	0·06	trace	0·01	trace
Nitrate of soda	—	—	—	—	—	—	—	—	—	—	—	trace	—	trace	trace	—	—
Oxide of iron	—	—	—	—	—	—	—	—	—	—	0·04	0·03	trace	0·06	0·03	0·08	trace
Alumina	—	—	0·018	—	—	—	—	—	—	—	0·06	0·07	trace	0·05	—	0·17	trace
Insoluble matter	—	—	—	—	—	—	—	—	—	—	0·19	0·17	0·15	0·10	trace	0·03	trace
Organic matter	—	—	—	—	—	—	—	—	—	—	—	—	—	—	—	trace	—
Water	2·901	1·364	2·155	—	—	—	—	—	—	—	6·10	6·20	1·30	1·10	0·60	2·30	4·60
	100·000	100·000	100·000	100·00	99·900	99·90	99·60	99·10	100·00	100·00	100·26	100·34	100·43	99·96	100·96	100·44	99·93

Of all these salts, chloride of magnesium is that which has the greatest influence upon the quality of the produce, both on account of its deliquescence in the air and its highly saline taste. For while pure chloride of sodium never attracts moisture from the air, it is well known how rapidly ordinary salt becomes wet in damp weather; this is still more evident when the salt has to be removed to a distance, and is proportionally more rapid when it contains a large quantity of chloride of magnesium. On the other hand, such salt is not unfrequently preferred in the kitchen to the purer kinds, as less of it is necessary for salting.

Berthier is the discoverer of a very ingenious method of getting rid of the chloride of magnesium during the soccage; slaked lime is to be added to the brine in the pan, until the whole of the chloride of magnesium is decomposed ($MgCl + CaO = CaCl + MgO$); on being further evaporated, the chloride of calcium formed, is decomposed with sulphate of soda into chloride of sodium and gypsum ($CaCl + NaO.SO^3 = NaCl + CaO.SO^3$). Of course the presence of a quantity of sulphate of soda equivalent to the chloride of magnesium, is here essentially necessary.

The requisite quantity of brine for each process of soccage, leaves, when the process is finished, a very impure solution of salt, which, however, is not so bad as to be at once rejected. A second, and sometimes even a third charge may be boiled down before the residue—*mother-liquor*—is removed, which is either used to produce an inferior kind of salt, or for the manufacture of Epsom salts or carbonate of magnesia. The following table shows the nature of the mother liquor.

Analyses of the Mother-Liquor.

Constituents.	Schöne-beck.	Stass-furth.	Halle.	Dürren-berg.	Kösen.	Reden-burg.	Moutier.	Unna, Westphalia.	
								Brande.	Liebig.
Chloride of sodium	15·057	13·623	6·494	18·384	15·632	5·32	20·30	7·45	2·21
,, potassium..	—	—	4·914	—	—	—	—	2·29	1·05
,, calcium....	—	—	5·350	—	—	—	—	9·98	21·78
,, magnesium	7·200	9·784	12·695	6·091	6·482	18·23	4·85	7·89	8·86
,, aluminium	—	—	0·042	—	—	—	—	—	—
Bromides and iodides ..	—	—	—	—	—	trace	—	0·10	0·16
Sulphate of potash	5·358	5·177	—	2·783	3·614	—	—	—	—
,, magnesia ..	3·522	1·596	—	0·798	4·759	5·44	9·50	—	—
,, lime	—	—	0·096	—	—	0·11	—	0·06	0·01
Ammonia	—	—	trace	—	—	—	—	—	—
Silica	—	0·005	0·006	—	—	—	—	—	—
Water................	68·863	69·885	70·401	71·944	69·513	70·30	64·85	72·23	65·91
	100·000	100·000	100·000	100·000	100·000	100·00	100·00	100·00	100·00

* In cases where the brine contains iodides and bromides, these,

from their high degree of solubility, will be found in the mother-liquor, and, as in Kreuznach, Unna, and Salzhausen, will communicate medicinal properties to it, which are particularly applicable to the cure of scrofulous diseases. The quantity of sulphate of soda can be increased by freezing the liquor, and then obtaining it from this, as from the scale (*Pfannenstein*) by crystallization. By evaporating the residual liquor, sulphate of potash may be separated, and lastly the chloride of magnesium is converted, by the addition of sulphate of soda and heating to 50° C. (122° F.), into Epsom salts.

Statistics of the Salt Trade.

We now give the statistics of the production of this important substance in this country from Mr. Hunt's report for 1860.

Cheshire.

The quantities carried down the River Weaver :—

White salt . . . . 26 cwt. to the ton,		759,486 tons.
Rock salt		71,043
By Bridgewater canal and by Railway from the Districts of Winsford and Northwich . . .		525,000

Worcestershire.

Stoke and Droitwich   197,000

Ireland.

Duncrue, near Carrickfergus   18,443

Total produce of the United Kingdom   1,570,972

Exports of Salt.

	Quantities.		
	1858.	1859.	1860.
	Tons.	Tons.	Tons.
To Russia	52,028	64,998	66,420
„ United States	227,195	219,678	268,083
„ British North America . .	86,985	76,931	76,138
„ „ East Indies	72,563	33,381	116,754
„ Other Countries	156,126	170,656	168,319
Total	594,897	565,644	696,714

The production in France, of salt from all sources, amounted to

330,210 tons in 1847.
465,435 „ 1848.
479,438 „ 1849.
495,183 „ 1850.
599,175 „ 1851.
428,037 „ 1852.

The annual production in Germany for 1846, the last year for which we can find any statement, was—

Holstein-Schleswig	2,000	tons.
Mecklenburg	3,750	„
Prussia	85,050	„

114,000 in 1855; 119,000 in 1856; 143,000 in 1857; 151,000 in 1858; 121,000 in 1859.

Hanover	18,428	„
Brunswick	1,200	„
Schwartzburg-Rudolstadt	3,000	„
Lippe-Detmold	1,400	„
Waldeck	225	
Saxe-Meiningen	4,100	„
Saxe-Weimar	1,500	„
Saxe-Coburg-Gotha	1,750	„
Reus-Schleiz	1,600	„
Hesse-Cassel	9,450	„
Hesse-Darmstadt	11,300	„
Wirtemberg	36,600	„
Baden	16,800	„
Bavaria	39,100	„

50,073 in 1856-7.

Austria, German Provinces	104,450	„
Total	339,703	„

The total production in the Austrian Empire was, on an average of 5 years—

German Provinces	104,450	tons.
Sclavanian „	167,333	„

And for Russia and Italy, the following quantities are given by Karsten:

The Russian Empire . . .	352,000	tons.
„ Italian States	247,000	„

SODA.

French, *Soude;* German, *Soda, Barilla;* Dutch, *Soda;* Spanish, *Barrilla;*
Portuguese, *Solda, Barrilha;* Italian, *Barriglia;* Russian, *Socianka;*
Arabic, *Kali.*

There is no substance in the entire range of chemical manufactures which occupies so important a place as *Soda,* whether
we regard the large amount of capital embarked in its production,
the number of workpeople employed, or the many purposes
to which it is applied.   The consumption of soda per head of the
population of any country, may very fairly be taken as an index
of the degree of civilization of the nation.   The principal form
in which it is found in commerce is the *Carbonate of Soda.*

In scientific language, the term soda denotes the oxide of
sodium, NaO, or its hydrate, NaO.HO, but technologically it is
applied to the *neutral carbonate of soda,* NaO.CO², which is the
compound of greatest importance in the arts.

Soda does not exist in nature in large quantities, or in any
degree of purity.   It was first produced in Egypt for purposes
of trade; but afterwards, Barbary had for a long time the
monopoly of the supply, and exported it to Europe, where it
was sold at a very high price.   After a time, deposits were found
in other localities, which deprived her of the monopoly.   Spain
then took the lead in supplying her *Barillas,* which have at
length given place to the artificial product.

Native Soda.

A mineral mass is found in several localities, consisting chiefly of
sesquicarbonate of soda, 2NaO.3CO² + 3 aq.; it is left in many
places as an incrustation, when the so-called "soda lakes" dry up
in the summer.   In Egypt it occurs in layers half an inch thick,
and is called *Trona* by mineralogists, and in South America,
where it is termed *Urao.*

The soda used in Egypt is brought thither from the interior of
Africa, and is made in the country itself, in the neighbourhood
of Memphis and Hermopolis, about a day's journey from Alexandria.   The small lakes from which it is obtained, are from 4 to
5 miles long, and from 1 to 1½ miles broad, but there are many

others which we should call *Ponds :* the depth is not great, that
called *Nehclé,* the deepest of all, being not more than 20 feet
deep. The heat of the summer months, dries up the smaller lakes
and ponds, and the more shallow parts of the larger ones, leaving
large masses of this salt behind, mixed with clay and other
substances ; these masses are called *Latroni* by the Egyptians.
The natives separate the impurities as much as possible by shovels,
and then spread out the soda in the sun to dry. It is afterwards
packed in baskets and carried to the Nile on camels, whence
it is conveyed to Cairo and Alexandria. The natives are in the
habit of throwing all kinds of substances into the water, as they
say, to promote the crystallization of the soda, which, as it
forms, floats on the surface ; hence the name *Natron,* from the
Arabic *Nether* or *Nathar,* to "rise up" or "start up."

Samples of this Egyptian soda have been analysed by

	Beudant.	Reichardt.
Carbonate of soda . . .	74·70	18·43
Sulphate of soda	7·50	31·11
Chloride of sodium . . .	3·10	45·77
Water	13·50	4·22
	98·80	99·53

Similar products are found in India, and one from Bengal,
analysed by Pfeiffer, was found to contain

Soda		22·59
Potash		2·65
Carbonic acid		16·00
Sulphuric acid		4·01
Chlorine		0·79
Water		17·59
Insoluble matters—		
Silica and sand . .	34·65	
Oxide of iron . .	1·08	
Aluminia . . .	0·26	36·45
Lime	0 16	
Magnesia. . . .	0·30	
		100·08

A native carbonate of soda has also been found at Clausthal,
as an efflorescence on the shale of a lead mine : it was analysed
by Kayser, and contained—

Carbonate of soda	92·07
Carbonate of iron	0·19
Carbonate of magnesia	3·32
Carbonate of lime	1·81
Water	1·85
	99·24

In Hungary, native soda occurs in the department of Bichar, near Mariatheresiopel; again in Lesser Kumania, near Shegedin, where there are several manufactories engaged in its preparation. The salt, there called *Széksó*, exudes as a snow-white crust upon the surface of the ground, and is collected as a greyish mass, mixed with earth, at an early hour of the morning, before sunrise, when it is said to be most abundant.

A specimen of this natural soda from Debreczin has been analysed by Wackenroder, who found—

Carbonate of soda	89·84
Chloride of sodium	4·35
Sulphate of soda	1·63
Phosphate of soda	1·46
Sulphate of lime	0·03
Carbonate of lime	0·24
Carbonate of magnesia	0·24
Silicate of iron	0·42
Silicate of soda	1·61
Silica	0·15
	99·97

And by Beudant, whose analysis corresponds with the formula $NaO.CO^2 + HO$.

Soda	43·20
Carbonic acid	30·40
Sulphate of soda	10·40
Chloride of sodium	2·20
Water	13·80
	100·00

This soda-earth is treated in the same way as *black ash*, and the ash of plants in the potash manufactories. It is lixiviated with water, and the lye is evaporated to dryness. The colouring matters are destroyed by heating it to redness.

Carbonate of soda also occurs in nature, combined with carbonate of lime, and has been named *Gay-Lussite*, by Boussingault, whose analysis is as follows :—

Carbonate of soda . . . . . . 34·5
Carbonate of lime . . . . . 33·6
Water . . . . 30·4
Clay : . . . . 1·5
———
100·0

which may be represented by the formula

$$NaO.CO^2 + CaO.CO^2 + 5HO.$$

This mineral is very abundant at Lagunilla, near Merida, in
Maracaibo, where the natives call it *clavos* or *nails*, in allusion to
its crystalline form.

Soda from Marine Plants.

Those plants which grow on the sea-shore or in the sea itself,
imbibe and assimulate the mineral ingredients of the sea-water.
Thus, they assimulate the soda of common salt, when some
function of vegetable life, as the saturation of an organic acid,
requires the presence of a base. Hence, without reference to the
plants exclusively inhabiting the ocean, as the *fucoïds*, we can
understand why the vegetation of the sea-shore, viz., certain species
of *Salsola, Triglochia, Salicornia, Atriplex, Statice, Batis,
Mesembryanthemum, Chenopodium, Reaumeria*, &c., should
be so constantly found in those localities which can only be
exchanged for the neighbourhood of salt-springs. When these
plants are incinerated, the compounds of soda with the organic
acids, remain in the ash as carbonate of soda. The ash contains,
in addition, sulphate and hyposulphite of soda, sulphide, chloride,
iodide, and bromide of sodium, ferrocyanide of sodium, and the
corresponding salts of potash and potassium. There are some
insoluble compounds, consisting of carbonate and phosphate of
lime, sulphide of calcium, magnesia, alumina, silica, sulphide of
iron, and particles of charcoal. In general, the ash of sea plants,
compared with those grown on the shore, is not so rich in
carbonate of soda, but contains a larger quantity of salts of
potash; there are exceptions; thus, Guibourt's analysis of the ash
of *Salsola tragus*, shows that it contains 29·04 per cent. carbonate
of potash, 17·89 chloride of potassium, 4·93 sulphate of potash,
40·26 carbonate of lime, and 7·80 phosphate of lime, and oxide of
iron.

The following table contains the results of analyses by Schweitzer, Forchhammer, Gödecheus, and James.

Composition of the Ashes of Sea-Weeds after deducting Carbon and Carbonic Acid.

Constituents.	Laminaria saccharina, North Sea.	Fucus serratus, North Sea.	Laminaria latifolia.	Furcellaria fastigiata.	Chondrus crispus.	Iridæa edulis.	Polysiphonia elongata.	Delesseria sanguinea.	Fucus distichus, Clyde.	Fucus nodosus, Clyde.	Fucus serratus, Clyde.	Fucus vesiculosus, Clyde.	Fucus vesiculosus, Mersey.	Fucus vesiculosus, North Sea.	Fucus vesiculosus, Denmark.	Fucus vesiculosus, Greenland.
Potash	24·77	16·02	16·91	21·36	18·00	23·43	21·23	13·72	22·40	10·07	6·61	15·23	—	17·68	5·08	17·86
Soda	1·94	6·01	—	24·77	19·47	16·06	5·06	21·34	8·29	13·80	21·13	11·16	18·16	8·78	7·78	21·43
Lime	6·50	7·05	8·28	6·02	7·11	10·23	2·92	2·80	8·79	10·98	11·09	8·15	16·77	4·71	21·65	3·31
Magnesia	8·13	7·59	6·80	11·04	11·30	—	30·56	5·94	7·44	10·93	11·66	7·16	13·19	6·99	10·96	7·44
Chloride of Sodium	33·72	38·72	26·92	—	—	1·66	13·34	—	28·29	20·16	18·76	25·10	9·89	24·38	3·53	25·93
Chloride of Potassium	—	—	10·18	—	—	—	—	—	—	—	—	—	—	—	—	—
Iodide of Sodium	4·70	·77	—	—	—	—	—	—	3·63	·54	1·33	·37	—	·15	—	—
Phosphate of Lime	8·41	4·95	12·90	2·96	·76	23·23	2·99	3·90	3·65	3·24	9·97	2·99	—	5·44	9·67	10·09
Phosphate of Iron	·75	·64	—	—	—	—	—	—	—	—	—	—	—	—	—	—
Oxide of Iron	—	—	—	—	—	—	—	—	—	—	—	—	—	—	—	—
Oxide of Manganese	—	—	—	·22	—	—	—	—	·63	·29	·34	·78	4·43	—	—	—
Sulphuric Acid	10·60	18·35	12·68	33·53	43·66	25·19	38·60	46·43	13·26	26·09	21·06	28·16	30·94	23·71	26·24	13·94
Silica	·58	·40	·69	—	—	—	2·33	12·38	1·56	1·90	·43	1·93	7·69	·28	11·04	—
	100·00	100·00	95·21	100·00	100·00	99·71	98·10	100·00	100·00	100·00	100·00	100·00	100·00	100·00	100·00	100·00
Percentage of ash in weed dried at 100° C.	9·78	25·83	13·62	18·92	20·61	9·86	17·10	12·17	20·40	16·19	18·53	16·99	13·22	20·56	—	16·23

Becker has analysed the ashes of the *Schoberia acuminata*, growing on the Steppes of Southern Russia, and found—

Sulphate of potash	2·77
Chloride of potassium	5·69
Chloride of sodium	53·69
Carbonate of soda	7·23
Phosphate of soda	0·50
Hyposulphate of Soda	0·42
Sulphide of sodium	trace
Insoluble matters	26·70
Loss by heat	3·00
	100·00

The saline plants which contained the most carbonate of soda, were the following: 100 parts of the ash of the

Salsola clavifolia	gave	45·99	per cent. carbonate of soda.
Salsola soda	„	40·95	„
Halimocnemum capsicum	„	36·75	
Salsola kali	„	34·00	
Kochia sedoïdes	„	30·84	„
Salsola brachiata	„	26·26	„

The following are the principal sources of supply of this kind of soda:

1. *The soda of Alicante, or Barrilla*, is procured from the *Salsola soda*, which is cultivated for this purpose. The plants are raised from seed, which is sown at the close of the year, and gathered in the September following. The plants are collected at low water, and dried in the sun, while those on shore are mowed down and dried in the same way. Pits are dug in the ground, and iron bars laid across these holes, on which the dry weed with layers of dry seeds are thrown, and the whole is then ignited. This is continued for some days, and the ash falling into the pits, which are 3 feet deep and 4 feet square, at last becomes so hot that it cakes into a semi-vitrified mass. The whole is covered over with earth and allowed to cool, and afterwards broken into pieces, in which state it is found in commerce. There are three qualities recognised in France:—The *Soude d'Alicante douce, barilla douce*, tolerably uniform, of a greyish-blue colour, which effloresces on the surface when exposed to the air. The second quality is termed the *soude or barilla mélangée*, darker than the former, very hard, and porous. The

third quality, *soude bourde*, consists chiefly of common salt and contains pieces of charcoal. These sodas are found in large masses of about 10 cwt., cased with broom, and covered with coarse cloth. The best qualities contain about 40 per cent. of carbonate of soda; but we can only give the composition of the third quality—*soude bourde*, analysed by Girardin, who found—

Carbonate of soda	2
Chloride of sodium	65
Sulphate of soda	30
Sand and charcoal	3
	100

2. *The Soda of Malaga and Carthagena*, which is obtained in the same way, and is equal in quality to the *Soude melangée* of Alicante. It occurs in trade, in large heavy lumps of a grey colour, mixed with white, greenish, and black patches, and contains about 14 per cent. of soda in the form of carbonate.

3. *The Soda of Teneriffe* is obtained in this island from the *Mesembrianthemum cristallinum.* It is in large irregular lumps of a dark grey colour, and contains about 20 per cent. of carbonate of soda.

4. *The Soda of Varec* or *of Normandy* is obtained from the sea-weeds, which grow in large quantities on the shores, chiefly near Cherbourg. It is found in rough masses, porous, and very irregular in quality. The following analyses by Girardin show the nature of Varec from—

	Rouen.		Villette.	Cherbourg.	Granville.		
Carbonate of soda	23·29	16·94	13·76	9·53	6·00	3·71	0·22
Chloride of sodium......	46·90	23·91	54·11	45·78	7·20	25·88	65·58
„ potassium .	—	—	10·53	16·00	—	19·64	15·70
Sulphate of potash	—	—	20·35	22·19	18·80	42·54	13·50
Oxysulphide of calcium	20·41	52·15					
Iodides	—	—	traces	traces	traces	traces	traces
Insoluble matters	8·40	6·00	—	1·50	—	0·23	—
Water	1·00	1·00	1·25	5·00	2·00	8·00	5·00
	100·00	100·00	100·00	100·00	100·00	100·00	100·00

5. *The Soda of Narbonne*, or *Salicor*, or *Salicote*, which is obtained by the incineration of the *Salicornia annua*, cultivated in the neighbourhood of Narbonne, on the shores of the Mediterranean. The following analyses have been made of this description of soda, from 20 hectogrammes.

	Chaptal.	Julia.
Soda	2·28	2·48
Sulphate of soda	—	0·40
Chloride of sodium	0·96	0·84
Sulphate of potash	—	trace
Sulphate and muriate of magnesia . .	—	trace
Insoluble matters	1·80	1·62
	5·04	5·34

6. *The Soda of Aigues-Mortes or Blanquette* is obtained by incinerating all the plants growing on the shores of the South of France, between Frontignan and Aiguemorte, consisting of *Salicornia Europœ, Salsola Tragus, Kali Atriplex portulacoïdes,* and *Statice Limonium.* The first named contains the most soda, and in former times chloride of sodium was often fraudulently added to this soda. Chaptal found from 20 hectogrammes—

Soda	1·28	0·84	0·210
Chloride of sodium	0·16	1·60	0·152
Sulphate of potash	0·20	0·02	0·501
„ and muriate of magnesia . .	traces	0·04	0·600
Insoluble matters	1·92	2·56	2·114
	3·56	5·06	3·577

7. *Kelp* was formerly prepared in large quantities on the coasts of Ireland, and on the Western coasts and islands of Scotland, but was inferior to the foreign sodas. It is still produced in considerable quantities in Scotland, for the manufacture of iodine and potash-salts, which will be described in detail in one of the following sections. The mother-liquors are boiled down to dryness and sold, as *kelp-salts*, containing from 7 to 14 per cent. of alkali, to the soda-makers for reducing their strong alkalies.

The quantity raised from all these sources is, however, insignificant compared with the enormous consumption of the present day, which is nearly all obtained from common salt.

The following were the quantities of barilla imported into this country in—

| 1850 . . . . | 34,880 cwts. |
| 1856 . . . . | 54,608 „ |

13*

Production of Soda from Common Salt.

HISTORICAL NOTICES.

In the beginning of the last century, Duhamel, and somewhat later Marggraf, heated nitrate soda with charcoal to obtain small quantities of carbonate of soda ; and a modification of this plan has been recently proposed, viz., to decompose fused nitrate of soda by metallic copper, and then separate the caustic soda from the oxide of copper by lixiviation.

In the year 1737, Duhamel proposed to reduce sulphate of soda by heating it with charcoal, and then to convert the sulphide of sodium into acetate by means of acetic acid. The acetate was converted into carbonate of soda by heat.

In 1778, Malherbes suggested the preparation of soda by heating sulphate of soda, iron, and charcoal. The sulphate was first reduced by the charcoal to sulphide, and then part of the iron was thrown into the melted mass. More charcoal was added, and after an interval, a further portion of iron, and so on, until the evolution of sulphuretted hydrogen ceased. The crude mass furnished caustic soda by lixiviation. The proportions employed were—

100 parts calcined sulphate soda.
20 „ charcoal powder.
11 „ red hot charcoal.
33 „ iron scales.

This process was in operation at Javelle, near Paris, before the Revolution.

Guyton and Carey were also manufacturing soda at Croisie before the Revolution, by a process founded on the discovery of Scheele, that caustic lime decomposed common salt, which had also been patented in this country with some modifications, in 1797, by a Mr. George Hodgson.

About this period, or in 1775, Scheele discovered that common salt is decomposed by oxide of lead, and this process was worked as a manufacturing operation in London, and at Walker by the late Mr. Losh—the compound of lead left in the process, being sold as a colour, under the name of " Turner's Patent Yellow."

In 1781, Constantini decomposed common salt by melting it with alum ; and in 1782 Guyton and Carey employed felspar for the same purpose.

In this year, 1781, a patent was secured by Dr. Bryan Higgins for decomposing common salt with sulphuric acid, and then reducing the sulphate to sulphide, by fusing it with coal in a reverberatory furnace. The sulphide was decomposed by iron, tin, or lead, and the mass being then raked out of the furnace, the patentee states it may be used in place of barilla.

Several other patents followed, but we must refer to our Appendix, where an abstract of all the patents relating to this manufacture, will be found.

Meyer, of Stettin, in 1784, decomposed the common salt directly by potash, and separated the chloride of potassium from the soda by crystallization.

This brings us to the time of the French Republic, previous to which continental nations were chiefly dependent upon Spain for a supply of soda, which was obtained from certain marine plants by incineration and lixiviation of the ash. These plants grow very luxuriantly on the coast of Spain, Sicily, and Teneriffe, and large supplies of *barilla*, then used in place of soda, were obtained from Alicante.

This country was more independent, for we obtained our supply from the ashes of sea-weed called *kelp*, produced in large quantities on the North and West coasts of Ireland, and later, on the Western shores and islands of Scotland, where at one time as much as 25,000 tons were annually made. Large quantities of potashes were also imported into this country from America and Russia, and were used for many purposes, in which they have been replaced by soda in the present day.

One of the first effects of the wars of the French Revolution was to cut off the supply of alkali into France, which led to the publication of an appeal to the French chemists by the Government of the Republic. Mr. Gossage has recently read a very interesting paper on this subject, before the Chemical Section of the British Association, and we now quote from his pamphlet the appeal and his observations, which follow.

" ' Considering that the Republic ought to extend the energy of liberty to all the objects which are useful in the Arts of first necessity—free itself from all commercial dependence—and draw from its own sources all the materials deposited therein by nature, so as to render vain the efforts and the hatred of despots, and should place equally in requisition for the general service all industrial inventions and productions of the soil, it is commanded that all citizens who have commenced establishments, or who have obtained patents, for the extraction of soda from common salt, shall make known to the Convention the locality of these establishments, the quantity of soda now supplied by them, the quantity they can hereafter supply, and the period at which the increased supplies can be rendered.'

" A commission was appointed in the first year of the French Republic, consisting of citizens Lelievre, Pelletier, D'Arcet, and Giroud, and they made their report in the following year, 1794. This report gives a summary of thirteen different processes, of which particulars had been submitted to the Commission, six of these commencing with the production of sulphate of soda by the decomposition of salt. The preference, in the judgement of the commission, was given to the operations devised by ' un Pharmacien,' (a term equivalent to our apothecary or druggist,) named Le Blanc, who had already erected a soda-manufactory near Paris, in conjunction with two of his countrymen named Dizé and Shée.

" It has been generally thought that the invention of Le Blanc's process was a consequence of the appeal made to their countrymen by the French Convention. It will be seen from the following particulars that this notion is erroneous. The Commissioners say in their report:—' Citizens Le Blanc, Dizé, and Shée (copartners), were the first who submitted to us particulars of their processes ; and this was done with a noble devotion to the public good. Their establishment had been formed some time previously at Françiade, but the consequences of the French Revolution, and of the war which followed it, having deprived them of funds, the works were suspended, and for some months past the manufactory had become a national establishment.'

" And further, ' This establishment had been erected entirely with the private funds of the partners. It would be difficult to collect together, in so moderate a space, more means and conveniences than are met with in this manufactory,—furnaces, mills, apparatus, and magazines, all arranged in the best order for the convenience of the service.'

" The report then gives a full description of the various processes which constitute Le Blanc's invention, which had evidently been perfected previously to the Revolution. These consist of—

" 1st.—The decomposition of common salt by means of sulphuric acid, and the consequent production of sulphate of soda.

" 2nd.—The decomposition of sulphate of soda by means of chalk and carbon, and the consequent production of black ash or rough soda, containing carbonate of soda and sulphide of calcium.

"3rd.—The separation of two constituents of the last product, by lixiviation with water, and the obtainment of carbonate of soda in a state fit for application in the arts.

"These are the processes now used for the manufacture of soda on the largest scale.

"In my recent perusal of this report of the Commissioners, I have been much struck with the completeness of Le Blanc's invention as therein described; in fact, as regards the main principles of the invention, and even the proportions of materials used, these are identically the same as those now in use in this country, as well as in France, after the lapse of seventy years from the date of the invention.

"Thus was given to the world, by an humble apothecary, an invention which has done more to promote the civilization of mankind than any other chemical manufacture, as well as affording employment to a larger number of workmen, and yielding wealth to their employers. I regret to add that the poor inventor met with the too common reward for the application of his talents to the public good. After a life of great privation, he ended his days in an asylum for paupers.

"The manufacture of soda from common salt having been thus made clear to the chemists of France, the requirements of that country for this alkali were speedily met by the erection of sundry establishments for its preparation. The chief of these were located near the Mediterranean Sea, a few miles from Marseilles, this locality being selected on account of its proximity to Sicily, from whence sulphur was imported for the manufacture of sulphuric acid, and a cheap supply of salt was obtained from the sea-water, while limestone was also found on the spot."

Mr. W. Losh, who has recently died at the patriarchal age of 91 years, commenced his experiments on the Tyne about the year 1800, when soda crystals were selling at £60 per ton (the present price being £4 10s.), and manufactured soda, as we have stated, by the decomposition of common salt by oxide of lead; but, having obtained a knowledge of Le Blanc's process about 1802, he introduced this plan at Walker in company with Lord Dundonald, and some other gentlemen, so that Mr. Losh may very properly be considered the father of the soda-trade in this country.

Until the repeal of the duty on salt, the soda-trade could scarcely be said to have taken its place as one of the manufactures of Great Britain. This repeal was effected in 1823, which Mr. Gossage calls the natal year of the soda-trade, and Mr. Muspratt commenced his large works in Liverpool in that year; since which time this manufacture has gone on extending, till it may fairly be called one of the staple trades of the Kingdom, having its main seats in Lancashire and Newcastle-on-Tyne, with large establishments at Glasgow, Birmingham, and Bristol, and other localities.

The history of the trade, short as it is, has been marked by more than one important epoch,—to which we may allude in passing: the discovery by Mr. Tennant of a solid bleaching agent, chloride of lime, by which, what was the *waste* muriatic acid, is rendered available—the condensing towers for this acid by Mr. Gossage—the substitution of sulphur ores for sulphur in the production of sulphuric acid, in consequence of the absurd and mischievous interference of the King of Naples with the export of the latter—the use of metal pans for the decomposition of the common salt by the late Mr. J. Lee and Mr. Gamble—and the introduction of Gay-Lussac's process for the saving of the nitrous gases of the chambers. We conclude our historical notice with Mr. Gossage's

Chronology of the Soda Trade.

Period.	Raw Materials used and Prices.	Quantity manufactured.	Prices.
1790—Barilla and kelp......................		Not known.	Not known.
1792—Le Blanc's process invented and applied in France	,,	,,	
1814—Crystals of soda, made from Bleacher's residua, and by Mr. Losh from brine............	} ,, {	Soda crystals £60 per ton.	
1823 & 1824 } Mr. Muspratt's Works commenced—using Common salt at ... 15s. per ton. Sulphur at .... £8 ,, Lime at .... 15s. ,, Coal at .... 8s. ,,	Probably 100 tons per week of Crystals and Soda ash.	Soda crystals £18 per ton. Soda ash £24 per ton.	
1861—50 Works in operation in Great Britain using Le Blanc's process. Raw materials in Lancashire costing— Common salt .... 8s. per ton. Sulphur from pyrites £5 ,, Limestone .... 6s. 8d. ,, Fuel .... 6s. ,,	5,000 tons per week.	Soda crystals £4 10s. per ton. Soda ash £8 per ton.	

1861—Annual value of produce, two millions sterling. Number of workmen employed in the manufactories, 10,000, exclusive of those engaged in mining for pyrites, limestone, and coal ; also those employed in navigation and other means of transport.

Manufacture of Soda.

The production of carbonate of soda as at present conducted, embraces several distinct operations which may be designated—

I.—The decomposition of the common salt or *Decomposing process.*

II.—The conversion of the sulphate into carbonate of soda, or the *Balling process.*

III.—The lixiviation of the black balls and evaporation of the liquors.

IV.—The carbonating and oxidising processes for the production of white alkali or soda crystals.

V.—The production of caustic soda.

VI.—The manufacture of bicarbonate of soda.

FIRST OPERATION.

The decomposition of Common Salt or the decomposing process.—Sulphate of Soda.

This salt was discovered by Glauber in 1658, and has since been known as *Glauber's salts*. It exists in almost all waters, and some mineral springs owe their purgative properties to its presence, such as those of Cheltenham and Leamington, in this country. It is deposited in large quantities from the hot springs at Carlsbad, and is so abundant in the spring of Paipa, in the Andes, that crystals of the salt are formed on the ground over which the water is thrown. When combined with 10 atoms of water, it is called by mineralogists *Mirabilite*, $NaO.SO^3 + 10HO$, and occurs in this form at Ischl and Hallstadt, in Austria. It is also found in Hungary, Switzerland, Italy, and Spain. A specimen found at Saint-Rambert, in France, has been analysed by Moisscuet, who found—

Soda	20·0
Magnesia	0·7
Sulphuric acid	26·0
Hydrochloric acid	trace
Water	53·3
	100·0

It is also found in large quantities in a cavern at Kailua in
Hawaii, Sandwich Islands. It occurs as an efflorescence, with
other salts, on the Siberian and Caspian Steppes, on the limestone
below the Genesee Falls and near the Sweetwater River, Rocky
Mountains, in the United States.

It also exists, combined with two atoms of water, as a volcanic
product from Vesuvius, which has received the name of *Exantha-
lose*, by Beudant, who found it to contain—

Soda	35·0
Sulphuric acid	44·8
Water	20·2
	100·0

In the anhydrous state it is known as *Thenardite*, and has
been found in large quantities in Spain and South America. It
has been analysed by Dick and Casaseca from

	Spain.	Tarapaca.
Sulphate of soda	99·78	97·48
Carbonate of soda	0·22	—
Insoluble matter	—	2·19
	100·00	99·67

Large beds of this salt have been discovered, within a recent
period, in the valley of the Ebro, on the confines of Navarre and
Old Castile, in Spain, principally, however, in the neighbourhood
of Lodosa. It is found lying between beds of clay and gypsum.
The beds are usually from 18 to 24 inches thick, but in one case
it is said to be no less than 18 feet thick. It has since been discovered in several other localities in Spain, Calatayud and Corvora,
in Catalonia, near Santandar, and Calmenar de Oreja, in the
province of Madrid. The quantity of this salt in the latter locality
has led to its being mined on a large scale, and works are now
being established there for its conversion into carbonate of soda.

Sulphate of soda also occurs combined with sulphate of magnesia, at Ischl, in Austria, and at Egra, in Bohemia; the former
has received the name of *Blœdite*, and the latter *Reussine*. A
similar substance is found near the salt lakes east of the mouth of
the Volga, to which Rose has given the name of *Astrakanite*.

These minerals have all been analysed with the following
results :

	Blædite, by John.	Reussine, by Reuss.	Astrakanite, by Göbel.
Sulphate of soda . . .	33·34	66·04	41·00
„ magnesia . .	36·66	31·35	35·18
„ lime. . . .	—	0·42	and sand 1·75
„ manganese .	0·33	—	—
„ iron . . .	0·34	—	—
Chloride of sodium . .	0·33	—	—
„ calcium . .	—	2·19	—
„ magnesium .	—	—	0·33
Water	22·00	—	21·56
	93·00	100·00	99·82

It is also found more widely distributed in combination with sulphate of lime, under the name of *Glauberite* (NaO.CaO) 2SO³. It exists disseminated in the salt of the mines of Villa-Rubia, in Spain, and of Vic, in France, and is found at Berchtesgaden, in Austria, and in the province of Tarapaca, in Peru. Glauberite consists of—

	Glauberite of		
	Varen-geuille, Pisani.	Villa-Rubia by Brogniart.	Berchtes-gaden, Köbell.
Sulphate of soda . .	50·50	51	48·6
„ lime . .	48·78	49	51·0
Chloride of sodium .	—	—	—
Sulphate of magnesia	—	—	—
Clay	0·40	—	—
Loss by heat	—	— ·	—
	99·68	100	99·6

	Atacama (Hayes).
Soda	21·32
Lime	20·68
Oxide of iron	0·14
Sulphuric acid	57·22
	99·36

Sulphate of soda is now manufactured in this country by decomposing common salt with sulphuric acid, and heat is applied to expel all the hydrochloric acid, until the mass is dry; it is

dissolved in water, and any excess of acid is neutralized by lime
or carbonate of soda. The liquor is then filtered through charcoal
and set aside to crystallize. This crystallization is either rapid or
slow, and either agitated or in quiet, according to the appearance
which it is intended the crystals should present. We regret to
add that in some instances the object is to produce a mass of
crystals having the appearance of Epsom salts, and they are
known in trade as *Mock Epsoms.*

Furnaces and Condensers.

The great bulk of the production is, however, intended for conversion into carbonate of soda. The decomposition of the common salt was formerly conducted in ordinary reverberatory
furnaces built of brick. A great loss of acid resulted and a heavy
expense for the repairs of the furnaces was incurred. The decomposing furnace was then lined with lead, and the sulphate was
afterwards dried on another bed of the same furnace, as shown in
Fig. 128.

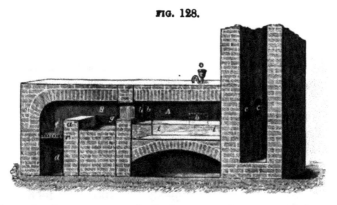

FIG. 128.

These leaden pans are built into reverberatory furnaces, or
rather the soles of the furnaces are lined with lead. Fig. 128
shows the manner in which a furnace of this kind is divided
behind the bridge a into two separate compartments, B and A.
In A, the hinder and less heated part, the decomposition of the
salt is effected, and in the front chamber B, which is considerably
hotter, all the free acid is expelled, the salt fritted, and the process completed. Both compartments are, therefore, in use at the
same time, the one occupied in decomposing the fresh charge,
the other in drying the previous one. The flame passes from
B through five apertures or slits, bb, into A, and thence

through three other wider ones, *cc*, into the chimney. Where several furnaces are worked at once, and on the same spot, one chimney serves for all, which must of course be large in proportion, and the smoke passes into it through subterranean flues. Below the grate *rr*, at *d*, is the ash-pit door; above it, at *e*, is the grate door; *g* and *h* are the working apertures. Apertures corresponding to *e*, *g*, and *h*, are made also on the opposite side, so that each compartment of the furnace is accessible from two sides. In order that the contents of the leaden pan, *ii*, may be easily removed, a part of this is cut away at *h*, which in the mean time is filled up with clay and bricks.

An important step was taken by the late Mr. J. Lee and Mr. Gamble, in the substitution of metal pans for the lead lining of the furnaces just described. The arrangement is as follows: Fig. 130 represents a section of the furnaces through the line *ab* (Fig. 129); in Fig. 130, *ccc*, are the three fireplaces for heating

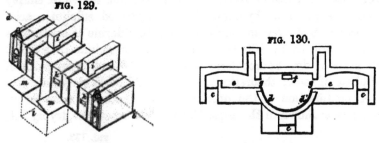

FIG. 129.

FIG. 130.

the metal pan, *dd*, and the two drying flats, *ee*, on either side. The charge of salt from 4 to 8 cwt. is thrown into the pan *d*, through the working door *k* (Fig. 129), the trap doors *mm* being closed during the charging. The proper quantity of acid at about 130° T., and heated to about 120° F., is run in from a concentrating pan, conveniently placed near the furnaces. The doors are closed, and the fire under the pan in the pit *l* set away. The muriatic acid which is evolved escapes through the flue *f*, to its condenser, and when the materials have been worked a few times, the pasty mass is removed by paddles through the openings *gg*, into one of the side drying furnaces, and a fresh charge of salt introduced into the metal pan. There are thus three charges in operation at the same time. The gas from the drying flats escapes by the flues *rr*, to their respective condensers, and when the sulphate has been properly furnaced, dry and as free from muriatic acid as possible, it is removed through the side working doors *kk*.

The salt cake, however, when removed from the furnace, always evolves more or less muriatic acid gas, which is offensive in some localities, and the plan shown in Fig. 131 is intended to obviate this evil. Instead of raking the charge out of the furnace, it is allowed to fall through an aperture l in the bottom—which at other times is closed by a leaden door — into the vault D. When it has sufficiently cooled, the operation is completed in B.

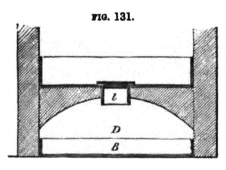

FIG. 131.

By the agency of the water in the sulphuric acid, hydrochloric acid is generated, and sulphate of soda (anhydrous) is produced. For $HSO^4 + NaCl = NaSO^4 + HCl$; sodium, therefore, simply takes the place of hydrogen in the hydrated sulphuric acid, whilst the hydrogen in combination with chlorine is evolved as hydrochloric acid; 100 parts of chloride of sodium thus afford 62 parts of dry hydrochloric acid, and 116 and more parts of sulphate of soda; theoretically there should be $121\frac{1}{2}$ parts.

A modification of the previous plan, adopted in some manufactories, is shown in the Figs. 132 and 133, where the drying flats

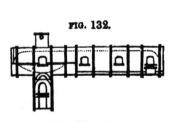

FIG. 132.

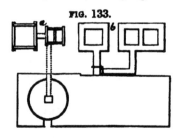

FIG. 133.

are on the side of the pan and connected together, the first charge being pushed forward on the flat next the fire-bridge, where the heat is greatest, and, as it dries, to make room for the damp charge from the decomposing pan.

The connection with the condensers a and b is shown here. The principal portion of the acid which escapes from the pan, is absorbed in the condenser a. This acid is the strongest, not only because the bulk of the acid is generated in the pan, but from the nature of the operation, this furnace partakes more of the character of a close furnace, with little or no admission of air, and consequently little or no dilution of the gas, which, when condensed, naturally produces a stronger acid.

The comparatively smaller quantity of acid from the drying flats, which is more diluted with air through the working doors being opened from time to time, to enable the workmen to stir the charge,—is condensed in the towers at b.

These condensing towers are shown in Figures 134 and 135.

The two towers, c and d, are built of large stone flags, well laid in cement and bound together by iron binders, e, and timber framing, f. c is filled with hard burnt coke, and supplied with water from a cistern, g. d is divided by a partition reaching nearly to the top. The gas enters at h, ascends and descends the smaller tower, d, for the purpose of cooling, and passes at i into the higher one, c, where the gas is so fully condensed, that the residual vapours are allowed to escape into the open air by the exit pipe, k.

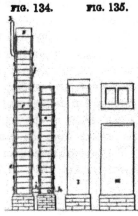

FIG. 134.   FIG. 135.

The other tower, the end view of which is shown in l, and the side view in m, is employed for the condensation of the gas from the drying flats; it is built of brick, divided in two parts by a middle wall, filled with coke and managed in the same way.

Another improvement has been adopted in the construction of the decomposing furnace, which consists in covering the metal pans with sheet iron, and we are indebted to the valuable work on Chemistry by Dr. Muspratt for the drawings and description of this arrangement.

Fig. 136 represents two decomposing pans, AA, in section, each weighing 5 tons, and set in the brick-work, so as to allow the fire to act on the bottoms and sides. The pans are covered with sheet iron, b, the joints being properly luted. E is the fire-place, and

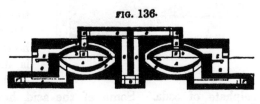

FIG. 136.

FIG. 137.

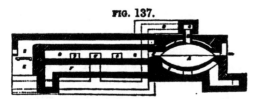

FF are the flues which pass over and under the pans, but not through them. The charge of salt is introduced at A and at K, exactly opposite, seen in Fig. 137. The gas escapes at CC, through a pipe composed of short lengths of earthenware tubing, carefully fitted and luted into each other, each length of pipe being soaked in tar while in a hot state, and allowed to cool before being used.

Two pans are worked together, and the gas brought into one main flue, as shown in Fig. 138. The *salt-cake* furnace or drying flat is built of fire-brick, and heated by the fire, H, the flues, FF, passing over and under, but not through it. G and E are side openings for stirring and withdrawing the charge. C is the tube for the escape of the acid vapours, and the charge is pushed into this furnace from the pan through K.

The general arrangement of the pans, furnaces, and flues, is shown in Fig. 138 ; AA being the pans, CC the acid flues, FF′

FIG. 138.

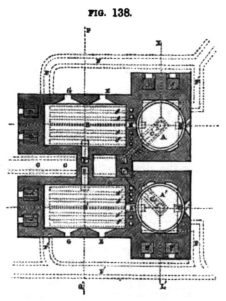

and *fff* indicating the direction of the smoke flues. The pan A is first heated to a proper temperature, and 16 cwt. of salt thrown in through the opening H. 1800 lbs. or 123¼ gallons of sulphuric acid of 1·450 specific gravity, heated to about 120° F., are run in, and the whole of the materials are well mixed with an iron rake. The door is closed, and the fire underneath is charged. In about an hour all the acid vapour is disengaged, and most of the salt converted into sulphate of soda. Some of the acid, however, exists as bi-sulphate of soda, and a greater heat is necessary to convert this and the undecomposed common salt into sulphate. This is accomplished by pushing the charge through the opening K into the salt-cake furnace, DD', when it is spread over the sole of the furnace, and frequently turned over while exposed to a lowered heat for about an hour. It is then withdrawn through the working door, G, and allowed to cool.

Two furnaces, as shown in the wood-cuts, will turn out 10 charges in a day, producing nearly 19 tons of sulphate of soda from 16 tons of common salt.

The muriatic acid which is evolved during the decomposition, is conveyed by the pipes to a tower, the interior construction of which is shown in section in Figure 139.

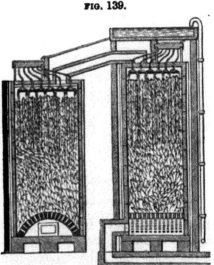

FIG. 139.

The gas enters at the bottom, passes upward through the descending shower of water, and what escapes condensation descends the second. The number or height of the towers must be regulated by the amount of salt decomposed, and the strength of the muriatic acid which the manufacturer intends to produce.

We here give drawings of the manner in which Messrs. Deacon and Robinson work the decomposing furnaces, by means of the heat generated in an ordinary cupola.

The fuel in the small cupola applied in combination with the furnace, is blasted with air, as is ordinarily practised when cupola furnaces are used for other purposes. The heat and products passing from the fuel in the cupola, may pass through the sides or over the top of the cupola, direct to the apparatus, or over or above the furnace with which the cupola is combined.

Figure 140 is a sectional elevation, and Figure 141 a plan of the same at the line AB, Figure 140. a is a cupola furnace or combustion chamber, constructed in the ordinary way, and covered with an earthenware slab, b, to prevent the heat passing

FIG. 140.

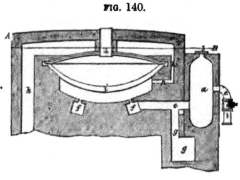

FIG. 141.

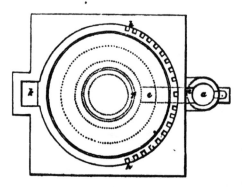

out of the top of the furnace; *c* is a blast pipe, which may be about six inches diameter, and fitted with a valve, *d*, for regulating the admission of the blast; *e* is a passage, along which the heat from the furnace is conveyed to the annular chamber, *f*, near the outlet from the cupola, and in the bottom of this passage is the cave or chamber, *g*, six inches wide at the top, and the full width of the passage, to receive the slag or cinders which may fall out of the furnace; at the bottom of this chamber a suitable door is placed, for convenience in removing the slag from time to time. In the top of the chamber *f* is a narrow annular opening or slit, which admits the heat from the chamber to the under side of the covered iron decomposing pot, *i*, which pot is of the description ordinarily used for the purpose. *hh* are small passages built in the brickwork surrounding the pot, to carry away the surplus heat from underneath; this heat is conveyed over the top of the pot, and by the flues *k* to the chimney or other outlet. With small coal as fuel, for a decomposing pot 9 feet 3 inches diameter, the cupola employed may be about 2 feet diameter and 6 feet 6 inches deep, the blast being about half a pound to the square inch. In other respects the working of the furnace during the process of decomposition, is the same as in an ordinary decomposing furnace.

The plan pursued in France differs in many details from the English process. The furnace arrangements are shown in Figure 142. A is the fireplace, supplied with coal or coke, the flame from which passes over the firebridge, *b*, into a reverberatory furnace, B, the sole of which is built of hard bricks, placed on edge, and very close, or of one or two pieces of stoneware, forming a hollow basin fixed in masonry.

A metal damper, *e*, can be raised or lowered at pleasure to form a connection, *e'*, with the second furnace, E, whose hollow sole is made either of metal lined with lead, or of hard stoneware. The damper *e* moves in a box luted with sand. This second furnace is heated by the products of the combustion from the first furnace, which escape through flues in the corners, *e'*, and pass under the basin. These vapours, which contain some acid gases, pass under the basin by a return flue and escape through the flues, *d'*, in the direction of two ranges of carboys, *l''*.

FIG. 142.

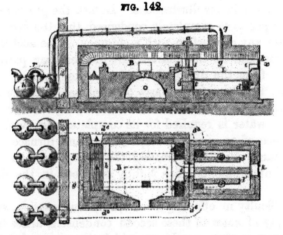

The second furnace is furnished with two pipes of stoneware, *gg*, which pass through the arch of the furnace, and are connected with the two sets of carboys, *hh*. A furnace, about 13 feet by 5 feet broad, requires from 72 to 78 carboys in the first two sets, and from 40 to 50 in the last two, each holding about 40 gallons.

The furnace E is charged with salt through the door K, the acid is run in through a syphon-funnel, and the door is shut, and well luted. The chemical action is promoted by the heat below, and the gas which escapes passes through the pipes *gg*, to the double range of carboys *hh*, one-third full of water, by which the hydrochloric acid is almost entirely absorbed. When the materials have become pasty, they are pushed into the first furnace through the opening *e*. The damper is again closed, and the

14*

charge is worked off in the usual way in the furnace B. The
gases and products from the furnace are conveyed, as we have
mentioned, to the second set of carboys, $i''i''$, but the condensation here is imperfect, and much acid ultimately escapes into the
atmosphere from the chimney with which they are connected.

Condensation of Hydrochloric Acid.

This is the confession of French authorities themselves, and
quite recently a report has been made to the Belgian government,
on the damage done to vegetation by the escape of acid gases
from their chemical manufactories. This cause of damage, which
formerly gave rise to much litigation in this country, has been
entirely removed by the adoption of Mr. Gossage's condensing
towers, as shown in our illustrations. In fact, the gas from the decomposing pan is so completely condensed, that the furnace does
not require any connection with a chimney. The acid from this
condenser is used for the manufacture of bleaching powder, &c.,
while that from the drying-flat or salt-cake furnace, which is
very weak, is either run to waste or employed in the production
of bicarbonate of soda. This acid is necessarily weak, as a large
quantity of water is required for its complete condensation, from
the presence of so much atmospheric air and furnace products.
Mr. Clapham's plan for increasing the strength of this acid is a
valuable improvement, which is fully described in the following
chapter.

The difficulty in effecting a complete condensation without a
large supply of water in these second condensers, arises from the
limit to the size of the condenser, which the power of draught of
the chimney imposes. This difficulty has been overcome by the
use of a steam jet instead of the chimney draught, patented by
Mr. Bell, and shown in the following woodcuts. a is an ordinary
reverberatory furnace for decomposing common salt (but the jet
is, of course, equally applicable to any other) connected by
means of a flue, b, with the lower end of a tall square tower
built of brick or stone, cemented by a mixture of clay and
tar. This tower is divided into two compartments, c and d, by a
partition wall, e, and the whole filled with small pieces of coke.
A supply of water is kept continually running through several
openings at the top of this tower, which is still further scattered
over the surface of the coke by impinging upon small rounded
pieces of stone or brick. At the lower end of the condenser, a

FIG. 143.

FIG. 144.

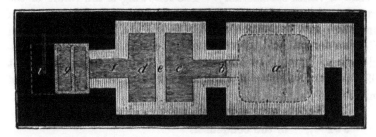

flue, *f*, leads to a cistern, *g*, partly filled with water, which is in
connection with an inverted cone or funnel, *h*, through which
a steam jet, *i*, blows into an underground flue. The drawing
only shows the steam escaping into the air as it was at first
applied. Fig. 143 shows a section of the arrangement, and Fig.
144 a ground plan.

The muriatic acid gas, as it escapes from the furnace, comes in
contact with the water falling through the coke in the usual way,
and is completely absorbed in its passage, running out at last by
openings, *kk*, at the bottom of the condenser. The traces of
liquid muriatic acid, which may be carried away by the mechanical
action of the draught, are retained by the water in the cistern,
g. The draught through the furnace and coke condenser is
maintained entirely by the action of the little jet of steam
blowing through the funnel or cones, and possesses considerably
more power for this purpose than the tallest chimney. The
steam is so perfectly free from acid, that the face may be exposed

to it without inconvenience. The use of the underground flue is simply to condense the waste steam, so that, at the open end, little or no aqueous vapour escapes into the atmosphere.

Instead of employing water for the condensing tower, Mr. Clapham pumps up the weak acid from the brick condensers, and by this simple arrangement secures all the hydrochloric acid in the form of a strong commercial acid.

These towers are generally filled with pieces of hard-burnt coke, but some manufacturers use flints. Others again employ soft lime-stones, when carbonic acid, which is harmless, is evolved and chloride of calcium is produced. Kuhlmann introduces carbonate of baryta, and makes use of the chloride of barium thereby produced, in the preparation of " Blanc fixé " or " Permanent white."

The proportion of acid obtained by the different plans for condensation varies considerably. The following is the result of the annual working at Risle, in 1854 : 74·5 per cent. of the weight of common salt decomposed (43·818 ctr.) was obtained in the form of commercial acid, corresponding to 23·18 per cent. of hydrochloric acid gas. Now 100 parts of the common salt produce about 109 parts of sulphate of soda, and 52·99 parts of this acid gas, of which—

23·18 parts were collected in the condensing vessels.

6·91   „   in the condenser.

22·90   „   were unaccounted for.

A carefully conducted trial of the English system of condensation by Mr. Clapham, yielded the following results :—

108 tons of common salt with 7 per cent. moisture and impurities, containing, therefore, 100·44 tons of chloride of sodium, and which should yield 61·99 tons of hydrochloric acid, were decomposed in the usual way, and the acid collected, which amounted to

172 tons of hydrochloric acid at 1·130 sp. gr. = 45·16 tons HCl.

167   „   „   „   1·040   „   = 13·49   „

Tons of real acid condensed . . . . .   58·65

„   „   lost or left as common salt in the sulphate . . . . . .   3·29

61·94 tons.

In another manufactory, where 250 tons of salt were decomposed per week, the result of the average working, calculating the produce per ton of salt, was as follows: 2240 lbs. of salt containing 7½ per cent. moisture, &c., yielded 2753 lbs. of hydrochloric acid at 28° T., or 1·140 sp. gr. = 730·86 lbs. HCl, and about 400 lbs. of weak acid from the dryers, varying in strength from 1·005 to 1·080 sp. gr. Hence the produce was nearly—

60 per cent. of the acid from the decomposers.

40 ,, ,, ,, ,, dryers and loss.

In a third manufactory, where nothing but weak acid was condensed, the following experiment was made: 860 lbs. of common salt, (containing 6·8 per cent. water, &c. = 801 lbs. NaCl = 494 lbs. HCl,) were decomposed, and yielded 8378 lbs. of acid, containing 5·75 per cent. HCl.

Hence HCl in salt = . . . 494
,, condensed = . . 481
,, lost = . . 13 494

OTHER PLANS FOR THE DECOMPOSITION OF COMMON SALT.

Chemists and manufacturers have devoted much time, and large sums of money have been expended, in trying various plans for decomposing common salt, which might prove more economical than that which we have just described.

These plans may be profitably studied, if we arrange the different suggestions under the following divisions:—

1. The direct action of sulphur ores.
2. The direct action of acids, or metallic and other oxides.
3. The double decomposition with metallic sulphates.
4. The double decomposition with earthy and other sulphates.
5. The double decomposition with carbonate of ammonia.
6. Electricity, &c.

1. *The direct action of Sulphur-ores*

has been tried by only one patentee, Mr. Longmaid, although the process had been suggested before the date of his patent.

2. *The direct action of Acids, or Metallic and other Oxides*,

has attracted much more attention, and the following proposals have been made for the use of—

Potash by Hagen in 1768.

Caustic baryta by Bergmann.

Caustic lime by Cohausen.

Oxide of lead by Scheele in 1775, and later by Chaptal and Bérard.

Oxide of zinc by Scheele in 1775.

Oxalic acid by Samuel in 1838, by Margueritte in 1855, by Kobell, and still later by Possoz.

Boracic and phosphoric acids by Margueritte in 1854 and 1855.

Sulphurous acid by Brooman in 1852.

Nitric acid by De Sussex in 1848.

Fluosilicic acid by Maugham and Chatelier in 1858.

Alumina by Tilghman in 1847.

Silicic acid by Blanc and Blazille in 1840, and since then by Anthon and Fritsche.

We have included alumina among the acids, as in Tilghman's process it plays the part of an acid.

3. *The double decomposition with some Metallic Sulphates,*

were among the first plans tried for the purpose—

Sulphate of iron having been proposed by Athenas in 1793, Dundonald in 1795, Phillip and Wilson in 1838, and Longmaid in 1847.

Sulphate of copper by Wilson in 1838, and Hunt in 1840.

Sulphate of zinc by Hunt in 1840, and Boulton in 1852.

Sulphate of manganese by Tennant, by De Sussex in 1847, and by Barrow in 1856.

Sulphate of lead by Margueritte in 1854.

Sulphurous acid and oxide of iron by Wöhler, and by Robb in 1853.

4. *The double decomposition with some Earthy Salts, &c.,*

among which the following must be mentioned :—

Sulphate of magnesia by Dundonald in 1795, and Payen and De Luna in 1855.

Sulphate of lime, &c., by Wilson in 1838, and more recently by Margueritte.

Sulphate of alumina by Constantini in 1781, by Dundonald in 1795, and by Cunningham in 1853.

Sulphate of ammonia by Poole, Payen, Persoz, Prückner, and Siemens.

5. *The double decomposition with Carbonate of Ammonia*

has attracted great attention, and plans for this process have been proposed by—

Hemming and Dyer in 1838.

Chisholm in 1852.
Gossage, Johnson, Turck, and Schloesing in 1854.
Belford in 1855.
Bell in 1857.
Schloesing again in 1858.
Bandiner in ?
Heerin in ?

6. *Electricity, &c.*

There have been many other proposals made, for which we must refer the reader to our abstract of the patents on this subject, as they cannot be classified, and, as far we can judge, are not likely to be carried into practice, even in the way of experiment.

We cannot of course attempt to give details of these numerous plans, many of which are very ingenious, and most of which show the great labour and attention which have been devoted to them by their authors. Even where the Patentees have claimed the use of the same materials, there are great differences in their respective modes of procedure, and we will select one or more plans from each of the divisions, to illustrate the general process indicated.

1. *Direct action of Sulphur-ores.*

This plan, resuscitated by Longmaid, consists in roasting common salt and pyrites at a low heat in a long reverberatory furnace with several beds, the materials being charged in at the far end, and the charge being moved forward on to the next bed as the decomposition progresses, each move bringing it nearer the fire-bridge, where the heat is greatest, and where the conversion of the common salt into sulphate renders the mixture more capable of sustaining a higher temperature without melting. This easy fusibility of the mass constitutes one of the difficulties of the process, as it necessitates a low heat, thus rendering the decomposition extremely slow and imperfect in proportion to the presence of fused particles. It is also found that the process is accompanied with a loss of sulphur, which escapes as sulphurous acid. It has on the other hand, some advantages, as the small sulphur ore, which is of much less value than the large, and ores of a low percentage of sulphur, can be used. The valuable metals, copper and silver, can also be extracted at a profit, when present in such minute quantities as not to pay for smelting.

The weak salt-cake obtained in this process, is brought up to strength by adding sufficient sulphuric acid to complete the decomposition.

Mr. Longmaid proposed to employ the chlorine evolved in making bleaching powder. He passed his gases through wooden flues immersed in water, to retain the hydrochloric acid and chloride of iron which always accompany the chlorine. The bleaching powder obtained was weak, but he then gave it an additional dose, in the usual way, to bring it up to full strength. The sulphate ash is lixiviated, and, in the insoluble residue, any tin present in the ore, is found as oxide, and may be recovered by washing, in consequence of its greater specific gravity. The copper is precipitated from the solution, either by iron or by milk of lime.

Messrs. Allen, who have worked this process on a manufacturing scale, have patented a plan for converting the salt and pyrites into sulphate of soda, which is shown in Figures 145, 146.

Figure 145 is a vertical section of a combined reverberatory furnace and kiln, and Figure 146 is a horizontal section of the furnace, and ground plan of the kiln. A is the reverberatory furnace with two beds, B and C. D is the fire, and EEE are outlets, or eye-holes, from the hinder end of the furnace to the flue, L, which leads to the chimney. The front bed, B, is level, and the other bed, C, rises in an inclined position to the hinder end of the furnace. Above or over the bed C, is the kiln, F, with its lower end open to the furnace, so that its contents rest on the bed C. The kiln is closed at the top by the door, G, through which fresh charges are

FIG. 145.

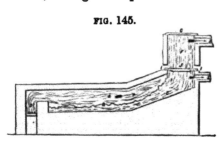

FIG. 146.

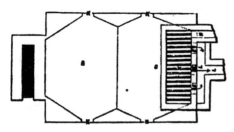

thrown in. H, is a flue leading from near the upper end of the kiln to the chimney. There are dampers in the flues L and H; II (Fig. 145) are holes on each side of the kiln, for the purpose of inserting the iron bars, J, from side to side, in order that they may bear up the contents of the kiln, and also to prevent their falling down on the hinder bed, C, of the furnace while the charge upon it is being moved forward to the front bed, B; KKKK (Fig. 146) are the doors or mouths of the reverberatory furnace. When the hinder bed C is emptied, the contents of the kiln are allowed to settle down by moving the bars J, and a fresh charge is put into the kiln.

The holes II are also useful for the admission of atmospheric air into the kiln, for the oxidation of the pyrites. The heat from the fire D passes over the beds B and C of the furnace, and up through the mass of materials in the kiln, thereby decomposing the pyrites, and forming peroxide of iron and sulphate of soda; the gaseous materials which are evolved, pass away by the flue H. When it is found that the whole of the heat from the fire is more than is required for the contents of the kiln, it is reduced by drawing open the damper in the flue L, leading from the back of the furnace to the chimney, the required distribution of the heat being regulated by the dampers with which the flues L and H are provided.

In some cases it will be found advantageous to have more than two beds to the reverberatory furnace, and in others it will be unnecessary to have more than one, in which latter case the charge, when finished, is drawn out of the furnace, from off the bed, underneath the kiln.

2. *The direct action of Acids or Metallic and other Oxides.*

Some of these processes will be described more in detail, to show in what way the inventors have endeavoured to realise their ideas. We commence with a notice of what may perhaps be regarded as the first manufacturing process for obtaining soda from common salt by means of oxide of lead. Kirwan mentions that soda was manufactured by this process in London so long ago as 1782, and the insoluble lead-salt was sold under the name of *Turner's yellow.* The same plan was adopted at Montpellier about the same time, and this was the first process employed at Walker, near Newcastle-on-Tyne.

50 parts of litharge, in fine powder, were mixed with 12½

parts of common salt dissolved in 55 parts of water. A portion of this solution was added to the litharge so as to form a paste, and allowed to remain untouched for a few hours. The surface acquired a white colour, and was then broken up, when another portion of the solution was added, and again allowed to rest. This was repeated until the whole of the solution had been used. At the end of two or three days the operation was complete, when the caustic soda was extracted by water, and the oxychloride of lead left behind :

$$NaCl + 2PbO = NaO + PbO, PbCl$$

The oxychloride of lead was melted, and a fine yellow product obtained, which was used as a pigment.

Oxide of zinc also acts in the same way ; but the decomposition is not complete under six to eight weeks, and six parts of the oxide are required for every part of common salt.

Blanc and Bazille use silicic acid to effect the decomposition, and they calcine the salt and sand in cast-iron cylinders placed in furnaces, so that the flame may pass round and raise them to a cherry-red heat. The axis of each cylinder is occupied by a large tube of the same length as the cylinder, and the tube is pierced with a multitude of holes. This tube is inserted at one end of each cylinder ; into the other end a large porcelain tube is adapted, through which the muriatic gas is allowed to pass off. Each cylinder is filled loosely with an intimate mixture of sea-salt and sand, and when the mass has arrived at a cherry-red heat, steam is made to pass very slowly into the perforated tube which traverses the cylinder, so as not to reduce the temperature below that of a cherry-red heat. The water is decomposed in the apparatus, and the transformation of the salt and sand into neutral silicate of soda goes on without risk. The proportions recommended by the patentees are—

Common salt . . . . . 280 parts,
Sand . . . . 200 „

Before proceeding to the transformation of the neutral silicate into sub-silicate, the mass is to be lixiviated, in order to extract the undecomposed portion of salt. The neutral silicate of soda obtained in this first operation being insoluble, it must be converted into a soluble sub-salt, by combining it with an additional proportion of soda. 60 parts of carbonate of soda and 100 parts of neutral silicate are calcined in a reverberatory furnace to form the sub-silicate of soda. At a cherry-red heat, the vitrification takes place, and the product in this state is soluble in hot water. The insoluble silicate, however, may be employed in the manufacture of glass with suitable additions of lime and alumina. The vitrified sub-silicate of soda is pulverized and thrown into hot water to effect its solution, which is submitted to the action of carbonic acid. The sub-silicate of soda is decomposed in this operation, carbonate of soda being formed, which remains in solution, and gelatinous silica is deposited. The operation is finished whenever the carbonic acid ceases to be absorbed by the solution. This silicic acid may be obtained as a white jelly, when the carbonic acid is free from smoke, in which state it is well adapted for the manufacture of fine glass.

De Sussex decomposes the chloride of sodium or potassium by nitric acid, in close vessels heated by steam to 212°. The decomposition proceeds very slowly, and the muriatic acid evolved is easily condensed. The alkaline nitrate is then mixed with its equivalent of lime, baryta, silica, or alumina, and exposed to a low red heat, when the nitric acid is expelled and condensed, ready for a fresh operation. The contents of the retort are in a caustic state, and by lixiviation the alkali is separated, leaving the earth fit for another decomposition.

In theory this process would leave nothing to be desired, for there would be no waste of materials, and two valuable products would be obtained, muriatic acid and caustic alkali, in a state of purity.

As we witnessed this plan in operation on a large scale near London, the extreme slowness of the decomposition, and the great danger of a loss of the nitric acid from too high a temperature, appeared to us to be the great difficulties in the way of its ever becoming a manufacturing process.

The leading idea of De Sussex, of obtaining the alkali with the ever-recurring materials for the decomposition of fresh portions of salt, without having recourse to sulphuric acid, appears to have influenced Margueritte, who employs phosphoric acid, but to obtain which he is obliged to use hydrochloric acid.

He decomposes phosphate of lead with hydrochloric acid, thus producing chloride of lead and phosphoric acid. The latter, having been separated from the precipitate, is mixed, evaporated, and calcined with a considerable quantity of common salt, which yields by distillation hydrochloric acid, and a fixed residue, viz., pyrophosphate of soda. This pyrophosphate of soda is dissolved, and then treated with caustic lime, until precipitation ceases. In this way a precipitate of pyrophosphate of lime is formed, while the solution contains caustic soda, which may be carbonated with carbonic acid. The phosphate of lime is next boiled with chloride of lead, thus giving chloride of calcium, and reproducing the phosphate of lead, with which the operation originated. These different reactions are represented by the following equations :—

$$2PbO.PO^{\ast} + 2HCl = 2PbCl + 2HO.PO^{\ast}.$$
$$2HO.PO^{\ast} + 2NaCl = 2HCl + 2NaO.PO^{\ast}.$$
$$2NaO.PO^{\ast} + 2CaO = 2NaO + 2CaO.PO^{\ast}.$$
$$2CaO.PO^{\ast} + 2PbCl = 2CaCl + 2PbO.PO^{\ast}.$$

Tilghman also attempted a solution of this problem of dispensing with the use of sulphuric acid and obtaining a constant supply of his decomposing agent in the following manner, which we believe was tried on a considerable scale in Glasgow.

He made a paste of common salt and alumina, and moulded it into small cylinders about 4 inches long and 1 or 2 inches in diameter, with a small opening in the centre from end to end of the cylinder. These cylinders were dried, placed in a kiln, and heated red hot. A current of steam was admitted, when the common salt was decomposed, hydrochloric acid being formed, which was evolved at the top and conveyed to a condenser. The cylinders retained their form, and were found to be composed of aluminate of soda with the salt which had escaped decomposition. They were then moistened with water and exposed to a current of carbonic acid and steam. On digesting the cylinders after this treatment, a solution of carbonate of soda was obtained, and the alumina left behind was ready for another operation.

We believe the imperfection of the decomposition—for our samples of the alkali only tested 32 per cent.—and the difficulty of recovering the light flocculent alumina—were practical difficulties which could not be overcome on a manufacturing scale. The decomposition is shown in the following equation :

$$\left.\begin{array}{ll}\text{Alumina} & . \ . \ Al^{2}O^{\ast} \\ \text{Common salt} & 3NaCl \\ \text{Water} & . \ . \ . \ 3HO\end{array}\right\}\text{produce}\left\{\begin{array}{l}3NaO.Al^{\ast}O^{\ast},\text{Aluminate of soda.} \\ \\ 3HCl,\text{Hydrochloric acid.}\end{array}\right.$$

The decomposition of common salt by oxalic acid, possesses more interest as a theoretical question than as a practical process, at least until some cheap plan for production of that acid has been discovered. Köbell states that the decomposition is as follows:—

Common salt 2NaCl ⎫
　　　　　　　　　 ⎬ produce ⎧ 2NaCO² Carbonate of soda.
　　　　　　　　　 ⎭　　　　 ⎨ 2HCl Hydrochloric acid.
Oxalic acid . C⁴H²O⁶ ⎭　　　 ⎩ 2CO Carbonic oxide.

3. *The double Decomposition by means of Metallic Sulphates.*

The idea embodied in this method was recognised long ago, and was one of the plans submitted to the French Commission. The decomposition, when sulphate of iron is employed, is as follows :—

$$NaCl + 2(FeO.SO^3) + HO = NaO.SO^3 + Fe^2O^3 + SO^3 + HCl.$$

Phillips proposed to employ the residue obtained by the oxidation of iron pyrites; but Thomas, Dellisse, and Boucard, distil over two-fifths of the sulphur from the sulphur ores, and leave the residuum to oxidation in the air. This is lixiviated, and a solution of common salt mixed with it, when the whole is exposed to a low temperature. The sulphate of soda crystallizes out and is freed from the oxide of iron by the addition of 1 or 2 per cent of milk of lime.

At Fahlun, in Sweden, the two solutions are boiled down to dryness, the residue exposed to a red heat, and the sulphate dissolved out in the usual way.

In other works, the two salts are boiled together, and the sulphate of soda salted out during the evaporation.

Instead of effecting the decomposition in a state of solution, the two salts may be mixed in the dry state and furnaced at a red heat, but Dumas says that a better produce of sulphate of soda is obtained by moistening the mixture and allowing it to remain for some days before calcining. Muspratt states that this process is employed in the North of France.

Sulphate of manganese was employed for many years at the celebrated works of Tennant, at St. Rollox, in the same way. The common salt was decomposed with sulphuric acid and peroxide of manganese, to obtain a supply of chlorine for making bleaching powder. A mixture of sulphate of soda, sulphate of manganese, and some free sulphuric acid, was left behind in the still, to which sufficient common salt was added to saturate the free acid and to furnish soda for the acid of the metallic sulphate. The whole was calcined in a reverberatory furnace, and the sulphate of soda dissolved out by water, or the mass was balled with lime in the usual manner.

A novel mode of employing sulphate of zinc has been proposed as a means of avoiding the evolution of the hydrochloric acid. Zinc and sulphuric acid are added to a solution of common salt, and the sulphate of soda separated by crystallization. The equation shows the nature of the change which takes place :

$$NaCl + Zn + HSO^4 = NaSO^4 + ZnCl + H.$$

Mr. Hunt substitutes the sulphates of copper and zinc, and effects the decomposition by mixing either of these sulphates in solution with a solution of common salt containing a slight excess of salt above the equivalent quantity necessary for mutual decomposition. He then separates the sulphate of soda by crystallization, and recovers the metals in the form of oxides by precipitation with lime. He deoxidizes his alkaline sulphate, and decomposes the resulting sulphide with his recovered metallic oxide. The sulphide is converted into sulphate by oxidation, and thus the circle of chemical change is never broken.

In theory this process, like many others, is perfect, but in practice there are great difficulties in effecting even a partial decomposition of the common salt, while the unmanageable character of bulky hydrated precipitates, and the impossibility of producing a pure alkaline sulphide, have more than once been sufficient to dishearten the manufacturer who has attempted to make these plans work well on a large scale.

4. *The double Decomposition with Earthy and other Sulphates.*

Margueritte's process combines the employment of sulphate of lead and an earthy sulphate with the common salt.

To carry out this application of sulphate of lead, he mixes equivalent quantities of this sulphate, prepared by roasting galena, and of common salt, and subjects the mixture to calcination at a strong and continuous red heat, by which it fuses and becomes quite clear and transparent. At the surface of the bath, thick vapours of chloride of lead are seen to rise, which evolution ceases as soon as the reaction between the chloride of sodium and the sulphate of lead is terminated, viz., when the common salt has been converted into sulphate of soda, and the sulphate of lead into volatile chloride of lead. At this stage of progress, the melted mass is run out, and, being treated with water, yields crystallized sulphate of soda, and leaves an insoluble residue of sulphate of lead, varying in quantity, according to the time during which the mass has been calcined. This sulphate of lead is to be used in the next operation. The chloride of lead which has been condensed, is pulverized and held in suspension in a solution of sulphate of magnesia, or agitated with water and sulphate of lime. The chloride of lead is converted into sulphate, giving rise to a soluble chloride which is removed by washing, and thus after each calcination the sulphate of lead that is required for the succeeding operation is reproduced. These reactions may be expressed by the following equations :—

$$NaCl + PbO.SO^3 = NaO.SO^3 + PbCl,$$
$$PbCl + MO.SO^3 = PbO.SO^3 + MCl,$$

in which MO represents the base of any soluble sulphate. When the chloride of lead is digested in water with sulphate of lime, magnesia, iron, alumina, &c., the whole of the sulphate is recovered, or the regeneration is complete, with the exception of a very small quantity, which is lost by a small portion of lead remaining in the liquor that is removed by washing. In the digesting process, rather dilute solutions must be used ; for, if sulphate of lead is brought in contact with a concentrated solution of chloride of potassium, sodium, magnesium, or calcium, the solution, being filtered and treated with an alkaline sulphide, yields an abundant precipitate of sulphide of lead ; whence it follows that the sulphate of lead, by way of double decomposition, is converted by either of the chlorides into a soluble chloride of lead, which, however, takes place only when the solution is concentrated. For this reason, sea-water cannot be employed in its natural state for this purpose : for the quantity of chloride of sodium and of magnesium which it contains, dissolves a proportionate quantity of lead, and this is avoided by diluting the sea water, and adding sulphate of lime.

The furnace bottom upon which the calcination of the mixture is made, must have as large a surface and as little depth as possible, for the purpose of facilitating the volatilization of the chloride of lead ; and the vault of the furnace must be rather low, so that the current of air carrying off the vapours of the chloride of lead from the surface, may be rapid and strong, with little volume, so as not to cool the melted mass. The furnace may, for instance, resemble a low gas-retort open at both ends, the lower sole of which is hollowed out some inches deep so as to contain the fused mass. At one end, there is an open door, through which the air is made to enter as required, and at the other end, a condensing apparatus in communication with a chimney, the object of these contrivances being to create a continuous current of air in the retort, for conveying the vapours of the chloride of lead into the condensing apparatus, where they are arrested. Instead of air, a current of steam may sometimes be used with advantage, as the mixture will thus be protected against the action of the air, and the steam will carry off the chloride vapours and condense along with them. The condenser must present a great number of surfaces which can either be in the dry state, or moistened by a jet of water or of steam, and these surfaces should be so constructed as to facilitate the collection of the condensed chloride of lead. The mixture of sulphate of lead and chloride of sodium is introduced into the retorts by means of a shovel or scoop, and the liquid mass being contained upon an incline, is discharged through an opening at the entrance of the retort, which is kept closed with loam or clay to the end of the operation.

We believe, without having witnessed this particular plan in operation, that an almost fatal practical difficulty in the way of its success, would be found in volatilizing and recovering the chloride of lead. We have always noticed that these volatile metallic chlorides undergo a decomposition in the volatilization, forming a mixed sub-chloride and oxide of the metal.

The decomposition of common salt by means of sulphate of magnesia has been introduced as a manufacturing process by Ramon de Luna, who employs the native sulphate of magnesia found in many parts of Spain, especially in the Province of Toledo, not far from Madrid. He heats a mixture of 1¾ parts of the lightly dried Epsom salts with 1 part of common salt, to dull redness. The hydrochloric acid is expelled, and sulphate of soda mixed with the magnesia remains behind. He employs water heated to 194° F. to dissolve the sulphate of soda, and separates any Epsoms that have escaped decomposition, by the addition of milk of lime, when sulphate of lime and magnesia are precipitated.

Balard employs the same plan for facilitating the separation of the potash-salts from his bittern liquors, adding Epsom salts to convert the chloride of potassium into sulphate, which easily crystallizes out from the chlorides of magnesium and sodium.

The same change takes place in the Yorkshire alum works,

when they use chloride of potassium to produce potash-alum, and the mothers are found to contain chloride of magnesium in place of the rough Epsoms which they ordinarily produce in such large quantities.

It is evident that this process is only adapted to localities peculiarly situated as to a supply of the two salts.

We have, under this division, to notice a process which is still employed in some manufactories of ammoniacal salts. It has been in use at Grenelle, in France, since 1795, and has been employed or suggested by Plückner, Persoz, Poole, and Siemens. It consists in dissolving 1 equivalent of sulphate of ammonia and $1\frac{1}{2}$ equivalent of common salt in hot water, and evaporating the liquor. The sulphate of soda produced by the double decomposition, gradually salts out, and is removed from time to time; when the liquor has been concentrated to such a point that a drop taken from the pan solidifies on cooling, it is run off, and the chloride of ammonium allowed to crystallize. An excess of common salt promotes the salting out of the sulphate of soda.

The main object of this process is, of course, to obtain the muriate of ammonia, but it is by no means certain that this is the most economical plan of producing this salt. The great loss of ammonia which arises in evaporating the solutions of most of its salts, shows that it would be more advantageous to produce the chloride, by distilling off the ammonia from gas-water, direct into hydrochloric acid, and fishing out the salt as it was formed, in the same way as is now pursued in manufacturing sulphate of ammonia.

5. *The double Decomposition with Carbonate of Ammonia.*

All the processes hitherto described, have had for their object the conversion of the common salt either into sulphate or some other form, preparatory to its employment in making the carbonate of soda; but this plan differs so far from the others, that this last salt is produced at once. As might be anticipated, it has, on this account, attracted great attention, and ought perhaps to have been described somewhat further on; but we introduce it here in connection with the decomposition of common salt. Messrs. Hemming and Dyer were the first to propose this process in 1838, and Schloesing has again made it the subject of a patent as late as 1858. We incline,

15*

however, to give Mr. Gossage's plan in detail, as his great experience in the manufacture of alkali is a guarantee that his suggestions have been made with a perfect knowledge of all
particulars necessary to the success of the process, if it be really
practicable on a large scale.

The chemical change is represented by the formula—

$$\left. \begin{array}{l} \text{NaCl} \\ \text{NH}^3 \\ 2\text{CO}^2 \\ 2\text{HO} \end{array} \right\} \text{ produce } \left\{ \begin{array}{l} \text{NaO.HO.CO}^2. \\ \text{NH}^4\text{Cl}. \end{array} \right.$$

The apparatus employed by Mr. Gossage consists of a cylindrical vessel, a, divided by partitions into chambers, a^1, a^2 a^3, &c.
This vessel is mounted on axles, b and c, supported on bearings.
The axle c is solid, and upon this axle a spur wheel, e, is fixed,
which is actuated by a pinion, f, on a shaft, g, moved by steam
power; and thus a slow rotatory motion is communicated to the
vessel a. The axle b is hollow and revolves in two stuffing
boxes, kk. A space exists between these stuffing boxes, forming
a chamber, m, with which two pipes are connected for the conveyance of gas. The hollow axle, b, is perforated with apertures

FIG. 147.

FIG. 148.

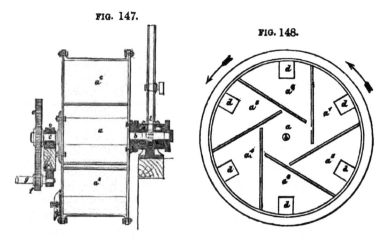

corresponding with the chamber m, through which gas is conveyed
from it to the vessel a. One of these pipes is used to convey carbonic acid, and the other ammonia. ddd are openings closed
with covers, by means of which the materials employed are
introduced into or withdrawn from the vessel a.

The vessel a is filled with the fluid mixture of saline compounds to the line of liquid marked *hj*. By reference to Figures 147 and 148 it will be seen that, as the vessel *a* is caused to revolve on its axles in the direction shown by arrows, each chamber will alternately be brought into the position of the chamber marked *a'*, in which position the fluid mixture of saline compounds is discharged from the chamber, at the same time that gases, contained in the vessel *a*, bubble through such fluid mixture, and by these means the gases contained in the vessel *a* are made to mix with the saline compounds, and are absorbed. For some time after beginning to work with a new charge of saline compounds in a mixing vessel, it is not necessary to subject the gases in this vessel to pressure; for so long as ammoniacal gas, as well as carbonic acid gas, is supplied to the mixing vessel, a condensation of the gases is effected without the aid of pressure; but when the total quantity of ammoniacal gas required for a charge has been introduced into the mixing vessel, it is then advantageous to introduce the remaining quantity of carbonic acid gas requisite for converting the sesqui-carbonate of soda produced, into bicarbonate of soda, by means of an air-pump, and to introduce the gas in such a manner as to occasion a pressure of not less than 10 lbs. per square inch.

During the condensation of the gases in the mixing vessel, a considerable evolution of heat takes place; it is, therefore, necessary to allow the mixture to cool down to a temperature not exceeding 100° F. before withdrawing the mixture from the mixing vessel.

Mr. Gossage produces his solution of sesqui- or bicarbonate of ammonia, by conveying a stream of ammonia to the middle of an ordinary condensing tower, while another stream of carbonic acid is entering at the bottom, and they both meet a shower of water falling from the top of the tower. Any ammonia escaping is made to pass into a second condenser, where it meets a supply of muriatic acid, so that no ammonia need be lost. He also heats the sesqui- or bicarbonate of soda in iron retorts, to obtain a supply of pure carbonic acid.

The decomposition of common salt in this process, is dependent on the relative solubility of the sal-ammoniac and sesqui- or bicarbonate of soda, which are produced; and as neither of them is insoluble in water, it is of great importance to use as little water as possible. Mr. Gossage recommends the use of three parts of water for every part of common salt present in the mixture.

When he uses a saturated solution of common salt, he either adds as much dry sesqui- or bicarbonate of ammonia, as contains 17 parts of ammonia tō every 240 parts of the saturated solution, or he adds the ammonia liberated from 54 parts of sal-ammoniac by lime, to 240 parts of the saturated solution, and then forces in the carbonic acid until a pressure of 10 lbs. per square inch is reached.

After completing the decomposition of a charge of salt and carbonate of ammonia in the mixing vessel, the mixture is transferred to another close vessel, which is designated a *filtering vessel*. This filtering vessel has a perforated diaphragm near its lower part, on which diaphragm is placed a linen or cotton cloth to serve the purpose of a filter, and thus prevent the passage of bicarbonate of soda through the perforations in the diaphragm. The filtering vessel is connected, by means of a pipe attached below the diaphragm, with a second vessel, also capable of being closed from the atmosphere, through which pipe the liquid can flow into the second vessel. The upper parts of the two vessels are connected by means of a pipe, so as to equalize the pressure of the gases when the vessels have been closed. The fluid is now allowed to drain off and contains sal-ammoniac, with some bicarbonate of soda in solution. A solution of bicarbonate of soda is introduced into the filtering vessel, and the solution allowed to drain through the mass of undissolved bicarbonate of soda until the bicarbonate is sufficiently freed from sal-ammoniac to render it suitable for sale. The bicarbonate of soda is either dried and ground for sale in this form, or heated to redness to obtain the carbonate for sale as alkali. The liquor containing the sal-ammoniac is run into a metallic still, and heated with lime to recover the ammonia for another operation.

We now notice some of the peculiarities of the other plans proposed to accomplish the same decomposition.

Chisholm purifies coal gas with a mixture of common salt and charcoal, and afterwards expels the ammoniacal salts by heat, leaving carbonate of soda behind. The ammonia is recovered by heating these salts with carbonate of lime, and conveyed through a mixture of common salt and charcoal, when the same process is repeated.

Mr. Bell, instead of passing the carbonate of ammonia through the dry salt and charcoal, causes his ammonia to act on a solution of the hydrochlorate or sulphate of soda, and collects the bicarbonate of soda as it precipitates.

Hemming and Dyer separated the bicarbonate of soda from the sal-ammoniac liquor, and dried it at the same time by pressure in a screw press, while Schloesing accomplishes the same objects in a centrifugal machine.

Schloesing and Rolland's plan of producing the bicarbonate of soda continuously, is worth a careful perusal, but it is too lengthy for us to print, after having devoted so much space to this particular process.

Mr. Parnell has pointed out the objection to which this process is liable, when carbonate of ammonia is used, and the disadvantages of employing sesquicarbonate of ammonia for the decomposition of the common salt. In the first case, the ammonia is not obtained as carbonate, but as sesquicarbonate, while part of the ammonia escapes in the free state; and in the latter, it is only the bicarbonate that has the power of decomposing common salt, so that, when sesquicarbonate is present, it is resolved into neutral and bicarbonate, the former of which has no action, and is very liable to be lost. In fact, the difficulty of preserving the ammonia from loss, is the great drawback to the adoption of the process.

SECOND OPERATION.

The Conversion of the Sulphate of Soda into Carbonate of Soda, or the Balling Process.

This is perhaps the most important stage in the manufacture, and consists in heating a mixture of sulphate of soda, carbonate of lime, and some deoxidising material, in a reverberatory furnace.

FIG. 149.

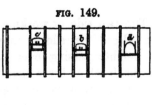

The furnaces are generally built in a range, and back to back, both on account of economy and increased strength. We give a sketch of one of these furnaces, in which it will be seen that it has two beds, the second being about 4 inches above the first, or that nearest the firebridge, which has an air course, *a*, through its length to keep it cool. Each bed has its own door, *bc*, but the fire-place, *d*, is banked up with coal, which is pushed over on to the grates from

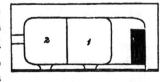

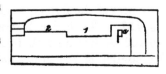

time to time, the great point in the management of the fire
being the regulation of the heat and the maintenance of a
deoxidising atmosphere inside the furnaces. The lining of the
furnace is, of course, constructed of good fire-bricks, and the
furnace strongly bound with iron binders and rods, as shown
in Figures 150 and 151, where two fire-places are shown, which
are charged at the end instead of at the side.

FIG. 150.

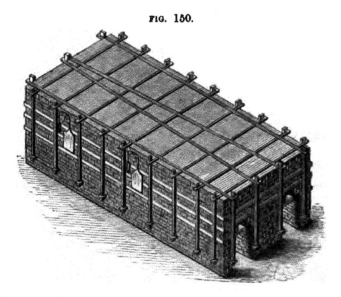

The two side or working doors
are closed by slides, supported
by weights passing over pulleys.
In the front of these openings, a
strong iron iron bar is suspended
from the top of the furnace, with
a bend in the middle, to carry
the heavy rakes while working
the charge.

FIG. 151.

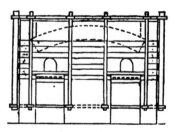

The balling furnaces are usually
constructed in this country with
a pan furnace, in which the tank liquors are boiled down by the
waste heat of the former, as shown in the accompanying woodcut,
Fig. 152. The fuel is thrown in through an opening, *a*, at the
end and banked up with fresh coal, so that no door is required.
The position of the grates and ash-pit is shown at *b* and *c*. The

FIG. 152.

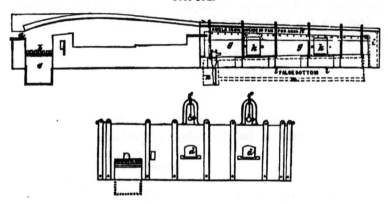

working doors, *dd*, are supported by balance weights, *ee*, over the pulleys, *ff*. The description of the boiling down pan is given further on.

The arrangements in France are somewhat different, and consist of an ordinary reverberatory furnace, Figs. 153, 154, and 155, with ash-pit and grate-bars, *ab*, fire bridge, *c*, which separates the fuel from the bed of the furnace, *dd*, of an elliptical form at the far end, from which two flues, *ee*, pass into the chimney, *f*. The four side openings, *gggg*, and that at the end, *h*, are fitted with iron bars, *iiii*, on which the work-

FIG. 153.

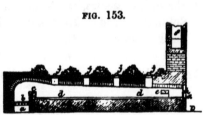

FIG. 154.

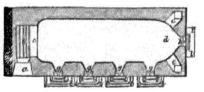

FIG. 155.

man rests his rake while working or withdrawing the charge of
materials. The mixture of sulphate of soda, chalk, and coal,
is dried on the arch of the furnace, and thrown in through the
openings, *jjjj*.

Leblanc recommended a mixture of—

             2000 parts of sulphate of soda,
             2000   „   carbonate of lime,
             1000   „   coal,

which are very nearly those now in use; but in France the
following are said to be the actual proportions employed for
making

	Crystals of Soda.	Alkali for Soap boilers.
Sulphate of soda . . .	200	190
Carbonate of lime . . .	250	220
Coal	90	100

We give below the mixtures employed in two of the largest
works in Lancashire, the stone being 14lbs.

Sulphate of soda	13 stones.	Sulphate of soda	16 stones.
Slacked lime .	18 „	Quick lime . .	17 „
Coal	9 „	Coal	10 „

And on the Tyne the following are used—

	Cylinder furnace.	Ordinary ball-furnace.	
		Ordinary alkali.	Caustic ash.
Sulphate of soda . . .	15	23	18 stones.
Chalk	15	23	21 „
Coals	10	13 to 15	11¼ „

The theoretical proportions being

      2 eq. Sulphate of soda, 2 $(NaO.SO^3)$ . .   142
      3 „ Carbonate of lime, 3 $(CaO.CO^2)$ . .   150
      9 „ Carbon   54
                                        ————
                                         346

Some manufacturers grind the materials, sift and mix them;
but this is not necessary so long as there are no large pieces, and
the whole are roughly mixed before being thrown into the
furnaces.

The following analysis by Mr. R. C. Clapham represents the
composition of the chalk delivered into the alkali works on the
Tyne:

Carbonate of lime	78·000
Sulphate of lime	0·244
Silicate of lime	0·325
Phosphate of lime	0·119
Chloride of sodium	0·163
Magnesia	trace
Alumina	0·220
Oxide of iron	0·200
Oxide of manganese	0·150
Silica	0·600
Water	20·600
	100·621

The furnace being brought to a bright red heat, the charge is thrown in, and spread evenly over the sole of the first bed, where it remains until it has become heated throughout its mass. It is then removed to the second or fluxing bed by means of a tool called a *slice*, or spatula, and exposed to the full action of the fire. In about fifteen or twenty minutes it begins to soften, and forms into roundish balls or clots, when it must be cautiously stirred to expose fresh surfaces and hasten the softening, until the whole mass appears to have the consistence of dough. The chemical changes commence at this point, with an evolution of gas, bubbles of carbonic oxide burst with a blue flame, and, rising with the other products of the decomposition of the coal, set the mass in motion, which must be constantly stirred by means of the slice and rake, so that all the materials may equally share in the decomposition. The mass at last melts, and from the evolution of gas appears to boil; this gradually ceases, and the whole at length assumes a tranquil fluid condition. The mass is now raked out into iron barrows, in which it solidifies on cooling, and is now termed a *ball;* hence the term *ball soda.*

Brown has made some interesting experiments on the effect of heat in promoting the decomposition, and the results confirm what the theory of the process would indicate. He exposed a mixture of 100 parts of sulphate ash, 100 parts of chalk, and 55 parts of charcoal, to six different degrees of heat, but in no instance allowed the materials to melt. The black ash was lixiviated, and the soluble salts analysed with the following results. The numbers 1 to 6 indicate the degrees of heat, the last being sufficient to soften copper.

CONSTITUENTS.	1	2	3	4	5	6
Carbonate of soda	10·8	15·4	29·0	26·0	19·0	14·4
Sulphide of sodium............	0·3	0·6	0·9	6·9	11·8	15·8
Hyposulphite of soda.........	—	—	0·4	0·6	0·9	1·6
Sulphate of soda	26·5	23·2	10·3	4·5	3·2	1·6

The weight of the charge is usually about 7 cwt., and a charge of this weight is worked off every hour. The success of this process depends very much on the judgment and skill of the workman, and two plans have been tried to supersede the manual labour by mechanical arrangements.

The first we must notice is that patented by Mr. W. W. Pattinson, so long ago as 1848, in which there is much skill and practical knowledge shown in contending with the serious difficulties in the way of such an operation. We believe that the wear and tear of the process induced the patentee to abandon his plan.

Figure 156 shows a longitudinal section of a ball furnace with two beds, but the description refers only to the working or fluxing bed, A. A vertical cast-iron shaft passes through the roof and bottom of the furnace, and rests in a foot-step, c. A horizontal cog-wheel, b, is fitted on the top, and receives motion from the prime mover. A piece of brickwork surrounds the vertical shaft where it passes through the furnace bottom, forming a sort of circular wall around it to prevent the materials in the furnace from coming in contact with the shaft. These brick pieces extend into the furnace bottom, and are securely fixed by being imbedded in the brickwork. The scraper, 3, nearest the centre, moves round with its ends nearly in contact with this pillar. Two horizontal arms of cast-iron, dd, extend from this vertical shaft within the furnace. The working bed, A, and also the heating bed, B, are made nearly circular, and these arms extend from the centre to near the circumference of each bed. These arms of cast-iron being exposed within the furnace to the direct action of the fire, would speedily get very hot, and indeed melt, but they are kept cool by a stream of water which constantly flows through them in a tube or passage, 1, provided for that purpose. The tube is an inch malleable iron pipe, round which the arm is cast. The cold water enters at the top of the shaft at ee, where a small cistern or furnace is fixed ; and after traversing both arms through the tube it issues at f, where it is delivered into an annular cistern, g, within which the shaft revolves. The quantity of water is so regulated that it issues into the cistern g at the temperature of about 160° F. which thoroughly protects the arms. The arms, dd, are placed at such a distance from the bottom of the furnace that they move round above the matters in the furnace; but there are attached to each arm three knives or stirrers, each kept in its place by the frame or eye of the stirrer being pushed over and along the arm, until it drops into a notch or groove cut on the upper surface. These are so formed, as alternately to push backwards and forwards the mass to be stirred as the arm revolves, and thus effect a thorough mixing and continual change of place and surface, the stirrers, 222, moving the mass towards the centre, and the stirrers, 333, moving it towards the circumference. Now the form of the stirrers and their position on the arms are such, that when the arms are made to revolve in the opposite direction, both stirrers push the mass outwards towards the circumference, and thus, when the ball is to be withdrawn, a slide, forming part of the furnace bottom, removed, the motion of the arms reversed, and the charge gradually expelled from the opening. In stirring the balls, one revolution per minute is sufficient, but when the motion is reversed to expel the ball, the speed, by suitable cog-wheels on the top of the vertical shafts, is increased to four revolutions per minute. During the time of working the ball in the fluxing bed of the furnace, as described, a ball is in process of heating in the heating bed B, and on removing the finished ball, a sliding piece of cast-iron, *h*, placed between the two beds to prevent any of the materials in the heating bed from being pushed over during the working of the ball, is withdrawn by its handle, the motion of the arms reversed, and the whole of the charge in the heating bed is pushed down into the working bed to be finished. Another charge is thrown into the heating bed to be ready in its turn for transference to the working bed, and in this way the operation proceeds continually. The mass to be acted upon is merely spread a few inches thick upon the bottom of the furnace, and is never fused, so that the revolving arms and knives or stirrers keep it constantly stirred, turned, and intermixed, with great ease and certainty.

The other plan consists in making the furnace itself revolve instead of the machinery, and this invention was patented by Elliott and Russell, in 1853. The invention, as worked by these parties, was not successful ; the balls could not be exhausted in the ordinary way, and the plan would have remained inoperative had not Messrs. Stevenson and Williamson,- of the Jarrow Chemical Works, introduced other and valuable improvements. It is now in successful operation at South Shields, and offers the following advantages :

1. The whole of the contents of the furnace are heated with such uniformity, that no portion ever reaches a temperature to cause a serious loss of soda from volatilization.

2. The atmosphere inside the furnace is always deoxidizing, as the furnace never requires to be opened during the mixing and fluxing.

3. Less skilled labour is necessary, from the larger amount of work done in a given time, and a more complete decomposition of the sulphate of soda is effected by the more perfect mixing of the materials.

4. There is less absorption of soda than in the ordinary beds of the balling furnaces, and less wear and tear, as no tools are used in the process.

We are indebted to Muspratt's work for a woodcut which shows the arrangement of this furnace.

a is a cylinder formed of a cast-iron case, or made of malleable iron plates, lined with fire-bricks, and having some portions in the shape of tiles projecting inwards from the brickwork, for the purpose of assisting the mixture of the materials during the revolution of the cylinder. Two cast-iron rings, *bb*, are fitted on to the cylinder, and they rest and run on strong iron rollers or wheels, *cc*, which are carried by proper bearings. The cylinder is made to revolve by means of a pair of pulleys and a system of wheels, shown in the woodcut. The cylinder is 13 feet long by 7 feet diameter, inside measurement. The furnace in which coal is burnt is shown at *d*, and the flame passes through a side opening and flue, corresponding with the circular openings, *e, f*, in each end of the furnace. After circulating in and around the inside of the furnace, the flame passes over the black ash liquors in the boiling down pan, *g*, to the chimney.

FIG. 157.

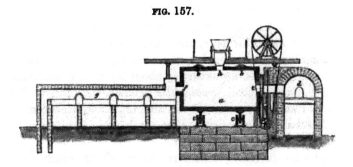

The cylinder is charged through an opening, *h*, which is brought under the hopper, containing a charge of materials. The progress of the operation is watched through an opening at the end, and when it is finished the position of the opening, *h*, is reversed. The cylinder is brought to a stand, and the opening being uncovered, the charge falls into a number of iron waggons, chained together, which are brought, one after another, under the opening.

In working the charge in this operation, the usual plan is to mix all the ingredients together and throw them into the balling furnace; but Messrs. Stevenson and Williamson keep the sulphate of soda back, until the chalk and coal have been heated together to a high temperature.

They introduce into the cylinder furnace as much undried chalk as is equal to 15 cwt. dry chalk and 7½ cwt. coal, while the cylinder is making one revolution in twelve minutes. In an hour's time, the cylinder is stopped, when a charge of 15 cwt. sulphate of soda and 2½ cwt. coals is added, and the cylinder is made to revolve at the same rate as before for half-an-hour, after which there is no tendency in the materials to be blown in dust into the flue. The speed of the cylinder is then increased to one revolution in a minute and a quarter for half-an-hour, when the operation is finished, and the charge is withdrawn.

The materials are not ground, but merely crushed and passed through an iron grating, the bars of which are 5 inches apart. It is not necessary to take great pains to mix the materials previous to introducing them into the furnace.

This modification in the manner of charging and working the balls is applicable to the ordinary furnace, and these gentlemen give the following instructions for a double-bedded ball furnace. The upper bed is to be charged every hour with about 2¾ cwts. of *dried* chalk and 1 cwt. of coal, previously *ground together*, and at the end of about an hour, these materials, having acquired a good red heat, are transferred to the working bed, on which they are evenly spread, and a mixture of about 2¾ cwts. of sulphate of soda and ¾ cwt. of coal, *previously ground together*, is introduced and intimately mixed with the highly heated coal and chalk. The balling operation is then conducted in the usual way and finished in about another hour.

Two cylinder furnaces, therefore, work off as much as five double-bedded furnaces.

Theory of the Decomposition.

Different views have been taken of the changes which occur in the balling operation, and we now give the opinions of those chemists who have directed their attention to this subject.

When a mixture of sulphate of soda and carbonate of lime is exposed to a high heat, sulphate of lime and carbonate of soda are formed, but they cannot be separated by water, as the first-named salts are reproduced. Again : when sulphate of soda is heated with carbon, sulphide of sodium is produced, and carbonic oxide, which escapes in the gaseous form. Again : if these two processes are combined, carbonate of soda and sulphide of calcium are obtained, but when the mass is lixiviated with water, a decomposition ensues, sulphide of sodium and carbonate of lime being formed ; or as represented in formulæ—

$$NaO.SO^3 + 4C = NaS + 4CO.$$
$$NaS + CaO.CO^2 = NaO.CO^2 + CaS.$$

Now, Dumas considers that sulphate of lime is first formed, and then reduced, becoming insoluble by combining with the extra equivalent of lime, while Gmelin adopts the previous view, the formulæ being modified to meet the actual proportions employed :—

$$3(CaO.CO^2) + 2(NaO.SO^3) + 9C = (CaO.2CaS) + 2(NaO.CO^3) + 10CO.$$

Liebig has, however, shown that the decomposition of sulphate of soda by carbon, is much more complex than that just mentioned, and explains his views by the following diagram :—

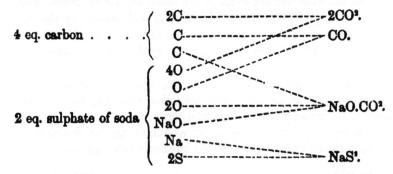

And if this be true, the subsequent decomposition of the balling operation must be much more complicated than either Dumas or Gmelin has represented.

We add, in the words of Mr. Gossage, the views he entertains on the nature of this decomposition :—" I pointed out in the year 1838, in the specification of a patent obtained by me for improvements connected with the soda-manufacture, that the undissolved residuum remaining from the lixiviation of black ash with water, consisted almost entirely of mono-sulphide of calcium and carbonate of lime ; in fact, when the lixiviation is effected with a large proportion of water, so as to produce a weak solution, the oxide of calcium may be converted entirely into carbonate of lime, which is then found *mixed* with the monosulphide of calcium ; and yet this conversion of oxide of calcium into carbonate of lime, and consequent disintegration of the supposed insoluble compound, is effected, and the supposed soluble sulphide of calcium is set at liberty, in the presence of solution of carbonate of soda ; and this takes place without occasioning the production of sulphide of sodium in solution. The fact is, that monosulphide of calcium is perfectly insoluble in water, and it is only to the extent to which polysulphide of calcium is formed, that solution takes place, and consequent production of sulphide of sodium. The real advantage obtained by the use of an extra proportion of lime arises from this affording a larger amount of surface for re-action, and thus expediting the fluxing operation, and thereby preventing the formation of polysulphide of calcium.

" In connection with this view of the subject, I pointed out that, when well-made black ash is digested in alcohol, there is no caustic soda dissolved, although the same black ash yields caustic soda abundantly by lixiviation with water. From this I inferred that caustic soda did not exist in the black ash ready formed, and that it was produced, during lixiviation with water, by the re-action of caustic lime on carbonate of soda, which could not have been effected without the disintegration of the supposed insoluble compound.

" Assuming these views to be correct, and omitting the minor products of the black ash operation, also assuming the decom-position of the sulphate of soda and formation of carbonate to be perfect, we may form this diagram as exhibiting the re-action of the materials employed, and of the products obtained :—

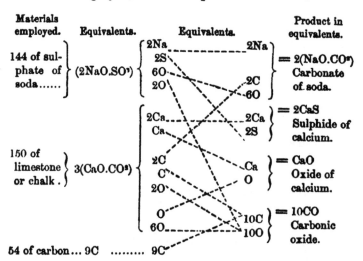

" In this formula I have assumed pure carbon to have been used, also that the gas evolved is entirely carbonic oxide, whilst, in fact, a large proportion of carbonic acid is produced. Small coal is also used, in practice, instead of pure carbon ; and it is desirable that the finished black ash should contain a considerable proportion of unconsumed pieces of coke ; therefore a proportion of small coal is used in the mixture, about double the quantity indicated for carbon."

These different opinions evidently hinge on the views which chemists hold regarding the nature of the insoluble lime-compound, of which there may be said to be four; for—

Unger represents it as 3CaS.CaO.
Dumas   „   2CaS.CaO.
Rose   CaS + CaO.HO.
Gossage   „   CaS.

We do not think that there is sufficient evidence at present to decide which, or if any, of these views is correct. The action of water and alcohol on the black ash would tend to indicate the presence of free caustic lime; but, whether it exists as such in the balls, or is a product of the decomposing action of water on the peculiar compound assumed to exist by the three first-named chemists, there are no facts to show.

This difference of opinion, of course, influences the analysts who have published the results of their examination of the black ash, and we give the following by Unger, Richardson, Kynaston, Brown, and Murphy, from—

	Cassel.	Glasgow.	Newcastle.
Sulphate of soda . . .	1·99	1·16	3·64
Chloride of sodium . .	2·54	1·91	0·60
Aluminate of soda . .	—	2·35	—
Carbonate of soda . .	23·57	35·64	9·89
Hydrate of soda . . .	11·12	0·61	25·64
Lime	—	6·30	—
Carbonate of lime . .	12·90	29·17	15·67
Oxysulphide of calcium	34·76	1·13	35·57
Sulphides of iron . . .	2·45	4·92	1·22
Silicate of magnesia .	4·74	3·74	0·88
Ultramarine	—	0·29	—
Carbon	1·59	8·00	4·28
Sand	2·02	4·28	0·44
Moisture	2·18	0·70	2·17
	99·86	100·20	100·00
	Unger, 1848.	Brown, 1848.	Richardson, 1845.

16*

Ringkuhl, in Cassel.

Carbonate of soda . . .	37·8
Chloride of sodium . . .	0·4
Oxysulphide of calcium .	40·0 ($CaO.2CaS$)
Lime	8·5
Magnesia	0·8
Soda	1·6
Silica	5·0
Alumina	1·2
Sulphide of iron . . .	1·2
Carbon	2·6

99·1

Unger, 1852.

Liverpool.

Carbonate of soda	36·879	41·489	28·89	
Chloride of sodium ...	2·528	1·308	3·07	
Sulphate of soda.........	0·395	0·748	0·82	
Silicate of soda	1·182	1·162	8·27	Caustic soda.
Aluminate of soda	0·689	0·392	0·40	Sulphate of sodium.
Sulphide of calcium ...	28·681	33·193	25·86	
Carbonate of lime	3·315	0·857	14·22	
Bisulphide of calcium .	0·435	—	—	
Hyposulphite of lime...	1·152	—	—	
Sulphite of lime	2·178	—	—	
Caustic lime	9·270	9·320	9·24	
Magnesia	0·254	—	2·03	Silicate of magnesia.
Sulphide of iron	0·371	—	—	
Sesquioxide of iron and phosphate of lime ...	2·658	3·020	6·23	
Alumina	1·132	1·020		
Charcoal	7·007	4·724	—	
Sand......................	0·901	2·259	—	
Ultramarine	0·959	—	—	
Moisture	0·219	—	0·99	

99·492	100·000	100·02

Kynaston. Murphy. Muspratt
and Danson.

The black balls or ash must be sufficiently porous to allow the water employed for lixiviation freely to penetrate the entire mass, and in fact to appear, except in colour, not unlike pumice-stone. It has a blackish grey colour, and is mixed with particles of coke or carbon ; the workmen soon learn from the external appearance to distinguish when their work has been well done.

Blythe and Kopp have patented the use of the different oxides and carbonates of iron, manganese, zinc, copper, and lead, instead of lime or its carbonate, in this balling process. They employ a charge consisting of—

> 2⅓ cwt. sulphate of soda.
> 1½ to 1¾ cwt. peroxide of iron.
> 114 to 120 lbs. small coal.

These materials are reduced to powder, mixed, and balled in the usual way; but they flux more readily, and become more liquid than the ordinary mixture. The decomposition is illustrated by the equation :—

$$
\left.\begin{array}{l} 2Fe^2O^3 \quad . \quad . \\ 3(NaO.SO^3) \\ 16C \quad . \quad . \end{array}\right\} produce \left\{\begin{array}{l} Fe^4Na^3S^3. \\ 14CO. \\ 2CO^2. \end{array}\right.
$$

When exposed to the air, the balls rapidly absorb carbonic acid, oxygen, and moisture, and fall down to a blackish powder, which, if exposed too long to the atmosphere, becomes hot from a further absorption of oxygen, and the product is deteriorated in quality.

This change is represented by the formula—

$$
\left.\begin{array}{l} Fe^4Na^3S^3 \\ 2O \quad . \quad . \\ 2CO^2 \quad . \end{array}\right\} produce \left\{\begin{array}{l} 2(NaO.CO^2) \\ \\ Fe^4S^2 + NaS; \end{array}\right.
$$

and when this mass is lixiviated with water, the insoluble residue by heating, furnishes sulphate of soda and sulphurous acid; thus—

$$
\left.\begin{array}{l} Fe^4NaS \\ \\ 14O \quad . \quad . \end{array}\right\} yield \left\{\begin{array}{l} 2Fe^2O^3. \\ NaO.SO^3. \\ 2SO^2. \end{array}\right.
$$

To obviate this and other inconveniences, the patentees place the balls on iron gratings in a close chamber, and admit a current of steam and carbonic acid, which promotes this *crumbling* process, the powder falling through the grates, and being removed from time to time to lixiviating tanks, where sufficient water is added, so that the liquor shall stand between 28° and 30° of Twaddell. The liquors may be boiled down to make soda-ash, or concentrated to produce crystallized carbonate of soda.

The residual sulphide of iron is drained and dried, and used in producing sulphurous acid for the manufacture of sulphuric acid, while the oxide of iron left in the burners can be again employed in the balling process.

Reiner has suggested the use of carbonate of baryta (Witherite) where it can be obtained sufficiently cheap. The tank refuse, in this case, consists of oxysulphide of barium (BaO.BaS), hyposulphite of baryta, and carbonate of baryta, which might be employed in the preparation of the salts of baryta used in trade.

Stromeyer has submitted Blythe and Kopp's process to a careful analytical examination, and concludes from his researches, that of every three equivalents of $NaO.SO^3$, two are reduced so as to form $2FeS$, and the soda is chiefly converted into carbonate, while the other equivalent of the $NaO.SO^3$ is only reduced to NaS, which combines with the two equivalents of FeS, producing a double sulphide, $NaS.2FeS$. When this compound is exposed to the air, it is rapidly decomposed, the alkaline sulphide yielding $NaO.S^3O^2$ and NaO, while the FeS is not altered. When treated with a limited quantity of water, the double sulphide becomes hydrated, and forms a black gelatinous mass, and with a larger quantity, an emulsion, which does not become gelatinous on standing. When it is treated with carbonic acid, sulphuretted hydrogen is evolved, while the soda is converted into carbonate, and the sulphide of iron loses its peculiar property of forming an emulsion when lixiviated with water.

These facts, which have been elicited by Stromeyer, point out the difficulties of this plan; and we know that, on the manufacturing scale, they have been so far confirmed as to explain the deficient produce of carbonate of soda. It is still a question how far the extra cost of labour and materials necessary to render available the soda, which becomes insoluble in the form $NaS.2FeS$, will militate against the commercial success of the substitution of metallic oxides for the lime in the balling operation.

We have now described the process for decomposing the sulphate of soda by the ordinary balling operation, and the modification of Blythe and Kopp; and we think this is the most appropriate place to introduce a description of some of the various plans which have been proposed to accomplish this decomposition without having recourse to Leblanc's method.

The number of these plans is so great, that we have adopted a classification, perhaps somewhat arbitrary, but which, at all events, will enable our readers to obtain a better idea of the whole subject.

New Plans for decomposing Sulphate of Soda.

They may be arranged under two classes: in the one the *sulphate* is the object of decomposition, and the *sulphide* in the second ; and each class may be subdivided, as the process may be said to be *direct* or *indirect*.

1. *Direct decomposition of the Sulphate of Soda.*

Samuel, 1838, by means of hydrate of baryta.
Claussen, 1852, by the same, and also carbonate of baryta.
Köhlreuter, 1828, by carbonate of baryta.
Lennig, 1851, by carbonate of baryta.
Wagner, ? , by bicarbonate of baryta.
Tilghman, 1855, by deoxidising agents.
Jacquemin, 1858, by carbonic oxide and steam.

2. *Indirect decomposition of the Sulphate of Soda.*

Samuel, 1838, by sulphide of barium, and the resulting sulphide of sodium by carbonic acid.

3. *Direct decomposition of the Sulphide of Sodium.*

Malherbe and Alban, 1788, by iron turnings.
Claufield, 1804, by lead or zinc, or their oxides.
Arrott, 1859, by oxides of iron.
Attwood, 1819, by carbonic acid.
Samuel, 1838, by the same.
Newton, 1852, by waste furnace gases.
Prückner and Persoz, ? , by oxide of copper.

4. *Indirect decomposition of the Sulphide of Sodium.*

Wilson, 1840, by bicarbonate of soda.
Werckshagen, 1853, by the same.
Clemm, 1853, by the same.
Bohringer, 1854, by the same.
Arrott, 1859, by the same, and carbonic acid.
Laming, 1859, by carbonate of ammonia.
Gossage, 1859, by carbonic acid and oxide of iron.
Wilson, 1859, by the same, and bicarbonate of soda.
Hunt, 1860, by the same, and oxide of iron.

We will select from the above lists such plans as afford the best illustrations of the general character of the class of improvements to which they belong. We cannot devote any space to the discussion of such plans as those of Samuel, Köhlreuter, and Claussen, partly because the decomposition by means of carbonate of baryta is very imperfect, partly because no one would think of employing so costly a decomposing agent as caustic baryta, and partly because it is obviously a waste of time and material to make sulphide of barium to produce sulphide of sodium, when the latter could be as cheaply made in the first instance.

It is possible that a large demand for pure sulphate of baryta might pay for the use of the carbonate of baryta, but it must always possess more or less of a crystalline form, which, *pro tanto*, diminishes its covering power as a paint.

1. *Direct decomposition of the Sulphate of Soda.*

We extract from Tilghman's pamphlet the following particulars of his process :—

The sulphate of soda may, to some extent, be decomposed by being subjected, at a high temperature, to the action of a current of heated steam, but owing probably to the volatile nature of the base at a high temperature, no large proportion can thus be obtained in the free state. To aid, therefore, the decomposing action of the steam, a substance is employed capable, when mixed with the sulphate highly heated and exposed to steam, of forming a combination with the alkaline base, which shall yet, when cold, give up the alkali to the action either of the water or of water and carbonic acid. Of the large class of substances possessing these properties, which are called combining substances, either alumina or the subphosphate of alumina answers, and the alumina may be prepared by strongly igniting sulphate of alumina or by any other well-known process. The subphosphate of alumina is prepared by mixing solutions of phosphate of soda and sulphate of alumina, and. adding to the solution a slight excess of ammonia. The alumina in the state of powder is mixed with an equal weight of the sulphate of soda, also in powder, and the mixture spread upon the hearth of a reverberatory furnace heated by a coke fire, the hearth being covered with a compact bed of native carbonate of magnesia 3 or 4 inches thick. Several clay steam-pipes are introduced through the roof of the furnace, so as to throw a current of heated steam over the whole width of the hearth ; these pipes are connected with a steam-boiler by a series of fire-clay tubes kept red hot.

A current of steam is then admitted from the boiler through the red-hot tubes upon the charge. The acid of the sulphate is carried off by the steam, and the acid vapours may be conveyed, together with the gases of the fire, into a leaden chamber to be combined into sulphuric acid by the usual means. The quantity of steam thrown upon the charge is kept at the point which produces the most rapid evolution of acid, and the charge is stirred occasionally so as to expose fresh surfaces to the action of the steam. As the contact of deoxidizing gases with the sulphate is injurious, such an excess of air is admitted, if necessary, by suitable openings above the fuel, as will render the atmosphere in the furnace oxidizing.

When a specimen of the charge shows by the usual tests that it contains no notable proportion of the sulphate undecomposed, the operation is completed. The charge is withdrawn and lixiviated with hot water, and when the clear solution of aluminate of soda thus obtained has become cold, an excess of carbonic acid is passed through it until no more precipitate of alumina is formed ; the clear solution of carbonate of soda is then drawn off and evaporated. The alumina thus recovered is again used as the combining substance.

The remarks we made on Tilghman's process for decomposing common salt by a similar plan are applicable here, and we believe they apply also to Jacquemin's proposal, which consists in passing steam and carbonic oxide over sulphate of soda heated to redness. The changes which take place in the latter process are represented by the formula—

$$NaO.SO^3 + 4CO = NaS + 4CO^2.$$
$$NaS + 2HO = NaO.HO + SH.$$

2 Indirect decomposition of the Sulphate of Soda.

For an account of Mr. Samuel's process, we refer our readers to the Abstract of Patents in the Appendix.

3 and 4. Direct and indirect decomposition of the Sulphide of Sodium.

In describing the processes patented by Laming and by Wilson, we shall have an opportunity of introducing some of the various suggestions of other inventors.

Laming employs what is usually termed *gas-water*, which he submits to a continuous process of distillation in some efficient form of apparatus, such, for example, as the square tower or column, known as the invention of Æneas Coffey, for the distillation and rectification of alcohol, and which, when made of iron and somewhat modified, is adapted for ammoniacal liquors. To modify Coffey's distillatory apparatus for this purpose, the small holes in its diaphragms should be suppressed, and one edge of each of them cut away to the extent of several inches at two opposite sides of the tower alternately, so as to open up a long serpentine or zig-zag passage for the fluid matters throughout the whole of the apparatus, instead of leaving one part of it divided into separate chambers. The edge of each diaphragm or plate where cut away may be turned up a little to keep back a shallow head of liquor, which by this arrangement is made to cascade from one plate to another until it flows out at the lower end of the series. The ammoniacal liquor distilled in such an apparatus, as it flows and falls in contact with steam passing in an opposite direction, gives out a constant supply of carbonic acid and ammonia, and generally also of sulphuretted hydrogen mixed with aqueous vapour,—while the steam which is perpetually entering below to effect the distillation, may be so regulated in quantity, as to become almost wholly condensed and thrown back by its contact with the worm through which the cold liquor is to be distilled ; it continually enters the apparatus at the upper part, so that very little of it passes out with the volatilized ammonia by the eduction pipe. The steam is in proper quantity, when the pipe just named has somewhat more than the lowest degree of heat that suffices to prevent the ammoniacal compounds from condensing within it, the rule being to maintain it at a temperature so high that the unprotected hand will hardly bear to remain upon it. As the vapours make their exit in a continuous stream, they pass into the vessel charged with sulphide of sodium, where the carbonic acid produces carbonate of soda ; the sulphuretted hydrogen as it arrives, mixes with that portion of the same gas which the carbonic acid liberates ; and the ammonia introduced, serves the purpose of promoting the liberation, by combining with the sulphuretted hydrogen.

The ammonia should leave the soda-vessel accompanied by not less than one and a half equivalents of sulphuretted hydrogen, which will generally be the case when the current is obtained by the distillation of liquors containing hydrosulphate of ammonia. The mixed gases are conducted by an induction pipe, into a close retort charged with pieces of magnesian, or ordinary limestone, or chalk, exposed to a full red heat, by which means carbonic acid is set free, and passes along with the ammonia, through an eduction pipe into an ordinary lead chamber, in which the two gases condense together as carbonate of ammonia. When the proportion of ammonia would otherwise be unduly large, the chamber in which the condensation takes place, may advantageously be maintained full of carbonic acid obtained from some additional source. Or the carbonate of ammonia as it leaves the retort may be dissolved in water by what is known in gas-works as a water-scrubber, or by other apparatus.

Or the ammonia may be recovered in an uncombined state, by receiving the gases from the soda-vessel in a dry gas-purifier or other appropriate vessel, charged with a revivifiable oxide of iron or manganese, made pervious by sawdust or other suitable matter, by which the sulphuretted hydrogen present is absorbed, and the ammonia set free. In this case, it is desirable, whenever the vapours leave the soda-vessel hot, to cool them down to near the atmospheric temperature, by causing them to pass, on their way to the metallic oxide, through an air or a water refrigeratory.

The free ammonia, as it leaves the oxide vessel in which it has been divested of sulphuretted hydrogen, may also be absorbed by cold water, and for this purpose a set of close scrubbers may be used, in the whole of which the same water is used consecutively until it is saturated with ammonia, beginning with that scrubber which is most distant from the oxide vessel, and communicates with the atmosphere at its upper part. To facilitate this absorption, the ammonia should be cold, arriving through a worm or other close refrigeratory, if necessary. The extent of absorbing surface can hardly be too great, especially if a large quantity of ammonia be required to be absorbed by a small quantity of water; and in case there should be any gas accompanying the ammonia not susceptible of absorption, a vent is left at the upper part of the apparatus for its escape.

The decomposition of sulphide of sodium by means of carbonic acid, either in the free state or combined with carbonate of soda, has long attracted attention, as the number of patents testify; and the industrial application of this idea has received an additional impetus from the discovery of the property which certain oxides of iron possess of absorbing sulphuretted hydrogen, and on subsequent exposure to the air undergoing decomposition, depositing the sulphur, and re-acquiring the property of absorbing fresh quantities of this gas : this has been termed the *revivifying* or *renovation* of the oxide of iron.

One of the practical difficulties of this plan is connected with the production of the sulphide of sodium. Arrott obtains the sulphide of sodium by roasting a mixture of 4 parts of sulphate of soda with 1 part of coal or coke, or by fusing the same mixture in a cupola or blast furnace, which must be kept supplied with a proper quantity of coke or coal as fuel. Laming obtains the sulphide by calcining a mixture of 2 parts of sulphate of soda with 1 part of coal or coke in a close vessel. Gossage heats a mixture of 2 parts of sulphate of soda with 1 part of coal or rather less coke, in a *blind furnace*, in which the heat is transmitted through the brickwork of the floor and roof, the heat employed being just insufficient to cause the sulphate of soda to become fluxed.

Whatever plan is employed for producing the sulphide, it must always be liable to more or less oxidation by atmospheric influences.

The details of these plans are very fully explained by Mr. Wilson, and we know that they have been very carefully worked out under the superintendence of Mr. Gossage.

He prepares a mixture of 4 cwt. of sulphate of soda, ½ cwt. of sulphate of baryta, and 2½ cwt. of small coal, or 2 cwt. of coke in coarse powder; fluxes this mixture in a reverberatory furnace in the same manner as mixtures of sulphate of soda with slack and lime are fluxed in the ordinary manufacture of carbonate of soda; and thus obtains sulphide of sodium, free, or nearly free, from sulphate of soda, and suitable for the manufacture of carbonate of soda by decomposition with carbonic acid.

Instead of using the ordinary reverberatory furnace to effect the fluxing, it may be performed still more advantageously, and with a still smaller proportion of sulphate of baryta in the mixture, by means of the rotating cylinder furnace. In using the rotating furnace, the decomposition of the mixture is effected without the necessity of introducing tools through openings into the furnace; and as the access of atmospheric air is thereby prevented, the gases evolved by the decomposition can be employed to supply carbonic acid for some of the succeeding processes.

When sulphide of sodium is prepared by causing sulphate of soda, in a state of fusion, to flow through a column of coke or coal maintained at a bright red heat, it has been found that the apparatus, as shown in the accompanying figures, is suitable for this purpose. Fig 158 is a vertical section of such apparatus; Fig. 160 is a transverse section, taken through the line ab, Fig. 158; and Fig. 161 is a transverse section taken through the line cd, Fig. 158. Fig. 159 is a detached view of one part of the apparatus.

This apparatus consists of a vessel, A, containing a column of coke or coal (when the latter material is used it is speedily converted into coke); this vessel is formed of iron plates, B, lined with fire-bricks, C. The column of coke is supported on the lower part of the vessel, B*, which is also coated with fire-bricks, C*. At a distance of about 2 feet from the bottom of the vessel, or at E, the space between the sides of the lining of fire-brick is diminished to about one-half of the space between the brick lining above and below this position; and this space is enlarged again gradually, so as to form a hopper of brickwork, extending from the wider part of the vessel to the narrower part at E. A similar provision is made in the upper part of the vessel at F. The object of these arrangements is to provide an unobstructed channel or passage for gases round the column of coke at each of the positions E and F. At D, there is an aperture, capable of being closed, communicating with the interior of the vessel A, through which sulphide of sodium, in a state of fusion by heat, can be run off. GG are pipes, communicating with the interior of the vessel A, below the part F. These pipes serve for the admission of air into the channel or passage surrounding the column of coke below the part F. It is better to have 8 or 12, or more, of these pipes. Each of these is furnished with a stop-cock for regulating the quantity of air admitted. H, is an opening provided with a door or stopper, J, for supplying coke or coal to the vessel A. K, is a flue or passage for the escape of gases from the vessel A. L, is an opening, furnished with a cover, M, for supplying sulphate of soda in the vessel A. This cover is placed in a channel or groove formed after the manner of a lute, which is made good to the iron casing of the vessel A and filled with sand.

In working with this apparatus, it is desirable that the supply

FIG. 158.

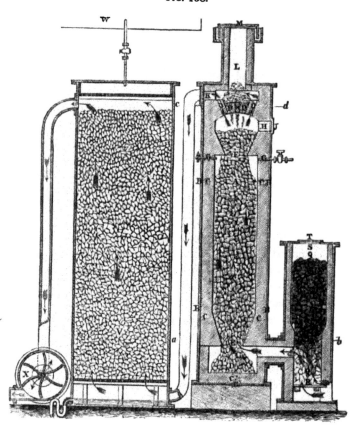

FIG. 159.

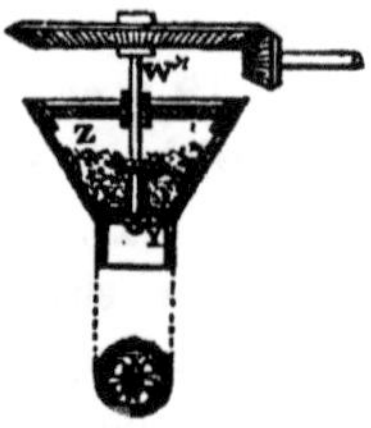

FIG. 160.

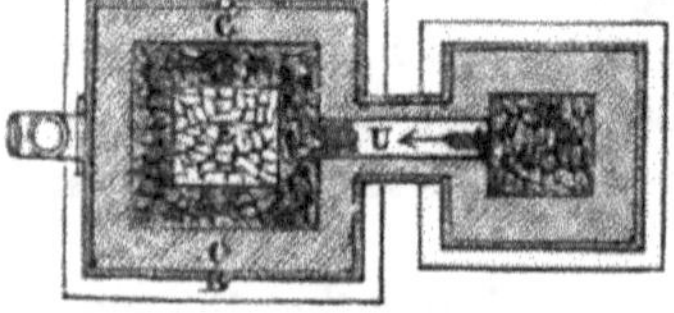

FIG. 161.

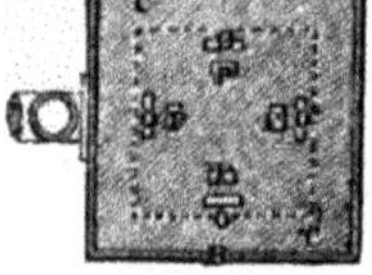

of fused sulphate of soda flowing into the column of coke, should be as uniform as can be conveniently arranged. It has been found that this can be effected by providing an arch or table of brickwork, N, in the upper part of the vessel A, on which supplies of about 2 cwt. at a time, of sulphate of soda can be thrown through the opening L, the cover M being immediately replaced. The arch or table N has four openings, OO, for the passage of heated gases up the vessel A; also four openings, PP, for allowing sulphate of soda, as it becomes gradually fluxed on the arch or table N, to flow into the column of coke in the vessel A. In place of using the arrangement described, with the moveable cover M, for supplying sulphate of soda to the vessel A, a hopper, Z (shown detached in Fig. 159), may be used for this purpose. This hopper is made of iron, and has a bottom plate, Y, through which there are two or more openings. The plate Y is covered with a plate X*, which has similar openings to those in the plate Y. A vertical spindle, W*, is attached to the plate X*, and supported by proper bearings. By means of this spindle, the plate X* may be so placed that its openings correspond with those in the plate Y, when sulphate of soda, contained in the hopper Z, will pass through the openings into the vessel A, and the spindle W* being turned so that the solid part of the plate X covers the openings in the plate Y, the passage of sulphate of soda into the vessel A is prevented. If the spindle W* be caused to revolve slowly, by some mechanical power, a regular supply of sulphate of soda can be passed from the hopper Z into the vessel A; and in this case the arch or table N may be dispensed with. Q is a furnace for burning coal or coke, constructed of a casing of iron plates, lined with fire-bricks. The coal or coke is supported on fire-bars, R, and there is an opening at S, furnished with a cover, T, for introducing them into the furnace. U, is a flue or passage through which the gaseous products of combustion pass from the furnace Q, to the vessel A, into the channel or passage surrounding the column of coke below the part E, and ascend from thence through such column of coke. The furnace Q and the vessel A are connected together by means of iron plates protected by fire-bricks. V, is a vessel formed of iron plates, and filled with siliceous pebbles or other suitable material. The object of using this vessel (which is connected with the vessel A by means of the flue K) is to cool the gases proceeding from the vessel A, and so render these gases capable of being used to furnish carbonic acid gas; also to arrest any compounds of soda which may pass over in vapour from the vessel A. W, is a cistern, from which water or solution of soda is caused to flow into the vessel V, and, becoming distributed over its contents, is brought into intimate contact with the gases. The same fluid, after having passed through the vessel V, can be elevated into the cistern W, and made to repass until it contains such a proportion of soda-compounds as to make it desirable that its contents should be obtained by evaporation. X is a centrifugal fan or blowing machine, the inlet passage of which is connected with the vessel V, while its outlet passage is connected with other apparatus, in which the carbonic acid of the gases produced in the vessel A is utilized. By means of this fan or blowing machine, atmospheric air is drawn into the furnace Q, causing the combustion of the coal or coke, and consequent production of highly heated carbonic oxide, which, passing through the flue V, enters the vessel A, below the part E, and ascends through the column of coke, which is kept at a bright red heat. The heated gases are drawn through the orifices in the table N, and pass from the vessel A by the passage K into the cooling vessel V, and thence through the fan or blowing machine X into other apparatus. As the fluxed sulphate of soda descends through the column of heated coke, it becomes decomposed, producing carbonic acid and carbonic oxide gases, which mix with the carbonic oxide gas proceeding from the furnace Q. A supply of atmospheric air is furnished through the pipes GG, below the part F of the vessel A, which causes the combustion of some of the carbonic oxide passing through the vessel A, and consequent production of an increased proportion of carbonic acid in the gases also causes an increase of heat in the upper part of the vessel A. This supply of air is regulated by the stop-cocks, care being taken that the quantity admitted shall not be sufficient to allow free oxygen to be present in the gases passing from the vessel A. A supply of coke, in small pieces, is thrown into the vessel A so as to keep up the column of coke. The quantity of sulphate of soda which may be decomposed by this apparatus, will depend upon the area and the depth of the column of coke, and the supply of sulphate of soda should be regulated accordingly. The sulphide of sodium produced, collects in a fluid state at the lower part of the vessel A, and should be withdrawn from time to time by opening the aperture D. This sulphide should be received in iron vessels, and immediately covered over with pounded coke to prevent oxidation by access of atmospheric air.

To obtain the bicarbonate of soda, a mixture is prepared, consisting of crystals of carbonate of soda with dry, or nearly dry, carbonate of soda, in the proportion of one part of the former to about three parts of the latter; and this mixture is placed in thin layers on shelves or trays contained in a closed chamber (such as is used in the ordinary manufacture of bicarbonate of soda from crystals of soda alone), and a supply of carbonic acid is conducted into the chamber. A supply of carbonic acid gas may be obtained by the calcination of limestone or chalk mixed with coke, or by the decomposition of sulphate of soda as already explained, or by other means hereafter described, and the gases being cooled, they are conducted to the chamber containing carbonate of soda.

In effecting the decomposition of sulphide of sodium by means of carbonic acid contained in bicarbonate of soda, it may be applied directly, or without being disengaged in the gaseous state from the bicarbonate, or indirectly, by being liberated in the gaseous form from the bicarbonate of soda. When the decomposition of sulphide of sodium is effected by the direct application of carbonic acid contained in bicarbonate of soda, a solution is obtained by lixiviating sulphide of sodium with water, in the same manner as black ash is ordinarily lixiviated. This solution is transferred to an iron vessel or boiler, capable of being closed, and arranged so that its contents may be heated, either by the application of fire or by the injection of steam, and having a pipe for conveying gas from it to other apparatus. Bicarbonate of soda is now added through an aperture provided for that purpose, which is immediately closed. The suitable proportions of the solution of sulphide of sodium and of bicarbonate of soda to be used, should be ascertained by a preliminary trial. Heat is then applied to the mixture, and as the carbonic acid of the bicarbonate becomes transferred to the base of the sulphide of sodium, sulphuretted hydrogen gas is liberated, and is conducted from the vessel employed to other apparatus, in which it is treated for the abstraction of sulphur by the means which will be described.

The application of heat to the mixture is continued until the whole, or nearly the whole, of the sulphur previously contained in the mixture has been expelled; the residual liquor now contains but little saline matter, except carbonate of soda, and by treating this liquor in the manner similar to that usually adopted with ordinary soda liquors, this substance may be obtained in the state of crystals or alkali.

When the decomposition of sulphide of sodium is effected by the action of carbonic acid in the gaseous state obtained from bicarbonate of soda, this salt is exposed to heat in iron retorts (similar to the iron retorts used in the manufacture of coal gas), and the carbonic acid gas thus evolved, is conducted to other apparatus containing the sulphide of sodium.

Carbonic acid obtained from other sources is equally applicable; thus, the pure carbonic acid gas obtained from earthy carbonates by the action of muriatic acid or other acids; also the impure carbonic acid gas, evolved when limestone or chalk is calcined with coke (proper care being taken to prevent the presence of free oxygen in such gas); the impure carbonic acid gas liberated during the decomposition of sulphate of soda by means of carbon, will likewise answer the purpose.

It may be applied to the decomposition of sulphide of sodium in solution in water, and, in this case, the solution should descend through a closed tower, nearly filled with small pieces of coke, over which the solution should be distributed as evenly as possible, at the same time that a supply of carbonic acid gas ascends through the interstices existing between the pieces of coke contained in the tower. The solution passing through the tower is thus made to absorb carbonic acid, producing carbonate of soda and liberating sulphuretted hydrogen gas, which is conducted to other apparatus, for the separation of sulphur by the oxide of iron. The liberation of sulphuretted hydrogen gas in this mode of producing carbonate of soda, is facilitated by introducing a supply of steam with the carbonic acid into the tower.

Carbonic acid gas may also be applied to the decomposition of sulphide of sodium in the solid form, and, in this case, a set of vats may be used, such as are used for the lixiviation of black ash in the ordinary manufacture of carbonate of soda, such vats being furnished with covers and pipes for the conveyance of gases to and from each other, and to and from other apparatus. The apparatus we have described in Gossage's process, page 25, will answer. In using this apparatus, the vats are charged with lumps of sulphide of sodium, and if they do not contain sufficient unconsumed coke to give suitable porosity to the sulphide of sodium when slaked to powder, some coke must be mixed with the lumps. A supply of steam is now introduced under the false bottom of each vat, and this is continued until the lumps of sulphide of sodium have become slaked to powder. A supply of carbonic acid gas, together with steam, is then conducted into the vats, and caused to pass through a series of these consecutively. By these means, carbonic acid becomes absorbed by the base of the sulphide of sodium, producing carbonate of soda, and sulphuretted hydrogen gas is liberated, which is conducted into other apparatus for the recovery of the sulphur. The application of carbonic acid gas to the sulphide of sodium is continued until it is found that the whole, or nearly the whole, of the sulphur previously contained in the sulphide of sodium has been expelled. The residual matter is then lixiviated with water, and a solution of soda produced, from which carbonate of soda may be obtained in a marketable form.

The separation of the sulphur by means of oxide of iron, may be conveniently effected by the use of an apparatus consisting of a series of chambers, built of bricks in Roman cement, and provided with a number of perforated floors, on which the oxide of iron employed, is placed. The gases proceeding from the apparatus in which the decomposition of the sulphide of sodium is being effected, are conducted into one of these chambers, and caused to percolate through the layers of oxide of iron placed on the perforated floors, and thence to other chambers also containing oxide of iron on perforated floors. By these means, the oxide of iron becomes highly charged with sulphur, and the last portions of sulphur can be abstracted from the gases. In place of using chambers for this purpose, it has also been found that a tower built of brickwork may be used, when the oxide of iron employed is in lumps. The tower may be about 6 feet square and 20 feet high, and should be provided with a grating at its lower part for supporting the lumps of oxide of iron with which the tower is filled, and through which the sulphurized oxide may be withdrawn. Gases containing sulphuretted hydrogen should be introduced into the tower below the grating; the gases ascending in the tower filter through the oxide of iron, and the sulphur is abstracted.

As the abstraction of sulphur from sulphuretted hydrogen gas progresses with the same oxide of iron, this oxide becomes loaded with sulphur, and gradually loses this power, which may be re-

17*

stored by causing atmospheric air to come in contact with the partially sulphurised oxide. Thus it has been found beneficial to cause sulphuretted hydrogen to pass through the oxide during a certain period, and then to shut off the supply of this gas from the chambers or tower and admit atmospheric air for some time. By these alternate applications of sulphuretted hydrogen and of atmospheric air, the oxide may be loaded with sulphur until its capability of abstracting a further quantity of sulphur from sulphuretted hydrogen gas is destroyed. Such sulphurised oxide is then withdrawn from the apparatus, and the sulphur set free, either by distillation in earthen retorts, when it is obtained as solid sulphur, or by combustion with atmospheric air, when it is obtained as sulphurous acid gas, applicable for the manufacture of sulphuric acid. The oxide of iron remaining after such separation of sulphur by distillation or combustion, can be used again in fresh operations.

The oxide of iron becomes so loaded with sulphur by the plan of alternate admissions of sulphuretted hydrogen and air, that it loses its power, more by the insterstices becoming filled up by the deposited sulphur than by any deficient chemical action, and it is therefore better to remove the sulphurized oxide to the open air, where, by being repeatedly turned over, the particles of finely divided sulphur adhere together, and the revivified or renovated oxide will be found to last much longer, and in the end to carry as much as 25 per cent. and upwards, of free or deposited sulphur.

Mr. Gossage mixes ground oxide of iron with the sulphide of sodium, and places this mixture, moistened with water, on trays, in apparatus similar to that employed in purifying coal gas. He then passes carbonic acid gas through the series of vessels until the sulphur has been transferred to the iron. He modifies this plan by using so much water with the materials, as to render the whole sufficiently fluid to permit the passage of the carbonic acid, which is driven into the apparatus by a force-pump. When he employs bicarbonate of soda as the source of the carbonic acid, he merely mixes a solution of the sulphide of sodium with the oxide of iron and bicarbonate in a close vessel, and applies heat while the whole is well agitated.

Hunt employs a different arrangement for carbonating the sulphide of sodium, and only uses the oxide of iron to facilitate the burning of the sulphur, so that none shall be lost in the process of converting it into sulphuric acid.

FIG. 162.

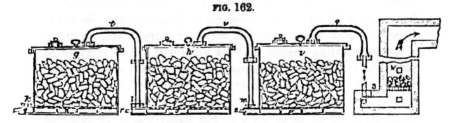

Figure 162 is a vertical section of the carbonating cisterns, and of the chamber where the sulphuretted hydrogen is burnt after being mixed with air.

The carbonic acid passes from the generators and enters the carbonating cisterns or vessels, g, h, i (Fig. 162), at the bottom openings marked k, l, m, underneath the perforated false bottoms, n, o, p, on which the sulphide of sodium is supported. Steam is at the same time introduced into each of the carbonating cisterns or vessels at the openings q, r, s, in the same direction as the carbonic acid. The carbonic acid gas passes out of the top of the first vessel by means of the curved pipe t, into the bottom of the second vessel h, and from the top of the second vessel h, by means of the curved pipe u, into the bottom of the third vessel, i, whence it passes by the pipe v, into the chimney w. The different junctions of the pipes with the carbonating vessels or chambers are made tight by means of water-joints. After the contents of the first vessel have been sufficiently carbonated (which is ascertained by taking a lump of the carbonated material out from time to time; when the lump is broken and carbonated crystals are formed inside, it may be considered finished), the gas is shut off from this vessel, and passed directly into the second vessel; and the first vessel, having been recharged with sulphide of sodium, the gas is made to pass through it last. Each vessel is thus made in turn to be the first of the series. These changes in the direction of the gas are made by means of a series of pipes. When the series of carbonating vessels is first set to work, the carbonic acid passes into the vessel g at k, and out of the vessel g at t into the pipe l, and thence through the vessels h and i into the chimney w. When the first vessel has been fully carbonated, and is recharged, the gas is made to take this course, namely, into the vessel h at 2, and from the vessel i through the pipe y into the vessel g, and from the said vessel g into the main supply pipe. By means of the pipes and openings.

represented, the gas can be passed in any desired order through
the vessels g, h, and i. The vessels g, h, and i, may be made
either of brick or of iron. A series of three, four, or more
of the carbonating vessels may be employed. The mixture of
carbonate of soda and carbonaceous matter, may either be lixiviated with water in the vessels g, h, i, so as to dissolve out the
carbonate of soda from the carbonaceous matter, which is afterwards removed to be re-used, or the mixture may be removed
and lixiviated with water in the usual way. As the sulphuretted
hydrogen, which the carbonic acid liberates from the sulphide of
sodium in the vessels g, h, i, passes to the chimney or chamber w,
it is mixed with atmospheric air, which enters the slit or opening 3, which slit or opening may be partially covered with a plate
or damper to regulate the quantity of air entering. The chimney
or chamber w is provided with two holes to look through, one
above the stratum 4, and the other below; the burning of the
gas can be observed, from time to time, through these holes. The
gaseous mixture is ignited under the stratum of oxide of iron,
pebbles, or broken brick 4, which becomes red-hot and rekindles
the gas, if from a temporary stoppage or otherwise the gas should
be extinguished. The sulphurous acid gas produced by the combustion passes into an ordinary oil of vitriol chamber, to be
converted into sulphuric acid in the usual manner. The sulphuretted hydrogen obtained when sulphide of sodium in solution
is treated with carbonic acid, may be burned in the way described,
and sulphuric acid obtained. The combustion chimney or chamber w is made of brick, lined with fire-brick, and of sufficient size
to create draught enough to draw the sulphuretted hydrogen from
the carbonating vessels.

THIRD OPERATION.

The Lixiviation of the Black Ash and Evaporation of the
Liquors.

1.—*Lixiviation.*

We now resume our description of the ordinary process in this
manufacture.

For some purposes, the black ash is employed, as it leaves the
furnace, and is an article of commerce, formerly known as British
Barilla. In such cases, the sulphate of soda employed contains
10—12 per cent. of common salt, which remains unchanged in
the black ash, and communicates to it the property of easily falling

to pieces in damp air, thus obviating the necessity of grinding. When used by soap-boilers, the presence of a certain quantity of common salt is desirable.

In some places, the balls are sprinkled with water in a tolerably hot furnace, where they soon swell up and fall to pieces in the moist atmosphere.

The oldest and most simple plan of lixiviation is that known in France by the technical term *Touillage*, for which we have no corresponding expression. It consisted in reducing the black ash to powder, sifting, and then mixing it with four times its weight of water in a cask. This mixture was left for a short time, and the liquor poured off, which was again mixed with a fresh portion of black ash, and this was repeated three or four times. The first portion was treated with another quantity of pure water, and afterwards passed through the other casks in the same way. This old method was of course very defective, and expensive from the amount of manual labour required for agitating the casks, &c.

This plan was replaced by a more methodical process. The balls, after standing for a few days, were broken to pieces and thrown into iron tanks, which were of all shapes and sizes, old steam-boilers, or cisterns, but which at last gave place to vessels constructed for the purpose. They were furnished with false bottoms, on which the black ash was thrown, and allowed to remain for some time in contact with the water. They were placed side by side as A, B, C, D (Fig. 163), on a solid foundation,

FIG. 163.

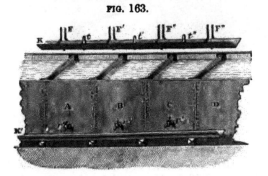

the false bottom being made of wood or iron pierced with small holes. A wooden spout K ran along the top, supported by iron holdfasts, FF', and supplied water or liquor to each tank by plugs t, t', t'. The liquor was run off from below by the cocks r, r', r'', into a spout K'. The liquors were supplied in rotation to the different tanks in proportion to the exhaustion of the black ash, the weaker liquors being run on the stronger ash, until they were brought up to the full strength.

The weak liquors were employed on fresh black ash until partially exhausted, when pure water was run on, and, lastly, water heated up to about 130° or 140° F. Each liquor was allowed to remain on the ash for eight hours, and the strength of these liquors varied from 1·007 to 1·200 sp. gr.

It is obvious, that while it was advantageous to use as little water as possible, to save the expense of evaporation, it was equally necessary to dissolve out all the soda ; and this method really necessitated the use of more water than was absolutely required, while the lixiviation of nearly exhausted ash invariably led to the decomposition of part of the carbonate of soda by reaction on the sulphide of calcium, which was specially apparent when hot water was used towards the end of the process.

An improvement on this plan, which is employed in some places on the continent, is shown in the Figure 164, and may be called *Lixiviation by Filtration*.

FIG. 164.

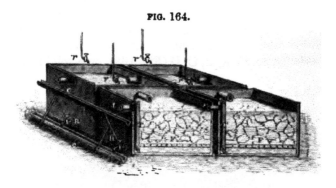

An iron pipe, T, open at both ends, with a spout near the top, is fixed in each tank, and the black ash is thrown on the false bottom, F', until the tank is nearly filled, when water is run in from the cock r''''. After a time, the liquor will stand at 1·600, 1·200, or even 1·280 ; and if fresh water is again run on from the same supply cock, it will drive the liquor up the pipe T, and discharge itself through the spout t', into another tank ; and by means of a plug at the upper end of the upright pipes, the overflow can always be regulated.

Another plan, more commonly adopted on the continent, is termed *lixiviation by displacement*; the principle on which it is founded is similar to that previously described.

Suppose that a certain quantity of water has taken up 8 per cent. from the first portion of ash; it will take up as much more from the second portion, and leave it containing 16 per cent.; it will be drawn off from the third portion with 24 per cent., and so on. Hot water dissolves more than its own weight, cold water half as much, still colder water at 8° C. (47° F.) dissolves about 23 per cent. of crystallized carbonate of soda. The arrangement (Fig. 165), shows the manner in which these principles have

FIG. 165.

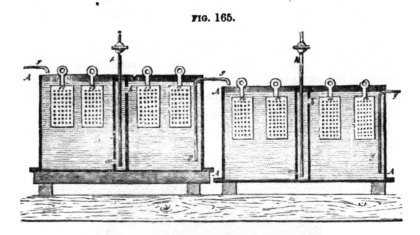

been practically carried out. Each of the iron lixiviating cisterns *AA*, is divided by a double partition, and the two halves are connected by an aperture *a* at the bottom, and another *b* at the top. In each compartment there are two sheet iron boxes, *nn*, pierced with holes in their sides like a sieve; these are filled with black ash, and suspended just below the surface of the water. The particles of fluid, as they become saturated and heavy, fall to the bottom, and make room for other particles, which become saturated in the same manner. The solution of the salt is thus effected much more rapidly than in the ordinary method, when the crude material collects at the bottom of the vessel, and such currents cannot take place. For the same reason (on account of displacement), the solution of a lump of sugar suspended just below the surface of water, is much more rapid than when it is placed at the bottom of the vessel; in the former case, the syrup is distinctly seen falling in streaks. A complete apparatus should have ten or twelve of these boxes placed side by side, and the one raised about 2 inches above the other placed immediately in front. In the same manner, the liquid is conducted from the first compartment through the lower aperture *aa* .. into the hollow partition, and thence through the upper opposite aperture, *bb*, into the second, from the bottom of which it enters the next lower box at the top through the pipes *gg* .. The steam pipes *hh* keep the temperature of the whole at about 40° (104° F.). The fresh water always enters the upper cistern first, whilst the lowest only is charged with fresh ash ; and while the water is traversing the whole series from top to bottom, the same space, but in an opposite direction, is traversed by the cases containing the ash, which from time to time are removed from one cistern to the other. The water is thus gradually converted into a stronger and stronger lye, until, on reaching the lowest cistern, it has attained the proper strength. On the contrary, the ash is gradually exhausted, and parts with the last portion of its soluble salts to the fresh water in the upper cistern.

The arrangement of a set of tanks on this plan is shown in Figure 166.

FIG. 166.

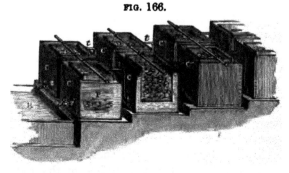

These two methods of *filtration and displacement* have been combined, and where there is space at command to permit the use of shallow tanks, it is very probable that less sulphide of sodium would be found in the soda-liquors : for the great heat generated by the slaking of the lime in the black ash, in the large quantities with which the deep tanks are filled, evidently favours the mutual decomposition of the carbonate of soda and the sulphide of calcium.

The arrangements are as follows : A series of shallow iron

tanks, ABCD, about 10 feet long by 7 feet wide, and 18 inches deep, are placed on benches, one rising above the other. They are filled with black ash ground to coarse powder, and when once in progress, the operation of lixiviation is conducted as follows. Fresh water is poured on the ash in D, and, after standing some time, is run off by the cock r'''' on the ash in lower tank, C, and so on, until it is run off from A by the cock r', a saturated liquor ready for boiling down. The partially exhausted ash is shovelled from each tank to the one below it, and the upper tank filled with fresh ash, the fully exhausted ash being removed from the lowest tank.

FIG. 167.

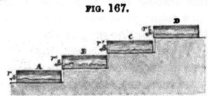

This plan undoubtedly costs more for labour, but it is a question whether this extra expense is not more than counterbalanced by the greater purity of the liquors.

The process of lixiviation by filtration was first used in the St. Rollox Chemical Works, and with several improvements which we will now explain, it is almost universally employed in this country.

Figure 168 is a plan of a set of four tanks about 8 feet square and 7 feet deep, adapted for lixiviating as much black ash as will yield 45 to 50 tons of soda ash per week. They are numbered 1 to 4, and a front view of them is shown in Fig. 169. They are fitted with the upright pipes previously mentioned, $a, a, a,$ which have each the arrangement of plugs b, b, b. The fresh water or weak liquor is supplied from a cistern, c, where it can be heated by a jet of steam and conveyed to any of the tanks by the pipe d. In the front view, two horizontal pipes are shown,

FIG. 168.

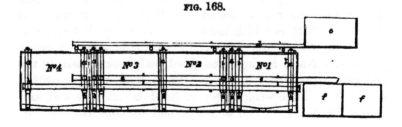

FIG. 169.

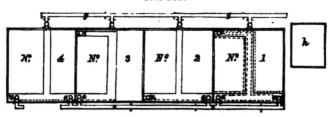

the upper one, *e*, communicating with each of the tanks, is intended to convey the strong liquor to the settlers *ff*, while the lower one, communicating only with Nos. 1 and 4 tanks, is intended to act as a return pipe, when these two tanks are to be placed in connection during the progress of the lixiviation. On the ground plan a third pipe *g* is shown, running off the tanks or conveying any weak liquors to the pump well *h*. The figures show how the connections are made, and require no further explanation beyond this, that the upright pipes *ii*, shown by dotted lines, are intended to represent those two pipes which are fixed at the back of the tanks Nos. 1 and 3.

We give the following practical directions for working the tanks.

Over the covers of the drains is placed about 3 inches of common furnace-slag, previously broken to pieces about 2 or 3 inches square and about 1 or 1½ inches thick, and on the top of this, 1 or 2 inches of the small cinders from the furnaces, about the size of a solid ½ inch, previous to filling the tanks with balls.

Fill the first three tanks with balls previously broken to pieces, about 56 lbs. weight; open the valves connecting Nos. 1 and 2 and Nos. 2 and 3; now run water on No. 1 till Nos. 1, 2, and 3 are filled, and allow all to rest till the liquor in No. 3 is 1·330 sp. gr. or 1·340 sp. gr.; then run off the liquor till it is reduced to 1·250 sp. gr., water running on No. 1 while the liquor is taken off No. 3, so as to keep the balls in all three tanks covered; again allow to rest and run as before, till the liquor in No. 1 is reduced to 1·010, at which time No. 1 is entirely shut off; and then run on No. 2, the valve connecting Nos. 3 and 4 being opened, and No. 4 having been filled with broken balls in the same manner as Nos. 1, 2 and 3. No. 4 tank is now treated like No. 3, and the weak liquor in No. 1 is run to a well, and pumped on No. 2, or run to waste, and at once cleaned out, taking care to examine the drainer and clear it if required. Now fill as before, and so soon as No. 2 is reduced to 1·010 sp. gr., work Nos. 3, 4, and 1.

In the larger manufactories, there are 3, 4, and 5 sets of these tanks, and they are arranged in double rows, as shown in the French plan.

The temperature of the water depends on the quantity of ash in the tank, the season of the year, and more especially on the *quality* of the black ash, as the balls are sometimes hard and dense, from having been kept too long in the furnace or exposed to too great a heat. The temperature of the water is thus found to vary at different times, according to the circumstances above named, from 90° to 140° F.

The time required to work off a tank, of course depends on the same causes; but on an average, those we have described are finished in 48 hours.

After the liquors have remained in the settlers long enough to deposit all the impurities held in suspension, they are pumped up to cisterns sufficiently elevated, to admit of their being conveyed by pipes to the pans and furnaces for concentration and boiling down to salt.

2.—*Evaporation.*

The enormous quantity of water to be evaporated and the different objects to which the soda-ash is to be applied, regulate the manner in which this operation is conducted.

In some localities, where an inferior white soda-ash is produced, the liquors are evaporated and the ash calcined in the same furnace. The furnace employed for this purpose is about 18 feet long by 9 feet wide, and the interior is lined with fire-bricks, between which and the wall of the furnace a partition of thin sheet iron is placed.

In other works, the waste heat of the balling furnace is employed for the evaporation of the tank-liquors, and the arrangements when the revolving balling furnace is working, are shown at p. 239. When the ordinary ball furnace is used, the plan of the salting furnace is shown at p. 233. The furnace is lined with an iron pan, *gg* (Fig. 152), the flame from the ball furnace passing over the surface of the liquors, *h*, while the steam and gaseous products pass to the underground flue at *i*. The salt is withdrawn through the side openings at *kk*, into an iron cistern or drainer, *ll*, placed alongside the furnace. This drainer is fitted with a false bottom, m, to allow the liquors which separate to run into a well n, from whence they can be removed by a pump, o.

Another system is pursued elsewhere by conducting the evaporation in iron pans, by what is termed *bottom heat*, which, as already explained, may be obtained from the balling furnaces. The pans are protected for some distance by a brick arch underneath, from the direct action of the fire and flame. As the liquors become concentrated, small crystals of monohydrated carbonate of soda, $NaO.CO^2 + Aq$, constantly fall to the bottom, which are raked to the other or farthest end of the pan, and are then collected with scoops, or *fished* out by means of perforated iron shovels : hence this system is called *fishing pans*.

In some alkali-works, the evaporation of these liquors is carried on in a range of three or four separate pans, or in one pan divided into three or four compartments. The liquor is brought to the boiling point in the first, and carried up to saturation in the next two, and finally boiled down to dryness in the last nearest the fire, or evaporated until all the carbonate of soda has salted out.

As might be expected, this operation of the soda process has received equal attention with the others, and many improvements have been introduced both to save time and fuel. We will, therefore, introduce a description of some of these plans, and commence with that of Kneller, which we believe, embodies an excellent idea.

Figure 170 represents an evaporating pan, AA, with a metal

FIG. 170.

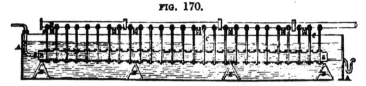

FIG. 171.   steam case, BB, through which superheated
                         steam or air can be driven. The steam-case
                         is fitted with a series of tubes, CC, which pass
                         through it at regular intervals, and are all
                         in connection with the main pipe, DD (Fig.
                         171). The steam case is supported on feet,
EE, which rest on the bottom of the evaporating pan. The liquors being heated by the steam or hot air in the case, BB, the evaporation is promoted by forcing currents of compressed air through the small tubes, CC.

We now describe the plans patented by Mr. Gossage for evaporating these solutions and for separating the saline compounds during the process. His pans are all placed on the same level. AA are two pans to which heat is applied by special fires or by waste heat from furnaces. These pans communicate with each other by a channel, *a*, and are connected with a cooling pan, B[1], by means of a channel, C. The pan B[1] is connected with a heating pan, E, through the channel D, and the pan E is connected with the cooling pan B[2] by the channel *e*, a similar arrangement being adopted with all the other pans in the range. Each of the cooling pans, B[1], B[2], &c., is provided with a vessel, F[1], F[2], &c., for distributing streams or bubbles of air through the liquor. These vessels, F[1], F[2], will be described hereafter, and they are connected with an air-pump by means of the pipes H[1], H[2], &c. I[1], I[2], &c., are cocks communicating with a reservoir

containing suitable liquor, through which it can be conveyed to the heating pans. Heat is applied to each of the heating pans by means of special fires, or by means of waste heat conducted from furnaces. In beginning to work with this set of apparatus, (all the pans having been supplied with liquors from which saline compounds are to be procured) heat is applied to each of the heating pans; and when the liquor contained in the pans AA has acquired such degrees of concentration and of temperature, that it will deposit salts, its temperature being reduced only a few degrees, a further supply of liquor is introduced into the pans AA, through the

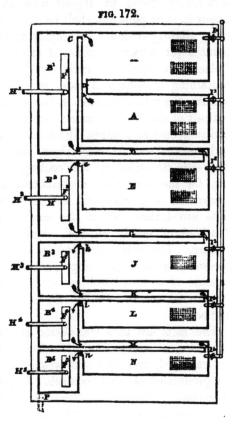

FIG. 172.

cocks II, which introduction occasions a flow of hot concentrated liquor through the channel C, into the cooling pan B';
at the same time, streams or bubbles of air are driven through
the liquor in the pan B', so as to accelerate the cooling of
the liquor and promote the evaporation of the water. Salts
deposit from the liquor in the pan B', and cooled liquor flows
from it through the channel D, to the heating pan E, where it
becomes heated and concentrated, and then flows into the cooling
pan B², where it is subjected to streams or bubbles of air passing
from the vessel F². The liquor follows a similar course through
all the other pans, depositing salts, and at last the remaining
liquor flows through the channel P into any convenient cistern.

By means of the cocks I' I², &c., such a supply of fresh liquor
is introduced into each of the heating pans (when required), that
the hot concentrated liquor flowing from such pans, shall not be
sufficiently strong to deposit salts at the temperature of the
liquor in such pans, but shall be in a condition to deposit salts on
being slightly cooled. In employing this series of pans and apparatus for obtaining different saline compounds of soda from
" black ash liquors " (previously deprived of sulphide of iron)
containing carbonate of soda, caustic soda, and ferrocyanide of
sodium, the salts deposited in the cooling pan B', consist almost
entirely of carbonate of soda; and these salts, after being drained
and subjected to a slight filtration of water or of a solution of
carbonate of soda through them, yield soda-ash of good quality.
The salts deposited in the cooling pan B², although consisting, for the most part, of carbonate of soda, contain a notable
proportion of caustic soda and ferrocyanide of sodium. These
salts are transferred to the cooling pan B', where they are
deprived of nearly all the caustic soda and ferrocyanide of
sodium. The salts produced in each succeeding cooling pan
become more and more impregnated with caustic soda and ferrocyanide of sodium, and the salts produced in the pan B³ are
transferred to the pan B⁴, and from the pan B⁴ to the pan B³,
and so on.

The salts are thus transferred through the different cooling
pans in succession, till they arrive at the pan B', whence they
are removed, and by slight washing are rendered suitable for
the production of good soda-ash. Caustic soda and ferrocyanide
of sodium (previously contained in the liquor employed) become
thus concentrated in the liquors flowing from the cooling pan B⁵;

and these liquors are subjected to processes which will be explained in the next chapter. When sulphate of soda and common salt are to be obtained separately, from solutions containing these salts, the deposition of each is effected by the relative proportion of the salts present in solution, and also by the temperature of the solution when the deposition takes place. Thus, when a solution containing 3 parts of sulphate of soda for 1 part of common salt, is evaporated by this system, the salts deposited in the cooling pan consist almost entirely of sulphate of soda; but when the solution contains nearly equal proportions of sulphate of soda and common salt, and the temperature of the solution in the cooling pan is reduced by streams of air or otherwise to about 92° F., the greater part of the salts deposited consists of common salt, and the solution or mother-liquors remaining, being concentrated by passing through a heating pan, and its temperature subsequently reduced to about 80° F., the greater part of the salts then deposited consists of sulphate of soda. By this mode of working, sulphate of soda can be obtained from solutions containing common salt, of sufficient purity to be suitable for the manufacture of alkali.

Mr. Gossage has carried these improvements a step further, by heating the air he draws through his liquors; in following the details of this later patent, we have endeavoured to connect it with the former by including the description of that part of the arrangements which are, more or less, common to both patents.

The apparatus, as shown in Fig. 173, consists of a pair of pans connected together at each end by means of open channels. In one of these channels is placed a paddle wheel, which being made to revolve, occasions a continuous circulation of the solution through the two pans. One of these is the boiling pan, to which heat is applied,—the other is the "salting pan," to which no heat is applied, and in which the salt is deposited. These pans are suitable for obtaining salts by the mere cooling and surface evaporation of the solution, as it flows through the "salting pan;" but it is found that the quantity of salt obtained, may be greatly increased by the evaporative action of streams of air passed through the solution.

An inverted vessel open at the bottom (called an "air-vessel") is placed in the "salting pan," with its open mouth dipping into the solution. A communication from the upper part of this vessel is made, by means of pipes, with an exhausting pump. The

action of this pump draws air from the "air vessel" and raises
the solution from the pan, until the notches at its lower edge,
being no longer covered with solution, admit the passage of air,
which bubbles through the column of solution elevated in the "air
vessel," and thus occasions a large evaporation of water from the
solution.

FIG. 173.

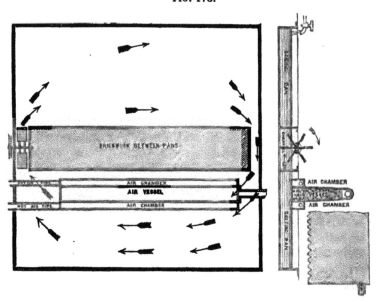

The pans are worked on the principle that no salt shall be
allowed to fall in the "boiling pan," to which alone heat is
applied, all the salt being deposited in the "salting pan" to
which no heat is applied. This is accomplished by so regulating
the quantity of fresh soda-solution supplied to the "boiling
pan," proportionately to the evaporation, that, by the time this
solution reaches the channel communicating with the "salting
pan," it shall not have quite arrived at the strength at which
salts are deposited from such solutions at a boiling temperature;—on flowing into the "salting pan," the solution becomes
cooled and immediately deposits salts. This deposition of salts
is greatly accelerated and increased by the evaporative and
cooling action of the bubbles of air, passing through the
column of solution contained in the "air vessel." The soda-
solution having deposited salts in the "salting pan" is drawn,

by the action of the paddle wheel, into the "boiling pan," where it becomes concentrated again, together with the supply of fresh solution, and again flows into the "salting pan" to deposit salts.

In the preceding description, it is supposed that air is drawn into the "air vessel" direct from the atmosphere (this vessel not having any air chambers). By adapting to this "air vessel" such "air chambers" as are shown in the figure and connecting these "air chambers," by means of pipes, with the flue of a "finishing furnace," or other sources of heated air, hot air can be caused to bubble through the column of solution in the "air vessel," and thus a much greater amount of evaporation is effected. Or, in place of causing air to be drawn through a column of elevated solution "by exhaustion," air can be driven through the solution by the forcing action of an "air pump"— thus causing the deposition of salts in the "salting pan" by the evaporation and cooling action of air forced through the solution.

As these arrangements prevent the deposition of salt in the "boiling pan" (to which alone heat is applied), they provide for the safe application of waste heat from black ash furnaces to these pans, without occasioning the destructive effects experienced when salts are deposited. By these simple means, an important economy of fuel and repairs is effected, and salts of very superior quality are obtained from black-ash liquors.

The two following plans derive their importance from the source of hitherto waste heat which they render available, not only for evaporating these liquors, but for the drying, roasting, carbonating and sulphating processes.

The first is that patented by Messrs. J. and W. Allen, represented in plan and section in figure 174, where AA are the coke ovens, B the boiling down pan, CC dampers in flue to shut off the ovens, DD flues connecting the ovens with the pans, EE doors for charging the ovens with coal, FF mouths of the pan, G draught flue to the chimney, H arch over the pan, K tile to allow surplus heat to escape when necessary, L arch of the coke oven.

Five ovens are attached to one pan, in order to keep the heat as uniform as possible, and to allow for the cooling of the ovens when drawing the coke. An oven charged each day maintains a pretty uniform heat, but in the other operations mentioned above, it is found that three ovens only are necessary for each furnace. When the oven is burnt off, the damper C is closed and the coke withdrawn. This waste heat can, of course, be applied

18*

either above the liquors or materials, or below the pan or furnace bottom.

FIG. 174.

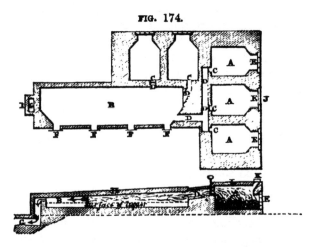

The other plan has been adopted at the Walker Chemical Works, under the charge of Mr. Clapham, and consists in making use of the waste gases from the neighbouring iron blast-furnaces, which are employed in the same way as is often practised in raising steam in iron works.

We have now to describe an ingenious plan patented by Messrs. Williamson and Stevenson, for applying the heat to the sides, instead of the bottom, of the evaporating vessel, and for preventing the encrustation of the salts on the heating surface of the boiler or pan. We have seen this plan at work and were surprised at the rapidity of the evaporation and of the salting out of the soda. The following is a description of the arrangements.

Figure 175 is a side elevation, Figure 176 a plan, and Figure 177 a vertical section, of an evaporating pan, arranged according to this invention. A is a circular pan with upright sides, heated by means of the fire-place B and the flue C, which is continued nearly all round the pan. Instead of a fire-place, the pan may be heated by the waste heat of a furnace. In the centre of the pan A, is placed a vertical axis E, to which are fixed horizontal projecting arms FFF, which nearly reach, but do not touch, the sides of the pan A. The top of the axis E, is fitted with a crown wheel G and pinion H, on the shaft of which is fixed a belt sheave J, whereby motion is communicated to the apparatus from a steam-engine shaft or other prime mover, and

FIG. 175.

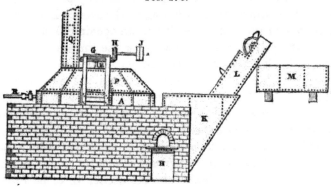

FIG. 176.

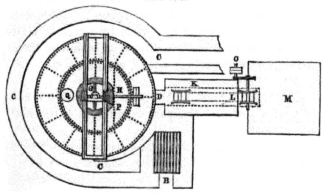

FIG. 177.

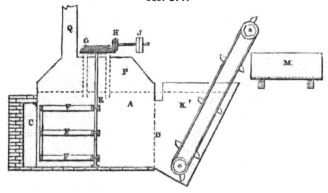

by the revolution of the arms F, a continuous circular motion is maintained in the contents of the pan A, thereby preventing the salts which are formed, from adhering to the sides. Between the fire-place B, and the exit of the flue C, an opening D, is made in the side of the pan A, commencing below the surface of the solution and extending down to the bottom of the pan, by which opening a communication is made between the pan A, and a settling vessel K. Through the opening D, the salts in suspension pass from the pan A, into the settler K, where they subside, and from which they are withdrawn by means of a dredging apparatus L, and discharged into a drainer M, provided with a perforated false bottom, from underneath which the liquor, drained from the wet salts, runs back into the settler K to be further evaporated in the pan A. The moving power is communicated to the dredging apparatus by a belt working on the sheave O. The pan A may be fitted with a close cover P, and the steam formed by the evaporation conveyed away by a pipe Q, and used for drying materials or heating solutions, or otherwise turned to account. R is a pipe furnished with a regulating stop-cock for supplying the solution to the pan.

A velocity of the current of the solution of about 6 feet per second round the sides of the pan is sufficient to prevent the adhesion of the salts.

Mr. Dale's plan contains an important principle which we think is overlooked in many localities where the latent heat of the aqueous vapour might render good service, especially in cases where, as we have seen it successfully applied, the concentration of weak liquors would not otherwise have paid for evaporation. He fills a boiler with the liquors and heats it in the usual way, and passes the steam through a coil of piping in the false bottom of an evaporating pan, also filled with liquor, and thus renders the sensible and latent heat available for the evaporation of the liquor in this second pan. Instead of an open pan, it may be a closed vessel or boiler, whence the steam may be conveyed to a third vessel, and so on, through a series of any practical number. The working pressure is, of course, less in each succeeding vessel, and proper cocks must be attached to the pipes or vessels to regulate this inequality of pressure.

The steam may also be employed for obtaining motive power by conveying it direct to an engine, or by applying it to generate steam in a boiler, or it may be conducted into a vessel, serving as a reservoir of steam, and there heat applied, so as to obtain superheated steam.

The calorific power of coals, or the number of lbs. of water which 1 lb. of coal can evaporate, varies from 11 to 14, but as this fuel is ordinarily burnt, barely the half of this work is obtained, and various plans, improperly called smoke consuming apparatus, have been invented to save this loss of power; but we know of only one attempt to apply it to the evaporation of these liquors, that, namely, which has been patented by Messrs. Deacon and Robinson. Respecting the application of this cupola furnace to the working of the pan for decomposing common salt we must refer to page 210. The modification necessary, when using it in connection with the evaporation of the black ash and other liquors, are so obvious as not to require any explanation.

Before proceeding to the next section, we may very appropriately introduce here the analyses which have been made of the ash obtained by boiling down these liquors, from Duisburg and Liverpool and Glasgow by—

	Mohr.	Kynaston.		Brown.	
Carbonate of soda	71·250	67·891	73·626	68·907	65·513
Hydrate of soda	24·500	14·245	13·600	14·433	16·072
Sulphate of soda	—	4·579	6·127	7·018	7·813
Sulphite of soda	0·102	—	—	2·231	2·134
Hyposulphite of soda ...	0·369	—	—	trace	trace
Aluminate of soda	—	1·214	—	1·016	1·232
Silicate of soda............	—	0·389	0·511	1·030	0·802
Chloride of sodium	1·850	6·061	4·431	3·972	3·860
Sulphide of sodium... ..	0·235	0·556	0·361	1·314	1·542
Cyanide of sodium	0·087	—	—	—	—
Alumina	1·510	—	—	—	—
Silica........................	0·168	—	—	—	—
Iron	traces	—	—	—	—
Insoluble matters	—	0·217	} 1·579	0·814	0·974
Water	—	5·005		—	—
Ultramarine...............	—	0·140	—	—	—
	100·071	100·287	100·235	100·735	99·945

FOURTH OPERATION.

The Carbonating and Oxidising Processes for the production of White Alkali and Soda Crystals.

An examination of the analysis of the ash obtained by boiling down the tank liquors, discloses the source of two defects, in the presence of caustic soda and sulphide of sodium, which are both prejudicial when pure carbonate of soda is wanted, either for the manufacture of glass or in the production of crystallised carbonate of soda.

The soda-salts obtained by any of the plans for evaporating the tank-liquors described in the last section, after having been allowed to drain, are mixed with small quantities of sawdust, and in some works with small coal, and calcined at a moderate heat in reverberatory furnaces, hence called *carbonating* furnaces. These furnaces are placed in ranges, with tanks fixed above for employing the waste heat in heating soda liquors.

The mixture is repeatedly stirred by means of *paddles*, or as they are called in some places *oars*, and rakes. As the sulphur is expelled and the carbon is being burnt off, converting the caustic soda into carbonate, the heat is increased to dull redness, and the operation continued until all the organic matter is destroyed. This operation requires great care, as, from the nature of the ash, too great a heat at first would cause a partial fusion, and the sulphur would not be expelled.

Messrs. Stevenson and Williamson, of the Jarrow Chemical Works, are engaged in introducing Mr. J. M. Napier's ingenious patent furnace for carbonating the crude soda.

The four figures, 178, 179, 180, and 181, represent this furnace. Fig. 178 shows a longitudinal section of No. 1, and Fig. 179 the same of No. 2 furnace. Fig. 180 is a plan in section of the furnaces, and Fig. 182 is an end view of both furnaces. A, Fig. 178, is a receiver into which the wet soda salt is thrown, B is a revolving rake, the speed of which is arranged to regulate the supply of the salt to the furnace. *aaa* are fixed shelves, made of fire-brick, and placed upon iron bed-plates, which pass from wall to wall across the furnace ; *bbb* are similar shelves, but attached to the vibrating frame, C, which is worked to and fro upon its wheels, D, by. the excentrics, EE, through the connecting rods, *dd.* By the action given to the alternate shelves, the salt is made to pass through the furnace, descending step by step until it is thrown upon the sieve, M, to which a quick vibrating motion is applied for sifting the salt. F, is a worm for collecting the fine portion of the salt which has passed through the sieve, and for conveying it to the elevator, G, which raises it to the distributing worm, L, by which it is passed into the hopper, T, of the furnace No. 2. H, is a rolling mill, into which the lumps fall from the sieve to be crushed. I, is a worm for collecting the crushed matter and passing it on to the elevator, J, which raises it to be returned to the furnace by the distributing worm, K.

The fires for heating the furnace are at NN, and the products of combustion, after passing over the salt, escape by the flue, e. O (Fig. 178), is an iron case containing a number of inclined shelves, down which the salt is made to pass by the action of a number of pins working through holes in the back plate of the iron case, and pushing the salt forward, which, as it lies upon the inclined shelves, is submitted to the action of the heat and products of combustion from the furnace, before it falls on to the first back of the furnace-bottom, to be progressively passed from shelf to shelf, as in furnace No. 1, until it descends from the last shelf, through the inclined box P, into the truck Q. Provision

FIG. 178.

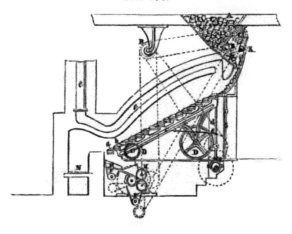

FIG. 179.

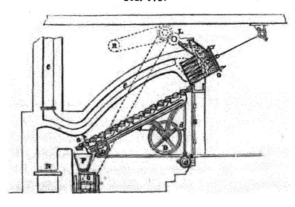

FIG. 181.

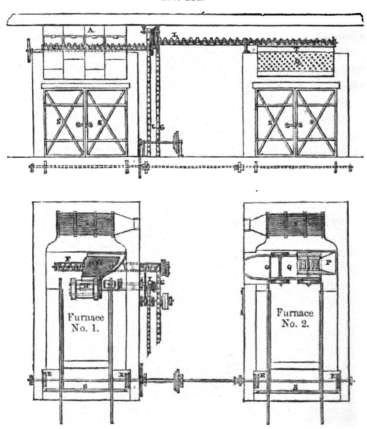

FIG. 180.

is made for closing the bottom of the inclined box P, while the
full truck is being replaced by an empty one.

R is the main shaft, through which the power is transmitted
to the various parts requiring motion. The vibrating frame c,
can be withdrawn from the furnace by the doors ss, when the
brickwork requires repairs.

No. 1 furnace is intended to be worked at a moderate heat,
but so as thoroughly to dry the material, which is afterwards
passed to No. 2 furnace to undergo there a much stronger heat.
The arrangement previously explained for returning the matter
crushed by the rolls, to pass again through the No. 1 furnace, is
introduced, to prevent the delay which would arise if the matter

were moved through the furnace slowly enough to permit the lumps to be completely dried.

The composition of the *soda ash* is shown in the following analyses:—

	English.		German.
	1	2	
Carbonate of soda . .	71·614	70·461	62·13
Hydrate of soda . . .	11·231	13·132	17·20
Sulphate of soda . . .	10·202	9·149	8·66
Chloride of sodium . .	3·051	4·271	3·41
Sulphite of soda . . .	11·170	11·360	0·35
Aluminate of soda . .	0·923	0·734	1·11
Silicate of soda . . .	1·042	0·986	2·56
Sand	0·313	0·464	0·62
Moisture.	—	—	3·96
	99·546	100·557	99·90
	Brown.		Unger.

White Alkali.

The alkali produced has a faint yellow tinge, and, after exposure to the air for some time, is dissolved in large vats by means of steam, a ton of water dissolving a ton of the ash, and the liquor standing at 54° Twaddell. The impurities are allowed to settle, and the liquor is either run into open pans *aa*, heated as before by the waste heat of the boiling down pan furnaces *bb*, or conveyed to the crystallising coolers.

These pan furnaces *bb*, are constructed of sheet iron, from 20 to 30 feet long, 7 feet wide, and about 2 feet deep. The fuel employed in the fires *cc*, is coke, in order to preserve the colour of the ash, and the steam and products of combustion pass off through the flue *d*, Fig. 182. The ash is removed through the doors *ee*, which project in front to facilitate the withdrawing of the product, now called *white alkali.*

FIG. 182.

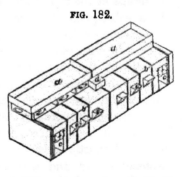

We give below the composition of different samples of alkali, No. 1 to 3, analysed by Brown, showing the effect of successive refinings, and Nos. 4 and 5 by Muspratt and Hanson, No. 4 being intended for soap-boilers, and No. 5 for plate glass.

CONSTITUENTS.	1		2		3		4	5
Carbonate of soda	79·64	80·92	84·00	83 76	84·31	84·72	77·08	78·55
Hydrate of soda	2·71	3·92	1·06	0 73	trace	0·28	4·88	4·15
Silicate of soda	1·23	1·32	0·98	0·73	0·41	0·32	2·40	0·25
Aluminate of soda	1·17	1·01	1·01	0·62	0·63	0·71	—	—
Sulphate of soda	8·64	7·43	8·76	9·49	10·26	0·76	5·11	1·70
Sulphite of soda	1·24	1·11	trace	0·38	trace	—	—	—
Hyposulphite of soda	trace	trace	—	—	—	—	—	—
Sulphide of sodium	trace	0·23	—	—	—	—	0·63	—
Chloride of sodium	4·13	3·14	3·22	2·29	3·48	3·14	7·13	5·62
Carbonate of potash	—	—	—	—	—	—	0·20	—
Carbonate of lime	—	—	—	—	—	—	0·32	0·33
Sulphide of calcium	—	—	—	—	—	—	0·20	—
Oxide of iron	—	—	—	—	—	—	0·32	0·27
Insoluble matter	0·97	0·77	0·71	0·84	2·25	0·50	0·66	0·48
Water	—	—	—	—	—	—	1·06	8·65
	98·83	99·85	99·74	98·84	99·34	90·43	99·99	100·00

Ward employs the carbonate of magnesia, known as *Greek stone*, along with sawdust in carbonating the crude ash, especially in localities where the fuel is loaded with sulphur, which he states withdraws the sulphur and yields a stronger and purer *white alkali*.

Various plans have been proposed and adopted for improving this operation. Mr. Shanks breaks the black ash into pieces, which are placed in close vessels, called *Carbonators*. The ash is moistened with water, and carbonic acid admitted, until the sulphide is decomposed, and the caustic soda is converted into carbonate. Instead of carbonating the black ash, a similar plan is applied to the tank liquors, which are allowed to fall over a mass of pebbles or flints, supported on an open arch of brick-work, while a stream of carbonic acid passes upward, and the waste gases escape at the top of the vessel or tower.

At a subsequent date, Mr. Wilson patented the use of bicarbonate of soda for the same purposes. He mixes the bicarbonate of soda with the liquors in the ordinary reverberatory furnaces, stirring the mixture until the mass becomes perfectly dry.

Mr. Brown proposed to employ nitrate of soda, but especially recommends its use with the drainings, or mother-liquors, obtained from the various evaporating plans, which are known as *red liquor*. He boils this red liquor, which consists chiefly of sulphide of sodium and caustic soda, in a metal pan, until it acquires the consistence of tar, and then adds the nitre, when the heat is increased to expel the remaining moisture, and flux the mass. Copious fumes of ammonia are evolved, and the mass froths, but gradually subsides down to a quiet flux, when it is run into moulds.

A plan commonly adopted in some works, and also depending on the principle of oxidation, consists in throwing some bleaching powder into the tank liquors; but in either case the presence of the sulphate of soda is perpetuated, which is very objectionable in many of the purposes to which white alkali is applied, and is certain, more or less, to disfigure the crystals of carbonate of soda.

Glauber's Salts on the surface of the former Salt.

Mr. Gossage endeavours to obviate this difficulty by the following patented processes, which have been adopted in several manufactories.

The black-ash liquors contain carbonate of soda, caustic soda, sulphide of sodium, sulphide of iron (held in solution by sulphide of sodium), and ferrocyanide of sodium. During the ordinary processes of evaporating black-ash liquors and calcining the salts, as already described, sulphide of iron and the iron of the ferrocyanide of sodium, become converted into oxide of iron, and this substance communicates an objectionable yellow colour to the soda-ash produced, rendering it unsuitable for certain applications.

Mr. Gossage finds that when the sulphide of sodium is converted by oxidation into hyposulphite, or sulphite, or sulphate of soda, the sulphide of iron is no longer soluble in these liquors, and consequently separates when allowed to subside. He oxidises the sulphide of sodium in the black-ash liquors by absorbing oxygen from air, passing them uniformly through a high tower, made of iron plates or other suitable material, and filled with small pieces of coke, supported on a grating near the bottom of the tower, while currents of air pass through the interstices of the pieces of coke. By means of this arrangement the black-ash liquors, spread in thin films over an immense number of surfaces, are presented to the currents of air passing through the tower, and the absorption of oxygen is complete. A circular tower, 6 feet diameter, and 20 feet high, filled with pieces of coke of the size of walnuts, is equal to the purification of as much black-ash liquors in a week as will produce from 30 to 40 tons of soda-ash.

When these purified black-ash liquors are used for the production of soda-crystals or of white alkali, the greater part of the caustic soda must be converted into carbonate of soda by the absorption of carbonic acid gas, and the carbonated liquor is concentrated in the usual way, or boiled down to dryness.

The mother-liquors obtained after the carbonate of soda has been crystallised out of the purified black-ash liquors, contain nearly all the ferrocyanide of sodium previously existing in the purified black-ash liquors. These mother-liquors are concentrated by evaporation until nearly all the carbonate of soda, previously contained in them, is deposited as salts, which are withdrawn from the concentrating apparatus at different stages of the operation. The first portions of salts contain but little ferrocyanide of sodium, the second portions contain more of this compound, and the last portions contain a considerable quantity of ferrocyanide, while the mother-liquors still remaining contain a large proportion of ferrocyanide in solution. The ferrocyanide of sodium is removed from the salts by washing with water, or by filtering a solution of carbonate of soda through them. A series of four vats, arranged on the system already explained, is employed, and the salts are placed on perforated false bottoms in the four vats. Water, or a solution of carbonate of soda, is introduced into the vat containing salts least impregnated with ferrocyanide, and the solution produced, containing carbonate of soda and ferrocyanide, flows into the adjacent vat, where it takes up a further portion of ferrocyanide, subsequently flowing into and through the two other vats, where it becomes still more impregnated with ferrocyanide. Water, or a solution of carbonate of soda is run into the first vat, until the salts are so far deprived of ferrocyanide as to produce white soda-ash of good quality. These washed salts are removed, and the vat is refilled with impure salts, this vat becoming the last in the series. By pursuing this mode of working, the salts contained in each vat successively become those which are least impregnated with ferrocyanide, and are first subjected to the action of the water or solution of carbonate of soda. The solutions so obtained, are mixed with the mother-liquors resulting from the previous concentrating operation, and if such liquors contain more than 10 parts of carbonate of soda for each part of ferrocyanide of sodium, they are concentrated by evaporation so as to obtain a further quantity of salts; and this concentration and separation of salts is continued until the mother-liquors then remaining, contain less than 10 parts of carbonate of soda for each part of ferrocyanide of sodium; but it is better that the proportion should be as 5 to 1. These liquors are mixed with sufficient water and caustic lime to convert the carbonate of soda into caustic soda, and the solution is concentrated, until the whole, or nearly the whole, of the ferrocyanide separates as salt, or it may be removed by crystallization. Or in place of converting the carbonate of soda into caustic soda, it may be converted into bicarbonate or sesquicarbonate of soda by impregnation with carbonic acid gas; a precipitated carbonate of soda of sparing solubility is then produced, from which the greater part of the ferrocyanide may be obtained by drainage, and by filtering water through the bicarbonate or sesquicarbonate, the operations above explained being applicable in this case.

The ferrocyanide of sodium obtained by either of the modes described, is refined until it is sufficiently pure to be suitable for use.

The salt obtained from the purified black-ash liquor in making alkali is, of course, impregnated with this ferrocyanide, which is removed in the manner already explained.

Mr. Stott runs his tank-liquor into close vessels, and pumps in the carbonic acid under pressure, either alone or after one or more of the oxides of iron, zinc, or manganese, in a fine state of division, have been mixed with the tank-liquor. We tried this plan many years ago on a large scale; but, while it decomposed the alkaline sulphide, the ash obtained still retained the yellow colour after calcining, the colour evidently arising from the presence of the ferrocyanide, which is not removed by this plan.

Mr. Deacon dispenses with the use of carbonic acid, and employs the metallic oxides or salts at specified temperatures, and with liquors of a definite strength. He prefers to use hydrated peroxide of iron, at a temperature of about 70° to 80° F., and with a liquor or ley about 70° Twaddell; but the lower the temperature and the weaker the lees, the more complete is the separation of the precipitate.

The application of heat while the metallic oxides, or salts, or sulphides remain in the lees, tends to render the lees of a darker colour; it ought to be used only when it is desirable to hasten the reaction of oxides or salts, which in the cold may, from some cause, be acted upon with difficulty. If the lees contain a large proportion of carbonated alkalies, the reaction is comparatively slow, and in order to obtain rapid and efficient action, a temperature of about 260° F. is desirable; but to obtain this temperature, the lees must contain much caustic alkali, or be heated under pressure, which latter operation is generally very inconvenient. The lees being raised to the necessary temperature and in a state of ebullition, which agitation facilitates the reaction, the sesquioxide of iron is added, and after the reaction appears to be complete, the lees are cooled down as quickly as possible to 80° or 90° F. Should the lees be stronger than 45° or 50° of Twaddell, they are diluted to about those strengths as rapidly, and with water as cold as possible, as the sudden change of temperature greatly facilitates the separation of the precipitate.

The metallic sulphides are separated from the lees as perfectly as possible, and the lees again agitated, but in contact with the air; the agitation, in the first instance, is principally of service when the lees are not boiled with the metallic salts or oxides, but the agitation and simultaneous exposure to atmospheric air are equally applicable in all cases, after as complete a separation as possible of the metallic sulphides has been effected, in order that their subsequent oxidation may be prevented.

The agitation which a man is able to effect, in any of the vessels ordinarily employed in the manufacture of alkali, is of very little service. The amount of agitation desirable, is at least equal to the agitation found requisite to obtain butter from milk or cream by churning, and the more rapidly the mixture of lees and metallic salts and oxides are agitated after admixture, the better, especially with the sesquioxide and sesquisalts of iron, which are otherwise apt to be partially reduced and to discolour the lees. The agitation should be continued until no more of the alkaline sulphide is decomposed. The metallic sulphides are then separated from the cool lees as described. The lees separated are again agitated with about the same violence, but also with free exposure to the air, and they should be as hot as possible during this operation. Any iron, or manganese, or alkaline or other sulphides in the solution, are thus oxidized, and the greater proportion, if not the whole, of the metallic oxides are rendered insoluble, and may be removed when the lees are not stronger than about 50° Twaddell, and are allowed to cool below 130° F. Any convenient mode of forcing or sucking air through the hot lees may be adopted, or a current of air may be made to pass through what may be termed " the churning vessel;" but Mr. Deacon gives the preference to Kneller's plan.

This exposure to air should be continued until no discoloration is produced by the addition of a salt of lead to a portion of the lees; and the precipitates produced must of course be separated as described.

Another part of his invention depends on the reaction of alkaline sulphides on ferricyanides, by which a precipitate containing sulphur and iron is formed. This reaction is hastened by the application of a moderate degree of heat not exceeding 130°, but a temperature of about 100° is best. The time necessary to effect these changes appears to be capricious, and often varies from twenty-four hours to a week or more. After a time, however, the reaction is complete, and a lye is obtained, which yields perfectly white crystals and alkaline salts; but the lye will contain nearly as much alkaline sulphides as at first, this method of purification not being intended for their removal.

Caustic lime or magnesia must now be added to the metallic sulphides produced, before washing out all the alkaline lees, as the oxidation of these sulphides is retarded by their being associated with an alkali or alkaline earth. In practice, an equivalent of magnesia or lime to an equivalent of the metallic sulphide will be found sufficient for such purpose. The lime or magnesia should be added before the washings are reduced below 15° Twaddell, but the lime or magnesia should not be added before the precipitate of metallic sulphides is produced, or a combination of the metallic oxides with the earths may be produced, which will prevent the reaction. If the lime or magnesia is added before, it will often assist the precipitate to subside.

The crude soda-ash, if washed with a saturated solution of carbonate of soda, is deprived of its caustic soda and ferrocyanide of sodium, which dissolve out first, and by continuing the operation, the common salt and sulphate of soda are at last removed. Ralston has proposed to perform this operation in the wrought iron tanks already described, and states that a soda-ash or white alkali may be obtained, containing 56 per cent. of soda.

M. Lestelle employs the following arrangement for carbonating the tank-liquors, which are run into a wrought iron vat, about 15 feet long, 4 feet wide, and 2½ feet deep. A grooved sheet of lead is supported near the bottom by metal rods, and as the carbonic acid arrives through the pipe T, it is diffused and retained among the liquor by this simple arrangement.

FIG. 183.

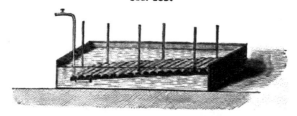

Margueritte has proposed to employ ammonia for obtaining the carbonate of soda in a pure state from these black-ash liquors. He brings this gas into contact with the liquors until they are saturated, when the carbonate of soda is precipitated in a crystalline and anhydrous state. He modifies this plan by washing the crude soda-ash with a saturated solution of ammonia, which he states will dissolve out the common salt, sulphate of soda, and sulphide of sodium, and leave the carbonate of soda behind, which when dried, is very pure.

Soda Crystals.

The soda-ash, or white alkali, obtained by any of the plans just described, is dissolved in the following manner, in sufficient quantity at one time, to produce from 2 to 4 tons of crystals. The dissolving pan is made of iron, of great depth, and filled with water to about two-thirds of its contents, which is heated nearly to the boiling point, when the alkali is thrown in, while a supply of steam keeps up the temperature until the solution has a specific gravity of 45° Twaddell, or 1·225 sp. gr.

The plan adopted in France is somewhat different. The ash is dissolved in a conical boiler A, fitted with a steam pipe C, and in which a small vessel D is suspended by pulleys. The boiler is filled to three-fourths with water from the pipe B, and the vessel D, containing the ash, is lowered into the water just below the surface. A jet of steam is admitted, which rapidly raises the temperature, and as the ash dissolves with an increase of volume of the liquid, the vessel D is raised from time to time. The strong liquor falls to the bottom of the boiler, while the water or weaker

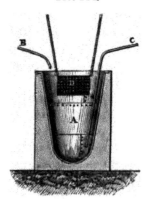

FIG. 184.

liquor rises and dissolves fresh portions of the ash until the whole acquires a density of from 1·27 to 1·30 sp. gr.

The solution is then filtered through bags of unbleached calico in some manufactories, or syphoned off into large iron settlers to deposit any insoluble matters, which is accomplished in about twelve hours. It is then run into boilers placed over naked fires, and maintained at the boiling point until it stands at the same strength as in the French plan. It is again run into settlers, and a small quantity of lime or bleaching powder, or both, mixed with water, is thrown in to assist in throwing down any remaining impurities and clarifying the liquor, which is allowed to settle, until the temperature falls to 92° F., when it is run into the crystallising coolers shown in Fig. 185.

FIG. 185.

The surface of the liquor is covered with bars of wood, or bars of iron are laid across the cones, which retain the crust of crystals adhering to them, and sustain the mass of crystals that grow beneath, termed *points* in the trade.

After the lapse of five to ten days, depending upon the season, the crystallisation is complete. The mother-liquor is run off and boiled down, producing a white alkali, but weak in strength. The crystals are allowed to remain a short time in the cones, then removed and placed on benches to drain and dry, when they are ready for packing.

In the crystallisation of carbonate of soda, simple as the process appears, great attention ought to be paid to the strength of the liquors, their freedom from sediment, the absence of sulphate,

19*

and the rate of cooling, in order to obtain good *points*, and to give to the solid mass of crystals what is technically called a *straight grain*. This latter character is of some importance to the retailer, as it enables him easily to divide the masses without loss.

The following woodcut of a beautiful block of crystals exhibited by Messrs. Chance, Brothers, & Co., in the International Exhibition of 1862, illustrates the above observations.

FIG. 186.

This crystallised salt has been analysed by Brown, with the following results, and the theoretical composition of the salt, $NaO.CO^2 + 10$ aq., with which it corresponds, is added.

	Analysis.		Theory.
	1	2	
Carbonate of soda . .	36·476	36·931	37·50
Sulphate of soda . . .	0·943	0·542	—
Chloride of sodium . .	0·424	0·314	—
Water	62·157	62·213	62·50
	100·000	100·000	100·00

When the water is expelled, the dry salt has the following composition :—

	1	2
Carbonate of soda	98·120	97·984
Sulphate of soda	1·076	1·124
Chloride of sodium	0·742	0·563
	99·938	99·671

Poggiale has determined the solubility of this salt in water at different temperatures, and has given the following table :—

100 parts of water at 32° F. dissolve 7·08 of carbonate of soda.

	,,	50	,,	16·66	,,
		86	,,	25·83	
	,,	86	,,	35·90	,,
	,,	219	,,	48·50	,,

In France, and sometimes in this country, the crystals are re-dissolved, in order to obtain an article of greater purity. The apparatus employed on the continent for this purpose, consists of a conical-shaped boiler, A, similar to the former, but heated over a naked fire, C, the flues, DD, from which pass round the boiler. The boiler is filled with crystals, and a small quantity of water is supplied from the pipe, B. Heat is now applied, and the whole speedily melts in the water of crystallisation, when the fire is withdrawn, and the boiler covered with a wooden lid. After standing for some time, the liquor is ladled into small coolers, about 18 inches in diameter, and set aside to crystallise, the crystallisation being completed in about eight days. The mother-liquor is run off, and the small cooler is placed in hot water, or in a leaden jacket and steam admitted. In either case, the crystals touching

FIG. 187.

the sides of the cooler, slightly melt, when the entire cake of crystals can be easily removed by the hands of the workman.

We have described in the previous section, Mr. Gossage's improvements for working the black-ash liquors, in which the portions of the salts first separated, consist almost entirely of crystals of carbonate of soda, but retain a portion of liquid caustic soda adhering to them, partly removable by washing. He has since patented a process for remedying this defect, by exposing them to the action of carbonic acid gas, or mixing them with bicarbonate or sesquicarbonate of soda.

The crystals obtained by the process above described, are composed of—

1 eq. soda .	. . . . .	32 or	22·22
1 carbonic acid .	. . .	22	15·28
10 water .	. . . .	90	62·50

$$NaO.CO^2 + 10 \text{ aq.} \quad . \quad 144 \quad 100·00$$

Their taste is sharp and slightly caustic, and when exposed to the air they effloresce, lose their water of crystallisation, and fall to powder, having the composition, $NaO.CO^2 + HO$. But if the efflorescence is very slow, they absorb carbonic acid; at the same time they lose part of their water of crystallisation, and form a salt having the composition $2NaO.3CO^2 + 4HO$.

When heated, they undergo the aqueous fusion, and lose all their water. The carbonate of soda is insoluble, but caustic soda is soluble in alcohol.

The solubility of the crystals in water depends on the temperature: thus at

Temperature.	Crystals dissolved by 100 parts of water.	Weight of water to dissolve 100 of crystal.
57° F.	60·4	165·7
97	133·0	12·0
100	1666·0	6·0
219	445·0	22·4

Thus, if a solution saturated with crystals at a temperature of 97° or 100° F. is heated up to 212° or 219° F., a large proportion of the salt precipitates, proving the importance of attention to the proper temperature on the part of the manufacturer.

The solution which has been saturated at a temperature of 97° F.

can be preserved in the liquid state for several days in a close tube, and even agitated without crystallisation. When the aqueous solution of the crystals is evaporated at a temperature of 212° or 219° F., the small crystals which precipitate, are composed of $NaO.CO^2 + HO$, and on exposure to a moist atmosphere they effloresce, absorbing 4 equivalents of water.

The carbonate which precipitates, either by the evaporation of the solution at 97° or by fusing the ordinary crystals at the same temperature, contains 5 equivalents of water, four of which are given off when the crystals effloresce in a dry atmosphere.

This completes what may be termed the ordinary operations of an alkali manufactory. The following table, exhibiting the composition of the salt and products of the soda manufacture by Mr. Brown, gives an excellent idea of the various chemical changes which occur.

Table exhibiting the composition of Salt and products of the Soda Manufacture.

Constituents.	Salt.	Sulphate of soda.	Soda-ball.	Waste.	Soda-ash before carbonating.	Soda-ash.	Carb. of soda before carbonating	Carbonate of soda.	Refined ash.	Crystallised carbonate of soda.	Carbonate of soda theory.	Ash made from cryst. carb. of soda.
Chloride of sodium	931·615	10·956	1·913	—	3·917	3·665	0·635	3·254	3·310	0·369	—	0·652
„ potassium	trace	—	—	—	—	—	—	—	—	—	—	—
„ magnesium	1·066	—	—	—	—	—	—	—	—	—	—	—
Sulphate of lime	10·098	9·731	—	4·281	—	—	—	—	—	—	—	—
„ magnesia	1·348	2·893	—	—	—	—	—	—	—	—	—	—
Carbonate of lime	1·500	—	—	24·220	—	—	—	—	—	—	—	—
Water	54·573	—	0·700	2·100	—	—	—	—	—	62·135	62·501	—
Sulphate of soda	—	962·170	1·160	—	7·415	9·676	8·036	9·027	10·012	0·742	—	1·100
Sulphite of soda	—	—	—	—	2·182	1·126	1·174	0·386	0·742	—	—	—
Peroxide of iron	—	—	—	—	—	—	—	—	—	—	—	—
Free acid	—	2·300	—	5·716	—	—	—	—	—	—	—	—
Sand	—	8·850	4·285	5·746	—	—	—	—	—	—	—	—
Carbonate of soda	—	3·100	35·640	1·309	67·210	71·037	80·279	83·881	84·517	36·703	37·499	98·052
Caustic soda	—	—	0·609	—	—	—	—	—	—	—	—	—
„ lime	—	—	6·301	—	—	—	—	—	—	—	—	—
Hydrate of soda	—	—	—	—	15·252	12·181	3·318	0·897	0·280	—	—	—
Aluminate of soda	—	—	2·350	—	1·124	0·828	1·095	0·816	0·674	—	—	—
Sulphide of sodium	—	—	1·130	—	1·428	—	0·230	—	—	—	—	—
Hyposulphite of soda	—	—	—	—	trace	—	—	—	—	—	—	—
Silicate of soda	—	—	—	—	0·915	1·014	1·276	0·882	0·366	—	—	—
Ultramarine	—	—	0·295	—	—	—	—	—	—	—	—	—
3CaS.CaO	—	—	29·172	20·363	—	—	—	—	—	—	—	—
Sulphide of iron	—	—	4·917	—	—	—	—	—	—	—	—	—
Silicate of magnesia	—	—	3·744	—	—	—	—	—	—	—	—	—
Carbon	—	—	7·998	12·078	—	—	—	—	—	—	—	—
Hyposulphite of lime	—	—	—	trace	—	—	—	—	—	—	—	—
Hydrate of lime	—	—	—	5·582	—	—	—	—	—	—	—	—
Bisulphide of calcium	—	—	—	3·583	—	—	—	—	—	—	—	—
Sulphide of calcium	—	—	—	8·527	—	—	—	—	—	—	—	—
Insoluble matter	—	—	—	—	0·894	0·390	0·870	0·781	0·374	—	—	—

FIFTH OPERATION.

The production of Caustic Soda.

This valuable material is becoming one of the trade-products of the soda manufacture, and, as we have already noticed, occurs in larger or smaller quantities in the black-ash liquors, according to the proportions in which the sulphate of soda, lime, and coal are employed. It promises, in fact, to be the most important article of the soda-trade, as it is in this form that soda is used in bleaching and soap-boiling. It is now manufactured in considerable quantities, in a state of great purity, by the Jarrow Chemical Company, and other manufacturers.

Caustic Soda from Black-ash and Soda-ash.

The liquors obtained by lixiviating black-ash are diluted with water to a sp. gr. of 1·09 or 1·12, or soda-ash is dissolved in sufficient water to form a solution of the same density. The soda is brought to the caustic state by means of carefully prepared hydrate of lime, and it is important to attend to the strength of the liquors, as lime does not completely decompose a strong solution of carbonate of soda.

The operation is performed in an ordinary cylindrical boiler, cut in two lengthways. An iron agitator is fitted to this boiler and made to revolve by steam power. The boiler is filled with liquor up to a given level, and the milk of lime added from time to time during the agitation. The decomposition is finished in about thirty minutes, when the liquor is run into large iron settlers to allow the carbonate of lime to deposit.

This *lime-mud* is transferred to a boiler similar to that just described and agitated with water, the insoluble matter allowed to deposit, and the clear liquor run off to be used, either in dissolving fresh portions of soda-ash, or in diluting the black-ash liquor. In other works the lime-mud and washings are filtered in the following manner : A large iron vessel is fitted with a false perforated bottom, in which is placed first a layer of coke, then another layer of pebbles, and upon this a layer of sand. The washings are either allowed to drain through these layers, or the filtration is hastened by creating a vacuum below the false bottom by means of an air-pump and by injecting steam. When the solution has passed through, fresh water is run on the lime, and this is repeated until the soda has been removed.

The liquors are now concentrated in ordinary steam boilers, the steam which is generated being employed in driving the engine power of the works, according to the patent of Messrs. Roberts, Dale, and Co. When the liquor has acquired a specific gravity of 1·24 and a temperature of 230° F., it is run into the boat-pan, when the subsequent treatment is similar to that which will be explained in detail in the following section.

Composition of the Lime-mud.

It has been found impossible to separate all the soda from this deposit, and it is generally believed that it is chemically united to the lime. Kynaston states that from 5 to 6 per cent. of soda remains in the wet mud, and Mr. Tait, to whose valuable paper we are indebted for much of our information on this subject, remarks that even in a carefully conducted laboratory experiment, he has found from 2¼ to 3 per cent.

This mud, when submitted to a dull red heat, parts with the soda, which can then be removed by washing with water. It is generally allowed to dry in a heap and afterwards used, instead of limestone, in the balling process. Muspratt suggests that, after being dried, it should be heated to expel the carbonic acid, and used again in the caustic state to prepare fresh lye, by which the soda present would be saved.

Filtration of the Lime-mud.

The principal drawback to this mode of producing caustic soda, is the loss of soda and the expense of managing the precipitated carbonate of lime. Mr. Ralston has proposed to employ the following arrangement for washing and filtering this precipitate. He builds a shallow tank a, Fig.188, of stone, water-tight, and raised from the floor. The bottom inclines towards one spot, from which a pipe b leads away the filtrate. A series of joints cc are ranged side by side on the bottom of the tank, with openings on their lower edge to admit the flow of the filtrate to b, and across these joints a number of supports dd are placed for carrying the porous back of which the filtering bed is formed; they are laid in cement to make the joints water-tight.

The precipitated carbonate of lime is run on to this bed and allowed to filter for some time, after which a few inches of water are gently run on the pasty mass, so as not to disturb the particles;

FIG. 188.

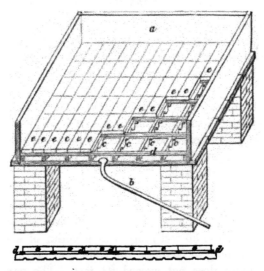

when this is properly managed, the water forces the strong lye out of the precipitate without much diminution in its strength.

We have seen an arrangement of this kind at work for filtering very flocculent precipitates; but, instead of porous bricks, the bed was formed of large stones covered with layers of gravel and fine sand, and it appeared to answer.

In our chapter on tartaric acid, full details will be found of some excellent arrangements for filtering similar precipitates.

Pure Caustic Soda.

Dr. Pauli describes the following method of making pure caustic soda at the Union Alkali Works, in Lancashire:—

Three tons of commercial caustic soda, containing an excess of water, alumina, and all the impurities which commonly occur in this substance, are fused in a cast-iron pot. During the evaporation nearly all the carbonate, and by far the larger quantity of the other salts, separate out on the surface as a scum, and can be easily removed. The liquid mass is then heated to dull redness, and kept at that temperature during the night. In the morning the mass appears perfectly transparent, the sides and bottom of the vessel being coated with cauliflower-shaped masses of crystals, consisting of silicate of alumina, with chloride of sodium, sulphate of sodium, and a little lime. The clear fused liquid is ladled off from these crystals, and, when cooled, is ready for use.

Caustic Soda from the Black-ash Liquors.

These liquors have a sp. gr. of about 1·30 and consist chiefly of hydrate of soda, with some carbonate and sulphate of soda, chloride of sodium, and a small quantity of ferrocyanide of sodium ; they owe their colour to a red compound of sulphide of sodium and iron.

In some manufactories, chloride of lime is added to these liquors to oxidize the sulphides and remove the iron, but nitrate of soda is generally preferred for this purpose. When the latter salt is used, the quantity necessary varies from ¾ of a cwt. to 1¼ cwts. per ton of caustic soda.

The operation is conducted in what is termed a *fishing* or *boat pan* from its peculiar shape. This pan is the invention of Mr. D. Gamble, and is made of ¼ inch iron plates riveted together. It is generally made 30 feet long with a width of about 8 feet. The sides are sloping to facilitate the collection of the salts, and its greatest depth does not exceed 2½ feet. The pan is set in brickwork and heated by a fire below, as shown in the woodcut.

FIG. 189.

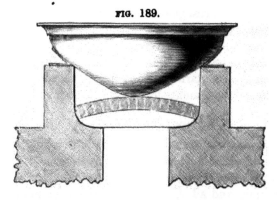

Mr. Gamble also introduced the use of centrifugal machines for separating the mother-liquors from the salts obtained from the black-ash liquors.

The liquors are concentrated in this pan up to a sp. gr. of 1·40, at a temperature of 260° to 270° F., and during the boiling some nitrate of soda is added. The salts which separate in this pan are fished out by a perforated ladle. The concentrated liquor is then run into settlers for a short time, when some salts are deposited during the cooling. The salts which separate, consist chiefly of carbonate of soda, with some sulphate of soda, and a small quantity of chloride and ferrocyanide of sodium : they are converted into soda-ash or into caustic soda by means of lime.

When the caustic liquor has settled, it is run into a hemispherical cast-iron pot. This pot has a diameter of about 5 feet at the top, to which is fitted a cast-iron ring about 2½ feet high, and the depth of the pot, including the ring, is 5 feet to 5½ feet. It is set in brickwork and heated by a fire below, or by the waste heat of another, and similar, pot used in the finishing process.

Nitrate of soda is added to the liquor in this pot, and the whole boiled from twelve to twenty-four hours, during which a further deposit of salts takes place, which are fished out by the workman. After being allowed to settle for some time, and when no more salts separate, the liquor is transferred to another pot, called the *finisher*, similar to the former, but always heated by its own fire.

The liquor is boiled in this pot until it contains 60 per cent. of soda, during which more nitrate of soda is added, while a thick scum, which forms on the surface, is skimmed off by the workman. When the concentration is complete the fire is slackened or withdrawn, and when the liquor has settled, it is run off into iron casks, where it becomes solid as it cools.

We will now describe the other processes which have been introduced for obtaining the caustic soda existing in the black-ash liquors.

We have already given an account of Mr. Gossage's improvements in treating black-ash liquors for the separation of the cyanogen and sulphur compounds, and we now add the account of his process for separating caustic soda from black-ash liquors, so as to obtain the caustic soda, in a useful form, without having recourse to the action of lime on carbonate of soda. He uses black-ash liquors which have been previously freed from sulphide of iron, and converts any hyposulphite of soda, which may be present in these purified black-ash liquors, into sulphite or sulphate of soda by oxidation, by the gradual addition of hypochlorite of lime. The black-ash liquors are then concentrated by evaporation, so as to separate the greater part of the carbonate of soda in the form of salts, more or less impregnated with caustic soda, and ferrocyanide of sodium, while the mother-liquors contain a large proportion of caustic soda, and some carbonate of soda, and ferrocyanide of sodium. The salts thus obtained are washed with water, or with a solution of carbonate of soda, by the filtration pre-

viously described, so as to obtain salts sufficiently free from caustic
soda and ferrocyanide, to be suitable for the production of white
soda-ash of good quality, and the liquors are added to the mother-
liquors previously mentioned. These mixed liquors are concentrated, either in the ordinary pans or by a process to be described
below, until nearly the whole of the carbonate of soda and
ferrocyanide of sodium are deposited as salts; the solution or
mother-liquor remaining, which has a specific gravity of about
1·450, contains scarcely any saline matter in solution, except
caustic soda. In this state, the caustic soda is particularly
suitable for the manufacture of soap, and for many other applications. By continuing the concentration of the solution of
caustic soda, in cast iron pans, it can be so far deprived of water,
as to become solid hydrate of soda when allowed to cool. The
salts obtained by the concentration of the mixed liquors, are
more or less impregnated with ferrocyanide of sodium and caustic
soda, and they are washed with water, or with a solution of carbonate of soda, by filtration, so as to separate the ferrocyanide of
sodium and caustic soda. The whole of the ferrocyanide is not
to be separated, but the washed salts are transferred from the
vats to a concentrating apparatus, in which black-ash liquors are
concentrated for the production of salts. If these liquors contain
more than 5 parts of carbonate of soda to 1 part of ferrocyanide
of sodium, they are concentrated, so as to obtain salts with a
larger proportion of ferrocyanide; but if the liquors contain less
than 5 parts of carbonate of soda to 1 part of ferrocyanide of
sodium, the carbonate of soda is converted into caustic soda by
means of lime, and the solution thus obtained is concentrated
until nearly the whole of the ferrocyanide is deposited as salt.

Mr. Gossage's plan of concentration consists in exposing the black-
ash liquors in films or sheets to the evaporative action of heated air.
An oblong vertical tower, about 6 feet long, 4 feet wide, and 20 feet
high, made of iron plates, is fitted inside with a number of iron
shelves or partitions, placed at an inclination of about 15°, so that
the liquor will flow down. One of these shelves is attached to
one end and to both sides of the tower, but does not extend to the
other end, thus leaving an open space of about 6 inches. The
next shelf or partition below this, is attached to the opposite end,
about 4 inches below the lower end of the upper shelf and to the
two sides of the tower, and is placed at a similar inclination, but
does not extend to the other end, leaving an open space of about

6 inches, and all the other shelves are similarly arranged. A supply of liquor being introduced on the upper part of the first shelf, it flows down to the next, and at last collects at the bottom of the tower. A current of hot air is drawn through the tower by means of a chimney draft, and thus comes into contact with the thin sheets of liquor as they flow over the shelves.

The hot concentrated liquor is conveyed from the bottom of the tower to iron pans, through which it flows, and becoming cooled it deposits salts, and the remaining liquors are pumped up to pass again through the tower. The liquors are passed through the tower until they are nearly deprived of carbonate of soda, when they may be used as a caustic soda lye.

Mr. Gossage has since introduced some further improvements for the same object. He employs different proportions of the materials for his balling mixture, so as to increase the percentage of the caustic soda: viz.,

	Usual mixture.	Mr. G.'s proportions.
Sulphate of soda	3 parts	3
Carbonate of lime	3 „	3
Small coal	1½ to 2 „	3

and adds a portion of the small coal to the mixture, as the fluxing proceeds.

He also adds a metallic oxide to the black-ash liquors when in a state of ebullition, so as to convert the sulphide of sodium into caustic soda, somewhat in the manner previously explained in the description of Mr. Deacon's plans for purifying these liquors.

Mr. Gossage has also proposed to convert the carbonate of soda obtained from the black-ash liquors, in the state of small crystals, and charged with caustic soda lye, into caustic soda by means of lime; and by the plan of concentration already described, he separates the sulphate of soda and other salts from the caustic soda.

Mr. Ralston modifies the treatment of the black-ash liquors so as to obtain a caustic soda nearly free from iron, and of great strength. He evaporates the liquors and separates the salts in the usual way; but instead of keeping the heat as low as possible, in the final evaporation of the lye, to prevent the discoloration of the product, he raises the temperature, until a state of dry fusion is reached, when the oxide of iron will be precipitated to the bottom, and the clear alkali can be ladled off. His plan is applicable whether the red liquors are used, or a lye obtained by

decomposing alkali with quick lime in the usual way. When
the stage of dry fusion is reached, he also forces jets of air
through the liquid to oxidise the sulphides. He pours the caustic
soda into casks or on to iron plates, so as to form slabs about half
an inch thick, which, when cold, he grinds to powder under edge-
stones.

He states that when the heat is raised in the final operation,
the iron separates as *hydrated* peroxide, but that when the dry
fusion commences, the oxide of iron becomes *anhydrous*, in which
state it settles to the bottom of the pan.

Mr. Ordway has described the following process for treating
black-ash liquors to obtain pure caustic soda. A liquor of 1·120
sp. gr. is raised to the boiling point, and heated with a milk of
lime containing 1 of lime to 6 of water, and the clear lye which
is run off is boiled up to about 1·470, and then removed to a
metal pan fixed over an open fire. A quantity of finely divided
metallic iron is thrown into the reddish-coloured lye, and the
evaporation continued, with constant agitation. The proportion
of iron must be so large, that the mass which remains must not
melt at a dull red heat. Large quantities of ammonia are evolved
during the drying, most likely arising from the decomposition
of the cyanogen-compounds. When the water is expelled, the mass
rapidly absorbs oxygen, and the colour passes from a black or dark
brown to a red. The fire is withdrawn, and the mass is occasionally stirred during the next two hours. It is then heated with
sufficient hot water to produce a lye of about 1·273 sp. gr., which
is allowed to settle and is then evaporated. In a short time, salts
begin to deposit, composed of sulphate, sulphite, and carbonate
of soda, which continue until the liquors have a specific gravity of
1·296, and as the density increases some common salt appears;
but when it has reached 1·346 sp. gr. no more salts are obtained.
Fresh supplies of liquor from other pans are run in, and the concentration continued until the strength of the lye reaches 1·428
sp. gr. The pan must be of sufficient capacity to allow room for
the frothing which takes place towards the conclusion of the process. The fire is now urged until the soda melts, and if it has a
reddish colour, arising from imperfect calcination, about 1 per
cent. of nitre is added to oxidise the sulphide of sodium. When
the dry fusion is fully reached, the mass is ladled into moulds
and allowed to solidify.

Caustic Soda from Sulphide of Sodium, &c.

Mr. Arrott employs the oxide of iron or manganese with sulphide of sodium. He prepares sulphide of sodium by fusing 4 parts of salt-cake and 1 part of coal in a cupola furnace, and adds the liquor obtained by lixiviating this product, to the metallic oxides above named, as long as they take up sulphur, by which he produces a caustic soda-lye.

Mr. Hunt produces sulphide of sodium by roasting sulphate of soda with coal or coke, lixiviating the mass when cold, and boiling the liquor with oxide of zinc or black oxide of copper.

Mr. Arrott also produces caustic soda by mixing an equal weight of alkali and peroxide of iron, or oxide of manganese, and exposing it for two or three hours to a full red heat in an ordinary iron gas-retort. He treats the mixture, after it is drawn from the retort, with hot water, which dissolves out caustic soda and leaves the metallic oxide behind in an insoluble state.

This process of Arrott's appears to be one of the most convenient for producing caustic soda.

Mr. Gossage has also patented the manufacture of caustic soda from silicate of soda, which he prepares by fusing soda-ash of 54 per cent. with its own weight of clean sand. He dissolves this soluble glass in water and adds to the solution 8 parts of slaked lime for every 100 parts of silica. The mixed solution of silicate and lime is well boiled and then allowed to remain quiet, to permit the silicate of lime to deposit, when the caustic soda lye is drawn off. Mr. Gossage proposes to employ the silicate of lime as a manure.

Wöhler has recently proposed the heating of nitrate of soda and peroxide of manganese in close vessels, in order to obtain pure caustic soda. No manganate of soda is formed, because, as Wöhler suggests, the temperature at which the nitrate is decomposed is too low for the formation of manganic acid.

When nitrate of soda is heated with sand, part of the soda combines with the silica, but the greater part, according to Malcolm, may be obtained in the form of caustic soda. Bishop Watson, in his Essays, mentions this process by which nitrous acid may be obtained; it has recently been proposed by Malcolm in his provisional specification, for the manufacture of nitric acid and caustic soda.

Commercial Caustic Soda.

There are three qualities of solid caustic soda, distinguished by their relative freedom from sulphides, and by the comparative smallness of the precipitate which they produce when dissolved in water. These qualities vary in strength up to 75 per cent., when obtained from the red liquors by Mr. Gossage's process. The common quality, containing 60 per cent, gives a coloured precipitate when dissolved in water. The stronger qualities are distinguished by possessing this property in a less degree, or by its entire absence. The best quality sometimes has a light blue colour, due, it is supposed, to the presence of a trace of manganate of soda, and proving the absence of sulphide of sodium or oxide of iron ; the sulphide would destroy the blue colour, and any oxide of iron becoming peroxidised, would gradually communicate a dirty olive-green tint.

The caustic soda is sometimes sold in the liquid state, under the name of *Soaper's lye*, which is prepared by concentrating the liquid up to a specific gravity of about 1·350, or 70° Twaddell.

It is also sold in the solid form, mixed with two-thirds of its weight of common salt, under the name of *Factitious Potashes*, and the red colour of the American potashes is communicated to it, by adding 1 per cent. of sulphate of copper, which is reduced to the state of red oxide by stirring the fused mass with a wooden pole.

Crystallized Caustic Soda.

Messrs. Gaskell, Deacon, & Co., to whom we are indebted for some of the above information, also manufacture another beautiful quality of caustic soda, viz, crystals of caustic soda, the process for making which they have patented. The concentration of the caustic soda-liquors is continued, until they are nearly strong enough to solidify on cooling, and they are then allowed to cool down to 120° F., after which they are run into suitable apparatus, kept at a gentle heat and protected from the air. The mother-liquors drain off, and the beautiful crystallized caustic soda is left behind.

These crystals are composed of—.

Soda	50·5
Hyposulphite of soda	0·5
Chloride of sodium	1·8
Water	47·2
	100·0

agreeing with the formula

$$3NaO.10HO \text{ or } 3NaHO^2.7HO.$$

Chemistry of this Process.

Referring the reader to the description of the process for making caustic soda from the red or tank liquors, we have further to observe that when nitrate of soda is added, the sulphide of sodium is oxidised, and the sulphide of iron is converted into peroxide, which is precipitated with the salts, and communicates the red colour they possess. It is obvious that the nitrate of soda must not be added until the temperature is sufficiently high, as some of it would escape decomposition, and fall down along with the salts.

The oxidation proceeds slowly in the boat-pan, where the temperature never rises higher than 270° F., the sulphide being converted into sulphate, and the nitrate into nitrite of soda. In the next vessel, where the heat reaches 290° F., the decomposition is more rapid, attended with a large evolution of ammonia. This ammonia arises partly from the decomposition of the cyanides, and partly from that of the nitrate of soda. Dr. Pauli, who has carefully examined this question, gives the following explanation of the change:

$$2NaS + NaO.NO^5 + 4HO = 2(NaO.SO^3) + NaO.HO + NH^3.$$

In the finishing pot, where the temperature rises above 310° F., free nitrogen is disengaged, as shown by the equation

$$5NaS + 4(NaO.NO^5) + 4HO = 5(NaO.SO^3) + 4(NaO.HO) + 4N.$$

Dr. Muspratt, looking to the sulphites and hyposulphites, &c., which are also present, represents the decomposition as follows:—

$$2NaS + NaO.S^2O^2 + NaO.NO^5 + 3HO = 4(NaO.SO^3) + NH^3,$$
$$\text{Or, } NaS + NaO.S^2O^2 + NaO.SO^2 + NaO.NO^5 + 4HO =$$
$$4(NaO.SO^3) + NH^4O.*$$

Neither of these series of equations represents all the changes which take place in so complex a solution, but it is highly probable that such decompositions do occur along with others, the principal oxidation being that represented in the equation of Dr. Pauli.

The separation of the iron in the form of peroxide, takes place chiefly in the settlers and coolers, but a portion is removed in the

---

* This equation is incorrect, the two sides not agreeing: it should perhaps be:

$$NaS + NaO.S^2O^2 + NaO.SO^2 + NaO.NO^5 + 4HO + O = 4(NaO.SO^3) +$$
$$NH^3 + HO.$$

scum which rises to the surface in the finishing pots. Another portion, however, remains dissolved in combination with the soda. Mr. Tait has found that precipitated peroxide of iron is dissolved by caustic soda, and even after dilution with water, a small quantity still remained in solution. He believes that part of it is merely suspended, owing to the finely divided condition of the oxide, a supposition confirmed by the fact, that when a clear bright solution of caustic soda is set aside in a stoppered bottle, the bottom bocomes coated in a few days with a deposit of peroxide of iron. There can also be no doubt that a portion of the iron exists in the caustic soda as ferric acid, and that the ferrate of soda is decomposed when the soda is dissolved in water.

Some of the caustic soda has also a delicate blue or greenish tinge, arising from the presence of manganese, which Mr. Murphy traces to the limestone used in the balling process.

This manganese and iron may also arise from the action of the caustic soda on the rust, or even the metallic iron of the boilers and pots, as suggested by Mr. Tait, for all these vessels undergo great wear and tear in these operations.

SIXTH OPERATION.

The Manufacture of Sesqui- and Bicarbonate of Soda.

The sesquicarbonate of soda occurs in nature, either in solution or in the solid form. The mineral waters of Vichy and Vals in France, of Bilin in Hungary, of Carlsbad in Bohemia, the Geyser springs in Iceland and in Teneriffe, and the Solfatara of Gaudeloupe, all contain this salt. It is found at the bottom of a rocky mountain in the province of Sukena, near Tripoli, and is called by the natives *Trona*, from which the terms *Nitrum* and *Natron* are probably derived. It forms thin crusts on the surface of the earth, and the walls of a fort, Cassar or Qaser, now in ruins, are said to be built of it.

It is also obtained from the bottom of a lake at Lagunillas, near Merida, in Venezuela, where it is called by the Indians *Urao*. It is collected every two years by the natives, who, by aid of a pole, plunge into the lake, separate the bed of earth which covers the mineral, break the urao, and rise with it to the surface of the water. It is then placed in boats and conveyed to a warehouse, and afterwards dried in the sun.

The Egyptian natron, deposited on the sides of the lakes to . the west of the Delta of Egypt, most probably contains the carbonate of soda in this form.

The plains on the shores of the Black and Caspian Seas, Persia, Arabia, India, Thibet, China, and Siberia, all yield this variety of carbonate of soda.

The following analyses show the composition of some of these natural products:—

Trona from Egypt, by

	Remy.	Fleischer.
Sesquicarbonate of soda . .	47·29	26·53
Carbonate of soda	18·43	2·37
Sulphate of soda	2·15	
Chloride of sodium . . .	8·16	
Silicate of soda	0·29	
Bicarbonate of lime . . .	0·20	
Bicarbonate of magnesia . .		71·10
Borate of soda	} traces	
Organic matter		
Water	19·67	
Insoluble matter	4·11	
	100·29	100·00

	Urao, Lagunilla.	Trona.				
		India.	Sukenna.	Agglomerated Crystals.	Radiating. Fibres.	Walls of Cassar.
Soda	41·22	32·99	37·00	37·428	38·62	32·67
Carbonic acid	39·00	34·01	38·00	39·274	40·13	33·15
Water	18·80	31·00	22·50	23·287	21·24	20·55
Sulphate of soda	—	—	2·50	—	—	1·96
Chloride of sodium	—	2·00	—	—	—	3·95
Earthy matters	0·98	—	—	—	—	2·33
	100·00	100·00	100·00	99·989	99·99	94·61
	Boussingault.	Reynolds.	Klaproth.		Beudant.	

Berthollet noticed that when chloride of sodium and carbonate of lime were found mixed, there was always an efflorescence of sesquicarbonate of soda, whence it has been supposed that its origin in nature is due to this reaction; but Dumas believes that the sulphate of soda is reduced by the organic matter which accompanies this salt and the chloride of sodium, to sulphide of sodium, which is converted into sesquicarbonate by the carbonic acid held in solution by the water.

Dufrenoy, on the contrary, does not believe that the trona found in the lakes in Fez, on the borders of the Great African desert, has been formed by a double decomposition, but that it has its origin from mineral sources, and is brought directly to these lakes, the waters of which keep it in solution. He adds, in support of this opinion, that MM. Boussingault and De la Rive find the same salt in Columbia in tertiary deposits.

Phillips and Rose state that it may be formed by boiling down and cooling an aqueous solution of bicarbonate.

Hermann found that when the carbonate and bicarbonate of soda are mixed and exposed to the air, this salt effloresces out on the surface. It is also formed when these two salts are heated to the point of aqueous fusion, but the most simple plan is to expose the bicarbonate to a moderate heat.

It is unalterable in the air, of which the ruined fort of Cassar is a remarkable illustration, and it is more soluble in water than the bicarbonate. It was formerly imported into this country and sold under the name of *tronse*.

Bicarbonate of Soda

is now a large article of commerce, and is principally manufactured on the Tyne.

It was formerly prepared by passing carbonic acid through a solution of the carbonate of soda, but this plan, being inconvenient and expensive, is now abandoned.

Where it is expensive to manufacture carbonic acid, that produced during fermentation is employed for the purpose, and in other localities, as at Vichy, the gas from the natural springs is made available in the following manner :—

A spring is surrounded by brick-work, in which an inverted bell-shaped receiver B is placed. An overflow pipe allows the

FIG. 190.

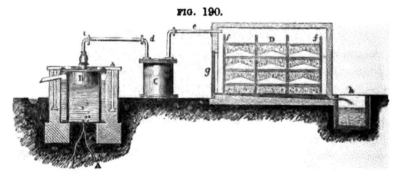

surplus water to escape, while the gas collects in the upper part of the receiver. It is allowed to escape through a pipe *cd*, and is washed in a small vessel C, whence it is conveyed through a pipe *e*, furnished with a stopcock, to a chamber or cistern D filled with carbonate of soda.

The crystals of soda are crushed and moistened with water, and spread over cloth to a depth of 2 or 3 inches. Each cloth is attached to a framework, which is supported between the uprights, and arranged in shelves inside the chamber D. As the water is expelled from the crystals, it is prevented from falling on the lower shelves by a roofing, as it were, of boards, which carry it to the two sides, where it escapes into a cistern, h, on the outside. Several chambers are connected together through which the gas passes, so as to prevent any loss. Sesquicarbonate is first formed, and then the bicarbonate, which retains only one equivalent of water, the other 9 eq. escaping, partly into the cistern H, while a portion remains behind in the interstices of the bicarbonate.

This damp bicarbonate is dried in similar chambers by means of carbonic acid from the same source, but previously deprived of moisture and heated up to about 100° F.

In some localities this manufacture is combined with the preparation of Epsom salts. Magnesian limestone is ground to powder, mixed with water, and introduced into a metal vessel A, lined with lead, through the man-hole A. An agitator B, isset in motion by a belt working round a sheave C, which can be stopped by shifting the belt on to a loose sheave C'. This keeps the limestone in suspension in the water, while chamber sulphuric acid is admitted through the bent pipe DD', the supply

FIG. 191.

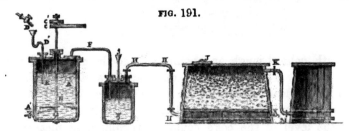

being regulated by the cock D'. The carbonic acid which is disengaged, escapes through the pipe FF to a vessel G, half full of water, where it is washed, and finally passes through another pipe HHH, to the bottom of the wooden cistern I, filled with crystals of soda, resting on a false bottom pierced with holes, to allow the water to escape by a cock, M.

The progress of the conversion of the crystals is examined, from time to time, through a manhole, J.

The excess of carbonic acid passes into a second cistern in connection with the first, and fitted up in the same manner.

The sulphates of magnesia and lime are withdrawn from the metal cistern A by a cock A'', and treated in the usual way for the manufacture of sulphate of magnesia.

During the famine in Ireland in 1847, when the demand for bicarbonate of soda became so urgent to mix with the Indian meal in baking the bread or cakes made with this flour, the supply of muriatic acid in some works was inadequate for the production of sufficient carbonic acid for the manufacture of larger quantities of this salt. In this emergency, one of us superintended the manufacture of carbonic acid from coke, with which large quantities of bicarbonate of soda were made. The arrangements were as follows:

A close furnace, A, was lined with fire-brick, and well bound with iron. The interior space, B, shown by dotted lines, was

FIG. 192.

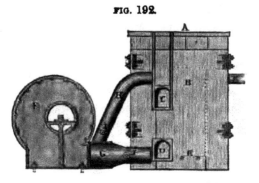

filled with coke through an opening C, and the ashes, &c., withdrawn at D. The coke rested on grate-bars E, and air was forced in by means of a fan-blast F. The pipes G and H were fitted with valves at I and J, so that the supply of air below or above the level of the coke could be regulated at pleasure, to convert any carbonic oxide that might be formed, into carbonic acid. The hot carbonic acid gas passed away through the iron pipe K, which was in a U form, to present sufficient cooling surface to the air. Just at the entrance of the cisterns containing the crystals of soda, two or three screens of fine wire-gauze were placed, which retained the dust carried away mechanically by the current. The other arrangements were the same as usual, except that the escape-pipe for the exit of the nitrogen, was fitted with a valve moderately loaded to increase the pressure inside, and thus facilitate the absorption of the carbonic acid.

Another method of producing carbonic acid has attracted attention, and is equally applicable to the manufacture of this salt. Meschelynk and Lionnet, and Dauglish, have published some observations on the process, and Blair has patented the following admirable plan for obtaining carbonic acid by this method. Blair's apparatus consists of fire-clay retorts, or pipes, 7 feet long by 15 inches diameter, open at both ends, to which metal mouth-pieces are attached, similar to those in coal gas works. Each mouth-piece is furnished with a metal branch, 2 inches in diameter, the one at the back for the admission of steam, and the front one for the escape of the gas and vapour. These retorts are built in ovens, and the fuel which is used for bringing the limestone up to a red heat, also heats a set of iron pipes imbedded in the brickwork between the ovens. The retorts are charged with 1½ to 2 cwts. of limestone broken small, and the doors luted with slacked lime. As soon as they are red hot, steam is admitted through the pipes between the ovens, at a pressure of about 10 or 12 lbs., when the carbonic acid is evolved. At the end of four hours, the charge is pushed towards the back end of the retort, and a fresh charge introduced. At the expiration of four hours, the first charge is removed, the second pushed back, and a third introduced. When necessary, the aqueous vapour may be removed by passing the products through ordinary coal-gas condensers.

The process now generally adopted, said to have been proposed by Schöffer or Smith, consists in decomposing chalk or limestone by means of hydrochloric acid. The apparatus consists of a large stone or brick cistern, laid in clay, and well bedded, or of metal tanks, lined with fire-bricks, and laid in suitable materials, as shown in Figure 193, EEE. They are filled with the chalk through a manhole, and the weak muriatic acid, as it flows from the condensers, is admitted by the pipe A, which descends to the bottom of the cistern, so that the acid, rising up through the chalk, disengages carbonic acid in its passage, while the solution of chloride of calcium overflows at C, and escapes by the pipe D. The carbonic acid passes off through the pipe B, whence it is conducted to the wooden or iron cisterns containing the soda crystals. These cisterns are filled through a door a, at the side, and are fitted with a small pipe to allow the water of crystallization to flow off, as the absorption of the carbonic acid progresses. Large masses of crystallized soda are piled one upon another

FIG. 193.

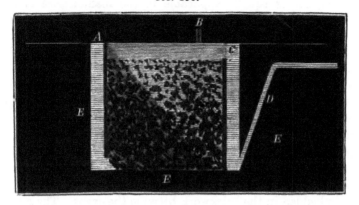

FIG. 194.

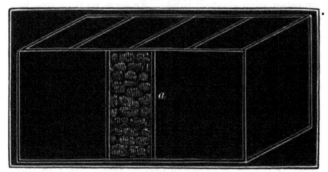

FIG. 195.

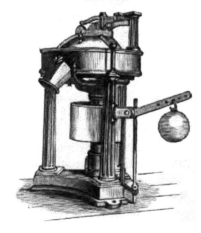

until the cistern is filled, as shown at *a*. Considerable heat is evolved during the operation, and at the end of ten days or a fortnight, the whole is converted into bicarbonate.

It is left, however, in a very damp state, and is removed to stoves, fitted with shelving, and dried by air gently heated in iron pipes. When perfectly dry to the touch, it is ground to a very fine powder between stones in a similar manner to flour, with this precaution, that the motion is not too rapid to cause such an evolution of heat as would injure the quality and appearance of the bicarbonate powder. This simple operation is of great importance in the manufacture of this article, as much of the facility with which it dissolves in water depends, upon the fineness of the powder.

A mill, called Bogardus's eccentric mill, represented in Figure 195, answers the purpose remarkably well. It is not expensive in its first cost, and works with rapidity and economy. Both plates revolve with nearly the same speed, in the same direction, on centres which are 1 inch apart. The centre or axis of the one revolves in a foot-step, and motion is communicated by means of belts or gearing, to the other plate. The circles which are cut in the plate, act like revolving shears by cutting every way, and when the mill is in operation, they cause a peculiar wrenching, twisting, and sliding motion, which is well adapted for every kind of grinding. The ground substance is freely delivered and never clogs the mill.

Mohr has proposed to mix the crystallised and anhydrous carbonates in such proportions, as to leave only a slight excess of water above that which is necessary for the water of crystallisation, when all the salt has been converted into bicarbonate. He admits the carbonic acid to the bottom of the mixture, by a tube which proceeds from a pneumatic apparatus similar to the ordinary gasometer.

Wilson has taken advantage of this suggestion in producing the bicarbonate which he uses in the processes described at page 257.

Artus states that the two salts absorb carbonic acid much more rapidly when they are mixed with recently calcined and powdered charcoal, made from soft wood, in the proportion of 2 parts of the salts to one part of the charcoal. The mixture is moistened and gassed for 24 hours, then removed, powdered, moistened, and gassed a second time. The materials are mixed with 8 parts of hot water, the solution filtered while hot, and the bicarbonate left to crystallise.

Schöffer originally employed the following process : he mixed 3 or 4 parts of soda-crystals with 1 part of the ordinary sesqui-carbonate of ammonia, and the semi-fluid mass, which evolves ammonia, was allowed to remain for 24 hours, and then exposed to the sun on a clear surface. The last portions of ammonia escape with the water, and the bicarbonate remains as a chalky mass. Schorz modified this process in order to save the ammonia, by distilling a mixture of—

50 parts of dry powdered carbonate of soda,
15   ,,   crystallised   ,,
41   ,,   carbonate of ammonia,

in a vessel immersed in water. The first nine-twelfths of the ammonia are easily distilled, the second two-twelfths come over with difficulty, and the last twelfth is removed by exposing the mass to the air.

This salt is composed of $NaO.2CO^2 + HO$, or rather $NaO.HO.2CO^2$, the water being an essential constituent. It has a slightly alkaline taste, and loses half its carbonic acid and all the water at a low heat. It is more soluble in hot than cold water, and when the solution is heated above 150° F. carbonic acid is said to be evolved, while at 212° it passes into sesquicarbonate, which soon changes to neutral carbonate of soda.

Poggiale has determined its solubility at different temperatures, and finds that 100 parts of water

At 32° F. dissolve 8·95 parts,
50   ,,   10·04 ,,
68   ,,   11·15 ,,
86   ,,   12·24 ,,
122   ,,   14·45 ,,
158   ,,   16·69 ,,

We have thus completed the description of this important manufacture in all its branches as at present conducted, and have endeavoured at the same time, to give some account of the various recent plans and modifications which have been either only suggested or carried into actual operation.

We propose now to give some details as to the cost of the manufacture and statistics of the trade, the refuse products which still wait some profitable application, and lastly, those plans which have been proposed for obtaining soda from other sources than common salt.

Cost of Manufacture.

Payen gives the following details of the cost of manufacture in one of the best situated localities in France, in 1848.

1. Cost of making *Sulphuric acid* per day at Marseilles:—

Sulphur, 1600 kil. at 15fr. . . . fr. 240

Nitrate of soda, 96 kil. at 50fr. . . 48

Coal, 24 hectolitres at 1fr. 66c. . . 39·84

Labour, 10 men at 3fr. 60c. (mean). 36

Incidentals, repairs, &c. . . . . 30·16

Produce, 4280 kil. acid at 1·850 sp. gr. fr. 394·00

Or, 8fr. 75c. per 100 kil. of oil of vitriol and 3fr. 58c. per 100 kil. of chamber acid of 1·555 to 1·590 sp. gr.

2. Cost of making *Sulphate of soda* in three double furnaces, each decomposing 1600 kil., or 4800 kil. in all, of common salt:

Common salt, 4800 kil. at 2fr. 48c. . . . . . fr. 119·05

Chamber acid, 1·555 sp. gr., 6420 kil. at 3fr. 58c. 223

Coal, 20 hectolitres at 1fr. 66c. . . . . . . 48·50

6 workmen at 2fr. 50c. (mean) 15

General charges . . . . 13·95

Produce, 4280 kil. of sulphate . . . . fr. 420 50

Or, 100 kil. cost 8fr.

3. Cost of manufacture of *Black ash* in four furnaces, working 32 charges:—

Sulphate of soda, 11,200 kil. at 8fr. . . . . . fr. 970·75

Ground limestone, 11,200 kil. at 50c. 56

Screened coal for the balling, 5,000 kil. at 2fr. 30c. 115

Ordinary fuel for furnaces, 72 hect. at 1fr. 65 . . 118·80

16 workmen and boys at 2fr. 20 (mean) 36

General charges . . . . 31·45

Produce, 16,800 kil. black ash . . . fr. 1328·00

which costs 7fr. 90c. per 100 kil., containing from 20 to 22½ per cent. alkali.

Mr. Gossage, in his recent pamphlet, gives the following statement of the quantities and cost of the raw materials required for the production of a ton of *soda-ash*, by the existing mode of working in Lancashire:

	£	s.	d.
1¼ tons of Irish pyrites	1	15	0
1 cwt. of nitrate of soda	0	12	0
1¼ tons of common salt	0	10	0
1½ tons of limestone	0	10	0
3½ tons of fuel	1	1	0
	4	8	0

to which must be added interest on capital, wages, repairs, packages, freights, &c., &c.

We are indebted to the kindness and liberality of Messrs. J. C. Stevenson and John Williamson, the able managing partners of the Jarrow Chemical Company, for the following interesting and instructive statement of the cost of a ton of refined *alkali* of 48 per cent., in 1860:—

Cwt.	qr.	lbs.		£	s.	d.		£	s.	d.
6	1	19	Sulphur at	7	17	9	per ton	2	10	7¾
0	2	8	Nitrate of soda	13	5	2*	"	0	7	7½
23	2	1	Salt	0	16	3¾	"	0	19	2¼
33	0	10	Chalk†	0	1	5½	"	0	2	5¼
67	2	7	Coal	0	2	11¼	"	0	9	11
			Wages					1	10	2
			Repairs					0	17	1½
			Interest on capital					0	6	4½
			Expenses of management, rates, taxes, insurance, &c.					0	5	8¼
			General stores					0	2	0¾
			Casks					0	10	0
			Net cost f. o. b. ship in the Tyne .					8	1	2¾

Or rather more than 2*d.* per cent. per cwt.

Statistics of the Trade.

The following return was drawn up by Mr. Williamson, of South Shields, for the year 1852, and is a model of a statistical account for manufacturers.

Statistical Reports of the Alkali Trade of the United Kingdom for the year 1852, drawn up by J. Williamson, Esq., Jarrow Chemical Company, South Shields.

From full returns by sixteen manufacturers, representing 46 per cent. of the trade.
From partial returns by eleven manufacturers, representing 35 per cent. of the trade.
No returns by sundry manufacturers, representing 19 per cent. of the trade.

Raw Materials consumed, in tons.

	Newcastle on Tyne.	Lancashire.	Glasgow and the Clyde.	Wales, Ireland, and South of England.	Midland.	Total.
Sulphur	7,580	none	3,000	none	940	11,520
Pyrites	33,750	40,220	9,000	12,000	5,292	100,262
Salt..................	57,905	40,152	19,120	12,000	8,370	137,547
Coals	232,020	136,400	80,000	36,500	34,500	519,420

* Net cost after allowing for the value of the resulting sulphate of soda.
† Damp chalk, containing 20 per cent. of water and upwards.

Quantity of Produce, in tons.

Alkali..............	23,100	26,343	12,000	7,000	2,750	71,493
Crystal soda	42,794	3,500	6,000	3,500	5,250	61,044
Bicarbonate of soda	4,046	1,200	none	none	516	5,762
Bleaching powder	5,000	1,250	5,000	1,850	none	13,100
Number of men employed	3,067	1,519	900	470	370	6,326
Amount expended in apparatus	£344,000	172,000	100,000	50,000	36,000	702,000
Amount annually expended in repairing the same	£69,500	23,000	20,000	11,000	6,200	129,700
Tonnage of shipping employed	189,100	71,200	70,000	35,000	8,000	373,300

71,193 tons alkali, at £10 . . . . . . £711,930
61,044 „ crystal soda, at £5 305,220
5,762 „ bicarbonate soda, at £15 . . 86,430
13,100 „ bleaching powder, at £10 . . 131,000

Total value of products . £1,234,580

Value of Materials imported from other Countries.

11,520 tons sulphur, at £6 . . . . . . £69,120
4,800 „ nitrate of soda, at £15 . . . 72,000
12,000 „ manganese, at £2 10s. . . . 30,000

£171,120

Amount contributed by the alkali trade to the } £1,063,460
annual income of this country . . . . . }

The following valuable table, prepared by Messrs. Henry Deacon, David Gamble, and John Hutchinson, shows that this trade has rather more than doubled in the last ten years :

Statistics of the Alkali Trade of the United Kingdom for May, 1862.

Annual value of finished products . . . £2,500,000
Weight of dry products, 280,000 tons.

Raw materials consumed per annum :

Salt	254,600	tons.
Coals	961,000	„
Limestone and chalk . . .	280,500	„
Pyrites	264,000	„
Nitrate of soda	8,300	„
Manganese	33,000	„
Timber for casks	33,000	„
Total . .	1,834,400	tons.

Capital employed in the manufacture:

In land £235,000
In plant, buildings, &c. 950,000
Working capital 825,000

Total capital . £2,010,000

Annual cost of materials for repairs:

Stone, bricks, slates, iron, timber, &c. £135,500

Labour, not including labour in transit:

	Number of hands.	Souls.	Annual amount of wages.
Directly employed	10,600	53,000	£549,500
Employed in getting casks	3,100	15,500	112,840
„ making salt	420	2,100	16,380
„ getting and breaking limestone	660	3,300	25,740
„ getting pyrites	4,030	20,150	157,150
„ felling and sawing timber for casks	330	1,650	10,140
Total labour employed in the manufacture and in the preparation of raw materials used in it	19,140	95,700	£871,750

Weight of raw materials and of finished products transported:

By inland navigation . . 481,350 tons per annum.
„ railway . . . . . . 1,154,300 „
„ coasting vessels . . . 722,300
„ foreign-going vessels . 182,800 „

Total . . 2,540,750 tons per annum.

About one-fourth of the raw materials and finished products are transhipped in transit.

Weight of Waste Products:

Alkali-waste, burnt ores, cinders, slag, debris, &c. . . . . . } 1,273,000 tons per annum.*

Waste acid, muriate of manganese, lime, &c. . . . . . . } 2,600,000 „

Total waste products . 3,873,000 tons per annum.

The export of this manufacture from this country has rapidly increased, as the following table proves:—

	Cwts.	£
1856	1,402,717	606,323
1857	1,534,405	759,426
1858	1,616,168	812,675
1859	2,027,609	1,024,203
1860	2,044,573	962,906

* Of this quantity about one-third, say 450,000 tons, is transported by railway and canals; the remainder is deposited in close proximity to the various works.

The next table gives the imports of this article into the Zollverein :—

1834	.	5,063 centner.	1849	.	95,228 centner.
1840	.	39,071 „	1850	.	120,000 „
1842	.	72,500 „	1851	.	126,986 „
1844	.	80,000 „	1852	.	134,000 „
1845	.	91,000 „	1853	.	86,885 „

A German chemist, Wagner, has recently discussed the question of the import of soda into the Zollverein, and says, that the power to meet the competition of England depends on—1st, the manufacturing processes; 2nd, the cost of labour; and 3rd, the price of fuel, salt, and sulphuric acid. He contends that the process is the same everywhere, viz., that of Leblanc, and that the application of the bye-products, hydrochloric acid and tank-waste, can exercise but little influence on the price of the main product. Further, that manual labour is somewhat higher in England than in the Zollverein, while fuel is one-third the cost in the former country. He considers that the sulphuric acid is not much cheaper in this country than in Germany, but adds that, thanks to the monopoly of salt in many of the German States, the price of salt is much higher. He gives the following figures as to the cost of salt in the different States :—

England	. . .	16 kreuzers per centner white salt.
Bavaria	. . .	1·0 florin	„	„
Austria	. . 1 to 2	„	„	„
England	. . .	9 kreuzers per centner rock-salt.
Wirtemberg	. .	30	„	„	„

He attributes the fact, that German soda is one-third dearer than the English, to the high price of fuel and salt in Germany, and that it can only compete with the latter by the protection of the cost of carriage, &c., and an import duty. He considers that there is not much to hope for, in obtaining cheaper fuel in Germany, but anticipates a lowering of the price of salt, from more productive sources of supply having been discovered ; and should the Zollverein be able to permit the native manufacturer to get salt at a small cost, he believes that Germany will not only be able to meet all home demands, but be in a position to export large quantities.

The principal seat of this trade on the continent is in the City of Marseilles, where the annual production is stated to be—

Alkali	. . . .	26,000 tons.
Soda-crystals	. . . . .	12,000 „
Bleaching powder	. . .	2,000 „

The exports of soda and alkali from France are yearly increasing, as is shown in the following table, which brings down the information to the latest date for which we can find any returns :—

1830	. . . .	606,853 kilogrs.
1835	. . . .	1,503,177 „
1850	. . . .	2,344,485 „
1856	. . . .	2,943,122 „

The next tables give the details of the imports and exports of the French trade for 1855.

Imports.

	Arrived.	Taken for consumption.
Sodas—		
Zollverein	52,876	2,179
England	76,117	1,553
Spain	37,896	96
Other countries . .	12,946	1,612
Kilgr.	152,835	5,440
Natrons—		
Belgium	20,771	20,771
Barbary	12,850	2,221
Other countries . .	19,506	—
Kilgr.	53,127	22,992

Exports.

Destination.	French and Foreign.	French.
Sodas—		
Belgium	94,217	94,217
England	41,314	41,214
Two Sicilies	117,275	117,275
Spain	962,524	858,646
Sardinia	604,547	531,407
Tuscany	327,251	327,251
Switzerland	370,786	307,689
Greece	52,075	52,075
Turkey	163,508	163,508
Rio de la Plata . .	61,274	60,097
Other countries . .	148,351	140,773
Kilgr.	2,943,122	2,964,152
Natrons— .		
Spain	5.000	—
Algeria	15,887	—
Kilgr.	20,887	—

The imports of sodas in 1856 were 235,312 kilgr., of which 183,419 kilgr. came from England ; and the exports were in the same year 3,442,941 kilgr., of which the following quantities were sent to

Spain . . . . 1,547,519 kilgr.
Sardinian States . . . . 858,524 „
Tuscany . . . . 256,409 „
Switzerland . . . . . 223,185 „

Refuse Products.

If we include the production of bleaching powder in our present manufacture, of which it is now an integral part, then these products are

Tank-waste,
Burnt sulphur-ore,
Chloride of calcium,
Chloride of manganese.

Several attempts have been made to turn the *tank-waste* to some economical use, and we must refer our readers to our account of some of those plans, viz., those of Losh, p. 185, part iv.|; Gossage, p. 29, part iii.; Townsend and Jullien, p. 39—42, part iii. ; and Bell, p. 72, part iii.

D'Arcet, some years ago, proposed to use it as a cement, a plan which has been recently revived by Kuhlmann and by Varren-traph ; but to our knowledge it has been employed for a similar purpose on the Tyne for upwards of the last 17 years. When used for making floors, it must be laid on the surface, when fresh, in thin layers, and well beaten down with a flat shovel. This application of the tank-waste is general in all the Alkali-works in this country.

Delanoue has proposed to boil the tank-waste with water and sulphur, by which a solution of bisulphide of calcium is obtained, which he says can be employed for the preparation of *milk of sulphur*, for sulphuring the vines as a protection against the vine disease, and in extracting cobalt and nickel from their ores.

The composition of this material is as follows :

	Brown.	Unger.		Musprett and Dawson.	
Carbonate of lime	24·22	19·56		41·20	23·42
Oxysulphide of calcium	20·36	32·80		—	—
Carbon....................	12·70	2·60		—	—
Silicate of magnesia ...	5·98	6·91		3·63	1·78
Sand	5·74	3·09		—	—
Oxide of iron	5·71	3·70	Phosphates and alumina	8·91	7·40
Sulphate of lime	4·28	3·69		2·53	4·59
Hyposulphite of lime .	trace	4·12		—	—
Hydrate of lime.........	5·58	11·79		8·72	12·03
Bisulphide of calcium .	3·58	4·67		5·97	0·62
Sulphide of calcium ...	8·52	3·25		25·79	36·70
Sulphide of sodium ...	—	1·78		1·44	2·87
Carbonate of soda	1·30	—		—	—
Water	2·10	3·45		1·73	10·59
	100·07	101·41		99·92 Fresh.	100·00 6 weeks old.

As regards the burnt sulphur-ores, we of course apply the term *waste* to those only which contain little or no copper, and the only application of them hitherto attempted, is that of Mr. Bell, to whose plan reference has already been made. The patentees have proposed to employ this refuse oxide of iron in some of the processes we have described for recovering the sulphur from the sulphides of sodium and calcium; and its application for a similar purpose has been proposed in purifying coal gas. Those burnt ores which contain sufficient copper are regularly smelted down to a regulus, from which the copper is obtained by the ordinary metallurgical operations. One of these plans is described at page 65.

Some portion of the chloride of calcium was, at one time, boiled down to dryness and employed in purifying coal gas; and another limited use, is its application to absorb and retain the water in distilling weak rough spirits. It may also hereafter be used for the manufacture of hydrochloric acid by Pelouze's process.

The chloride of manganese is also sometimes boiled down to dryness and employed in the purification of coal gas. It is also used for the recovery of the manganese by processes which will be described in the next chapter.

Kuhlmann has patented the application of this chloride of manganese, with chloride of calcium, for the manufacture of chloride of barium or strontium.

He first neutralises the excess of acid by the carbonate of baryta or strontia, and runs the solution into a reverberatory furnace at a red heat, previously charged with a mixture of sulphate of baryta and ground coal, in the following proportions:

320 to 400 parts of liquid chloride of manganese and barium.

100 parts of sulphate of baryta.

30 to 35 parts of coal in powder.

The materials are heated, when they become a pasty mass like black-ash, emitting flames of carbonic oxide; after three or four hours the charge is withdrawn. The chemical change is expressed by the equation—

$$MnCl + BaO.SO^3 + 4C = MnS + BaCl + 4CO.$$

Kuhlmann also decomposes the chloride of manganese with lime, recovering the oxide of manganese, and employing the chloride of calcium from this or any other source, in the following proportions:

Sulphate of baryta	233	parts.
Coal in powder.	48	„
Chloride of calcium (calculated dry). . .	111	„
Lime, 28 parts, or chalk	50	„

and operates in the same way .The decomposition which ensues is explained by the equation—

$$2(BaO.SO^3) + 8C + 2CaCl + CaO = 2BaCl + 2CaS.CaO + 8CO;$$

or he combines the two processes, mixing

100 parts of ground sulphate of baryta,

40 „ of coal,

170 „ of saturated liquid chloride of manganese,

13 „ of lime or chalk.

The chloride of barium is removed by lixiviation, crystallized, and employed for various purposes.

This material and the previous one, after being mixed with lime to neutralise the free acid and precipitate the manganese, are boiled down to dryness, and calcined until all the metallic oxides are rendered insoluble in water. The dry mass is then dissolved and filtered, and from the pure solution of chloride of calcium, sulphate of lime is precipitated by sulphate of soda. The sulphate of lime is washed, and then subjected to pressure to remove as much of the water as possible. The soft silky white sulphate of lime is then sold, as a *pearl-hardener*, to paper makers, for which application Mr. Jullien has secured a patent.

Notwithstanding all these efforts and applications, the great bulk of all these materials are still refuse products.

There are three sources besides common salt, from which it has been proposed to obtain soda; viz.,

> Felspar,
> Cryolite,
> Nitrate of soda;

from the first, we have the two plans to describe, which have been patented in the names of Ward and of Newton.

Felspar.

Ward proposes to subject orthoclase as well as albite, and, in fact, pumice-stone, lava, or any natural silicate containing an alkali, to his treatment. He mixes the mineral with fluoride of calcium and carbonate of lime, all in fine powder, and frits the whole in an ordinary reverberatory furnace. The frit is then lixiviated, when the solution will be found to contain the caustic alkali. He regulates the proportion by supplying each equivalent of silica and alumina with three equivalents of lime, and adding 7 or 8 per cent. of the fluoride of calcium to the mixture. There are minor details, but having indicated the general nature of the process, we must refer to the Repertory of Arts for 1858, p. 221, for fuller information.

The other plan, patented in the name of Mr. Newton, consists in heating either the soda or potash felspar with phosphate of lime and lime, in the following proportions:

> 100 parts of felspar,
> 50 „ phosphate of lime,
> 300 to 400 „ lime.

The materials are ground to powder, and calcined in a reverberatory furnace at a low red heat for about two hours. A double decomposition ensues, by which silicate of lime and phosphate of the alkali are produced. The silicate of lime combines with the silicate of alumina, and when the mass is lixiviated with water, the lime decomposes the phosphate of the alkali, reproducing phosphate of lime, while the alkali remains in the solution in the caustic state.

While Ward employs his residuum as a pozzolano material

suitable for the manufacture of hydraulic cement, Newton recommends his to be used as a manure.

Cryolite.

Spilsbury first proposed to obtain soda from this mineral, to which Moigno has since recalled the attention of chemical manufacturers.

The composition of cryolite has been determined by—

	Hermann.	Chodnew.	Durnew.
Aluminium	18·69	16·43	13·41
Sodium	23·78	26·54	32·31
Potassium	—	0·50	—
Calcium	—	—	0·25
Magnesium	—	0·93	—
Yttrium	—	1·04?	—
Fluorine	57·53	53·61	53·48
Oxides of iron and manganese	—	—	0·55
Loss by heat	—	0·95	—
	100·00	100·00	100·00

from which the formula has been calculated by

Hermann . $3NaF + 2Al^2F^3$, and
Chodnew . $2NaF + Al^2F^3$.

The cryolite is ground to powder, and mixed with its own weight of lime, and either calcined, or boiled with as much water as will form a milk. The boiling must be continued for several hours. When the mixture has been calcined, the mass must be exhausted with water.

The hydrofluoric acid, alumina, and lime, form an insoluble compound, while the caustic soda remains in solution. The following equation shows the change :

$$Al^2F^3.3NaF + 6(CaO.HO) = 3(NaO.HO) + Al^2O^3 + 6CaF + 3HO.$$

Spilsbury proposes in addition, to fuse wolfram with half its weight of cryolite, and exhaust the melted mass with water. The solution will contain tungstate of soda, from which the tungstic acid may be recovered as tungstate of lime, by adding one part of lime to six parts of the former salt, leaving the caustic soda in solution.

Nitrate of Soda.

We refer to pages 310 and 317 of Part IV., for a full description of the processes which are now pursued for decomposing this salt by carbonate or caustic potash, by which saltpetre, and either carbonate of soda or caustic soda, is obtained.

We have thus brought our account of this important manufacture to a close, and without venturing to predict what may be the course of its future development, we think that the peculiar renovating property of the peroxide of iron, points out its application to the recovery of the sulphur, and that sulphide of sodium will be the form from which the soda will hereafter be obtained.

CHLORINE AND ITS BLEACHING COMPOUNDS.

Historical Notices of their Introduction.

BLEACHING is a very ancient art, and was doubtless discovered soon after mankind had found out how to clothe themselves with textile materials. We read in the Scriptures of "fine linen, white and clean," and both the Greeks and Romans were familiar with many materials used for this purpose.

The ancient plan of washing and exposure to the air, was pursued until very recent times, and for a long period Holland enjoyed the reputation of having the best process, so that the brown linen of Scotland used to be sent to that country to be bleached, and when returned, was called *Hollands*, a name still retained in trade. The Dutch methods were at last introduced into Scotland in 1749, and the first attempts at improvement consisted in employing sulphurous or sulphuric acid, diluted with water, in place of sour milk. The great advance in the art of bleaching, however, was the observation of Berthollet in 1785, that an aqueous solution of chlorine, which had been discovered by Scheele in 1774, possessed the power of destroying vegetable colours. Berthollet showed the process to Watt at Paris in 1786, and Watt, on his return to Scotland, tried the plan on 1500 yards of linen in the bleach-fields of his father-in-law near Glasgow. It was brought under the notice of the bleachers at Aberdeen by Professor Copeland, and was introduced at Manchester, about the same time, chiefly through the labours of Dr. Henry. It was soon found that the texture of the goods was injured by the action of the chlorine, and that the workmen were much affected by the gas. Berthollet obviated these difficulties, in a great measure, by adding potash to the water, and Henry followed by substituting lime. About this period, the eminent manufacturer, Mr. Tennant, to whom we are indebted for the adaptation of this valuable agent to all the purposes of bleaching, took out his first patent for the manufacture of a saturated solution of chloride of lime, which he superseded by another patent in 1799, for the dry chloride of lime, or *bleaching powder*, as it has since been termed.

It is interesting to notice how soon the application of Scheele's discovery was made to the practical purposes of life, and that within a century, the manufacture of chlorine and its bleaching compounds has gone on increasing, until the annual commercial value exceeds a quarter of a million sterling, and has given rise to a trade in manganese from Germany and Spain, of great importance.

There are numerous substances and mixtures which possess the power of bleaching, such as chromic, nitric, nitrous, nitrous-sulphuric, sulphurous, and manganic acids; a mixture of a salt of tin and chlorate of potash, evolving chlorous acid; a mixture of chromate of potash, chromate of lead or chlorate of potash, and hydrochloric acid, evolving chlorine; a mixture of peroxide of manganese and hydrochloric acid also evolving chlorine; a mixture of peroxide of manganese and sulphuric acid evolving oxygen; and air ozonised through the action of phosphorus or electricity, &c., &c. Our present object is limited to an account of chlorine and its bleaching compounds.

CHLORINE

is met with in nature only in a state of combination. It occurs combined with hydrogen, as free hydrochloric acid, in some mountain torrents in South America, and as a gaseous exhalation from active volcanoes, both free and combined with ammonia; with potassium and sodium, in sea-water, plants, and animals, and with the latter metal in those enormous deposits of common salt found in so many localities; with magnesium and calcium, in several brine springs; and with lead, mercury, silver, and copper, in the mineral kingdom.

As an illustration of the variety of its combinations in nature, we give Bunsen's analyses of two samples of minerals collected from the fumaroles of Iceland :—

Chloride of ammonium . . .	81·68	74·32
$4Fe^2O^3$, Fe^2Cl^3	5·04	6·75
$4Al^2O^3$, Al^2Cl^3	3·73	0·28
Chloride of magnesium . . .	1·69	5 45
Chloride of calcium	0·53	4·63
Chloride of sodium	1·73	2·33
Chloride of potassium	0·53	0·70
Silica	0·95	0·25
Water and insoluble residue . .	3·12	5·29
	99·00	100·00

Chlorine gas is obtained by the mutual action of peroxide of manganese and hydrochloric acid, or of peroxide of manganese, common salt, and sulphuric acid. In the former case, the colour of the liquid becomes dark brown by the formation of bichloride of manganese, $MnO^2 + 2ClH = MnCl^2 + 2HO$, but as the heat increases, the bichloride is decomposed into MnCl and Cl, which escapes as a gas. The products by weight are as follows:

Raw Materials.		Products.	
Peroxide of manganese, 1 eq. $= 43·6$		Chlorine . . . . . 1 eq. $= 35·5$	
Hydrochloric acid . 2 eq. $= 73·0$		Chloride of manganese 1 eq. $= 63·1$	
Water . . . . . 12 eq. $= 108$		Water . . . . . 14 eq. $= 126.0$	
	224·6		224·6

In the latter case, the decomposition is represented by the equation

$$2HSO^4 + NaCl + MnO^2 = MnSO^4 + NaSO^4 + 2HO + Cl.$$

In practice, however, more sulphuric acid is necessary than the two equivalents according to theory : for the sulphuric acid has a tendency to form bisulphate of soda, which, although it yields chlorine with a mixture of salt and peroxide of manganese, does so only at a high temperature ; moreover, the decomposition is not perfect unless the salt is dry. Hesse states that $2\frac{1}{2}$ eq. of acid are necessary, Beringer 3 eq., and Dobereiner even says that 4 eq. must be added to obtain all the chlorine.

The apparatus for preparing the chlorine will be fully described in the following pages. The absorption of this gas is easily effected by water, in a series of carboys or Woulfe's earthenware jars. The chief point in the absorption is the temperature, which requires great attention, as the solubility of chlorine in water varies with the temperature:

100 cm. of water at 0° C. absorb 145 cm. of chlorine.

,,	.	8	,,	304	,,
		17		237	
		35		160	
,,		50	,,	109	,,

Hence it appears that water at 8° C. possesses the power of absorbing the largest quantity of this gas, and this temperature is near the point when the hydrate of chlorine, Cl.10HO, is decomposed, so that a saturated aqueous solution of this gas is really a solution of a mixture of hydrate of chlorine and free chlorine.

When chlorine is passed into water kept at a temperature of between 0° to 1°, till the liquid is converted into a thick paste, and then thrown on a filter to remove the excess of water, the hydrate may be obtained tolerably pure. It is formed of dark yellow crystalline plates, and must not be allowed to cool below 0° C., for at —4° it freezes fast to the filter. It is composed of—

Chlorine . . . . 27·7
Water . . . . 72·3
_____
100·0

At 7° C. it liquefies, while the gas escapes.

When chlorine, as usually prepared, is passed through strong sulphuric acid, it is deprived of all moisture, and rendered perfectly dry.

Chlorine is nearly 2½ times as heavy as atmospheric air, so that it falls towards the lowest parts of the buildings where it is manufactured or employed.

Its compound with water, and other weak combinations, as the hypochlorites, easily part with their chlorine, which acts on a number of substances, combining with the hydrogen and liberating oxygen, to which its bleaching and disinfecting properties are partly, if not wholly, due.

Chlorine, even when largely diluted with air, cannot be inhaled without producing great irritation and a choking sensation. It is a bad supporter of combustion, but many bodies take fire spontaneously when introduced into the gas.

These properties, and the difficulty of managing so powerful a liquid, have, in most cases, led to the disuse of chlorine and its aqueous solution, although it is sometimes employed for bleaching rags for the manufacture of paper.

As we have already stated, the principal form in which chlorine is now made available is in combination with lime, improperly called *chloride of lime*.

When chlorine is brought in contact, in the gaseous form, with alkaline or earthy bodies, different products are formed, depending on their state of hydration, the degree of dilution, or the temperature, but in every case accompanied by an oxidation of the chlorine.

Thus, when it is made to act on lime, one equivalent of the lime is decomposed, the oxygen going to form hypochlorous acid, which reunites with another equivalent of lime, while the calcium combines with a second equivalent of chlorine, thus:

$$2Cl + 2CaO = CaCl + CaO.ClO.$$

A similar reaction takes place with a weak solution of potash or soda in the cold;

$$2Cl + 2KO = KCl + KO.ClO;$$

but if the solution is concentrated, and the temperature increased, the oxidation is more complete, and chlorate of potash is produced—

$$6Cl + 6KO = 5KCl + KO.ClO^5.$$

The same changes take place with the lime, but at a higher temperature; in this case oxygen is evolved.

This introduction leads us to the

CHLORIDE OF LIME.

This substance has been supposed by Schafhäutl and Schönbein, to exist in a fluorspar found in Bavaria, but Schrötter believes that this mineral contains ozone and not hypochlorous acid.

The materials employed in the production of this compound are hydrochloric acid, lime, and manganese ores, which we will first describe, and then go into the special details of the manufacture.

I. HYDROCHLORIC ACID.

This acid has been found in an uncombined state, in some mountain streams in South America, of which we have given analyses at page 46. Humboldt found it in the thermal springs of Chucandiro, Guinche, and several other localities between Valladolid and the Lake of Cusco, in Mexico. It is also present in the gases and vapours evolved from volcanoes, especially in those from Vesuvius. Bunsen has also found it among the gases of the fumaroles of Iceland, one of his analyses of which we quote:

Nitrogen . . . . 81·81
Oxygen . . . . 14·21
Carbonic acid . . . . 2·44
Sulphurous acid . . . . . . 1·54
Hydrochloric acid, not determined.

Bunsen ascribes the presence of this acid among these gases, to the decomposition of the chlorine-compounds in the lava by the action of silicates and steam at a high temperature. He even found from 0·246 to 0·447 per cent. of hydrochloric acid in the lava which flowed from Hekla in 1845.

Hydrochloric acid is chiefly obtained from common salt by the action of sulphuric acid, but other plans have been proposed and carried into practice in some places.

1. *Preparation from Common Salt and Sulphuric Acid in Furnaces.*

We have already described how this acid is procured in such enormous quantities in the manufacture of sulphate of soda, and the apparatus employed for its condensation. When a stronger acid is wanted, a different system of condensation is adopted. Instead of conveying the gas from the decomposing pans through flues, it is carried by means of earthenware pipes to a series of carboys or Woulfe's jars, as shown in Fig. 196. These earthenware pipes, previous to being fitted up, are heated red-hot, and immersed in large boilers filled with hot tar containing a large proportion of pitch.

FIG. 196.

The pipes *aa* are connected with the arch *b* of the decomposing pan at one end, and convey the gas to a series of 30 or 60 carboys partially filled with water, where it is said, that upwards of 90 per cent. of the whole acid is obtained in a concentrated form.

Mr. Clapham has patented a plan for converting the weak into strong acid, which has been noticed at page 214. His apparatus consists of a metal vessel *a*, lined with gutta-percha, receiving weak acid from a cistern *b* by the pipe *c*; *d* is a pipe leading from the air-pump of an engine to the vessel *a*, and *e* is a pipe leading the weak acid, forced from the vessel *a*, to the cistern *f*, at the top of a condenser marked No. 1. Nos. 1, 2, and 3 are ordinary condensers, Nos. 2 and 3 being surmounted by water-cisterns, *g* and *h*; *i* is a pipe leading to the decomposing pans from No. 1 condenser, *k* is a pipe leading from No. 2 condenser to the drying furnace, and *l* is a pipe leading the waste gas from No. 2 to No. 3 condenser; *m* is a pipe leading weak acid to the cistern *b*.

FIG. 197.

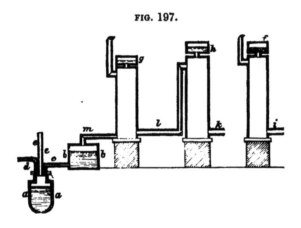

The weak acid solutions, as they usually run from the last or weak condensers, are cooled, and are then, instead of water, passed down what is called No. 1 condenser, or the strong acid condenser commonly used in decomposing common salt; or the weak solutions are exposed to a stream of weak or waste muriatic acid gas, till they acquire sufficient strength for use in the manufacture of bleaching powder. The apparatus employed resembles in principle, that used in working Gay Lussac's apparatus for saving nitrate of soda, and consists of a series of cooling cisterns made of Caithness flags, secured with suitable cement. From thence a gutta-percha pipe leads the solutions to an egg-shaped vessel made of metal or wood, and carefully lined with india-rubber or gutta-percha, or it may be made of thick sheets of gutta-percha. This latter vessel is attached to an ordinary air-pump, by which the liquid contents are forced through a gutta-percha pipe, or a metal pipe lined with india-rubber, to a cistern placed on the top of No. 1 condenser, from which a regular supply may be run into the condensers by the usual apparatus.

In some localities in France, in the Vosges for example, advantage is taken of the occurrence of suitable stones for making condensers, instead of the carboys. The stone flags are made into rectangular cisterns, to hold about 70 cubic feet, and are filled one-third full with water. They are ranged one above another,

FIG. 198.

AA′A″. The gas from the furnace enters the lowest through an earthenware pipe, T, ascending upwards through the series by the connecting elbow-pipes T′T″, while the water for condensation descends in the opposite direction, the communication between the different cisterns being maintained by the small lead pipes tt′t″, which are placed at the level of the liquid. The strong acid flows off by the pipe r, into the carboy B.

2. *Preparation in Metal Cylinders.*

This method is adopted in many localities. The apparatus consists of metal cylinders, AA, arranged in pairs, built in an oven of a furnace-range CD, which may contain 10 or 20 such ovens. The oven is constructed with a strong arch D, which reflects the heat, while the flame from the fire circulates all round the cylinders. The smoke, &c., escapes at e, into a flue fg, which leads to the chimney H. The cylinder is closed at both ends by metal lids, or built up with bricks, and the gas escapes through an earthenware pipe fixed in the lid, which is luted to a connecting piece j. This pipe j, passes into the elbow of the first row of

FIG. 199.

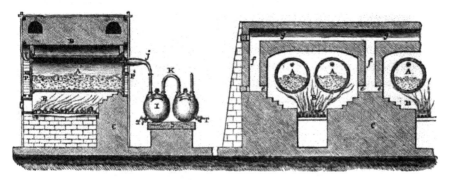

carboys I, which are connected with a second row by a double
elbow-pipe K.   There are six rows of carboys, all connected
together by similar elbow-pipes LL'L", and the last communicates with the chimney H by a pipe, O.

The front lid is furnished with a handle P, and it is luted to
the cylinder by means of clay.   An opening in the upper side of
this lid is made to admit an elbow-syphon S, through which the
acid is poured on the salt.

The carboys in the first range are empty, in order to collect the
impure acid which passes over, more or less mixed with sulphuri
acid, but the other carboys are all filled one-half full of water.

The cylinders are about 2 feet in diameter, and $5\frac{1}{2}$ feet long,
and are intended to hold about 3 cwt. of salt, which must be
spread evenly over the bottom of the cylinder.   The acid being
changed, the fire is gradually urged until all the gas has been
driven off, which is ascertained by the cooling of the connecting
pipe j.

The next woodcut (Fig. 200) shows more of the details of the
condensing apparatus, where, A and B, repre-

FIG. 200.

sent two carboys connected by the elbow-pipes
CD, the gas passing from D to C towards the
chimney : there is also a communication below
the level of the liquid, by means of thick tubes
of gutta-percha EE', FF', and if the carboy is
filled with water through the funnel GH, the
current will flow from A to B, in the opposite direction of the
gas, by which means the water becomes gradually saturated.

Each cylinder yields about 450 lbs. of acid, containing nearly

40 per cent. of real acid, which is drawn off by cocks fixed in the carboys nearest the cylinder.

The sulphate of soda is then removed from the cylinder and sold to alkali makers under various names.

The acid vapours corrode the upper part of these cylinders, and after a comparatively short time, it is neces-

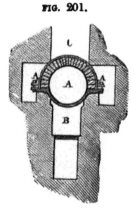

FIG. 201.

sary to reverse them, so that the under part on which the materials have lain, is, in its turn, exposed to the gas. In order to remedy this defect, the following arrangement has been adopted with great advantage. The cylinder A, is cast in two halves, B and C, as shown in Fig. 201, and bolted together on the outside. The upper half is lined with bricks d to protect the metal from the acid vapours. The setting of a single cylinder is represented in the figure.

3. *Preparation in Glass Retorts.*

This method, as followed in France, consists in using glass retorts C (Figs. 202, 203), coated with clay or cement, and placed in connection with large glass receivers C', or with receivers shaped as shown at D, each of them being connected with a bottle E, and this again communicating with another by an elbow pipe.

The retorts being filled with salt, and the acid poured in through a bent funnel C" (Fig. 202), they are placed in two rows, to the number of 4, 6, 8, or 10, and heated by a furnace A, placed in front, and covered with an arch BB'. The flame passes through the flue B, among the retorts, which are protected from its direct action by walls of brick, &c., and then escapes to the chimney GG'. The same plan is adopted in Germany, but the glass retorts are placed in metal pots, heated by a common fire in a galley furnace (*Galeerenofen*). The retorts are covered with clay, and the neck enters an empty receiver to retain the impure acid. The pure gas escapes from this first condenser through a lead pipe into a second, third, &c., partially filled with water, where the acid is absorbed.

In preparing hydrochloric acid by either of the two methods just described, it is important to attend to the proportions of the materials. When equivalent quantities are used, that is to say,

FIG. 202.

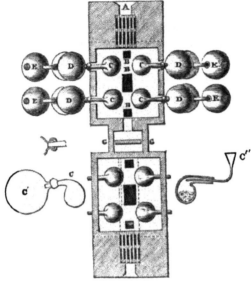

FIG. 203.

FIG. 204.

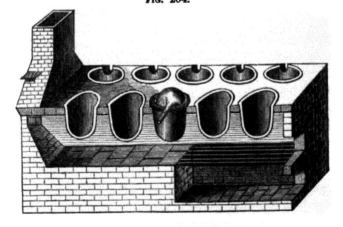

2 eq. common salt (= 117) to 2 eq. sulphuric acid [2 (HO.SO^s) = 98], only half of the salt is decomposed at a moderate temperature, thus,—

Common salt, 2NaCl
Sulphuric acid 2(HO.SO^s)
} give {
Bisulphate of soda, NaO.HO2SO^s,
Hydrochloric acid, HCl,
Common salt, NaCl ;

and at a higher temperature, the bisulphate of soda and common salt act upon each other in such a manner as to produce hydrochloric acid and neutral sulphate of soda :

NaO.HO.2SO^s
and
NaCl
} give {
2(NaO.SO^s)
and
HCl.

1 equivalent sulphuric acid to 1 equivalent common salt is, therefore, sufficient to convert the whole of the chlorine into hydrochloric acid ; but the decomposition is effected much more easily and with considerably less expenditure of fuel, when a double proportion of sulphuric acid is used, that is to say, 2 equivalents, HO.SO^s to 1 equivalent NaCl. The whole of the soda is then converted into bisulphate, and the whole of the chlorine obtained as hydrochloric acid at a moderate heat :*

$$NaCl + 2(HO.SO^s) = HCl + NaO.HO.2SO^s.$$

---

* The manner in which the reaction takes place is much more clearly seen when sulphuric acid is regarded as a bibasic acid, and represented by the formula H^{2}S^{2}O^s (atomic weight = 98). In that case, we see at once that the first reaction between this acid and the chloride of sodium must necessarily be the formation of bisulphate of soda and hydrochloric acid, as shown by the equation

$$NaCl + \left.\begin{matrix}H\\H\end{matrix}\right\} S^2O^6 = \left.\begin{matrix}Na\\H\end{matrix}\right\} S^2O^s + HCl ;$$

and, consequently, when the materials are mixed in this proportion (98 sulphuric acid to 58·5 common salt) the whole of the chlorine is at once obtained as hydrochloric acid. If, on the other hand, an excess of chloride of sodium is present, the reaction begins in the same way as before ; but when the heat is increased, the excess of chloride acts on the bisulphate of soda in such a manner as to produce hydrochloric acid and neutral sulphate :

$$NaCl + \left.\begin{matrix}Na\\H\end{matrix}\right\} S^2O^s = HCl + \left.\begin{matrix}Na\\Na\end{matrix}\right\} S^2O^s.$$

Similar remarks may be applied to every case of the preparation of a volatile monobasic acid, e.g., hydrobromic, hydrocyanic, nitric, by decomposing its soda or potash-salt with sulphuric or other bibasic acid. The reaction always takes place with the greatest facility, and at the lowest possible temperature, when the proportions of the materials are such that the residue consists of an acid sulphate or other salt.

22*

4. *Preparation without Sulphuric Acid.*

When common salt and pyrites are carefully roasted, with free access of moist air, hydrochloric acid is expelled, mixed however with more or less chloride of iron and sulphurous acid, so that, although this process has been combined with the manufacture of bleaching powder, the presence of these impurities and of a large excess of atmospheric air was fatal to its success.

Similar objections are applicable to those plans which are based on the decomposition of weathered sulphate of iron and common salt by heat in contact with the air, the hydrochloric acid evolved being then accompanied by considerable quantities of chlorine.

The process of Ramon de Luna has been attended with more success. He heats two parts of crystallized sulphate of magnesia with one part of common salt, in metal cylinders connected with the ordinary condensing apparatus. The common salt is decomposed and is converted into sulphate of soda, while hydrochloric acid is freely disengaged, and the magnesia is left behind mixed with sulphate of soda.

Barrow and De Sussex have proposed to heat chloride of magnesium in metal cylinders or fire-clay retorts, and condense the hydrochloric acid which is evolved; but the great heat necessary to effect this decomposition must entail a very serious cost in the wear and tear of the apparatus.

De Sussex has also patented the use of the chloride of iron and aluminium, and chloride of lead, mixed with iron filings or any other metal capable of decomposing water.

Pelouze has proposed to mix chloride of calcium with sand, to prevent too great a fluidity in the mass, and to expose the mixture at a high heat to the action of steam to obtain the hydrochloric acid. The same chemist has since found that chloride of calcium, exposed to the action of a high heat and steam, is almost wholly decomposed with an evolution of hydrochloric acid.

Kuhlmann employs a process, which, though it includes the use of sulphuric acid, may with propriety be mentioned here, as the acid is not used in the same manner as when common salt is decomposed. He decomposes a solution of chloride of barium with sulphuric acid and obtains a weak hydrochloric acid of 1·0448 sp. gr. without having recourse to distillation. He then concentrates this acid up to 1·1111 sp. gr.

Impurities in the Commercial Acid, and Purification.

These impurities are derived, partly from the presence of foreign substances in the materials used, and partly from the process. The yellow colour of the acid may arise from the presence of sesqui-oxide of iron, chlorine, or organic matter, the first coming from the corrosion of the apparatus or the nature of the condensing arrangements, the other two from impurities in the materials.

Sulphuric and sulphurous acids are generally present, the first being carried over mechanically during the ebullition caused by the heat and the evolution of the gaseous acid, and the latter originating, either by the reducing agency of the organic impurities, or from the action of sulphuric acid on the iron at the high heat employed.

Chlorine and nitrous acid arise from the presence of nitric acid in the sulphuric acid used in decomposing the salt.

The most serious impurity is arsenic, coming from the same source and probably present in the form of chloride of arsenic. Various statements are made as to the quantities of this impurity present.

Wackenroder found	$\frac{1}{1000}$	of metallic arsenic.
Dupasquier ,	$\frac{1}{100}$	,
Wittstein ,	$\frac{1}{10}$	
Reinsch ,	$\frac{1}{500}$	,

Common salt, lime, and other fixed substances are also found, the first being often present, carried over during the process, and the others arising, in part from the water employed for condensation, and in part from the action of the liquid acid on the condensing apparatus.

The arsenic may be precipitated by sulphuretted hydrogen, but the acid must be filtered through asbestos or some other material, before it is redistilled.

Dumont states that the sulphurous acid and chlorine may be separated by passing a stream of carbonic acid through the impure product of any of the processes.

The sulphuric acid is removed by a salt of barium, and the other impurities by redistillation, collecting only the intermediate distillate.

Properties of the Pure Acid.

Hydrochloric acid gas condenses with the moisture always present in the air, forming clouds more or less dense. It is rapidly absorbed by ice, but still more by water, which at common temperatures takes up 480 times its own volume, or nearly its own weight, of the gas. Roscoe and Dittmar (Chem. Soc. Qu. J. xii. 128) have carefully examined the relations existing between the quantities of real hydrochloric acid, HCl, contained in the saturated aqueous acid at different temperatures, and believe that the law may be empirically expressed by saying, that, for each temperature there exists a definite hydrate of hydrochloric acid, the existence of which depends essentially upon that particular temperature.

The specific gravity of the gas is 1·247; the concentrated aqueous acid, represented empirically by the formula, HCl.6HO, has a density of 1·210.

When the gas is mixed with $\frac{1}{2}$ vol. of oxygen, and an electric spark passed through the mixture, water and chlorine are produced. The same decomposition ensues when spongy platinum is heated in this mixture of the gases.

Hydrochloric and sulphurous acids do not act upon each other in a state of solution; but when mixed in the dry gaseous state, they are resolved into sulphur, chlorine, and water.

A mixture of 2 volumes of hydrochloric acid gas and 1 volume of hypochlorous acid gas, forms 1 volume of aqueous vapour and 2 volumes of chlorine gas, or in equivalents, $HCl + ClO = HO + Cl^2$.

Many salifiable metallic oxides act on this acid gas, producing water and anhydrous metallic chlorides: under these circumstances baryta and strontia become red-hot, and lime disengages considerable heat. The dioxides of copper and mercury, and the oxides of lead and silver, act in the same manner with the aqueous acid. With the metallic peroxides, both the liquid and gaseous acids yield a chloride of the metal and free chlorine:

$$MnO^2 + 2HCl = MnCl + 2HO + Cl.$$

The concentrated acid fumes in the air, and boils at a temperature which is lower in proportion to its strength, giving off a portion of the acid in the gaseous form. The acid, which has a constant boiling point under the ordinary atmospheric pressure, contains 20 per cent. of hydrochloric acid gas, which corresponds to the formula HCl.16HO ; it distils over unchanged at 110° C. ; but Roscoe and Dittmar (*loc. cit.*) have shown, with regard to hydrochloric as well as other acids, that the composition of aqueous acids of constant boiling point varies with the pressure under which the ebullition takes place : whence they infer, that such aqueous acids are not really definite hydrates, but only mixtures of acid and water, whose constituents pass into vapour at the boiling temperature, in the same proportion as that in which they occur in the liquid.

The liquid acid has a pungent, purely acid odour, but it acquires a saffron smell when iron is present. It absorbs carbonic acid in small quantities.

Applications of Hydrochloric Acid.

The impure acid obtained from the furnaces or cylinders, is chiefly employed for the manufacture of chlorine and the hypochlorites ; but it has recently been largely used in the production of chloride of zinc, which is now applied to the preservation of timber, cordage, &c., and for disinfection, especially in France. It is also employed for the manufacture of chloride of ammonium (sal-ammoniac) ; in dissolving out the earthy matters from bones in the manufacture of glue, &c. ; for the purification of sands containing iron ; for dissolving incrustations in boilers, pipes, &c. ; in the production of freezing mixtures, when mixed with crystals of sulphate of soda; and in decomposing galena for the manufacture of a white pigment, Pattinson's oxychloride of lead. The weak impure acid is also used for the production of carbonic acid from chalk, for the manufacture of bicarbonate of soda.

The pure acid is employed to make aqua regia for dissolving gold, platinum, and various alloys ; in the production of the chlorides of tin, antimony, &c. ; for softening ivory for the preparation of various surgical instruments ; for washing linen and other tissues impregnated with chloride of lime ; in the preparation of sealing wax, &c., &c.

II. THE OXIDE OF MANGANESE.

1. Ores of Manganese.

Manganese occurs in nature combined with sulphur, carbonic, silicic, and phosphoric acids, and with oxygen, under various mineralogical forms; but only a small number of the oxygen-compounds have found their way into commerce.

There are several distinct combinations of manganese and oxygen ; viz.,

The Protoxide or Manganous oxide . . . . . MnO.
Sesquioxide or Manganic oxide . . . . Mn^2O^3.
Binoxide, Peroxide, or Black oxide . . . MnO^2.
Manganoso-Manganic oxide, or Red oxide Mn^3O^4.
Manganic acid . . . . MnO^3.
Permanganic acid . . . . Mn^2O^7.

But the only one that will occupy our attention is the binoxide or black oxide.

As Bischof states, the ores of manganese present a great similarity to those of iron, both as regards their deposits and form. Those which are crystallized, occur in veins in both the primary and secondary formations, and the amorphous forms are found in irregular masses lining the cavities of the rocks, in various shapes, botryoïdal, stalactitic, and compact.

The formation of these mineral masses has been attributed to sublimation, to solution and subsequent deposition, and to electro-chemical action ; the most recent explanation by Deville ascribes it to the decomposition of nitrate of manganese. Some recent researches of Deville and Debray seem to confirm this opinion. They have discovered that by boiling 500 grammes of black oxide of manganese with 10 grammes of carbonate of potash, and eva-porating the filtered solution to dryness, a residue was obtained, composed as follows :

Sulphate of lime . . . . 1·03
Chloride of calcium . . . . . 2·05
      ,,   magnesium . . . . 0·84
      ,,   sodium . . . . . 1·74
Nitrate of soda . . . . 3·53
      ,,   potash . . . . . . 6·29
                         —————
                         15·48

And they further add the fact, that when the neutral or acid nitrate of the protoxide of manganese is heated with water up to 150° C. in a close vessel, a black peroxide of manganese is deposited, very like many of the natural ores, both in form and appearance.

The ores which will be described are never found in nature in the pure state, but more or less combined or mixed with water, baryta, silica, the oxides of iron, potash, lime, magnesia, and some-times with the oxides of copper and cobalt. Those employed in this manufacture are chiefly found in Germany, Spain, France, Belgium, and Holland; and smaller quantities are raised in this country, Piedmont, and America.

The following mineral species, more or less mixed with each other, constitute the commercial ores of manganese.

1. *Hausmanite* is one of the ores of manganese containing the least available oxygen, and corresponds exactly to the artificial manganoso-manganic oxide, or, as it is more familiarly known, red oxide. It is found associated with other manganese ores, with porphyry near Ilmenau in Thuringia, and at Ihlefeld in the Hartz. It has been analysed by Turner and Rammelsberg, who found

	Turner.	Rammelsberg.
Red oxide of manganese . .	98·902	99·44
Oxygen	0·215	0·05
Baryta.	0·111	0·15
Silica	0·337	—
Water	0·435	—
	100·000	99·64
	Ihlefeld.	Ilmenau.

2. *Pyrolusite* is the most abundant ore of manganese, and one of the richest in oxygen. It occurs in various forms, in amorphous masses, in stalactites, in radiated fibrous masses, &c., sometimes dull, and sometimes with a metallic lustre. It is soft and stains the fingers with a dark grey or black colour.

It is extensively mined at Eglersburg, Ilmenau, and other places in Thuringia; also at Vorderehrensdorf in Moravia. It also occurs in considerable quantities at Brandon and other places in Vermont, and in other localities in the United States.

Analyses of Pyrolusite.

	Turner.			Scheffler.	Arfvedson.	Dufresnoy.	Riegel.
Red oxide of manganese	85·617	84·05	85·62	87·0	83·56	72·5	86·00
Oxygen	11·699	11·78	11·60	11·6	14·58	9·8	11·45
Sesqui-oxide of iron	—	—	—	1·3	—	14·2	0·40
Alumina	—	—	—	0·2	—	—	—
Baryta..........	0·665	0·53	0·66	1·2	—	—	—
Lime	—	—	—	0·3	—	—	CaO trace
Silica	0·553	0·51	0·55	0·8	—	1·4	0·71
Water	1·566	1·12	1·57	5·6	1·86	1·6	1·40
	100·000	97·99	100·00	98·2	100·00	99·5	100·16
	Devon-shire.	Eglers-burg.	Ihlefeld.	Ilme-nau.	Unde-naes.	Tarn.	Krettnich.

3. *Braunite* corresponds with the so-called manganic or sesqui-oxide. It is generally crystalline, harder than felspar, and does

not soil the fingers. It is of a dark brown colour, with sub-metallic lustre.

It occurs both massive, and crystallized in veins, traversing porphyry near Ilmenau in Thuringia, at Ihlefeld in the Hartz, at St. Marcel in Piedmont, at Vizianagram in India, and in Vermont in the United States.

Analyses of Braunite.

	Turner.	Tonnager.	Bechi.	Scott.	Damour.
Protoxide manganese ...	86·94	—	88·31	—	19·17
Red oxide „ ...	—	86·40	—	—	—
Deutoxide „ ...	—	—	—	73·79	67·37
Oxygen	9·85	—	3·08	1·86	—
Peroxide iron	—	1·57	4·75	12·91	1·45
Baryta	2·26	—	1·03	—	—
Lime	—	—	—	—	1·22
Magnesia	—	—	—	2·34	—
Silica........................	trace	9·84	0·75	8·30	10·43
Water	0·95	1·98	2·08	0·54	—
	100·00	99·79	100·00	99·74	99·64
	Eglersburg.	Tellemark.	Elba.	Vizianagram.	Marceline.

4. *Manganite* differs from Braunite in containing water, the latter being the anhydrous sesquioxide of manganese, and the former the hydrated oxide. It has a dark steel-grey or iron-black colour, and occurs both in a crystalline and fibrous form. It has about the same hardness as carbonate of lime. The large percentage of water, chemically combined, is its most distinguishing feature.

The *Varvacite* of Phillips and the *Newkirkite* of Thomson, both belong to this species of manganese ore. Varvacite is an altered form, consisting largely of pyrolusite. It is found in veins traversing porphyry at Ihlefeld in the Hartz, at Ilmenau in Thuringia, at Christiansand in Norway, and at other places. It often contains fluorspar, sulphate of baryta, peroxide of iron, and variable quantities of peroxide of manganese. Some pure samples have given the following results when analysed by

	Arfvedson.	Gmelin.	Turner.
Sesquioxide of manganese ..	89·92	—	—
Manganese	—	62·86	62·68
Oxygen	—	27·64	27·22
Water	10·08	9·50	10·10
	100·00	100·00	100·00
	West Gothland.	Ihlefeld.	

5. *Psilomelane* has a dull metallic lustre, and a dark steel grey colour. Its ordinary form is amorphous, but it is sometimes met with in botryoïdal and stalactitic masses. It has about the same hardness as fluorspar. It is a common ore of manganese, found in Devonshire and Cornwall, at Ihlefeld in the Hartz, at Ilmenau, and other places. It is met with in mammillary masses at Chittenden and other localities in Vermont in the United States.

The following analyses of this ore have been published.

Analyses of Psilomelane.

	Turner.		Rammelsberg.	Fuchs.	Scheffier.	Berthier.
Red oxide of manganese	69·80	70·97	81·36	81·8	73·3	64·10
Oxygen	7·36	7·26	9·18	9·5	9·8	7·50
Sesquioxide of iron	—	—	1·43	—	0·3	6·80
Protoxide of copper	—	—	0·96	—	—	—
„ cobalt	—	—	—	—	—	—
Alumina	—	—	—·	—	2·1	—
Lime	—	—	0·38	—	1·8	—
Magnesia	—	—	0·32	—	—	—
Baryta	16·36	16·69	—	—	5·8	4·60
Potash	—	—	3·04	4·5	—	—
Silica	0·26	0·95	0·53	—	1·7	10·00
Water	6·22	4·13	3·39	4·2	4·3	7·00
	100·00	100·00	100·59	100·0	99·1	100·00
	Schneeberg.	Romanéche.	Horhausen.	Baireuth.	Ilmenau.	Thiviers.

	Herter.	Ebelmen.	Rammelsberg.	Clausbruch.	Heyl.	Bahr.
Protoxide of manganese	74·61	70·60	70·17	77·23	68·00	64·64
Oxygen	16·06	14·18	15·16	15·82	13·62	17·16
Sesquioxide of iron	—	0·77	—	—	—	—
Protoxide of copper	0·46	—	0·30	0·40	0·36	—
„ cobalt	—	—	0·54	—	—	0·03
Alumina	—	—	—	—	—	—
Lime	1·84	—	0·60	0·91	0·20	0·61
Magnesia	0·64	1·05	0·21	—	0·53	0·29
Baryta	2·40	6·55	8·08	0·12	8·59	16·04
Potash	0·92	4·05	2·62	5·29	0·27	0·29
Silica	—	0·60	0·90	0·52	2·18	—
Water	2·70	1·67	1·42	—	3·95	—
	99·63	99·47	100·00	100·29	97·70	99·06
	Elgersburg.	Gy, Haute Saone.	Heidelberg.	Ilmenau.	Langneberg.	Skidberg, Sweden.

6. *Wad* is the native peroxide of manganese; but under this name are included mixtures of different oxides, which cannot be regarded as chemical compounds. They occur in amorphous and reniform masses, generally earthy and amorphous, but in some

cases compact. Wad is very soft and soils the fingers, and is rapidly acted upon by hydrochloric acid. It is very light, and often loosely aggregated. The colour is dull black and bluish-black. It is found in the counties of Columbia and Duchess, New York; at Vicdessos in France; and at different localities in England, Sweden, &c.

The following analyses indicate a great variety in the composition of these ores, which have probably been produced by the decomposition of other ores, partly oxides and partly carbonates.

Analyses of Wad.

	Klaproth.	Turner.		Berthier.		
Red oxide of manganese	68·0	79·12	—	—	—	—
Protoxide „	—	—	—	69·80	62·4	58·5
Peroxide „	—	—	38·59	...	—	—
Oxygen	—	8·82	—	11·70	12·8	10·4
Sesquioxide of iron	6·5	—	52·34	—	6·0	5·7
Protoxide of copper	—	—	—	—	—	—
„ cobalt	—	—	—	—	—	—
Alumina	—	—	—	7·00	—	10·7
Baryta	1·0	1·40	5·40	—	—	—
Lime	—	—	—	—	—	—
Magnesia	—	—	—	—	—	—
Potash	—	—	—	—	—	—
Silica, &c.	8·0	—	2·74	—	3·0	1·8
Water	17·5	10·66	10·29	12·40	15·8	12·9
	101·0	100·00	109·36	100·90	100·0	100·0
	Claus-thal.	Devon-shire.	Derby-shire.	Vicdessos.	Groro-dite.	Siegen.

	Schef-fier.	Rammels-berg.	Igel-ström.	Beck.		Bahr.	Kus-sin.
Red oxide of manganese	—	—	—	—	—	66·16	—
Protoxide	66·5	67·50	—	—	—	—	64·40
Peroxide	—	—	82·51	68·50	58·50	—	—
Oxygen	12·1	13·48	—	—	—	—	7·37
Sesquioxide of iron	1·0	1·01	0·77	16·75	22·00	2·70	11·12
Protoxide of copper	—	—	—	—	—	0·02	—
„ cobalt	—	—	—	—	—	—	—
Alumina	—	—	6·30	—	—	0·75	—
Baryta	8·1	0·36	—	—	—	15·34	—
Lime	—	4·22	1·91	—	—	0·59	—
Magnesia	—	—	0·69	—	—	0·28	—
Potash	—	3·66	—	—	—	0·28	—
Silica, &c.	2·5	0·47	1·43	3·25	2·50	0·92	—
Water	9·8	10·30	5·58	11·50	17·00	12·07	14·10
	100·0	101·00	99·19	100·00	100·00	99·11	96·99
	Ilmenau.	Rübeland.	West-goth-land.	Hills-dale, N.Y.	Anster-litz, N.Y.	Skid-berg.	Krum-man.

Most of the ores found in commerce are by no means so pure as those represented by the analyses given in the preceding tables. The ores generally contain as much of the higher oxides as is equal to 60 or 70 per cent. of the peroxide. The differences of the ores in their natural state are very great, as the following table indicates, in which the quantity of chlorine is given, which the same weight of each ore produces; the ores named are those used in France.

Pure manganese ore	796
Ore from Crettnich	752
Calvéron	765
Périgneux	517
Romanêche	469
Laveline	464
Pesillo, Piedmont . . .	442
Saint-Marcel, Piedmont .	278

Some ores contain, moreover, so much oxide of iron, carbonate of lime, or water, as really to be of no commercial value; nevertheless, if the impurities were removed, a residue suitable for manufacturing operations would be left.

When the quantity of water is great—and we have known cases where the species called Wad, have contained upwards of 50 per cent.—it is easily expelled by applying a regulated heat, not exceeding 100° C., under roughly constructed floors, taking the precaution of frequently turning the mineral.

Mr. Cook has patented a process for treating ores containing the other impurities, so as to remove them without destroying the higher oxides of manganese present in the mineral. He heats such ores with hydrochloric acid of 8° Twaddell, or with sulphuric acid of 12° Twaddell, or with the waste acid liquor from the bleaching stills, either in the hot or cold state, and effectually removes the oxide of iron and earthy carbonates, without injury to the peroxide of manganese.

We have hitherto confined our attention to the natural sources of the oxide of manganese, but various processes have been proposed to recover the manganese from the waste liquors of the bleaching still, which will be explained more fully hereafter. This is, however, the more appropriate place to describe and discuss some of these plans for the preparation of, what may be termed, the artificial peroxide of manganese.

2. *The Artificial Oxides of Manganese.*

We will describe, in one of the following sections, the ordinary process for making chlorine by the action of hydrochloric acid on the mineral peroxides of manganese. The residual liquor which is run off from the stills, contains protochloride of manganese, more or less mixed with sesquichloride of iron, small quantities of alumina and silica in a soluble form, with different compounds of lime, magnesia, baryta, potash, and soda, and the insoluble gangue of the mineral.

A number of suggestions have been made to turn this waste product to some commercial use. We have already described Kuhlmann's process at p. 325, and noticed, at the same place, some of the other proposals. We confine ourselves at present, to an examination of those plans which have for their object, the recovery of the manganese in a form capable of being again employed for the production of chlorine.

All these plans may be said to aim at the production of, either

    The red oxide of manganese, or

    The black oxide of manganese ;

and the following, somewhat arbitrary, classification of these processes will assist our explanation :

1. Decomposition of manganese salts by heat.
    Walters, in 1843, the sulphate by heat.
    Glover, in 1844, the chloride by heat.
    Elliott, in 1856, the chloride by heat and steam.

2. Precipitation of the oxide and exposure to heat and air.
    Binks, in 1839, by soda, lime, &c., and heat.
    Balmain, in 1855, by gas water and heat.
    Pattinson, in 1856, by lime and heat.

3. Separation of the lower oxides by acids.
    Binks, in 1839, by weak hydrochloric and sulphuric acids.
    De Sussex and Arrott, in 1844, by hydrochloric acid.
    Elliott and Pattinson, in 1856, by the same.

4. Decomposition of the carbonate of manganese by heat.
    Dunlop, in 1855, by lime and carbonic acid.
    Gossage, in 1856, by magnesia and carbonic acid.

5. Oxidation by chemical reagents.
    Binks, in 1839, by nitrate of soda with the oxides.
    De Sussex and Arrott, in 1844, by manganic or waste nitrous acids.
    Gatty, in 1857, by nitrate of soda with the manganese salt.

We will select one patent for full details, as an illustration of the process of each class, and conclude with a description of Muspratt and Gerland's plans for obtaining the more valuable metals, such as copper, cobalt, &c., when present in the residual liquors.

1. *The decomposition of the Manganese-salts by heat.*— Elliott precipitates the iron present in the still-liquor by means of chalk, and separates the chlorides of manganese and calcium from the oxide of iron, &c. He then throws down the protoxide of manganese from the purified solution by means of lime in close vessels, in order to prevent as much as possible the access of air. This protoxide of manganese, freed from the chloride of calcium, is mixed with fresh still-liquor to precipitate the iron it contains, and this purified solution of chloride of manganese is boiled down to salt in pans or reverberatory furnaces.

This salt is now spread over the surface of a bed of a reverberatory furnace, constructed with four or five beds. After remaining twenty-four hours on the bed farthest from the fire, it is moved on to the adjoining bed, where it remains for about the same period, and so on, till it arrives at the bed nearest the fire, where it is exposed to the greatest heat. The mass is occasionally stirred, and a current of steam may be admitted into the furnace, as it facilitates the decomposition of the chloride of manganese. The hydrochloric acid, which is disengaged during the operation, is condensed in the usual way. The oxide of manganese thus obtained is a mixture of protoxide and peroxide, and may be used for generating chlorine. Mr. Elliott's peculiar claim is for the purification of the chloride of manganese from iron by means of the protoxide, and we have described his process, in preference to the others, partly on this account.

Walters precipitates the oxides of manganese from the still-liquor by means of iron, obtaining the sulphate of iron at the same time, as he applies his process to the liquor which results from using a mixture of hydrochloric and sulphuric acids and peroxide of manganese. He also heats the sulphate of manganese in close vessels, and conveys the sulphurous and sulphuric acids which are disengaged, to ordinary chambers.

2. *The precipitation of the oxide and exposure to air and heat.*—Binks, who was the first to adopt this plan, precipitates the manganese from the still-liquors by means of soda, sulphide of sodium, or lime; or he mixes the materials in the moist state, and

furnaces the mixture until the oxide is formed, and then separates the oxides of manganese by washing.

Balmain employs the ammoniacal liquor of gas-works to precipitate the manganese, and throws the sediment on a bed of sand or cinders, to be washed, and to allow the water to run off by drainage.

When Binks precipitates the manganese as a sulphide, he decomposes it by an acid, and burns the sulphuretted hydrogen as a source of sulphurous acid in making sulphuric acid.

In whatever manner the oxide of manganese is obtained in this first operation, it is afterwards exposed to a red heat in a reverberatory furnace with free access of air, until, as Balmain says, it ceases to burn like tinder. Pattinson advises the use of a regulated temperature, considerably below the melting point of tin.

3. *The separation of the lower oxides by acids.*—This plan was first proposed by Binks, who recommends that the mixed oxides should be boiled with the dilute acids, while Elliott states that it should be done in the cold. Binks employs sulphuric acid, varying in strength from 1·100 to 1·400 sp. gr., and hydrochloric acid of 1·030 to 1·100 sp. gr. The proportions of acid for the red oxide, should be at least one-half its weight of sulphuric acid of 1·840, and one-half its weight of real hydrochloric acid, calculated for the strengths of the acids as recommended.

In all cases the protoxide of manganese which is dissolved, is treated in the same way as the original still-liquor.

4. *Decomposition of the Carbonate of Manganese by heat.*— Of all the plans which have been proposed, none have been found to pay on a manufacturing scale, except that patented by Dunlop. It consists in precipitating the manganese in the form of carbonate by carbonate of ammonia, or by lime or carbonate of lime, and passing a current of carbonic acid through the liquid.

The plan is carried out at St. Rollox in the following manner: The still-liquor, when drawn off, is allowed to stand to permit all insoluble matters to subside, or a milk of lime is cautiously added to throw down the sesquioxide of iron, alumina, and silica, until the precipitate ceases to show the characteristic red colour of sesquioxide of iron. The clear solution of the chloride of manganese is lifted by machinery from the underground tank into large shallow cisterns, where it it is intimately mixed with finely ground chalk. This thick milky fluid is then run into large boilers, 9 feet in diameter and 80 feet long, where the decomposition is completed. An iron shaft passes through these boilers, fitted with metal arms, which are intended to agitate the contents, and the whole is kept in motion by a steam-engine. Steam is then admitted under a pressure of 2 to 4 atmospheres, and by the combined action of the heat and pressure, the decomposition of the chloride of manganese by the carbonate of lime, is finished in about four hours. The carbonate of manganese is then allowed to subside, the solution of chloride of calcium is run off, and the precipitate carefully washed. The carbonate of manganese is thrown up in heaps on an inclined surface to drain.

This partially dried material is then placed in small low waggons of sheet iron, supported on rollers, and slowly drawn through the calcining furnace by means of a chain. The furnace holds 48 of these little waggons, and this operation resembles the mode of annealing articles made of flint glass. The furnace is 50 feet long, 12 feet wide, and 10 feet high. A fire-brick flue runs down the centre of the bottom of the furnace, and is connected at the far end with two return metal pipes, which lie on each side of the flue. A uniform heat of about 315° C. is thus easily maintained in the furnace, on which four lines of rails are laid for the small waggons to traverse.

The half-dried material loses all its water and part of its carbonic acid, as the waggons pass along the first line of rails, and as they return down the second line, where the heat is greater, a further escape of carbonic acid ensues, and ultimately the expulsion of all the acid, and the peroxidation of the manganese, is completed on the third and fourth lines. This operation lasts about 48 hours, and the colour of the material gradually changes from brown to black. The ends of the furnace are closed by loose hanging doors, so that a sufficient supply of air is always present. The fire-place is situated below the floor of the furnace, and requires the greatest attention.

Reissig has carefully examined the action of heat on the carbonate of manganese, and has found that at a temperature of 300° C. all the carbonic acid is expelled, and a mixture or compound left behind, consisting of 3 equivalents of manganese and 5 equivalents of oxygen, or $2MnO^2 + MnO$. The following table

shows the action of the heat during a given time, and the extent
of the peroxidation :—

Temperature.	Duration.	Per cent. of peroxide of manganese.
220° C. . .	2 hours . . .	37·26
240 . . .	1 ,, . . .	53·59
240 . . .	2 ,, . . .	57·22
240 . . .	3 ,, . . .	60·12
250 . . .	1 ,, . . .	54·68
250 . . .	2 ,, . . .	61·58
250 . . .	3 ,, . . .	63·15
260 . . .	1 ,, . . .	56·04
260 . . .	2 ,, . . .	63·32
260 . . .	3 ,, . . .	65·39
270 . . .	1 ,, . . .	58·21
270 . . .	2 ,, . . .	64·87
270 . . .	3 ,, . . .	66·36
280 . . .	1 ,, . . .	64·35
280 . . .	2 ,, . . .	67·54
280 . . .	3 . ,, . . .	72·50
290 . . .	1 ,, . . .	66·83
290 . . .	2 ,, . . .	67·89
290 . . .	3 ,, . . .	72·80
300 . . .	1 ,, . . .	68·76
300 . . .	2 ,, . . .	71·43
300 . . .	3 ,, . . .	73·91
300 . . .	3 ,, . . .	72·96
300 . . .	3 ,, . . .	73·14

We do not know, in the whole range of the applications of
chemistry to manufactures, a more beautiful illustration of the
shrewdness of an inventor in availing himself of a delicate
chemical decomposition.

Gossage employs the same principle, but substitutes magnesia
for lime, and having mixed his sulphate or chloride of manganese
with carbonate of magnesia, he forces a supply of carbonic acid
among the materials under a pressure of 30 lbs. to the inch.
The other operations are the same as those already described.
He employs the resulting chloride of magnesium for acting
on caustic magnesian lime, in order to obtain a supply of magnesia. He also claims the use of carbonates of iron, manganese,

or zinc, obtained by any of these processes, for decomposing sulphide of sodium in the tank-liquors of the soda-manufacture, to obtain carbonate of soda and the metallic sulphide.

5. *The oxidation by Chemical Reagents.*

Binks first proposed this method of oxidising the lower oxides of manganese. He adds nitrate of soda to the still-liquor, and heats the mixture till the decomposition is complete, obtaining sulphate of soda, nitrous acid gas, and peroxide of manganese. He, however, prefers to use the materials in the dry state, and has given the following proportions as the best adapted for carrying out his plan :

Nitrate of soda . . . . 1 part
Dry sulphate still-residue . . . . 2 ,,
Inferior oxides or carbonate of manganese from the chlo-⎱
    ride still-residue prepared by any of his previous pro-⎰ 3 ,,
    cesses . . . .

Or,

Nitrate of soda . . . . 1 ,,
Dry sulphate still-residue . . . . 8 ,,
Hydrate or carbonate of lime . . . . 3 ,,

He conducts the operation in a close iron retort, and brings the nitrous acid gas or binoxide of nitrogen which is evolved, in contact with his inferior oxide or carbonate of manganese at a low red heat in order to peroxidise them.

Gatty's plan is very similar : he concentrates the still-liquors, and for every 72 lbs. of dry chloride, or 84 lbs. of dry sulphate, of manganese present, he adds 85 lbs. of nitrate of soda. This mixture is dried at a gentle heat, and afterwards exposed to a low red heat in an iron retort. The nitrous acids are either condensed, or conveyed to an ordinary leaden chamber, to take part in the manufacture of sulphuric acid.

The residue obtained by Binks or Gatty is afterwards lixiviated to remove the soluble matters, and the peroxide of manganese is fit for further use.

De Sussex and Arrott convert the manganese in still-liquors into red oxide or carbonate by any of the plans already mentioned, and mix one part of their product with three parts of caustic or carbonate of soda or potash. This mixture is fused on the hearth of a reverberatory furnace with free access of the air, and afterwards run or ladled out of the furnace. When cold, it is dissolved in hot

23*

water, forming a solution of chameleon-mineral, which is exposed
to the air, or to the action of carbonic acid, or alkaline bicarbonates, by which a peroxide of manganese and an alkaline
carbonate are obtained.

They also peroxidise the lower oxides or carbonate of manganese, by exposing them to the action of the waste nitrous
compounds, which are evolved in the manufacture of oxalic acid,
nitrate of lead or copper, &c.

Having completed this account of the processes specially intended to produce peroxide of manganese, we have to describe
Muspratt and Gerland's plan for separating nickel, cobalt, and
copper, when present, from the still-liquors, previous to the
oxidation of the manganese. The patentees first separate the
iron by means of lime, or by heating the dry residue, and
then throw down the copper by sulphuretted hydrogen. This
being separated, they precipitate the nickel and cobalt by a
solution of the tank-waste, which is added till the flesh-
coloured sulphide of manganese appears. The small quantity of
sulphide of manganese is removed from the nickel and cobalt by
hydrochloric acid, and their further separation is effected by any
of the plans adopted by manfacturers. The manganese left in
solution can be recovered by one or other of the processes already
described.

Applications of the Oxide of Manganese.

Payen states that the still-liquors have also been employed by
the potters as a source of their black colours, and that when
thickened with gum or dextrin, they have been applied for
printing textile goods, immersing the cloth in a solution of soda
to precipitate the oxide, and then exposing it to the oxidising
action of bleaching liquor to bring out the black colour.

The purer and richer qualities of oxide of manganese are
employed in the manufacture of the finer kinds of glass to
oxidise the iron, and so produce an article without colour.

Large quantities, as free from sulphur as possible, are also used
in the manufacture of homogeneous iron and steel by the processes
of Heath, Uchatius, Mushet, and others.

Statistics of the Manganese Trade.

This country was first supplied from the mines in Devonshire,
and afterwards by importations from Germany and Belgium, but
latterly the demand has been met by the ores raised in Spain.

After the produce of the mines in Devonshire had fallen off, the demand was met chiefly by the ore raised in the Duchy of Nassau, which rose from 496 centner in 1828, to 670,192 centner in 1857, and then fell off to 349,887 centner in 1859, in consequence of the competition from Spain. The production of these mines was—

In 1828	. . . . .	496	centner.
1837	. . . . .	8,011	,,
1847	. . . . .	213,679	,,
1857	. . . . .	670,192	,,
1858	. . . . .	463,502	,,
1859	. . . . .	349,887	,,

The total production of manganese-ores in Germany in 1857 was as follows :—

Nassau	670,192	centner.
Hesse-Darmstadt . . .	145,269	,,
Saxe-Gotha	40,000	,,
Prussia	22,384	,,
Schwarzburg-Sondershausen	8,742	,,
Saxony	2,927	,,
Hesse-Cassel	2,570	,,
Black Forest	1,200	,,
Hanover	1,000	,,
Austria	882	
Bavaria	500	
	895,666	,,

The annual production in Spain is said to be nearly 500,000 centner.

Muspratt states that the total consumption in this country is about 25,000 tons per annum, of which from 17,000 to 18,000 tons are used in the manufacture of bleaching powder.

The following details of our imports for 1858 are given in Ure's Dictionary of Arts, Manufactures, and Mines:

From the Hanse Towns	. tons	262
,, Holland	,,	23,280
,, France	,,	239
,, Other parts . . .	,,	390
	Tons	24,171

The next table shows the imports of this ore into the port of Newcastle-on-Tyne for the last five years:

	1857.	1858.	1859.	1860.	1861.
Holland	6,820	7,507	7,427	2,759	3,375
France	—	3	—	213	180
Belgium	—	—	150	—	250
Norway	1	—	—	—	—
Germany	126	27	121	19	—
Spain	98	—	4,566	5,964	2,240
	7,045	7,537	12,264	8,955	6,045

Mangin gives the following statement of this trade in France:

		Imports in 1857.	1858.
From Belgium	kil.	1,486,086	1,172,783
„ Holland	„	580,776	586,392
„ Germany	„	501,935	
„ Spain	„	154,543	586,392
„ Other countries	„	21,025	
	Kils.	2,744,365	2,345,567

III. CHOICE AND PREPARATION OF THE LIME.

The selection of a suitable lime is very important, as the presence of iron and manganese gives a yellow colour to the bleaching powder, and it is stated that the oxides of the latter metal occasion a spontaneous decomposition of the hypochlorite of lime. A limestone or chalk should be chosen, therefore, as free as possible from metallic oxides, silica, alumina, and other foreign matters. It is said in Germany that magnesian limestone answers better than ordinary limestone; but this is opposed to the generally received opinion in this country, where that which gives the heaviest lime is preferred, the Irish being heavier than the French.

French and Irish limestones are used in this country for this manufacture. The limestone is burnt in kilns specially constructed for the purpose, as shown in Figure 205. A represents one kiln complete, B the other kiln, with the front and end wall removed, to show the position of the flues a and b, with exit for the gaseous products at the top, c and d. The charge is thrown in through E, against the back wall, C, and slopes to about the point e, in the front. The fires are worked through the openings ff, and the direction of the heat is regulated by managing the draught through a and b at the bottom,

FIG. 205

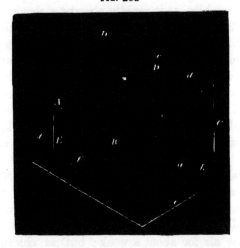

and D at the top. The open door E is roughly built up, and when the calcination is complete, the quicklime is removed through the same opening.

The slaking of the lime is even more important than the selection of the limestone. Dingler states that an excess of water renders the absorption of the chlorine more difficult, and it, of course, reduces the strength of the bleaching powder. If, on the other hand, too little water be employed, the lime is converted chiefly into chloride of calcium and chlorate of lime, without any hypochlorite being formed. Houton-Labillardiere therefore recommends that it should be slaked with a large quantity of water, brought to the consistence of a cream, and the excess of water afterwards expelled by heat.

According to theory, 100 of lime combine with 32·14 parts of water to form hydrate of lime, $CaO.HO$, which should absorb 126·6 parts of chlorine. The following table is given to show the importance of a proper hydration of the lime :

100 parts of lime slaked.	Chlorine absorbed.	Chemist.
With 16 parts of water............	32	Morin
With 32 to 64 parts of water ...	63	,,
Carefully and completely	86 to 100	Ure
„ „	113	{ Houton-Labillardière
Ordinary practice	64	

The quicklime ought to be slaked as soon as possible after it is taken from the kiln, as it rapidly absorbs carbonic acid. It is spread over the floor of the building to the depth of 6 to 16 inches, according to different authorities, and the water is applied through the rose of a watering-can, evenly over the surface, while a workman occasionally turns it to bring the whole in contact with the liquid. Wagner states that 100 lbs. of quicklime require 30 lbs. of water; Muspratt advises 5 gallons for every cwt., which is nearly in the ratio of 45 lbs. to the 100 lbs. of lime; while Barreswil and Girard give as much as 60 of water to the 100 of lime.

After the slaking is finished, the lime is carefully sifted through fine wire-gauze or hair-cloth to remove any lumps, the interior of which cannot be reached by the chlorine. Wagner says that the hydrate of lime absorbs the chlorine better after it has been exposed to the air for about a week; but as chlorine has no action on carbonate of lime, which must be more or less mixed with the hydrate after such exposure, it is difficult to understand how this can take place.

Labarraque recommends the addition of 5 per cent. of common salt to the hydrate of lime, by which, he says, the absorption of the chlorine is much facilitated.

MANUFACTURE OF BLEACHING POWDER.

The production of this compound involves the use of hydrochloric acid and manganese to obtain a supply of chlorine, and of lime to absorb this gas. The apparatus consists, therefore, of two sets of vessels: the one in which the chlorine is generated, called *stills*, and those which contain the lime for its absorption, called *chambers*, *receivers*, or *condensers*. These vessels are constructed of different materials, and of various shapes, sometimes of stoneware, strong sheet lead, cast iron, and of stone. We will first describe some of these stills and chambers, then give the details of the process as generally followed, and lastly, an account of the plans which have been proposed to dispense with the use of manganese.

I. THE APPARATUS.

1. *The Still.*

In the early stages of this manufacture, the chlorine was obtained from a mixture of common salt, peroxide of manganese and sulphuric acid; and the apparatus then employed, which is still in use where this process is followed, consists of two parts, as shown in Figure 206, the lower one being generally made of lead, and enclosed in a casing of cast iron, and heated by steam from the pipe R, which circulates in the intermediate space.

FIG. 206.

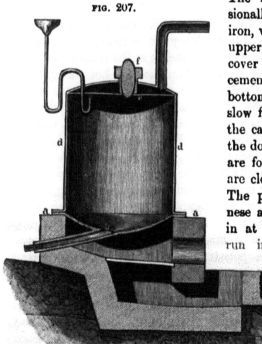

FIG. 207.

The lower half is occasionally formed of cast iron, with a groove in the upper part, to which the cover is luted with good cement. In this case, the bottom is heated by a slow fire, placed beneath the cast-iron bottom. In the dome of the still there are four openings, which are closed by water lutes. The peroxide of manganese and salt are thrown in at M, and the acid is run in through a bent leaden funnel S; the chlorine escapes through a small lead pipe T; and an iron shaft, covered with lead, passes through A, by which the agitator is occasionally turned round to bring the materials into contact with each other. When the operation is finished, the waste products are run off through the pipe K.

In Germany a still is used of a different construction, represented in Fig. 207. It consists partly of iron and partly of lead. The lower part is made in the shape of a flat iron basin with a projecting rim *aa*, with holes for screws. The heat is applied to this basin from below. The upper part is formed of a cast leaden cylinder *dd*, with a flange and holes corresponding with those in the iron rim of the basin. An iron ring covers the lead flange, and a cement of albumen and lime being laid between the iron and lead, the whole is firmly bolted together by screws.

The top is fitted with an elbow-funnel, and an opening *f*, for charging the solid materials. The waste products escape at *b*, and the chlorine through the bent pipe, *c*.

Morfit first proposed to substitute an earthenware vessel, for the lead lining of the still, which itself was all made of lead ; and Mr. Gamble made the next improvement, which is explained below.

A, A, A, A (Fig. 208), are four stills, the bottoms made of earthenware, the tops of lead. They are surrounded by cast-iron casings, leaving a space of three inches at the bottom and sides. The stills are heated by the circulation of hot water, saline solutions, or steam from a boiler, G. B, B, B, B, are branch pipes leading from the stills to a main-pipe G. At the end of the pipe G, there is a shorter pipe D, which can be made to move alternately to the receivers E, E, by means of water-lutes. A small door, about 10 inches square, is made on the side of each receiver E, at F, to withdraw the chloride when the process is completed, and they must be made air-tight by luting. These chambers or receivers are adapted to the manufacture of bleaching powder ; but the stills are applicable to the manufacture of an aqueous solution of chlorine, or of solutions of the chlorides of lime, potash, or soda. Any suitable receiver may be adapted to these processes, care being taken that the water-lutes shall be made so deep as to overcome the pressure in the receivers. Fig. 208 is in part a section, and in part an elevation, of the same apparatus. A, shows the earthenware still bottom, and the manner of attaching it to the leaden top and iron casing. The bottom is an inverted cone, 2 inches

FIG. 208.

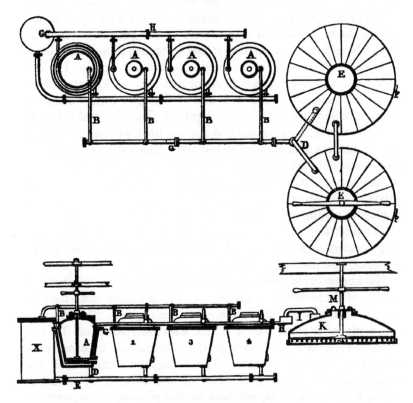

thick, 6 inches of the upper part being made conical. To these 6
inches, the top is fitted with fat lute, and firmly bound round
by an iron hoop. The lower part of the lead is prolonged
below the hoop, and turned up with a flange, so as effectually to
close in the space for hot water, and extend over the flange of
the iron casing, to which it is firmly attached by screws, and
made water-tight with cement. For the purpose of discharging
the still, a pipe proceeds from its bottom, 5 inches long, 3 inches
internal diameter, and 2 inches in thickness. It is made to pass
through another pipe in the outer case, 2 inches long, and 8 inches
internal diameter. The space between these pipes must be made
secure from leakage, by lead or cement. When the still is at
work, the pipe will be secured by a plug. The branch pipes,
B, B, B, B, which convey the water within the casing, pass
through the flange of the leaden top at C; D, is a branch pipe

to convey the cooled water to the main pipe, which returns it to the boiler. F, is an agitator for stirring the materials in the still, and is secured to the two beams in such a way, by nuts and collars, as to prevent it from touching the bottom of the still. From the still A, the main pipe, joined by branch pipes from stills 2, 3, 4, is connected with the receiver K, by the moving pipe L. The bottoms of the receivers are made of cast iron, with a ledge about 2 inches deep, the sides and top of wrought iron. They are made circular, about 1 foot in depth at the side, and 2 feet at the centre. M, is an agitator to stir the lime, working in a pivot at the bottom and a collar at the top. The stills and receivers have water-lute lids, through which they are charged. This kind of still is peculiarly adapted to the use of hydrochloric acid, or a mixture of hydrochloric and sulphuric acids, which observation applies to all the modifications to be described.

The next step in improving the apparatus in which the chlorine is generated, was taken by Mr. Lee, who employed stone as the material, and heated the materials from above.

His apparatus is shown in Figure 209, which is a longitudinal section of three retorts or ovens. a, a, a, are three stone troughs or vessels in which the chlorine is generated. The troughs are made of a stone which is not materially acted on by the process, such as is now used in the construction of condensers and flues of alkali works. The troughs are supported by brick, enclosed by walls of brick b, and covered by arches c, through which the heat, requisite for the process, is transmitted, thus forming separate ovens. A door in front is provided for the

FIG. 209.

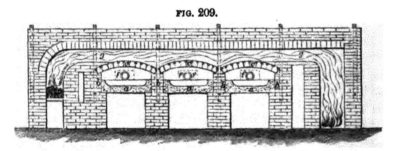

introduction of materials, and made air-tight when at work by any of the usual lutings, and at the other end there is a pipe f, of lead or earthenware, for conveying away the chlorine. The heat necessary for the operation is communicated to the materials in the chambers or ovens *a*, through the arches *c*, *c*, by the heat of the fire passing over them, the flame and heat of the fire being carried over the arches *c* along a horizontal flue *g*. In order to render the arches *c*, *c*, tight, they are covered with a cement of 4 parts of fire-clay, 2 parts of ground fire-brick, and 1 part of common salt, decreasing or increasing the common salt, as the cement is required to bear a greater or less heat. The process is commenced by introducing a quantity of manganese in lumps into the troughs or ovens *a*, *a*, *a*; the doors are then closed and luted; a quantity of muriatic acid is afterwards gradually introduced through a glass or earthenware pipe fixed in the front wall, and passing into the troughs.

The manufacturers at length found that the stills might be made entirely of stone, and we give a section of one of these stone stills, which also shows Mr. Pattinson's improvement in the mode of working them.

a, *a*, *a*, represent the still; *b*, *b*, its cover; *c*, an opening in the cover for introducing or withdrawing the materials; *d*, an earthenware pipe for the passage of the chlorine; *i*, *i* is a stone shelf or a false bottom, laid across the still, and supported by upright pieces of stone, at a height of from 5 to 10 inches from the true bottom of the still; *e*, *e* are two openings in the cover, one for running in the acid, and the other for guaging the depth; *j* is another opening in the lid for the insertion of a stone pipe *h*,

FIG. 210.

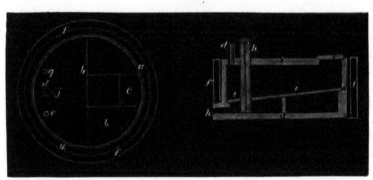

which reaches to the true bottom of the still, and through which steam can be admitted among the materials; *f*, *f* is the intermediate space between the stone still and its double iron casing, to heat the apparatus externally by steam.

Mr. Pattinson found, in working this apparatus, that great inconvenience and loss arose from the breaking of the stone vessel or still, and the consequent escape of more or less hydrochloric acid into the space between its outer surface and the inner surface of the iron case in which it is placed, as well as great delay, inconvenience, and loss, from the necessary frequent removal of the stone vessels themselves. He avoids this great

FIG. 211.

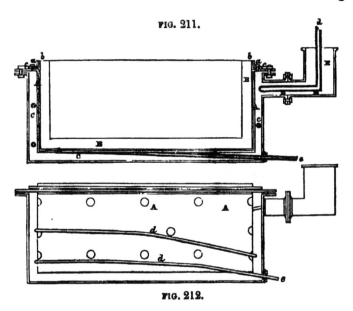

FIG. 212.

loss by permitting the use of the stone vessels, after they may have become cracked or broken, in such a way as would render them quite useless in the old mode of working. The method by which this is accomplished, consists in providing a case made with an external and internal iron vessel, shown in Figs. 211 and 212, similar to his former steam case, except that the inner iron case, A (Figs. 211 and 212), instead of being whole, as formerly, is perforated with a number of holes, as shown in the figures. Into this perforated iron plate, the stone vessel B is inserted, but the upper edge of the stone vessel, rises above the edge of the iron pan about 5 or 6 inches, as shown in the figure. Around this projecting portion of the stone vessel or still is placed a ring of iron a, a (Fig. 211), a little larger than the stone, so as to leave a space of from $\frac{1}{4}$ to $\frac{1}{2}$ an inch all around; b, b, its flange, resting upon the flange of the inner and outer pans. The inner iron pan and this ring are firmly bolted together, and rendered water tight by cement. The bottom of the space between the upper edge of the stone vessel and the iron ring is covered with a strip of vulcanized india-rubber c, c. A layer of fire-clay half an inch thick is pressed upon this, and the remainder of the space b, b, filled with melted lead, which is then caulked so as to render the whole space c c, between the outside of the stone vessel and the outer iron pan, steam- or water-tight under considerable pressure. Communicating with this space by the pipe D, is a small iron cistern E, 10 or 12 inches diameter and 3 feet high, its upper edges being about 2 feet above the upper edge of the stone vessel, and in this cistern is a lead pipe d through which steam may be passed so as to heat its contents when necessary; and this pipe is continued and coiled two or three times quite round the inner iron pan, in the space between it and the outer one, as at d d, Fig. 212, to where it issues at e, at which point it has passed through the outer iron case and the opening secured by a flange. On beginning to work, heated coal-tar (which has been previously thickened by boiling to such a degree as to be stiff when cold, but not brittle,) is run into the iron cistern E, until the whole of the space C, between the outer iron pan and the stone, and also the pipe, is completely filled in, and until the iron cistern is nearly full.

These stone stills have now superseded every other kind. On the Tyne, they are formed out of solid blocks of stone, so that there are no joints or external framework necessary. When the dressing of the stone is finished, the still is heated to expel all moisture, and boiled for 24 hours in a large iron pan filled with coal tar, which penetrates the outer skin of the stone; when cold it is found to be covered with an almost impervious coating.

In Lancashire, a different plan is adopted in constructing these stills. They are formed of stone flags 6 to 8 inches thick, 10 feet long by 5 feet wide, and 3 feet deep inside. Figures 213 and 214 show the way in which these stones are put together, the joints being made with a cement of china-clay and tar. Along each side of the flag is placed a range of fire-bricks, a a, on which the ends of narrow stone flags b, b, rest. These flags are placed so as to leave open spaces c, c, between them. d is the stone pipe for admitting the current of steam; e, e, are man-hole doors for charging the materials; f the escape pipe for the chlorine; g the stone pipe for removing the waste still-liquors; and h, a small

FIG. 213.

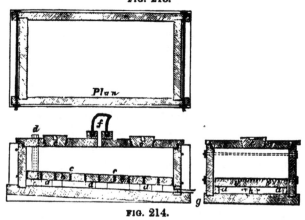

FIG. 214.

man-hole door for cleaning out the refuse which accumulates on the bottom, under the stone flags b, b, on which the manganese is spread. These stills are sometimes cased in stone, the intermediate spaces being used for the admission of steam for heating the inner stonework.

Some of the stills used in Germany differ from those now in use in this country : they are made of stoneware. The following description of this kind of still is given by Gentele. A, Fig. 215, is the stoneware still, a the

FIG. 215.

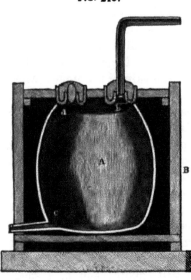

opening for charging the materials, b the exit pipe for the chlorine, c the escape pipe for the waste liquors, B the cistern between which and the still, steam is admitted. The necks a and b project above the top or cover, and are surrounded with a rim, in which a cement of clay and linseed oil can be pressed down to make an air-tight lute. Several of these stills are placed in the same cistern, which is made of wood or stonework. The other kind approaches the English in character. It consists of an octagon-shaped stone vessel, the stone flags being partly bound together by the iron rings *c c*, and partly by the metal bed-plate, *a a.* The edges of the flags are grooved, and when brought together the groove is filled with cement, so as to make a tight joint. A small fire below heats the bed plate *a a*, which is protected from the action of the acid by a double course of fire-clay quarls or flags set in fire-clay. The stone cover *m n*, is also coated with cement. It is fitted with an opening, *o*, for charging the manganese, with a pipe, *p*, for the passage of the chlorine, and with a second bent funnel-pipe, *q*, for running in the

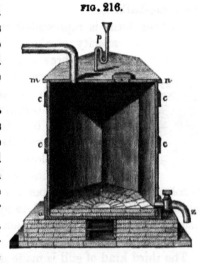

FIG. 216.

hydrochloric acid. The waste liquor is allowed to escape through the stoneware pipe and cock *a z*.

This stone still is sometimes heated by steam, in which case it is cased with an external iron or stone jacket, with an intermediate space between the still and its jacket.

Three forms of still are employed in France, one of which is shown in Fig. 217. A is a boiler made of thick sheet lead; B a vessel of stoneware to contain the hydrochloric acid; C an inner stoneware vessel to contain the manganese; D the gasholder, made of stoneware; E the exit pipe for the chlorine; F the furnace for heating the water-bath A; G G the smoke flue; R

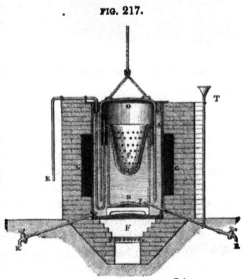

FIG. 217.

a stoneware pipe and cock to empty the vessel B; T T a lead pipe to run in a supply of muriatic acid without the necessity of raising the gasholder; and R' another stoneware pipe and cock to empty the water-bath.

Another form is represented in Fig. 218. The carboy, with a capacity of about 22 gallons, is made of stoneware with three openings; the largest, in the centre, admits a stoneware tube or vessel pierced with holes, A, the top of which can be covered by a stoneware lid, made tight with cement. The opening on the left is intended for the introduction of the hydrochloric acid, and that on the right is fitted with a lead pipe for the escape of the chlorine. The stoneware tube is filled with the manganese, and as the manganese is gradually dissolved by the acid, the portion in the upper part falls down within reach of the acid. A range of these carboys is placed in cisterns filled with a saline solution or sand, heated from below by a fire.

FIG. 218.

The third kind of still is made of the stoneware of the Vosges, Rive-de-Gier, or Volvic, c, c', c , built in brickwork. Each still has a capacity of about 35 cubic feet, and openings, P, P', P'', in front, which are closed by metal plates. The manganese and acid are charged through these openings, and the chlorine escapes by the lead pipes t, t', t''. The stills are heated by fires placed directly under them.

FIG. 219.

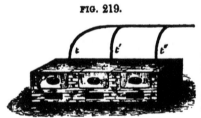

2. The Chambers, Receivers, or Condensers.

These vessels are constructed of lead, stone, or brick, and of different forms and dimensions. In some localities lead is preferred, as the cooling action of the thin metallic sides assists in preventing the temperature from rising too high; but they are more costly in the first instance, and more liable to wear and tear than chambers built of stone flags. The lead receivers a, a, are miniature acid chambers, the sheet lead being supported by an external framing of timber b, and the top sheets suspended from joists c, resting on the framework of the sides. Figure 220 (1) shows the side view of a still *d*, and the receiver *a*; Figures 220 (2) and 220 (3) are a section *e*, and a view of the bottom *f*, of the still; Figure 220(4) represents the stone flag-floor *g*, of the receiver. Large manhole doors are made in the sides of the receiver for the workmen to enter and spread the lime on the floor. These doors have glass windows to watch the progress of the operation, and there are openings near the bottom, through which the lime may be turned over by rakes.

In Lancashire these chambers are made with stone bottoms, the sides and top of lead. They are 30 feet long, 10 feet wide, and 6 feet high.

In other places, these receiving chambers are built of stone flags, and are arranged in rows in a large, lofty, and well-ventilated building, "as figured in Fig. 221. A is the passage between the two rows of chambers B, B, B; and D, D, D are open spaces between each pair of chambers, in order to keep them as cool as possible by a free current of air.

The stills are placed in an adjoining building, and the gas is conducted to the chambers by the pipes *a, a, a*. The gas, before entering the chamber, passes through small leaden purifiers *b, b, b*,

FIG. 220.

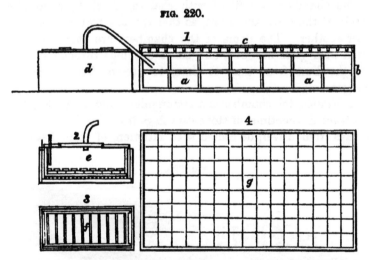

filled with hard burnt coke or flints, down which a small stream of strong sulphuric acid falls from the pipes *c, c,* which branch off into each purifier. This acid dries the gas, and flows away through the pipes *d, d, d.* The gas enters the chambers by the pipes *e, e, e.*

24*

FIG. 221.

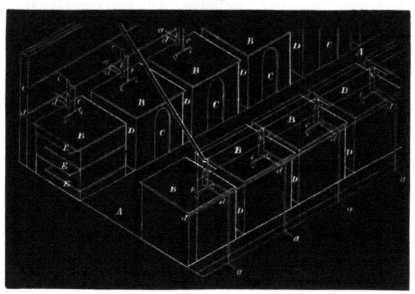

The chambers are fitted with shelves, E, E, E, open alternately at the back and front, so that the gas traverses one shelf
after another. The doors of the chambers are shown at c, c, c,
fitted with glass windows, and small openings through which to
stir the lime from time to time, to present fresh surfaces to
the gas.

In France, the chamber is a rectangular building M (Fig. 222),
constructed sometimes of stoneware flags from the Vosges, of flags
of lava from Auvergne coated with bitumen, of stone flags from

FIG. 222.

Rive-de-Gier, or of bricks carefully laid in cement. They are from 13 to 20 feet long, about 3 to 5 feet broad, and 2 to 3 feet high. At the front end, there is a small channel formed by a partition wall, in which any condensed liquid collects and escapes below by an overflow pipe. On the top at the farther end there is a safety tube. A door, K, is made at the far end for charging the lime. which is spread over the floor by a shovel; in some works it is thrown in through openings in the top or roof, and in other places small waggons of sheet iron, K (Fig. 223), are employed, which are pushed towards the partition wall at the fore end of the chamber by means of the long handles; the loose bottom of the waggon is drawn out by the rod

FIG 223.

shown in Fig. 223, and the lime is allowed to fall evenly over the floor to the depth of 6 or 8 inches.

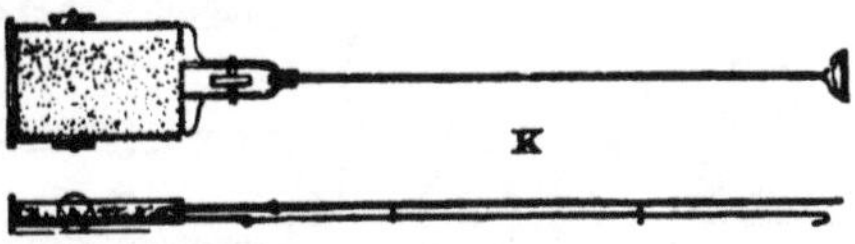

In large manufactories, the chambers are built with three or four shelves, the stone flags being supported by an external timber framing. The chamber A, A', shown in Fig. 224, is about 23 feet long, 10 feet high, and 3 feet broad. The interior is

FIG. 224.

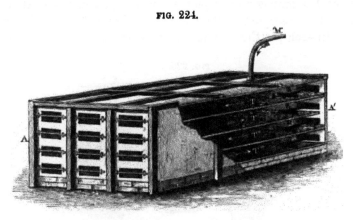

fitted with three shelves made of the same material as the chamber, or of an inferior quality of glass. They communicate by open spaces left at alternate ends, so that the gas passes in rotation over the floor and shelves T, T', T'', T''', after entering through the pipe M.

This form of chamber is modified in France by giving it much larger dimensions. The arrangements of still, purifier, &c., are the same as already described. The chamber is built of stone, but the shelves are made of a resinous wood, and wide apart, in order to prevent the over-heating of the materials, as represented in Fig. 225.

FIG. 225.

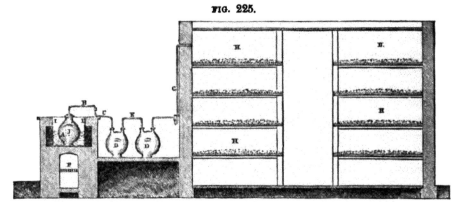

Each end of the chamber is furnished with doors for each shelf, through which the lime can be spread over its surface, and stirred during the progress of the operation.

Booth has proposed to employ a range of pots instead of chambers, for the absorption of the chlorine. Figure 226 explains his plan. In each pot there is an inverted leaden funnel, I, with a

FIG. 226.

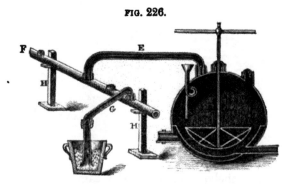

waved outline to facilitate the passage of the gas. The slaked lime is thrown loosely round the funnel. The pot is closed with a wooden cover, and the chlorine arrives from the still through a pipe E, which communicates, by means of water-lutes, with a supply pipe F, supported on stands H, H. The pots are placed in connection with this pipe by branch-pipes, G, to the number of from twenty to forty. These pots are placed in tanks and surrounded with water to keep down the temperature, which never exceeds 110° F., to which the greater density of the bleaching powder made in this way, is probably due.

In some manufactories revolving cylinders are employed to contain the lime. This was the plan first adopted in Germany, and was proposed by Oberkampf and Widmer as early as 1816. They filled a cylinder or barrel, A, two-thirds with the hydrate of lime, and the chlorine was admitted through a perforated pipe, d. The gas was conveyed from the still by a smaller pipe, e, which was luted to the larger one by means of a cement. The cylinder was supported by uprights, a and b, and was made to revolve by a handle or otherwise at e.

FIG. 227.

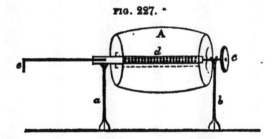

A similar arrangement is in use in this country, as represented in Figs. 228 and 229. A, A, A, are cylinders of ¼ inch sheet iron, about 5 feet diameter and 14 feet long, coated inside with cement. They are fitted with manhole doors, B, B, B, at one end, and are fixed on a strong timber framing, at such a height above the floor as to admit of placing a cask below them, to receive the bleaching powder, which is raked out into a sieve. The powder is then sifted direct into the cask. The axles, c, c, c, are hollow and perforated, to allow the gas to escape into the cylinders. The gas is supplied from the stills by the pipes D, D, D. Cogged or toothed wheels, E, E, E, are attached to each cylinder, and they are driven by worms, F, F, F, fixed on the driving shaft G. These worms are made to slide along the driving shaft, so that any cylinder may be thrown in or out of gear when necessary. The motion is very slow, being only from 12 to 15 revolutions per hour. Each cylinder is charged with 14 cwt. of a lime at a time.

FIG. 228.

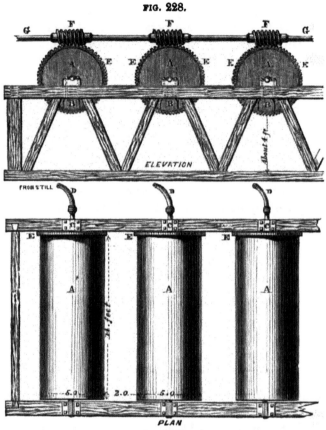

ELEVATION

FROM STILL

PLAN

FIG. 229.

II. THE PROCESS WITH MANGANESE.

There are three plans now followed for generating the chlorine: the one consisting in the use of common salt, peroxide of manganese, and sulphuric acid, in which case sufficient acid must be employed as will leave bisulphate of soda and sulphate of manganese behind in the still. The theoretical proportions are as follows :

Common salt, NaCl	58·5		Chlorine	35·5
Peroxide of manganese, MnO²	43·6		Bisulphate of soda, NaO.HO.2SO²	120·0
Sulphuric acid, 3(HO.SO³)...	147·0	=	Sulphate of manganese, MnO.SO²	75·6
			Water, 2HO	18·0
	249·1			249·1

Or, instead of the common salt, some makers substitute hydrochloric acid, when the proportions become—

Peroxide of manganese, MnO^2	43·6		Chlorine, Cl	35·5
Hydrochloric acid, ClH	36·5	=	Sulphate of manganese, NaO.SO^3	75·6
Sulphuric acid, HO.SO^3	49·0		Water, 2HO	18·0
	129·1			129·1

And, lastly, when hydrochloric is the only acid employed.—

Peroxide of manganese, MnO^2	43·6		Chlorine, Cl	35·5
Hydrochloric acid, 2HCl	73·0	=	Chloride of manganese, ClMn	63·1
			Water, 2HO	18·0
	116·6			116·6

The actual proportions differ, however, from the above theoretical quantities, as the oxides of manganese of commerce are never pure, containing lime, baryta, and oxide of iron, and only from 60 to 75 per cent. of peroxide of manganese, all of which combine with the acid and necessitate the use of a proportionally larger weight.

As regards the first plan, the additional equivalent of sulphuric acid is required, from the circumstance long since pointed out by Mitscherlich, that when sulphuric acid acts on common salt, it forms bisulphate of soda, which is only afterwards decomposed by a high heat: thus—

$$\left.\begin{array}{c} 2NaCl \\ 2(HO.SO^3). \end{array}\right\} \text{ yield } \left\{\begin{array}{l} NaO.HO.2SO^3. \\ NaCl + HCl. \end{array}\right.$$

And hence the proportions previously given. (See page 339.)

The materials are thrown into the still, the fire lighted and urged, until the temperature of the interior rises to 180° F., and the agitator occasionally turned round to raise the manganese and bring all the materials into contact.

When hydrochloric acid is employed, the manganese in coarse powder is thrown on the stone shelf of the still, and the acid run in until it reaches the guage. Steam is then admitted into the outer case and allowed to act for 36 to 48 hours, during which time the temperature of the materials rises to 180° F. Mr. Pattinson employs the manganese in lumps, and when the interior temperature has reached 180° F., which it does in about 18 hours, he introduces steam at 20 to 25 lbs. pressure, through the steam-pipes, among the materials. The steam is not blown in continuously, as the chlorine would be evolved too rapidly, but at intervals of half-an-hour, on and off, until the temperature of the materials has reached 212° to 220° F. This lasts about 6 hours, when the charge is worked off, or about 24 hours after first starting.

The advantages of this mode of working are, that the production of chlorine from the same weight of materials is increased, while the process is expedited. The grinding of the manganese is also rendered unnecessary, and there is no occasion for any mechanical agitation of the materials, for the steam produces sufficient movement to free the lumps of manganese from any sediment deposited upon them, and thus continually presents a fresh surface to the action of the acid.

The steam must not be blown among the materials until they have reached a temperature of 180° F., to prevent a reduction of the strength of the acid, and to avoid the danger of explosions.

The chlorine is dried by passing through the purifiers, as explained at p. 371, before being allowed to enter the chambers.

By using steam of this pressure, the same necessity does not exist, as in other plans, for the external application of heat, as there is less risk of the acid being weakened by the condensation of the steam.

In working Mr. Pattinson's admirable plan, it is essential to keep his small iron cistern, E (Fig. 211, p. 366), pretty full of thickened coal-tar in a melted state; for this obvious reason, that the pressure of the coal-tar being into the still, any cracks in the stone are always filled up with the tar, and thus any leakage of acid into the iron case is prevented.

We have already noticed the importance of a proper hydration of the lime to insure the absorption of chlorine, so as to produce the full equivalent of hypochlorite; and we have now to draw attention to the equally important precaution of a proper temperature in the chambers. If the absorption be too rapid, a great development of heat ensues, Morin having found the temperature to rise to about 246° F. under such circumstances. Dingler states that when this occurs, oxygen is evolved, which statement, however, is not confirmed by Morin. It has been found that when the temperature ranges from 90° to 240° F., the lime absorbs the same quantity of chlorine as in the cold; but the product possesses only two-thirds of the bleaching power, and one-third of the' hypochlorite of lime is converted, partly into chloride of calcium, but chiefly into chlorate of lime.

We therefore see the necessity of conducting the operation very slowly, the advantage of low chambers with a thin layer of lime, and every facility for withdrawing the heat evolved by the chemical action.

The stills are usually charged with from 3 to 5 cwt. of manganese, and three charges are worked off in about 90 hours. Thus, in one manufactory, 14½ cwts. of lime are spread over the shelves of the chamber, and two stills are charged with 5 cwt. of 64 per cent. manganese, which, when worked off, is followed by a second charge of 4 cwt., and this by a third charge also of 4 cwt., or in all 26 cwt. of manganese. The produce varies from 24 to 26 cwt. of bleaching powder, of 35 to 38 per cent. strength. The steam is blown in four hours after charging, and withdrawn during the last hour, during which period there is a danger of some hydrochloric acid, chloride of manganese, and chloride of iron passing over; hence the necessity of washing the chlorine in the way we have described, before it is admitted to the chambers.

This is perhaps the proper place to state, that in another manufactory, where 350 tons of salt were decomposed weekly, and the strong hydrochloric acid employed in making bleaching powder, the following were the actual working results, reduced to a standard of 20 cwt. of salt.

Common salt . . . . 2240 parts.
Hydrochloric acid, at 28° Tw.   . . . . 2753 „
Manganese, at 60 per cent. . . . . . 448 „
Bleaching powder obtained of 39 per cent. . 416 „

The large stills in Lancashire are worked in the following manner: eight stills supplying five chambers with gas. Each still is charged with 7 to 8 cwt. of manganese, and 90 cubic feet of hydrochloric acid of 28° Twaddell. The mode of working the still is the same as we have described, and the charge is finished in 24 hours. Each chamber is filled with 43 cwt. of lime, and requires a week for saturation, when the produce is about 64 cwt. of bleaching powder of 35 per cent. strength.

The manganese is spread on the stone flags, and as it dissolves, the insoluble matters are washed off its surface, by the agitation of the liquid caused by the current of steam, and they fall through the open spaces c, c, below the false bottom, whence they are removed by a rake through the manhole h, as often as the accumulation renders it necessary.

In France, with the apparatus previously described, the chlorine produced from 8 carboys, saturates 137 kilgr. of hydrated lime, which yields 200 kilgr. of ordinary bleaching powder.

Mr. Burnett has very clearly pointed out some of the disadvantages attending this mode of operating, by which it is necessary so often to recharge the stills. If the chlorine is generated too rapidly, or if the lime is nearly saturated, the stills being connected with only one chamber, and the absorbing power of the lime being diminished, there is great danger of a loss of chlorine by leakage. To obviate this risk, in some measure, the materials are very gradually mixed, by which much time is lost, and another detention occurs in the discharging and recharging of the stills. The chlorine, as well as the chambers, varies in temperature in this system of working, and the chlorine may reach the lime in a moist state, thereby injuring the quality of the bleaching powder. He proposes to remedy these defects in the present plan, by working the stills and chambers in common, collecting the chlorine as it is produced in the stills into one main pipe, and conveying it by branch pipes to the chambers. He adds that greater efficiency would be given to the apparatus; the chlorine might be generated more rapidly than is safe at present; a uniform pressure could be maintained; the absorption by the lime would be regular and uninterrupted; a larger produce and a better quality of bleaching powder, would be obtained; and there would be less escape of gas, so prejudicial to the health of the workmen.

This suggestion has been, in some measure, anticipated by the arrangement of stills and chambers patented by Mr. Gamble. described at p. 363, in which the lime, already partially saturated with chlorine, is exposed to the strongest gas, and what remains is brought in contact with a surface of fresh lime.

The time occupied in saturating the lime varies, of course, according to the quantity, &c., from 36 to 48 hours; but when the layer of lime is thin, and the apparatus cool, it is sometimes completed in 24 to 30 hours. During the absorption, the surface of the lime is renewed from time to time by raking it through the openings left for this purpose.

The workmen are exposed in all the manipulations of this process to the action, either of the vapours of the hydrochloric acid, or of the chlorine, which unavoidably escapes at different times. It is advisable to have at hand, either milk or eggs beaten up with sugar and water, which the men can drink whenever they are seriously affected by this vapour or gas.

Payen also recommends the following arrangements, which are more applicable in France than in this country, whenever the carboys are to be emptied, or the acid transferred from one to another carboy. The glass syphon, CC (Figure 230), is fitted with a neck of gutta-percha, B, which, when pressed down, closes the neck of the carboy, and suspends the working of the syphon. By blowing air through the small side tube D, the acid in the carboy is forced up the shorter limb of the syphon, which is set in operation ; the rapidity of the flow can be regulated at pleasure, by pressing the little tube between the fingers.

FIG. 230.

FIG. 231.

Hydrate of lime does not absorb all the chlorine which theory indicates it ought to do, and the stability of ordinary bleaching powder has been ascribed to the presence of this excess of lime, while a similar stability of the solution has been supposed to depend on the excess of water. The following table shows the relation they bear to each other.

Composition of Chloride of Lime

In powder.	In solution.
Chlorine . . $2Cl$ = 71	Chlorine . . $2Cl$ = 71
Lime . . $4CaO$ = 112	Lime . . $2CaO$ = 56
Water . $4HO$ = 36	Water . . $4HO$ = 36
219	163
Or,	Or,
Chlorine . . . . 32·4	Chlorine . . . . 43·6
Lime . . . . . 51·1	Lime . . . . . 34·3
Water . . . . . 16·5	Water . . . . . 22·1
100·0	100·0

Modifications of the Process.

Some chemists and manufacturers have proposed to obtain the chlorine, when employing manganese, by plans differing from those already described which are in ordinary use. They may be classified under the two following divisions:

1. Hydrochloric acid through the manganese.
   Maugham, 1836.
   Seybel, 1842.
   Clement, ?
   Monod, 1856.
2. Earthy and metallic chlorides.
   De Sussex, 1847, chlorides of magnesium and aluminium, with manganese, with or without sulphuric acid.
   Binks, 1853, chlorides of manganese, zinc, &c., with manganese and sulphuric acid.
   Barrow, 1856, dry air over chloride of manganese.
   De Luna, ? sulphate of magnesia and common salt, with manganese.

1. Hydrochloric Acid through Manganese.

Maugham was the first to propose to employ the hydrochloric acid gas, as it was generated during the decomposition of the common salt by sulphuric acid, to act directly on the manganese. He filled an upright cylinder with the peroxide of manganese in lumps, which he kept moistened by allowing a small stream of water to fall over the lumps. The cylinder was fitted with a casing or jacket for the admission of steam, to maintain the materials in the cylinder at a temperature of 130° F. The hydrochloric acid gas entered at the top of the cylinder and passed down through the manganese, and the gases escaping at the bottom were washed in a cistern containing water. The chlorine was then conveyed to the chambers.

Seybel and Monod both employ the hydrochloric acid as it is generated, but they pass it up through the manganese, which is covered with water. Monod places his manganese on a perforated stone shelf, while Seybel keeps his manganese in motion by means of an agitator.

We give the details of Seybel's plan (Fig. 232), to illustrate the nature of this modification. a, a, represents the leaden vessel or retort, in which the process of manufacturing sulphate of soda is performed by decomposing common salt by sulphuric acid; b is a man-

FIG. 232.

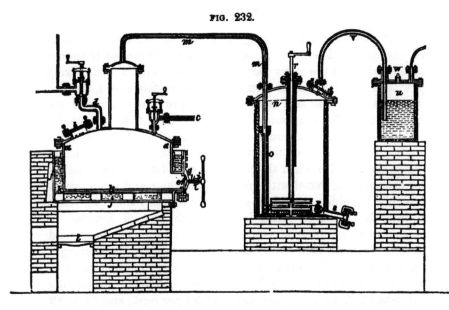

hole, with a cover, by which the charge of common salt is thrown
into the vessel or retort *a a*, the cover being securely fixed
by clamp-screws during the process of decomposition; *c* is a pipe
leading to the chimney, with a valve to close it while the process
of decomposition is being performed; *d* is a pipe leading to a
reservoir or vessel containing sulphuric acid, of specific gravity
1·71, with a valve to close it when the proper quantity has
been run into the vessel or retort *a a*; *e* is an outlet for running
off the charge, with a cover of lead, *f*, retained by the iron
frame *g*, which is screwed in its place by the screw *h* passing
through a fixed arch bar *i*. The retort or vessel *a a*, is placed
in an iron vessel *j j*, which contains oil, or other suitable
material, to act as the heating medium, and the vessel *j* has
false bottom or frame *k*, which is perforated so as to allow the
heating medium to come in contact with the bottom of the
vessel or retort *a, a*. The vessel *j* is heated by a fire or furnace
at *l*; and there should be a damper in the flue leading to the
chimney, to regulate the heat of the retort, which should not
exceed 330° F. 20 cwt. of salt is thrown into the retort, and
30 cwt. of sulphuric acid, of 1·71 run upon it. The hydrochloric
acid passes off through the pipe *m m*, into the vessel *n*, containing the manganese. This vessel is lined with firebrick or

earthenware tiles, for resisting, as much as possible, the action of the acid. The pipe m is connected with an earthenware pipe o, which at the lower part is connected with a hollow ring p, with several small holes around the inner surface at q, q, through which the muriatic acid is conducted below the water and the manganese. The manganese is agitated, from time to time, by an agitator z, made of iron covered with lead, working through a stuffing box r. s is an outlet for removing the materials remaining in the vessel n when the process is complete; t is a manhole for charging the vessel with ground manganese and water, and v is a tube by which the chlorine is conveyed away from the vessel n into the vessel u, which is lined with tiles of fire-clay or earthenware, where it is washed by passing through water, and thence into a similar vessel, again to be washed, there being a small quantity of manganese, 10 lbs. in each of the vessels u. The chlorine may then be admitted to the chambers. In charging the vessel n, 7 cwts. of manganese of 62 per cent., and 11 to 12 cwts. of water, are thrown in, and the manhole is closed. These charges will be worked off in about fifteen hours, and care should be observed to keep the process, during that time, as equal as possible. w is a plug in the vessel u, for drawing off the water by a syphon, after the process has been completed, which water is used on charging the vessel n. The workman can also ascertain at the plughole, towards the end of the process, whether chlorine is coming over; if not, he should stir with the agitator, in order to cause the manganese to float in the water. When chlorine ceases to come over, the process is complete, and the workman, after a little practice, with care, will judge, by listening at the pipe or tube v, whether the process is going on regularly; if he finds, by the quick bubbling of the vapours in the vessel u, that the materials in the vessel n are boiling too freely, he will cease stirring, and if that does not retard the action sufficiently, then he should damp the fire.

It is almost unnecessary to point out the many practical difficulties of these plans. The difficulty of regulating the relative supplies of gas and manganese, the necessity of greater heat than could be managed in the vessels with the manganese, the danger of hydrochloric acid passing over with the chlorine, and many other objections, are obvious; and we believe that no attempt has ever been made to carry them into practical operation.

2. Earthy and Metallic Chlorides.

De Sussex proposed to replace the common salt, in the process of making chlorine already described, by the chloride of magnesium or aluminium, the process being in other respects the same. He afterwards boiled the still-liquors to dryness, and heated the resulting dry sulphates in earthenware retorts, to recover the sulphuric acid. Or, he precipitated the manganese by magnesia when using chloride of magnesium, and peroxidised the manganese by one of his methods explained at p. 356, while the sulphate of magnesia was recovered.

Binks added to these chlorides, that of manganese obtained from the still-liquors, as well as those of zinc and calcium.

De Sussex further proposed to mix any of the above chlorides with peroxide of manganese or any salt of manganic acid, all in a perfectly dry state, and to expose them to a red heat in an earthenware retort. He mixed 48 parts of dry chloride of magnesium or 46 parts of dry chloride of aluminium with 80 parts of dry peroxide of manganese, so that the hydrochloric acid evolved from the chloride, meeting the nascent oxygen of the peroxide of manganese, was converted into chlorine and water.

Barrow's plan consists in boiling down the still-liquors to dryness and heating the dry residue to redness in a retort, while a current of dry air was passed over it, by which he claimed to obtain a supply of chlorine for making bleaching powder, and recovered an oxide of manganese fit for further use.

Ramon de Luna has also suggested that chlorine may be obtained by highly heating a mixture of common salt, sulphate of magnesia, and peroxide of manganese ; thus,

$$
\left.\begin{array}{l}
2(MgO.SO^2 + HO) \\
\\
2NaCl . \quad . \quad . \quad . \\
\\
MnO^2 . \quad . \quad . \quad .
\end{array}\right\} \text{produce} \left\{\begin{array}{l}
2(NaO.SO^2). \\
2MgO. \\
MnCl. \\
Cl. \\
2HO.
\end{array}\right.
$$

None of these plans have been conducted on a manufacturing scale.

III. THE PROCESS WITHOUT MANGANESE.

The proposals which have been made to dispense with the manganese have all failed in practice, either from some difficulties in the manufacturing operations or from being too expensive in working, with the exception of that patented by Dunlop.

All these plans may be conveniently arranged in the following way, as respects the sources of the chlorine.

1. From hydrochloric acid and other gases.

    1839, BINKS, hydrochloric acid by carbonic acid, &c., and heat.

    1845, OXLAND, hydrochloric acid, air, and heat.

    1846, JULLION, hydrochloric acid and oxygen through heated asbestos or spongy platinum.

2. From metallic chlorides.

    1845, LONGMAID, roasting pyrites and common salt, or heating metallic chlorides with air.

    1847, LONGMAID, heating metallic sulphates and common salt with air.

    1855, THIBIERGE, air over heated protochloride of iron.

3. From chromates and hydrochloric acid.

    1847, McDOUGAL and RAWSON.

    1857, SHANKS.

These processes differ from each other in the methods pursued for restoring the chromate.

4. DUNLOP'S process.

1. *From Hydrochloric Acid and other Gases.*

Binks has proposed to pass hydrochloric acid gas, obtained from the usual source, along with the gases produced by heating nitrate of soda to redness in an iron retort, through a red-hot porcelain or earthenware tube, in the proportion of 4 volumes of the former to 1 volume of the last-named gases. He obtains a mixture of chlorine, nitrogen, and water, which he washes in water to remove any undecomposed hydrochloric acid, and conveys the purified gases to an ordinary chamber filled with lime.

He has also proposed to decompose hydrochloric acid gas by mixing it with carbonic acid, in the proportion of 1 volume of the latter to 2 volumes of the former, and passing the mixture through red-hot earthenware tubes, by which he obtains chlorine, carbonic oxide, and water. The resulting gases are cooled, washed with water, and applied to the production of bleaching powder.

Oxland's plan consists in passing a mixture of 1 volume of dry hydrochloric acid gas and 2 volumes of atmospheric air, through an air-tight reverberatory furnace, filled with pumice-stone kept at a red heat. The gaseous products containing chlorine are cooled, washed with water, and conveyed to the usual chamber.

Jullion proposes a similar plan, but substitutes asbestos or spongy platinum for the pumice-stone.

In these processes, the oxygen of the nitrous compounds, of the carbonic acid, and of the atmospheric air, oxidises the hydrogen of the hydrochloric acid, to form water, thus liberating the chlorine; but, simple as they appear in theory, the practical difficulties attending the methods proposed to carry them into execution, would deter any manufacturer from undertaking them.

2. From Metallic Chlorides.

Longmaid obtains chlorine by passing dry air over a roasted mixture of pyrites and common salt, or over a mixture of a metallic sulphate and common salt, or over a metallic chloride; and Thibierge produces chlorine from dry protochloride of iron by passing dry air over this salt when heated in perforated clay vessels.

These three patents are substantially the same as far as relates to the production of the chlorine, which, in each instance, comes ultimately from the oxidation of the metallic chloride by the oxygen of the air, and in this respect are analogous to Barrow's process.

In Longmaid's first process, the roasting of the mixture of the pyrites and common salt results in the evolution of sulphurous acid and the formation of sulphate of iron, so that from this point it resembles his last-patent. His processes, therefore, as well as that of Thibierge, become practically identical with that of Oxland in their resulting products, with the addition of sulphurous acid in the plan of Longmaid.

The reaction is the same as takes place in the roasting of silver ores preparatory to amalgamation. The mixture of sulphate of iron and common salt, when heated, undergoes the change shown in the equation,

$$Fe^2O^3.3SO^3 + 3NaCl = 2(FeO.SO^3) + NaO.SO^3 + 2NaCl + Cl;$$

and by the admission of the air the latter undergoes this decomposition:

$$2(FeO.SO^3) + NaO.SO^3 + 2NaCl = Fe^2.O^3 + 3(SO^3.NaO) + 2Cl.$$

Longmaid employs a rectangular furnace about 20 feet long by 7 feet wide, the chamber of the furnace being made as air-tight as possible, and heated by flues above and below. The dry air is driven into the chamber of the furnace by means of a fan, and the gases generated are carried through a wooden flue 40 feet long by 1 foot square, which is immersed in water, to remove any hydrochloric or sulphurous acids or metallic chlorides carried off with the chlorine.

We have seen this process in operation, but, from the large admixture of nitrogen and atmospheric air, it was impossible to bring the bleaching powder above 20 per cent.; and the same difficulty will be encountered whenever any of these plans are tried, in which little or no pressure can be exerted in the receiving chamber, from the large proportion of residual gases arising from the use of the atmospheric air as the oxidising agent.

3. From Chromates and Hydrochloric Acid.

McDougall and Rawson employ any of the chromates or bi-chromates with hydrochloric acid in a free or nascent state, whereby chlorine is evolved, and a chloride or sesqui-chloride of chromium is formed together with a chloride of the base of the chrome-salt. They treat these latter products with nitric acid and distil off the hydrochloric acid, thus obtaining the nitrate of chromium and of the base of the chrome-salt. They then heat these nitrates and condense the lower oxides of nitrogen which are given off, in the following manner:

The gases are conveyed into a vessel containing water, by a pipe dipping below the surface of the water. Air is permitted to enter into this vessel and to mix with the gas which bubbles through the water. This mixture of air and nitrous gas is drawn, by a pneumatic apparatus, through a series of vessels similar to the first, each containing a tube dipping into the liquid, and another tube or pipe attached to its top to connect it with the next vessel, thus making the nitrous gas to pass alternately into air and water, by which means it is converted into nitric acid. Thus, when $3NO^4$ is passed into water at a temperature of 100° F. or upwards, $2NO^5 + NO^2$ results, the $2NO^5$ (i.e., two atoms of nitric acid) remain in solution, while the NO^2, which is uncondensable gas, bubbles through the liquid and mixes with the air in the vessel above the liquid; it instantly takes two atoms of oxygen from the air, and becomes NO^4, which, passing again through the liquid, becomes nitric acid and nitrous gas as before, and thus nearly the whole of the nitrous vapour or gas is reconverted into nitric acid. When the temperature of the liquid into which the nitrous gas is passed, is much under 100° F., a slightly different but equally beneficial reaction ensues.

The residue, after heating the nitrates, consists of the chromate or bichromate first employed, and can again be used. The patentees prefer to use the chromate of lime.·

Shanks's process consists in placing a quantity of chromate of lime in an ordinary stone still, and adding as much hydrochloric acid as will decompose it, which point is easily ascertained by the liquid becoming of a beautiful grass-green colour. The charge in the still works very well and yields a gradual and steady supply of chlorine, until more than one-half has been expelled; it is then advisable to apply heat by steam or otherwise, either internally or externally, as we have already explained when describing the bleaching powder stills, in order to drive off the last portions of chlorine.

The contents of the still, a solution of the chlorides of calcium and chromium, is then run into a large open stone tank, a quantity of hot water is added, and the whole mixed together. Lime previously slaked, in the form of milk of lime, is added to the liquors in the tank, while being well stirred with a paddle, until the free acid is just neutralised. As much more milk of lime is now added as will, first of all, precipitate the oxide of chromium, and afterwards a further quantity, sufficient to combine with the chromic acid which is formed in the next operation.

The precipitate is allowed to subside, and the resulting solution of chloride of calcium is run off from the precipitate of oxide of chromium and lime, which is placed on a drainer. The precipitate having become a little stiff, is thrown into a close furnace, with apertures for the free admission of atmospheric air into the bed or hearth on which the material is placed. The heat of this furnace is kept at low redness, when the oxidation of the materials goes on rapidly, so that the whole mass is transformed into chromate of lime, which can be employed for another operation.

Mr. Shanks informs us that they have made many tons of chloride of lime by the above process.

4. Dunlop's Process.

This beautiful chemical process is described by the late lamented and talented inventor in the following general terms:—" I produce chlorine by effecting mutual decomposition between the following substances :—muriate of soda, or any other muriate, nitrate of soda, or any other nitrate, muriatic acid, nitric acid. The assistance of sulphuric acid is in some cases required. I generally prefer employing it, in order to obtain for residuum, sulphate of soda, suitable for the manufacture of soda, &c. The above substances may be employed either altogether or only two of them ; as, for instance, a nitrate with a muriate (in which case, of course, sulphuric acid must be employed) ; or a muriate with nitric acid, or a nitrate with muriatic acid ; or muriatic and nitric acids may be used together."

The apparatus consists of large cast iron cylinders, which are laid horizontally and built over a furnace : both ends of the cylinders are clear. They are from 7 to 8 feet diameter and 6 to 7 feet long, and the front end is fitted with a manhole for charging the mixture of nitrate and muriate of soda, among which the sulphuric acid is run in, through a pipe descending from a cistern placed above the level of the cylinders. The gases escape by an earthenware pipe at the other end of the cylinder, and pass through five Woulff's bottles, made of lead, which are filled with sulphuric acid to a depth of 24 inches. This large quantity of acid is necessary to moderate, by the pressure it exerts, the violent ebullition which ensues when the materials are mixed in the cylinders. All parts of the apparatus require to be made very tight, in consequence of this great pressure.

Each cylinder is charged with 12 cwts. of nitrate of soda, and the equivalent quantities of common salt and sulphuric acid of about 150° Twaddell. The quantity of acid used is sufficient to form bisulphate of soda, which, after the decomposition is complete, is run off in a liquid condition, into a furnace containing common salt, where a good furnace sulphate-ash is produced for the manufacture of alkali.

The chemical decomposition is explained by the equation,

$$\left. \begin{array}{l} \text{Common salt} \quad . \quad 2NaCl \\ \\ \text{Nitrate of soda } NaO.NO^5 \\ \\ \text{Sulphuric acid } 6(SO^2.HO) \end{array} \right\} = \left\{ \begin{array}{l} \text{Bisulphate of soda} \\ \qquad\qquad 3(NaO.HO.2SO^3) \\ \text{Chlorine} \quad . \quad . \quad . \quad . \quad 2Cl \\ \text{Nitrous acid} \quad . \quad . \quad . \quad . \quad NO^3 \\ \text{Water} \quad . \quad . \quad . \quad . \quad . \quad . \quad 3HO. \end{array} \right.$$

And when nitric acid is substituted for the nitrate of soda, the products are the same :

$$4(SO^3.HO) + NO^5 + 2NaCl = 2(NaO.HO.2SO^3) + 2Cl + NO^3 + 2HO.$$

The charges are worked off in about 36 hours, at a temperature of 400° to 500° F.

The nitrous acid is absorbed by the sulphuric acid as the gases pass through the condensers; the hydrochloric acid is removed by water in an ordinary tower; and the chlorine is conveyed to the usual chambers.

The nitrous-sulphuric acid is employed in the manufacture of sulphuric acid in the manner previously described.

Mr. Dunlop also proposed to employ this nitrous-sulphuric acid for another purpose. He expelled the nitrous compounds and received them into a suitable chamber, into which a current of sulphuretted hydrogen was streaming at the same time, by which he obtained sulphur and sulphate of ammonia, thus—

$$4SH + NO^4 = NH^4O.SO^3 + 3S.$$

MANUFACTURE OF BLEACHING LIQUIDS.

We have already pointed out that when chlorine is brought in contact with lime, in the presence of an excess of water, the chloride of lime obtained in this case is much richer in chlorine than in the bleaching powder; and inasmuch as the latter is always dissolved in water before being used, many parties prefer making a liquid chloride of lime. Several other liquid chlorides are also manufactured, the details of which will now be explained.

1. *Liquid Chloride of Lime.*

The apparatus employed at Mulhouse for the production of this bleaching liquid, consists of a range of glass vessels *a a a*, in which the chlorine is generated from a mixture of hydrochloric acid and peroxide of manganese.

FIG. 233.

FIG. 234.

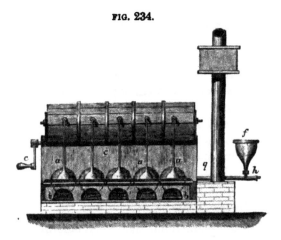

The glass globes are placed in separate sand baths, the whole forming a kind of galley furnace (Galeerenöfen), $b\,b$, which is made of cast iron, lined with firebrick. Each vessel is heated by its own fire, and the smoke escapes through the pipe $q\,q$.

The vessel c, which contains the lime and water, is made of sandstone, of a cylindrical shape, and covered with wood d, which is coated with coal-tar varnish, so as to be perfectly air-tight. This cover fits into grooves in the walls of the cylindrical stone tank, and the joint is made fast by cement. A spindle passes through this vessel, to which a number of arms with boards are attached, and the whole is turned by a winch. This spindle with its agitators is shown in Figures 235 and 236.

FIG. 235. **FIG. 236.**

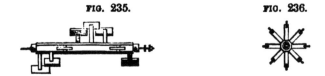

The milk of lime is run in through a funnel f, and the liquid chloride of lime is drawn off at h.

The cylinder is filled with 3 lbs. of lime and 80 lbs. of water, and the chlorine, after being washed, is introduced just above the level of the liquid through stone or lead pipes. The contents are agitated during this period, which serves the double purpose of presenting a fresh surface of lime to the gas, and of preventing any excessive elevation of temperature.

Another arrangement consists in employing a range of carboys *c*, of peculiar construction, as shown in Fig. 237.

The lime and water are introduced through the wide opening E, until the liquid rises high enough to shut off the access to the air. The chlorine enters through the pipe B in a very slow stream, during which a workman stirs up the lime from the bottom by a rod T. Any chlorine not absorbed, escapes through the pipe D. At the end of about 12 hours the saturation is complete, when the contents are allowed to stand that the liquid may become clear, when it is syphoned off for use.

FIG. 237.

Schlieper has made some valuable experiments on this liquid chloride of lime, and has found that strong solutions evolve oxygen when heated to 212° F., while weak solutions do not give off any, even when boiled for a length of time at a temperature of 230° F., but are decomposed into chlorate of lime and chloride of calcium,

$$3(CaO.ClO) = CaO.ClO^5 + 2CaCl$$

2. *Liquid Chloride of Magnesia.*

Claussen prepares this liquid by adding an equivalent quantity of sulphate of magnesia to the liquid chloride of lime, and after the sulphate of lime has fallen down, he decants off the clear supernatant solution, which he states is more efficacious in bleaching flax and hemp than the other compounds of chlorine.

3. *Chloride of Soda, or Liquid of Labarraque.*

This solution may be prepared from liquid chloride of lime by double decomposition with carbonate or sulphate of soda, or by passing chlorine into a solution of caustic soda, thus—

$$2Cl + 2(NaO.HO) = NaCl + NaO \cdot ClO \, 2HO,$$

or into a solution of carbonate of soda, thus—

$$4(NaO.CO^2) + 2Cl = (NaCl + NaO.ClO) + 2(NaO.2CO^2).$$

In the first case, about 20 lbs. of good bleaching powder are dissolved in about 26 gallons of water, and the whole filtered. A solution of 40 lbs. of soda crystals in about 9 gallons of warm water is added, when carbonate of lime is precipitated. This is filtered, and the solution constitutes the liquid of Labarraque.

In the other mode of preparation, about 40 lbs. of soda crystals are dissolved in about 20 gallons of water, and a stream of chlorine is passed through the solution until it has a specific gravity of 1·06. When the density rises above this point, it is an indication of the formation of chloride of sodium and chlorate of soda.

4. *Chloride of Potash, or Water of Javelle.*

This substance, named from *Javelle*, near Paris, where it was originally made in 1792, was the first artificial bleaching agent prepared by manufacturers.

It may be made by any of the processes described for manufacturing the chloride of soda, and was replaced by this substance after Labarraque proved that they possessed the same properties. The colour of the first products was more or less of a rose tint, arising from the accidental mixture of a small quantity of a salt of manganese; but this colour is now an essential point with purchasers, and to meet the effect of long custom, a small quantity of the residue from the still is added to the pure colourless solution of chloride of potash.

1 part of potash is dissolved in 8 parts of water, and chlorine passed through the solution in the cold, when the chloride of potassium separates in fine crystals.

When chlorine is passed through a strong solution of ordinary carbonate of potash, silica separates, but little or no carbonic acid is expelled, as it forms a bicarbonate with the undecomposed salt, and the evolution of carbonic acid commences only after the passage of the chlorine has been continued for some time. The solution must not be supersaturated with the chlorine, as the hypochlorite undergoes decomposition in the presence of an excess of this gas.

When chlorine is passed through a mixture of 24 parts of potash and 1 part of water, a dry compound is obtained, composed of hypochlorite of lime, chloride of potassium, and bicarbonate of potash, which is a very good bleaching agent.

5. *Liquid Hypochlorite of Alumina.*

Mons. Orioli has patented the application of this substance to the purposes of bleaching and disinfection.

He prepares it by mixing equivalent quantities of hypochlorite of lime and sulphate of alumina. The sulphate of lime immediately falls, and the solution contains this bleaching agent.

It is a very unstable compound, and hence acts very rapidly on all organic colouring matters, in regard to which it may be compared with a mixture of hypochlorite of lime and an acid, but possesses the great advantage over all such mixtures in remaining neutral during the whole of the action. When these mixtures are used, it has been found impossible to remove the last traces of hydrochloric acid which is formed by the action of the chlorine on the hydrogen of the colouring matter. Fabrics bleached in this way often grow more or less brown and friable, and this defect of the ordinary methods of bleaching has led to the introduction of the various so-called "Anti-chlors," sulphites or hyposulphite of lime, soda, and alumina. Pelouze has explained the cause of the defect by proving the solubility of the cellulose in hydrochloric acid. The hypochlorite of alumina, by the energy of its action, and its conversion into chloride of aluminium, by parting with its oxygen while bleaching, meets all the objections to the use of such bleaching agents.

The chloride of aluminium formed in the act of bleaching, being an antiseptic salt, has the further advantage, when used in the manufacture of paper, of preserving the pulp from fermentation, so that it may be preserved for months without undergoing any change.

It is also applicable to the preservation of animal matters, embalming bodies, &c., as it forms very stable compounds with the protein and gelatinous tissues, and destroys the fermenting matters with which they are impregnated.

Orioli also states that it can be used as a mordant instead of acetate of alumina. At the heat of the bath, all the hypochlorous acid is disengaged, while the alumina remains in combination with the material.

6. *Liquid Hypochlorite of Zinc.*

Varrentrap and Sace have proposed to employ this compound on account of the energy of its bleaching action. When solutions of bleaching powder and sulphate of zinc are mixed, sulphate of lime and oxide of zinc are precipitated, leaving hypochlorous acid in solution, of which advantage might be taken in many cases. When solutions of bleaching powder and chloride of zinc are mixed, no precipitate is formed, as the chloride of calcium remains dissolved. The energy of the bleaching action of this solution renders it equal to that of the ordinary bleaching powder and a mineral acid, while there is no danger of destroying the fibre, as happens when some of the previous compounds are used. When it is employed in the manufacture of paper, the presence of the sulphate of lime and oxide of zinc would add to its value, as they remain mixed with the pulp as hardeners.

Waste Product of the Stills.

This waste product, consisting chiefly of the chloride of manganese, has attracted much attention, with the object of reconverting the manganese into peroxide. Several processes suggested for this purpose have already been described, and we introduce the following plans in this place, as they embrace other objects in addition to the recovery of the manganese.

The first is that recently proposed by Mons. F. Kuhlmann *fils*, but which had already been made the subject of two patents, by Binks in 1839, and by Gatty in 1857. The latter chemist, however, connected this plan with the manufacture of sulphuric acid, and the former with the peroxidation of protoxide of manganese, while M. Kuhlmann employs it for the production of nitric acid. It consists in heating nitrate of soda with chloride of manganese, when a mixture of nitrous acid and oxygen is disengaged, and common salt with oxide of manganese remains behind in the apparatus. The decomposition of the two salts commences below a temperature of 446° F., which heat must be carefully maintained during the operation. The reaction which ensues is represented by the equation—

$$5MnCl + 5(NaO.NO^5) = Mn^5O^6 + 5NaCl + 5NO^4 + 2O.$$

The oxide of manganese supposed to remain, might be viewed as a compound of $3MnO^2 + 2MnO$, but M. Péan de Saint-Gilles, as will be explained, has found it to contain chlorine.

The two salts are heated in a stone retort, and the gases which are evolved are passed through a series of vessels containing water. Nitric acid is formed, and some binoxide of nitrogen is disengaged, which, coming in contact with the air in the vessels, absorbs oxygen, reproducing a further quantity of nitric acid.

Kuhlmann states that 100 parts of nitrate of soda thus treated yield 125 to 126 parts of nitric acid of 1·320 sp. gr., which is nearly that obtained by the usual process.[*]

---

[*] The chlorides of calcium, magnesium, and zinc, furnish similar results, common salt and an oxide of the metal being formed. Sulphate of manganese likewise decomposes the nitrate, furnishing the same quantity of nitric acid and sulphate of soda. Sulphate of lime may also be employed,

M. Péan de Saint Gilles has examined this process, and finds that the decomposition commences at 428° F., while the evolution of nitrous gases continues from eight to ten hours. The residue, after being washed, presents the appearance of peroxide of manganese, but is in reality an oxychloride of this metal. He found it to consist of—

Manganese	63·8
Chlorine	11·3
Oxygen	23·5
Water	0·9
Loss	0·5
	100·0

represented by the formula $3Mn^2O^3.MnCl$, and, of course, when it is employed in the manufacture of chlorine, the hydrochloric acid necessary for its decomposition must be regulated accordingly.

The other proposal is that patented by Mr. Haeffely, which consists in exposing burnt cupreous pyrites to the action of the still-liquor. The apparatus consists of a series of stone vessels, arranged after the plan of the lixiviating tanks described in the previous chapter. These tanks are filled with the poor copper sulphur-ores, after the sulphur has been burnt off, and the still-liquor is run into the first tank, whence it flows through the whole series. The copper is dissolved out in about a week, when the exhausted ore is withdrawn and replaced by a fresh charge.

The solution is then acidulated by the addition of a little still-liquor, and a definite quantity of tank-waste is afterwards added to throw down the copper as sulphide, which is filtered and washed, when it may be reduced or converted into blue vitriol.

This plan has the recommendation of working up, so to speak, the three waste products of alkali works: viz., burnt pyrites, still-liquor, and tank-waste.

but a higher temperature is necessary, and the yield of nitric acid is only 90 per cent. of the nitrate used. In this case the residue consists of a mixture of sulphate of lime and soda.

THE PROPERTIES OF THE BLEACHING COMPOUNDS.

The composition of chloride of lime may be represented thus :

Lime, CaO	28·0	38·6
Water, HO	9·0	12·4
Chlorine, Cl.	35·5	49·0
	72·5	100·0

Or,

Chloride of calcium, CaCl . .	55·5	38·3
Hypochlorite of lime, CaO.ClO	71·5	49·3
Water, 2HO	18·0	12·4
CaCl + CaO.ClO + 2 eq. .	145·0	100·0

But that manufactured on the large scale never contains more than 41 per cent., and the great bulk only from 32 to 37 per cent., of which 2 per cent. has no bleaching power, being present in the form of chlorate of lime.

The composition of commercial bleaching powder is assumed by Brande and Phillips to be as follows :

Hypochlorite of lime, 1 eq., CaO.ClO	32·65
Lime, 2 eq., 2CaO	25·57
Chloride of calcium, 1 eq., CaCl . .	25·34
Water, 4 eq., 4HO.	16·44
	100·00

Or,

Trihypochlorite of lime, 1 eq., 3CaO.ClO .	58·22
Chloride of calcium, 1 eq., CaCl	25·34
Water, 4 eq., 4HO	16·44
	100·00

Fresenius has recently made some experiments to ascertain the nature of the chemical constitution of bleaching powder. He analysed a sample, and gives the following results—

		Equivalents.
Hypochlorite of lime .	26·72	1·00
Chloride of calcium . .	25·51	1·23
Lime	23·05	2·20
Water	24·72	
	100·00	

which may be arranged thus :—

		Equivalents.
Hypochlorite of lime	26·72	1·00
Chloride of calcium (in combination with the above as CaO.ClO + CaCl)	20·72	1·00
Chloride of calcium in excess . . .	4·79	0·23
Hydrate of lime	30·46	2·20
Combined water and moisture . . .	17·31	
	100·00	

He verified this composition by treating a weighed portion of the chloride of lime with a known volume of hydrochloric acid, the difference in the strength of the acid before and after the experiment, indicating the quantity of acid saturated by the lime. The reaction is as follows :—

$$(CaO.ClO + CaCl) + 2HCl = 2CaCl + 2HO + 2Cl.$$

He found that 100 of the bleaching powder had combined with 57·206 of anhydrous acid. Now the

26·72 hypochlorite of lime and 20·72 chloride of calcium, forming 47·44 pure chloride of lime, would require of hydrochloric acid . . . . 27·24

30·46 hydrate of lime would require ditto . . . . . . 30·01

57·25

while the experiment gave 57·206.

Again, he treated, in two series of experiments, 50 grammes of bleaching powder with 10 successive, gradually increasing portions of water. 80 grammes of water were used in the first lixiviation, filtered, and 20 grammes of filtrate were obtained. A second lixiviation produced 30 grammes, a third 100, a fourth 120, a fifth 150, a sixth and a seventh somewhat more, an eighth above 200, a ninth and tenth above 300.

These solutions were analysed as soon as filtered, and the following numbers were calculated from the analytical results :—

Solution.	Hypochlorite of lime.	Chloride of calcium.	Equivalents of CaCl for 1 eq. of CaO.ClO.
1	4·5371	15·9437	4·5279
2	5·2254	7·6451	1·8851
3	3·6968	2·6064	0·9084
4	1·6004	0·9897	0·7968
5	0·4672	0·2971	0·8193

Solution.	Hypochlorite of lime.	Chloride of calcium.	Equivalents of CaCl for 1 eq. of CaO.ClO.
6	0·1600	0·1198	0·9647
7	0·0602	0·0487	1·0422
8	0·0128	—	—
9	0·0060	0·0067	1·4319
10	0·0039	0·0043	1·4521
Residue . .	0·1389	0·0321	0·2981

The conclusions drawn by Fresenius (as condensed by Mr. Crookes), are—

1. The first portion of water dissolved all the chloride of calcium, as the last portions contained very little.

2. The hypochlorite of lime only gradually entered into solution.

3. All the free chloride of calcium was dissolved at the end of the third lixiviation ; for afterwards the proportion between the CaO.ClO, and the Ca.Cl, was nearly regular.

4. As the chloride of calcium dissolves completely at once, while the hypochlorite only dissolves principally after the third lixiviation, it follows that these two compounds are in combination, or that the water decomposes the chloride of lime into hypochlorite and chloride of calcium.

5. That the hydrate of lime is combined with the chloride of calcium, forming a basic chloride, which would explain why 4 eq. of hydrate of lime take up only 2 eq. of chlorine ; and this opinion, he thinks, is supported by the fact that the crystalline compound, $3CaO.CaCl + 16HO$, is converted into hydrate of lime and chloride of calcium by water

6. That the formula of chloride of lime is
$$CaO.ClO + CaCl.2CaO + 4HO.$$

7. The hydrate of lime separated, when the chloride of lime is treated with water, retains a certain proportion of hypochlorite with such force, that repeated washings cannot remove it. It is necessary, therefore, in accurate estimations of the hypochlorous acid in chloride of lime, to value the deposit along with the solution.

Bleaching powder, when properly manufactured, is a perfectly white uniform powder, without being gritty or lumpy. It is dry when first made, and becomes moist, only very gradually on exposure to the air, but finally deliquesces, evolving oxygen slowly, and forms at length a liquid consisting of chloride of calcium, which possesses no bleaching properties. It emits a faint chlorine odour, in reality that of hydrochlorous acid.

Heat converts chloride of lime into a mixture of chloride of calcium and chlorate of lime, with evolution of oxygen, and sometimes of chlorine, the compound at the same time losing its bleaching power. If the heat be gradually applied, the chlorine is evolved before the oxygen. The same amount of oxygen is evolved by heat, whether the bleaching powder retains its full bleaching power, or whether it has been reduced one third by too high a temperature during its manufacture. The decomposition by heat is explained by the equation

$$9CaCl + 9(CaO.ClO) = 12O + 17CaCl + CaO.ClO^{\ast}.$$

The remarkable circumstance attending the effect of heat is, that at a low temperature one-third of the hypochlorite of lime is changed into chloride of calcium and chlorate of lime, while the remainder is unaffected; but at a higher heat, this residue is resolved into oxygen and chloride of calcium.

An excess of chlorine has the same effect as heat, decomposing the hypochlorite into chloride and chlorate, with evolution of oxygen.

The aqueous solution of chloride of lime is a transparent colourless liquid, with a slight odour of hypochlorous acid, and an astringent taste.

When the solution is boiled, the same change takes place as when the powder is heated, with this difference, that pure oxygen only is evolved; and the same fact has been noticed by Morin as to the quantity of oxygen given off, whatever the strength of the bleaching powder employed in making the solution. This decomposition also occurs when the solution is preserved in close vessels at the ordinary temperature, especially if exposed to light.

Gay Lussac found that the hypochlorite, when exposed to the direct rays of the sun, is converted into chlorite of lime, $CaO.ClO^{\ast}$.

Acids, including the carbonic, convert chloride of lime into chlorine and a lime-salt.

$$CaCl + CaO.ClO + 2SO^{\ast} = 2(CaO.SO^{\ast}) + 2Cl;$$

but if added in smaller quantities, they disengage hypochlorous acid. When carbonic acid is passed through the solution, its bleaching power is destroyed, chlorine is expelled, carbonate of lime is precipitated, and chloride of calcium, with some bicar-

bonate of lime, remains dissolved in the water. Atmospheric air
acts in the same way, only more slowly ; but if it be deprived of
its carbonic acid, no change is produced. In the open air the
liquid slowly evolves chlorine, which effect ceases, however, when
the surface becomes covered with the crust of carbonate of lime,
and then the same change takes place as in close vessels, chloride
of calcium and oxygen only being produced. This circumstance
is of great importance when the solution is used for disinfecting
purposes ; in such cases the crust ought to be constantly broken up.

The bleaching process probably consists, in most cases, in the
decomposition of water or other oxygen-compounds by the
chlorine, and the evolution of nascent oxygen, which is the real
bleaching agent ; sometimes, however, a substitution of chlorine
for hydrogen takes place, as in the action of chlorine and water
on indigo :

Indigo   Hydrochloric
blue.   Water. Chlorine.   Isatin.   acid.
$$C^{16}H^5NO^2 + 2HO + 2Cl = C^{16}H^5NO^4 + 2HCl,$$

                                                     Hydrochloric
Isatin.   Chlorine.   Chlorisatin.   acid.
and : $C^{16}H^5NO^4 + 4Cl = C^{16}H^5NO^4Cl^2 + 2ClH.$

The solution of bleaching powder destroys organic colours and
the odour of decomposing organic matter in the same way, but
only in the presence of an acid to liberate the chlorine. Thus
tincture of litmus is not immediately decolorised by the solution,
but only after a time, when the carbonic action of the air has
effected a partial decomposition. This reaction is employed to
ascertain when sufficient chlorine has been passed into the liquids
in manufacturing the solutions of chloride of potash and soda ;
as soon as the colour of the litmus disappears, the supply of
chlorine must be suspended.

When 1 part of bleaching powder is mixed with 3 parts of
water, and a strong solution of sugar is added to the mixture, a
violent reaction ensues, with great increase of temperature.
Mr. Bastick, in making this experiment on one occasion, had his
apparatus dashed in pieces in a few moments. The results of the
decomposition are formate of lime, lime, chloride of calcium, and
water.

                Hypochlorite   Formate   Chloride
Sugar   of lime.   Lime.   of lime.   of calcium. Water.
$$C^{12}H^{11}O^{11} + 6(ClO.CaO) + 6CaO = 6C^2HCaO^4 + 6CaCl + 5HO ;$$

but when pure hypochlorite of lime is employed, the results of the decomposition are carbonic acid, chloride of calcium, and water.

The solution of bleaching powder acts on starch, cotton, linen, and all vegetable fibres, in the same manner. It has no sensible action on camphor, the essential oils, and similar bodies. Organic compounds of animal origin are similarly affected, with formation of ammonia.

When the *alkaline hypochlorites* are heated, the same decompositions take place as those already described, and hence, in making solutions of these salts, 6 atoms of chlorine must be added to 7 of the alkali, in which case 3 atoms of chloride are formed, leaving 3 of hypochlorous acid to 4 of the alkali. If chlorine is passed into a solution of carbonate of potash, previously saturated with chloride of potassium, this salt is precipitated; and the same observations apply to the corresponding soda-salts.

When hypochlorite of potash, containing 1 atom of base and acid, is kept at 212° F. for some time, it evolves 13 per cent. of oxygen of the acid; but if there be 4 atoms of base to 1 atom of acid, the quantity of oxygen disengaged amounts to 36 per cent. Powdered manganese added to the solution has the same effect, while the liquid acquires a pink colour, from formation of permanganic acid. These chlorides, when an excess of alkali is present, may be evaporated to dryness at 122° F. without being resolved into chloride and chlorate, the residue retaining considerable bleaching power.

The action of acids on the alkaline hypochlorites is the same as that which they exert on the corresponding lime-compounds, and similar remarks apply to their bleaching properties.

The aqueous solutions of the hypochlorites exercise the same kind of oxidising action as the solution of the acid itself, being themselves converted into chlorides at the same time. They convert phosphorus into phosphorous and phosphoric acids; sulphur into sulphurous and sulphuric acids; arsenic into arsenic acid; tin and copper into oxychlorides, with evolution of a portion of the chlorine and oxygen; precipitated sulphides into sulphates, &c., &c.

Chloride of lime precipitates hydrated peroxide of manganese from the sulphate of the protoxide. With nitrate of lead, a white precipitate of chloride of lead is first formed, which changes to yellow, and ultimately to brown, as the chloride passes into peroxide of lead, with an evolution of chlorine caused by the action of the hypochlorite left in solution : thus, in the first instance,

$$(CaCl + CaO.ClO) + PbO.NO^5 = PbCl + CaO.ClO + CaO.NO^5,$$

and afterwards,

$$CaO.ClO + PbCl = PbO^2 + CaCl + Cl.$$

If the nitrate of lead is in excess, the change to brown is effected by the action of the hypochlorite of lead in solution, which, when filtered from the precipitate, becomes turbid, and deposits peroxide of lead, with evolution of chlorine:

$$PbO.ClO = PbO^2 + Cl.$$

Chloride of lime precipitates calomel from a solution of nitrate of protoxide of mercury (mercurous nitrate), and the calomel quickly changes to red oxychloride, while the supernatant liquid loses its bleaching power, and contains corrosive sublimate in solution.

APPLICATIONS OF CHLORINE AND THE HYPOCHLORITES.

These compounds have proved of incalculable benefit in manufactures, for sanitary purposes, and in medicine; but they do not all possess the same value, the dry chloride of lime far exceeding the others in importance.

We may enumerate their application to the bleaching of cotton, linen, and hemp, and of the rags for paper; to the discharging of colours in printing processes; to the formation of the double chloride of sodium and aluminium in the manufacture of the latter metal; to the preparation of chlorate of potash; to the extraction of iodine; to the production of the red prussiate of potash; in the commercial valuation of indigo and other colouring matters; to all the varied purposes of disinfection; and to medical purposes.

As regards the latter application, we copy the statement of Pereira in his " Materia Medica."

" These two properties—viz., that of destroying offensive odours, and that of preventing putrefaction—render the alkaline hypochlorites most valuable agents to the medical practitioner. We apply them to gangrenous parts; to ulcers of all kinds attended with foul secretions; to compound fractures accompanied with offensive discharges; to the uterus in various diseases of that viscus, attended with fetid evacuations. In a word, we apply them in all cases accompanied with offensive and fetid odours. As I have already remarked, with respect to hypochlorite of soda, their efficacy is not confined to an action on dead parts, or on the discharges from wounds and ulcers; they are of the greatest benefit to living parts, in which they induce more healthy action, and the consequent secretion of less offensive matters. Furthermore, in the sick chamber, many other occasions present themselves in which the power of the hypochlorites to destroy offensive odours will be found of the highest value; as to counteract the unpleasant smell of dressings or bandages, of the urine in various diseases of the bladder, and of the alvine evacuations. In typhus fever, a handkerchief or piece of calico, dipped in a weak solution of an alkaline hypochlorite, and suspended in the sick chamber, will be often of considerable service both to the patient and the attendants.

"The power of the hypochlorites to destroy *infection* or *contagion,* and to prevent the propagation of epidemic diseases, is less obviously and satisfactorily ascertained than their capability of destroying *odour.* Various statements have been made by Labarraque and others, in order to prove the disinfecting power of the hypochlorites with respect to typhus and other infectious fevers.

"But, without denying the utility of these agents in destroying bad smells in the sick chamber, and in promoting the recovery of the patient by their influence over the general system, I may observe that I have met with no facts which are satisfactory to my mind as to the chemical powers of the hypochlorites to destroy the infectious matter of fever. Nor am I convinced by the experiments made by Pariset and his colleagues, that these medicines are preservative against the plague.

"Considered in reference to medical police, the power of the alkaline hypochlorites to destroy putrid odour and prevent putrefaction, is of vast importance. Thus, chloride of lime may be employed to prevent the putrefaction of corpses previously to interment, to destroy the odour of exhumed bodies during medico-legal investigations, to destroy bad smells, and prevent putrefaction in dissecting-rooms and workshops in which animal substances are employed (as cat-gut manufactories), to destroy the unpleasant odour from privies, sewers, drains, wells, docks, &c.; to disinfect ships, hospitals, prisons, and stables. The various modes of applying it will readily suggest themselves. For disinfecting corpses, a sheet should be soaked in a pailful of water containing a pound of chloride, and then wrapped around the body. For destroying the smell of dissecting-rooms, a solution of the chloride may be applied by means of a garden watering-pot. When it is considered desirable to cause the rapid evolution of chlorine gas, hydrochloric acid may be added to the chloride of lime.

" Hypochlorite of lime (or of soda) is the best *antidote* in poisoning by hydrosulphuric acid, hydrosulphuret of ammonia, sulphuret of potassium, and hydrocyanic acid ; it decomposes and renders them inert. A solution should be administered by the stomach, and a sponge or handkerchief, soaked in the solution, held near the nose, so that the vapour may be inspired. It was by breathing air impregnated with the vapour arising from chloride of lime, that the late Mr. Roberts (the inventor of the miner's improved safety lamp) was enabled to enter and traverse with safety the sewer of the Bastile, which had not been cleansed for 37 years, and which was impregnated with hydrosulphuric acid. If a person be required to enter a place suspected of containing hydrosulphuric acid, a handkerchief moistened with a solution of chloride of lime should be applied to the mouth and nostrils, so that the inspired air may be purified before it passes into the lungs."

NATURE OF THE BLEACHING COMPOUNDS.

Chemists are not agreed as to the true character of the constitution of these compounds. Some have supposed that they are formed by the simple absorption of chlorine by the base, the chlorine entering into combination with the oxide, CaO.Cl, while Raab regarded them as mixtures of the metallic chlorides with peroxides of the metals and hydrate of chlorine.

Miller, Muspratt, and others consider that they are analogous to peroxides, part of the oxygen being replaced by chlorine.

Balard's opinion, which is that generally adopted, regards them as compounds or mixtures of a metallic chloride with a hypochlorite,

$$ClCa + ClO.CaO ;$$

but several facts militate against this view ; for example, chloride of calcium is deliquescent and soluble in alcohol, whereas bleaching powder, when properly made, does not deliquesce ; nor does alcohol dissolve any appreciable quantity of chloride of calcium out of this compound. Again, when chloride of lime is agitated with half its weight of water, Dingler states that the filtrate contains 4 per cent. of chlorine, that the first, second, and third washings yield solutions of the same strength, but that afterwards the filtrate becomes poorer in chlorine, which would not have been the case had this compound contained chloride of calcium ready formed, for then the first extract would have been the richest in chlorine. Further, Muspratt states that when bleaching powder is saturated with chloride of calcium, chlorine is evolved. He further proved that anhydrous lime does not absorb chlorine; while the addition of another atom of water to hydrate of lime increased the absorption of this gas, which increase also took place when such bodies as the chlorides of barium and potassium, the sulphates of baryta, lime and soda, caustic lime, &c., were mixed with the hydrate of lime.

This question also remains unanswered: Why does bleaching powder always contain so large a proportion of lime uncombined with chlorine, however long the hydrate is exposed to the action of the gas? It has been supposed that this lime formed a basic compound with the chloride of calcium, assumed to exist in bleaching powder. Rose has proved the existence of a similar compound, composed of $CaCl + 3CaO.16HO$, which is sometimes found in the stills when ammonia is prepared by the decomposition of sal ammoniac and lime. If such a combination existed in bleaching powder and resisted the action of chlorine gas, much of the difficulty attending Balard's view would be removed. This is not, however, the case: for Bolley and Lohner have submitted this very compound to the action of chlorine, and obtained hypochlorite of lime. Fresenius has recently published some experiments upon the action of water on bleaching powder, from which he infers that this basic chloride of calcium does really exist under the form of $CaCl.2CaO + 4Aq$. These experiments have been described at p. 399.

The peculiar energy of chlorine and hypochlorous acid has been explained by assuming that the oxygen is liberated in the form of ozone; but this idea, as well as the nature of the bleaching compounds, are questions which must be left to future research for a satisfactory answer.

Cost of the Manufacture of Bleaching Powder.

We regret we cannot give such admirable details of the cost of this manufacture as in that of alkali. The quantities of the materials have been already given in the explanations of the processes, to which may be added, that about 20 cwts. of coal are used for every ton of bleaching powder produced, and that the charges for labour, casks, and repairs are estimated at about £2 5s. per ton of bleaching powder.

The following statement is given by Payen as the cost of making 1000 kilgr. of bleaching powder per day, which can be obtained from a chamber about 20 feet by 23 feet, each square foot of whose floor is covered with about 3 lbs. of hydrate of lime. The necessary acid and manganese are worked up in 8 sets of generators, each set consisting of 4 carboys.

				Francs.
Hydrochloric acid	2,500 kilgr. at 6 per cent.			150·0
Peroxide of manganese consumed.	700	„	15 „	105·0
Caustic lime	600	„	5 „	30·0
Coal, 8 hectr.				27·5
8 Workmen				24·0
Rent, interest, management, repairs, packages, &c. . .				48·5
				———
900 kilgr. of bleaching powder, at 32 per cent. . . .				385·0

METHODS OF TESTING THE OXIDES OF MANGANESE, BLEACHING POWDER, AND HYDROCHLORIC ACID, TO ASCERTAIN THEIR COMMERCIAL VALUE.

A. Tests for the Oxides of Manganese.

The value of these oxides to the manufacturer, depends upon the quantity of chlorine which they can furnish when exposed to the action of hydrochloric acid, this quantity varying according to the percentage of those oxides present which contain more oxygen than the protoxide. This is shown by the following equations, which represent the action of hydrochloric acid on the binoxide, sesquioxide, and red oxide of manganese :

$$MnO^2 + 2HCl = MnCl + 2HO + Cl,$$
$$Mn^2O^3 + 3HCl = 2MnCl + 3HO + Cl,$$
$$Mn^3O^4 + 4HCl = 3MnCl + 4HO + Cl.$$

But in practice the proportions of some of the impurities in the manganese of commerce are, so far as they are present, of equal importance; such, for example, as oxide of iron, the carbonates of lime, baryta, &c., which not only consume the hydrochloric acid, without yielding chlorine, but evolve carbonic acid which combines with the lime to the deterioration of the bleaching powder.

The necessity therefore of knowing the percentage of the higher oxides, and the nature and quantity of impurities in the manganese of commerce, have led chemists to devise various rapid, but more or less accurate, methods for determining these points.

All these testing operations may be classed under three divisions : 1st, The amount of oxygen disengaged by a red heat or by sulphuric acid ; 2nd, The oxidation of oxalic acid ; and, 3rd, The evolution of chlorine from hydrochloric acid.

In selecting a sample which shall fairly represent the bulk, it is advisable to arrange to do so during the weighing, and on a scale sufficiently large to avoid any error. Some cwts. having been collected, according to the quantity for sale, it ought to be ground to powder under edge-stones, swept and cleaned beforehand, and from this powder a sample must be carefully taken for analysis.

The first step is to determine the moisture present, which is done by drying a weighed quantity at a temperature of 230° or 240° F., and then to proceed immediately with the testing to avoid any absorption of moisture by the powder.

I. Heat or Sulphuric Acid.

The plan of testing manganese ores by the loss they sustain at a red heat, from its liability to give erroneous results, is no longer used ; but the process proposed by Gay-Lussac, to determine the quantity of oxygen they give off by the action of sulphuric acid, is very simple and convenient.

It depends upon the change represented by the equation,

$$MnO^2 + HO.SO^3 = MnO.SO^3 + HO + O.$$

A small retort a is employed, and 3 grammes of the ore is introduced together with 25 cc. of strong sulphuric acid.

The retort a (Fig. 238) is connected with a narrow tube b, which is carried under the graduated test-glass c. This test-glass is placed in a jar d, containing an alkaline liquid to absorb any carbonic acid that may be evolved from the impurities of the manganese.

The contents of the retort are heated to the boiling point, which heat is continued until no more gas is disengaged. The tube is then withdrawn, and the apparatus is allowed to cool, when the oxygen gas is measured, and the volume corrected for pressure, temperature, and the tension of the aqueous vapour.

One volume of oxygen is equivalent to two volumes of chlorine,

FIG. 238.

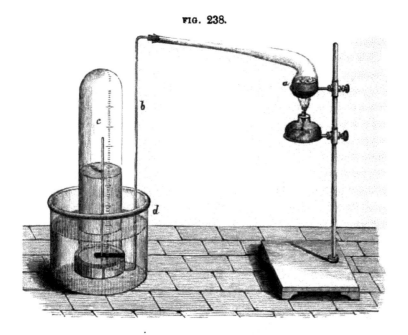

so that this process furnishes a ready method of ascertaining the volume of chlorine which a given sample of ore will disengage when treated with hydrochloric acid.

II. *Oxidation of Oxalic Acid.*

The plan generally adopted is that devised by Frezenius and Will, founded on this principle as first applied by Berthier and Thompson.

The principle of this process depends upon the oxidation of oxalic acid or an oxalate, when the oxygen of the peroxide is disengaged by the action of an excess of sulphuric acid, as explained by the equation,

$$2MnO^2 + 2(HO.SO^3) + C^4H^2O^8 = 2(MnO.SO^3) + 4HO + 4CO^2.$$

Each equivalent, therefore, of binoxide of manganese = 44, gives rise to the evolution of 2 equivalents of carbonic acid = 44. Thus, with a sample weighing 50 grs., the weight of carbonic acid evolved multiplied by 2 will give the percentage of peroxide of manganese present in the ore.

The process is conducted in the following manner :

The flask B is half filled with sulphuric acid, and a weighed portion of the manganese introduced into A, after which twice this weight of oxalate of soda is added, and about 2 oz. of water are poured into the same flask.

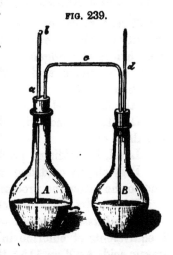

FIG. 239.

The tube a is closed with a little wax stopper at b, and the corks are inserted in the flasks. The whole apparatus is then weighed. A small quantity of air is withdrawn from B, by means of suction applied to a caoutchouc tube at d, when a small quantity of sulphuric acid passes over into the flask A. An evolution of carbonic acid immediately commences, and when it slackens, a further quantity of acid is drawn over as before, and this is repeated until the ore is fully decomposed, which ensues in about five minutes if the sample has been well pulverized. The cessation of the evolution of carbonic acid on the addition of the last portion of sulphuric acid, and the disappearance of the black colour of the manganese, prove that the ore has been all decomposed.

Sufficient sulphuric acid is now drawn over into A to heat the contents and expel the carbonic acid dissolved by the liquid; the wax stopper is removed from b; and air is drawn through the apparatus at d, until it no longer tastes of carbonic acid. The apparatus is allowed to cool and reweighed. The loss in weight is carbonic acid, which gives the weight of peroxide in the sample of manganese.

The simpler and more convenient apparatus described under "Acidimetry," may also be used for the process just described, with equally accurate results.

Those ores which contain carbonates, require a previous treatment, or a modification of the process. It is conducted as above described, but without the addition of oxalic acid; and the loss, deducted from that previously found, gives the quantity of carbonic acid produced from the oxidation of the oxalic acid by the action of the sulphuric acid on the binoxide of manganese.

III. *The evolution of Chlorine from Hydrochloric Acid.*

Different processes have been proposed for estimating the quantity of chlorine evolved by the action of hydrochloric acid on these oxides of manganese. One class depends on the oxidation either of arsenious acid or of a salt of iron, and the other on the liberation of iodine.

To the former belong the processes of

Gay-Lussac and Price, by oxidizing arsenious acid,

and those of

Levol, by oxidizing protochloride of iron,

Poggiale, by oxidizing sulphate of iron and ammonia.

And to the latter the process of

Bunsen, by liberating iodine from iodide of potassium.

1. *Gay-Lussac's process.*—Arsenious acid, $AsO^3 = 99$, requires 2 equivalents of chlorine in order to effect its conversion into arsenic acid, $AsO^5 = 115$; thus,

$$AsO^3 + 2Cl + 2HO = AsO^5 + 2HCl ;$$

or, in other words, 100 parts of peroxide of manganese are required for the conversion of 113·6 parts of arsenious into arsenic acid.

The process is conducted in the following manner :

The weighed sample of manganese ore is placed in a small flask d (Fig. 240), holding only 30 to 40 cc., and strong hydrochloric acid

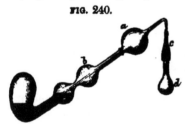

FIG. 240.

added until the flask is two-thirds full. The connecting pipette a is then made fast to the neck of the flask by means of a caoutchouc tube c. The end of the pipette is slightly bent upwards, and it is inserted into the receiver b.

The neck of this receiver has two bulbs blown in it, to allow the liquid to rise during the pressure of the gas as it collects, and to prevent any risk of the liquid being driven out by the sudden evolution of the chlorine. The receiver is filled with a solution of potash.

The flask is gently heated until all the manganese ore is decomposed, and afterwards boiled to drive over all the chlorine into the receiver. When no more chlorine comes over, the liquid begins to run into the pipette, which must be removed and washed with distilled water. These washings are poured into a beaker, to which the contents of the receiver are transferred, together with its own washings.

The potash-solution is now ready for being tested by the

normal solution of arsenious acid. The measures of the test-solution used, correspond with the quantity of peroxide of manganese in the sample.

Price employs a solution of permanganate of potash in connection with this process. The manganese is dissolved in a normal solution of arsenious acid in hydrochloric acid, and the arsenious acid on which the chlorine has not reacted, is determined by means of permanganate of potash.

The normal solution of arsenious acid is prepared by dissolving 113·6 grms. of arsenious acid in potash, which corresponds to 100 grms. of pure peroxide of manganese, and then adding pure hydrochloric acid, so as to form a litre.

To obtain the test-solution of permanganate of potash, 50 cc. of the arsenious liquor are diluted, which corresponds to 5 grms. of peroxide of manganese, and the number of cubic centimetres necessary to convert the arsenious into arsenic acid is determined.

. The process is performed as follows : 10 grms. of the sample are placed in a small flask, 100 cubic centimetres of arsenious liquor added, and a bulb-tube adapted to it (Fig. 241), filled with a solution of caustic potash.

Heat is applied, and when the solution is complete it is allowed to cool, that contained in the flask is decanted, and the liquid in the bulb-tube is added to it. The solution is diluted with water, and the quantity of arsenious

FIG. 241.

acid which has not been oxidised by the chlorine, is determined by permanganate of potash, from which the value of the sample of manganese is ascertained.

When oxide of manganese is dissolved with heat in the arsenious liquid, chloride of arsenic, which is volatile, is formed ; but a loss of chlorine is prevented by employing a weak arsenious solution, and adapting the bulb tube to the flask.

2. *Levol and Poggiale's process.*—This process depends on the conversion of a known weight of a proto-compound of iron into a sesqui-compound, and on the circumstance that proto-chloride of iron dissolved in hydrochloric acid absorbs chlorine

perfectly so long as it is not changed into the sesquichloride
The change which ensues under these circumstances is as follows :

$$MnO^2 + 2ClH + 2FeCl = MnCl + 2HO + Fe^2Cl^3.$$

Levol gives the following instructions for performing the
operation : 4·858 grms. of pianoforte wire are placed in a flask
holding 3 decilitres, with a short and rather a wide neck,
closed with a cork, in which a straight funnel
tube is fixed, with its lower end drawn out to
a point. 100 grms. of pure hydrochloric acid are
poured in, the cork being loosely inserted in
the neck to allow the hydrogen to escape, and the
whole is gently heated. When the iron is dissolved,
3·980 grms. of the manganese ore, wrapped up in
paper, are added, the cork is made tight, and the
flask is shaken. The contents are boiled for a few
minutes, allowed to cool, and a slip of litmus paper
suspended, as shown in Figure 242.

FIG. 242.

As the manganese of commerce is never pure,
the quantity of the sample employed cannot convert all the iron
into sesquichloride, so that another operation is necessary to
determine how much protochloride is left in the solution. This
is managed by using a normal solution of chlorate of potash, and
the theory of its action depends on the fact that when treated
with hydrochloric acid and heat, it disengages as many equivalents of chlorine as it contains equivalents of oxygen,

$$KO.ClO^5 + 6HCl = KCl + 6HO + 6Cl.$$

And

$$KO.ClO^5 + 12FeCl + 6HCl = KCl + 6HO + 6Fe^2Cl^3.$$

An aqueous solution of this salt is made in such proportions,
that 100 grms. shall contain 3·170 grms. of chlorine,] with
which a burette is filled. This normal solution is then poured,
drop by drop, into the funnel of the tube in the cork, until the
litmus-paper changes colour, proving that the chlorine is in
excess, and that all the iron has passed to the state of sesquichloride. The quantity of the normal solution which has been
used is noted, and the strength of the sample is found thus :—
The strength of the manganese + the quantity of normal
solution used being 100
And the quantity of the latter being, say 15

Leaves the strength of the manganese 85

If the quantity of normal solution necessary to discharge the colour of the litmus be found to be a cubic demicentimetre, this must be deducted from the above to give the real strength of the manganese.

Instead of chlorate of potash, the normal solution may be made with permanganate of potash, which undergoes reduction to convert the protoxide into sesquioxide of iron, thus :—

$$10FeO + KO.Mn^2O^7 = 2MnO + KO + 5Fe^2O^3.$$

The strength of this normal solution is determined by the methods of Margueritte, Mohr, and others, and it is then used in the same way as that made with chlorate of potash ; but the solutions must be employed quite cold, especially when hydrochloric acid has been employed.

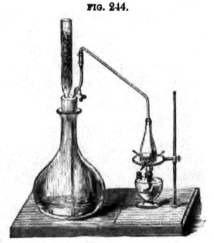

FIG. 243.

Weighed quantities of pure sulphate of protoxide of iron, or of sulphate of protoxide of iron and ammonia, may be substituted for the metallic iron.

A very convenient form of burette is that invented by Geissler, a modification of Gay-Lussac's, in which the narrow tube is placed inside the wide tube, Fig. 243.

3. *Bunsen and Mohr's processes.*—Bunsen's plan consists in treating the ore in the way described at page 109, in the same apparatus, but filled with a solution of iodide of potassium, when the chlorine displaces the iodine, which is estimated by means of sulphurous acid, as described at page 33, and from this the quantity of peroxide of manganese is easily calculated.

FIG. 244.

Mohr employs the apparatus, Fig. 244, in applying his process to the examination of peroxides like those of manganese. The ore is boiled with the hydrochloric acid in the small flask, and the large one should have a capacity of about 1 litre.

The wide mouth of the large flask is accurately fitted with a cork, through which two glass tubes pass. The larger tube is open at both ends, and filled with pieces of glass; the small tube dips into the liquid with its end drawn to a small point. A small ring or clamp hangs loosely on this tube, which rises almost straight from the flask, in order that the acid liquid, which condenses, may flow back into the flask.

Two solutions are required: the one of iodine dissolved in iodide of potassium, which is prepared as already described at page 32; and the other of arsenite of soda, which is made by boiling 5 grms. of pure arsenious acid with 10 grms. of bicarbonate of soda until the solution is complete, when water is added to bring the volume up to 1 litre.

An excess of this solution is poured into the flask, and a sufficient quantity of a solution of carbonate of soda is added through the wide tube. The connection being made by the caoutchouc tube, the contents of the small flask are boiled until the glass tube below the connection becomes hot, and an effervescence commences in the large flask from the presence of some hydrochloric acid having been driven over. The caoutchouc tube is closed by the clamp, and the apparatus left to cool, the tube being disconnected which leads from the small flask. The pieces of glass in the wide tube and the gas tube in the large flask are washed with water, and the washings added to its contents.

Starch-paste is now added, and then the iodine-solution, until a permanent blue colour is produced, when the strength of the same is read off from the burette.

It is scarcely necessary to remark that there must be no smell of chlorine perceptible in either flask when the distillation is finished.

B. *Tests for Chlorine, or Chlorimetry.*

The value of bleaching powder depending upon its available chlorine, various processes have been devised for determining its strength, and the term *chlorimetry* has been given to this operation.

An object of so much importance in a trade dealing with so valuable an article, has led to an unusual number of suggestions to combine every facility in the manipulations with the greatest accuracy in the results.

In order to give as clear and succinct a description of those various processes as possible, we will adopt the following arrangement:—

1. Descroizilles and Welter . . . Indigo.
2. Gay-Lussac, Penot, and Mohr . Arsenious acid.
3. Duflos, Fordos and Gélis, Nöllner, and Mohr . . . . . } Hyposulphite of soda.
4. Bunsen . . . . { Iodine and sulphurous acid.
5. Runge . . . . { Protochloride of iron and copper.
6. Dalton, Graham, and Otto . . Protosulphate of iron.
7. Gay-Lussac . . . . Ferrocyanide of potassium.
8. Balland de Toul and Marezeau. Subchloride of mercury.
9. Henry and Plisson, and Schencke. Ammonia.

Fresenius gives some useful and necessary instructions as to the preparation of the sample of bleaching powder to be tested, which are applicable whatever process is adopted.

He recommends the trituration of the samples with the water, added by degrees, in a mortar, in the usual way, and when the first portion of the liquid is poured into the flask or beaker, the renewal of the process with the powder left in the mortar, until the whole powder has become intimately ground up with the water. The flask is always to be shaken before using the contents, as Fresenius states that this turbid solution gives a higher strength than the clear liquid which is so often used in these processes. In one instance he found 22·6 chlorine in the clear fluid, 25 in the residuary mixture, and 24·5 in the uniform turbid solution.

1. *Welter and Descroizilles' process.*—This is still much used by bleachers, but is very unsafe, as the indigo itself is of so uncertain a composition, that aqueous chlorine is now actually employed to determine its value.

The process adopted is to mix 1 part of the indigo with 9 parts of concentrated sulphuric acid, and digest the mixture at 100° C. for 6 or 8 hours. The solution is diluted with 990 parts of water, which constitutes the test-liquid. A sample of the bleaching powder is weighed, mixed with water, and the test liquid added until the colour disappears. The number of degrees of the test-liquid used, are then read off, and the strength of the chloride of lime thence determined.

2. *Gay-Lussac's process.*—This plan depends on the oxidation of arsenious acid by chlorine, $AsO^3 + 2Cl + 2HO = AsO^5 + 2HCl$.

The solution of arsenious acid is made of such strength that 10 cc. are equivalent to 0·1 of chlorine. For this purpose, 13·96 grms. of acid are dissolved in a solution of soda in a flask, the liquid is then diluted with water strongly acidulated with hydrochloric acid, and the volume made up to a litre with water.

10 grms. of the bleaching compound having been prepared as described, and the bulk of the liquid made up to 1 litre, as marked in the flask C (Fig. 245) ; a graduated burette (Fig. 247), is filled

FIG. 245.　　　FIG. 246.　　FIG. 247.　　　FIG. 248.　　　FIG. 249.

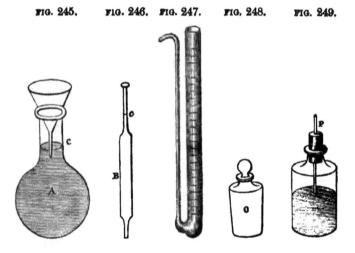

with a portion up to 0°. This burette, devised by Gay-Lussac, is divided into 200°, 100° being equal to 10 cubic centimetres. 10 cc. of the arsenious solution are drawn up into the pipette B, and transferred to the bottle O (Fig. 248), to which a few drops of the indigo solution are added from the bottle P (Fig. 249), to give a blue colour to the liquid. The bleaching liquid is then poured into the bottle O, until the blue colour disappears. A few drops of the indigo are again added, followed by fresh portions of the bleaching liquid, until the fluid suddenly entirely loses its colour, which the addition of a small drop of the indigo solution fails to restore.

As the bleaching liquid contained 0·1 grm. of chlorine, the strength of the sample will be obtained by dividing 1000 by the number of cubic centimetres which have been used.

Penot's process.—The difficulty of the previous process consists in ascertaining, with sufficient delicacy, the exact point when the blue colour really disappears, and the object of Penot's modification of Gay-Lussac's process is to obviate this objection.

He employs an alkaline solution of arsenious acid, and colourless iodised paper, which turns blue on the addition of a trace of free acid.

The paper is prepared by heating a mixture of—

    1 grm. of iodine
    7   „   crystallised carbonate of soda
    3   „   starch
    $0\frac{1}{4}$ litre water

until it becomes colourless, and the volume is then made up to $\frac{1}{2}$ a litre. White paper is dipped into this solution and dried ready for use.

The test liquid is prepared by dissolving with heat—

    4·4 grm. of arsenious acid,
    13   „   crystallised carbonate of soda,
    $0\frac{3}{4}$ litre of water,

and making up the volume to 1 litre by a further addition of water.

The operation is conducted in the same way as before, but the burette is filled with the test-liquid in this case, and cautiously added to the bleaching liquid until a drop of the mixed liquids ceases to cause a blue colour in the iodised paper.

Mohr's process is a modification of the above, but by no means so simple.

3. *Fordos and Gelis's process.*—These chemists substitute hyposulphite of soda for the arsenious acid in Gay-Lussac's process.

The test-liquid is made by dissolving 2·77 grms. of the hyposulphite of soda in one litre of water, which is exactly equivalent with Gay-Lussac's arsenical normal liquid, and neutralises its own volume of chlorine.

10 cc. of the test liquid is poured into a beaker and diluted with 100 parts of water, and the whole is rendered slightly acid and coloured blue with a few drops of the solution of indigo. The sample of bleaching liquid is then added, and the operation conducted as the celebrated French chemist has described.

The blue colour is but slowly destroyed, which gives· time to the operator more carefully to observe the change than is possible with some of the other processes.

The advantages of this plan arise from the fact of dispensing with the use of so dangerous a liquid as the arsenical solution, and from the circumstance that the chlorites, which are sometimes present from the alteration of the hypochlorites under the influence of light, decompose this salt, while they do not act upon the arsenious acid. It has, however, this disadvantage, that the test-liquid must be prepared each time it is required. The danger of precipitating sulphur on the addition of the acid, is avoided by the previous dilution of the test liquid.

Nöllner's process.—Duflos first drew attention to the conversion of sulphurous and hyposulphurous acids into sulphuric acid by chlorine, as the basis of a chlorimetrical process. Nöllner has since applied it for this purpose, and precipitates the sulphuric acid by a salt of barytes. The change is shown in the equation,

$$NaO.S^2O^2 + 3HO + 4Cl = 2SO^3 + NaCl + 3HCl.$$

1 gramme of the bleaching powder is pounded with 2 grammes of hyposulphite of soda in a mortar, and the mixture is transferred with water to a flask, and warmed on a vapour-bath. The excess of hyposulphite is decomposed by hydrochloric acid, until the liquid becomes clear and the precipitated sulphur has collected into drops. The whole is filtered, and the sulphuric acid is precipitated in the usual way.

116·5 of sulphate of baryta represent 71 of chlorine.

The dangers of this process consist in the possible presence of some sulphate in the hyposulphite, and of chlorine in the hydrochloric acid, while the tedious operations of washing and weighing the precipitate are drawbacks for manufacturing purposes.

Mohr's application of this salt depends on the estimation of iodine by means of a test-liquid. The iodine is liberated by the action of chlorine in an acid solution of iodide of potassium, and the precipitated iodine is determined by a test-liquid of hyposulphite of soda, thus:—

$$2(NaO.S^2O^2) + I + HO = NaI + NaO.S^4O^5 + HO.$$

The process is conducted in the following manner : 10 grammes of the bleaching powder are mixed with water and some small pieces of broken glass, in a strong flask, which is shaken, until the bleaching powder is thoroughly and minutely subdivided. The milky fluid is then introduced into the litre-flask, and water added to bring up the volume to 1 litre. As much more water is added as corresponds to the bulk of the broken glass, which does not amount to a cubic inch. 100 cc. of this liquid corresponds with 1 gramme of the bleaching powder.

The solution of iodide of potassium is made by dissolving 10 grammes of this salt in a decilitre of water.

The solution of the hyposulphite of soda is prepared by dissolving 24·8 grammes in 1 litre of water, so that 1 cc. corresponds with 0·0127 of iodine and 0·00355 of chlorine.

100 cc. of the milky bleaching liquid is mixed with 25 cc. of the iodine liquid, and weak hydrochloric acid is added while the whole is being stirred, until the fluid has a faint acid reaction. The liquid remains clear, but becomes of a dark brown colour, and the test hyposulphite liquid is added until the colour disappears.

The number of cc. of the test-liquid required for decolorizing the mixed fluid + ·355 will give the strength of the sample.

The formation of the tetrathionate of soda exerts no influence on the working of the process.

The iodine fluid may be collected and the iodine precipitated, when it can again be used for making iodide of potassium.

One peculiarity of this plan is, that the liquid containing the iodine after the conclusion of one operation, can be employed for a second and third trial.

4. *Bunsen's process* depends on the mutual action of iodine and sulphurous acid in the presence of water, when the latter is not present beyond 0·05 per cent. in the liquid : the change then takes place as follows : $I + HO + SO^2 = HI + SO^3$.

The process is managed as follows : 10 cc. of the bleaching liquid are poured into a beaker, and about 6 cc. of the solution of iodide of potassium are added. The mixture is diluted with 100 cc. of water, acidulated with hydrochloric acid, and the iodine liberated is determined in the way already described at p. 35.

One equivalent of iodine corresponds with one of chlorine, which renders the calculation very easy. (See pp. 31—36 of this volume.)

5. *Runge's process.*—This plan depends upon the conversion of protochloride of iron into perchloride by chlorine, and upon its

reduction back to protochloride by the action of copper, as explained by the equation

$$Fe^2Cl^3 + 2Cu = 2FeCl + Cu^2Cl$$

2 grammes of the bleaching powder are introduced into a flask with a long neck, and 30 grammes of water are added. A solution of 0·6 grammes of pianoforte wire in hydrochloric acid is poured in, and the whole agitated for a short time. An excess of hydrochloric acid is added, and a strip of sheet copper, weighing exactly 4 grammes, is placed inside the flask. The contents are then boiled until the dark colour of the liquid changes to a pale green, when the copper is withdrawn, washed, dried, and weighed. A loss of weight in the copper of 64 = 2Cu represents 36 of chlorine.

6. *Process of Dalton, Graham, and Otto.*—Dalton first suggested the application of the oxidation of the protoxide of iron by chlorine, as a means of determining the quantity of the latter in a bleaching compound. Graham worked out the process, which was introduced into Germany by Otto.

When these substances are mixed together, the following reaction ensues :

$$2(FeO.SO^3) + (SO^3.HO) + Cl = Fe^2O^3.3SO^3 + Cl ;$$

therefore, 2 equivalents of crystallised sulphate of iron = 278, correspond with 1 equivalent of chlorine = 36, or 1 of chlorine oxidises the iron in 7·722 of the copperas.

The test-liquid is prepared by dissolving 3·96 grammes of the salt of iron in 200 cc. of water, and acidulating the solution with hydrochloric acid.

5 grammes of the bleaching powder are dissolved in 20 cc. of water, and introduced into the burette up to zero. This is poured cautiously, drop by drop towards the end of the operation, into the test liquid, until the oxidation of the iron is complete.

This point is ascertained by saturating pieces of bibulous paper with ferrocyanide of potassium and acid, touching them from time to time, with the rod from the beaker holding the test liquid, to which the bleaching liquid has been added. As soon as no blue colour is produced, the number of degrees of the burette are read off, each of which will correspond with 0·5 of chlorine.

7. *Gay-Lussac's process.*—The method of employing ferro-

cyanide of potassium in estimating the value of a bleaching compound depends on this reaction :

$$2K^2FeCy^3 + Cl = K^3Fe^2Cy^6 + KCl$$

35 grammes of the ferrocyanide of potassium are dissolved in 1 litre of water, which forms the test liquid. This solution is rendered acid, and then takes a yellow colour in the presence of the bleaching liquid. A few drops of a solution of indigo are added, which communicates a fine green colour to the solution.

The solution of bleaching powder is now cautiously added, until this green colour suddenly disappears, succeeded by the yellow, when the degrees of the burette being read off, the strength of the sample is ascertained.

8. *Process of Balland de Toul and Marezeau.*—This plan depends on the oxidation of the protochloride or protoxide of mercury, and was first suggested by Balland de Toul.

He poured nitrate of protoxide of mercury into a solution of chloride of lime, until a precipitate was formed, which did not disappear on agitation ; but this plan is liable to great errors.

His plan, as modified by Marezeau and Gay-Lussac, is conducted in the following manner. The chemical change is explained thus :

Subchloride of mercury . . .	Hg^2Cl =	235·5
Chlorine	Cl =	35·5
Chloride of mercury	$2HgCl$ =	271·0

The proof liquid is made by dissolving nitrate of the protoxide of mercury in water, so that each cubic centimetre shall contain 0·036 grms. of mercury ; 2·5 cc. will, therefore, require 0·005 litre of chlorine to convert the mercury into protochloride or subchloride, and as much more to form perchloride.

2·5 cc. of this proof liquid is poured into a beaker, diluted with water, and hydrochloric acid added while the solution is being stirred, until no further precipitation takes place.

5 grms. of the bleaching powder are dissolved in $\frac{1}{2}$ litre of water, and a burette graduated into half cubic centimetres filled with the solution.

The bleaching liquid is then added to the proof or test-liquid until the precipitate is redissolved, when the number of degrees of the solution in the burette used, will give the percentage of chlorine in the sample.

9. *Process of Henry and Plisson.*—This method of estimating chlorine differs from the previous processes, inasmuch as it consists in measuring the nitrogen which is produced when ammonia is decomposed by chlorine. Thus

$$4NH^3 + 3Cl = 3(NH^3.HCl) + N ;$$

hence 1 equivalent of nitrogen represents 3 of chlorine.

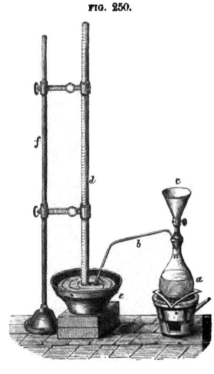

FIG. 250.

The apparatus consists of a small flask, a, holding from 300 to 400 cc., fitted with a cork, in which a glass funnel c and a bent glass tube b are fixed. The tube d, graduated into cubic centimetres, receives the gas, and is kept in its position by the support f; this tube holds 80 centimetres.

10 grms. of the bleaching powder, dissolved in 250 cc. of water, are introduced into the flask, and 1 decilitre of ammonia, diluted with its own weight of water, is added. The whole is now gradually heated, and the gases collected in the tube, filled and standing in an alkaline liquid. When the disengagement of the gas has ceased, the apparatus is filled with water through the funnel, to drive over the gaseous contents of the flask e.

The air which filled the apparatus at the commencement of the operation, is mixed with the nitrogen from the ammonia, and its volume is determined by absorbing the oxygen by any of the usual plans. The residual gas is the nitrogen due to the process, from which the chlorine can be determined.

This process, although very accurate, requires some dexterity in the manipulation, and is, moreover, not adapted for manufacturing purposes.

C. Estimation of the Chlorine in Hydrochloric Acid, Common Salt, &c.

There are two volumetrical processes by which the estimation of the chlorine in these compounds can be made, one proposed by Levol, and the other by Mohr.

1. *Levol's process.*—This plan is based upon the decomposition which salts of silver undergo when mixed with alkaline chlorides. When a solution of common salt is poured upon a precipitate of phosphate of silver, an alkaline phosphate and chloride of silver are produced. Now, if nitrate of silver is added to a mixture of a phosphate and an alkaline chloride, there will be no interchange of constituents between the nitrate and phosphate, until all the chlorine has been precipitated, these facts depending upon the difference of the solubility of the salts. In other words, the *yellow* precipitate of phosphate of silver will not appear until all the chlorine has fallen in combination with silver.

The process is conducted as follows: 1 grm. of the substance containing the chlorine is dissolved in 50 parts of water, about 5 grammes of a cold saturated solution of phosphate of soda are added, and the liquid, if acid, is neutralised by carbonate of soda; the test-solution in a burette, graduated in cubic centimetres and tenths, is cautiously added until the precipitate becomes slightly yellow. The first drops of the solution of nitrate of silver produce a yellow precipitate which disappears on agitation.

The test-liquid is made to contain 0·0305 grm. of silver per cubic centimetre, which are equivalent to 0·010 grm. of chlorine, so that 100 cc. contain 1 grm. of chlorine. The result is, therefore, given directly in hundredths.

2. *Mohr's process.*—Mohr's plan consists in employing chromate of potash, which is much superior to the phosphate of soda, as the bright red of the chromate is sooner recognised than the faint yellow of the phosphate of silver.

The standard solution of nitrate of silver must, as in the previous case, be perfectly neutral, as the coloured silver salts are soluble in acid.

The process is managed as follows : The substance containing the chlorine is weighed, placed in a beaker, and dissolved in water. A few cc. of the solution of chromate of potash are added, and then the standard solution of nitrate of silver is cautiously poured in from the graduated burette, until the red colour of the chromate of silver appears. The result is obtained as in the previous process.

This process gives the most accurate results; but Fresenius states that it requires 0·1 cc. of the silver solution in excess, to produce the red colour, which must be deducted from the total quantity employed to give the real strength.

This process is applicable to the determination of chlorine in urine, and in the potashes and alkali of commerce.

When the chlorine in the chlorides of barium, calcium, magnesium, mercury, &c., is to be determined, these compounds must be previously decomposed by carbonate of soda, a few drops of chromate of potash added, and the standard silver-solution applied in the way already explained.

POTASH.

French, *Potasse*; German, *Potasche*; Danish, *Potaske*; Italian, *Potassa*; Russian, *Potasch*; Polish, *Potass*.

HISTORICAL NOTICES.

The name of this substance, in each language, is derived from its origin, viz, the evaporation of the solution of *ashes* of plants in iron *pots*. The term *alkali*, applied by the Arabians to the carbonate of soda obtained from the ashes of marine plants, was afterwards extended to the carbonate of potash found in the ashes of land plants, and to carbonate of ammonia.

At an early date it must have been found that the addition of lime rendered these substances caustic: hence the terms *mild* and *caustic* alkalies. These caustic alkaline solutions were probably known to the Greeks and Romans, as Pliny tells us of soap being made in his time from wood-ashes and tallow; and although he does not mention the use of quicklime in connection with the ashes, he was acquainted with it, as he describes a mixture of goat's tallow and quicklime in the same account.

This is still further confirmed by the circumstance that Pliny describes an Egyptian nitre, made in his days, which he says must be kept in close vessels, else it becomes liquid. In a work ascribed to Galen, the process of increasing the strength of a lye of ashes, by the addition of quicklime, is described. Paulus Ægineta and Geber also describe a similar process. According to Julius Pollux, the *Konia* mentioned by Aristophanes and Plato, was used for washing, and was made from a lye of ashes.

Under the term *Borith*, which is very similar to the Arabic *Boruk*, used in the Scriptures, it is probable that both soda and potash were included. It first occurs in Jeremiah ii., 22, " For though thou wash thee with nitre, and take thee much soap (*borith*), &c. ;" and again in Malachi iii.,.2. And although there is nothing here to indicate the origin of these substances, it is probable that some difference existed, or both would not have been mentioned in the same passage, unless they had possessed different cleansing properties. Six different kinds of *boruk* are mentioned by Persian authors ; and as the Arabians describe the use of the ashes of the poplar, and Sanscrit authors,˙ that of the *Butea frondosa*, it is clear that potash had long been employed before the distinction between it and soda was pointed out.

Black first explained, in 1756, the action of lime on these lyes, by withdrawing the carbonic acid.

The older chemists distinguished ammonia as the *volatile alkali*, while the others were called *fixed alkalies*. When the difference between the two latter was pointed out by Duhamel in 1736, and Marggraf in 1758, the potash was named *vegetable alkali*, because it was chiefly found in plant-ashes, and soda received the name of *mineral alkali*, as it existed in rock-salt ; which terms are commonly used in the early patent specifications to distinguish these two substances. Klaproth afterwards proved that potash existed in many widely-diffused minerals, and in consequence proposed that the term *Kali* should be applied to it, while the French and English adopted the names of *potasse* and *potash*.

Sources of Potash.

Potash occurs in a great variety of forms in the mineral kingdom, as *sulphate* in alum, alum-stone, polyhalite, and arcanite or glaserite; as *silicate*, united with various other bases, in apophyllite, harmotome, chabasite, pearl-stone, pumice-stone, felspar, leucite, mica, lepidolite, schorl, haüyne, pinite, latrobite, nephelin, —in nearly all the clays,—in smaller quantities in limestones of the most various formations,—in native oxide of manganese, specular iron ore, &c.; as *chloride of potassium*, sublimed from volcanoes, and sparingly in rock-salt; as *sulphate of potash*, and as *chloride, bromide*, or *iodide of potassium*, in sea-water, salt springs, and different mineral waters; as *nitrate*, in various soils in tropical countries; and in organic bodies, combined with *carbonic, phosphoric, sulphuric, hydrochloric, nitric*, and *organic acids*, but more abundantly in vegetables than in animals.

Although potash is not found in nature in such large quantities as soda (including common salt), its importance in manufactures may be appreciated by the following statement of the quantities and values of its different salts extracted or imported annually into this country, in the form of

Nitrate of potash	tons	12,000 =	£480,000
Potashes and pearlashes . .	„	7,500 =	225,000
Muriate of potash	„	2,230 =	37,910
Sulphate of potash from kelp	„	670 =	5,360
Do. do. argols	„	30 =	5,400
Cream of tartar	„	1,000 =	120,000
			£873,670

Presence of Potash in Animals and Plants.

Extensive as the trade in potash appears, far more importance must be attached to these salts in their relation to vegetable and animal life.

Potash appears, from the admirable researches of Liebig, to be as indispensable to the flesh as soda is to the blood, and the absence of potash in the food induces that terrible malady of seamen, the scurvy. It is an equally essential and important constituent of milk. Dr. Garrod has examined the various kinds of food in reference to this point, and the following table contains the results of his investigation.

TABLE I., *showing the quantity of Potash in grains in one ounce (avoirdupois = 7000 grs.) of the following alimentary substances* (Garrod).

Baker's bread, London, City	0·259
Do. do. West-End	0·257
Home-made bread (probably containing potato flour)	0·262
Best white flour	0·100
Bran	0·609
Rice	0·005
Rice	0·011
Oatmeal	0·054
Split peas	0·529
Raw beef	9·599
Salt beef, raw	0·394
Do. boiled (slightly salted)	0·572
Boiled mutton	0·637
Dutch cheese	0·230
Boiled potato (large size)	1·875
Raw potato (small)	1·310
Boiled potato, without peel and well done, water containing much potash	0·529
Onion (small)	0·333
London milk (1 fluid ounce)	0·309
Orange (not ripe) including septa	0·675
Lemon-juice (1 fluid ounce)	0·852

which indicates at once what may be termed the *scorbutic* and *antiscorbutic* foods.

All the potash in animals, is derived from the vegetable kingdom. The two following tables show how universally this substance is present, and, at the same time, indicate that no more important ingredient can be added to restore the exhaustion of the soils of our cultivated lands, than potash, in some form which may be retained by the soil and assimilated by the plant.

TABLE II, *showing the per centage of Potash and Soda, and their Chloride in the Ashes* (1) *of the following Plants* (Way and Ogston, and others); *of Fruits and Vegetables* (Richardson).

Plant or parts of the Plant.	Potash.	Soda.	Chloride of Potassium.	Chloride of Sodium.
Wheat, grain..................	30·02	3·82	—	—
Do. straw	17·98	2·47	—	—
Do. chaff	9·14	1·79	—	—
Barley, grain	21·14	—	5·65	1·01
Do. straw	11·22	—	—	2·14
Oats, grain	20·63	—	1·03	—
Do. straw	19·46	1·93	2·71	4·27
Do. chaff	6·23	3·93	—	0·24
Rye, grain..................	33·83	0·39	—	—
Do. straw	17·20	—	0·30	0·60
Maize, grain..............	28·37	1·74	—	trace
Do. stalks and leaves	35·26	—	—	2·29
Buckwheat, straw	31·71	—	7·42	4·55
Peas (gray), seed	41·70	—	3·62	1·24
Do. straw	21·30	4·22	—	—
Beans (common field), grain	51·72	0·54	—	—
Do. straw	32·65	2·77	—	11·54
Tare, straw	33·82	—	3·27	4·03
Red clover	25·60	—	9·06	6 02
Lucerne	27·56	—	11·64	1·91
Flax, seed..................	34·17	1·69	—	0·26
Do. straw	21·53	3·68	—	9·21
Rape, seed..................	16·33	0·34	—	0·96
Do. straw..............	16·63	10·57	—	2·53
Potato, tuber	43·18	0·09	—	7·92
Do. stem	29·53	3·95	—	20·42
Do. leaves	17·27	—	4·95	11·37
Turnip, seed..............	21·91	1·23	—	—
Do. bulb	23·70	14·75	—	7·05
Do. leaves	11·56	12·43	—	12·41
Mangold Wurzel, root	21·68	3·13	—	49·51
Do. leaves	8·34	12·21	—	37·66
Chicory root	34·64	—	8·92	2·98
Rice, grain	20·21	2·49	—	—
Tobacco leaves..............	36·37	—	2·50	2·51
Hop flowers	19·41	0·70	—	—
Cactus	7·83	28·19	—	14·87
Cocoa-beans	37·14	1·22	—	9·65
Tea (Souchong), extract	47·45	5·03	—	3·62
Coffee (Java), extract............	51·45	—	1·96	—
Chenopodium maritimum, flowers and young shoots	4·40	2·30	—	71·30

Fruit or Vegetables.	Potash.	Soda.	Chloride of Potassium.	Chloride of Sodium.
Pear, whole	54·69	8·52	—	trace
Apple, do.	35·68	26·09	—	—
Cucumber, do.	47·42	—	4·19	2·06
Gooseberry, do.	38·65	9·27	—	1·23
Celery	22·07	—	33·41	—
Carrot..................	32·17	5·00	—	9·62
Parsnip	36·12	3·11	—	5·54
Lettuce	46·01	5·29	—	7·82
Spinach	9·69	34·96	—	7·93
Endive	37·87	12·12	—	trace
Fig, whole..................	28·36	24·14	—	4·02
Walnut, kernel	31·11	2·25	—	trace
Chestnut	39·36	19·18	—	4·82
Asparagus	6·01	34·21	—	12·94
Onion	32·35	8·04	—	4·49
Broccoli, head	47·16	—	6·22	trace
Do. leaves	22·10	7·55	—	—
Cauliflower, head	34·49	14·79	—	2·78
Radish, root	21·16	—	1·29	7·07
Do. leaves..............	5·05	11·09	—	8·50
Kidney bean..............	36·83	18·40	—	2·80
Plum, fruit	54·69	8·72	—	0·62
Do. shell	21·69	7·69	—	trace
Do. seed	26·52	1·94	—	0·49
Cherry, whole	51·85	1·12	—	2·02
Strawberry, do.	21·07	27·01	—	2·78
Pine apple, do..................	49·42	—	0·88	17·01
Do. tops..................	19·66	31·11	—	2·42
Orange, whole..................	38·72	7·64	—	trace
Rhubarb, stalk..................	50·59	0·46	—	8·84
Do. leaves..................	14·47	31·77	—	trace
Cow cabbage, head..............	40·86	2·43	—	—
Do. stalk..............	40·93	4 05	—	2·06

The following table (from Watts's "Dictionary of Chemistry") exhibits the degree of variation in the composition of the Ash of Cultivated Plants, and will also serve to illustrate some remarks on this point in the following pages.

TABLE III. *Composition in 100 parts of the Ash of Bean Seed and Bean Straw.*

Locality.	Potash.	Soda.	Lime.	Magnesia.	Sesqui-oxide of Iron.	Sulphuric acid.	Silica.	Carbonic acid.	Phosphoric acid.	Chloride of potassium.	Chloride of sodium.	Ash, per Cent. In substance undried.	In substance dried at 212° F.	Analyst.
I. SEED.														
Faba Vulgaris.														
Common field bean from Holland	20·8	17·6	7·3	8·9	1·0	1·3	2·4	—	38·0	—	2·4	—	—	Bichon
Do. Alsace	46·3	—	5·3	9·0	—	1·6	0·5	1·9	33·7	1·5	—	—	—	Bousingault
Do. Giessen	33·9	13·0	4·9	6·3	0·7	—	0·5	—	40·5	—	—	—	—	Buchner
Do. England	51·7	0·6	5·2	6·9	trace	3·0	0·4	3·4	36·7	—	3·3	2·37	2·65	Way and Ogston
Masagan bean (seed sown)	36·7	0·1	12·1	6·0	0·6	4·3	1·5	1·6	33·7	—	—	2·85	3·43	Do.
Do. raised from it on clay soil	43·4	1·3	5·8	5·7	0·1	3·1	0·4	3·4	36·7	—	—	2·68	3·01	Do.
Do. do. sandy soil	45·7	—	13·3	6·5	0·6	3·1	0·4	0·8	26·9	0·9	1·6	2·48	2·07	Do.
Heligoland bean (seed sown)	43·9	1·6	7·7	7·7	0·3	5·1	2·2	2·6	29·9	—	—	2·54	2·90	Do.
Do. raised from it on clay soil	43·5	2·4	4·8	5·6	0·1	6·2	0·7	2·8	30·6	—	3·2	2·53	2·94	Do.
Do. do. sandy soil	40·7	—	3·2	7·7	0·3	5·3	0·04	0·3	33·3	1·2	3·2	2·90	3·83	Do.
Phaseolus Vulgaris.														
Haricot bean from Worms	38·9	11·3	8·9	9·0	0·1	2·5	0·4	—	31·3	—	0·5	—	—	Levi
Do. Alsace	51·0	—	6·0	11·9	—	1·8	1·0	3·3	28·4	0·2	—	—	—	Bousingault
Do. Kurhessen	22·1	21·4	5·5	7·5	0·3	2·3	1·5	—	25·5	—	3·4	—	—	Thon
Do. England	36·8	18·4	7·7	6·3	2·8	4·0	4·1	—	17·0	—	3·8	0·68	—	Richardson
II. STRAW.														
Common field bean	18·3	13·6	29·3	7·1	3·0	3·1	11·3	—	13·1	—	0·4	4·97	3·56	Hertwig
Do.	33·9	3·8	19·6	3·5	0·8	1·4	2·6	20·3	0·6	—	11·5	5·17	3·81	Way and Ogston
Masagan bean on clay soil	18·7	18·9	18·9	8·1	0·7	1·4	2·2	24·4	6·5	—	10·0	4·64	5·05	Do.
Do. sandy soil	25·6	4·1	22·4	4·7	0·5	5·4	4·6	22·6	3·3	—	6·9	6·47	7·24	Do.
Heligoland bean on clay soil	19·6	—	18·3	4·9	0·4	3·9	1·5	25·7	11·1	2·6	11·0	6·05	6·69	Do.
Do. sandy soil	21·1	0·3	23·6	6·9	2·0	3·1	7·3	18·1	8·4	—	8·3			Do.

In fact, certain plants may be, chemically, as much distinguished
by the potash and soda found in their ashes, as some are by the
presence of an excess of lime and magnesia, and others by silica.
Liebig has classified plants according to this marked characteristic
of their ashes, as in the following table.

TABLE IV.

General Composition of the Ashes of three classes of Plants.

	Plants.	Salts of Potash and Soda.	Salts of Lime and Magnesia.	Silica.
Silica Plants.	Oat-straw with seeds............	34·00	4·00	62·00
	Wheat-straw	22·00	7·20	61·05
	Barley-straw with seeds	19·00	25·70	55·03
	Rye-straw	18·65	16·52	63·89
Lime Plants.	Tobacco (Havannah)..........	24·34	67·44	8·30
	Do. (Dutch)..................	23·07	62·23	15·25
	Do. (Artificial soil).........	29·00	59·00	12·00
	Pea-straw......................	27·82	63·74	7·81
	Potato-herb	4·20	59·40	36·40
	Meadow-clover	39·20	56·00	4·90
Potash Plants.	Turnips	81·60	18·40	—
	Beet-root......................	88·00	12·00	—
	Potatoes (tubers)	85·81	14·19	—
	Jerusalem artichoke............	84·30	15·70	—
	Maize-straw	71·65	6·50	18·00

The exhaustion of a soil, therefore, may as easily arise from a
deficiency of the alkalies, as from a want of phosphoric acid, to
which the attention of agriculturists is, we fear, too exclusively
directed at the present time.

The previous tables explain, what experience has proved, that
certain plants grow better on one soil than on another; thus leafy
shrubs are weak and sickly on siliceous and calcareous soils,
deficient in potash, while vigorous and thriving on soils derived
from the disintegration of granite, &c.; and, on the other
hand, pines, which require very little potash, flourish on sandy
soils.

The quantities of potash and other mineral substances removed
from the soil in a four years' course of wheat, barley, turnips, and
hay has been calculated by the late pains-taking Prof. Johnstone;
and the following table contains his results, with the totals of
which we add the results of the experiments of Boussingault, by
way of comparison.

TABLE V. *Johnstone's Four Years' Course.*

	Grain.	Straw.	Bulbs.	Tops.	Hay.	Total.
Potash	14·39	32·73	142·66	88·82	38·22	316·82
Soda	7·05	1·21	17·31	16·76	12·05	54·38
Lime	2·24	27·62	46·24	72·14	44·45	192·69
Magnesia	7·60	12·14	18·16	9·58	7·09	54·57
Oxide of iron, with a little oxide of manganese and alumina	1·11	5·96	4·35	2·67	0·58	14·67
Phosphoric acid	35·76	10·56	25·77	28·80	15·12	116·01
Sulphuric acid	0·12	13·15	46·24	38·81	9·20	107·52
Chlorine	0·02	3·55	12·24	49·75	4·06	69·62
Silica......................	14·71	233·08	27·03	2·67	78·23	355·72
	83·00	340·00	340·00	310·00	209·00	1282·00

Boussingault's Five Years' Course, once Manured.

1st year, crop of potatoes (tubers without herb). 246·8 lbs.

2nd do. do. wheat (straw and corn) . . . 371·0 „

3rd do. do. clover . . . . 620·0 „

4th do. do. { wheat . . . . 488·0 „

{ or fallow turnips . . . . . 108·8 „

5th do. do. oats (straw and corn) . . . . 215·0 „

In what manner the potash and other mineral constituents are assimilated by the vital agency of the plant, or what part they play in the vegetable organism, is, for the present, veiled in mystery. It is only known, and for our immediate purpose it is of the greatest importance, that a considerable portion of the saline bases are in combination with the vegetable acids in the plants. Thus potash, in the vine, is combined with tartaric acid, in wood-sorrel, with oxalic acid; and lime in other plants is united with malic acid, &c.

Presence of Potash in Soils.

Among the sources of potash named at p. 428, that derived from minerals is the most important for the growth of plants and trees. Among these minerals felspar (orthoclase) contains 12 to 17 per cent., mica 3 to 5 per cent., basalt and clinkstone 0·75 to 3 per cent., clay slate 2·75 to 3·31 per cent., clay and loam 1·5 to 4 per cent. of potash. Liebig has calculated, from these data, the quantity of potash which would exist in one Hessian acre (26,910 square feet) in a layer of soil formed by the disintegration of the following rocks, to a depth of 20 feet.

TABLE VI.

Felspar		1,152,000	lbs.
Clinkstone . .	from 200,000 to	400,000	„
Basalt . . .	„ 47,500 „	75,000	„
Clay-slate . .	„ 100,000 „	200,000	„
Loam . . .	„ 87,000 „	300,000	„

The oxidising action of the air, and the disintegrating power exerted by water under changes of temperature, both help to break up the hardest rocks, while the experiments of Struve have proved that water impregnated with carbonic acid decomposes rocks containing alkalies, which are dissolved and carried away. It is also certain that plants themselves, by producing carbonic acid during their decay and by means of the acids exuded from their roots while living, assist in destroying the cohesion of mineral masses. Thus air, water, and the alterations of the seasons, all promote the disintegration of rocks, and help to render their potash available for plants.

The analyses of orthoclase by Crasso place the nature of this change in a very intelligible form.

TABLE VII.

	Original Mineral.	Decomposed Mineral.	Constituents removed.
Silica	65·21	32·50	32·71
Alumina . . .	18·13	18·13	—
Potash	16·66	2·80	13·86
Soda	—	0·25	—
Lime	—	0·35	—
Magnesia . . .	—	0·27	—
Oxide of iron . .	—	0·73	—
Water	—	5·11	—
	100·00	60·14	46·57

Where the two silicates of potash and soda exist side by side in the same mineral mass, the latter is much more rapidly decomposed, and the alkali removed, than is the case with the former: thus Struve found the relative proportions of these alkalies in minerals, in their natural and decomposed condition, to be as follows:

TABLE VIII.

In a phonolite from Rothenberg—

	Potash.	Soda.
Natural condition . . .	3·45	9·70 per cent.
Decomposed do. . . .	5·44	3·26 „

.In a basalt—

	Potash.	Soda.
In natural condition . .	1·35	7·35 per cent.
Decomposed do. . . .	2·62	2·31 ,,

which remarkable fact is even more clearly shown by the analyses of phonolite by Meyer and Rammelsberg.

Locality	As a whole.		Soluble part.	
	Marienberg.	Whisterschan.	Marienberg.	Whisterschan.
Silica	56·652	54·090	43·244	41·220
Alumina	16·941	24·087	21·000	29·231
Protoxide of iron	3·905	1·248 Peroxide	7·816	2·427
Peroxide of manganese	—	0·319	—	0·638
Lime	1·946	0·687	2·986	1·034
Potash	9·519	4·244	0·035	3·557
Soda	2·665	9·216	7·112	12·108
Magnesia	1·697	1·379	—	1·261
Oxide of copper	—	0·012	—	0·025
Water	4·993	3·279	13·325	6·558
	98·318	98·561	95·518	98·136

There appears, therefore, to be a provision in nature by which the more rare and valuable potash shall be stored up in the soil for the plant.

Extraction of the Potashes.

The method employed depends on the nature of the source whence the crude material is obtained ; and as the ashes of both land and marine plants are employed for this purpose, as well as the bye-products from manufactories of beet-root sugar and tartaric acid, &c., while in other localities sea water, brine springs, and minerals are used, we will adopt the following arrangement in describing the various processes for operating on these different materials.

I. From the ashes of land plants—

    1. As hydrate and carbonate of potash, from the ashes of timber, plants, &c.

    2. As hydrate and carbonate of potash, in the form of a bye-product from the manufacture of beet-root and cane sugar.

    3. As sulphate and carbonate of potash, as bye-products in the manufacture of wine and tartaric acid.

II. From the ashes of marine plants—

    As sulphate of potash and chloride of potassium, from kelp and barilla, or varec.

III. From sea water and brine springs—
As sulphate of potash and chloride of potassium, in combination with sulphate of magnesia and chloride of magnesium.

IV. From felspar, &c.

V. From wool.

I. THE ASHES OF LAND PLANTS.

1. *Potashes.*

This term was originally applied to the salts obtained by lixiviating the residue left after burning the timber in clearing the land ; but, attention having been called to the subject, much information has been obtained in reference to the

Nature of the Ash.

When vegetable substances, wood, &c., are destroyed by fire, all the mineral constituents taken from the soil remain in a fixed state, after the volatilisation of the combustible matters, formed from the carbon, hydrogen, &c. This residue is called *the ash,* whatever substances it may contain. All the vegetable acids, which were combined in the plants with mineral bases, are burnt, and leave *carbonates* in their place in the ash, as the bases are of that class which do not lose their carbonic acid by heat, as is the case with the earths. The mode of combination in which the several constituents are found in the ash, gives, however, a very inadequate and incorrect idea of their actual state in the living plant, as the temperature at which they are burnt gives rise to violent decompositions and new combinations. This has been very clearly proved by Professor Rose in his memoir on the incineration of organic bodies.

The following bases have been found in the ash : *potash, soda, lime, magnesia, alumina, oxide of iron* and *manganese ;* and the acids are *carbonic, sulphuric, phosphoric, silicic,* which are accompanied by *chlorine.* Oxide of copper, and other bases and acids, have been found, but evidently from some exceptional circumstances connected with the locality, as at Mansfeldt.

The extraordinary differences in the quality and quantity of ash obtained under different circumstances, arise from various causes.

Although differences of this kind in different species of plants would not, perhaps, be thought singular, yet it is a remarkable fact, that the ashes of two specimens belonging to one and the same class, are seldom perfectly alike in composition, and are often very different. The researches which have been made upon this subject leave no doubt that plants, in the absence of one constituent of the soil, can take up some other, and that the bases can, to a certain extent, replace each other and act conjointly with reference to the functions of the plants. Now, as the action of a base is dependent upon the amount of oxygen which it contains, and is, indeed, measured by that quantity, the bases of such ashes —independent of that portion which is in combination with the mineral acids—must together contain one and the same quantity of oxygen, which, indeed, appears proved by the close approximation obtained by experiment. In the ash of a fir, Saussure found 3·6 carbonate of potash, 46·34 carbonate of lime, and 6·77 carbonate of magnesia. In the ash of the same kind of tree, but from a different locality, 7·36 carbonate of potash, 51·19 carbonate of lime, and no magnesia. In both cases, the oxygen of the acid in these bases is 9·0. The ash of a tobacco-plant from the neighbourhood of Debreczin, in Hungary, contained in combination with carbonic acid, 43·91 potash, 3·40 soda, 41·79 lime, and 10·9 magnesia; that of another kind of tobacco, from the same neighbourhood, 42·03 potash, 46·97 lime, and 11·00 magnesia, without any soda. The total amount of oxygen in the former case is 24·27, in the latter 24·57 (Will and Fresenius).

On the other hand, Way and Ogston, from their numerous analyses of the ashes of cereals and root-crops, came to the conclusion that the composition of the ash of a plant is independent of the nature of the soil, that the preponderance of any constituent, lime or silica, for example, in the soil, by no means leads to a preponderance of the same in the plant, and that an abundance of soda in the soil does not cause that alkali to take the place of potash in the plant. These conclusions, however, are not confirmed *to their full extent* by the researches of other chemists.

Malaguti and Durocher, from the analysis of a large number of plants growing wild in Bretagne, conclude that the constitution of the soil does influence that of the plant-ash to a certain extent, and in particular, that plants growing on calcareous soils are much richer in lime than those growing on clay soils. (Ann. Ch. Phys. [3] liv. 257.)

Will and Frezenius found, that barley grown in the interior of Germany, contained less soda in proportion to the potash, than the same plant grown in the maritime districts of the Netherlands.

Daubeny, in 1852 (Chem. Soc. J. v. 9), made a series of experiments on the growth of barley in soils manured with carbonate of potash, carbonate of soda, and chloride of sodium, respectively, and found that the presence of a large quantity of soda in the soil did increase the proportion of this alkali in the plant, as compared with the potash, *but only to the extent of about 8 per cent.*, whence he concludes that the power of soda to replace potash is confined within very narrow limits, and that potash is the alkali essential to the growth of the plant. He refers also, for confirmation of this conclusion, to the well-known fact that marine algæ, though growing in a medium containing a very large quantity of soda, and very little potash, nevertheless contain as much of the latter as of the former: whereas, if the replacing power existed to the extent which has sometimes been asserted, these plants ought to contain very little potash. Daubeny likewise points out that all soluble constituents of the soil, not absolutely poisonous, are taken up by the roots and circulate in the sap through the plant; that some of these constituents are taken up by the plant, and incorporated in its tissues, while others are again excreted; and that if any such non-essential substance is present in the soil in large excess, it will also circulate through the vessels of the plant in increased quantity, and when the plant is cut down and incinerated, will be found in considerable quantity in the ash, though by no means essential to the plant.

The latter researches of Dr. Cameron lead to the same conclusion. He gives the following analyses of the ashes of (whole) plants grown in (artificial) soils, containing a small proportion of potassium compounds.

TABLE IX.

Constituents.	Barley.	Oats.	Peas.	Poa trivialis.
Potash and chloride of potassium......	4·76	6·04	5·22	7·16
Soda and chloride of sodium............	11·62	10·14	14·01	20·46
Lime..	10·40	8·05	40·90	11·48
Magnesia	10·11	10·75	10·42	4·60
Sesquioxide of iron......................	0·96	1·22	0·82	0·40
Sesquioxide of manganese..............	0·10	trace	0·16	trace
Silicic acid	46·61	50·42	9·20	38·20
Sulphuric acid............................	4·16	5·12	6·10	5·04
Phosphoric acid	11·28	8·26	13·17	12·66
	100·00	100·00	100·00	100·00

He is of opinion that his experiments tend to prove :

1. That barley, oats, peas, and *poa trivialis*, cannot be perfectly developed on a soil destitute of potassium.

2. That sodium is capable of replacing potassium, in those plants, to a limited extent.

3. That when sodium predominates over potassium in those plants, they are stunted in development, flower with difficulty, and produce very inferior seed.

4. That there is every reason to infer, from the result of his investigation, that similar researches on other plants would give similar results.

From all these observations, and from the well-known fact that certain inorganic substances are essential to the devolopment of particular organs of plants—as phosphoric acid to the grain of cereals, silica to their stems, &c., as also the sickly growth of some plants on some soils, which yet thrive on others, and *vice versâ*, we must conclude that this supposed replacing power of different bases for each other in the plant does not exist to any important extent.

But even in the same specimen, as might have been anticipated, the nature of the ash varies with the part of the plant from which it was obtained. Different vital functions being performed by different organs, these have to transform and assimilate other kinds of food, and in variable quantity, and even this again is not uniform throughout the different periods of development and changes of season. Thus, Saussure found in the leaves of the oak, gathered on the 10th of May, 5·3 per cent. of ash, on the 27th of September, 5·5 ; in the wood he found 0·2, in the bark 6·0, in the wood of the branches 0·4, in the bark of the same 6·0, and in the inner bark 7·3 per cent. of ash. So in the wood of the white beech 0·6, in the sap-wood 0·7, in the bark 0·7 per cent. of ash. Again, in the leaves of the poplar gathered on the 26th of May, 6·6, on the 12th of September, 9·3, in the wood of the stem 0 8, and in its bark 7·2 per cent. of ash. The analyses of Staffel and Wolff, at pages 447 and 448, also illustrate the same point.

Rammelsberg has published some very curious results upon the power of selection exerted by different organs of the same plant. He found potash, but not a trace of soda, in the seed of rape and peas, while the straw yielded both the alkalies, but by far the larger portion was soda. The soil, upon which the plants were grown, contained both alkalies.

The analyses of Rammelsberg, Hertwig, and Fresenius and Will, exhibit the extent of these variations.

TABLE X.—*According to Rammelsberg.*

	Rape,		Peas,	
	Seed.	Straw.	Seed.	Straw.
Potash	25·18	8·13	43·09	8·20
Soda	—	19·82	—	12·50
Lime	12·91	20·05	4·77	30·53
Magnesia	11·39	2·56	8·06	6·93
Peroxide of iron	0·62			
Phosphoric acid....................	45·95	4·76	40·56	9·21
Sulphuric acid	0·53	7·60	0·44	7·01
Carbonic acid	2·20	16·31	0·79	17·36
Muriatic acid......................	0·11	19·93	1·96	7·15
Silicic acid.......................	1·11	0·84	0·33	0·62

TABLE XI.

Hertwig.				Will and Fresenius.		
	Fir Wood.	Fir Bark.	Needles of Fir.		Rye	
					Grain.	Straw.
Carbonate of soda...	7·42	2·95	29·09	Potash	32·72	17·19
„ potash	11·30			Soda	4·44	—
Chloride of sodium .	trace	—		Lime	2·91	9·06
Sulphate of potash .	—	—		Magnesia	10·12	2·41
Carbonate of lime...	50·94	64·80	15·41	Oxide of iron	0·82	1·36
„ magnesia	5·60	0·93	3·89	Phosphoric acid	47·22	3·84
Phosphate of lime...	3·43	5·03		Chloride of sodium ...	trace	0·56
„ magnesia	2·90	4·18		„ potassium	0·41	0·25
Protoxide of manganese	trace	—	38·36	Sulphuric acid........	1·46	0·82
Basic phosphate of iron	1·04	1·04		Silica	0·17	64·50
„ „ of alumina .	1·75	2·24				
Silica	13·37	17·28	12·36			

Lastly, the influence of an extraordinary soil is sometimes shown in the composition of the ash. Amongst others, Böttinger found in the ash of the wood of a fir-tree, which certainly was sickly, and growing upon a doleritic soil (in the neighbourhood of a manganese mine), 15 per cent. of protoxide of manganese, whilst in healthy wood of the same tree, only 2·7 per cent. was obtained. In like manner, Saussure obtained from flowering vetches on arable land, 12 per cent., and from the same plant grown in distilled water, only 4 per cent. of ash.

In general, the herbaceous plants yield much more ash than shrubs, and these, again, more than trees. The trunk (wood) in these plants, contains little ash, the branches and twigs considerably more ; but the bark and leaves afford the largest quantity. Saussure found in the peeled branches of oak 29 times as much ash as in the wood, in the bark 30 times, in the inner bark 36 times, in the sap-wood twice, and in the leaves 36 times; so in the white beech, the sap-wood contains 1·1 times, the bark 22 times as much ash as the wood. According to Hertwig, the quantity of ash from the bark of beech is 19 times that from the wood; the bark of the fir produces 5·4 times, the needles 5·8 times as much ash as the wood. These relations are more clearly seen in the following tables, which contain all the determinations of special interest for the preparation of potashes.

TABLE XII. *Quantity of Ash in* 100 *parts*: TREES.

Species of tree.	Part burnt for ash.	Karsten.	Berthier.	Mollerat.		Saussure.	Winkler.	Hertwig.	Will and Fresenius.
Oak	Wood	young 0·15 / old 0·11	3·30	1·40	1·97	0·2	—	—	—
	Branches	—	—	—	—	0·4	—	—	—
	Bark	—	6·00	—	—	6·0	—	—	—
	Leaves	—	—	—	—	5·5	—	—	—
	(Charcoal)	—	2·50	—	—	—	0·75	—	—
Red beech	Wood	young 0·37 / old 0·40	—	0·612	0·58	—	—	0·38	—
	Bark	—	—	—	—	—	—	6·62	—
	(Charcoal)	—	3·00	—	—	—	1·25	—	—
Hornbeam	Wood	young 0·32 / old 0·35	—	1·14	1·12	0·6	—	—	—
	Sap-wood	—	—	—	—	0·7	—	—	—
	Bark	—	—	—	—	18·4	—	—	—
	(Charcoal)	—	2·65	—	—	—	—	—	—
Alder	Wood	young 0·35 / old 0·40	—	1·39	—	—	—	—	—
Birch	Wood	young 0·25 / old 0·30	1·00	1·07	—	—	—	—	—
	Charcoal	—	—	—	—	—	0·80	—	—
Scotch fir (P. picea)	Wood	0·15	—	1·68	—	—	—	—	—
	Seeds	—	—	—	—	—	—	—	4·47
	(Charcoal)	—	—	—	—	—	1·11	—	—
Fir (P. abies)	Wood	young 0·22 / old 0·25	—	—	—	—	—	—	—
	Bark	—	—	—	—	—	—	1·78	—
	Needles	—	—	—	—	2·90	—	2·31	—
	(Charcoal)	—	1·24	—	—	—	1·44	—	—
Pine (P. sylvatica)	Wood	young 0·12 / old 0·15	0·83	1·80	—	1·19	—	—	—
	Needles	—	—	—	—	2·60	—	6·25	—
	Seeds	—	—	—	—	—	—	—	4·98
	(Charcoal)	—	—	—	Branches	1·50	1·38	—	—
Lime	Wood	0·40	5·00	1·45	—	—	—	—	—
	(Charcoal)	—	—	—	—	—	3·55	—	—
Poplar	Wood	—	—	1·306	1·24	0·80	—	—	—
Elm	Wood	—	—	2·28	—	—	—	—	—
Ash	Wood	—	—	2·30	—	—	—	—	—

TABLE XIII. *Ash* in 100 parts. SHRUBS.

Elder	1·64	Berthier
Do.	1·39	Mollerat
Hazel	0·50	Saussure
Wortleberry	2·60	,,
Do.	0·68	Mollerat
Barberry	0·71	,,
Juniper	1·84	,,
Wild rose	0·71	,,
Green-weed or whin	1·62	,,
Heckle	1·66	,,
Blackberry	0·76	,,
Broom	1·48	,,
Sumach	1·71	,,

Ash in 100 parts. HERBS.

Potato straw	15·00	Berthier
Do. leaves	1·15	Mollerat
Peas-halm	5·05	Hertwig
Do. do.	11·30	Boussingault
Do. do.	8·10	Saussure
Oat-straw	5·10	Boussingault
Wheat do.	4·40	Berthier
Do. do.	7·00	Boussingault
Do. do.	4·30	Saussure
Rye do.	0·30	Karsten
Do. do.	3·60	Boussingault
Maize do.	12·20	Saussure
Nettle	10·67	Pertius
Thistle	4·03	,,
Rushes	4·33	,,
Cane-stem	1·70	Karsten
Fern	2·75	,,
Do.	2·90	Mollerat
Do.	5·00	Pertius

Quantity of Potash in Trees, &c.

The following tables give the quantity of potash found in 1000 parts of the following trees, shrubs, or plants, according to Vauquelin, Pertius, Kirwan, and De Saussure.

TABLE XIV.

Pine	0·45
Poplar	0·75
Beech.	1·45
Do. bark	6·00
Oak	1·53
Do. bark of branches . .	4·20
Box	2·26
Willow	2·85
Elm	3·90
Vine	5·50
Thistles	5·00
Cotton-grass	5·00
Rush	5·08
Fern	6·26
Wheat-straw	3·90
Do. before flowering . . .	47·00
Barley-straw	5·80
Maize-stalk.	17·50
Bean-stalk	20·00
Stinging nettles	25·03
Vetch	27·50
Wormwood	73·00

Höss: Analyses from 1000 parts.

	Ash.	Potash.
Pine wood	3·40	0·45
Beech do.	5·80	1·27
Ash do.	12·20	0·74
Oak do.	13·50	1·50
Elm do.	25·50	3·90
Willow do.	28·00	2·85
Vines	34·00	5·50
Ferns	36·40	4·25
Wormwood.	97·40	73·00
Fumitory	219·00	79·90

Abbene and Blengini: Analyses from 1000 parts.

	Ash.	Potash.
Dahlia with blossoms and leaves	79·92	19·98
Do. stems after flowering time	44·57	3·60
Do. bulbs	99·16	13·44
Do. branches	23·05	2·56
Acacia-branches	24·59	2·56
Grape-stems	88·88	41·66
Vine	46·66	12·73

	Ash.	Potash.
Skins of pressed grapes . . .	72·91	14·88
Stems of a cluster of grapes . .	—	39·81
Grape-stones	—	9·50

Constituents of the Ash.

The constituents which are found in the ash have already been noticed in a general way. The bases are united with the acids in such a manner, that the whole appears to be a mixture of alkaline, earthy, and metallic salts—a mode of combination which, as it exists in the ash, is probably more dependant on the heat employed in incineration, than on the vital agency of the plant. If the ash is treated with water, an important separation results ; all the sulphates, chlorides (some silicate), and the carbonates of the alkalies in particular, are dissolved—the really valuable portion, therefore, of the alkaline constituents—while the carbonates, phosphates, and silicates of the earths remain insoluble. In examining ashes, therefore, the relation of the soluble to the insoluble portion,—i. e. the value of the ash to the manufacturer of potash,—ought always to be examined. The best among the older determinations, are those of Berthier, and of the more recent, those of Fresenius, Hertwig, Böttinger, and others.

TABLE XV. *According to Berthier.*[*]

Constituents of 100 parts of ash.	White beech wood.	White beech charcoal.	Red beech wood.	Oak wood.	Oak charcoal.	Oak bark.
Soluble in water. Carbonic acid.........	...	4·43	3·65	2·88	4·39	1·45
Sulphuric acid	...	1·30	1·19	0·97	0·90	0·37
Hydrochloric acid ...	...	0·83	0·85	0·01	0·62	0·04
Silicic acid	...	0·18	0·16	0·02	0·15	0·05
Potash..................	...	9·12 }	10·45	8·11	9·43	4·33
Soda	...	2·14 }				
Together............	19·22	18·00	16·30	12·00	15·50	6·25
Insoluble in water. Carbonic acid........	26·92	24·43	27·53	34·99	26·91	37·22
Phosphoric acid......	8·11	7·22	4·77	0·71	6·27	—
Silicic acid	4·05	3·20	4·85	3·36	1·52	1·03
Lime	31·31	35·75	35·66	48·41	39·95	47·78
Magnesia..............	6·33	5·70	5·86	0·53	7·15	0·75
Oxide of iron........	1·30	0·08	1·25	—	0·09	—
Oxide of manganese .	2·76	5·70	3·77	—	2·60	6·98
Together............	80·78	82·08	83·70	88·00	84·50	93·75

[*] In these tables, the entire quantity of the soluble constituents, added to that of the insoluble, amounts in each case to 100 ; but as there is

Constituents of 100 parts of ash.	Lime wood.	Birch wood.	Alder wood.	Fir wood.	Fir charcoal.	Pine wood,
Soluble in water.						
Carbonic acid	2·96	2·72	—	7·76	7·34	2·89
Sulphuric acid	0·81	0·37	1·24	0·80	3·75	1·67
Hydrochloric acid	0·19	0·03	0·06	0·08	—	0·92
Silicic acid	0·17	0·16	—	0·26	1·09	0·18
Potash	6·55	12·72	—	16·80	{ 15·32	4·41
Soda					22·55	3·53
Together	10·80	16·00	18·80	25·70	50·00	13·60
Insoluble in water.						
Carbonic acid	35·75	26·04	25·17	17·17	10·75	32·77
Phosphoric acid	2·51	3·61	6·25	3·14	0·90	0·91
Silicic acid	1·80	4·62	4·06	5·97	6·50	4·19
Lime	46·53	43·85	40·76	29·72	13·60	38·51
Magnesia	1·97	2·52	2·03	3·28	4·35	9·56
Oxide of iron	0·09	0·42	2·92	10·53	11·15	0·09
Oxide of manganese	0·54	2·94	—	4·48	2·75	0·36
Together	89·20	84·00	81·20	74·30	50·00	86·40

According to Berthier.

Constituents of 100 parts of ash.	Mulberry tree.	Nut tree wood.	Elder wood.	Wheat straw.	Potato straw.	Fern.
Soluble in water.						
Carbonic acid	5·82	3·11	7·71	trace	0·26	4·35
Sulphuric acid	2·09	0·78	2·06	0·20	0·97	1·62
Hydrochloric acid	1·01	0·08	0·13	1·31	0·50	3·19
Silicic acid	—	0·08	0·06	3·53	—	—
Potash	13·16	} 11·27	21·54	5·05	2·47·	19·84
Soda	2·91					
Together	25·00	15·40	31·50	10·10	4·20	29·00
Insoluble in water.						
Carbonic acid	31·75	32·33	22·06	—	—	17·96
Phosphoric acid	1·36	4·19	5·83	1·08	—	5·68
Silicic acid	2·19	3·67	2·25	73·36	36·4	15·48
Lime	34·85	37·06	34·57	5·72	—	30·39
Magnesia	3·48	3·84	1·76	—	—	0·50
Oxide of iron	0·38	3·50	0·08	2·42	—	0·50
Oxide of manganese	0·98	—	1·26	{ 7·25 Potash	—	9·50
Together	75·00	84·60	68·50	89·90	95·80	71·10

invariably a loss in separating by analysis, the different substances composing the ash, the addition of the numbers against each constituent does not always make up that sum; the deficiency, therefore, exhibits the amount of loss. In some cases, however, the deficiency arises from the whole of the constituents not having been estimated.

TABLE XVI. *According to Hertwig.*

Constituents of 100 parts of ash.		Beech wood.	Beech bark.	Pine needles.	Bean straw.	Pea straw.	Potato straw.
Soluble in water.	Carbonate of potash	11·72	...	} 10·72 {	13·32	4·16	... ... } 4·1?
	„ of soda	12·37	...		16·06	8·27	... ...
	Chloride of sodium.	trace	...		0·28	4·63	Sulphate of soda
	Sulphate of potash.	3·49	...	1·95	3·24	10·75	Chloride of sodium 2·2˙
	Silicate of potash...	—	...	3·90	—	—	
	Together	27·77	3·02	12·70	32·91	27·82	6˙·?
Insoluble in water.	Carbonate of lime...	49·54	64·76	63·32	39·50	47·81	... ... 4˙˙˙
	Do. magnesia	7·74	16·90	1·86	1·92	4·05	... ... 3·7?
	Phosphate of lime .	3·32	2·71	} 6·35 {	6·43	5·15	} ... ... 5·73
	Do. magnesia	2·92	0·66		6·66	4·37	
	Basic „ of iron...	0·76	0·46	0·88	} 3·49 {	} 0·90 {	... ... 1·3?
	„ „ alumina.	1·51	0·84	0·71		} 1·20 {	... ... 2·7?
	„ „ manganese	1·59	—	—	—	—	... ... —
	Silica.................	2·46	9·04	10·31	7·97	7·81	... ... 29·˙l
	Together	72·23	96·98	86·30	65·97	72·18	93·03

Wittstein has published the following analyses of the wood and bark of some trees grown in Bavarian forests.

TABLE XVII.

Constituents of Ash.	Birch.		Red Fir.		White Fir.	
	Wood.	Bark.	Wood.	Bark.	Wood.	Bark.
100 parts dried at 100° C. } left ashes	0·2930	1·2835	0·2460	2·8140	0·2773	3·2986
Potash	9·6370	0·5886	3·6457	3·5219	22·5529	5·2859
Soda	3·2340	4·7754	18·9483	2·7591	4·8113	2·4028
Lime........................	41·6054	39·4420	33·5218	41·5123	33·0493	46·0615
Magnesia	5·8999	5·5446	4·3650	3·1286	6·1686	1·9991
Alumina	0·0517	0·5467	0·0532	0·3886	0·1923	0·2629
Protoxide of manganese......	2·8106	6·0687	2·1344	2·3220	3·2357	1·1786
Oxide of iron	0·9246	0·7106	0·6510	0·3795	0·4053	0·5458
Oxide of copper	—	—	—	0·2281	—	0·0300
Chlorine	0·3865	1·3764	0·2454	0·1270	0·1913	0·7766
Sulphuric acid.................	0·4328	0·2394	2·1344	0·6799	3·7051	1·0338
Phosphoric acid	6·9723	5·8213	3·5432	1·6959	5·0404	1·6152
Silica	2·9822	13·6935	1·4087	10·4204	0·9240	5·5647
Carbonic acid	25·0630	21·1928	29·3489	32·8367	19·7238	33·2431
	100·0000	100·0000	100·0000	100·0000	100·0000	100·0000

The three following tables contain the analytical results of Staffel, which illustrate the previous remarks as to the influence of the season on the composition of the ashes of horse-chestnut and nut trees; of Wolff, which show how far the composition ot the ashes varies in different parts of the same tree; and of Sprengel and others, which exhibit the effect of different soils on the nature of the ash of beech-wood and its leaves.

TABLE XVIII

Staffel.	Horse Chestnut.					
	Wood.		Bark.		Leaves.	
Constituents.	Spring.	Autumn.	Spring.	Autumn.	Spring.	Autumn.
Potash............................	57·57	17·54	54·96	22·61	46·38	14·17
Lime	5·92	50·99	9·24	61·34	13·17	40·48
Magnesia	4·08	5·17	4·36	3·99	5·15	7·78
Alumina	—	0·23	—	0·18	0·41	0·51
Oxide of iron	0·31	0·63	1·66	0·31	1·63	4·69
Oxide of manganese	—	trace	—	—	—	—
Sulphuric acid	0·82	—	—	1·05	2·45	1·69
Silica	1·80	0·71	0·67	1·06	1·76	13·91
Phosphoric acid...............	19·02	21·73	19·54	6·95	24·40	8·22
Chloride of potassium	10·47	2·98	9·56	2·50	4·65	8·55
	99·99	99·98	99·99	99·99	100·00	100·00
Ash per cent. in dry material	10·908	3·380	8·680	6·570	7·690	7·520
,, fresh ,,	1·198	1·693	1·342	3·171	1·376	3·288

Staffel.	Nut Tree.					
	Wood.		Bark.		Leaves.	
Constituents.	Spring.	Autumn.	Spring.	Autumn.	Spring.	Autumn.
Potash............................	40·78	14·88	44·52	11·06	42·04	25·48
Lime	22·24	55·92	18·37	70·08	26·86	53·65
Magnesia	8·92	8·09	7·25	10·55	4·55	9·83
Alumina	—	—	—	0·29	0·18	0·06
Oxide of iron	2·71	2·23	0·85	0·40	0·42	0·52
Oxide of manganese	—	trace	—	—	—	trace
Sulphuric acid	4·94	3·15	4·45	0·15	2·58	2·65
Silica	2·41	2·86	2·67	0·71	1·21	2·02
Phosphoric acid...............	14·89	12·21	19·94	5·85	21·12	4·04
Chloride of potassium	3·10	0·65	1·94	0·90	1·04	1·73
	99·99	99·99	99·99	99·99	100·00	99·98
Ash per cent. in dry material	10·028	2·987	8·748	6·403	7·719	7·005
,, fresh ,,	0·899	1·476	1·381	3·683	1·092	2·570

TABLE XIX.

E. Wolff. Constituents	Horse Chestnut.								
	Young wood.	Young bark.	Young leaves.	Leaf stalk.	Leaves of blossom.	Flower stalk.	Pollen stamen.	Calyx.	Young fruit
Potash	13·24	5·23	18·45	24·93	44·09	47·15	44·80	45·59	44·16
Lime	32·58	47·70	25·48	18·04	10·68	7·65	10·81	9·54	8·37
Magnesia	3·70	1·06	2·27	2·53	3·01	1·08	2·43	4·57	1·99
Sulphuric acid	—	—	7·92	3·13	—	2·90	—	2·90	2·08
Silica	1·97	0·70	4·27	0·87	1·13	0·60	0·58	1·31	0·64
Carbonic acid	24·16	37·92	12·89	16·78	21·58	17·64	21·47	22·17	18·75
Phosphoric acid	14·58	3·75	19·48	12·32	13·31	14·07	15·33	12·94	17·55
Chloride of potassium..	9·67	3·64	9·24	21·40	6·20	8·91	4·58	3·88	5·46
	100·00	100·00	100·00	100·00	100·00	100·00	100·00	100·00	100·00
Per cent. of ash	1·08	7·85	7·62	13·36	6·10	11·36	6·56	6·65	4·29

The analyses a, g, h are by Sprengel; b, c by Witting; b of wood grown on sandstone, near Marburg; c of wood grown on the muschelkalk, in Kurhessen; d, e, f by Heyer and Vonhausen, the wood was grown on a basaltic hill near Giessen.

TABLE XX.

Constituents	Wood.			Wood with bark.			Leaves.	
				Stem.	Large branches.	Twigs.		
	a	b	c	d	e	f	g	h
Potash	24·9	10·9	6·9	13·1	12·5	11·8	—	5·1
Soda	1·1	1·2	0·2	3·1	1·7	1·8	0·7	1·0
Lime	27·4	13·5	43·6	39·8	37·8	40·2	51·7	37·7
Magnesia	6·6	12·0	5·4	10·1	13·4	9·0	6·1	7·9
Alumina	2·2	0·06	—	—	—	—	1·1	—
Oxide of iron	—	—	0·6	0·5	0·3	0·6	0·8	0·4
Oxide of manganese	7·4	3·4	trace	0·9	1·0	0·6	4·1	2·4
Sulphuric acid	7·1	1·0	0·6	0·4	0·5	1·0	1·9	1·2
Silica	5·2	6·2	2·1	6·2	5·5	8·2	27·0	28·5
Carbonic acid	—	26·2	28·2	19·6	17·4	16·2	—	10·5
Phosphoric acid	—	5·6	7·5	6·0	9·6	10·2	6·6	4·6
Phosphate of iron ..	16·1	—	—	—	—	—	—	—
Chloride of sodium ..	2·6	6·7	0·6	0·1	0·8	0·1	—	0·2
Charcoal	—	2·1	—	—	—	—	—	—
Sand	—	10·7	3·7	—	—	—	—	—

The concluding table, compiled by Hohenstein, is of great value, containing, as it does, the results of manufacturing operations in Russia; those of Simon's manufactory, and Hohenstein's own works; the returns made by the Prussian inspector Hartig; and the statements published by Hermann and Leuchs.

TABLE XXI.

No.	Common Name	Botanical Name
1	Acacia, foliage	Robinia pseudacacia
2	Alder	Betula alba, Trunk
	Do.	Do. Branches & Leaves
	Do.	Do. Roots
3	Ash	Fraxinus excelsior, Trunk
	Do.	Do. Foliage
	Do.	Do. Roots
4	Aspen	Populus tremula
5	Alder	Alnus salicifolius
6	Barberry	Berberis vulgaris
7	Beans	Vicia faba
8	Beech	Fagus sylvatica
	Do.	Do. Foliage
	Do.	Do. Fresh Foliage
9	Beetroot	Beta vulgaris, Leaves
10	Bilberry	Vaccinium myrtillus
11	Birch	Betula albo
	Do.	Do. Foliage
12	Blackberry	Rubus fruticosus
13	Boxwood	Buxus sempervirens
14	Broom	Spartium scoparium
15	Buckthorn	Rhamnus catharticus
	Do.	Do. Wood

No.	Name of Plant. Common Name.	Name of Plant. Botanical Name.	Hermann's Results of the Yield of Potash, 1000 lbs. give: Ashes. Pounds.	Hermann's: Raw. Potashes. Pounds.	Hermann's: Calcined. Potashes. Pounds.	Hermann's: 1000 lbs. of Ashes give of Potashes.	Hermann's: 100 lbs. of the Plant contain in Potash.	Hermann's: 100 lbs. of the Plant contain in Soda.	1000 lbs. of Ashes give of Crude Potashes according to Simon's manufacturing Results.	Hohenstein's Results. 1000 lbs. yield. Black. Potashes	Hohenstein's: 100lbs. black. Pure. Potashes	Hohenstein's: (column M)	Hartig's Results. 1000 lbs. yield. Black. Potashes	Hartig's: 100lbs. black. Pure. Potashes	Hartig's: (column P)	Results from the Russian Works. 1000 lbs. yield. Black. Potashes. Pounds.	Russian Works: 100lbs. black. Pure. Pounds.	Russian Works: Pure. Pounds.	Russian Works: Yield Potashes.	E. F. Leuch's Results. 1000 lbs. yield. Ashes.	Leuch's: Crude Potashes.	Leuch's: Calcined Potashes.	Leuch's: 100 parts of the dry Plant contain Potash.	Leuch's: 100 parts of the dry Plant contain Soda.	Leuch's: 1000 of Ashes yield lbs. of Potashes. Lbs.
16	Buckwheat	*Polygonum fagopyrum*	—	—	—	418¾	0·332	0·062	—	—	—	8	—	—	—	—	—	10½	—	—	—	—	0·332	0·062	418¾
	Do.	Do. *aviculare*	—	—	—	—	0·642	0·049	—	—	—	—	—	—	—	—	—	—	—	—	—	—	0·642	0·049	—
	Do.	Do. Straw	—	—	—	295	—	—	—	—	—	—	—	—	—	—	—	—	—	—	—	—	—	—	295
17	Cabbage leaves	*Brassica Oleracea var. Capitata*	—	—	—	—	2·370	1·154	—	—	—	—	—	—	—	—	—	—	—	—	—	—	2·370	1·154	—
18	Carraway	*Carum carvi*	—	—	—	—	0·825	0·072	—	—	—	—	—	—	—	—	—	—	—	—	—	—	0·825	0·072	—
19	Carrot	*Daucus carota*, Leaves	—	—	—	—	3·236	0·921	—	—	—	—	—	—	—	—	—	—	—	—	—	—	3·236	0·921	—
20	Chestnut	*Castanea*, Fruit	32·4	19·6	—	—	—	—	—	—	—	—	—	—	—	—	—	—	—	32·2	19·5	—	—	—	—
21	Horse-Chestnut	*Æsculus Hippocastaneum*	—	—	—	—	—	—	161·2	—	—	—	—	—	—	—	—	—	—	—	—	—	—	—	—
	Do.	Do. Ashes	—	—	—	720	—	—	—	—	—	—	—	—	—	—	—	—	—	—	—	—	—	—	720
	Do.	Do. Fruit	23·5	16·7	—	—	—	—	—	—	—	—	—	—	—	—	—	—	—	23·5	16·7	—	—	—	—
22	Common Celandine	*Chelidonium majus*	—	—	—	250	—	—	—	—	—	—	—	—	—	—	—	—	—	—	—	—	—	—	250
23	Common Mugwort	*Artemisia vulgaris*	96·42	72	—	748¾	0·604	0·154	745·64	—	—	—	—	—	—	—	30	—	—	97·44	78	—	—	—	748¾
	Field Southernwood	Do.	—	—	—	—	0·370	0·030	—	—	—	—	—	—	—	—	—	—	—	—	—	—	—	—	—
24	Common Wild Cabbage	*Brassica campestris*	—	—	—	—	0·883	0·550	—	—	—	—	—	—	—	11½	1½	2¾	—	—	—	—	—	—	—
25	Corn Marigold	*Chrysanthemum segetum*	—	—	—	—	1·234	0·244	—	—	—	—	—	—	—	—	—	—	—	—	—	—	1·234	0·244	—
26	Corn Spurrey	*Spergula arvensis*	—	—	—	—	0·930	0·420	—	—	—	—	—	—	—	—	—	—	—	—	—	—	0·930	0·420	—
27	Corn Thistle		106	19·8	—	—	—	—	—	—	—	—	—	—	—	—	—	—	—	105	19·6	—	—	—	—
28	Cudweed	*Gnaphalium arenarium*	120·5	22·5	—	—	—	—	—	—	—	—	—	—	—	—	—	—	—	120·7	25	—	—	—	—
29	Dog Rose	*Rosa canina*	—	—	—	—	—	—	—	—	—	—	—	—	—	6¼	0¼	2¼	—	—	—	—	—	—	—
30	Dogwood	*Cornus sanguinea*	—	—	—	—	—	—	—	—	—	—	—	—	—	6	0¾	3½	—	—	—	—	—	—	—
31	Dutch Peat		509·57	1·04	—	—	—	—	—	—	—	—	—	—	—	—	—	—	—	509·37	1·04	—	—	—	—
32	Dyer's Broom	*Genista tinctoria*	—	—	—	—	0·450	0·052	—	16¼	2	2¾	16	2	—	16¼	1¾	2¾	—	—	—	—	0·450	0·052	—
33	Eagle Fern	*Polypodium Pteris aquilina*	40·0	6·4	—	211	0·058	0·370	—	27	15	—	—	—	—	29	14¼	12½	—	40·0	6·2	—	0·058	0·370	—
34	Elder	*Sambucus niger*, Trunk	20	4·06	3·7	—	—	—	—	—	—	—	—	—	—	13¾	1⅛	2¼	—	20	4·06	3·7	—	—	—
	Do.	Do. Roots	16·5	2·83	2·43	—	—	—	—	—	—	—	—	—	—	—	—	—	—	16·4	2·81	2·42	—	—	—
35	Elecampane	*Inula salicina*	—	—	—	—	0·774	0·168	—	—	—	—	—	—	—	—	—	—	—	—	—	—	0·774	0·168	—
A	B	C	D	E	F	G	H	I	J	K	L	M	N	O	P	Q	R	S	T	U	V	W	X	Y	Z

A	B	C	D	E	F	C	H	I	J	K	L	M	N	O	P	Q	R	S	T	U	V	W	X	Y	Z
36	Elm	*Ulmus campestris*	23·5	3·9	—	166	0·791	0·370	164	23	2½	2¾	—	—	—	23	3	3¼	—	23·5	3·9	—	0·791	0·370	166
	Do.	Do. Leaves	—	—	—	—	0·579	0·160	—	—	—	—	—	—	—	—	—	—	—	—	—	—	1·579	0·160	—
	Do.	Do. Branches & Twigs	—	—	—	—	—	—	167·2	—	—	—	—	—	—	—	—	—	—	—	—	—	—	—	—
	Do.	Do. Effusa	—	—	—	—	—	—	—	23¼	2¾	3	23	2¾	—	23	3	3¼	18	—	—	—	—	—	—
37	Feather Reed		43	5	—	—	—	—	—	—	—	—	—	—	—	—	—	—	—	43	5	—	—	—	—
38	Field Bean	*Vicia faba*, Straw	—	—	—	—	1·656	0·050	—	—	—	—	—	—	—	—	—	—	—	—	—	—	—	—	—
39	Foxtail Grass	*Alopecurus pratensis*	—	—	—	330	—	—	—	—	—	—	—	—	—	—	—	—	—	—	—	—	—	—	330
40	Fumitory	*Fumaria officinalis*	218·5	79	—	360	—	—	358·31	—	—	—	—	—	—	219	79	9	—	219	79	—	—	—	360
41	Golden Rod	*Solidago virga aurea*	—	—	—	—	0·552	0·076	—	—	—	—	—	—	—	—	—	—	—	—	—	—	0·552	0·076	—
42	Hair Moss	*Polytrichum jaccæ folium*	—	—	—	—	trace	trace	—	—	—	—	—	—	—	—	—	—	—	—	—	—	trace	trace	—
43	Hazel Nut	*Corylus Avellana*	30	2·8	2·28	—	—	—	—	—	—	—	—	—	—	—	—	—	—	30	2·7	2·27	—	—	—
44	Heath	*Erica vulgaris*	—	—	—	—	0·094	0·200	—	14½	1½	3	14	1½	—	14	1¾	3	—	—	—	—	0·094	0·200	—
45	Honeysuckle	*Lonicera xylosteum*	—	—	—	—	—	—	—	—	—	—	—	—	—	10½	1	2½	—	—	—	—	—	—	—
46	Hornbeam	*Carpinus betulus*, Wood	11·21	1·45	—	—	—	—	—	—	—	—	—	—	—	—	—	—	—	11·21	1·45	—	—	—	—
	Do.	Do. Twigs	25	3·3	3	—	0·061	0·013	219·9	12½	1¼	1¾	11	1¼	—	12½	1¼	2¾	—	25	3·3	3	0·061	0·013	—
	Do.	Do. Foliage	—	—	—	—	1·215	0·689	221·2	—	—	—	—	—	—	—	—	—	—	—	—	—	1·215	0·689	—
47	Ivy	*Hedera helix*	39·6	2·7	—	—	—	—	—	—	—	—	—	—	—	15¼	1¾	3½	—	—	—	—	—	—	—
	Do.	Do. Leaves, autumn	39·2	19·71	—	—	—	—	—	—	—	—	—	—	—	—	—	—	—	54·21	19·73	—	—	—	—
	Do.	Do. do. February	94·3	19·62	—	—	—	—	—	—	—	—	—	—	—	—	—	—	—	94·2	19·60	—	—	—	—
48	Jointed Glasswort	*Salicornia herbacea*	—	—	—	—	0·170	3·688	—	—	—	—	—	—	—	—	—	—	—	—	—	—	0·170	3·688	—
49	Juniper	*Juniperus communis*	—	—	—	—	—	—	—	18	1½	2¾	18	1½	—	18½	1½	2¾	—	—	—	—	—	—	—
50	Lady's Mantle	*Alchemilla vulgaris*	—	—	—	—	0·506	0·040	—	—	—	—	—	—	—	—	—	—	—	—	—	—	0·506	0·040	—
51	Larch	*Pinus larix*, Trunk	—	—	—	—	—	—	144·1	—	—	—	—	—	—	—	—	—	—	—	—	—	—	—	—
	Do.	Do. Roots	—	—	—	—	—	—	116·24	—	—	—	—	—	—	—	—	—	—	—	—	—	—	—	—
52	Lime Tree	*Tilia parvifolia*, Leaves	—	—	—	—	1·552	0·198	—	—	—	—	—	—	—	—	—	—	—	—	—	—	1·552	0·198	—
	Do.	Do. Wood	—	—	—	—	0·056	0·035	—	—	—	—	—	—	—	—	—	—	—	—	—	—	0·056	0·035	—
53	Do.	Do. *cordata*	—	—	—	—	—	—	—	14¼	1	1¾	14	¾	—	14½	¾	1½	—	—	—	—	—	—	—
54	Lucerne, green	*Medicago sativa*	—	0·73	—	—	0·362	0·166	—	—	—	—	—	—	—	—	—	—	—	—	0·75	—	0·362	0·166	—
55	Maize	*Zea mays*, Straw	88	17·05	—	—	0·189	0·004	—	—	—	—	—	—	—	—	—	—	—	88	17·05	—	0·189	0·004	—
	Do.	Do. Ashes	—	—	—	193	—	—	—	—	—	—	—	—	—	—	—	—	—	—	—	—	—	—	193
56	Maple, trunk	*Acer Pseudoplatanus*	27·9	4·6	—	—	0·115	0·005	—	—	—	—	—	—	—	11¼	1¼	2¾	7	27·9	4·6	—	0·115	0·005	—
	Do. leaves	Do.	—	—	—	—	0·931	0·580	—	—	—	—	—	—	—	—	—	—	—	—	—	—	0·931	0·580	—
57	Meadow Trefoil		—	0·78	—	—	—	—	—	—	—	—	—	—	—	—	—	—	—	—	0·78	—	—	—	—
58	Milfoil	*Achillea millefolium*	—	—	—	—	0·425	0·152	—	—	—	—	—	—	—	—	—	—	—	—	—	—	0·425	0·152	—
59	Moss Stems	*Hypnum cuspidatum*	—	—	—	—	0·140	0·178	—	—	—	—	—	—	—	—	—	—	—	—	—	—	0·140	0·178	—
60	Nightshade, woody	*Solanum dulcamara*	—	—	—	270	—	—	—	—	—	—	—	—	—	7	1¾	6½	—	—	—	—	—	—	270
61	Walnut Tree	*Juglans regia*	—	—	—	140	—	—	—	—	—	—	—	—	—	—	—	—	—	—	—	—	—	—	140
	Do.	Do. Trunk	—	—	—	—	—	—	149·64	—	—	—	—	—	—	—	—	—	—	—	—	—	—	—	—
	Do.	Do. Branches & Twigs	—	—	—	—	—	—	150·22	—	—	—	—	—	—	—	—	—	—	—	—	—	—	—	—
	Do.	Do. Roots	—	—	—	—	—	—	98·80	—	—	—	—	—	—	—	—	—	—	—	—	—	—	—	—
62	Oak	*Quercus robur*	15·5	1·5	—	111	0·066	0·014	111	15	1⅔	3¼	14	1¾	—	13½	1½	3	10	—	—	—	0·066	0·014	—
	Do.	Do. Foliage	30	5·1	—	—	0·710	—	—	—	—	—	—	—	—	—	—	—	—	—	—	—	0·710	—	—
	Do.	Do. Branches & Twigs	—	—	—	—	—	—	112·16	—	—	—	—	—	—	—	—	—	—	—	—	—	—	—	—
	Do.	Do. Trunk	23·5	2·2	1·98	—	—	—	—	—	—	—	—	—	—	—	—	—	—	23·5	2·2	2	—	—	111

No.	Common Name	Botanical Name	Ashes	Raw	Calcined	1000lbs. of ashes give of Potash	Potash	Soda	1000 lbs. of ashes give of Crude Potash according to Simon's manufacturing	Black 1000 lbs.	Pure	Black 1000 lbs.	Pure	Black 1000 lbs.	Pure	Black 1000 lbs.	Pure	1000lbs. black	Yield Potash	Ashes	Crude Potash	Calcined Potash	Potash	Soda	1000 of ashes yield lbs. of Potashes
	Oak	Quercus robur, Roots	24·33	2·5	2·35	—	—	—	—	—	—	—	—	—	—	—	—	—	—	24·32	2·5	2·35	0·710	—	—
	Do.	Quercus pedunculata, Green foliage	—	—	—	—	—	—	—	—	—	—	—	—	—	—	—	—	—	—	—	—	0·700	trace	—
	Do.	Do. Ripe do.	—	—	—	—	—	—	—	—	—	—	—	—	—	—	—	—	—	—	—	—	0·870	0·002	—
63	Oat Straw	Avena sativa	—	—	—	410	0·870	0·002	—	—	—	—	—	—	—	—	—	—	4	34	4·7	—	0·235	—	410
64	Urache	Atriplex, Ashes	28	4·7	—	256¾	0·235	—	182·08	—	—	17·2	—	16½	2	—	3	—	—	—	—	—	0·440	0·196	256¾
65	Pea	Straw	—	—	—	182	0·440	0·196	114·54	—	—	17·2	—	17¾	2	—	3	—	—	12	—	0·7	—	1·640	2·260	182
66	Pine	Pinus picea, Needles	—	—	—	—	1·640	2·260	—	—	—	—	—	—	—	—	—	—	—	—	—	—	0·846	0·443	—
	Do.		12	0·7	—	—	0·846	0·443	—	—	—	—	—	—	—	—	—	—	—	—	—	—	0·641	0·307	—
67		Pinus abies, Roots	—	—	—	—	0·641	0·307	45·1	18¾	1	18	2	12	—	—	4	—	—	—	—	—	—	—	—
68	Plume-thistle	Cnicus arvensis	—	—	—	333	0·138	0·009	—	1½	1¼	—	—	11½	9½	1¼	2	—	—	—	—	—	0·138	0·009	333
69	Poplar	Populus dilatata, Leaves	—	—	—	—	0·819	0·009	—	16½	2¼	16·1½	—	61·13½	5¼·15	1½	1	—	—	—	—	—	0·819	0·009	338
	Do.	Do. tremula	—	—	—	—	—	—	—	—	—	—	—	9½	—	—	—	—	—	—	—	—	0·138	0·086	—
	Do.	Do. Leaves	—	—	—	—	0·735	0·049	—	—	—	—	—	—	—	—	—	—	—	—	—	—	0·923	0·188	—
70	Potato	Helianthus tuberosus, Green leaves	—	—	—	—	0·419	0·111	—	—	—	—	—	—	—	—	—	—	—	—	—	—	0·419	0·111	—
	Do.	Do. before flowering	—	—	—	—	—	—	—	—	—	—	—	—	—	—	—	—	—	—	—	—	—	—	—
	Do.	Do. after do.	39	7	—	—	1·000	0·040	—	—	—	—	—	—	—	—	—	—	—	39	7	—	—	—	—
71	Prickly Sedge	Carex muricata	—	—	—	—	0·494	0·103	—	—	—	—	—	17¾	3	—	3	—	—	—	—	—	0·494	0·103	—
72	Privet	Ligustrum vulgare	—	—	—	—	0·006	0·007	—	—	—	—	—	—	—	—	—	—	—	—	—	—	0·006	0·007	—
73	Red Clover	Trifolium pratense	—	—	—	—	—	—	—	—	—	—	—	—	—	—	—	—	—	—	—	—	—	—	—
74	Rest-harrow	Ononis spinosa	—	—	—	—	—	—	—	—	—	—	—	—	—	—	—	—	—	—	—	—	—	—	—
75	River Sedge		—	—	—	—	—	—	—	—	—	—	—	—	—	—	—	—	—	—	—	—	—	—	—
76	Rush (soft)	Juncus effusus	—	—	—	—	—	—	—	—	—	—	—	—	—	—	—	—	—	—	—	—	—	—	—
77	Saintfoin	Hedysarum Onobrychis	—	—	—	—	—	—	—	—	—	—	—	—	—	—	—	—	—	—	—	—	—	—	—
78	Scotch Fir	Pinus sylvestris	—	—	—	—	—	—	—	—	—	—	—	—	—	—	—	—	—	—	—	—	—	—	—
A	B	C	D	E	F	G	H	I	J	K	L	M	N·O·P	Q	R	S	T	U	V	W	X·Y	Z

A	B	C	D	E	F	G	H	I	J	K	L	M	N	O	P	Q	R	S	T	U	V	W	X	Y	Z
	Scotch Fir	Pinus Sylvestris, Bark	127·75	28·3			0·297	0·264												127·8	25·4		0·297	0·264	
	Do.	Do. Needles	56·88	6·1																56·95	6·12				
79	Soapwort	Saponaria																							
80	Sow Thistle	Sonchus arvensis, Do. Roots					0·261	0·163						104½		25	6			107·5	25		0·261	0·163	
	Do.	Do. oleraceus					0·436	0·276															0·436	0·276	
81	Spring Cinque-foil	Potentilla verna	108	25			0·493							17		9½	4						0·493		
82	Stinging-nettle (larger)	Urtica dioica			232½	232½																			232½
83	Sugar Cane				110	110																			110
84	Sumach	Rhus coriaria	57	20		409								404		34	3½			67·02	20				409
85	Sunflower	Helianthus	30	6·7										105		19½	4½			40	5·9				
86	Thistle	Carduus, 1st sort																								
	Do.	Do. 2nd sort	220	16·12		67½															220	18·12				67½
87	Tobacco-stalks	Nicotiana tabacum													16½		2	3½								
88	Tormentil	Tormentilla erecta													88½		17½	4½								
89	Traveller's Joy	Clematis vitalba	104	27·4			0·897	0·632												34	27·5		0·897	0·632	
90	Turkish Wheat		34	8·07		162½																			162½
91	Vetch	Vicia sativa									11	3 11			33½		5½	4½			34	3·05				
	Do.	Do. graeca										3 11			16		1	1½								
92	Vine	Vitis, Branches					0·268								18½		1½	2						0·268		
	Do.	Do. Leaves																								
93	Gueldar Rose	Viburnum opulus																								
94	Wayfaring Tree	Do. lantana	67·3	11·83							11½	3½ 11			10		1½	1			67·26	11·80				
94	Wild Cammomile	Matricaria Parthenium													13		1½	1½								
95	Wild Service-tree	Crataegus torminalis																								
	Whitethorn	Do. oxyacantha													28	3	3	2¼			28					
96	Willow	Salix alba	28	2·63		102	1·655	0·605	101·79													2·85		1·655	0·605	109
	Do.	Do. vitellina																								

Grabner, in his work on Natural Philosophy for Foresters, says, in reference to this subject, that 1,000 parts of the most productive timber do not yield, on an average, more than from 2 to 4 parts of potashes, bark from 4 to 6 parts, while many garden vegetables furnish from 20 to 40 parts, and wormwood rises as high as 70 to 80 parts.

PRODUCTION OF POTASH ASHES.

In the case of forest culture, the removal of the timber off the land does not impoverish the soil, because the yearly disintegration of its mineral matter yields a supply of potash sufficient to meet the annual growth of the trees. But when the produce of the soil is carried away in larger quantities than can be restored by the operations of nature, such a soil must, sooner or later, become more or less barren, unless the deficient mineral matter is returned in the artificial form of manures. The whole profit, therefore, to be derived by extracting potash from the ashes of weeds grown on arable land, would be simply so much deducted from the fruitfulness of the soil.

Hermbstädt tried the plan of planting wormwood, which grows on very poor soil, for the production of potashes. From his recorded experiments, one Magdeburg acre (18,000 square feet) yields three crops in summer, weighing 20,000 lbs. of dry plant, which leave 2,364 lbs. of ash, from which 936·6 lbs. of potashes were obtained, or 11·8 per cent. of ash, and 4·7 of potashes.

Beichos has proposed to grow tansy for the same object, and states that the acre yields 1250 lbs. of potashes.

More recently, Fresenius has suggested the growth of the common marigold (*Crysanthemum segetum*) for this manufacture. This plant is found in enormous quantities in the Westerwald. Bangert has examined it in reference to this question and has found that the

Fresh plant yields .	1·61 per cent. ash,	
Air-dried ditto „ .	6·87	„
Dried at 100° C. .	8·52	„

which consists of—		
Soluble portion . . .	63·31 per cent.
Insoluble portion . .	36·69 „
	100·00

The ash contains—	
Potash	24·86
Soda	6·21
Chloride of sodium	16·10
Sulphuric acid	5·12
Carbonic acid	12·36
Phosphoric acid	6·16
Silica	4·68
Lime	14·08
Magnesia	6·96
Oxide of iron	1·02
Oxide of manganese, sand, &c. .	3·06
	100·61

From our previous remarks, it must be obvious that the locality must be of a most exceptional character, before any plant can be grown for the sole purpose of extracting potashes from its ash.

It is very different in extracting the potash from the ashes of fire-wood, as we have pointed out how the operation of nature provides for the annual supply of potash to the land, and the ultimate exhaustion of such a soil is so very remote as not to occasion any fear with regard to such an issue of forest culture ; a single cubic foot of felspar is sufficient to supply an oak copse, covering a surface of 26,910 square feet, with potash for five years.

In some thinly-populated countries, as in the north of Europe, the Western States of America, and the British Provinces of North America, wood is so abundant and cheap, that it can be burnt solely for the sake of the ash. In Russia, where manure is of very little value, the straw and weeds are burnt for this purpose. Yeast, and the lees of wine, in Burgundy and other wine districts, the residue from the brandy distilleries at Valenciennes, and the stems and leaves of the indigo plant in Java; are all employed for the production of potashes.

Maillard has also patented a process for lixiviating the ashes of turf and peat to procure the potash-salts, and Barruel, after distilling cotton-seed for illuminating gas and lubricating oil, extracts the potash from the charred residue left in the retorts.

The Preparation of the Ash

from forest timber, is principally carried on in North America, Russia, Sweden, Germany, and Tuscany, on an immense scale, also in France, in the district of the Vosges.

In many of the States of the American Union, the incineration is managed in dry pits, sunk in the ground to a depth of 3 to 4 feet. The timber and plants are thrown in from time to time and burnt, until the pit is nearly full of ashes ; this plan is said to be superior to burning them in pyramids, on which the combustion goes on too rapidly, and there is less protection from the wind.

The ashes from the ordinary hard wood from $2\frac{1}{2}$ acres of land, are stated to be sufficient to make a barrel of potashes, worth, if of second quality, £7 10s., after paying all expenses, and often going a long way, says Morris, towards enabling the settler in Canada to meet the first cost of his land.

On the Continent, similar plans are adopted; the dry pit is covered with flat stones to prevent any contact with the ground; the depth is between 3 and 4 feet, with a length and breadth of 6 and 4 feet. The plants and timber are partially air-dried before they are burnt. A small lot is set on fire in the pit, and fresh quantities are added to keep up the combustion until the pit is nearly full of ash. In other localities, a flat surface is chosen and covered with stones, on which the trees and plants are arranged in piles, which are set on fire, fresh quantities of timber being added as before, until a heap of ashes is obtained. A rough shelter from the wind is constructed of branches placed round the fire, to prevent, as far as possible, a loss of ash during the stirring of the heap, which is required to render the combustion perfect.

This latter plan produces better ashes than the former, where, the access of air being limited, a quantity of charcoal always remains unconsumed.

In former times, in Germany, the ashes obtained by burning the trees and shrubs in the forests were called *Waldasche* (forest ashes) in contradistinction to those procured as a bye-product from furnaces, which were termed *Brennasche*. The former were lixiviated in a rough manner by the woodmen in the forest itself, dried and calcined, and in this form sold to the potash manufacturers, who refined them in the usual manner. This intermediate product, between the crude ash and the potashes of commerce, is called *Okras* or *Ochras*, and is produced in the neighbourhood of Dantzig, in the east of Prussia, and in Sweden. For the preparation of this ash, old decayed trees are selected. The wood, split up into billets, is burnt to ash on the spot, in a low situation protected from the wind, and with as little fire as possible; the ash having been separated from the charcoal, half-burnt wood, and stones, is then collected together in a shed. When a sufficient quantity has been brought together, it is submitted to a kind of crude calcination. This operation is commenced by the workman transforming the whole mass, by the gradual addition of water and constant agitation with rakes, into a dough or stiff mortar, which is then arranged in alternate layers with pinewood, so as to form a sort of pile. Upon a series of billets, arranged on the ground, a layer of this dough is deposited, then another series of billets in a cross direction, another layer of the ash, and so on. After igniting this pile, which is several feet high, the fire is urged to its greatest intensity, until the

ash becomes red-hot and fused; when this state has become
general throughout, the pile is destroyed and the fused mass
stirred about with poles, to which it attaches itself, for the
greater part, on cooling. It has the appearance of a slag, and
is of a bluish colour.

The crude ash is treated with water, to extract the soluble
portion; the lye so obtained is evaporated, and the residue
heated to redness.

The Lixiviation of the Ashes.

The process of lixiviation is similar in principle to that
practised in soda works, but is carried out in a very crude
manner. The tubs used for this purpose are generally merely
tar-barrels sawn in two, a number of these being furnished with
two cross beams upon which rests a false (sieve-like) bottom
covered with straw, and below this is a plug, a; these are called
the *ash-tubs*, Fig. 251. In order to facili-
tate the descent of the lye into the space
between the two bottoms, a tube, b, is introduced at the commencement for the escape
of the air. The ash cisterns are arranged in
a double row, one above the other; the first
upon a scaffolding, and the last set upon
wooden stands. Parallel with these, and
in great part sunk in the earth, is a third
front row of empty cisterns, "the *wells*."

FIG. 251.

Before charging the ash cisterns, it is necessary to sift the ash
from coal and charred wood (which are thrown upon the fire),
and to moisten it thoroughly, that the water may pass through it
of uniform strength. When all the cisterns in both rows are
charged with moist ash, which is firmly pressed down upon the
straw, that no interstices may be left, the lixiviation begins, and
the first cistern in the uppermost row is supplied with a quantity
of fresh water. In taking up the soluble portion from the ash, this
becomes converted into a strong lye, and on leaving a, and being
in a fit state for boiling, it is immediately brought into the pan
for evaporation. In the meantime, a second portion of water
is poured upon the ash in the first cistern, which is not exhausted; a second, much weaker, lye is produced, which is
allowed to run into the first ash cistern of the second row, whence
it also issues in a fit state of saturation for boiling. The third

portion of water is still too weak on leaving the first and second cisterns, and is only fit for boiling after having passed through a third cistern (the second of the upper row); thus every fresh portion of water comes at last in contact with a fresh portion of ash. While the last ash cisterns are at work, those first used can be emptied and refilled.

In this manner, some of the cisterns are always ready charged for bringing up the lye to the proper state of saturation (it should contain from 20 to 25 per cent. of salt); and as water is poured upon the ash as long as anything is dissolved (which is estimated by the hydrometer), no portion of the soluble matter, or no potash, is lost. By treating the ash with cold water, nearly all the sulphate of potash would remain behind; but partly in order to increase the weight of the potash, partly because the potash in that salt is made available in many of its applications, it is thought preferable to use hot water, at least towards the end of the operation.

The simple and suitable process just described, which is that practised at Blansko in Moravia, is not so generally followed as it might be. In the Odenwald and Hinterland (Grand Duchy of Hesse), the imperfect method of lixiviation gives rise, with other inconveniences, to a great waste of fuel. In the latter district, the lye with which the pan is supplied is so weak, that the fire must be kept up for 60 hours, in order to obtain from 70 to 100 lbs. of crude potash by evaporation.

Hohenstein describes the following arrangement for lixiviating the ashes, as it is practised in some parts of Russia and Germany. The ash-tubs A and B are made of pine or fir-wood, and bound with three iron hoops. A is a front view of one, and B is shown in section. The depth is three feet, the diameter at top, 3 to 5 feet, and that of the bottom, $2\frac{1}{4}$ to $4\frac{1}{4}$ feet. The tubs are supported

FIG. 252.

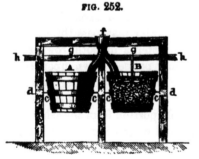

by iron bolts, c c c, let into the middle iron hoop, and resting in sockets, d d d, in the upright beams, e e e, round which the tub can be made to turn, and thus facilitate the emptying of the exhausted ash. The interior fittings are the same as those ·

already described. The wet or damp ashes are filled into the tube through the funnel f, and the weak lye or water is run on through the cocks g g in a supply pipe h h.

In France the water is gradually added to the cisterns or tubs until it rises above the level of the ashes, where it is left for twelve hours, and then run off direct to the evaporating vessels. Three or four washings remove all the soluble portion of the ash. Instead of cutting the tubs in two, they are used whole; and in other works large iron cisterns are employed, capable of holding from 20 to 30 cwt. of the ash.

Payen states that they are lixiviated in a series of tanks A, B, C, D, E, &c., the highest of which is twice the size of the others. The ashes are thrown into the smaller iron cisterns, a, b, c, d, &c., which are pierced with small holes. These cisterns are suspended in the tanks A, B, &c., which are filled with water;

FIG. 253.

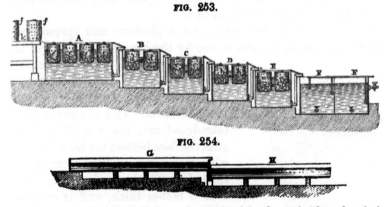

FIG. 254.

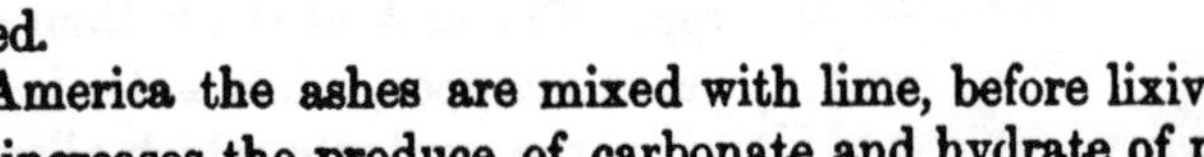

as the water in A becomes charged with the soluble salts, it is run into B, and so on through the whole series, until it arrives in the settlers F F, whence the evaporating pans G H are kept supplied.

In America the ashes are mixed with lime, before lixiviating, which increases the produce of carbonate and hydrate of potash, probably by decomposing the sulphate and silicate of potash. This addition of lime has also been recommended by M. Stœcklin in France and Germany.

The ashes are collected as pure as possible and then sifted in a drum (Fig. 255) to separate the charcoal and other impurities. This drum-sieve is a wooden frame-work cylinder, A, covered with wire-gauze, the meshes of which are small or coarse, according to

FIG. 255.

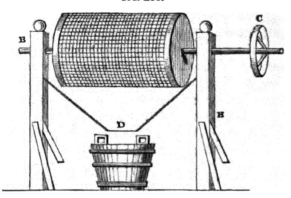

the degree of fineness to be given to the ashes. It is mounted on uprights, B, and can be made to revolve by hand, or by means of a shaft and pulley, C. The shelf or hopper, D, guides the fine ashes into tubs as it leaves the sieve.

The sifted ash is heaped up on a platform and spread out, leaving an opening in the centre for the lime, which is added in the proportion of 50 to 80 lbs. for every 1000 lbs. of ash. The lime is slaked and then thoroughly mixed with the ashes. This mixture is then introduced into the vat (Fig. 256), after the false bottom has been covered with straw to prevent the stoppage of the holes. The layer of ashes is made 6 or 8 inches thick, and is covered with straw, followed by another layer of ashes, and so on, until the vat is filled to within 8 or 12 inches of the top. The cock at the bottom of the vat is left open, while water is run in upon the contents, which allows the air to escape without disturbing the materials. When the liquid makes its appearance below, the cock is closed, and the water allowed to dissolve the soluble salts. When saturated it is run off by the cock, and fresh portions of water are added, until the runnings appear white, which is an indication that the ashes are exhausted.

FIG. 256.

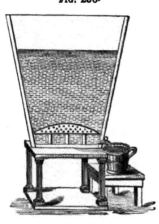

The insoluble residue left by the German plan of lixiviation, has been analysed by Wolff, after exposure to the air for about twelve months. It contained

Carbonate of lime . . .	41·55
Phosphate of lime . . .	11·30
Soluble silica	3·25
Magnesia	2·55
Potash	0·74
Sand and clay	36·16
Organic matter	4·61
	100·16

It is used in Germany as a manure, in the manufacture of green bottle glass, and in the saltpetre plantations.

In the analysis, Wolff has represented the silica and potash as not in combination, which is probably quite correct, for the carbonic acid of the air decomposes the silicate of potash left after lixiviation. This circumstance has been applied in practice in some parts of Germany, in a process termed " Keimen," which consists in spreading the sifted and moistened ash upon a level stone floor, where it is exposed to the air for several months, during which it is frequently moistened, and turned over.

Evaporation of the Lye.

The colour of the lye is dark brown, which arises from the carbonate of potash dissolving part of the humus left by the imperfect incineration of the plants.

The lye is evaporated in flat iron pans with corrugated bottoms, which present a greater surface to the fire, or in cast-iron boilers which expose a large evaporating surface. More space is thus afforded to the workmen for breaking up and removing the mass of salt, while the bottom can be supported in several places. The same arrangement for applying heat as that employed with the salt pans (p. 170) is the most appropriate, and the flame in escaping is advantageously used for warming a feeder, whence the evaporating pan receives its supply. The evaporation goes on quickly at first, and the evaporated water is constantly replaced by lye of the proper strength, until the contents of the pan become thick and syrupy, and a hot portion of the lye quickly solidifies to a crystalline mass on cooling. No further addition is then made, and the whole is evaporated to dryness at a moderate heat.

In this manner a saline mass is left, containing about 6 per
cent. of water, and firmly attached to the bottom. The colour is
derived from empyreumatic substances, which give a yellowish-
brown appearance to the lye. The *crude potash*, or the *flux*
(*Fluss*), has now to be broken away from the pan by means of
chisels or axes, and the evaporation is then commenced anew.
To avoid this operation, the lye may be evaporated to dryness
with constant stirring, which prevents the salt from adhering to
the sides or bottom of the pan, but is by no means a less laborious
process, and affords a salt containing at least twice as much
water; the pans, however, are not so liable to injury.

The arrangement of the pans for boiling down these liquors, as
generally adopted in Germany, is shown in Figs. 257 and 258, in
section and ground plan. The lyes are run into an iron pan, *a*,
where they are warmed by the waste heat from the other pans,

FIG. 257.

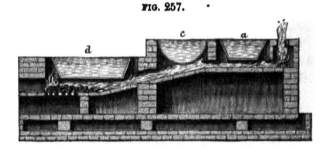

FIG. 258.

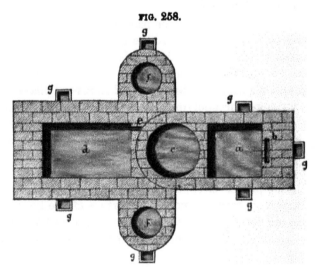

regulated by a damper, *b*. These lyes are afterwards boiled in an adjoining copper pan, *c*, which is 5½ feet diameter. The warm lyes or partially concentrated liquors are run into the large iron pan, *d* by a pipe *e*, which is connected with the previous pans. This pan is 10 feet long by 6 feet wide, and is heated directly by the fire below. The supply of liquor being kept up while the concentration is going on, an economy of fuel is thus effected. When the potash-lyes have become thick, they are transferred to the metal pans *f f*, where they are heated to dryness previous to calcination. The flues, *g g g*, are furnished with dampers for regulating the course of the heated air and gases from the fires under the iron pan *d* and the metal pans *f f*.

In Russia two kinds of potash are at present manufactured; namely, that from *wood-ashes*, more particularly in the districts beyond the Wolga, in the Goub, Neschnei-Nowogrod and Kasan; and that from the ash of *straw*, and the *vegetation of the Steppes*, which is brought from the south, and is only about 10 per cent. inferior to the former. The production of ash constitutes one of the fiscal duties imposed by the landlords upon the peasantry in those countries; the method of preparation is consequently carried on upon a large scale on the estates, but in a crude and in a somewhat different manner. The ash (obtained from fire-hearths, and by burning the weeds) is three times extracted, and the lye is evaporated in flat copper pans heated by a straw-fire, not to dryness, but, in the first instance, only to the point of crystallization, when it is drawn off into wooden cisterns. The pure carbonate of potash deposits in brown crystals, which, after being drained, are calcined in a muffle furnace. The mother-liquor is returned to the wells. In these districts, the whole value of the straw and brush-wood, which is used as fuel, consists in the ash they produce.

In some works in Germany, the sulphate of potash is separated by evaporating the lye to such a strength, that while no carbonate of potash will crystallize out, the whole liquor may be run into wooden cisterns and allowed to cool while it is occasionally stirred. The sulphate of potash crystallizes, and the liquor is boiled to dryness. The sulphate of potash is used in the manufacture of alum.

In France both sheet-iron pans and metal boilers are employed, heated by a naked fire, and the sulphate of potash is separated in the manner we have described.

Calcination, or Preparation of Pearl-ash.

Crude potash, as already observed, contains water and empyreumatic substances, which must be destroyed by a red heat, by "*calcination*." The calcining furnace for potash (Fig. 259), differs from an ordinary reverberatory furnace in having a double

FIG. 259.

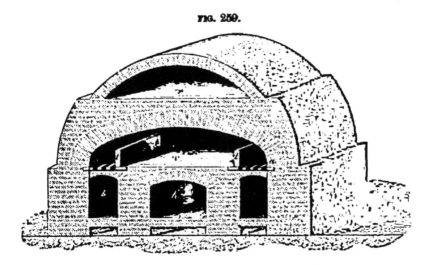

firing (with the grates *a a*, and the ash-pits *b b*), which is separated by two narrow iron plates (fire bridges), *t t* from the working or *calcining hearth*, the latter being from 3 to 4 feet wide. The space *h* is left open, that the sole *c* may be kept more dry, and for economy in the construction. The arch *e* is not necessary, and may be used for any other purpose. The air which feeds the fire, has access through the ash-pits *b* and the grates *a* to the fuel, and produces a flame which plays above the fire bridges *t t*, over the calcining hearth.

The ground-plan of a similar furnace is shown in Fig. 260, where the sole is represented at *a*, with the charging door at *b*. The grates *c c*, extend the whole length of the furnace, and are fired with the wood through the doors *d d*. Instead of iron fire-bridges, hollow brick walls, *e e*, are constructed between the grates and the sole of the furnace.

Dry wood in small billets is used, in order to obtain a long flame to play over the materials. Before introducing the charge, the furnace should be heated until it is perfectly dry, and it is advisable not to allow the furnace to cool between

FIG. 260.

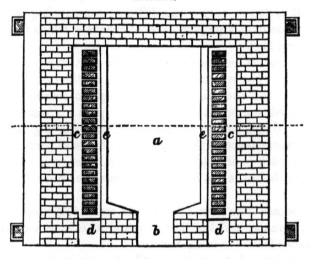

two consecutive operations, as the reheating is a source of expense. As soon as the furnace has nearly attained a red heat, about 3 or 4 cwts. of the crude potash is broken up and spread carefully over the calcining hearth. The mass now assumes a different appearance, depending upon the greater or lesser quantity of water of crystallisation which it contains, and this is evolved with more or less frothing; when too much is present, the operation is impeded, as the heat must then be raised very gradually. By constant agitation, the water is driven off in as uniform a manner as possible. After the lapse of about an hour, all the water has been expelled, and the heat is increased, that the whole mass may become red hot, and that the air, in passing through the furnace, may consume the combustible and colouring matters. At this period, the potash takes fire, and burns with a flame, becoming black by the carbonization of the combustible matter; it soon, however, becomes clearer and clearer, until, at length, a sample appears quite white, and free from carbon. An incrustation generally attaches itself to the sole of the furnace, and to remove this, the flame must be allowed to play upon it, when it softens and is easily loosened.

When the heat is not carefully managed, the mass fuses before the combustible matters are destroyed, and they are covered over with the fluid salt; they are decomposed, but leave their charcoal mixed with the potash, which cannot afterwards be burnt to appear white and clean. The carbonate of potash, fortunately, is difficult of fusion, so that this difficulty can only happen when the ash contains a large proportion of chlorides, which fuse at an incipient red heat.

If the operation has been successful, the next step is to granulate the salt. The workman breaks down all the lumps with an iron paddle, and the small pieces are kept in constant motion while exposed to a bright clear fire. The pure white salt is then ready to be drawn, and to it the name of pearl-ash is given, from a fancied resemblance to the bright white lustre of pearls.

The pearl-ash is allowed barely to cool, when it is packed in close casks, before it has had time to absorb moisture from the air.

Crude potash loses from 15 to 20 per cent. of its weight in the calcining furnace, and the time required to calcine a charge of 3 cwt. varies from five to six hours.

In some manufactories the muffle furnace is constructed to work off charges of upwards of 20 cwt., of which Fig. 261 represents a front view, the chimney K, in front, carrying off the gases, &c.

FIG. 261.

Calcined potash is a hard, porous, and granular saline mass; the white colour sometimes merges into pearl grey, or becomes yellow or blue. Single pieces often exhibit blue and red spots on the broken surface. The red colour is caused by oxide of iron, the grey by the mixture of particles of carbon, and the blue by the action of the alkali on oxide of manganese, giving rise to the formation of a minute quantity of manganate of potash.

Refining of Pearl-ashes.

In America a similar plan is pursued, but in some cases the *pearl-ash* is redissolved, the liquor concentrated, and then allowed to cool. The more insoluble salts crystallise out, and the clear liquor being drawn off into clean flat iron pans, it is evaporated to dryness, during constant stirring towards the end. The potash forms white lumpy granulations, and as thus treated, is called *salt of tartar*.

In France the process differs from the above; instead of dissolving the whole of the pearl-ashes in hot water, they are washed several times with cold water, which gradually takes up all the carbonate of potash, leaving the most of the sulphate of potash and chloride of potassium, these salts being much less soluble than the former in cold water, and especially in a cold solution of carbonate of potash.

The solutions obtained by this plan have a specific gravity of about 1·4735, and are evaporated in pans which gradually diminish in depth, arranged as shown in Figs. 262, 263. The liquors are

FIG. 262.

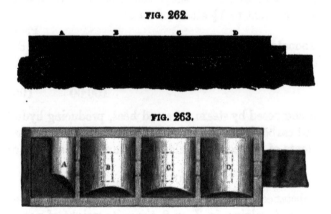

FIG. 263.

run into the deep pan A, heated directly over the naked fire, and as the evaporation progresses, fresh portions of the solutions are added until it is full of concentrated liquor. This is transferred by ladles to the adjoining pan, and the process continued in the same way with each pan, the stronger solutions being always transferred into the more shallow pans, where the boiling is assisted by means of a flat iron poker. By this plan of boiling down the liquors, the escape of the steam is much facilitated, the danger of

30*

the scattering of the carbonate is avoided, and a cleaner product
is obtained. The dry salt is left in D, and by constant careful
stirring during the drying it is procured in that granular form
which is preferred in commerce.

Carbonate of potash has a specific gravity of 2·2643. It is
fusible at a bright red heat and volatile at a white heat. It has
a strong alkaline taste, but is only slightly caustic. It is very
soluble in water, and when the solution is boiled to 1·26 sp. gr.,
and allowed to cool, crystals are deposited, containing two
equivalents of water, $KO.CO^2.2HO$, composed of

Potash	54·13
Carbonic acid	25·23
Water	20·64
	100·00

In its ordinary state, as met with in commerce, it is deliquescent, and rapidly absorbs moisture, yielding a dense solution,
forming the *oleum tartari per deliquium*. It is insoluble in
alcohol.

The granulated salt of the shops contains 16 per cent. of water,
which is equivalent to $1\frac{1}{2}$ atoms of water.

		Phillips.
Carbonate of potash, $KO.CO^2$ . . .	83·64	84
Water, $1\frac{1}{2}$ HO	16·36	16
	100·00	100

It is decomposed by steam, at a red heat, producing hydrate of
potash and carbonic acid gas, and when heated to whiteness with
2 equivalents of charcoal, it yields potassium and carbonic oxide.

CAUSTIC POTASH.

This substance may be prepared from the refined pearl-ash in
the cold, by dissolving in 7 or 8 times its weight of water, and
adding 60 parts of quicklime for every 100 parts of the potash
salt; the whole is occasionally agitated, and after some hours
the decomposition is complete. The lye is siphoned off from
the precipitated carbonate of lime and evaporated in iron pans
nearly to the consistence of a syrup, whence it is ladled into metal
pots, where it is heated until it undergoes the igneous fusion.
In this state it is ladled into polished iron trays, or on to plates
of tinned copper or of silver.

It may also be prepared with the aid of heat. An iron vessel, with a closely fitting cover, is employed. One part of pearl-ash is heated with 12 parts of water till it boils, and a mixture of ⅓ parts of quicklime with 3 parts of warm water, in a fine cream, is added by degrees. The mixture is boiled for a few minutes after the addition of each portion of slaked lime, to allow the precipitated carbonate of lime to acquire greater density, so as to fall more readily to the bottom. The whole is boiled for about 15 minutes, after all the lime has been added, with the cover on, and then left for the precipitate to subside. The caustic solution should not effervesce on adding an acid, nor give any precipitate with lime-water. The lye is now rapidly concentrated in covered iron pots, and if it becomes turbid, it must be set aside in close vessels to deposit the precipitate which has formed. The clear liquor is then rapidly boiled down in metal or, better, in silver basins, till the oily hydrate begins to evaporate in white clouds.

It may be allowed to cool in the basin and afterwards broken up into small pieces, or as is more commonly done, it is run into cylindrical moulds (Fig. 264). These moulds are formed in two

FIG. 264

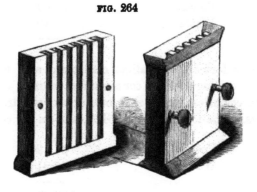

halves, which can be screwed together to form a tight joint. The fused potash is poured in at the top, and when solid, the sticks are removed by unscrewing the mould. These sticks constitute the *potassa fusa* of the shops.

It is necessary that the lime be well slaked, and not merely ground to powder, to make it effective in this process, otherwise a larger quantity must be used, which, *pro tanto*, increases the difficulty of removing the lye from the carbonate of lime. A large quantity of water is also absolutely necessary, as Liebig found that a strong caustic lye deprives carbonate of lime of its carbonic acid, and 1 part carbonate of potash mixed with 4 parts of water undergoes no decomposition; even with 10 parts of water, the separation of the carbonic acid is not complete.

As usually sold in the shops, it contains the impurities of the pearl-ashes and oxide of iron from the process, from all of which it may be separated by treatment with alcohol and evaporation to dryness.

The sticks of potash are more or less coloured from these impurities brown, grey, and blue; but when pure, the potash is a white, hard, brittle substance, often with a fibrous texture. It melts below redness, forming a clear oily liquid, which volatilises at a fullred heat in white pungent vapours. It is very deliquescent and dissolves in water with great rise of temperature. It creates a peculiar soapy feeling between the fingers, from dissolving the cuticle. It dissolves and decomposes the organic tissues, whence its use in surgery, and its name of *caustic* potash. It has no odour, but a caustic urinous taste. It unites with the fat oils to form soap, whence its aqueous solution was called *soap-lye*. It consists of $KO.HO$, and contains in 100 parts—

Potash 84
Water 16
                    ———
                    100

Potash also forms a crystalline compound with water, which separates on cooling from a concentrated solution of potash, in transparent, colourless, acute, rhombohedrons, with truncated terminal edges. The formula is $KO.5HO$, and its per centage composition—

Potash . . . . . . 51·19
Water . . . . . . 48·81
                    ———
                    100·00

COMMERCIAL POTASHES.

In commerce the various kinds of potashes are classified, either according to the locality whence they are imported, or the place where they are made, or the route by which they arrive : thus there are American, Russian, Turkey, German, Moselle, Illyrian, Saxon, Bohemian, Dantzig, Heidelberg, &c., potashes.

The *American* comes by way of New York and Philadelphia. It is of a reddish colour, sometimes grey and violet; the pieces are hard, but very deliquescent. That from New York is preferred to that from Philadelphia. There are three kinds : 1st. The red potashes, mixed with white pieces, standing 54 to 58 per cent.; 2nd quality, equally caustic with the preceding, but standing only 48 to 52 per cent., and the interior of the pieces are darker in colour; 3rd quality stands only 30 to 45 per cent., of a dark colour. They come in oak barrels holding about 5 cwts.

The *White American*, or pearl-ash, is in a granular condition, with a faint blue tint. There are three kinds : the New York, standing 55 to 58 per cent., and those from Boston and Philadelphia, 25 to 40 per cent. They arrive in barrels holding from 3 to 4 cwts.

The *Russian*, called also *St. Petersburg* or *Odessa*, because forwarded through those cities; and also *Kasan*, as manufactured near this last locality. These potashes are in light, friable, irregular masses of a bluish colour, and arrive in barrels of poplar, holding from 7 to 12 cwts. The strength is about 50 to 52 per cent.

The *Riga* potashes are similar to the Russian, but more variable in quality, standing from 30 to 50 per cent.

The *Polish* or *Podochinsky*, also called *Potasse de Paille* because made from straw, is denser and harder than the Russian.

The *Dantzig* resembles pearl-ashes, but is more friable, and stands from 50 to 60 per cent. The barrels contain about 4 cwts.

The *Tuscan* is in powder, mixed with large and small pieces of different colours, whence the three qualities—1. *Grey*, friable, with a decided blue shade of colour, standing 60 per cent.; 2. *White*, nearly colourless, harder, but standing only 50 to 55 per cent; and 3. *Blue*, of about the same strength as the last. They arrive in barrels holding from 6 to 10 cwts.

The other potashes are all of a similar character, the above being sufficient to illustrate the character of the commercial article. These potashes are never completely soluble in water; sometimes a considerable refuse is left, and at other times a mere trace. This arises either from an imperfect filtration through the straw in the ash cisterns, or from a portion of the ashes of the fuel being carried over the fire-bridge and mixing with the potashes. The caustic potash is found only in American, and arises from the use of lime in preparing the lye. The sulphide of potassium sometimes found, is formed by the action of the organic matters on the sulphate of potash. The presence of the other impurities is explained by a reference to the analysis of the ashes of the plants from which the potash is made.

The following analyses of the different kinds of potashes, by different chemists, convey a very good idea of their composition.

I. Hermann's Analysis of Potashes from Kasan.

Insoluble portion.

Lime	0·054
Alumina	0·012
Manganic acid . . .	0·013
Silica	0·132—0·211

Soluble portion.

Carbonic acid	27·790
Potash	47·455
Soda	2·730
Sulphate of potash . .	17·062
Chloride of potassium .	3·965
Bromide of potassium .	trace
Phosphate of potash . .	0·443
Silica	0·344—99·789

100·000

II. Pesier's Analysis of Potashes of Commerce.

Constituents.	Tuscany.	Russia.	America.		Vosges.
			Red.	Pearl-ashes.	
Sulphate of potash	13·47	14·11	15·32	14·38	38·84
Chloride of potassium...	0·95	2·09	8·15	3·64	9·16
Carbonate of potash ...	74·10	69·61	68·07	71·38	38·63
Carbonate of soda	3·01	3·09	5·85	2·31	4·17
Insoluble matters	0·65	1·21	3·35	0·44	2·66
Moisture	7·28	8·82	not es-timated.	4·56	5·34
Phosphoric acid, lime, silica, &c., and loss... }	0·54	1·07		3·29	1·20
	100·00	100·00	—	100·00	100·00

III. Bley's Analyses of Illyrian Potashes.

Carbonate of potash . . .	78·75	82·85
Carbonate of soda } Sulphate of soda . } . . .	12·50	12·50
Insoluble matter	8·75	4·65
	100·00	100·00

IV. *Van Bastelaer's Analyses of American and Russian Potashes.*

Constituents.	American Potashes.			Russian Potashes.
	1	2	3	
Carbonate of potash	21·47	30·43	12·96	50·84
Caustic potash	4·46	9·33	21·71	—
Sulphate of potash	20·08	25·05	23·70	17·44
Chloride of potassium	7·55	4·20	7·89	5·80
Carbonate of soda	22·99	23·90	17·07	12·14
Moisture	9·37	3·11	0·81	10·18
Insoluble matters	14·08	3·98	15·86	3·60
	100·00	100·00	100·00	100·00

V. *Meyer's Analyses of American Potashes* (New York).

Constituents.	Best Quality.		First Quality.	Second Quality.		Third Quality.
Carbonate of potash .	43·68	24·57	56·01	15·07	53·15	38·47
Hydrate of potash . .	49·68	44·43	5·61	38·69	4·49	—
Sulphate of potash .	4·07	16·14	27·70	19·76	21·30	53·34
Chloride of sodium .	1·64	4·40	10·49	6·60	5·37	0·62
Carbonate of soda . .	—	4 27	—	4·70	14·01	6·03
Insoluble matter .	0·72	6·19	0·19	15·86	1·69	1·54
	99·79	100·00	100·00	100·68	100·01	100·00

The importance of the trade in potashes may be seen from the following table of the imports into this country since 1841.

Imports of Pearl and Pot-ashes.

	Cwts.			Cwts.
1841 . . .	91,844	1851 . . .		184,043
1842 . . .	122,518	1852 . . .		199,911
1843 . . .	146,951	1853 . . .		151,944
1844 . . .	164,485	1854 . . .		155,739
1845 . . .	181,574	1855 . . .		?
1846 . . .	?	1856 . . .		105,941
1847 . . .	?	1857 . . .		140,833
1848 . . .	?	1858 . . .		150,432
1849 . . .	158,541	1859 . . .		155,663
		1860 . . .		141,087

The exports from Canada, in 1859, were 25,598 barrels of potashes, and 12,221 barrels of pearl-ashes, of the aggregate value of £221,000.

2. *Potashes as a Bye-Product from the Manufacture of Beet-root and Cane Sugar.*

Beet-root belongs to the potash class of plants, but in some few instances a large proportion of soda exists in the ash; thus Boussingault found in the two kinds—

	Potash-ash.	Soda-ash.
Potash	48·9	30·1
Soda	7·6	34·2
Lime	8·8	3·1
Magnesia	5·5	3 2
Chlorine	6·5	18·5
Sulphuric acid	2·0	3·8
Phosphoric acid . . .	7·6	3·5
Silica	10·0	3·0
	96·9	99·4

When the manufacture of sugar from this plant was introduced into Europe, at the commencement of this century, the value of the ash as a source of potash soon attracted attention. Mathieu de Dombasle was the first manufacturer who attempted to combine the production of sugar with the extraction of potash. He proposed to strip the beet-root of its leaves and incinerate them; for he had found that 100 parts of dry leaves left about 10 per cent. of ash, of which about one-half consisted of good potashes. It was soon discovered that the mode of cultivation which produced the largest yield of sugar was not adapted to give an abundance of leaves rich in potash, and his plans were abandoned.

No further steps were taken until, about 20 years ago, Dubrunfaut suggested that the uncrystallisable sugar in the molasses might be converted into alcohol, and the potash-salts extracted from the residue.

This proposal was first carried into practice at the distillery of Messrs. Serret, Hamoir, Duquesne, & Co., and has since been adopted on a large scale at Valenciennes, Lille, and Rocourt in France, and at Magdeburg and other places in Germany. The large beet-root sugar manufactory at Waghäusel, in Baden, produces about 300 tons of these potashes, of which an analysis is quoted at p. 480.

This new and extending manufacture depends, therefore, on inducing a fermentation in the molasses, and then separating the alcohol by distillation,—evaporating the residual liquors to dryness, calcining the solid mass, and treating the pure salts in a similar manner to that described for the ordinary potashes.

Molasses.

The raw beet-root sugar has been analysed by Moinier, who found that the uncrystallisable sugar present varied from 0·12 to 3·40 per cent. His analyses of two samples furnished the following results :

	1	2
Pure sugar	95·28	95·23
Uncrystallisable sugar	0·12	0·80
Colouring matter	0·20	0·25
Ashes	1·10	1·02
Water	3·20	2·70
	100·00	100·00

But this uncrystallisable sugar, in some molasses, rises up to 90 per cent. Payen, Poinsot, and Brunet have made a detailed analysis of this material, and found

Water	21·740
Organic matter containing 1·47 of nitrogen	65·680
Carbonates of potash and soda	9·699
Sulphate of potash	0·280
Chlorides of potassium and sodium	1·578
Carbonate of lime	0·874
Carbonate of magnesia	0·118
Silica	0·009
Alumina and oxide of iron	0·022
	100·000

The mineral matter left by the incineration of the molasses, has been analysed by Krocker, who found the portions soluble and insoluble in water, to contain—

Constituents.	Soluble.	Insoluble.
Potash	47·88	1·70
Soda	2·34	0·17
Lime	—	5·29
Sulphuric acid	1·53	—
Silica	0·85	0·22
Carbonic acid	22·39	3·79
Phosphoric acid	—	0·29
Phosphate of lime, magnesia, and oxide of iron	—	0·63
Chloride of sodium	12·92	—
	88·11	12·09

With these analytical details the processes which are pursued will be rendered more intelligible.

The molasses are reduced to a specific gravity of 1·085 by the addition of water, and sufficient sulphuric acid added to render the liquid slightly acid. This liquid is then mixed with 2½ per cent. of the yeast of beer, and run into cisterns holding from 5000 to 6000 gallons, where it is exposed to a temperature of 68° F., and allowed to ferment for 5 or 6 days. The specific gravity falls to 1·022, and it is submitted to distillation to separate the alcohol, which amounts to 4 or 5 per cent. of the liquid. The residue from the still, which now stands at 1·029, is called *Vinasse*, and contains all the salts withdrawn from the soil by the plant.

Vinasse.

This liquid is mixed with chalk to saturate any free acid, and the clear liquid is decanted. This addition of chalk or lime has a most beneficial action on the composition of the salt produced, as the following comparative analyses plainly indicate :

Composition of the Potashes.

	Without Lime.	With Lime.
Carbonate of potash	42	52
Carbonate of soda	29	30
Sulphate of potash	10	1
Sulphide of potassium	3	—
Chloride of potassium	16	17
	100	100

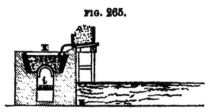

FIG. 265.

The liquid is concentrated in iron pans with convex bottoms, as shown in Fig. 265. The upper sides of the pan are wide and spreading, so that the froth which forms during the boiling may not overflow. De Massy has proposed to concentrate the vinasse by causing it to fall over a series of shelves in a tower, up which the waste heated gases of the furnaces of the manufactory are drawn by means of an ordinary chimney draught.

The evaporation is continued until the liquid acquires a syrupy consistence and a specific gravity of from 1·217 to 1·372. It is then run into tanks to allow the sulphate of lime to precipitate, when the calcination of the clear brown liquor is effected in the reverberatory furnace (Fig. 266). It has three compartments;

FIG. 266.

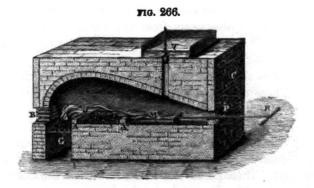

the fuel is charged in the first, and the adjoining one, M, receives the solid residue left by the evaporation of the vinasse, in the third, M'. In the side wall, openings are left corresponding with each division of the furnace, for working the materials, and another door, P, at the end of the furnace, serves a similar purpose. The flues are shown at C, and a grating of bricks at B in the side-wall of the furnace G, to admit air. The draught must be very powerful to burn off the organic matter, and the admission of air through B, passing over the burning fuel, very much facilitates this part of the calcination.

The clear liquid is run into a pan, V, from which it is admitted into the compartment M', through a pipe passing through the arch of the furnace, and capable of being closed by a plug, T. Each compartment is separated from the adjoining one by bridges, A A'. The liquid is rapidly concentrated and becomes pasty. Great care is then necessary to break up the large pieces which form, by means of rakes worked through the door P and that in the front of the furnace. When the mass is dry, it is removed to the bed M, and a fresh quantity of vinasse run into M'. The dry mass on the bed next the furnace takes fire, and the combustion is also facilitated by the presence of a quantity of nitrate of potash in the vinasse. The heat at this point must not be too great, so as to prevent the dry material forming a hard mass, which is difficult to lixiviate, and to avoid the danger of the carbon reacting on the sulphates to produce an alkaline sulphide, which cannot afterwards be removed. As soon as the carbon is burnt off, and a portion, taken from the furnace, gives a clear colourless solution with water, the charge is withdrawn, allowed to cool, and packed in casks.

In some localities the mass is withdrawn from the furnace as soon as the flame disappears from the surface, and thrown up into heaps in the air, when a slow combustion continues until the carbon is burnt off.

The potashes obtained in this process vary very much in composition. Those obtained from districts where the cultivation of the beet-root has been recently introduced, are much richer in potash than those coming from soils long cultivated and exhausted. These latter potashes contain much more soda, and are of proportionally less value, though showing the same alcalimetrical strength. These potashes are called *Salin* on the Continent: a sample analysed by Ducastel contained—

Carbonate of potash.	31·68
Chloride of potassium	12·28
Sulphate of potash	1·33
Carbonate of soda	1·76
Carbonate of magnesia . . .	2·13
Carbonate of lime	16·15
Oxide of iron and alumina . .	7·35
Carbon and silica	7·75
Sulphide of calcium	7·12
Water	10·68
	——
	98·23

In addition to the impurities in the above analysis, considerable quantities of prussiate of potash are obtained, and in some samples sulphocyanide of potassium is found. The strong evolution of ammonia which takes place on lixiviating some of the raw products with water, evidently arises from the presence of these cyanogen-compounds.

The raw *Salin*, or beet-root potashes, contains on an average,

Sulphate of potash, from	3 to 5 per cent.
Chloride of potassium „	20	„
Carbonate of potash „	30 to 35	„
Carbonate of soda „	18	

Refining of the Raw Salin.

This crude material is lixiviated by any of the methods already described, and the last liquors, which have a very low specific gravity, are employed instead of water, with fresh portions of the salin.

The liquors are, in some manufactories, boiled down to dryness, and Ducastel has analysed a sample of potashes so obtained, which consisted of

Carbonate of potash	67·20
Sulphate of potash	2·91
Chloride of potassium	26·09
Carbonate of soda	3·80
	100·00

In other localities, the strong liquors, having a sp. gr. of 1·399 are concentrated up to 1·473, during which, and while the liquor is cooling in crystalising vats, a quantity of sulphate of potash separates, which is fished out. After standing some time, a large crop of crystals of chloride of potassium forms on the bottom and sides of the coolers. The mother-liquors are pumped up into the pan and concentrated up to 1·555 sp. gr., when they are again run into the coolers. In three or four days a mass of beautiful crystals is formed, consisting of a double salt of potash and soda, $KO.CO^2 + NaO.CO^2 + 12HO$, of which two-thirds are $KO.CO^2$. The mother-liquor is now very rich in carbonate of potash, and is boiled to dryness, in a similar apparatus and with the same precautions as we have described at p. 462.

The double salt above named is either redissolved in water and recrystallised, when a double salt is formed containing less carbonate of potash; or the former salt is treated with a small quantity of warm water, which leaves a great portion of the carbonate of soda behind in an insoluble state, and the latter double salt obtained by crystallisation. This double salt is then melted in a metal pan, in its water of crystallisation, and boiled, when a carbonate of soda with 1 equivalent of water, $NaO.CO^2 + HO$, separates, leaving the mother-liquor rich in the potash-salt, which is obtained in the manner previously described.

All these different salts of potash and soda are washed with water to remove the adhering liquor, and dried. The various washings and mother-liquors are mixed with corresponding original liquors, so that the loss of any salt is avoided as far as possible.

It is obvious that this system of cultivation and manufacture must end ultimately in an entire exhaustion of the soil, but meanwhile large quantities of potashes from this source are finding their way into commerce.

The following table contains some of the analyses of these potashes.

Locality or description.	Waghäusel.		Raw Potashes.		Refined Potashes.	Soda Salt.	Polnisch-Weistritz.
Carbonate of potash........	94·392	88·73	29·05	53·90	76·440	7·410	85
Carbonate of soda	trace	6·45	13·63	23·17	16·330	59·000	10
Sulphate of potash	0·282	2·27	9·16	2·98	1·197	0·896	
Chloride of potassium	2·409	1·01	18·82	19·69	4·160	10·400	
Iodide of potassium	0·111	0·03	—	—	—	—	
Silica	0·700	0·12	—	—	—	—	5
Phosphoric acid	—	trace	—	—	—	—	
Water	1·764	1·39	12·13	—	0·600	21·400	
Insoluble matters	0·183	—	17·01	0·26	1·149	0·894	
	99·841	100·00	100·00	100·00	99·876	100·000	100
Chemist	V. Groningen.	Keller.	V. Bastalaer.		Pesier.		Schward.

Mons. Billet employs a totally different process for treating the vinasse, and somewhat similar to that which Mr. Stanford has proposed for sea-weed, fully explained in the following pages.

The apparatus of Billet is shown in Figures 267 and 268, and

FIG. 267.

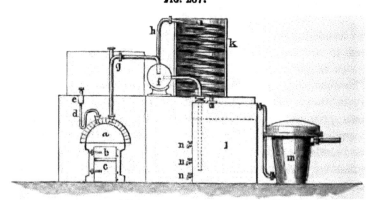

consists of an ordinary gas retort, a with the fire place and ash pit, b, c. The retort is charged by a bent pipe, d and funnel, e, which can be made air-tight by a plug. The products of the distillation pass into a small vessel, f, through a bent pipe, g. Part of the tar condenses in f, and the vapours pass on through another pipe, h,

FIG. 268.

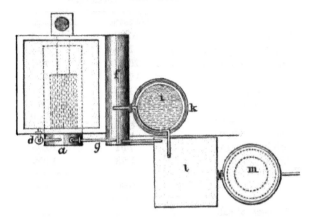

into a worm, i, kept cool in a vessel, k, full of water. The condensed products are conveyed to a reservoir, l, while the gas
passes through a purifier, m, to a gasometer. The various condensed products are drawn off at different levels by the cocks
n, n, n. When the retort is filled with charcoal, the lid is removed and the contents are withdrawn. The potash is extracted
by lixiviation with less loss than in the previous process, and in a
state of much greater purity.

We are, unfortunately, unable to give any information as to the
quantity of potash-salts produced from the beet-root manufacture ;
but the aggregate of all Europe must be very large, and it is
daily increasing, as we have heard that a system of collecting the
crude salin from the smaller sugar manufactories, has been
organised in Germany, by parties who treat the whole in works
specially built for refining this material. Wagner states that
1,000 parts of molasses yield on an average, from

100 to 110 parts of carbonate of potash, and

30 to 40 „ { sulphate of potash and chloride of potassium.

The application of the processes just described, to similar waste
products in the manufacture of cane-sugar, might be attended
with some advantages in the English sugar-producing colonies.
The analyses of the ashes from the raw sugar and molasses, as
imported into this country, and of the molasses from Louisiana,
confirm the correctness of this supposition.

	Richardson.	
	Molasses.	Raw Sugar.
Potash	36·23	19·42
Chloride of potassium . .	1·58	8·03
Chloride of sodium . . .	25·87	15·46
Lime	12·72	14·67
Magnesia	11·14	10·72
Oxide of iron	2·62	6·55
Oxide of manganese . . . }	trace	{ 0·71
Oxide of copper . . . . }		{ trace
Sulphuric acid	7·91	10·85
Silica	1·93	13·59
	100·00	100·00
Per cent. of ash	3·60	1·33

Composition of the Saline contents of 2¼ *gallons of Molasses.*

	Grammes.
Acetate of potash	208·31
Chloride of potassium	113·63
Sulphate of potash	84·46
Acid phosphate of lime . . .	51·01
Acetate of lime	15·18
Phosphate of copper	0·21
Silica	22·85
Gummy matter	66·28
	561·93

A process has, in fact, been patented in this country in the name of Mr. G. Seymour, communicated by Mons. Leplay of Pont-de-Madron, near Toulouse, by which he recovers the potash and soda. The patentee treats the saccharine juices from canes or beet-root with a solution of caustic baryta or sulphide of barium; precipitates the saccharate of baryta, which is washed with a weak solution of caustic baryta to remove the impurities; and collects all the liquors together in a convenient apparatus. The baryta is removed, either as hydrate by evaporating these liquors, or precipitated as carbonate by forcing in carbonic acid, and the mother-liquor is then boiled down to obtain the alkaline salts.

3. Sulphate and Carbonate of Potash as Bye-Products in the Manufacture of Tartaric Acid.

Previous to the application of tartaric acid in manufacturing operations, the deposits obtained during the vinous fermentation, the lees of wine, &c., were employed for the production of a very pure carbonate of potash.

The bitartrate of potash, which exists in the juice of the grape, is only slightly soluble in a mixture of alcohol and water, and is separated during fermentation as a crust on the sides of the casks. This crust is an article of commerce under the name of *Crude Tartar* or *Argol*, which is called *white* or *red*, according to the wine from which it is obtained. When tartar is dissolved in hot water, and the saturated solution allowed to cool, the surface is soon covered with a coating of fine crystals of the bitartrate, which was called, and is still known as, *Cream of Tartar*.

The argols of commerce are found in crystalline cakes of a reddish colour, and are composed of bitartrate of potash, tartrate of lime, sometimes biracemate of potash, and organic matters.

The best descriptions come from Italy, Spain, Marseilles, Montpellier, Bordeaux, &c. Brescius gives the following average composition of different samples.

Bitartrate of potash	82·7
Tartrate of lime	7·6
Impurities	9·7
	100·0

And Scheurer-Kestner has analysed that from Tuscany, with the following results :

Tartaric acid, hydrated . . .	73·67
Potash	22·13
Glucose	0·62
Lignine	0·88
Silica	0·32
Oxide of iron	0·26
Magnesia	1·39
Colouring matters	0·73
	100·00

He also gives the following statement of the percentage of bitartrate of potash and tartrate of lime in different samples.

31*

From	Quality.	Bitartrate of Potash.	Tartrate of lime.
Alsace	White tartars	77·50 ; 84·95 ; 85·10	4·6 ; 7·3 ; 9·9
Switzerland . . .	Do. do.	73·50 ; 85·00	7·7 ; 18·3
Burgundy . . .	Red do.	32·10	46·25
Tuscany	White do.	84·50 ; 85·20 ; 88·50	
Hungary	Do. do.	67·30	9·20
Spain	Red do.	24·20	45·20

When these tartars are heated, they are decomposed, swell up, and leave a black mass, which is called *Black Flux;* when deflagrated with saltpetre, the residue is white, and is called *White Flux.* These fluxes are much used in assaying metallic ores.

The tartars were formerly heated in contact with the air till the residue was white, and then sold under the name of *Woad-ash,* which originated from - the circumstance that they were selected by dyers of blue colours to supply the *woad* tubs.

The lees of wine, which are obtained from the *racking* or *defecation* process, as well as the *yeast* of wine, are now used for the production of carbonate of potash.

After the wine has been fermented and drawn off, the yeast or lees are collected into one vessel and allowed to settle down, in order to separate the remaining portion of the wine, which is used for making vinegar. A thick mud remains behind, which is placed in bags holding about 40 lbs., and submitted to pressure. The cakes, when they leave the press, are bent in a curved shape in order to dry more quickly and completely. When brittle and pulverisable, they are ready for combustion.

The combustion is performed upon a kind of plastered threshing floor, which is surrounded by loose brick walls, to keep the ashes together. The height of this wall is increased as the bulk of the ashes enlarges.. A bundle of twigs in the middle of the floor is first ignited, and a few cakes are placed around it, to which others are added as soon as the former have become red-hot, and this operation is continued until all the stock has been exhausted. The chief care is to regulate the fire in such a manner as completely to burn off the organic matter, without causing so strong a flame as to carry off the ashes or volatilise the alkali. 1,000 cakes, weighing 60 cwt., yield on an average about 10 cwt. of ash of a superior quality, which is very light and porous, white, with patches of blue or green, arising from the presence of traces of manganese. These ashes are called in France, *Cendres gravelées.*

A similar product is obtained from the residue, or *vinasses*, in the manufacture of brandy, which has been analysed by Porter, who found—

Potash	38·52
Soda	4·47
Lime	5·19
Magnesia	7·33
Oxide of iron	1·50
Sulphuric acid	6·10
Silica	2·84
Carbonic acid	12·27
Phosphoric acid	16·78
Chloride of sodium	4·00
	99·00

The stalks of raisins, when burnt, also leave an ash containing potash, but it is not so pure as the above, and requires the same treatment as the ordinary ashes of plants.

In the manufacture of tartaric acid as at present conducted, the argols are dissolved in boiling water, and lime is added, which, combining with 1 equivalent of the tartaric acid, is precipitated as tartrate of lime. The potash remains in solution in combination with the other equivalent of the acid, and is either decomposed by chloride of calcium, sulphate of lime, or dilute sulphuric acid, when chloride of potassium or sulphate of potash are obtained by evaporating the clear solution.

Gatty has taken out a patent for an improvement on this mode of treating the solution containing the neutral tartrate of potash, by which the alkali is obtained in the form of a carbonate. He employs a cylindrical wooden vessel, capable of holding 400 gallons, and fitted with an air-tight lid. This vessel is filled with a mixture of 300 gallons of the solution of neutral tartrate, of 1·025 specific gravity, and 34 gallons of a milk of lime, containing 1 lb. of lime per gallon. An agitator, worked by a shaft passing through a tight stuffing-box, serves to mix the carbonic acid, which is forced into the liquid until it is saturated. An insoluble tartrate of lime is formed, and bicarbonate of potash remains in solution, from which it is obtained by evaporating the liquid to dryness. It is calcined in a reverberatory furnace when not sufficiently pure, redissolved, and the solution boiled to dryness.

Wagner has proposed to substitute carbonate of baryta for the lime. He treats the crude tartar with the usual quantity of water, and adds 9·8 parts of carbonate of baryta to every 19·8 parts of the tartars. The clear solution of neutral tartrate of potash is run off from the insoluble tartrate of baryta, and mixed with 8·5 parts of hydrate of baryta dissolved in water. The whole is then heated up to the boiling point, and a current of carbonic acid passed through until all the baryta is precipitated, and the caustic potash converted into carbonate. On boiling down the solution, a pure carbonate of potash is obtained, but Wagner recommends the addition of a little sulphate of potash during the concentration, to remove any baryta that may happen to be present.

II. The ashes of Marine Plants.
Kelp and Varec.

Under the general term *Seaweed*, are included the marine algæ, which are extensively used for various purposes, as food and medicine, manure, in the production of kelp and varec, and some parties' have proposed to employ them in certain manufactures. This arrangement will be adopted in the following description of their varied uses. For much of the information, in some of these applications, we are indebted to the valuable and interesting communications made to the Society of Arts, by Messrs. E. C. C. Stanford, and P. T. Simmonds, and Dr. Macgowan.

Sea-weed as Food and Medicine.

The applications of the algæ for these purposes are not numerous, and in fact are very limited in this country. These plants possess nutritious and demulcent properties : they contain a peculiar gum, sugar, starch, a little albumin, and a comparatively large proportion of nitrogen. The species chiefly used are the following :—

Laminaria digitata.—Tangle or sea-girdles ; the leathery flat fronds, when young, are used as food for both man and cattle. It is known under different names, *tangle* in Scotland, *redware* in the Orkneys, *sea-girdles* in England, and *sea-wand* in the Highlands.

The tender fronds of the young stems are eaten by the Scotch, and are boiled and given to cattle in Nordland. The frond of this species, no doubt, constitutes the *goître-leaf*, chewed in the Himalayas, as the whole plant abounds in potash and iodine.

The stems of this plant make excellent handles for knives, files, &c.; if the blade be inserted when the stem is fresh cut, and the weed then allowed to dry, it contracts, holding the blade with great firmness, and presents a brown wrinkled appearance, something like buckhorn. The handle, when varnished, is very durable. The dried stalks are used as fuel in the Orkneys and in Bretagne.

Laminaria saccharina.—Sweet tangle, called *sea-tape* in China, where it is eaten. It is also highly esteemed in Japan, where it is extensively used as an article of diet, being first washed in cold rain-water, and then boiled in milk or broth. This plant has been found by Dr. Stenhouse to contain large quantities of mannite, a peculiar unfermentable crystalline sugar, usually procured from manna; and likewise occurring in the lilac and privet plants, which belong to the same natural order (*Oleaceæ*) as the manna ash, *Fraxinus rotundifolia.* Mannite contains $C^{12}H^{14}O^{12}$, and is formed when ordinary cane-sugar is submitted to the lactic fermentation. This sea-weed contains from 12 to 15 per cent. of mannite, which may often be seen on the dry frond as an efflorescence. Dr. Stenhouse found it in *L. digitata*, and in the fuci, *Aliaria esculenta* and *Rhodomenia palmata.* The *Halidrys siliquosa* contains 5 to 6 per cent.

The *Kambon* of the Japanese, and the *Sea-cabbage* of the Russians, is the *Fucus saccharinus*, which is found in great abundance on the islands and shores of Eastern Asia. It is spread out on the sand to dry, then collected in heaps resembling haycocks, and covered with matting, until the time arrives for loading the vessels which carry it from the Kurile Islands to the ports of the southern islands of Japan. It is used in soup, or wrapped round fish, when both are boiled and eaten together. It is often broiled on the fire, salt strewed over it, and eaten without any other dressing. On some parts of the coast of Niphon it is gathered, dried, and roasted over the coals, rubbed down to a very fine powder, and eaten with boiled rice or in soup.

Laminaria potatorum furnishes the natives of Australia with a proportion of their instruments, vessels, and food, while other species of the same family constitute an equally important resource to the poor on the west coast of South America, to the

Fuegians and inhabitants of the country near the Straits of
Magellan Under the name of *bull-kelp* it is largely eaten in New
Zealand and Van Diemen's Land, where it is very common. It
has a thick stem and a flat oval-shaped leaf, and is about the thickness of sole-leather. Mr. Howie, who has frequently lived entirely
upon it for days, describes it as very nutritious and fattening.
It is picked from the rocks, dried in the sun and afterwards
roasted The natives place it on the embers of a wood fire, and
turn it until all the parts are equally cooked. It is then soaked
in fresh water for ten or twelve hours, and is fit for eating.
When hung up and dried, it may be preserved for months. The
natives carry it about with them on their journeys, and use it as
we do bread, eating it by itself or with animal food.

Fucus vesiculosus, bladder-wrack, lady-wrack, *Quercus marina*, or black-tang, is used in Gothland under the name of
swine-tang, boiled and mixed with flour for feeding hogs. In
Iceland, it passes under the name of *Doughlaghman*. It is
employed as winter food for cattle and sheep in the Western
Hebrides. The cattle in the Isle of Lewis eat it with apparent
relish, and thrive well on it. It has been much employed, externally, as a friction in glandular enlargements, and the juice has
been given by Dr. Russell internally with advantage in the same
complaint. Carbonised in close crucibles, it constitutes the
Æthiops vegetabilis, formerly used in scrofula.

Porphyra laciniata, laver, sloke, or slokaun, is pickled with
salt, and eaten with pepper, vinegar, and oil. It is stewed as a
luxury in England, and sold under the name of *laver*, in Ireland
as *sloke*, and in Scotland as *slaak*. It makes a wholesome and
palatable dish. The thin purple and green fronds are only
gathered during the winter months, being too tough in the
summer. After being cleaned, they are stewed with a little
butter, to prevent them from acquiring a burnt flavour, and the
substance is brought to Belfast, where it is sold for 5d. per quart.
Before using, it is again heated with butter, and is eaten with
vinegar and pepper. The pepper dulse (*Laurencia pumatifida*),
distinguished for its pungent taste, is often used as a condiment
when other seaweeds are eaten. There was a time when the cry of
" Buy dulse and tangle " was as common in the streets of Edinburgh and Glasgow, as that of " Water-cresses " now is in London.

P. vulgaris is eaten in China as a relish with rice, under the
name of Tang-Tsai (purple vegetable).

Alba latissima, oyster-green, or green laver or sloke, is inferior to the above, but is used in much the same way, or seasoned with lemon juice.

Rhodomenia palmata, dulse, dylisk, or dillish, known among the Highlanders as dullisg or water-leaf, has an odour of violets, a slightly acrid taste, and is nutritious but sudorific, for which property it is employed in the Isle of Skye. The dried plant was formerly much chewed in Scotland and Ireland. When washed it is almost tasteless, and has a mucilaginous taste in the mouth. The Icelanders wash it thoroughly in fresh water, and then dry it in the air, when it becomes covered with a white powder, which is sweet and palatable. It is then packed in casks, and preserved for eating. In this state it is used with fish and butter, but the higher classes boil it with milk, with the addition of rye-flour. In Kamschatka it is employed to produce a fermented liquor. It is also consumed in considerable quantities throughout the maritime countries of the north of Europe, and in the Grecian Archipelago, being a favourite ingredient in ragouts, to which it imparts a red colour and a thick consistence. Cattle are very fond of this seaweed, and sheep and deer eat it with so much avidity, that they are occasionally drowned by going too far from land in search of it at low water.

Iridœa edulis, called dulse in the S.W. of England, after being pinched between hot irons, is eaten by the fishermen, and is said to taste like roasted oysters.

Aliaria esculenta grows on rocks in deep water, is known by the local names of bladder-locks and hen-ware in Scotland, honey-ware in the Orkneys, and murlins, &c., in other places. The stout midrib is eaten when the olive-green frond is removed. It forms part of the simple food of the poorer classes on the Irish and Scotch coasts, Denmark, Iceland, and the Faroe Islands. This fucus is of rapid growth, and will attain the length of six feet in as many months. It is said to have the hardness of the raw carrot, with the fishy and coppery flavour of an oyster.

Chorda filum, sea-lace, or sea-catgut in Orkney, is collected as food for cattle in Norway, and when twisted and dried, it possesses a strength and toughness that adapts it for fishing-lines.

Durvillœa utilis is used as an article of food in Valparaiso,

and by the poorer inhabitants on the west coast of South
America.

Eucheuma speciosum and *Gigartina speciosa* are edible sea-
weeds eaten in Australia.

Sargassum bacciferum is the celebrated gulf-weed of the
Atlantic, the stems of which are said to constitute the *goitre-
sticks* chewed in South America, where that disease is prevalent,
but they more probably come from the stems of the *Laminaria
digituta.*

Chondrus crispus and *C. mamillosus*, Irish moss, carrigeen
moss, pearl moss. The fronds of these plants, when bleached,
enter to some extent into commerce, and the market supply is
obtained from Clare and the west coast of Ireland. They afford
demulcent and nutritive jellies, and of all the algæ, are the most
used in England for food and medicine. They contain a large
proportion of gelatin, and have to some extent replaced isinglass,
and are sometimes employed in Ireland as a substitute for size.
The *C. crispus* contains 79 per cent. of a peculiar gelatinous
principle called *carrigeenin*, which differs from ordinary gum in
not being precipitated by alcohol, from gelatin by affording no
reaction with tincture of galls, and from starch-jelly by giving no
blue colour with tincture of iodine. It is converted into grape-
sugar by boiling with dilute sulphuric acid. It consists of
$C^{11}H^{10}O^{10}$ according to Schmidt, but Mulder gives the formula
$C^{21}H^{19}O^{19}$. Dr. Davy's analysis represents it as composed of
three substances, one analogous to gum, another to isinglass, and
the third of a peculiar character. He gives the following as the
proportions in which they exist :—

Gummy matter	28·5
Gelatinous matter	49·0
Insoluble matter	22·5
	100·0

Carrigeen moss is coming into use for feeding cattle, and is
employed for stiffening silks, and for sizing pulp in the manufacture
of paper. A fucus allied to this, found at the Cape of Good
Hope, is boiled into jelly, and, mixed with sugar and the juice of
oranges or lemons, makes a very agreeable dish. In Bavaria,
carrigeen moss is employed in clarifying beer, and has been
used in this country for the same purpose. It is also employed
in pulmonary complaints, and in making bandoline.

Plocaria helminthocortos, which has several synonyms, viz., *Gracilaria lichenoïdes* or *Spærococcus*, *Gigantina lichenoïdes*, Corsican moss, &c., has been used by the inhabitants of Corsica, as an anthelmintic for many centuries. It is found in other parts of the Mediterranean, is nutritious and strongly scented.

The *Plocaria candida*, called *agar-agar* by the Malays, and *bulung* in Java, is imported into this country as commercial *Ceylon moss*. It is a small and delicate fucus of white colour, and contains about 70 per cent. of starch and vegetable jelly. When washed with water, and pressed, to remove the salt and some mucilage, it is employed as a preserve.

The edible birds'-nests esteemed as a delicacy in China, are probably constructed from this plant by a species of swallow, the *Hirundo esculenta*. Mr. Crawfurd states that it forms part of the cargo of every junk leaving the Archipelago, and the price on the spot varies from 5s. 8d. to 7s. 6d. per cwt.

The agar-agar is prepared at Malacca in the shape of a clear jelly. The plant is found on the rocks at Pulo Ticoos, and on the shores of the neighbouring islands. It is blanched in the sun until it is quite white. It is also obtained on the submerged banks in the neighbourhood of Macassar in, Celebes, by the Bajoulaut, or sea gypsies, who send it to China. When boiled with sugar, it forms a sweet jelly, and is highly esteemed both by Europeans and natives for the delicacy of its flavour. The bamboo lattice-work for lanterns in China is covered with paper which is rendered semi-transparent when saturated with this gum. It is also used in the manufacture of silk and paper, and is preferable to flour for making paste, as it is avoided by insects.

Gelideum corneum is the *algue de Java* which is made into an iced jelly, and sold in Ningpo under the name of *Nin-mau* (ox-hair vegetable). Payen has recently examined it and extracted a peculiar principle from it, amounting to 27 per cent. He calls it *gelose*, and states that it has ten times greater gelatinising power than the best isinglass. Its reactions are similar to carrigeenin. Mr. Stanford examined a large number of the British algæ, but found this compound in very few of the species. Taking the *Laminaria saccharina* as a type of the membranous species of algæ, he found it to contain, besides mannite, 34 per cent. of a gum soluble in water, and resembling dextrin. When boiled with a mixture of equal parts of nitric acid and water, this seaweed yielded 22 per cent. of oxalic acid in fine crystals, which probably came from the decomposition of the gum.

Laurencia papillosa (tan shwui). This, as well as the previous species, is used in the preparation of *tang-tsai,* or Chinese moss, the method of preparing which the Chinese have recently learnt from the Japanese.

Gracillaria crassa, (*ki-tsai,* hen-foot vegetable,) is cooked with soy or vinegar, and is also used as a bandoline for the hair by the Chinese ladies.

Hai-tai (sea tape) is sent into the interior of China wherever coal is used, as it is considered a corrective for the deleterious exhalations from this fuel. It is usually boiled with pork. It comes from the Shantung province.

Nostoc edule is also eaten in the East, and *Nostoc arctium* in the Arctic regions. Mr. Frere sent to Bombay some specimens of *Nostoc collinum* from Shikarpoor, which were found thickly scattered over the space of two or three square miles, after a shower of rain, in May, 1855. The natives described it as a shower of *gosht,* or meat, the germs of which, Mr. Simmonds states, might have been carried through the air, and after germinating, taken a visible form, which, having no earthly origin that the Scindians could conceive, and coming in tangible masses, after a storm, they naturally concluded, must have been showered down from the heavens.

There are many other algæ used exclusively as food in China and Japan, and we may probably learn much regarding these edible plants from the Chinese, who are evidently the algæ-consuming nation.

Mr. Simmonds states that the *Maratheum feniculaceum,* which resembles the finer seaweeds, and grows in most of the rivers of Veraguas, is used as one of the culinary vegetables on the Isthmus of Panama. It is highly esteemed by the inhabitants, and the young leaf-stalks, when boiled, have a delicate flavour, not unlike that of French beans.

Dr. Davy and Professor Apjohn have examined some of these plants, which appear to be deserving of the reputation they possess among the poor. The results of their analyses are given in the following tables.

PLANTS.	Water.	Dry Matter.	Nitrogen per cent. in dry Matter.	Protein in dry Matter.
Chondrus crispus, bleached	17·92	82·08	1·534	9·587
„ unbleached, Ballycastle	21·47	78·53	2·142	13·387
Gigartina mammillosa „	21·55	78·45	2·198	13·737
Laminaria digitata, or dulse tangle.	21·38	78·62	1·588	9·925
„ black tangle.	31·05	68·95	1·396	8·725
Rhodemenia palmata, or dylisk	16·56	83·44	3·465	21·656
Porphyra laciniata, or levre	17·41	82·59	4·650	29·062
Iridœa edulis, Ballycastle	19·61	80·39	3 088	19·300
Alaria esculenta, or murlins	17·91	82·09	2·421	15·150

The water is, of course, less than when fresh from the sea, but the very large proportion of nitrogen is equal, and even larger, than in most other vegetable esculents, with which Dr. Apjohn contrasts them in the next table—

Proportion of Nitrogen in 100 *parts of various Edible Substances dried at* 212° *F.*

Potatoes . . . . 0·541
Flour of first quality . . . . 1·817
Beet-roots . . . . 1·848
Mangolds . . . . 1·781
Swedish turnips . . . . . . 1·843

That scourge of seamen, the scurvy, is due to the insufficient supply of potash in the salt meat, the juice of which has diffused into the brine; and the juices of limes and lemons are largely imported for its cure, which act, as is supposed, by virtue of the potash they contain. The marine algæ, which would be a far better source, might be procured fresh from the sea, and cooked as vegetables.

SEA-WEED AS MANURE.

The value of seaweed as manure is most appreciated in the Channel Islands; the "varec" or "vraic," as the weed is there called, is considered so valuable that special laws are enforced for its regular collection and fair distribution amongst the agriculturists, many of whom use no other manure. "Point de vraic, point de hangard" ("no seaweed, no corn-yard"), has passed into a local proverb. The seaweeds are of two kinds—"vraic venant," and "vraic scié." The former is the drift weed cast up by stormy seas on their sandy but rock-bound shores. This is the most valuable, consisting chiefly of *Laminaria digitata* and *L. saccharina*, very rich in iodine and salts of potash; it is allowed to be raked up and collected all the year round, from sunrise to sunset, the time being prolonged during the winter to 8 P.M. This is the constant employment of the cottagers on the coast both of Guernsey and Jersey, and the collection is at its height in stormy weather. The work is very laborious, the large wooden rakes used, being often torn out of the hands of the "vraiqueurs" by the waves. The beautiful sandy bays, which abound in these islands, are the scenes of their toil. The weed is either thickly spread on the land, and ploughed in fresh with a deep plough, or dried on the beach, and burnt on the cottagers' hearths as fuel; certainly not on account of the cheerful appearance of the fire, or its pleasant odour, but because the charred ash thus produced sells at 6*d*. per bushel for manure. The fire smoulders quietly, it is never extinguished, but constantly renewed, and the whitest of all smoke ascends night and day from the rude chimneys of these humble dwellings. The "vraic scié" is the "cut weed," cut off the rocks at low tide, consisting principally of *Fucus vesiculosus, F. serratus, F. nodosus;* the time of cutting it is fixed by law, at Guernsey, from July 17th to August 31st, and at Jersey twice a year, commencing March 10th and June 20th, and lasting about ten days each time. The summer cutting is made a regular holiday in both islands, and to the young "vraiqueurs" of both sexes it is an occasion of great festivity.

It is computed that about 30,000 loads of vraic of all kinds are annually obtained from the rocks and bays of Guernsey and the adjacent small island of Herm, valued at £3,000 or 2*s* per load; this is a mere nominal price, if the value of the potash and iodine alone be taken in consideration. The quantity of vraic collected in Jersey is, probably, quite equal to that obtained in Guernsey, but the vraic used is generally not so rich, and thus more is burnt and less ploughed in than at Guernsey. The best seaweeds are stacked in Jersey; a dozen stacks of "vraic venant" thatched over, are common objects in a farm-yard, and small barns are given up to its storage when dry. The value set on vraic may be judged of by the fact, that the inhabitants of Sark, having none on the island, import it in fishing boats from Herm, five miles distant; fifty Guernsey and Sark boats may be seen at once at Herm engaged in this traffic, and those who are acquainted with the precipitous nature of the rocks of Sark and its dangerous currents, will appreciate the value of vraic in that island.

"Drift weed" is also largely used in Ireland, as the only manure for the potato crop; this is interesting, because the potato requires a considerable supply of potash. This alkali can hardly, however, be required in the Channel Islands, as the granitic sub-soil would, on disintegration, furnish it in abundance; it is probably the earthy phosphates that render the weed so fertilising there; this is borne out by the fact that the lixiviated seaweed ash, from which the alkalies have been removed, meets with a ready sale in Guernsey, and is esteemed indeed richer, no doubt on account of the increased per centage of phosphates. The residual seaweed ash from the iodine factories in France, is highly valued as a manure, and constantly carried a distance of thirty miles from the factory.

The agriculture in the Western Islands is also enriched by this manure, and some of the tangle is brought into Oban by fishermen in boats, and sold at 1s. per' load. On the S.E. coast of Fife it is laid on the stubble at the rate of twenty cartloads an acre, and ploughed in; the clover crop never fails, and this is a crop requiring much phosphate of magnesia, an important constituent of seaweed ash. In the Isle of Lewis, twenty tons of seaweed are considered ample manure for a Scotch acre.

The agricultural produce of the Isle of Thanet, in Kent, is said to have been tripled by the use of this manure, and the farms on the Lothian coasts let for 20s. to 30s. more rent per acre where the tenants have a right of way to the sea coast, where the weed is thrown ashore. In England, however, seaweed is, for the most part, little valued by agriculturists as an actual manure, and appears to be regarded rather as an economical and useful covering to protect turnips and other roots from winter frosts. Farmers object to its bulk and expensive carriage, particularly now so many portable artificial manures are offered for sale.

Mr. Simmonds, in his valuable work on waste products, states that a deposit of seaweed of a million tons or more, nearly equal to the best guano, has been going on for centuries at the Isle of Foulney, the outlet of the tidal river ten miles from Ulverstone; although little used generally, it has been successfully employed on a farm there for the last ten years.

Seaweed is used largely as manure in Cornwall and the Isle of

Man. It is rotted in heaps, covered with ditch and road scrapings, weeds, &c. It is not deemed a powerful manure, for it contains a large proportion of water, and its mineral constituents are chiefly soda. Mr. Simmonds suggests that, dried and charred, it would be useful to disintegrate the waste fish for manure. On the Cornish coast it is extensively used as a manure for the growth of early potatoes, especially near Penzance.

Professor Way gives the following analysis of decomposed and concentrated seaweed:

Organic matter .	65·62
Soda	13·65
Soluble salts . .	16·00
Nitrogen . . .	3·23 = 4 of ammonia
	98·50

Mr. Scott, writing on Mr. Stanford's paper, says Great Britain, irrespective of Ireland and the Scottish Isles, possesses a sea-coast line of about 7,000 miles, from which he believes that, on an average, 3,000 tons of seaweed per mile per annum might be collected.

SEA-WEED AS APPLIED TO SOME MANUFACTURES.

Acetic Acid.

A process has been recommended by Dr. Stenhouse (*Philosophical Magazine*) for the manufacture of acetic acid from wet weeds by fermentation. His experiments were conducted with some of the fuci; these were mixed with lime, and kept moist at a temperature of 90° F. He obtained by distillation with sulphuric acid, an average of 1·5 per cent. of anhydrous acetic acid; it was contaminated with butyric acid. This might however be separated and turned to account in the manufacture of butyric ether or essence of pine-apple. The best method would be to ferment the fuci in pits with lime or chalk, at the ordinary temperature in the summer, leaving each portion for two or three months, and supplying its place by a fresh load until the lime was saturated; the liquid would then be pumped out, evaporated to dryness, the residue sold as crude acetate of lime, and the weed carried to the manure heap. The whole process might be roughly and economically carried on by an agriculturist near the sea.

Silver.

The ashes of several fuci have been found by M. Malaguti to contain silver, as well as lead and copper, in minute proportion. The silver in the ash of *Fucus serratus* is estimated at $\frac{1}{10000}$ of its weight, or about $3\frac{1}{2}$ oz. to a ton. Other algæ probably contain this metal, as it has been found universally present in sea-water; but this would not be an economical source, and it is probable that the 2,000,000 tons, calculated by M. Tuld to be dissolved in sea-water, might be better extracted by the copper sheathing of ships, in which the copper is gradually replaced by the silver; the sheathing thus becomes, after a time, rich in silver. It has been found in the sheathing to the extent of 17 oz. per ton.

Paper.

Two patents were obtained in 1855, the first dated June 20th, by Martenoli de Martinoi and others, of San Francisco, and the second dated November 29th, by Charles Maybury Archer, for the employment of seaweed in the manufacture of paper, and for the production of textile fabrics. Neither of these was proceeded with; the true algæ, in fact, having no fibres in their structure, would appear to be singularly inappropriate for the strength required in paper. Another patent was recently obtained by Ebenezer Hartnall, of Ryde, dated May 31st, 1861, for the application of the grass wrack, *Zostera marina*, to the manufacture of pulp for paper, to be used alone or in combination with other fibrous materials. This appears more practicable, as this plant is not one of the algæ, and it does contain fibre resembling the grasses. This patent, however, was not proceeded with.

Another patent was obtained in 1858, dated August 5th, by Donald M'Crummen, "for the application, use, and treatment of marine plants, heath, or heather, and other vegetable productions, as well for the manufacture of paper as for the production of alkaline and other salts." According to his specification, the plants are crushed or bruised, and exposed to the action of boiling water or steam until the soluble salts are removed; the insoluble residuum, after treatment with various agents for disintegration, is to be employed in the manufacture of paper, millboard, and papier maché. The solutions obtained are evaporated to dryness to recover the salts. This is the only patent which has been proceeded with.

Colours.

Mr. W. L. Scott writes, that although the natural colours of many seaweeds are very beautiful, especially among the *Rodosperms*, no attempts have been made to extract them.

Several varieties of *Griffithsia* yield a brilliant crimson colour to pure water, which is precipitated again on the addition of certain soluble chlorides. This substance (which might be very appropriately called *Rubalgine*) appears to combine with alumina and other metallic oxides. Again, on many shores, masses of a fine yellowish-brown alga may be observed, which, after a few hours' exposure to the sun's rays, change to a deep red hue. The colour may generally be developed from these weeds, (*Plocamium*) artificially, under the influence of an oxidising agent, such as nitric acid or formate of potassa.

Substitute for Horn, &c.

Another application of seaweed has been patented by Mr. Chislin for the manufacture of various useful articles for which horn, shell, whalebone, indurated leather, &c., have hitherto been employed. One of his processes consists in immersing the weed, from which all extraneous matters have been removed, in a hot lye of caustic lime for three hours, and then steeping it in very dilute sulphuric acid. It is then placed in a solution of common salt, afterwards washed in pure water, and dried, after which it may be shaped in any form desired. Mr. Chislin also bleaches the material he obtains, by the usual agents, and dyes it by means of picric acid, logwood, alum, &c. He produces the variegated brown, so desirable in imitating horn, by rubbing off the external surface of the natural protuberances by means of fine emery paper, and then applies a varnish composed of shellac dissolved in methylated spirit. Copal varnish may sometimes be advantageously used, and at other times gold or bronze powder, ultramarine blue, or other colours in fine powder.

Mr. Chislin employs another mode of operating to obtain a material with the appearance of carved wood. He grinds the dry material to powder, which is then mixed with glue, to which some alum and powdered resin are added. The whole is thoroughly incorporated until it acquires the consistence of putty. Coal-tar, or bitumen in solution, may be substituted for the substances previously mentioned, in which case the articles

made from the mixture, must be baked at a temperature of 300°
to 500° F. to harden them. When the mass has become quite
uniform and of the proper consistence, it is rolled out or submitted to pressure in moulds, and the articles, after drying, are
brushed over with oil, and polished.

Substitute for Soap.

The Chevalier Claussen found that some common seaweeds
entirely dissolved in the alkalies, and formed a compound
which could be employed in the manufacture of soap. Mr.
Simmonds states that the Brazilians use a malvaceous plant,
sida, for washing, instead of soap, and the Chinese employ
the flour of beans in the scouring of their silks.

Production of the Kelp or Varec.

The value of marine plants, as a source of alkali, has long been
known, but the systematic production of a material from seaweed, at least in these islands, is of comparatively recent origin.
It appears to have been first regularly pursued in Ireland, whence
it was introduced into Scotland about the year 1730 by Mr.
McLeod.

It was at first carried on for the carbonate of soda it contained,
which was unduly enhanced in value by the high war duties
levied on barilla and salt. At the beginning of the present
century, Highland kelp sold for £20 or £22 per ton, and Maccul-
loch states that the kelp shores of the island of North Uistralone
let for £7000 per annum. The total production of Scotland
about this period was 20,000 tons per annum, which quantity has
never since been reached. When the duty was removed from
foreign produce, and the manufacture of soda from common salt
commenced, the value of kelp rapidly fell, and as a consequence,
its production as an article of trade. The discovery of iodine in
1812 opened up a new demand for kelp, and the production is
now much increased, but still far below the quantity manufactured previous to the peace. Kelp and varec are now produced
solely on account of the iodine and potash-salts which they
contain, but Mr. Stanford's plans open up a wider field, of which
full details will be given in the following pages. The composition

32*

of the ash of these marine plants is, therefore, of great importance, and the following tables of analyses are given to indicate how widely the different plants differ in this respect.

COMPOSITION OF THE ASH, ETC., OF SEA-WEEDS.

The first table gives the results of Dr. Anderson's analyses, which include not only the composition of the ash, but that of the organic matter.

Dr. Anderson's Analysis.

Constituents.	Fucus nodosus.	Fucus vesiculosus.	Laminaria digitata. Collected in autumn.		Stem and frond collected in spring.	Mixed weeds in the state in which they are actually used.
Water	74·31	70·57	88·69		77·31	80·44
Albuminous compounds	1·76	2·01	0·93		8·32	2·85
Fibre, &c.	19·04	22·05	4·92		10·39	6·40
Ash	4·89	5·37	5·46		8·98	10·31
	100·00	100·00	100·00		100·00	100·00
Nitrogen	0·28	0·32	0·15		0·53	0·45
The ash consisted of—			Stem.	Frond.		
Peroxide of iron	0·25	0·35	0·20	0·50	0·45	2·35
Lime........................	9·60	8·92	7·21	7·29	4·62	18·15
Magnesia	6·65	5·83	2·73	5·91	10·94	6·48
Potash	20·03	20·75	5·55	11·91	12·16	12·77
Chloride of potassium...	—	—	58·42	26·59	25·83	9·10
Iodide of potassium ...	0·44	0·23	1·51	2·09	1·22	1·68
Soda	4·58	6·09	—	—	—	—
Sulphide of sodium	3·66	—	—	—	—	—
Chloride of sodium	24·33	24·81	15·29	80·77	19·34	22·08
Phosphoric acid	1·71	2·14	2·42	2·66	1·75	4·59
Sulphuric acid............	21·97	28·01	2·23	8·80	7·26	6·22
Carbonic acid	6·39	2·20	4·11	2·49	15·23	13·58
Silicic acid	0·38	0·67	0·33	0·99	1·20	3·00
	100·00	100·00	100·00	100·00	100·00	100·00

The first four analyses give the composition of the seaweeds, after they have been separated from all foreign substances; the last, that of the mixture taken from the heap just as it is used in Orkney; and its value is then enhanced by small shells and marine animals adhering to the plants, which increase the amount of phosphoric acid and nitrogen.

The next table embraces the analyses of the ash of certain fuci, of which only one detailed analysis has been published, with the exception of the *Fucus digitatus* and *F. nodosus*.

Composition of the Ash of certain Fuci.

Constituents of the Ash.	Laminaria saccharina, North Sea.	Fucus serratus, North Sea.	Laminaria latifolia, Hoffmannsgave.	Furcellaria fastigiata, Cattegat.	Chondrus crispus, Cattegat.	Iridæa edulis, Cattegat.	Polysyphonia elongata, Hoffmannsgave.	Delesseria sanguinea, Cattegat.	Fucus digitatus from the mouths of the Clyde.	Fucus nodosus from the mouths of the Clyde.	Fucus serratus from the mouths of the Clyde.
Potash	24·77	16·02	16·91	21·36	18·00	23·42	21·23	13·72	22·40	10·07	4·51
Soda	1·84	6·01	—	24·77	19·47	16·06	5·06	21·34	8·29	15·80	21·15
Lime	6·50	7·05	8·98	6·02	7·11	10·23	2·92	2·30	8·79	10·98	11·09
Magnesia	8·13	7·59	6·80	11·04	11·80	—	20·56	5·94	7·44	10·93	11·66
Chloride of sodium	33·72	38·72	26·92	—	—	1·66	13·85	—	28·39	20·16	18·76
" potassium	—	—	10·18	—	—	—	—	—	—	—	—
Iodide of sodium	4·70	0·27	—	—	—	—	—	—	3·62	0·54	1·33
Phosphate of lime	8·41	4·95	12·80	3·96	0·76	23·23	2·99	3·90	5·63	3·34	9·67
" iron	0·75	0·64	—	—	—	—	—	—	—	—	—
Oxide of iron	—	—	—	—	—	—	—	—	0·62	0·29	0·34
" manganese	—	—	—	0·22	—	—	—	—	—	—	—
Sulphuric acid	10·60	18·35	12·63	32·63	42·86	25·19	28·66	40·42	13·26	26·69	21·06
Silica	0·58	0·40	0·69	—	—	—	2·83	12·38	1·56	1·20	0·43
	100·00	100·00	95·21	100·00	100·00	99·79	98·10	100·00	100·00	100·00	100·00
Percentage of ash in weed dried at 212° F.	9·78	25·83	13·62	18·92	20·61	9·86	17·10	13·17	20·40	16·19	15·63
Chemists	Schweitzer.	Schweitzer.		Forchhammer.					Godechens.		

This table shows the great difference which exists in the percentage of ash left by these plants, which may partly depend upon the species, and partly upon the season of the year at which they were collected, as Dr. Anderson's analysis of the digitata in some measure confirms. The manner of collecting the sample will also, as Mr. Stanford remarks, have some influence on the result; the fronds being always more or less covered with zoophytes, which largely increase the amount of ash, without adding to the soluble salts.

The third table exhibits the results of five analyses of the *Fucus Vesiculosus* from different localities, which give some idea of the extent to which the ash of the same species is liable to vary in its composition.

Composition of the Ash of the Fucus vesiculosus from different localities.

Constituents.	Godeschen, Month of the Clyde.	Month of the Mersey.	Schweitzer, North Sea.	Forchammer, Denmark.	Forchammer, Greenland.	Average.
Potash................	15·23	—	17·68	9·03	17·86	11·96
Soda	11·16	15·10	5·78	7·78	21·43	12·25
Lime	8·15	16·77	4·71	21·65	8·31	10·92
Magnesia	7·16	15·19	6·89	10·96	7·44	9·53
Chloride of sodium .	25·10	9·89	35·38	3·53	25·93	19·82
Iodide of sodium ...	0·37	—	0·13	—	—	0·25
Phosphates of lime and iron............	2·99	—	5·44	9·67	10·09	5·64
Oxide of iron.........	0·83	4·42	—	—	—	0·95
Sulphuric acid	28·16	30·94	23·71	26·34	13·94	24·62
Silica	1·85	7·69	0·28	11·04	—	4·06
	100·00	100·00	100·00	100·00	100·00	100·00

The preceding tables show that the

Potash and soda vary from	.	15 to 40 per cent.
Lime	„ .	3 to 21 „
Magnesia	„ .	7 to 15 „
Common salt	„ .	3 to 35 „
Phosphate of lime	„ .	3 to 10 „
Sulphuric acid	„ .	14 to 31 „
Silica	„ .	1 to 11 „

It still remains to be determined, how these variations depend on the age of the plant, the part of the plant incinerated, and other circumstances to which no attention has yet been paid.

C. Knauss has found in the ash of seaweed from the White Sea, after deducting 13·5 per cent. of charcoal and 42·6 per cent. of quartz-sand, the following constituents.

Sulphate of potash	19·43
Sulphate of soda	13·01
Sulphate of lime	7·02
Chloride of sodium	25·66
Bromide of sodium.	1·32
Iodide of sodium	0·46
Carbonate of lime and magnesia, with oxide of iron, alumina, and phosphoric acid . . . . . . }	27·27
Silica	5·83
	100·00

The following tables contain the results of Mr. Stanford's investigation, and presents them in a form more adapted for commercial purposes. The first five columns are the present kelp-bearing species; the sixth and seventh, the *Zostera marina* and *Rhodomela pinastroïdes*, have not yet been employed for this manufacture. The water was driven off at a temperature of 212° F., and the estimations of the iodine were made with weeds carefully carbonised in a covered porcelain crucible, at a low red heat.

TABLE I. *Wet Sea-weeds, Centesimal.*

Constituents.	Laminaria digitata.	Laminaria saccharina.	Fucus vesiculosus.	Fucus serratus.	Fucus nodosus.	Zostera marina.	Rhodomela pinastroïdes.
Water	82·000	81·000	71·000	78·500	58·000	82·000	83·000
Organic matter	12·710	12·920	23·280	16·220	35·030	14·720	12·878
Soluble ash	4·170	4·810	4·100	3·360	4·880	2·580	2·940
Insoluble ash	1·120	1·270	1·620	1·920	2·090	0·700	1·182
Total	100·000	100·000	100·000	100·000	100·000	100·000	100·000
Yield of kelp	5·900	6·600	7·100	6·100	7·630	3·320	6·368

TABLE II. *Dry Sea-weeds, Centesimal.*

Constituents.	Laminaria digitata.	Laminaria saccharina.	Fucus vesiculosus.	Fucus serratus.	Fucus nodosus.	Zostera marina.	Rhodomela pinastroïdes.
Organic matter	70·112	67·646	80·358	75·079	83·412	81·796	65·748
Soluble ash	23·560	25·598	14·079	15·859	11·614	14·319	17·297
Insoluble ash	6·328	6·756	5·563	9·062	4·974	3·885	16·955
Total	100·000	100·000	100·000	100·000	100·000	100·000	100·000
Yield of kelp	33·335	35·112	24·381	28·782	18·159	18·444	27·460

Composition of the Soluble Ash.

	Laminaria digitata.	Laminaria saccharina.	Fucus vesiculosus.	Fucus serratus.	Fucus nodosus.	Zostera marina.	Rhodomela pinastroïdes.
Total amount of sulphuric acid	2·135	1·627	4·165	4·620	2·927	1·498	2·619
Total amount of alkalies expressed as chlorides }	21·526	23·886	11·400	14·396	10·733	12·209	15·910
Total amount of potash	6·893	8·958	2·043	4·408	2·546	2·940	6·075
Total amount of iodine	0·4788	0·1583	0·00985	0·0382	0·0422	0·0157	0·0378

Collection of the Sea-weed and Manufacture of the Kelp.

This manufacture is at present confined, in this country, to the western and northern islands of Scotland, the north-west coast of Ireland, and Guernsey. In France, the principal localities are in Normandy and Brittany, and in the chapter on Soda, the chief seats of the Spanish trade are mentioned.

Scotland.

Collecting the Weed.—In Scotland, the kelp harvest lasts from June to September, during which full employment is given to the inhabitants on the sea coast, in collecting the weed and burning it into kelp.

This kelp is known in commerce under the terms *cut-weed* and *drift-weed;* the former being prepared solely from weeds growing upon the shores, alternately immersed and left dry by the flow and ebb of the tides; the other is made from weeds or *wrack* which is torn from the rocks and thrown on the beach during storms. These deep-sea weeds differ from the more landward plants, and are always covered with salt water.

The cut-weed kelp, which is that chiefly prepared in the Highlands, is made from two plants of the same order, known under the technical terms *yellow wreck* and *black wreck.* The former floats in the water when cut, and is more easily managed than the other, which does not float; and it is more commonly used in the Highlands for this reason. It is more of a land plant, so to speak, and yields less iodine and potash than the drift-weed or deep-sea plants.

The mode of cutting and collecting this yellow wreck, or *Fucus nodosus,* called *bladder-wrack* by the Irish, is conducted in the following manner. A company of six or eight men are provided with a small boat, furnished with anchor, painter, and oars, three or four common hay pitchforks, some shovels, and two or three wheel-barrows. Each man has also a reaping hook and about 30 yards of stout rope. The cutting of the weed is carried on only during *spring* tides, when the largest quantity of weed is exposed during the ebb. A short time before low water, the men station themselves along the shore, about 6 or 8 yards from each other, and cut the weeds just as wheat is reaped, throwing the weeds behind them with the left hand. The ropes are then laid along the water's edge, the ends being tied together. Each man wraps

a portion of the weed round the rope, so as completely to incase it
and to give the rope sufficient buoyancy to rise with the weeds
when they float. The men then turn round and cut the weed
lying between them and the shore. The tide advancing, floats
the weeds enclosed within the swaddled rope, the farther ends of
which are drawn up as far as possible on the shore, and made fast
to a rock. When the tide is full, the men in the boat and on
shore, haul the rope with the wrack into *port*, as they say.
When the tides have carried the wrack as high on the beach as
possible, it is spread out to dry in the sun, and burnt during the
neap tides. It is important to finish these operations before the
weeds suffer from rain, when they are apt to ferment.

The *black-wrack*, or *Fucus serratus*, is also cut with the hook,
but as it does not swim, it is collected in large boats holding 4 or
5 tons. The boats are moored near the rocks left bare by the
receding tide, and the weed, as it is cut, is wheeled into the boats
and landed on the shore at the succeeding full tide.

The *drift-weed*, or *Laminaria digitata*, is in many cases collected by women, who, when they have collected a lapful, carry it
to the shore, or lay it in heaps on the rocks, whence it is transferred to the *kelp-cart* and conveyed to the beach beyond tide
mark.

Burning the Weed.—When it can be managed, the weeds are
spread out in thin layers on the ground and turned over at times,
until quite dry. Two days of fine weather are quite sufficient for
the purpose; and to prevent the weeds getting wet by the heavy
dews, they are collected in the evening into little heaps, called
quoils, which are spread out again on the next day.

The kilns for burning the weed, are built on level ground,
generally about 14 to 16 feet long by 2 feet broad, surrounded
with a stone wall 8 to 10 inches high. They are placed at right
angles to the direction of the prevailing winds, that the smoke
may prove of the least annoyance. Two men attend each kiln,
which is kindled with a layer of dry heath or straw, and on which
the dried seaweed is placed. When this is nearly consumed,
fresh weed is added, and this operation continues all the day.
The *burner* spreads the weed carefully over the burning mass
with his pitchfork, leaving the ends lying over the walls of the
kiln, to prevent the fresh weed crushing the burning mass below,
and to allow the air to obtain freer access. The weed is never
allowed to burst into flame, which wastes the salts, and the

burner also prevents the mass *falling into holes*, which is apt to cause a partial fusion, by closing the outside of the walls of the kiln with grass sods, so as to regulate the admission of the air. The burning is divided into two periods termed *floors ;* when this operation has been in progress from six to eight hours, the burner levels the surface, throws in all the half-burnt material, and allows the whole to remain ten to fifteen minutes. In the meantime, a portion of the end walls is pulled down, and by means of *clats* or rakes, called *corag* in Gaelic, the men begin to knead and work up the porous ash, when it melts and runs together. The whole of the ash being brought into this state, it is spread over the bottom of the kiln, and forms a cake from 3 to 6 inches thick, which forms the *floor*. The burner then adds fresh weed and proceeds as before, until the end of the day, when a second *floor* is formed in the same way. The number of floors varies ; in Collonsay, two is the usual number, but in the West Islands, four and even six are made.

In some places circular enclosures are used, and during the season, columns of smoke and flame ascend like beacon-fires, which are answered by hundreds of others all along the coast.

The fused mass is broken up by throwing water upon it, and the kelp so made is a semi-vitreous mass, in the best specimens of which, the carbonised stems of the *Laminaria digitata* may be recognised. It is of this weed,—called *bardarrig* in the " drift-weed,"—that the most valuable kelp is made, which, when pure, will yield from 10 to 15 lbs. of iodine per ton. The " cut-weed " is called *slaten-varra*, and its produce of iodine is not half of that yielded by the former.

Ireland.

The kelp in the north-west of Ireland is made chiefly from the " drift-weed," consisting of *sea rods*, or *tangle*, or *staffa*, as the Laminaria are called in Gaelic. It comes in during April and May, and is burnt in small heaps, or in holes 18 inches deep, and at a lower temperature, so that the kelp is more porous and much richer than that from the Western Islands. Some of the drift-weed is stacked in the winter for burning in the summer. A portion of Irish kelp is also made from cut-weed, principally from the fuci.

Guernsey.

In Guernsey, twice a year, a day is set apart for gathering the seaweed, when persons of all ages and both sexes strive to collect as much as possible of the coveted treasure; and so violent is the scramble, that, Sir G. Head states, peace officers are summoned to regulate the proceedings. That which is cut or scraped from the rocks is called *vraic scié*, which is considered superior to that washed ashore by the tide, and called *vraic venant*,— the term *wraic* in Scotland having obviously the same origin. We have seen how carefully each little farmer stacks his stock of weed and uses it as fuel for domestic purposes, the ashes being carefully preserved and employed as manure. Mr. Arnold, of Guernsey, is the only lixiviator of kelp on this island, and the local Government have more than once tried to stop the lixiviation altogether. Professor Graham first drew attention to the richness of the Guern-sey ash in iodine, but the peculiar laws of the island prevent any great extension of its manufacture. The probable origin of this superior weed in the Channel Islands, arises from the quantity torn up from the Atlantic, as it is chiefly found on the western coasts of Guernsey and Jersey.

A large quantity of this drift-weed finds its way up the Chan-nel, and is washed in and out of the harbours, and thrown on the flat coasts; this is seen especially in Brading Haven, in the Isle of Wight, and many thousand tons of seaweed of various kinds are deposited on the shores of Sussex, and along the east coast of this country. Except in the northern parts of this east coast, little or no use is made of this seaweed, even for purposes of manuring the adjoining lands.

France.

On the west coast of France, in Normandy and Brittany, especially on the shores of La Manche, this trade has existed from time immemorial. The *drift* weed is collected all the year round, but the *cut* weed only in March and October.

When the weed has been gathered or cut, it is spread out in the sun to dry, and afterwards stacked and covered, to protect it from the weather. These stacks stand until July or August, when the weed is burnt. Dry weather is selected for this opera-tion, which is performed in a sheltered place. Round or rectan-gular pits are dug in the ground; the first being from 2 to 3 feet

in diameter, and 16 to 20 inches deep ; and the other 6 to 7 feet
long, by about 2 feet wide and 1 foot deep. The sides and bottom are protected by stones, bricks, or metal plates. The pit is
filled with dry weed and set on fire. A dense white smoke is
produced, of a disagreeable smell, from the empyreumatic and
ammoniacal vapours which are disengaged. As the burning proceeds, fresh weed is added, and in time the heat becomes so great
as to soften and ultimately to fuse the ashes. A workman occasionally stirs the mass, to bring all the organic matter in contact
with the air, to ensure as perfect a combustion as possible. When
the pit is full of well-fused salts, it is left to cool. The fused mass
takes the shape of the pit, and sometimes weighs as much as 9 to
10 cwts.

Composition of the Kelp and Varec.

The great difference in the composition of the ash of seaweeds,
even of the same species from different localities, might lead one
to expect corresponding differences in the kelp and varec. These
differences do exist in these substances as manufactured in different localities ; but from the enormous quantities burnt at one
time, 20 to 24 tons of weed yielding only 1 ton of produce, there
is great similarity in the kelp or varec from the same place, year
after year, so much so, that the buyers can rely with tolerable
certainty on a fair uniformity in the composition. The differences,
however, in the produce from different localities, are very great.
Thus, one kind will yield 1 of iodine per 100 of kelp, another 1
per 1,000, and some none at all. And the same remark applies
to the salts : thus, in Golfier-Besseyre's analyses, the sulphate
of potash varied from 11 to 44 per cent., and in one case fell as
low as 2 per cent. ; the chloride of potassium, from 12 to 35 per
cent., and in one instance was only 0·36 per cent. ; the sulphate of soda, from *nil* to 35 per cent. ; the carbonate of soda,
from *nil* to 17 per cent. ; and the common salt from 9 to 70 per
cent.

The following analyses represent the composition of drift
and cut-weed from Scotland, Irish and Scotch kelp, and the
French varecs, by Golfier-Besseyre, who states that those samples which had no iodine, contained appreciable quantities of
bromine.

Brown's Analysis of a sample of Orkney Kelp made from Drift-weed.

INSOLUBLE PORTION :—

Carbonate of lime . . .	2·59
Phosphate of lime . . .	10·55
Sulphide of lime	1·09
Silicate of lime	3·82
Carbonate of magnesia. .	6·55
Sand.	1·57
Alumina	0·14
Carbon	0·92
Hydrogen	0·14
Nitrogen	1·15
Oxygen	0·66—29·18

SOLUBLE PORTION :—

Sulphate of potash . . .	4·53
Sulphate of soda. . . .	3·60
Sulphate of lime	0·28
Sulphate of magnesia . .	0·92
Sulphite of soda	0·78
Hyposulphite of soda . .	0·22
Sulphide of sodium . . .	1·65
Phosphate of soda . . .	0·54
Carbonate of soda . . .	5·31
Chloride of potassium . .	26·49
Chloride of sodium . . .	19·33
Chloride of calcium . . .	0·23
Iodide of magnesium . .	0·32
Bromide of magnesium .	trace
Water	6·80—71·00

100·18

Lamont's Analysis of a sample of Kelp, made from Cut-weed from the Hebrides.

INSOLUBLE PORTION :—

Sulphide of calcium . .	2·09
Phosphate of lime . . .	6·44
Carbonate of lime . . .	11·17
Magnesia	5·70
Alumina	0·23
Oxide of iron	1·42
Silicate of lime	2·77
Sand	1·65
Carbonaceous matter . .	3·30
Loss	1·33—36·10

SOLUBLE PORTION :—

Sulphate of potash . . .	7·74
Chloride of potassium . .	19·86
Chloride of sodium . . .	3·47
Iodide of sodium . . .	0·20
Sulphide of sodium . . .	2·92
Hyposulphite of soda . .	1·98
Sulphite of soda	2·71
Sulphate of soda	15·10
Phosphate of soda . . .	0·95
Carbonate of soda . . .	5·15
Loss	0·82—60·90
Moisture in sample.	3·00
	100·00

Irish Kelp (Richardson).

SOLUBLE IN WATER :—

Sulphate of potash . . .	17·76
Sulphate of soda . . .	14·79
Chloride and iodide of sodium	· 14·32
Chloride of manganese. .	0·20—47·07

Soluble Portion, brought forward . .   47·07

SOLUBLE IN HYDROCHLORIC ACID :—

Soda	2·57
Lime	0·70
Magnésia	3·10
Sulphuric acid	0·53
Phosphoric acid	2·06
Silica	1·85
Carbonate of lime . . .	22·35
Sulphide of iron	2·07—35·23

INSOLUBLE PORTION :—

Charcoal	2·40
Sand	11·02—13·42
Water and loss	4·28

100·00

Scotch Kelp (Richardson).

SOLUBLE IN WATER :

Sulphate of potash . . .	24·52
Sulphate of soda	12·29
Sulphide of sodium . . .	8·57
Chloride of sodium . . .	18·28—63·66

SOLUBLE IN HYDROCHLORIC ACID :—

Lime	1·00
Magnesia	3·96
Peroxide of iron	0·60
Phosphate of iron . . .	2·56
Silica	1·32
Sulphur	0·20
Carbonate of lime . . .	13·40—23·04

INSOLUBLE PORTION :—

Charcoal	2·22
Sand	7·08—9·30
Water and loss	4·00

100·00

Analyses of French Varec, by Golfier-Besseyre.

Department	Locality of the Varec	Composition of the Varec in 100 parts in		Composition of the Rough Salts.							Relation between the Salts of Potash and Soda converted into	
				Salts of Potash.				Salts of Soda.				
		Salts	Insoluble matters	Sulphate of potash	Chloride of potassium	Iodide of potassium	Chloride of sodium	Carbonate of soda	Sulphate of soda	Hyposulphate of soda	Chloride of potassium	Chloride of sodium
Pas-de-Calais	D'Andreselles, near Boulogne	25·00	75·00	29·95	4·08	traces	47·72	—	18·25	10 77	30·53	59·47
Seine-Inférieure	No name given	20·50	79·50	30·66	14·60	traces	18·89	—	25·08	traces	43·26	54·74
	Ménil-à-Caux, Commune of Criel	61·00	89·00	27·37	17·05	traces	55·57	4·68	35·53	—	40·44	59·56
	Eletot	28·73	71·25	43·59	6·47	—	9·56	4·97	29·17	—	49·47	50·53
Manche	Jourville, Capitainerie of Marcouf	57·00	43·00	17·90	18·65	—	29·81	13·95	29·07	—	35·79	64·21
	Gatteville — Barfleur	62·50	37·60	17·77	32·36	traces	16·85	—	—	0·45	40·00	60·00
	Fermanville	39·00	61·00	35·79	2·76	—	61·07	5·74	—	2·30	36·84	63·16
	Rhétonville	48·00	52·00	30·47	18·54	0·0008	42·96	0·75	—	—	49·47	50·53
	Querqueville	40·00	60·00	14·94	24·45	—	34·64	1·26	25·22	—	38·95	61·05
	Becquets — Cherbourg	44·00	56·00	38·62	14·56	—	50·06	0·70	—	—	45·26	54·74
	Anderville — Beaumont	45·50	54·50	35·55	15·01	0·0061	48·74	0·27	—	—	47·37	52·53
	Ammonville	41·25	58·75	34·43	19·86	—	43·45	3·20	—	—	51·58	48·42
	Dielette — Dielette	44·20	55·80	30·54	24·02	—	42·24	1·92	—	—	53·68	48·42
		67·00	33·00	32·00	24·26	0·0048	41·93	3·40	—	—	53·68	46·22
Finistere	Granville, sold as salts	59·50	47·50	34·10	8·70	—	53·80	3·40	—	—	38·95	61·05
	Coasts of La Manche, without name	—	61·50	41·47	8·47	—	50 66	2·40	—	—	40·92	59·08
	Isle of Beniquet	38·50	58·75	42·50	4·14	0·0220	53·36	—	—	—	43·16	56·84
	Labrevais, Arrondissement of Brest	41·25	38·75	31·87	27·08	0·0104	41·06	—	—	—	56·84	43·16
	Point du Raz	54·50	45·50	29·71	8·78	0·0100	61·51	—	—	—	51·58	48·42
	Primelin, Arrondissement of Quimper	39·00	61·00	29·06	0·36	0·0066	67·99	2·99	—	—	26·31	73·69
	Audierne	41·50	58·50	31·01	18·88	—	60·11	—	—	—	38·95	61·05
		50·00	50·00	24·66	23·45	0·0090	35·96	2·13	—	—	47·37	52·53
		43·00	57·00	8·86	35·03	0·0260	35·95	12·90	—	7·95	43·16	56·84
	Plouhinec	32·00	68·00	10·94	18·93	—	40·10	1·16	26·54	2·38	31·58	68·42
	Esquibien	49·25	50·75	24·81	18·64	0·0140	53·78	2·77	—	—	41·05	58·95
	Plozevet	76·87	23·13	20·96	32·47	0·0080	44·22	2·34	—	—	51·58	48·42
	Isle of Glénans	41·00	59·00	28·81	6·72	0·0090	54·13	0·34	—	—	32·63	67·37
	Isle of Quémènes	48·50	51·50	18·64	20·06	0·0160	52·04	1·87	30·99	—	38·95	61·04
Vendée	Noirmoutiers — 1st quality	45·50	34·50	23·52	10·05	0·0060	41·99	6·57	—	—	55·79	44·21
	" 2nd "	57·50	42·50	9·64	29·37	—	29·90	9·49	—	14·96	36·84	63·16
	" 3rd "	39·50	60·50	11·07	31·36	—	24·67	16·70	—	14·39	45·26	54·74
Var	Salt Works of Hyères	43·25	56·75	26·44	19·49	—	35·53	12·76	—	20 14	47·37	52·63
		72·50	27·50	2·37	12·19	—	69·93	16·51	—	10·99	12·63	87·37

Some years ago, Professor Rose pointed out the great disadvantages of the plan generally adopted in the incineration of plants, and showed to what an extent the nature and composition of the ash was affected. His remarks apply with far greater force to the description above given, of the mode of preparing kelp and varec, and Mr. Stanford has very clearly stated the objections to the present system, the principal of which are the following:

1st. The high temperature developed during the burning, by which much of the iodine and some of the potash are volatilised. The loss of iodine, on an average, may be said to be equal to the present yield of this substance from kelp, as at present manufactured.

2nd. The high temperature also enables the carbon to deoxidise the alkaline sulphates, reducing them to sulphites, hyposulphites, and sulphides. These accumulate in the mother-liquors and necessitate the consumption of a large quantity of sulphuric acid to reconvert them into sulphates. Dr. Wallace, in a paper read before the British Association, at Aberdeen, September 14, 1859, estimates the cost of extracting all the salts and the iodine from kelp, at from 25s. to 28s. per ton, of which from 11s. to 13s. are expended in oil of vitriol; the disadvantage is therefore obvious. And the addition of the oil of vitriol to the mother-liquor, sets free large volumes of sulphuretted hydrogen, from the decomposition of these sulphur-compounds. To such an extent is this deoxidation carried, that kelp produced from the *fuci*, which are rich in sulphates, gives off sulphuretted hydrogen by simple solution in water.

3rd. The crude manner in which the weed is burned in a rough clay pit, or in heaps on the beach, whence the kelp becomes mixed with the clay, sand, or stones, which are sometimes also employed as adulterations to the extent of 50 per cent.

4th. The general neglect of the winter supply, on account of the difficulty of drying the weeds, at that season, in this country. This consists principally of the deep-sea algæ, torn up by storms: it occurs then in the greatest quantity, while it is also much the richest in iodine and potash, and is almost entirely lost. Even in the summer, during a wet season, large quantities of drift weed collected by the kelpers, are rotted by the rain, and rendered useless for burning; and this, says Mr. Paterson, "was the fate of many thousands of tons during the kelp season" in 1861.

5th. The entire loss of the heat produced, and of the products of combustion.

6th. The dense smoke and unpleasant odour evolved in burning, banishes the kelp-burner to desolate shores, and, consequently, increases the distance of the lixiviating factory; the ash in its long transit suffers much by exposure to rain, and the residuum of lixiviation, a valuable manure in the country, is a bye-product in a city. It may be urged, also, that the poor kelpers, in their arduous work, receive but a small part of the real value of their labour.

Various plans have been proposed to remedy the present system. Some parties have suggested charring the weed in reverberatory and other forms of furnaces or kilns.

Mr. Lamont has tried the following plan on a small scale with marked success. The weeds were boiled with a small quantity of water, evaporated to dryness, and the residue charred. The mass, treated with water, gives a colourless solution, and Dr. Wallace has analysed a sample of kelp prepared in this way, chiefly from the *Laminaria digitata*, with the following results:

Sulphate of potash	14·35
Chloride of potassium	42·49
Chloride of sodium	36·47
Sulphate of soda	3·90
Iodide of sodium	1·78
Carbonate of soda	1·01
	100·00

Dr. Kemp has suggested a process, which consists in selecting the stems of those algæ which are the richest in iodine, and crushing them; they are then placed in a tank for a few days, and the most soluble salts, including the iodides, extracted by cold water acidified with hydrochloric acid. The solution is then treated with chloride of lime, and the iodine thus set free is precipitated by amido-acetate of lead. The cakes of weed left after pressure, are dried and used as fuel, the ashes being preserved *on account of the iodine they still retain*, as well as the other salts. This process has not been adopted, probably because the manipulation, down to the burning of the fuel into ashes, is a complicated labour superadded to the ordinary process, and with little advantage, for the ashes have still to be worked for iodine and potash. The crushing of the weeds, and the drying of the

crushed cakes, present great practical difficulties, which those who
have had any experience in compressing fresh vegetables will
fully appreciate. Dr. Wallace's suggestion, in the paper before
referred to, is more simple, and he states that it is rather given to
call attention to the subject, than as a perfect process. He wishes
to supersede the present fused kelp by a loose charred ash. He
recommends the weed to be charred rather than burnt, and as
this ash, from its bulk, would be inconvenient for carriage, it
should be lixiviated with a small quantity of water, and the
solution roughly evaporated to dryness. He says, "By this
treatment a very pure salt would be obtained, the iodine would
be wholly preserved, while the cost of working would be more
than counterbalanced by the saving of vitriol that would be
effected by the absence of the sulphur-compounds."

Dr. Wallace gives the following results of some analyses of the
quantities of potash per cwt., and of iodine per ton, in salts
prepared by his plan, in the ashes of the plants, and in the two
varieties of ordinary kelp :

		Laminaria digituta.	*Fucus serratus.*	*Fucus nodosus.*
Pure salt	Potash . . .	34	31	23 per cwt.
	Iodine . . .	38	15	10 per ton.
Ash . .	Potash . . .	26	19	14 per cwt.
	Iodine . . .	28	9	6 per ton.

		Very fine drift.	Cut-weed.
Kelp .	Potash . . .	24	15½ per cwt.
	Iodine . . .	18	4 per ton.

This idea meets some of the difficulties of the question, but
not all ; the saving of iodine would to a great extent be effected,
and it would make the difference which now exists between the
best Highland kelp and that of the Channel Islands,—the
former a vitreous mass, with more than half the iodine volatilised,
the result of igneous fusion,—the latter a loose charred ash, the
result of slow combustion at a low temperature, and containing
much more of the iodine. The difficulty, however, of completely
charring such a substance as seaweed on open slabs, is considerable, and the salts resulting from its lixiviation would be coloured
by any portion not perfectly carbonised. The Guernsey kelp gives
a coloured solution and dirty salts, which require re-crystallisation
before they can be brought to market. Of Mr. Standford's plan,
full details will be given in a distinct section.

33*

Another very feasible plan has been proposed by Mr. Meldrum, and his arrangements are seen in Fig. 269. A is a kiln, into which dried seaweed is thrown through the epening, C. There

FIG. 269.

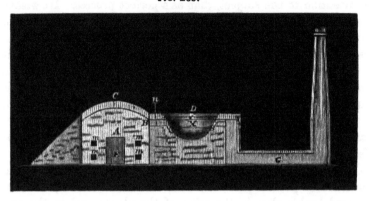

are two doors, E, one on each side, opposite one onother, for the purpose of removing the kelp. When at work, they are built up with loose stones, to allow a current of air to pass in among the seaweed. There are additional openings at *m m* for the same object, and also for stirring the materials. B is a circular flue round the pan, D. The flue B being closed by a damper, *n*, the damper in the flue leading to the chimney, is opened; or when there is more than one pan, the heat is conducted to another of the pans. The opening C is fitted with a door. If there were only two pans, fire-places might be built in front at X, and fed with peat, if there was not sufficient dry seaweed for both the kiln and furnaces. The seaweed may be dried on the roof of the kiln and on the arch of the flue G.

The kiln being charged with dry weed, it is ignited, and may be burnt at any degree of heat, by regulating the draught by means of the dampers. The ashes are lixiviated, and the liquors boiled and salted in the pans, while the mothers are evaporated to dryness. The two kinds of salt are then sent to the manufacturer to separate the potash-salts and extract the iodine, while the insoluble residue might be used to manure the neighbouring lands.

Statistics of the Kelp Trade.

Mr. W. Paterson furnished Mr. Stanford with the following particulars of this trade. The prices named do not include

freight and charges to the potash and iodine manufactories. The
average is taken on the years 1860 and 1861 :

Western Islands.

1,800 tons cut-weed kelp, at £2 . . .	£3,600
800 „ drift-weed kelp, at £4 . .	3,200
400 ,, „ at £6 . .	2,400
3,000	£9,200

Orkney and Shetland Islands.

1,200 tons drift-weed kelp, at £6 . .	£7,200
150 „ cut-weed kelp, at £2 10s. . .	375
1,350	£7,575

Ireland.

2,500 tons of drift-weed kelp, at £4 .	£10,000
500 „ „ at £3 10s.	1,750
1,000 , at £5 .	5,000
500 , at £6 .	3,000
200 , „ at £6 5s.	1,250
80 ,, „ at £7 .	560
1,300 „ of cut-weed kelp, at £2 13s.	3,445
6,080	£25,005

Total Amount of Kelp, and average Price, for the Years 1860—61.

Scotland . .	4,350 tons, at about £3 17s., valued at	£16,775
Ireland . .	6,080 „ „ £4 2s. „	25,005
	10·430		£41,780

In round numbers, the average yield of British kelp may be
taken at 10,000 tons, giving, at £4 per ton, an annual income of
£40,000.

This quantity of kelp represents about 200,000 tons of wet
weed, an amount which, large as it may seem, is insignificant
compared to the immense masses of seaweed annually deposited
on the coasts of Great Britain and Ireland.

Messrs. Tissier & Son estimate the total annual production

of kelp in France at 25,000,000 kils, or upwards of 24,000 tons,
so that this manufacture is more than twice as large as the yield
of Great Britain.

Extraction of the Salts.

This manufacture is carried on to the greatest extent in
Glasgow and its neighbourhood, where there were no less than
nine establishments in 1854, the largest, by far, being that of
Mr. W. Paterson, who consumes from 7,500 to 8,000 tons of kelp
per annum; one at Borrowstownness, belonging to a most
scientific manufacturer, Mr. F. R. Hughes; two in Ireland, near
Lough Zwilly, in the centre of the kelp districts; one in Guernsey;
and several in France—the largest belonging to Messrs. Tissier,
& Son, who work up from 4,000 to 5,000 tons per annum.

The separation of the three principal salts contained in kelp, is
based on three facts, viz.:

1st. That common salt is nearly as soluble in cold as in hot
water; 2nd. That chloride of potassium is more soluble in hot
than in cold water; and 3rd. That sulphate of potash is only
slightly soluble in cold water.

The following table exhibits the extent of the solubility of
these salts in pure water at 54° and 212° F.

Salts.	Quantities dissolved by 100 parts of water	
	at 54° F.	at 212° F.
Sulphate of potash .	10·5	27 0
Chloride of potassium	32·0	59·4
Chloride of sodium .	35 5	40·0

It must not, however, be overlooked, that the *first solutions* of
the chloride of potassium render the sulphate of potash *less
soluble*, while the chloride of sodium renders sulphate of potash
more soluble in the *second lye* than in pure water.

The methods of separating the salts can, of course, be varied,
but whatever plan is adopted, the liquors gradually become
stronger and stronger, until the mothers acquire a density of
1·555 to 1·647, when they require a special treatment, to be
hereafter more fully detailed when the extraction of iodine and
bromine is described.

Lixiviation.

In Scotland the kelp is broken up, by large hammers, into small pieces of the size of the stones used for repairing our roads. It is then thrown into large vats, made of cast-iron, arranged as in Fig. 270. A filter made of straw, dried seaweed, or small stones, covered with a mat, is placed in each vat, at m and B. Assuming the vats to have been at work, and A 4 just filled, the plugholes m being closed, as well as the communication, B C, between A 1 and A 4, water is run on to A 1. The water falls through the kelp in that vat, and overflows by the pipe B C into A 2, thence to A 3, and lastly into A 4. As soon as the last vat is full, the supply of water is shut off, and the whole allowed to stand for 12 hours. The liquor is then run off by the opening m, into the spout G M, which conveys it to the underground receiver, K whence it is pumped up

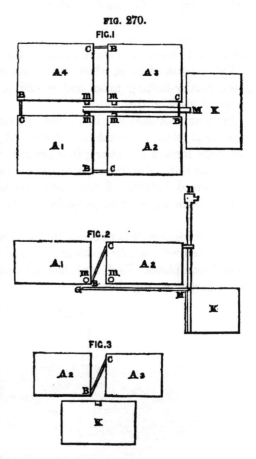

FIG. 270.

into the boiling pans. When the liquors in A 1, fall to 5° Twaddell, it is run off through the opening m, into the receiver K; the vat is emptied and refilled with fresh kelp. The communication between the strong liquors of A 1, and the weaker lye of A 2, is closed, and the weak liquor in the receiver is pumped back upon A 2, from which it flows round through A 3, and A 4, on to A 1. Another twelve hours are allowed to elapse, and the liquor is again drawn off at m, and this process is repeated, each vat being

filled and emptied alternately, all liquors below 20° T. being
pumped back on the fresh kelp. At first, the liquors stand at
50° T., but by the time the last vat is exhausted, they fall to
35° T., below which strength no liquors are evaporated, in order
to save fuel. The filters are renewed at every fifth operation,
and the pipes B, C, must be straight, to admit of their being
easily cleaned.   The end C, of the pipe must also be placed
about 4 inches below the top of the vats, as some pressure is
necessary to force the liquors round to the fourth vat.

The exhausted mass, which still contains some kelp, is a dark
green earthy-looking mass, and is sold to the manufacturers of
common glass bottles.

In some manufactories in France, the varec, after being broken
into small pieces, is crushed under an edge-stone rolling over a
bed-plate formed of a strong iron grating, the bars of which are
about one-third of an inch apart ; and in other works, the crushing
is performed between fluted metal cylinders, similar to those
used in bone mills, or in preparing chalk for the alkali manufacture.

The lixiviation is conducted in wooden tanks, about 3 feet
deep, fitted with a false bottom, F, made of wood, or of perforated iron plates. Each tank has two pipes, with three openings
to connect them together.

FIG. 271.

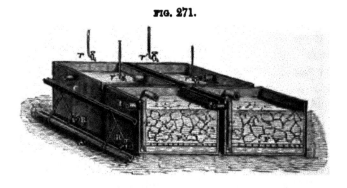

The lixiviation is conducted very methodically, the first washing being intended to extract the chlorides of potassium and
sodium, while the sulphate of potash is obtained in the second lye.

Fresh water or a weak lye is run on the fresh kelp, and after
standing some time, it is conveyed to tanks containing kelp
which has already been treated with water, and thence to other

similar tanks, until the density reaches 1·12 to 1·16. These lyes contain the chlorides of potassium and sodium in large quantities, while the sulphate of potash is present in only a small proportion, and they are afterwards specially treated for the separation of the two alkaline chlorides.

The residual mass is now submitted to a similar washing until the lyes stand at 1·06, and principally contain sulphate of potash.

The kelp-liquor, from 35° to 50° T., is pumped up into the evaporating pans, and boiled down to 60° T., during which a quantity of common salt and sulphate of soda salt out, which is regularly fished up and thrown into a wooden vessel with a perforated bottom, to allow the drainings to fall back into the pan. Figs. 272, 273, and 274, represent a side view, front view,

FIG 272.   FIG. 273.

FIG. 274.

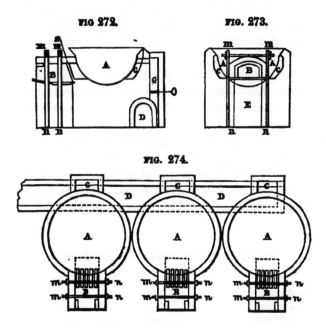

and plan of the top of these pans, in which A shows the pans, B the fire-places, E the ash-pit, C the circular flue, and G the branch flue leading to the main flue, D. The iron binders $m\,n$, support the arch of the fire place. The furnaces are purposely placed at the sides of the pans, as a bottom heat would have a tendency to cause the salt to adhere to the metal, which, becoming red-hot, would, sooner or later, crack.

The salt which falls during the evaporation of the liquid, is an

impure sulphate of potash, which is drained as already explained.
The liquor, when boiled up to 60° T, is allowed to settle for an
hour, and run off into a cooler, where the chloride of potassium
crystallises. It requires five days·to cool, and the crystallisation
is known to be finished by the scum on the surface falling to the
bottom. The mothers are run into a receiver, whence they are
pumped into an evaporating pan. The chloride of potassium is
thrown into a basket, and allowed to drain into the cooler for 24
hours. The second liquor is now boiled up to 68° T., during
which sulphate, muriate, and carbonate of soda salt out, which are
drained in the same way as before. The hot liquid is allowed to
settle for an hour, run into a cooler, and another crop of chloride
of potassium obtained. The mothers are run off again, treated
as before, and boiled up to 74° T., during which process the same
salts of soda separate, and are fished out. The liquor, run into
coolers, furnishes a third crop of chloride of potassium, which is
drained in the same way.

During the evaporation of the first liquor, the pan is occasionally fed with small quantities of kelp-liquor, but during the
boiling of the second and third liquors, the pan is supplied only
with their corresponding mother-liquors.

The salts which separate during the boiling of the second and
third mothers, are sold under the name of *kelp-salts*, and contain
from 5 up to 10 and 14 per cent. of alkali; but they vary very
much in composition, some actually containing only 1 per cent.
of carbonate of soda, as the following analyses prove.

Analyses of Kelp-salts (Richardson).

	1	2	3	4
Carbonate of soda	1·08	6·23	15·04	23·68
Chloride of sodium	80·65	69·97	64·69	41·17
Sulphate of soda	10·03	4·70	15·27	17·91
Sulphate of potash	7·47	10·43	traces	trace
Chloride of magnesium	—	—	traces	—
Insoluble matters	—	—	traces	—
Water	1·60	8·50	4·74	18·00
	100·83	99·83	99·74	100·76

The first crystallisation contains about 86 to 90 per cent. of
pure chloride of potassium, and the rest is chiefly sulphate of

potash. The second and third crops are more pure, and contain 96 to 98 per cent. of the chloride of potassium. The fourth crop contains some sulphate of soda.

The mother-liquor has a specific gravity of 1·33 to 1·38, and contains sulphate of soda, alkaline sulphides and hyposulphites, carbonate of soda, and the compounds of iodine and bromine.

Sulphate of Potash.

The sulphate of potash which is obtained from the kelp-liquors, is found in commerce in two forms : one, which is in powder and granular, is called *soft* or *granulated*, and the other, which is hard, and in flat cakes, is call *plate* sulphate of potash.

The *granulated* sulphate is obtained, as already mentioned, by the simple evaporation of the liquors, during which it falls as a heavy granular salt, which is fished out of the boiling liquor by perforated ladles. It is a very impure substance, containing sulphate, carbonate, and muriate of soda, with some sulpho-cyanide of potassium, iodide of sodium, and insoluble matter. The amount of the sulphate of potash varies considerably : cut-weed kelp seldom gives a salt containing more than 32 per cent. ; a good drift-weed kelp, a salt with from 44 to 65 per cent. It is used for making *plate*-sulphate ; and when poor in sulphate of potash, it is sold for agricultural manures.

The plate-sulphate is made, either from the granulated salt, or directly from kelp liquors. The process is technically called *plating*.

The granulated sulphate is either dissolved in weak kelp-lye, or in water, with heat, until the liquor stands at 44° T. when it is boiled up to 48° or 50° T., and then run into coolers to crystallise. When the crystallisation is complete, the mothers are drawn off and saturated with the granulated salt as before. The crystallised sulphate is not removed as usual, but left to form, as it were, the basis of a fresh crop. The mothers, saturated as explained, are again brought up to strength, and run into the same cooler, where a second crop of crystals forms on those already lining the cooler. This operation is repeated four, six, or even seven times, yielding as many crops of sulphate, in the form of a hard thick cake.

In plating directly from kelp-liquors, they are boiled up until sulphate of potash begins to salt out. The fires are damped, and after allowing impurities to subside, the liquor is run into a cooler to crystallise. In a few days, a crop of plate-sulphate has formed. The mothers are drawn off, and another boil of fresh liquors run on the first crop, when a second crystallisation forms, and this operation is repeated until the plate has acquired sufficient thickness.

This plate-sulphate is, of course, not pure, and contains small quantities of carbonate of soda, common salt, and insoluble matters.

A curious phenomenon accompanies the crystallising of this salt. On looking into a vat, when the temperature is about 100° Fahr., an emission of vivid sparks or scintillations is seen, as the crystals separate from the liquor, analogous to the phosphorescence of arsenious acid in the act of crystallising from its solution in hydrochloric acid.

Dr. Penny, to whom we are indebted for this account of plating, has analysed this salt. It never contains more than 75 to 78 per cent of sulphate of potash, as found in commerce. The mean of twelve analyses of pure specimens, gave Dr. Penny the following composition :

Potash	42·22
Soda	49·54
Sulphuric acid	48·24
	100·00

which corresponds very closely with the formula

$$3(KO.SO^3.) + NaO.SO^3.$$

The following analyses of the sulphate of potash of commerce were made on samples taken from parcels of upwards of 100 tons :

		Richardson.	
		Irish.	Scotch.
Sulphate of potash .	77·43	75·28	83·06
Sulphate of soda . .	21·31	20·89	14·89
Sulphate of lime . .	—	0·80	—
Sulphate of iron . .	trace	—	—
Chloride of sodium .	0·76	0·54	0·67
Insoluble matter . .	trace	1·04	—
Moisture	0·59	1·55	1·44
	100·09	100·10	100·06

which agree very closely with Dr. Penny's results.

It is more fusible than pure sulphate of potash, and its density and solubility in water are greater than the pure salt. 100 parts of water at 207° Fahr. dissolve

29 parts of sulphate of potash, and
40·8 parts of plate-sulphate.

The solution of plate-sulphate furnishes salts of a great variety of composition, when submitted to successive crystallisations in the usual way. The following is a list of these various crops of salts :—

1st crop, sulphate of potash, with a little plate-salt.
2nd crop, plate-salt, with short prisms of sulphate of soda.
3rd crop, chiefly sulphate of soda in large and bold crystals.
4th crop, layer of plate-salt, with sulphate of soda above.
5th, 6th, and 7th crops, sulphate of soda.
8th, 9th, and 10th crops, almost wholly plate-salt.
11th crop, sulphate of soda.

The Processes followed in France for separating the Chloride of Sodium.

These depend on the same principle as the Scotch plan, but as the lixiviation in the former country is different, it is modified in the manner now to be explained. The lye containing the alkaline chlorides is concentrated by evaporation until the solution stands at 1·333, during which operation the common salt is continuously deposited, and is fished out by the workman, and thrown into wooden hoppers, arranged above the pan, so that the drainings may fall back into the liquor. This method of removing the salt is expensive and tedious, and has given place to the following plan. A round iron vessel, c, with one or more series of small holes in the upper part of the sides, and resting on three feet, is suspended over the pan, B B, heated by a naked fire, A. This vessel hangs, usually, just below the level of the liquor, but can be lowered to the bottom of the pan when necessary.

The salt, as it precipitates during the concentration, is

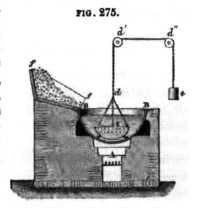

FIG. 275.

thrown into this vessel by the ebullition of the liquid. When
full of salt, it is lifted out of the liquor by depressing the balance
weight e, which is attached by a chain, d, working over the
pulleys d d''. Two bars of iron are laid across the edge of the
pan, on to which it is lowered, and allowed to drain. The salt is
thrown on a leaden shelf, f f, inclining towards the pan, where
the drainage is completed.

A workman keeps stirring the boiling liquor all the time, to
prevent the salt attaching itself to the pan, and also to facilitate
its collection in the vessel C.

The salt thus obtained contains chloride of potassium, and
even sulphate of potash, which are removed in the following
manner: The salt is thrown into wooden boxes, fitted with false
bottoms, and while the lye from a previous operation is run upon
it, a jet of steam is admitted below the false bottom. The salt
is stirred by the workman for about an hour, and the whole
allowed to stand for some time to settle, when the liquor is
drawn off. This operation is repeated again, but with fresh
water, and in about an hour the solution stands at 1·27, when it
is run off, and employed, as mentioned, for another portion of
salt.

The potash-salts are almost entirely removed by this treatment, and the purified salt is left to drain on shelves, and afterwards dried at a gentle heat in a reverberatory furnace. It is
used for curing fish; but in some manufactories it is redissolved,
and salted out during the boiling of its solution; the mother
liquor is found to contain the 2 per cent. of chloride of potassium
and the small quantity of iodide of potassium which the salt
had retained.

Separation of the Chloride of Potassium.

The lye from which the common salt has been removed, and
which stands at 1·333 sp. gr. is left in the pan for an hour, the
fire being previously damped. It is then syphoned off into rectangular iron coolers, about 18 inches deep, and the chloride of
potassium crystallises, as it cools, in the form of small white
crystals. In some works the mothers are again treated in the
same way; but when the lixiviation has been carefully conducted,
all the chloride of potassium is obtained in one operation. The
mothers, in each case, contain the iodine. The chloride of potas-

sium always contains some common salt, which is removed by washing it with lye in the cold, and the former salt is dried in a stove or in an ordinary furnace.

Separation of the Sulphate of Potash.

This salt is obtained from the second washings from the kelp, as previously explained, which vary in strength from 1·06 to 1·147 sp. gr. This lye is evaporated in flat iron pans, until the density reaches 1·25 or 1·27, during which the sulphate separates in small microscopic crystals, which are removed by ladles and thrown into wooden hoppers to drain.

The mothers are run into coolers, when some chloride of potassium and sodium are obtained by crystallisation, and added to those obtained in the preceding operation.

The sulphate of potash is washed with lye from previous operations, and then with cold water; the washings being added to the different lyes used in working.

The chlorides still remaining in the various mother-liquors, are removed by repeated and successive concentrations and crystallisations. During the concentrations, chloride of sodium salts out, and on cooling the liquors, fresh quantities of chloride of potassium crystallise. As soon as a density of 1·634 is reached, these two salts are obtained mixed, when they are redissolved and treated as previously explained, to separate them. In each case of salting or crystallisation, the products are carefully washed and the washings added to the mothers, as these liquors become more and more rich in iodine. When the chlorides have been separated, the last mothers are reserved for a special treatment for the extraction of the iodine and bromine, which will form the subject of the next chapter.

Conversion of the Potash-salts into Nitrate of Potash.

The processes employed for this purpose are fully explained in another chapter, but now that Mr. Hill's patent has expired, this use of the kelp potash-salts has acquired increased importance; and M. Theroulde has described how this extraction from kelp can be carried on simultaneously with their conversion into nitrate of potash. This plan is adopted to a large extent by some of the manufacturers in France, and was proposed many years ago by Mr. Rowlandson of Liverpool.

The apparatus consists of ten or more rectangular vessels of

sheet iron, arranged side by side, and in two rows; they are
each provided with a false bottom of perforated sheet metal or
wood. These vessels are connected together by pipes passing
from the upper part of one vessel to the space under the false
bottom of the next vessel of the series. These vessels are charged
in succession, with about 2 tons of the ashes to be treated. The
water is introduced under the false bottom of the vessel containing the ash most nearly exhausted, and percolates upwards
through the contents of the vessel. From the top of this vessel,
the liquid passes by a pipe to the space under the false bottom of
the next vessel of the series, and in this manner the liquid passes
on from vessel to vessel, until it traverses the ash in the vessel
last charged, and from the top of this vessel the concentrated
solution escapes to a large tank. The potash in solution being
determined, a sufficient quantity of nitrate of soda is added, to
convert it into nitrate of potash. In order to facilitate the
solution of the nitrate of soda, a frame is fitted into the upper
part of the tank before mentioned, and the surface of this frame
is covered with wire cloth; the nitrate of soda is spread on the
platform thus formed, and the platform being below the surface
of the liquid in the tank, the nitrate of soda quickly dissolves.
When the whole of the nitrate of soda is dissolved, the liquid is
run into suitable boilers and evaporated, to separate common salt
and other soda-salts, which are removed by a rake as they fall.
When the liquid is concentrated to 50° to 55° Beaumé, the evaporation is stopped, and it is set aside to crystallise; the salts
which fall during the evaporation are placed to drain in a vessel
of convenient size, containing a perforated false bottom, on which
the salts rest; they consist, almost entirely, of chloride of sodium
and other soda-salts, but still contain small proportions of nitrate
of potash; this is separated by washing, steam being blown
through the salts while they are draining. The steam is introduced under the false bottom of the draining vessel; this causes
the nitrate of potash to redissolve and filter away, as it is very
soluble in hot water; the same process also separates the iodide
and bromide of sodium.

The salt which crystallises out on cooling, when the evaporation is stopped, is drained from the mother-liquor; it consists
almost entirely of nitrate of potash, but contains also small
quantities of chloride, iodide, and bromide of sodium, and other
soda-salts, and these are separated by washing. The mother-liquor,

from which the nitrate of potash has crystallised, is again evaporated, together with the liquors resulting from washing the salts, and the chloride of sodium and other soda-salts are separated and treated as before described. When the liquor is concentrated to the point before mentioned, nitrate of potash is again crystallised out, which also is treated as before, and the operation is repeated.

Mr. Stanford's Improvements in treating Sea-weeds.

The value of this gentleman's researches induces us to give fuller details than would have been allowable under other circumstances. The experiments he made when submitting these marine plants to destructive distillation in iron retorts, the products he obtained, and their bearing on the value of this branch of our national industry, are described in the following pages. The products were the same in all cases, and differed only in the proportions obtained from the various species. They were:

An inflammable gas;

Water, containing carbonate and acetate of ammonia, and a kind of naphtha;

A fluid tar containing a volatile oil, some of which also floated on the water.

A light charcoal remained in the retort; this was lixiviated with water, and treated as ordinary kelp; the solutions obtained were colourless, and the salts perfectly white and pure. The mother-liquor formed a striking contrast to those usually seen; it was nearly colourless, and contained mere traces of sulphur-compounds; the amount of iodine yielded was unusually large.

The products of distillation were then examined. The gas burnt exceedingly well, giving a flame of little luminosity at first, but becoming very luminous towards the end of the distillation. A portion collected at first, gave 41·66 per cent. of carbonic acid. Another portion, collected towards the end of the distillation, gave only 13·3 per cent. of carbonic acid, and contained olefiant gas. The gas given off in the distillation of wet weed is also inflammable, but the flame is occasionally extinguished by the vapour of water simultaneously generated; a portion of this gas gave 50·84 per cent. of carbonic acid.

The condensed liquid was very alkaline, and yielded ammonia in abundance, on distillation with lime. The ammonia appears to exist in the liquid as bicarbonate, as crystals of this salt were

found in some of the gas tubes. The chloride of ammonium in
the crude state, prepared by neutralising with hydrochloric acid,
has an odour of picoline, an alkaloid which is also found in the
volatile oil. The naphtha distilled from this liquid burns with a
slightly luminous flame. Rectified over lime, its specific gravity
is 0·826 ; this begins to boil at 155° ; it is not a definite compound, but appears to be a mixture of acetone and methylic
spirit ; by repeated rectification, the boiling point of the first
portion is reduced below 140° Fahr., but no definite separation of
it could be effected, either by treatment with chloride of calcium
or by repeated fractional distillation. The presence of acetone
considerably improves its value as a solvent of sandarac and
shellac. It has also, in the crude state, a peculiar odour, from
the presence of a small quantity of some ethereal body.

The residuum in the still, after the distillation of the ammonia
and naphtha, was found to be a solution of acetate of lime ;
distilled with hydrochloric acid, it furnished acetic acid, the
last portions of which also contained butyric acid.

When the tar was distilled with water, a volatile oil of very
peculiar odour was obtained, similar to that of burning seaweed.
In the crude state, it contains about 1·2 per cent. of the naphtha,
and about 1·6 per cent. of a body analogous to creosote. Dilute hydrochloric acid removes from it about 3·7 per cent. of a
mixture of volatile alkaloids, containing picoline, &c., and a red
pitchy colouring matter is separated. It also contains pyrrol.
After purification with oil of vitriol and caustic soda, the oil presents
the appearance and properties of a mixture of pure hydrocarbons
analogous to those from coal-tar naphtha. Its specific gravity is
0·841, and its boiling point rises from about 180° to above 400°
Fahr. This oil must not be confounded with that obtained by
Dr. Stenhouse, in distilling some species of fucus with sulphuric
acid, and named by him *fucusol*, and from which he prepared
fucusamide and fucusine. The oil here treated of is a pure
hydrocarbon, and has not been hitherto obtained from this
source. It may be applicable to many of the uses for which
coal-tar naphtha is so much in demand. The colouring matter,
separated by an acid from the oil, is a product of decomposition.
It is soluble in oil of vitriol, glacial acetic acid, alcohol, and
methylic spirit. It is insoluble in water, either cold or boiling,
and is separated from all its solutions by its addition ; it is also
insoluble in chloroform, ether, benzole, bisulphide of carbon, oil

of turpentine, the fixed oils, and dilute acids. It is unaffected by boiling hydrochloric acid. Nitric acid has no effect on it in the cold, but decomposes it when boiling, forming oxalic acid. Boiling oil of vitriol carbonises it. It is decolorised by chlorine. The caustic alkalies partially decolorise, but do not dissolve it; the colour is restored by an acid. It is, however, chiefly remarkable in containing iodine, which it holds in such close combination, that a boiling solution of potash does not remove it, and it is only by fusion with hydrate of potash, that the iodine can be detected. It is thus converted into oxalate of potash and iodide of potassium; it was found to contain 0·35 per cent. of iodine. The tar, after the volatile oil had been removed, presented the appearance of Stockholm tar, which it might well replace; it was then distilled alone, and a fixed oil thus separated; this contains on an average 5·2 per cent. of a crude acid, analogous to creosote. The oil, after this has been removed by potash, is purified with oil of vitriol, and it then resembles the paraffin oil of commerce.

In the following tables, the tar is that left after distilling in the presence of water. The volatile oil, having thus been removed, is stated separately, its weight being included in the aqueous portion. The paraffin oil is distilled from this tar; the paraffin in it was not estimated. The pitch left behind in the distillation, was not weighed. The naphtha, volatile oil, and paraffin oil were in all cases measured, as (in the crude state) impurities tell more on the weight than on the measure.

The products obtained from the residual charcoal are also shown in the tables, the alkalies, for convenience, being expressed as chlorides. The pure charcoal is the total amount of pure carbon yielded, and shows the value of each weed for heating purposes. The ash is pure, containing 20 per cent. of earthy phosphates; unadulterated kelp, as usually obtained, contains about double the quantity of ash there stated, the difference being unconsumed carbon; as, however, it contains the same amount of phosphates, it would be worth no more than the quantity here indicated.

Products obtained by the destructive distillation of Dry Sea-weeds. Centesimal.

Products.	Laminaria digitata.	Laminaria saccharina.	Fucus vesiculosus.	Fucus serratus.	Fucus nodosus.	Zostera marina.	Rhodomela pinastroides.
Charcoal, including ash......	43·750	47·040	38·040	45·450	35·166	43·272	56·610
Tar	1·861	2·123	3·835	2·474	3·116	0·805	1·078
Aqueous portion	24·509	27·151	31·988	24·978	24·434	25·341	29·931
Gas	29·880	23·686	26·137	27·098	37·284	30·582	12·381
Total...........	100·000	100·000	100·000	100·000	100·000	100·000	100·000

Products from the Water and Tar.

	Laminaria digitata.	Laminaria saccharina.	Fucus vesiculosus.	Fucus serratus.	Fucus nodosus.	Zostera marina.	Rhodomela pinastroides.
Volatile oil	0·707	0·708	0·989	0·825	not estimated	0·312	0·319
Paraffin oil	0·698	0·796	1·438	0·928	1·168	0·302	0·404
Naphtha	0·450	0·427	0·616	0·315	not estimated	0·355	0·480
Ammonia............	0·609	0·965	0·736	0·862	0·929	0·537	0·611
or as Sulphate of ammonia	2·362	3·713	2·794	3·348	3·606	2·086	2·373
Acetic acid	0·194	0·245	0·620	0·652	not estimated	0·181	0·272
or as Acetate of lime	0·259	0·326	0·816	0·469	...	0·241	0·363
Colouring matter	0·041	0·093	0·127	0·077	...	0·028	0·029
Volatile iodine..........	0·00724	0·007039	not estimated	0·003128	...	not estimated	not estimated.

Products from the Charcoal.

	Laminaria digitata.	Laminaria saccharina.	Fucus vesiculosus.	Fucus serratus.	Fucus nodosus.	Zostera marina.	Rhodomela pinastroides.
Pure charcoal	13·862	14·695	18·398	20·529	18·578	25·088	22·358
Chloride of potassium ...	10·831	14·104	3·216	6·940	4·008	6·203	9·693
Chloride of sodium.........	10·695	9·782	8·184	7·456	6·725	7·006	6·217
Insoluble ash	6·328	6·756	5·563	9·063	4·974	3·885	16·955
Iodine	0·4788	0·1582	0·00985	0·0382	0·0422	0·0457	0·0378.

Products obtained by the destructive distillation of Dry Sea-weeds, per ton.

Products.	Laminaria digitata.	Laminaria saccharina.	Fucus vesiculosus.	Fucus serratus.	Fucus nodosus.	Zostera marina.	Rhodomela pinastroides.
	Cwt. lbs.	Cwt. lbs.	Cwt. lbs.	Cwt. lbs.	Cwt. lbs.	Cwt. lbs.	Cwt. lbs.
Charcoal, including ash......	8 84	9 46	7 68	9 10	7 4	8 73	11 36
Tar	41·8 lbs.	47·5 lbs.	86 lbs.	55·4 lbs.	69·8 lbs.	18 lbs.	24·1 lbs.
Aqueous portion......	55 galls.	61 galls.	72 galls.	56 galls.	54 galls.	57 galls.	67 galls.
Gas (approximative)	1,205 c. ft.	956 c. ft.	1,052 c. ft.	1,089 c. ft.	1,504 c. ft.	1,234 c. ft.	500 c. ft.

Products from the Water and Tar.

Products.	Laminaria digitata.	Laminaria saccharina.	Fucus vesiculosus.	Fucus serratus.	Fucus nodosus.	Zostera marina.	Rhodomela pinastroides.
Volatile oil	253·4 oz.	253·7 oz.	354·5 oz.	295·7 oz.	not estimated	111·8 oz.	114·3 oz.
Paraffin oil	250·2 oz.	285·3 oz.	515·4 oz.	332·6 oz.	418·6 oz.	108·2 oz.	144·8 oz.
Naphtha	161·3 oz.	153 oz.	220·8 oz.	112·9 oz.	not estimated	127·2 oz.	172 oz.
Sulphate of ammonia......	529 lbs.	83·2 lbs.	62·6 lbs.	75 lbs.	80·8 lbs.	46·7 lbs.	53·2 lbs.
Acetate of lime	5·8 lbs.	7·3 lbs.	18·3 lbs.	10·5 lbs.	not estimated	5·4 lbs.	8·3 lbs.
Colouring matter	0·92 lb.	2 lbs.	2·8 lbs.	1·7 lbs.	...	0·63 lb.	0·65 lb.
Volatile iodine	0·1622 lb.	0·1576 lb.	not estimated	0·07007 lb.	...	not estimated	not estimated

Products from the Charcoal.

Products.	Laminaria digitata.	Laminaria saccharina.	Fucus vesiculosus.	Fucus serratus.	Fucus nodosus.	Zostera marina.	Rhodomela pinastroides.
	Cwt. lbs.	Cwt. lbs.	Cwt. lbs.	Cwt. lbs.	Cwt. lbs.	Cwt. lbs.	Cwt. lbs.
Pure charcoal	2 86	2 105	3 76	4 12	3 80	5 7	4 52
Chloride of potassium	2 19	2 92	0 72	1 43	0 90	1 27	1 105
Chloride of sodium......	2 15	1 107	1 71	1 55	1 38	1 44	1 27
Insoluble ash	1 30	1 39	1 13	1 91	0 111	0 87	3 43
Iodine	10·72	3·54	0·221	0·856	0·845	1·024	0·847

The following table exhibits the loss of the valuable products in the present mode of manufacturing kelp. Mr. Stanford states that the quantity of iodine is under, rather than overstated.

Products now obtained from a ton of Kelp, and those which would be added by the New Process of Manufacture.

NOW OBTAINED.

Products.	Laminaria digitata.		Laminaria saccharina.		Fucus vesiculosus.		Fucus serratus.		Fucus nodosus.		Average.	
	Cwt.	lbs.	Cwt.	lbs.	Cwt.	lbs.	Cwt.	lbs.	Cwt.	lbs.	Cwt.	lbs.
Chloride of potassium ...	6	56	8	20	2	72	4	96	4	46	5	35
Chloride of sodium	6	45	5	53	6	78	5	24	7	41	6	25
Insoluble ash	3	89	3	103	4	63	6	38	5	53	4	92
Iodine	0	12·77	...		...		...		...		0	5

ADDED BY THE NEW PROCESS.

Products.	Laminaria digitata.	Laminaria saccharina.	Fucus vesiculosus.	Fucus serratus.	Fucus nodosus.	Average.
Volatile oil	$4\frac{3}{4}$ galls.	$4\frac{1}{2}$ galls.	$9\frac{3}{5}$ galls.	$6\frac{1}{2}$ galls.	not estimated.	$6\frac{1}{4}$ galls.
Paraffin oil	$4\frac{3}{4}$ galls.	$5\frac{1}{3}$ galls.	$13\frac{1}{5}$ galls.	$7\frac{1}{3}$ galls.	$14\frac{1}{3}$ galls.	9 galls.
Naphtha	3 galls.	$2\frac{3}{4}$ galls.	$5\frac{2}{3}$ galls.	$2\frac{1}{2}$ galls.	not estimated.	$3\frac{1}{4}$ galls.
Sulphate of ammonia......	1 cwt. 46 lbs.	2 cwt. $17\frac{1}{4}$ lbs.	2 cwt. $32\frac{1}{2}$ lbs.	2 cwt. $38\frac{1}{2}$ lbs.	3 cwt. $108\frac{1}{2}$ lbs.	2 cwt. 48 lbs.
Acetate of lime	$17\frac{1}{2}$ lbs.	21 lbs.	75 lbs.	$36\frac{3}{4}$ lbs.	not estimated.	37 lbs.
Colouring matter	$2\frac{3}{4}$ lbs.	$5\frac{3}{4}$ lbs.	$11\frac{1}{2}$ lbs.	6 lbs.	...	$6\frac{1}{2}$ lbs.
Pure charcoal	8 cwt. $35\frac{1}{3}$ lbs.	8 cwt. 59 lbs.	15 cwt. 10 lbs.	14 cwt. 41 lbs.	20 cwt. 49 lbs.	13 cwt. 39 lbs.
Gas (approximative)	3615 cubic feet	2771 cubic feet	4313 cubic feet	3811 cubic feet	8272 cubic feet	4456 cubic feet
Iodine	19·39 lbs.	...	...	...	...	5 lbs.

The next table shows the comparison between the products of the destructive distillation of peat and sea-weed. The peat results are taken from the valuable report of Messrs. Kane and Sullivan, which have been fully detailed in our work on fuel and its applications, Part I., p. 368 to 382.

Products obtained in the destructive distillation of Dry Sea-weeds, compared with those at present obtained in the destructive distillation of Peat.

Products.	Average Yield of Sea-weed. Present Kelp-bearing Species, Nos. 1, 2, 3, 4, and 5.		Average Yield of Peat. By Kane and Sullivan.	
	Per ton.	Per 100 tons.	Per ton.	Per 100 tons.
Volatile oil	289·3 oz.	181 galls.	320 oz.	200 galls.
Paraffin oil	360·4 oz.	225 galls.	157·7 oz.	98·6 galls.
Naphtha	162 oz.	102 galls.	66·3 oz.	42 galls.
Sulphate of ammonia	70·9 lbs.	3 tons 3 cwt. 34 lbs.	22·4 lbs.	1 ton.
Acetate of lime	10·5 lbs.	9 cwt. 42 lbs.	6·8 lbs.	6 cwt. 8 lbs.
Pure charcoal	3 cwt. 49¼ lbs.	17 tons 4 cwt. 22 lbs.	4 cwt. 103 lbs.	24 tons 11 cwt. 108 lbs.
Gas (approximative)	1,161 cubic feet.	116,100 cubic feet.	1,475 cubic feet.	147,500 cubic feet.
Colouring matter	1·85 lbs.	1 cwt. 73 lbs.	—	—
Chloride of potassium	1 cwt. 63 lbs.	7 tons 16 cwt. 28 lbs.	—	—
Chloride of sodium	1 cwt. 79¼ lbs.	8 tons 10 cwt. 110 lbs.	valueless	valueless
Insoluble ash	1 cwt. 34¼ lbs.	6 tons 10 cwt. 90 lbs.	103 lbs.	4 tons 11 cwt. 108 lbs.
Iodine	3·26 lbs.	326 lbs.	—	—
Paraffin	not estimated.	not estimated.	3 lbs.	300 lbs.

Mr. Stanford then draws attention to the improvements he propose, which are as follows .—

1. Seaweeds of all kinds are to be stored in sheds ; they keep perfectly well under cover ; they should first be collected in heaps, to drain off the superficial water, and then laid out in drying sheds ; in summer, advantage may be taken of the sun's rays. Seaweeds, when laid out thin, are not so difficult to dry as is generally supposed, and when dry, keep perfectly well under cover.

2. The seaweeds, thus dried, are then to be compressed into cakes by hydraulic or other pressure ; this is not essential, but the cakes occupy less room in stowage, and if the charcoal obtained is to be used for fuel, this treatment improves it.

3. The cakes, or the unpressed weed, are then to be distilled at a low red heat, in iron retorts ; the tar and aqueous products to be collected in suitable condensers, and the gas in a gasometer. The gas may be employed for heating the stills used for rectifying the products, for heating the drying sheds, or even for lighting the factory ; it might even be treated according to Pettenkofer's method of superheating, and used as a means of lighting a district. The best kind of retorts for the purpose, are cylindrical vessels of wrought iron, heated externally, and placed vertically, having the base and the upper end conical, the former furnished with an air-tight iron damper plate, for withdrawing the charge when the carbonization is complete, and the latter provided with a movable iron hopper for introducing the charge.

4. The charcoal is to be lixiviated and treated as ordinary kelp, and then thrown out in heaps to dry in the air. When raked from the retorts, it should be allowed to fall into the lixiviating water, or into iron boxes, to protect it from the air ; if the latter plan is adopted, their heat may be rendered available for drying the weeds by wheeling them into the drying shed. The lixiviation will require larger tanks than those at present employed, on account of the greater bulk of the charcoal ; it has the advantage, however, of floating on water, and as the water in contact with it, when saturated, sinks to the bottom, it is quickly replaced by a fresh portion, and the solution is thus rapidly effected. The solution should be roughly evaporated to dryness, and the salt, thus obtained, sold to the potash and iodine manufacturers.

5. The washed and air-dried charcoal is to be used for heating the retorts and evaporating the solutions of the salts. Should, however, peat be very abundant in the neighbourhood, this charcoal may be manufactured into manure, by treating it with the ammoniacal liquid; or applied to some of the many uses of charcoal, and the peat employed as the fuel. The ash from the charcoal is a valuable manure; it usually contains over 20 per cent. of earthy phosphates. A few samples gave the following results :—

	Laminaria digitata.	Laminaria saccharina.	Highland kelp.	Zostera marina.
Phosphates of lime and magnesia . .	24·16	18·50	22·83	15·90
Carbonates of lime and magnesia . .	16·10	20·11	18·38	16·37
Sand, &c.	56·76	57·78	56·56	65·36
Charcoal	2·98	3·61	2·23	2·37
	100·00	100·00	100·00	100·00

The phosphates of magnesia predominate, and these are partially soluble in water. The proportion of phosphate is about equal to that in Peruvian guano, and if the crude ammoniacal salt, obtained by distillation, were added in the proportion of about 40 per cent., a manure would be obtained worth from £10 to £12 per ton, of which from 3 cwt. to 4 cwt. would be sufficient for an acre of land. The phosphate of magnesia it contains, points to its special application to beet-root and clover. Mixed with about 5 per cent. of the chlorides of potassium and sodium, it would be equally beneficial to other root and cereal crops.

6. The products of distillation ought to be treated as follows: —The tar is syphoned off, and distilled with an equal measure of water in an iron tar still; the light volatile oil passes over with the condensed water, on which it floats. This is decanted, and treated with dilute sulphuric acid, which removes picoline and other oily bases, and the red colouring matter is deposited. This substance is washed and dissolved in spirit, and the solution deposits it on evaporation. The light oil is then agitated with from 5 to 10 per cent. of oil of vitriol, washed with water and caustic soda, and finally re-distilled. The residual tar, from which the light oil has been removed, is then pumped into another iron still, and a stronger heat applied. The paraffin oil is thus obtained, and purified by oil of vitriol, caustic soda, and re-

distillation. The residual pitch may be employed for the manufacture of patent fuel, &c., or pumped while hot into brick ovens provided with an iron pipe to carry off the heavy vapours, and subjected to a red_ heat, by which a further portion of paraffin oil is obtained, and a good coke left in the still, commercially valuable to iron-founders on account of its freedom from sulphur. The liquid in the condensers, being separated from the tar, which sinks to the bottom, is mixed with excess of lime, and distilled in a capacious iron still, provided with a suitable condenser. Ammonia and naphtha pass over, and are received into hydrochloric acid. The solution of acetate of lime remaining in the still, is run out, evaporated to dryness, and the black impure acetate thus obtained, purified by charring, re-crystallisation, distillation, or conversion into acetate of soda. The ammoniacal distillate, which has been neutralised by hydrochloric acid, is re-distilled till the specific gravity of the distillate rises nearly to that of water. This is best distilled by the agency of steam. The first portion which comes over is the naphtha; this is collected separately, the weaker liquor subsequently distilled being returned to the still with the next charge. Redistillation over quicklime yields it in a state of purity. The solution of chloride of ammonium, remaining in the still, is run out, evaporated, crystallised and the crystals sublimed, according to the ordinary method of making sal-ammoniac of commerce.

This process offers the following advantages :—

1. Retention of the whole of the iodine.

2. Easy and rapid lixiviation, colourless solutions, and pure salts.

3. Absence of sulphur-compounds in the mother-liquor, great saving of oil of vitriol, and no evolution of poisonous gases.

4. Factory to a great extent self-supporting, having its own means of heat and light, the fuel being extracted from the weed itself.

5. Manufacture continuous, affording employment to the kelpers all the year round, and at a higher rate of remuneration.

6. Extension of the manufacture, as this process allows a much larger margin for profit, and admits of the lucrative working of the commonest weeds.

Statistics of the trade in the Potash-salts.

Taking the average annual production of kelp in these Islands to be 10,000 tons, and the average produce of salts per ton of kelp as follows :—

500 lbs. of chloride of potassium,
150 do. sulphate of potash,
300 do. kelp-salts ;

and valuing these salts at the following prices of the average of the last three years :

Chloride of potassium . . . £17 0s. 0d. per ton.
Granulated sulphate of potash . 8 0 0 „
Kelp-salts . . . . 1 5 0 „

The total annual value and production in round numbers will be, viz.—

2,230 tons chloride of potassium £37,910
670 tons sulphate of potash 5,360
1,340 tons kelp-salts 1,675
———
£44,945

MM. Tissier furnished Mr. Stanford with the following statement of the products annually extracted from varec, by the seven principal factories in France.

Factories.	Iodine and Iodide of potassium.	Bromine and Bromide of potassium.	Chloride of sodium.	Chloride of potassium.	Sulphate of potass.	Nitrate of potass.
	kilos.	kilos.	kilos.	kilos.	kilos.	kilos.
Le Conquet .	20,000	1,500	800,000	500,000	200,000	200,000
Granville . .	20,000	800	800,000	—	—	1,000,000
Cherbourg. .	5,000	—	200,000	180,000	80,000	—
Montsarac .	4,500	—	180,000	150,000	75,000	—
Pont-l'Abbé .	4,000	200	150,000	140,000	70,000	—
Portsall . .	4,000	—	150,000	140,000	65,000	—
Quatrevents .	2,500	—	100,000	90,000	50,000	—
	60,000	2,500	2,380,000	1,200,000	540,000	1,200,000

III. FROM SEA-WATER AND BRINE SPRINGS.

1. Sea-Water.

Potash, in some form or another, is a constant constituent of nearly all sea-waters and brine springs, and amounts, on an average, to 0·257 parts, as chloride of potassium and sulphate of potash, in 1000 parts of the latter.

Within the last 10 years, its extraction from the water of the Mediterranean has been systematically pursued in the South of France, where a long line of level shore, a tideless sea, a powerful summer's sun, and a sufficient degree of cold in winter, have all combined, as natural advantages, to lend success to the processes.

The observation of Green, that chloride of sodium and sulphate of magnesia undergo a mutual decomposition when in solution, followed up by the facts noticed by Berthier, that this decomposition not merely takes place, but is actually facilitated by a low temperature, led Balard to investigate the question more in detail, and he was rewarded by finding that the winter temperature of the South of France was low enough to render the decomposition and extraction of the salts possible as a commercial operation. We have already noticed, at page 183, how the reverse decomposition of these salts, at a different temperature, is employed to remove the chloride of magnesium from the brine-springs of Rodenberg.

The mean composition of sea-water is as follows :—

Water		96·470
Chloride of sodium	2·700	
Chloride of potassium	0·020	
Chloride of magnesium	0·360	
Sulphate of lime	0·140	
Sulphate of magnesia	0·240	3·530
Sulphate of potash	0·005	
Carbonate of magnesia . . . . . . } 0·004		
Carbonate of lime . . . . }		
Bromides, iodides, and organic matters .	0·011	
Carbonic acid, essential oils	0·050	
		100·000

But M. Usiglio has proved, that the water of the Mediterranean contains a smaller proportion of magnesian salts than is found in general in the waters of the ocean.

The following table exhibits the results of his analyses of the water of this sea, according to weight and measure.

Constituents.	Weight obtained from 100 grammes of sea-water.	Weight obtained per litre of sea-water.
Oxide of iron	0·0003	0·003
Carbonate of lime	0·0114	0·117
Sulphate of lime	0·1357	1·392
Sulphate of magnesia	0·2477	2·541
Chloride of magnesium	0·3219	3·302
Chloride of potassium............	0·0505	0·518
Bromide of sodium	0·0556	0·570
Chloride of sodium	2·9424	30·182
Total fixed constituents	3·7655	38·625
Water	96·2345	987·175
Weight in grammes...............	100·0000	1025·800

The processes of obtaining salt from sea-water, and especially that followed near Marseilles, have been described in the chapter on common salt (page 157). Mons. Usiglio has submitted the water of the Mediterranean, in reference to this question, to a careful investigation, and examined the products obtained at different stages during the evaporation of the water. His results are shown in the following table :

Density of the solution in degrees of Beaumé.	Volume of the mothers each time, Litre.	Salt deposited, in grammes.						
		Oxide of iron.	Carbonate of lime.	Hydrated sulphate of lime.	Chloride of sodium.	Sulphate of magnesia.	Chloride of magnesium.	Bromide of sodium.
3·50	1·000	—	—	—	—	—	—	—
7·10	0·533	0·003	0·064	—	—	—	—	—
11·50	0·316	—	trace	—	—	—	—	—
14·00	0·245	—	trace	—	—	—	—	—
16·75	0·190	—	0·053	0·560	—	—	—	—
20·60	0·144	—	—	0·562	—	—	—	—
22·00	0·131	—	—	0·184	—	—	—	—
25·00	0·112	—	—	0·160	—	—	—	—
26·25	0·095	—	—	0·051	3·261	0·004	0·008	—
27·00	0·064	—	—	0·148	9·650	0·013	0·036	—
28·50	0·039	—	—	0·070	7·896	0·026	0·043	0·073
30·20	0·030	—	—	0·014	2·624	0·017	0·015	0·036
32·40	0·023	—	—	—	2·272	0·025	0·024	0·052
35·00	0·016	—	—	—	1·404	0·538	0·027	0·062
Totals in grms.		0·003	0·117	1·749	27·107	0·623	0·153	0·223

The composition of the saline matters in the mother-liquors at 25°, 30°, and 35° Beaumé, was as follows :

Constituents.	25° Beaumé = 1·1968 sp. gr.	30° Beaumé = 1·2459 sp. gr.	35° Beaumé = 1·2992 sp. gr.
Sulphate of lime................	2·07	—	—
Sulphate of magnesia.........	22·64	78·76	114·48
Chloride of magnesium	29·55	101·60	195·31
Chloride of potassium	4·90	18·32	32·96
Bromide of sodium............	5·23	14·72	20·39
Chloride of sodium............	268·90	212·80	159·79
Totals	333·29	426·20	522·93

When the density of the sea-water has risen to 35° B. the separation of the salts depends, no longer on the mere evaporation of the water, but on the alternation of temperature during the day and night. Thus, salts which crystallise out during the night are partially redissolved during the day, and the normal state of the fluid varies with the temperature. Usiglio pursued his researches, under these circumstances, by decanting the mothers from each crop of crystals when formed.

The sea-water at 35° B. was exposed in shallow vessels to the air, and the following successive crystallisations obtained. During the *first night*, a crop of nearly pure crystals of sulphate of magnesia was formed. The *first mothers*, by evaporation during the next day, gave a mixture of common salt, Epsom salts, and occasionally some chloride of potassium. The *second mothers*, exposed during the *second night*, again yielded a crop of sulphate of magnesia, tolerably pure. The *third mothers*, having a specific gravity of 1·308 to 1·320, during the evaporation on the following day, gave a crop of very mixed salts, composed of common salt, Epsom salts, chloride and bromide of magnesium, and a double sulphate of magnesia and potash, $MgO.KO.2SO^3 + 6HO$. The *fourth mothers*, during the *third night*, left another double salt, $2MgK.3Cl + 12HO$, which was often mixed with the previous double salt. The *fifth mothers*, during the forenoon of the day following, had a specific gravity of 1·359, but yielded no crystals. At noon, the double chloride appeared, as also during the night; and when the temperature fell, either rapidly or very low, some Epsom salts were mixed with the double chloride. The *sixth mothers*, with a specific gravity of 1·372, deposited the most of their salts, but still retained some common salt and Epsoms with a large quantity of chloride of magnesium; the latter salt crystallised freely in autumn when the temperature fell to 40° F.

Guided by similar results in his own researches, Mons. Balard was led to the conclusion that the double decomposition of these salts of sea-water might be facilitated by attending to these two principles, viz. :

1st. That a mixture of two salts with different bases and acids, increases the solubility of each salt.

2nd. That a mixture of two salts of the same base or of the same acid, diminishes the solubility of each salt.

Manufacturing Operations.

The extraction of the salt is continued during the summer in the manner already described at p. 158; and as the mother-liquors become more dense, the common salt obtained is more and more impure, as the following table shows: the weight of the salt is given in grammes :—

Strength, Beaumé.	10 litres, after each deposit, become.	Sulphate of lime.	Chloride of sodium.	Sulphate of magnesia.	Chloride of magnesium.	Bromide of sodium.
26·25	0·950	0·508	32·614	0·040	0·078	—
27·00	0·640	1·476	96·500	0·130	0·356	—
30·02	0·302	1·444	26·240	0·174	0·150	0·358
32·04	0·230	—	22·720	0·254	—	0·518

These mothers are concentrated up to 35° B., during which two crops of salts are obtained, consisting of Epsom salts and common salt, in about equal proportions.

Separation of the Sulphate of Potash and Chloride of Magnesium.

The mothers from these crystallisations are then passed through two series of shallow ponds, made tight by clay mixed with hair or some fibrous material, and well beaten down. A crop of common salt is obtained during the day, and during the night, a mixture of sulphate of magnesia with a double salt of sulphate of magnesia and potash. The mothers from these salts are run into other ponds, where another crop of crystals is obtained, consisting chiefly of a double chloride of potassium and magnesium,

$$2MgCl.KCl + 10HO.$$

The mothers from these salts stand at 39° B., and at one time were thrown into the sea, but are now boiled up to 44° B. in iron pans, and become solid on cooling. This mass of crystals of chloride of magnesium is then exposed to a red heat in clay retorts, through which a current of superheated steam at 300° C. is passed, and the hydrochloric acid condensed by the French plan. The magnesia left in the retorts, is worked up with clay into very refractory fire-bricks, and might be employed in building the hearths of blast-furnaces, the basic property of magnesia eminently fitting it to resist the action of the silica in the slags and metal.

The mixed salt of sulphate of magnesia and double sulphate of magnesia and potash, is redissolved in pure water and evaporated, when the double salt, $KO.SO^3 + MgO.SO^3 + 6HO$, is obtained, and the mother, containing only the Epsom salts, is stored up for the operations in winter.

This double salt of potash and magnesia is used in the manufacture of alum, and for the production of carbonate of potash by the same process as is used for converting sulphate into carbonate of soda.

The Summer Operations.

The crop of mixed salts, sulphate of magnesia, and common salt, is redissolved in pure water and stored away in deep ponds : this solution is called *Eau pur sang*.

The mothers standing at 29° to 31° B., and which have not been worked up, are also preserved in large deep store-ponds.

In a third set of store-ponds or reservoirs, which, like the others, are very deep, another set of mothers, standing at 25° B., is also preserved. These mothers are called *Eaux vierges*.

Thus the operations in summer, besides yielding the common salt, produce three classes of mother-liquors.

1. The *Eau pur sang*, rich in sulphate of magnesia, but free from chloride of magnesium.

2. The *Eaux mères*, saturated with sulphate of magnesia and common salt, but containing other salts.

3. The *Eaux vierges*, saturated with common salt and nearly free from other salts.

The Winter Operations

consist in taking advantage of the cold weather, to effect the decomposition of the sulphate of magnesia by the common salt.

The three kinds of mothers are run into shallow ponds, in such proportions that these two salts shall be present with an excess of half an equivalent of common salt, in order to render the sulphate of soda which forms, less soluble in the mother-liquor.

The decomposition is complete in the one or two following nights, the sulphate of soda crystallises out, and the mothers must be immediately run off, to avoid a re-solution of part of the sulphate of soda during the higher temperature of the succeeding days.

Balard's Direct Process.

The mothers standing at 32° to 35° B. yield a crop of Epsom salts and common salt,—No. 1; and the mothers from this mixed salt are kept for the cold weather. They are then exposed, in shallow ponds, to a temperature of 2° or 3° C., when a large crop of sulphate of magnesia is obtained,—No. 2.

The mothers from this salt, are preserved in deep covered reservoirs, for treatment in summer, when they are exposed in shallow ponds and yield a double chloride of potassium and magnesium,—No. 3.

This double salt is dissolved in water heated by steam, and on cooling, deposits the chloride of potassium,—No. 4. The mothers from this salt, mixed with these from No. 3, are then boiled up to 44° B., and on cooling yield a mass of chloride of magnesium, —No. 5.

The salts No. 1 and No. 2 are dissolved at a temperature of 35° C. (95° F.) in water containing common salt, in the proportion of $1\frac{1}{2}$ equivalent of the latter for each equivalent of sulphate of magnesia in the mixture of salts No. 1 and No. 2. The temperature being reduced to zero, a double decomposition ensues, and three-fourths of the sulphate of magnesia is converted into sulphate of soda, which is obtained in the form of crystals.

The low temperature is produced artificially, by an improvement introduced by Mons. Merle. He originally employed a refrigerating machine worked by the evaporation of ether; but this is now replaced by the highly ingenious apparatus of M. Carré, in which cold is produced by the rapid volatilisation of liquefied ammonia-gas.[*] The salt-gardens or lakes in La Camarque, of upwards of 6,000 acres, were worked on this system last June, and are to be extended to 20,000 acres.

* See Jury Reports of the International Exhibition of 1862.

Produce of a Salt Lake.

The salt works of Baynas, managed on the system first described, and which cover a surface of 150 hectares, or about 370 acres, yield as follows :—

200,000 cwts. of common salt.

1,200,000 cubic feet of mother-liquors at 31° B.

20,000 cwts. of summer salts . . $\left\{ \begin{array}{l} \text{11,000 cwts. of crystallised sulphate of magnesia.} \\ \text{9,000 cwts. of double salts of} \left\{ \begin{array}{l} \text{potash.} \\ \text{soda.} \\ \text{magnesia.} \end{array} \right. \end{array} \right.$

1,400,000 cubic feet of . . . $\left\{ \begin{array}{l} \text{Eaux mères} \\ \text{Eaux pur sang} \\ \text{Eaux vierges} \end{array} \right\}$ = 90,000 cwts. of sulphate of soda.

This manufacture is rapidly extending in France, and several cargoes of the impure Epsom salts have already arrived in the Tyne for the purpose of being refined. With our extended sea-coast and, on many points, almost valueless adjoining land, and cheap fuel, these processes might be introduced into this country with advantage.

2. *Brine Springs.*

At pages 163 and 185, some analyses are given of brine springs and of the mother-liquors which are obtained after the extraction of the salt ; but up to this time, we believe, no attempts have been made to recover the potash they contain. M. Margueritte has proposed a plan for this purpose, which is also applicable to the mothers produced by treating sea-water.

His process is thus described in his specification. A solution of common salt is made, into which a current of hydrochloric acid gas is admitted, each bubble of which occasions an agitation as the salt precipitates, and this agitation continues until the liquid is saturated with the acid, and has lost all its salt. The salt is then drained and spread on the sole of an oven or stove-room, suitably heated, where the adhering acid is driven off.

Chloride of potassium is precipitated by this acid as easily as common salt, so that it may be separated from the mother-liquors of salt-marshes, and from the chloride of magnesium and the sulphate of magnesia, which are not precipitated by this acid. Hence, instead of crystallising the double chloride of potassium and magnesium, which must be dissolved again in order to separate the chloride of potassium, a current of hydrochloric acid gas will precipitate the potash salt.

The sulphates of soda and of potash, and generally all the soluble salts of these two bases, are partly converted into, and precipitated as, chlorides of sodium and potassium. When hydrochloric acid gas is passed through the concentrated waters of the Mediterranean Sea, the whole of the soda and potash is precipitated in the state of chlorides. When a solution of the double sulphate of potash and magnesia is submitted to the action of hydrochloric acid, the potash is thrown down in the form of chloride of potassium.

The passage of the hydrochloric acid gas through the liquid, may give rise to some obstructions, which are easily removed by means of a rod, so adjusted in the tube as to be capable of moving from above, as shown in Fig. 276. But instead of the muddling process, the hydrochloric acid gas may be brought up to the surface of the liquid by means of two plates hung in the same manner as common scales, which being alternately immersed, keep the liquid in a state of continual agitation, as represented in Fig. 276, in which A shows the entrance of the gas; B the entrance and level of salt water; C C the plates or scales; D, E, D, the beam; F the fulcrum; K K the stuffing box for the rods; N the level of the salt deposited, and also of the efflux of the liquid acid.

FIG. 276.

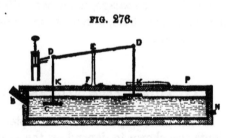

Hydrochloric acid gas is admitted through the tube A; the plates C C are moved by any suitable power, and being immersed alternately in the liquid, keep the surface in a state of constant agitation. The absorption of the acid by the salt water, causes the salt to precipitate, and the operation is stopped when the liquid is saturated with acid. The salt having deposited, the lid P is lifted, and the salt collected by means of a shovel.

Payen has also shown that a solution of chloride of magnesium, standing at 44° B., precipitates the alkaline chlorides from mother-liquors of 25° to 26° B.

Lillo & Co. have taken advantage of Green's discovery of the action of lime on salts of magnesia, to obtain the chlorides of

35*

sodium and potassium in a pure state, from the mother-liquors of salt-works. A milk of lime is prepared with a strong solution of salt, and added to the mother-liquor in sufficient quantity to decompose the sulphate of magnesia. The whole is heated up to the boiling point, and then run into vats, to allow the sediment to fall to the bottom. The common salt is obtained from the clear liquor in the usual salting pans, and when the chloride of potassium begins to appear, the liquor is run into coolers. The common salt in suspension first subsides, and a small quantity of chloride of sodium and potassium crystallises out as the liquor cools. When the temperature falls to 140° to 160° F., the liquor is run off into another set of coolers, where the chloride of potassium crystallises. The mother-liquors from this last crop of crystals and the first crop of the two salts, are then mixed with fresh mothers, and submitted, as before, to the *soccage* process as pursued in salt works. The salts are again separated, and the process repeated, until the residual liquors contain too much chloride of magnesium for further treatment.

The chloride of potassium is placed in baskets to allow the chloride of magnesium to drain away, and the common salt obtained in the soccage, is washed with a warm saturated solution of chloride of sodium, to remove the magnesian salt which adheres to the crystals.

IV. Potash from Felspar, and other Minerals.

In the former part of this chapter, the origin of potash in plants was shown to depend on the existence of this substance in the mineral matter forming part of the soil. The chief source of this potash is in the granite, gneiss, and other felspathic rocks which form so large a portion of the framework of the globe. It is true, there are some minerals which may, with more propriety, be termed potash-minerals—such as polyhalite, picromerite, arcanite, and carnallite,—but which, with the exception of the latter, occur in such small quantities as scarcely to deserve notice in a work devoted to manufactures. In the deposit of salt at Stassfurth, this mineral occurs more or less mixed with other salts. The upper layer of this deposit, nearly 200 feet thick, contains what is termed in Germany *Abraumsalz*, or dung-salts, a sample of which, analysed by Peters, was found to contain—

Chloride of potassium . . 19·16
Chloride of sodium . . . 32·84
Chloride of magnesium . . 17·08
Sulphate of magnesia . . 15·09
Sulphate of lime . . . . 2·04
Sand . . . . 2·71
Water, &c. . . . . . . 11·08

100·00

An examination of this deposit and salt, was made by a friend of one of the authors, who came to the conclusion that its peculiar composition, the high price and restrictions imposed by the Prussian Government, its distance from a place of shipment, &c., do not, at present, render it probable that it can become a commercial source of potash. Deposits of these minerals may, however, be hereafter discovered, and the following table exhibits their composition from different localities, as determined by various chemists.

Polyhalite.

	Ischl. Strome-yer.	Aussee. Rammels-berg.	Gmund. Joy.	Hallstatt. V. Hauer.	Ebensee. V. Hauer.	Vic. Jenzsch.
Sulphate of potash	27·70	28·10	28·11	14·81	19·12	25·87
Sulphate of soda	—	—	0·75	—	—	1·69
Sulphate of lime	44·74	45·43	42·78	56·41	61·18	44·11
Sulphate of magnesia ...	20·03	20·59	19·05	11·04	13·53	19·78
Sulphate of iron	—	—	0·36	—	—	—
Chloride of sodium	0·19	0·11	1·75	12·16	0·23	0·24
Peroxide of iron	0·34	0·33	—	—	0·41	1·01
Silica, &c.	—	0·20	—	—	—	0·52
Water	5·95	5·24	6·41	5·58	6·05	6·16
	98·95	100·00	99·21	100·00	100·52	99·38

Arcanite.

	Vesuvius.
Sulphate of potash	71·4
Sulphate of soda	18·6
Chloride of sodium	4·6
Chlorides of ammonium, copper, and iron	5·4
	100·0

Picromerite.

	Vesuvius. Scacchi.
Potash	23·43
Magnesia	9·94
Sulphuric acid	39·78
Water........................	26·85
	100·00

Carnallite.

	STASSFURT. Oesten.
Chloride of potassium	24·27
Chloride of sodium	4·82
Chloride of calcium	2·82
Chloride of magnesium	30·98
Sulphate of lime	1·05
Oxide of iron	0·14
Water, &c.	35·92
	100·00

Kremersite.

	VESUVIUS. Kremers.
Potassium	12·07
Sodium	0·16
Ammonium	6·17
Iron	16·89
Chlorine	55·15
Water	9·56
	100·00

Rock-salt from Vesuvius.

	1822.	1850.	
	Laugier.	G. Bischof.	Scacchi.
Chloride of potassium	13·90	53·84	37·55
Chloride of sodium	83·10	46·16	62·45
Sulphate of soda	1·60	—	—
Sulphate of lime	0·07	—	—
	98·67	100·00	100·00

Bischoff and Nöggerath have also noticed a saline efflorescence in the Valleys of Brohl and Tonnestein, which contained potash, and of which they have published the following analysis—

Sulphate of potash	18·901
Carbonate of potash	43·872
Chloride of potassium	18·273
Carbonate of soda	20·612
	101·658

It also occurs in the form of potash-alum, as an efflorescence on argillaceous minerals, more particularly alum-shale. Whitby in Yorkshire, Hurlet and Campsie in Scotland, and Cape Sable in Maryland, are some of the localities where this efflorescence has been found. It has also been found at the volcanoes of the Lipari Islands and Sicily.

Another mineral containing this alkali, in considerable quantities, is known to mineralogists, under the name of *Alunite*, or *alum-stone*. It is found at Tolfa, near Civita Vecchia; at Musay and Bereghszasz in Hungary; on Milo and other islands in the Grecian Archipelago; at Pic de Saucy, Mont Dore, in the Auvergne, in France; Gleichenberg in Stiria; and at Zabrze in Upper Silesia.

The following analyses of this mineral have been made—

Constituents.	Tolfa.	Tuscany.	Hungary.	Mi o.	Mont Dore.	Gleichenberg.	Zabrze.
Alumina	39·65	40·00	26·00	30·00	31·80	19·06	34·53
Potash	10·02	13·80	7·30	9·40	5·80	3·97	10·45
Sulphuric acid	35·49	35·60	27·00	31·00	27·00	16·50	36·06
Peroxide of iron	—	—	4·00	—	1·44	1·13	—
Zinc, magnesia, &c.	—	—	—	—	—	1·40	—
Silica	—	—	26·50	19·00	28·40	50·71	—
Water	14·83	10·60	8·20	10·60	3·72	7·23	18·96
	99·99	100·00	99·00	100·00	98·16	100·00	100·00
Chemists	Cordier	Descotils.	Berthier.	Sauvage.	Cordier	Fridau.	Löwig.

Potash also exists in the green-sand deposits of New Jersey, U.S., and other places, and H. Wurtz has proposed to extract it by fusing this sand with chloride of calcium, when, by lixiviating the fused mass, he states, all the potash is obtained in the form of chloride of potassium.

It is, however, to the potash-felspars that chemists have long looked as a source of potash, if any cheap industrial process could be found for extracting the alkali from these minerals.

There are four kinds of felspar, in only one of which is potash found in any quantity. The potash-felspar is termed *Orthoclase*, and soda is found in *Albite*, and in *Oligoclase* or *Soda-spodumene*.

The following table shows the composition of these potash and soda felspars :—

Analyses of Potash-felspar.

	St. Gothard. Abich.	Siberia. Abich.	Freyberg. Kersten.	Norway. Gmelin.	Ceylon. Brouchiart & Malaguti.
Silica	65·69	65·32	65·52	65·18	64·00
Alumina	17·97	17·89	17·61	19·99	19·43
Potash	13·99	13·05	12·98	7·03	14·81
Soda	1·01	2·81	1·70	7·08	—
Oxide of iron	—	0·30	0·80	0·63	—
Oxide of manganese	—	0·19	—	—	—
Lime	1·34	0·10	0·94	0·48	0.42
Magnesia	—	0·09	—	—	0·20
	100·00	100·00	100·00	100·39	98·86

	Oligoclase.			Albite.	
	KIMITO.	SCHLESWIG HOLSTEIN.	ARRILGE.	ARENDAL.	ST. GOTHARD.
	Chodnew.	Wolff.	Laurent.	G. Rose.	Thaulow.
Silica...................	63·80	64·30	62·60	68·46	69·00
Alumina	21·31	22·34	24·60	18·30	19·43
Soda	12·04	9·01	8·90	9·12	11·47
Potash	1·98	—	—	—	—
Sesquioxide of iron	—	—	0·01	0·28	—
Lime...................	0·47	4·12	3·00	0·68	0·20
Magnesia	—	—	0·20	—	—
	99·60	99·77	99·31	97·84	100·10

Fuchs was the first chemist who attempted to solve the problem of extricating this alkali from the felspars and micas. He calcined them with lime, and then lixiviated the frit, when he obtained 19 per cent. from the former, and 15 per cent. from the latter. Since his researches were made, numerous processes have been published for the same purpose.

These plans aim at obtaining the potash in the form of—

1. Potash, or carbonate ⎰ as in the processes of Fuchs, Laurence,
   of potash. . . . . ⎱ Meyer, Ward, and Hack.
2. Sulphate of potash and ⎰ in the processes of Wurtz and Tilgh-
   chloride of potassium . ⎱ man
3. Potash-alum . . . ⎰ in the processes of Sprengel, Turner,
                       ⎱ and Muspratt.

1. As Potash or Carbonate of Potash.

Laurence heats the mineral to redness, throws it into water, and grinds the produce to fine powder. This powder is mixed with damp sawdust, then piled up in heaps with alternate layers of straw, which are watered from time to time with urine or other nitrogenous liquids. These heaps remain in this state for six months, during which the fermentation and the chemical action which follows, facilitate the subsequent treatment. The materials are mixed with a thick cream of lime, and made into bricks, which are calcined at a high temperature. The residue is lixiviated, and the silicate of lime which has been formed, liberating the potash, remains behind while the alkali dissolves.

Hack's process does not differ from Meyer's, of which the following account contains the details. It consists in making an intimate mixture of 100 parts of felspar, and from 139 to 188

·parts of lime (in the form of hydrate or chalk), which is exposed for some hours to a temperature between a bright red and white heat. The product ought not to contain any free lime nor carbonate of lime. It is ground to powder, and heated with water for two to four hours, under a pressure of eight atmospheres, when a caustic lye, free from lime, is obtained, which contains all the soda in the felspar, and a quantity of potash amounting to about 9 to 11 per cent. of the weight of the mineral. This lye is saturated with carbonic acid, and evaporated, during which alumina and silica are first precipitated, after which the carbonate of soda salts out, leaving the carbonate of potash in solution. The residual mineral mass can be used in the preparation of Portland cement.

Ward's plan consists in using fluorspar along with lime, and in adding a clay rich in alumina, when it is intended to employ the residuum of the lixiviation in the preparation of cement.

The materials are prepared in the same manner as in the preceding plans, and when the alkaline mineral is quartzose, a quantity of pure clay is added in such proportions that each equivalent of silica and alumina shall meet three equivalents of the earthy base. The proportion of fluoride of calcium is recommended to be about 7 or 8 per cent. of the mixture.

The mixture is calcined in a reverberatory furnace at a heat ranging from red to bright red, so as partially to fuse or *frit* the ingredients. This is accomplished in about 30 to 60 minutes, and the mass is lixiviated in the usual manner. .

The alkaline liquor is boiled down, and yields a crude hydrated alkali, containing about 25 per cent. of silica. The liquor, treated with carbonic acid, is freed from the silica, which precipitates, and furnishes carbonate of potash when boiled down.

The residue, insoluble in water, is simply calcined when hydraulic cement is wanted, but clay is added previous to the calcination, when a puzzolana is required.

2. *As Sulphate of Potash or Chloride of Potassium.*

Wurtz's treatment of green-sand has been briefly noticed, the principle of which had been previously adopted by Tilghman, in operating on felspars containing potash.

The proportions recommended by Tilghman are calculated for a mineral containing 16 per cent. of potash.

To prepare sulphate of potash, he uses a potash-felspar, lime, or its carbonate, and the sulphate of either lime, baryta, or strontia, and afterwards lixiviates the mixture with water. The materials are ground to fine powder and intimately mixed, in the proportion of two parts of felspar, one part of lime, or an equivalent quantity of carbonate of lime and one part of sulphate of lime. The mixture is thrown upon the hearth of a reverberatory furnace, and heated to bright redness for about eight hours, turning the charge from time to time, so that all parts may receive equal heat. Although the formation of sulphate of potash is most rapid at a high temperature, the heat is not raised so high as to cause the fusion of the mass, as the subsequent extraction of the salt by water is then rendered more difficult. The presence of a deoxidising atmosphere in the furnace being injurious to the formation of the sulphate of potash, a sufficient quantity of air is admitted above the fire to preserve the atmosphere of the furnace in an oxidising condition.

When the charge has remained at a high red heat from eight to ten hours, it is withdrawn from the furnace and repeatedly lixiviated with hot water, as some of the salt obstinately adheres to the sulphate of lime. The solution of sulphate of potash is then evaporated, the sulphate of lime which deposits during the evaporation being continually removed. When the sulphate of baryta or strontia is used in place of the sulphate of lime, the same process is pursued, but the extraction of the sulphate of potash by water is more easy, owing to the less solubility of the sulphates of baryta and strontia. When a cheap and abundant supply of sulphurous acid gas can be obtained, as from the roasting of sulphurous ores, the use of the sulphate of lime or other sulphate in the mixture may be dispensed with, by doubling the quantity of lime or of carbonate of lime before prescribed, and exposing the charge, while at a red heat in the furnace, to a current of sulphurous acid gas and air, with frequent stirring ; sulphate of lime is thus formed during the process, and the formation of sulphate of potash takes place as in the former case.

To obtain chloride of potassium, a potash-felspar and the chloride of either sodium, calcium, or iron, is heated at a temperature above the fusing point of the chloride employed. When common salt is used, it is mixed with an equal weight of finely ground felspar. The mixture is well dried and introduced into a horizontal iron cylinder, with an opening only at one end, which is closed with an iron door or cover, and luted tight. For the escape of any gas that might tend to burst the cylinder, a small hole is made through the upper part of this door, which is closed at pleasure by a loosely-fitting plug. The cylinder is covered on the outside with fire-brick, to protect the iron from the action of the fire. The cylinder and its contents are heated to bright redness for about six hours; the heat should be above the fusing point of the chloride employed, but it ought to be kept below the temperature at which the felspar would melt, as the charge would then be more difficult to remove from the cylinder. The cover is then taken off, and the charge raked out as quickly as possible into an iron pot, which is immediately covered, and kept closed until the mass is cool. The soluble salts are extracted by water, and the chloride of potassium obtained, either by boiling down the liquor to dryness, or by evaporation and crystallisation in the usual manner.

3. *As Potash-alum.*

Sprengel was the first chemist to propose extracting the potash from these minerals, in the form of alum. His researches were commenced in 1830, and the process he found best adapted for this purpose, consisted in mixing the finely-ground mineral with strong sulphuric acid to the consistency of a thin paste, which was left for two months. The mass was then lixiviated with a large quantity of water, and on evaporating the solution, alum was obtained in the usual crystalline form, so pure as not to need recrystallisation.

Turner patented his process in 1842, and employed sulphate of potash instead of sulphuric acid. The ground felspar is mixed with its own weight of sulphate of potash, and thrown on the upper part of the inclined bed of the reverberatory furnace (Fig. 277), which has previously been brought to a white heat. When a glass has been formed, which flows down the inclined bed of the

FIG. 277.

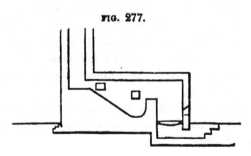

furnace, carbonate of potash, equal in weight to the sulphate used, is gradually added to the glass at the lower end of the furnace. This operation is repeated, until the *sack* of the furnace is filled with glass. The carbonate of potash must never be added until the sulphate is completely decomposed. When this glass is boiled with water, two-thirds of the silica in the felspar, in combination with part of the potash, is dissolved, and a porous mass is left behind, having the composition of elæolite. This porous mass is boiled with dilute sulphuric acid of 1·20 specific gravity in leaden pans, in the proportion of 160 lbs. of dry acid for every 285 lbs. of the felspar employed. This solution, when concentrated, yields crystals of alum, and the mothers are boiled down to dryness to render the silica it contained insoluble. The dry residue is heated with water to recover the alum, which is obtained by concentrating the solution and crystallisation.

The silicate of potash liquors from the elæolite, are either treated with carbonic acid to precipitate the silica, or filtered through a bed of caustic lime, which retains the silica. In the former case, carbonate of potash is obtained, and in the latter, a caustic potash lye.

This process was tried on a manufacturing scale on the Tyne, but abandoned on account of the loss of potash and excessive wear and tear, from the great heat necessary in the operation, as well as from the slow progress of the lixiviation of the glass, which would have required a large plant to have carried on an extensive trade.

Muspratt mentions that the potash may be extracted from felspar, in the form of alum, by the following process : 2 parts of felspar and 3 parts of fluorspar are mixed with sulphuric acid and exposed to a low heat, until the vapours of hydrofluosilicic acid cease to be evolved. The sulphuric acid first disengages the hydrofluoric acid, which then decomposes the silicates, forming hydrofluosilicic acid, which is evolved, as shown in the equation,

$$Al^2O^3.3SiO^2 + KO.SiO^2 + 12CaF + 4(HO.SO^3) =$$
$$Al^2O^3.3SO^3 + KO.SO^3 + 12CaO + 4(HF.SiF^2).$$

The sulphate of lime formed in the first stage of the process, is afterwards decomposed, and the sulphuric acid is transferred to the potash and alumina.

The mass is lixiviated with water, concentrated, and after the separation of some sulphate of lime, a crop of alum, in a nearly pure state, is obtained.

V. POTASH FROM WOOL.

Vauquelin has shown that the grease taken from the wool of sheep, is a complex mixture of different organic acids combined with potash. His analysis gives the following details,—

Potash-soaps, forming the principal portion.
Carbonate of potash, in small quantities.
Acetate of potash, in considerable quantities.
Chloride of potassium.
Lime in combination with an animal acid.
Animal matters.

MM. Maumené and Rogelet have established a manufactory at Rheims to recover the potash from this grease. Their process consists in placing the wool in casks, so arranged that the washing may be conducted in a methodical manner. The wool is pressed down in the casks as much as possible, and cold water poured over it. When cold water is used, no greasy particles escape with the brown-coloured solution which is formed, and all sand, &c. is retained by the wool, which acts like a filter. The solutions have a specific gravity of 1·10.

They are boiled down to dryness, and the so-called *suintate* of potash presents the appearance of baked molasses. It is broken into lumps and calcined in retorts, producing tar, ammonia water, and gas. The residue in the retorts is lixiviated, and the first liquors boiled up to 30° or as far as 50° B. The chloride of potassium and sulphate of potash crystallise out on cooling, and the mother-liquor is boiled down to dryness, when a carbonate of potash, *free from soda*, is obtained.

1,000 lbs. of raw wool generally yield from 140 to 180 lbs. of dry suintate of potash, which produces about half its weight of pure carbonate of potash, and 5 or 6 lbs. of sulphate and chlorides.

IODINE.

French, *Iode;* German, *Iod;* Italian and Spanish, *Iodio;* Arabic, *Iud.*

HISTORICAL NOTICES.

THE discovery of this substance is interesting, and illustrates what Herschell calls "Residuary Phenomena." A clever French chemical manufacturer, Courtois, was treating, in 1811, the potash-salts from kelp, with a view to convert them into nitrate of potash, and found his apparatus, copper pans, &c., so much corroded, as compelled him to seek the cause of so serious a loss. After incredible labour, he was rewarded by the discovery of this hitherto unknown substance. He introduced his discovery through Clement-Désormes, to the French Institute, and its investigation was immediately taken up by Gay-Lussac and Sir H. Davy, who confirmed all that Courtois had found. It is melancholy to think that Courtois, to whom society is under such obligations, died in poverty, and that his widow was a pensioner of the Pharmaceutical Society of Paris.

SOURCES OF IODINE.

This substance does not occur in nature in a free state, but in combination, generally with potassium. It is found in small quantities, but widely distributed through all the three great kingdoms of nature, mineral, vegetable, and animal.

It has been found in a cadmiferous zinc ore of Silesia, and in the white lead ores of Catorce, in Mexico. It also forms two mineral species with silver and mercury; the latter was discovered by Del Rio, and called *Coccinite;* the former, termed *Iodargyrite,* has been analysed and contains,

| | LOS ALGODONES, CHILI. | | | DELIRIO'S MINE. |
	Domeyko.	Damour.	Smith.	Field.
Iodine	53·75	54·03	52·93	54·02
Silver	46·25	45·72	46·52	45·98
	100 00	99·75	99·45	100·00

These analyses correspond with the formula AgI—

1 at. Iodine, 15·86 = 54·03
1 at. Silver, 13·50 = 45·97

100·00

The former occurs in veins of steatite at Abarradon, in Mexico, and at Delirio, in Chili. It has also been found at Guadalajara, in Spain. The latter has only been found at Cavas Viejas, in Mexico.

Rivier and Fellenberg have found iodine in the dolomite of Wallis, in Saxony, and 1,000 parts of the rock, boiled with water, yielded—

Iodine	1·560
Chlorine	0·015
Sulphuric acid	traces
Lime	0·738
Magnesia	0·272
Soda, with traces of potash	0·862

3·447

Lembert has also discovered iodine in the limestone near Lyons and Montpellier, and Genteles has detected it in the shale of Latorp, in Sweden.

Bussy found hydriodate of ammonia, mixed with sal-ammoniac, in the sublimate from the burning coal-mine of Commentry; and Mène has discovered iodide of ammonium in gas water, while the coke, left in the gas retorts, also contained iodine.

Fischer found iodine in the coals of Silesia, and Riegel has noticed it in the ashes of coal, peat, &c. H. Straub found it in the peat of Hofwyl and Klobach, and in the peat-ashes of Gifhorn, in Hanover.

Lamy, Fehling, Reuss, and Macadam have each found iodine in the potashes of commerce, and it exists in considerable quantity in the nitrate of soda of Peru. Reichardt has analysed the mother-liquors left in treating the *Caliche* or native salt, and found in 100 parts

Chloride of magnesium	1·121
Chloride of sodium	8·594
Sulphate of magnesia	2·214
Nitrate of soda	23·300
Iodate of soda ($NaO.IO_5$)	0·440
Loss of water at 212° F.	57·406
Water in chemical combination and loss of analysis . . . . }	6·925

100·000

Iodine also exists in combination with the alkali-metals, calcium, and magnesium, in numerous brine-springs, among which may be enumerated Sülze, in Mecklenburg; Kolberg, in Pomerania; Salzuffeln and Königsbronn, near Unna; Rehme, near Minden; Halle, in Saxony; Salliez, in France; Bolechow and Drochobyez, in Gallicia; Kenahwa, in North America; Guaca, in the province of Antiogena, in New Granada.

It also gives a medicinal value to the following mineral waters :

Name of Spring.	Country.	Chemist.
Aix	Savoy	Cantù
Ascoli	States of the Church	Egidi
Bath	England	?
Bonnington	Scotland	Turner
Busko	Poland	Heinrich
Castel Nuovo d'Asti	Piedmont	Cantù
Castellamare	Naples	Sementini
Challes	Savoy	Henry
Chatenois	France	Schaenfelle
Cheltenham	England	Daubeny
Colberg.	Prussia	Kruger
Cusset	France	Henry!
Eleusis	Greece	Landerer
Friedeischhall	Saxony	Riecker
Hall	Austria	Von Holger
Hassfurt	Bavaria	Von Bibra
Hauterive	France	Henry
Heilbrün	Bavaria	Vogel
Hombourg	Hesse	Liebig
Ischl	Austria	?
Iwonicz	Gallicia	Torosiewicz
Carlsbad	Bohemia	Berzelius
Kreuznach	Hesse Darmstadt	Osarnu
Leamington	England	Daubeny
Lipso	Greece	Landerer
Lippspring	Prussia	Witting
Luièche	Switzerland	Morin
Luhotschowitz	Moravia	Planiava
Néris	France	Henry
Popayan	South America	?
Pré-Saint-Dizier	Piedmont	Abbene
Pyrénées	France	Henry
Sales	Italy	Angelini
Salzschlurt	Germany	Fresenius and Will
Seidschutz	Bohemia	Berzelius
Tambangan	Java	Fresenius
Tarasp	Switzerland	Casselmann
Toeplitz	Bohemia	?
Unna	Westphalia	Liebig
Vichy	France	Henry
Wildegg	Switzerland	Buchner

Chatin has found iodine in the waters of various streams and rivers, and concludes from his examination, that the iodine increases with the proportion of iron present, and that the waters of volcanic districts are richer than those from the oolitic and limestone formations. According to this chemist and Fourcault, the drinking waters of those districts where scrofula and cretinisin prevail, contain no iodine.

Chatin also states that he has found iodine in rain-water in different localities, Paris, Pisa, Florence, &c., but none in the rain-water of mountainous districts. He also states that it exists in the air, at times, in some localities; but Grange, Lohmeyer, and Macadam failed to detect its presence, although Chatin alleges having found $\frac{1}{10}$ milligramme in 4,000 litres of the air of Paris.

Large quantities exist in the ocean, but Davy, Gaultier, Fyfe, and Sarphati could not detect it. Balard, however, found it in the water of the Mediterranean, and Pfaff in that of the Baltic.

It is, however, in marine and littoral plants that the most iodine is found. The following quantities of iodine have been found in 100 parts of the dry plants which have been examined by,—

	Sarphati.	Davy.	Balard.	Stanford.
Fucus Filum	0·0894	—	—	—
Laminaria digitata . . .	0·1350	—	—	0·4788
Do. saccharina . .	0·2300	—	—	0·1583
Fucus vesiculosus . . .	0·0010	—	—	0·00985
Do. nodosus	very little	—	—	0·0422
Do. saccatus	0·1240	—	—	—
Do. siliquosus	0·1420	—	—	—
Zostera marina	0·0005	—	same	0·0457
Rhodomela pinastroides .	—	—	—	0·0378
Uloa linza, Pavonia . .	—	0·059	—	—
Lactuca	0·0550	—	—	—

Gaultier and Fyfe, and the chemists already mentioned, have found iodine in a number of other marine plants. It has been stated that only those plants contain iodine, which are fed by the Gulf Stream, which is considered to be the carrier of the iodine from the Gulf of Mexico; but Mr. Stanford, who has devoted so much attention to this subject, says that although nearly all contain iodine, only a few species of algæ contain it in large quantities.

The ashes of those Fuci, &c., which yield the kelp of Scotland, and the varec of France, are rich in iodine; but, according to Davy and Fyfe, the ashes of the Salsolæ and other shore plants, from which the Spanish barilla and the Roman and Italian sodas are made, contain little or none.

Some land plants, tobacco, &c., a species of Salsola, which grows in the floating gardens on the fresh-water lakes near the City of Mexico, and, as we have seen, various peats, also contain iodine.

Various marine animals contain iodine ; the common sponge ; the horse-sponge ; various species of Sertularia and Tubularia ; various kinds of Rhizostoma and Cyana ; various species of oysters, Doris, and Venus ; and the oil from the liver of the cod (*Gadus Morrhua*) ; 100 parts of the light-brown oil contain, according to

$$\text{Gräger} \quad . \quad . \quad . \quad . \quad . \quad . \quad 0.0846 \text{ iodine.}$$
$$\text{Wackenroder} \quad . \quad . \quad . \quad . \quad 0.1620 \quad \text{do.}$$

And according to De Jongh, the three oils contain—

$$\text{Pale oil} \quad . \quad . \quad . \quad . \quad 0.03740 \text{ iodine per cent.}$$
$$\text{Pale brown oil} \quad . \quad . \quad . \quad 0.04060 \quad \text{do.} \quad \text{do.}$$
$$\text{Brown oil} \quad . \quad . \quad . \quad . \quad 0.02950 \quad \text{do.} \quad \text{do.}$$

Lastly, various articles of food contain iodine, such as salted Scotch herrings, wine and cider, &c.

The iodine of commerce is, however, almost entirely derived from kelp or varec, and Mr. Patterson states that the present yield of iodine from good drift-weed kelp, varies from 8 to 16 lbs., but that similar kelp from the islands of North and South Uist, and the County of Donegal do not produce more than 4 to 6 lbs. per kelp ton of 22½ cwts. Mr. Stanford's improved mode of preparing kelp increases the yield of iodine, one sample of *Laminaria digitata* having produced 32 lbs. of iodine per 20 cwt. of kelp.

The following plans are employed or have been proposed for the extraction of iodine from the various mother-liquors obtained by treating kelp and varec, as explained at pp. 523 and 527, from mineral waters ; and from the mothers of the caliche of Peru.

1. *From Kelp.*

The mother-liquors, which have a specific gravity of 1·33 to 1·38, contain, in addition to the iodides and bromides, a quantity of alkaline sulphides, sulphites and hyposulphites, which would be oxidised by the action of the iodine, unless they were previously decomposed. Sulphuric acid is therefore employed for this purpose, which precipitates some sulphur and disengages sulphurous acid and sulphuretted hydrogen. The quantity of acid necessary for this decomposition differs, of course, with the varying character of the kelp, and must be determined, each time, by adding a little to a test sample, until there is no effervescence. In general, about 14 or 15 per cent. of its bulk is sufficient ; but it is desirable to avoid using an excess, as it may, under the influence of the heat which is generated, cause a loss of iodine, either in the free state or as hydriodic acid. The sulphuric acid should not be stronger than 1·700.

The acid is added to the kelp-mothers in a wooden tub by means of a siphon, and the liquor is stirred while the acid is running in. A violent effervescence ensues, and the liquor is allowed to stand for a day or two, during which the sulphur collects on the surface. This sulphur is removed, washed, dried, and sold. About the same quantity of sulphur is obtained as of iodine. Some free chlorine would also appear to escape during the decomposition, as the mother-liquors in the vicinity of the escaping gases, are covered with a thin coating of a violet colour, which can only arise from some iodine being liberated by chlorine.

The saturated liquor is then run into the iodine retort until it is about three-quarters or four-fifths full.

The apparatus for disengaging the iodine, consists of a cast iron pan, A (Figs. 278, 279), seated over a fire-place, G, from which the flues, O, pass round the pan ; X X X, are three leaden pipes for conveying the iodine-vapours to the eighteen condensers, S.

FIG. 278.

FIG. 279.

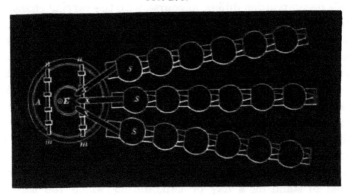

FIG. 280.

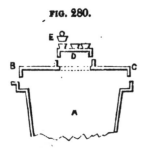

This pan is covered with a leaden top, shown in Fig. 280; A is the pan; B, C, the lead cover with a man-hole in the centre to admit of cleaning the pan; D the cover of the man-hole, with four openings, three of which are fitted with lead pipes; the manganese is charged through the fourth which is afterwards stopped by a lead plug: *m, n*, in the previous figures, represent two iron bars laid across the pan, having leaden straps passed round them and soldered to the lead cover, to prevent its sinking. All the joints of the retort and receivers are made tight with clay or cement.

In other works, the retort is made of a different shape, but built over a naked fire, F, in the same way as the pan. This form of retort is shown in Fig. 281, and consists of a metal cylinder,

FIG. 281.

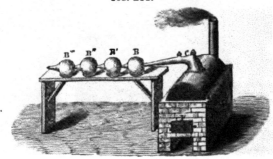

about 4 feet in diameter, the upper portion being usually made of lead, with two openings, which can be closed with lead stoppers. Two stone capitals, or heads, conduct the iodine-vapours to the condensers B, B; lead is also employed for this part of the apparatus, which is fitted with an opening, C, on the top of the neck, through which the progress of the operation can be watched.

The receivers or condensers vary in number, and sometimes only two rows are employed. These receivers are made of glass or stoneware; in the latter case, they are 2 feet in one direction by 1 to 1½ in the other, with necks at each end, about 4 inches in diameter. They are fixed only loosely together, and a damp cloth is placed on the opening of the last of the series. Each receiver is fitted with a small hole, in the under side, for the droppings of the aqueous vapour carried over during the distillation. These receivers are also sometimes made of two halves, as in Fig. 282.

FIG. 282.

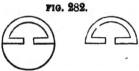

Another form of condenser, A, B, C, D, is represented in Fig. 283, which has many advantages over the previous system

FIG. 283.

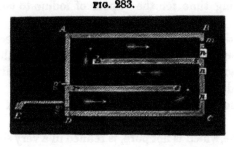

of carboy-receivers. It is made of stone flags with two shelves, r, s, and h, s, grooved into the sides, thus presenting a large condensing surface to the iodine vapours, passing in the direction of the arrows. E is the pipe from the retort; m an opening for the escape of the air; n, n, n, three openings for the removal of the iodine, which are closed, during the operation, with stone flags, luted with clay. The end, B C, of the condenser is raised a few inches higher than the other, A D, to allow any condensed water to run off through the openings g, g.

The liquor in the retort, is then raised as nearly as possible to the boiling point; finely ground manganese is introduced through the openings E (Fig. 280), or C (Fig. 281), in small quantities of 6 to 8 ounces at a time; and the stoppers are immediately replaced, luted with a little of the manganese. On each addition of the manganese, the iodine is disengaged and condensed in the receivers. The small holes C, and near C (Fig. 281), are opened before adding the manganese, to ascertain whether the evolution of the iodine, from the action of the previous charge, has ceased. The contents of the retort must also be frequently stirred through the opening E. When the violet vapours of iodine cease to appear on the addition of manganese, a little of the liquor is drawn up and tested in a glass tube with sulphuric acid, and then with manganese, and at last boiled, to ascertain the condition of the contents of the retort, and the operation is then completed according to the reactions of the test sample. The last portions of the hydriodic acid are difficult to decompose, and the liquor must be boiled towards the close of the operation.

The contents of the still are run off each night, to prevent any action on the metal, and to allow all the parts of the apparatus to cool. The advantage of this alternate working and cooling, consists in preventing the condensers becoming too hot, and thereby giving time for the vapour of iodine to condense in the receivers.

A distillation lasts fourteen days, the receivers becoming coated with a lining of iodine, from $\frac{1}{4}$ to $1\frac{1}{2}$ inches thick, which is sometimes so hard and strong as to offer some difficulties in its removal, and acting on the eyes of the men.

The iodine is packed in glass bottles or in oak casks holding 1 cwt.

This iodine, which is not pure, is refined in a very simple manner. It is placed in large stoneware bottles, fixed in a cistern full of water kept at the boiling point, and a leaden pipe conducts the iodine vapours from the bottle to condensers similar to those above described.

This process is generally pursued in Scotland; but Mr. Whytelaw, of Glasgow, modifies the process in adding only one measure of sulphuric acid carefully, and in small portions at a time, to 8 parts of the mother-liquors; carbonic acid and sulphuretted hydrogen escape at once, and after exposure to the air for a day or two, sulphurous acid is evolved, accompanied by a deposition of sulphur. The liquor is then run off from the sulphate of soda which has crystallised out, and conveyed into a leaden cylinder placed horizontally in a sand-bath, where the contents are gradually heated up to 212° F., but not higher, as at 244° F. chloride of iodine begins to distill over. Cyanide of iron also sometimes makes its appearance in the form of white needle-shaped crystals, in the last receiver. The presence of chlorine, iodine, and bromine in the nascent state, induces also the formation of a very volatile triple compound. A little iodine is also lost in the form of iodide of lead, which remains behind in the liquid in the still.

The chemical change in the still is represented by the following equation—

$$KI + MnO^2 + 2(SO^3.HO) = KO.SO^3 + MnO.SO^3 + HO + I;$$

or the chlorine may be first formed, which then displaces the iodine : $KI + Cl = KCl + I.$

2. From Varec.

The process employed by Courtois, consisted in distilling the mother-liquors obtained, as explained in the previous section, with sulphuric acid, when

$$KI + 2SO^3 = KO.SO^3 + SO^2 + I;$$

but the loss of iodine was very great, as the sulphurous acid acting on the iodine and water, produced sulphuric acid and hydriodic acid, which escaped.

This plan has been succeeded by a process introduced by Barruel, which consists in neutralising the mother-liquor, with the usual precautions, evaporating it down to dryness, and mixing the residue with 10 per cent. of its weight of peroxide of manganese. This mixture is then heated to low redness in an iron vessel, until a sample no longer evolves sulphuretted hydrogen, or deposits sulphur, when treated with sulphuric acid. The mass is dissolved in water, so as to form a solution standing at 36° B., when the subsequent operations are the same as those pursued in some French establishments, where this preliminary preparation of the mothers is dispensed with, except as regards their neutralisation with sulphuric acid.

In each case, the solutions containing the iodine are poured into large stoneware jars holding about 45 gallons. Each jar is made with three openings, one being larger than the other two.

FIG. 284.

A glass or lead tube from the opening on the right side, communicates with a chlorine still; the centre is fitted with a waste pipe, and is left open. The chlorine, in traversing the mother-liquors, liberates the iodine, and any excess escapes through the waste-pipe, but at the same time, exerts sufficient pressure on the surface of the liquid to cause it to rise and lute the third opening. The iodine separates in the form of a black powder, which gradually accumulates at the bottom of the jar. The uid is occasionally stirred with wooden rods, and the iodine is moved by wooden shovels or rakes. The above quantity of uid requires to be treated with chlorine for several days, before the iodine is liberated.

An excess of chlorine must be avoided, as it might form a latile chloride of iodine, and moreover precipitate the bromine soon as all the iodine has been separated. The condition of e liquor is easily ascertained in this way: a few drops of a ution of iodide of potassium are added to a test-sample, when line is immediately thrown down, if an excess of chlorine present; or a few drops of red nitric acid added to another test-mple, will equally precipitate iodine, if there is a deficiency of lorine.

When the operation is complete, the precipitated iodine is aced in stoneware funnels and washed with pure water, to re-ove all the adhering mother-liquor, then allowed to drain, and essed between folds of paper to dry it as completely as possible.

Sublimation of the Iodine.

This operation purifies the iodine, and transforms it into a ystalline mass. It is sometimes performed by placing the ude iodine on a plate heated from below by a water-bath, and vered by an inverted bell-glass; but the following is the plan ually adopted. The crude iodine is placed in large retorts, , C, arranged in a sand-bath, B, heated from below. These torts are connected with receivers, R, R, with small openings the side to allow the aqueous vapours to escape. The lower urt is fitted with a perforated false bottom to permit the con-nsed water to fall below. They are arranged to the number of 10 or 12 in a galley-furnace, which is sometimes heated by side flues. Six or seven sublimations are conducted successively, in order to obtain the iodine in larger crystals and as dry as possible. The retorts are charged each time with about 45 lbs. of crude iodine, and they must be entirely covered with the sand, to prevent any condensation of iodine in the necks, which, for this reason, are made as short as possible, and the receivers placed close alongside the sand-bath.

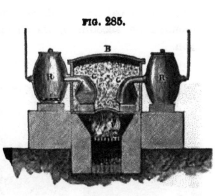

FIG. 285.

Theroulde's Process.

This chemist combines the manufacture of nitrate of potash with the extraction of the salts of kelp and varec, as described at page 528. After three crops of salts have been taken from the liquors, he proceeds to separate the iodine from the mothers by the following process. He manufactures oxalic acid by acting on some organic substance with nitric acid, and conveys the nitrous acid which is evolved, through his kelp mother-liquors. The iodine is precipitated, separated, washed, and sublimed, in the manner already described.

3. From Mineral Waters.

This process had its origin in the offer of the Academy of Fine Arts at Florence, to adjudge one of its prizes for a cheap and manufacturing process for extracting iodine from mineral waters. Bechi proposes to employ charcoal for extracting this substance from the spent artificial iodised baths and from the natural saline springs which contain very little of this substance. He found that iodine is so completely separated by charcoal that it may be estimated by the increased weight of the latter. The charcoal completely precipitates the iodine from its solutions on the surface of its particles, where it remains fixed in the same way that a piece of cloth retains its colour. The mere application of a strong heat will not expel the iodine, nor has chlorine any power, even when assisted by warmth. Cold and hot water and alcohol do not dissolve the smallest trace.

Animal charcoal acts most powerfully, but lamp-black is re-commended on account of its cheapness and its state of fine division.

FIG. 286.

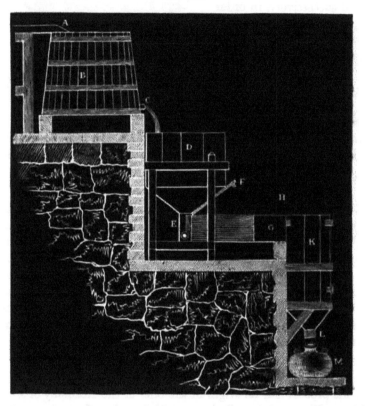

A represents the pipe which conveys the original water, and B the cistern, where the iodine-compound is decomposed by a mixture of 1 part sulphuric acid and 2 parts nitric acid; an excess of the acid is not found to interfere with the subsequent operations.

C is the pipe which conveys this acid liquid to a filter D, filled with charcoal, which rests upon a flannel or other material, to prevent the fine particles of charcoal passing off along with the water through the waste pipe E. F is a foot-board, by which the iodised charcoal is removed to a tank G, and thence, by a slide H, into another filter K, where it is mixed with the hydrated protoxide of iron. A sufficient quantity of water is allowed to filter slowly through this mixture, until all the soluble iodide of iron is taken up. This solution, which is kept as strong as possible, passes through some flannel at L, into a receiver M. The mixture of charcoal and oxide of iron is washed with weak muriatic acid, so as to render it useful for further operations.

Bechi prefers the use of the hydrated oxide to caustic alkali, as the iodide of iron is very soluble and easily separated from the charcoal, while the caustic alkali has a tendency to form an iodate of potash, which, being but slightly soluble, necessitates the use of a much larger quantity of water. The iodide of iron is also easily converted into an alkaline iodide by Baup's method. The iodide of iron may be obtained directly from the solution, by evaporating it in a tubulated retort, through which a stream of hydrogen gas is made to pass, to prevent the formation of any basic iodide of iron.

This iodide of iron can also be evaporated to dryness in an iron pan, and the residuum distilled with peroxide of manganese and sulphuric acid in the usual way to obtain the iodine ; or the method of Serullas may be employed.

4. From the Caliche of Peru.

An analysis of the mother-liquor of caliche is given at page 559. It will be observed, that there is more iodine in this liquor than in ordinary kelp-liquors, a fact indirectly confirmed by Krafft, who found, by recrystallising some commercial nitre, that it contained no less than 0·59 grammes of iodine per kilogramme.

Grüneberg has investigated this question, and having found that the mother-liquor from the caliche, contained part of the iodine in the form of iodide of magnesium, which was decomposed into magnesia and hydriodic acid during the evaporation, he proposes to avoid this loss by adding caustic soda to the mothers. He adds iron filings to the mothers, which are then gently warmed, while sulphate of copper is added, by degrees, until the solution indicates the absence of iodine and iodic acid. The presence of the former is ascertained by treating a test sample with starch and nitric acid, and the latter by starch and hydrochloric acid. All the iodine of the iodide of sodium combines with the copper, while the iodine of the iodic salts combines with the iron and copper, forming a mixed precipitate, in which some basic chloride of copper is also found. This precipitate is washed and dried at a gentle heat, and distilled with half its weight of peroxide of manganese, and as much sulphuric acid.

The receiver contains crystals of iodine and chloride of iodine. The latter is heated with carbonate of potash, when all the iodine is separated with the formation of chloride of potassium and chlorate of potash.

Jacquelin and Faure have patented the following plan for extracting the iodine from the mother-liquors of caliche and from commercial nitre. Their process depends on the action of sulphurous acid on compounds of iodic acid, as illustrated in the equations—

1. $5SO^2 + IO^4 = 5SO^3 + I.$
2. $5SO^2 + MgO.IO^4 = MgO.SO^3 + 4SO^3 + I.$

A solution of the nitre is made to stand at 36° or 37° Beaumé, of which 1 litre is taken and mixed with a solution of sulphurous acid, cautiously added and with constant stirring, until all the iodine is precipitated, when the quantity of sulphurous acid used is noted, from which data the manufacturing operations are managed.

Any convenient quantity of the solution of the nitre is run into a cistern built of stone or fire-bricks, and enclosed with a cover, the under side of which is coated with glass plates. The cistern is fitted with an agitator, having arms made of stoneware, and worked from above through the cover. This agitator is kept in motion while the solution of sulphurous acid is being mixed with the liquid. The iodine which precipitates is collected, washed, and sublimed by any of the plans already described.

When the iodine is present in the form of an iodide and an iodate, both chlorine and sulphurous acid are to be employed, the necessary quantities of these precipitating agents having been previously ascertained by test samples. If the iodate is present in the largest proportion, the aqueous chlorine is first employed and then the aqueous sulphurous acid, but when the iodide is the chief ingredient, this treatment is reversed.

Modifications of the Chemical Treatment.

Various plans have been proposed to obtain the iodine from these mother-liquors, and we are induced to notice some of them, as it is possible one process may be better suited to one locality than another.

Grüneberg's treatment of the mother-liquors of the caliche of Peru, is substantially the same as that proposed by Soubeiran.

This chemist employs dilute solutions, to which he adds sulphate of copper as long as any precipitate is formed,

$$2KI + 2(CuO.SO^3) = 2(KO.SO^3) + Cu^2I + I.$$

The liquid containing the free iodine is then decanted, and the washings of the diniodide of copper being added, they are mixed with sulphate of copper and iron filings, until the smell of iodine disappears.

$$I + 2(CuO.SO^3) + 2Fe = Cu^2I + 2(FeO.SO^3).$$

The precipitated diniodide of copper must be quickly separated from the liquid and iron filings, and washed before the sulphate of iron has time to oxidise. The two precipitates of diniodide of copper are dried at a gentle heat, as, at a higher temperature, the presence of some disulphate of peroxide of iron would cause a decomposition, and iodine would be disengaged and lost. The diniodide is then mixed with two or three times its weight of peroxide of manganese and as much sulphuric acid as will form the whole into a paste, and sublimed in an ordinary apparatus—

$$Cu^2I + 2MnO^2 + 4SO^3 = 2(CuO.SO^3) + 2(MnO.SO^3) + I ;$$

or, if the acid is omitted and a stronger heat applied,

$$Cu^2I + 3MnO^2 = 2CuO + Mn^3O^4 + I.$$

Berzelius and Liebig have proposed to replace the iron filings by the protosulphate of iron, and Schwarz by the anhydrous sesquichloride of iron—

$$Fe^2Cl^3 + KI = 2FeCl + KCl + I,$$

while Duflos dispenses with the iron and its salts altogether, and saturates the liquor with sulphuric acid.

Liebig modifies Soubeiran's process by treating the diniodide of copper with caustic potash, in order to manufacture the iodide of potassium, as will be explained at length in the following pages.

Soubeiran's principle is excellent, but its cost, compared with the other plans in use, and the bulky precipitate formed, which subsides very slowly, have prevented its introduction into the manufacture of this substance.

M. Tissier, the large French manufacturer, adds nitric acid direct to the mother-liquors, but Berzelius states that this plan causes a loss of iodine. The reaction is shown in the equation—

$$KI + 2NO^5 = KO.NO^5 + NO^4 + I.$$

Stefanelli and Doveri propose to heat the salts containing iodine, with sulphate of lime and free admission of air, or with sulphate of lime and peroxide of manganese in a close vessel.

Ubaldini has also found that the iodine is liberated when the iodide of potassium is heated with any of the following substances, in a current of air:—nitrate, sulphate, oxalate, carbonate or muriate of ammonia; sulphate, nitrate, phosphate, or borate of soda; chloride of sodium, potassium, or calcium; sulphate of potash or magnesia; nitrate of lime; boracic or silicic acid.

Wagner recommends the sesquichloride of iron, but dissolves the liberated iodine in bisulphide of carbon, which can be distilled off at a temperature of 50° C., leaving the iodine behind. Schwarz has drawn attention to the fact that the bromides are not decomposed by this compound of iron.

Schönbein has described the action of several substances capable of effecting the decomposition of the alkaline iodides, such as the metallic oxides and acids containing 2, 3, or 5 equivalents of oxygen, the most remarkable being the sesquioxide of iron, oxide of copper, arsenious and chromic acids. Of these the most desirable appears to be the chromic acid. When three parts of bichromate of potash are mixed with two of iodide of potassium, and distilled dry in a stoneware retort, the whole of the iodine is liberated, leaving behind a greenish mass of chromate of potash and sesquioxide of chromium. The decomposition which takes place is as follows,—

$$5(KO.2CrO^3) + 3KI = 8(KO.CrO^3) + Cr^2O^3 + 3I.$$

The iodine yielded by this process is very pure, and no loss in the materials is sustained, as the chromate and oxide left behind are both valuable materials.

Mr. Horsley, of Cheltenham, had discovered, as early as 1853, that iodine might be precipitated in the crystalline form, by using bichromate of potash and sulphuric acid, a process which has since been introduced, for estimating the quantity of iodine, by Penny and Bunsen.

Luchs has also recently given full details for the management of this process.

Properties of Iodine.

Iodine has a black-grey colour, with a metallic lustre, not unlike that of micaceous iron ore; transmits a red light in thin laminæ; crystallises usually in elongated octahedrons with a rhomboidal base; the finest crystals are obtained from its solution in ether. It is solid at ordinary temperatures, but melts at 107° C. (224·6° F.); forms a black-looking liquid at 175° C.

(347° F.) and boils at 180°C. (356° F.), and sublimes in the form
of a beautiful violet vapour, whence its name; it solidifies from
the liquid, or condenses from the state of vapour, in laminated
masses. It volatilises slowly, even at common temperatures,
emitting an odour of chlorine or of chloride of sulphur. It attacks
all the animal membranes and acts as a violent poison. It is very
slightly soluble in water, but more so when other salts are present;
soluble in alcohol, ether, and bisulphide of carbon. It has a weak
affinity for oxygen, but very great affinity for hydrogen; stains
the skin and paper with a brown colour, from the formation of
hydriodous acid; and is characterised by the blue colour it produces with starch-paste, which, according to Stromeyer, is so
delicate as to give a perceptible blue tinge, when not more than
$\frac{1}{470000}$th part of iodine is present in the liquid. Iodine forms
the heaviest of all known vapours, its specific gravity being 8·716.

BROMINE.

WE take up the discussion of this substance here, as many of the
statements in the following sections apply both to bromine and
to iodine.

Sources of Bromine.

Bromine has been found in small quantities in the zinc ore of
Silesia and in rock-salt from some few localities. It also exists in
combination with silver, sometimes as pure bromide (bromargy-
rite) containing 42·55 per cent. bromine and 57·45 silver, but
more frequently as chlorobromide, in the mines of Huelgoeth, and
in Mexico and Chili. Two mineral species from the latter country
have been found to contain—

	Megabromite CHILI.		Microbromite CHANARCILLO.		COPIAPO	
	Richter.	Field.	Field.	Field.	Plattner.	R. Muller.
Bromine	26·49	33·82	16 84	19·82	20·09	12·40
Chlorine	9·32	5·00	14·92	13·18	13·05	17·56
Silver	64·19	61·07	68·22	66·94	66·86	70·04
	100·00	99·89	99·98	99·94	100·00	100·00

The chlorobromide is also termed *Embolite.*

Bromine is found, accompanying iodine, in nearly all the salt springs and many of the mineral waters mentioned in the previous chapter. It is present in the waters of the Mediterranean, Gulf of Trieste, North Sea, Baltic, and Dead Sea. It has likewise been found in numerous marine plants and animals; but the quantity is always very small, and it is not extracted in most of the manufactories, even where the iodine is obtained.

Bromine was discovered by Balard in 1826, while engaged in his researches connected with the extraction of salt from the waters of the Mediterranean at Marseilles.

Extraction of Bromine.

From Bittern.—Those mother-liquors from which the iodine has been previously removed, are freed from as many of the crystallisable salts as can be separated by evaporation, and chlorine is passed through the liquid as long as the yellow colour increases in intensity. An excess of chlorine would combine with the bromine, and thus interfere with the subsequent treatment. The yellow liquor is poured into a large glass globe having a neck fitted with a stopcock. Ether is added, and the whole shaken, when the yellow is replaced by a hyacinth-red colour, from the absorption of the bromine, which was set free by the decomposition of the bromide by the chlorine. The stopcock is now opened and the aqueous solution cautiously poured out until the ethereal solution reaches the neck of the vessel, when the stopcock is closed. An aqueous solution of potash is added and the contents again shaken. The ether is decanted to serve for another operation, and the potash solution is evaporated to dryness. The residue, consisting of bromide of potassium and bromate of potash, is ignited to destroy the bromide of carbon, which is probably formed by the action of the bromine on the ether. The ignited mass is mixed with one-third of its weight of peroxide of manganese, and distilled with one part of oil of vitriol diluted with half its weight of water. The chemical change is represented by the equation—

$$NaBr + 2SO^3 + MnO^2 = NaO.SO^3 + MnO.SO^3 + Br.$$

The distillate is collected under water and redistilled with chloride of calcium, to render it anhydrous.

Another and simpler plan consists in heating the mother-liquors, prepared as before, with peroxide of manganese and hydrochloric acid,—or when a sufficient quantity of chlorides are present in solution, with the manganese and sulphuric acid. The bromine obtained is purified in the way already described.

From the Mothers of Brine-springs.—At Schönebeck, 240 parts of the mother-liquors are distilled in a glass retort with 3 parts of peroxide of manganese, and 4 parts of sulphuric acid diluted with half its weight of water, and the distillate is collected in a receiver containing a solution of potash. The liquid in the receiver is evaporated to dryness, and the bromine is obtained by distilling the residue with peroxide of manganese and sulphuric acid. Another and better plan consists in heating the mothers with sulphuric acid, to destroy the chlorides, and separating the sulphates by crystallisation, when the remaining liquid is distilled in the usual way, to obtain the bromine it contains.

The mothers from the Kreuznach springs are treated in a different manner. They are boiled down to one-third of their bulk in an iron pan, and after standing some days, the liquid is drawn off from the crop of salts which have crystallised. This liquid is diluted with water, and the lime present is precipitated with sulphuric acid. The clear liquid, separated from the gypsum by straining and pressure, is evaporated to dryness. The residue is treated with its own weight of water, separated from some insoluble gypsum, and distilled with peroxide of manganese and hydrochloric acid, with the usual precautions.

Le Chatelier, of Paris, has proposed to apply his aluminate of soda to the separation of the magnesian salts from the brine-waters of sea-water, salt lakes, and salt springs. He especially alludes to the mother-waters of M. Balard's process, which contain in 1,000 parts at 37° B.

Chloride of magnesium	319·65
Sulphate of magnesia	30·51
Chloride of sodium	4·16
Bromide of sodium	7·75
	362·07

This liquor is mixed with a solution of aluminate of soda, containing a quantity of soda equivalent to the acids combined with the magnesia, when the latter base is precipitated along with the alumina. The solution, by evaporation and salting, furnishes a liquor rich in bromine, which is then obtained by any of the processes already described.

The aluminate of magnesia is treated with a weak acid, whereby the alumina is recovered, and may be reconverted into aluminate of soda in the usual way.

From the Mothers of Varec.—These mothers, from which the iodine has been precipitated by chlorine, as described at p. 568, are distilled in glass retorts, connected, without any lute, with a tube-funnel and receiver, and again with a glass cylinder by means of a bent tube. The mothers are first concentrated, and a test-sample is treated with a few drops of chlorine-water, to ascertain that all the iodine has been precipitated in the previous operation ; 156 parts of the mother-liquor, 4 parts of peroxide of manganese, and 3 parts of oil of vitriol are then boiled until the red vapours of bromine cease to appear. Fresh portions of manganese and acid are added to the contents of the retort, if a test-sample has shown that any bromine has been left, and the distillation is renewed until all the bromine has been obtained.

The bromine condensed in the first receiver is driven over by a gentle heat into the glass cylinder, which is kept cool by a mixture of ice and salt.

Balard collects the bromine in a leaden receiver filled with pieces of iron, and extracts it from the bromide of iron which is formed.

In some establishments, the bromine is saturated with potash, evaporated to dryness, and calcined. This mass is treated with manganese and sulphuric acid in the usual way, and the bromine collected under sulphuric acid of 1·84 specific gravity.

In Theroulde's process, the mothers from his iodine operation, are mixed with peroxide of manganese, and chlorine is passed through the liquid, which is afterwards treated in the usual manner.

The last mothers are then boiled up to 1·60 specific gravity during which the chloride of sodium and other salts, which fall, are fished out, and the liquor is set aside to crystallise, when another crop of nitrate of potash is obtained.

Properties of Bromine.

Bromine is liquid at ordinary temperatures, with a very intense orange-red or brown colour by reflected light, and a hyacinth-red colour by transmitted light. It has an extremely offensive odour which adheres to substances that have been saturated with it, and a very sharp, burning, astringent and nauseous taste. Its vapour, mixed with a large quantity of air, may be taken into the lungs with impunity, but larger quantities produce oppression, increased secretion of the mucous membrane, and headache. It colours the skin yellow and then brown, producing violent itching, while the stains are only removable with the skin itself. It discharges vegetable colours like chlorine, and corrodes organic substances. It extinguishes flame, first however tingeing it red at the top and green below. It volatilises rapidly in the air, and boils at 47° C.; freezes at from −19° to −25° C. into a yellow-brown, brittle, laminated mass, covered with bluish-grey . spots. It is a disinfectant, and resembles iodine in its medicinal action.

MANUFACTURE OF THE IODIDE AND BROMIDE OF POTASSIUM.

These two compounds are the most important, and are employed in the preparation of most of the other combinations of iodine and bromine. Various plans have been proposed for preparing them, but the simplest is that in which caustic potash is employed.

Powdered iodine, or that which is precipitated by chlorine, after being washed with water, is dissolved in a lye of caustic potash of specific gravity 1·25, until the blood-red colour of the liquid gradually disappears or becomes only slightly yellow, when a few drops of a weak solution of potash are added, to render the liquid perfectly colourless. Two compounds of iodine are formed in this process—

$$6KO + I^4 = KO.IO^5 + 5KI;$$

and the solution containing them, is boiled down to dryness. The dry mass is introduced into a covered metal crucible and submitted to a gentle fusion. This operation requires care, for too great a heat might volatilise part of the iodide, and too little might leave some iodate undecomposed. Freundt, Orfila, and Scanlan have recommended the use of charcoal-powder to obviate both the above dangers. Freundt advises that 10 per cent. of the weight of the iodine should be employed. Some disadvantages attend the use of charcoal, for while it deoxidises the iodic acid, it has a tendency to prevent the union of the iodine with the potassium, and when the caustic lye contains sulphate of potash, this latter is reduced to sulphide, which combines with part of the iodine.

37*

The change which takes place in either case, results in the formation of iodide of potassium—

$$KO.IO^5 = KI + 6O \; ; \; \text{or,}$$
$$KO.IO^5 + 3C = KI + 3CO^2.$$

The fused mass is treated with water, the solution filtered, and
when concentrated up to 1·75, it is set aside to crystallise.

Alcohol may be substituted for water, and produces a purer salt,
as it leaves most of the impurities behind.

Some manufacturers employ an excess of caustic alkali, in order
to give the crystals a certain degree of opacity, which has become
a fashion in the trade. Others dip the crystals in a solution of
carbonate of soda, and afterwards dry them in a stove.

Another process, which has long been pursued in France, was
proposed by Baup of Vevey, and Caillot of Paris. 30 parts of
iron filings and 500 parts of water are put into an iron pan, and
gently heated, when 100 parts of iodine are cautiously added
in small quantities at a time, and when the dark colour of the
solution has disappeared, the whole is filtered. A solution of
carbonate of potash is added to the filtrate, and the precipitate
which forms is separated and washed. The solution is evaporated
to dryness, dissolved in water, evaporated up to strength, and set
aside to crystallise.

Liebig has modified this plan by adding only 20 parts of the
iodine to the iron, and dissolving 10 parts in the potash-solution,
by which he states that the decomposition is more rapidly and
completely effected, when the two solutions are mixed.

Pypers has proposed another, which appears to be more practical than any of the preceding. He gently heats a mixture of—

130 parts of iron,
100   „   iodine,
75   „   carbonate of potash,
144   „   water,

dries the mass thoroughly, treats it with water, filters, evaporates
to dryness, and then completes the process in the usual way.

Criquelion-Hurant replaces part of the iron by lime, which is
said to facilitate the washing of the bulky precipitate obtained
from the iron.

Dorvault has substituted sulphate for the carbonate of potash,
with great success. A quantity of sulphate of potash, equal in
weight to the iodine employed, is dissolved in water and added
to the iodide of iron, prepared as already explained. Quicklime

is added to the mixed solutions, and the whole heated by steam. Sulphate of lime and oxide of iron are precipitated. The operation is performed in a large metal pan, furnished with cocks at different levels, to run off the clear solution as the precipitate subsides. The washing of the precipitate is conducted in the same vessel and in the same way. The solution is evaporated to dryness, and the other operations are the same as before. This process is said to produce a purer salt than other plans.

FIG. 287.

Le Royer, Girault, and Dumas, propose to substitute zinc, and Serullas, antimony, for the iron in the previous processes.

Thevenot of Dijon employs a plan differing from any of the preceding. He uses the following materials :—

Sulphate of baryta .	500 parts.
Lamp-black. . . .	100 „
Iodine.	400 „
Sulphate of potash .	266 „
Water. . .	2,000 to 3,000 „

The two first-named substances are intimately mixed and exposed to a strong heat in a covered crucible for three hours. The mass is pounded and treated with the water. The iodine is added to the solution by degrees, and stirred each time. Any colour in the solution, from an excess of iodine, is removed by an addition of a little of the solution kept in reserve. The whole is filtered and washed; and the solution being gently heated, the sulphate of potash, in powder, is added by degrees until all the barytes is thrown down. The solution is separated by filtration, and treated in the same way as those obtained in the previous processes.

Dr. Squire gives the preference to a beautiful process recently proposed by Baron Liebig. It consists in covering 1 part of phosphorus in a basin, with 40 parts of hot water, to which 20 parts of iodine are gradually added with frequent agitation. A violent action ensues, and a large proportion of the phosphorus is converted into the amorphous variety, which acts, however, equally well in reducing the iodine. The colourless liquid contains phosphoric and hydriodic acids:

$$P + 5I + 5HO = PO^5 + 5HI.$$

It is poured off, and milk of lime added until the mixture becomes alkaline. The whole is filtered to remove the phosphate of lime, and the filtrate containing iodide of calcium is boiled down with 12 parts of sulphate of potash. After being reduced to half its bulk, the whole is allowed to cool, and filtered. A small quantity of pure carbonate of potash is added, to throw down any lime, and the filtered solution yields, on evaporation, crystals of pure iodide of potassium. Dr. Squire states that this salt, prepared by Liebig's plan, has a peculiar pinkish hue, which is completely removed by fusing the salt before crystallisation.

The crystals of iodide of potassium obtained, are beautifully white, in the form of cubes or octahedrons, soluble in three-fourths of their weight of water, in six parts of alcohol of 0·85 specific gravity, at ordinary temperatures. The salt melts at a temperature below redness, and forms pearly crystals on cooling.

It is often adulterated with carbonate, sulphate, and muriate of potash, which can be easily detected by the ordinary chemical re-agents. Iodate of potash is also sometimes present, from imperfect manufacture. This salt is detected by adding an acid, which liberates hydriodic acid, and this re-acts on the iodate, setting iodine free :

$$KO.IO^5 + 6HI = KI + 6HO + I.$$

The manufacture of the bromide of potassium is precisely similar, but requires greater expedition, to avoid the dangers arising from the vapour which is so rapidly evolved from the bromine.

Iodine and bromine are both employed in photography to a considerable extent. Iodine is also largely used in medicine for the treatment of goître, scrofula, &c. It is also employed in the production of the beautiful red colour, *iodide of mercury*.

STATISTICS OF THE TRADE IN IODINE AND BROMINE.

Mons. Tissier estimates the production of iodine and iodide of potassium in the seven principal manufactories of France at 60,000 kilogrammes, and Payen states that 9,000 kilogrammes of iodine are made. From the best data we can obtain, the annual produce in France may be assumed to be—

Iodide of potassium	54,000 kilogrammes.
Iodine	6,000 ,,
Bromine	700 ,,

M. Mercier estimates that 90 per cent. of the iodide of potassium is consumed in medicine, pharmacy, and for chemical purposes, and only 10 per cent. in photography.

We are not aware that bromine is manufactured in this country to any extent; but as regards the production of iodine, if we take the data furnished by Mr. Patterson, the quantity extracted will not fall far short of 100,000 lbs. The consumption of late years has been increased by the application of the iodides of potassium, ammonium, and cadmium in the art of photography.

ESTIMATION OF IODINE.

As in the case of chlorine, chemists have devised a number of different processes for determining the quantity of iodine in any given substance, chiefly, however, by *volumetric* methods, many of which might, with equal propriety be termed *colorimetric*. In attempting to classify these processes, any principle adopted in the arrangement must be, to some extent, arbitrary.

In the following classification, the volumetric processes are sub-divided into those in which the coloration depends on the use of starch, and those in which no starch is employed. In this way there are four classes—the gravimetric, the volumetric, and the colour-volumetric, with or without starch. Those processes which are best adapted for the manufacturer, or which are applicable in certain special cases, will be described.

1. *Gravimetric Processes.*

Wallace and others . . . by silver and ammonia.
Lassaigne, Thompson . . by palladium.

2. *Volumetric Process.*

Kersting . . . . . . by palladium.

3. *Colour-volumetric Processes—without Starch.*

Penny, Bunsen . . . . { by bichromate of potash and hydro-chloric acid.

Mayer . . . . { by sulphates of copper and iron and permanganate of potash

Luca, Dupré, Stein . . . { by tin, chlorine, or bromine, with bisulphide of carbon or chloroform.

Moride . . . . by nitric acid and benzole.

Rabourdin . . . . . . by potash, chloroform, and acids.

Berthet . . . . by iodate of soda and sulphuric acid.

Duflos, Schwarz . . . . { by sesquichloride of iron, with ar-senite of soda or permanganate of potash.

4. Colour-volumetric Processes—with Starch.

Kersting . . . . by chloride of mercury and bromine.
Schwarz. . . . . by hyposulphite of soda.
Price. . . . . { by nitrite of potash and hydrochloric acid.
Grange . . . . by hyponitric acid.
Bunsen . . . . by sulphurous acid.
Mohr. . . . . by arsenious acid.
Golfier-Besseyre, Reiman . by chlorine.
Hempel. . . . . . . { by sulphuric acid and sesquichloride of iron, with or without chloroform.

1. The Gravimetric Processes.

There are only two substances employed for the quantitative determination of iodine : viz., silver and palladium. The latter metal has been generally substituted for the former, as in the materials met with in commerce, chlorine is always present to some extent, and the estimation by means of silver is liable to error from this source, notwithstanding that, Wallace, Pisani, and Field state that ammonia effects a perfect separation of the chloride and iodide of silver.

Lassaigne first introduced the use of palladium, and this process was improved by Dr. R. D. Thompson. The solution containing the iodine is slightly acidulated with hydrochloric acid, and protochloride of palladium in solution is added until no more precipitation takes place. The mixture is allowed to stand in a warm place for one or two days, and the russet-black precipitate is collected on a weighed filter and washed, first with water, then with alcohol, and lastly with ether, in order to facilitate the drying. The heat must be cautiously regulated, as at 212° F. the precipitate is liable to a partial decomposition. The weight is taken when the precipitate loses no more on repeating the weighing. Rose has shown that the quantity of iodine may be estimated from the palladium left on igniting the precipitate in the ordinary manner in a porcelain or platinum crucible. 53 parts of palladium answer to 126 of iodine.

2. The Volumetric Process.

There is only one plan by this process properly so called, that of Kersting, which also consists in the use of palladium. He employs a solution of pure iodide of potassium containing 1 part of iodine in 1,000 parts of water, which is prepared by dissolving 1·308 grammes of the pure salt in 1 litre of water. An acid solution of protochloride of palladium, containing 1 part of the metal in 2,370 parts of the liquid, which is prepared by dissolving 1 part of the metal in nitro-hydrochloric acid, evaporating the solution to dryness at 212° F., adding 50 parts of strong hydrochloric acid, and diluting the solution with 2,000 parts of water, any insoluble matter being allowed to deposit. These solutions are employed to ascertain that the exact quantity of the metal is present in a given volume of the test-liquid.

The substance under examination is dissolved in water, and when the quantity of iodine has been approximately ascertained, the solution is diluted to contain about 1 part of iodine in 1,000 of water.

The probable quantity of iodine having been ascertained by a preliminary trial, the solution of the palladium is added to the solution under examination until no further precipitate is formed, and the number of degrees used is read off from the burette. A check analysis is made, reversing the use of the solutions, and the quantity of the liquid noted which is required to throw down the metal from a given measure of its standard solution.

3. Colour-volumetric Processes—without Starch.

Several of these methods give very exact results, and are distinguished for the facility with which they can be performed.

Penny, Bunsen, and Horsley's Method.—This method depends upon the following re-action :

$$3KI + KO.2CrO^3 + 7HCl = 4KCl + Cr^2Cl^3 + 7HO + 3I.$$

10 grains of pure bichromate are dissolved in 70 measures of the burette, which holds 1,000 grains of water up to zero, and the remainder is filled with strong hydrochloric acid up to 0°. A weighed quantity of the salt is dissolved in 1,000 grains of water, and the test-liquor from the burette added until all the iodine has been precipitated. This point is ascertained by adding a drop to a mixed solution of freshly-prepared protochloride of iron and sulphocyanide of potassium until a red colour is observed, when the degrees on the burette are read off. As every 10 divisions of the burette contain 1 grain of bichromate, which corresponds with the precipitation of 2·437 grains of iodine, the quantity of this substance in the solution is at once ascertained.

Luca and MM. Dupré's Method.—M. Luca's plan depends

on the liberation of the iodine by the action of bromine, and the colour which the former substance communicates to bisulphide of carbon.

4 grammes of bromine are dissolved in 4 litres of distilled water, and 40 cc. of this solution are diluted with water to make a litre. Each cubic centimetre of this test-liquid contains the hundreth of a milligramme of bromine.

Two graduated burettes are filled, the one with the test bromine liquid, and the other with bisulphide of carbon. The solution of a weighed quantity of the substance to be examined, is placed in a stoppered bottle with some bisulphide of carbon from the burette, to which a few measures of the test bromine liquid are added, and the whole shaken. The iodine, disengaged by the bromine, dissolves in the bisulphide of carbon with a rose colour. This is removed and replaced by a fresh quantity, until the bisulphide ceases to acquire any colour on the addition of the last portion of the test-liquor.

The measures of the test-liquor required, give by calculation the quantity of iodine present.

MM. Dupré's method is superior to the former, as the point of saturation depends on the disappearance of the violet colour, without the necessity of removing the ethereal solution. This method is based on the action of chlorine on a metallic iodide, in which the first equivalent of iodine liberated combines with 5 of chlorine to form pentachloride of iodine.

The iodine solution is placed in a stoppered bottle with chloroform, and the chlorine test-solution is added until the violet colour disappears, when 6 equivalents of chlorine correspond with 1 of iodine.

Rabourdin's Method is adapted to the determination of iodine in organic substances, and especially in cod-liver oil.

A mixture of 50 grammes of cod-liver oil, 5 grammes of caustic alkali, and 15 grammes of water, is heated until the organic matter is destroyed. The residue is washed with water and filtered. To the filtrate, 10 drops of nitric acid, 1 or 2 grammes of sulphuric acid, and 4 grammes of chloroform, are added. In time the chloroform separates with a violet colour.

The test-liquor is prepared by dissolving 1 centigramme of iodide of potassium in 100 cc. of water, and adding the above quantities of nitric and sulphuric acids and chloroform. This test-liquor is added to water until the shade of the coloration corresponds with the previous liquid, and the number of cc. required indicates the quantity of iodine in the oil.

Duflos and Schwarz's Method.—When a metallic iodide is heated with sesquichloride of iron, the iodine escapes with the aqueous vapour, and protochloride of iron is produced—

$$Fe^2Cl^3 + 1K = 2FeCl + KCl + I.$$

The iodine may be condensed in a solution of iodide of potassium, and the quantity determined by Mohr's method, and as a control, the quantity of protochloride of iron may be ascertained by the permanganate of potash; 2 equivalents of the protochloride correspond with 1 equivalent of iodine.

4. *Colour-volumetric Processes—with Starch.*

Price's Method.—This delicate test for iodine consists in mixing the liquid for examination with starch-paste, and sufficient hydrochloric acid to acidify the solution. A solution of nitrite of potash is then added, and if iodine is present, a dark blue colour is immediately produced. The acid disengages the iodine in the form of hydriodic acid, which is, in its turn, decomposed by the acid of the nitrite, and the free iodine then furnishes the characteristic re-action with the starch.

Dr. Price illustrates the delicacy of his test by the following experiment A thin transverse sectional slice of the stem of the *Fucus laminaria digitata* being moistened with starch-paste and dilute hydrochloric acid, and a drop of a solution of the nitrite added, the formation of the blue compound may be distinctly seen with the aid of a microscope.

Mohr has recommended the following process for preparing the solution of starch. The starch is pounded with a little water and a concentrated solution of chloride of zinc at ordinary temperatures. A paste is formed which yields, on the addition of water, a turbid solution of starch. The zinc is precipitated by carbonate of soda, and filtered. The filtrate is a clear solution of starch, which can be used without the necessity of warming, and possesses great delicacy in its re-action with iodine.

Fresenius has drawn attention to the influence of the temperature in promoting the re-action between the starch and iodine. The solution of iodide of potassium and nitrous-sulphuric acid indicated the following quantities of iodine at the different temperatures:—

From $\frac{1}{110000}$ to $\frac{1}{155000}$ iodine at 0° C., and only

$\frac{1}{110000}$	,,	13
$\frac{1}{135000}$	,,	20
$\frac{1}{155000}$	,,	30

he therefore recommends, that in the examination of very weak solutions of iodine compounds, they should be cooled down to 0° C.

Bunsen's Method.—This elegant and accurate process is fully described in the chapter devoted to sulphurous acid (p. 33).

Mohr's Method.—This process is based on the oxidation of arsenious acid, thus—

$$AsO^3 + 2\,NaO + 2I = AsO^5 + 2NaI.$$

The preparation of the test solution of arsenite of soda has been already described (p. 419) 0·5 gramme being brought up to a volume of 250 cc., 50 cc. is transferred to a beaker to which 10 to 20 cc. of a cold saturated solution of bicarbonate of soda are added with a little starch-paste. The iodine liquid is now mixed until the blue colour just begins to appear : 99 of arsenious acid represent 253·76 of iodine.

ESTIMATION OF BROMINE.

Three methods have been proposed to determine the weight of bromine in solutions, one of which is colorimetric and two volumetric.

Heine's Method.—The bromine is liberated by chlorine and absorbed by ether. The colour of this solution is compared with test coloured solutions of different strengths, and an estimate formed of the correspondence between two of the standard solutions with that under examination.

Figuier's Method depends on the yellow colour communicated by bromine to the liquid, which disappears on boiling.

For analysis, 2·5 grammes of bromide of sodium are dissolved in 500 cc. of water, so that a burette with 100 cc. of this solution, contains 5 decigrammes of this salt. The strength of the normal solution of chlorine is ascertained by adding a few drops of hydrochloric acid at 100 cc. of the above solution of bromide of sodium, and then the chlorine solution from an alcalimetric burette, loosely closed with a cork. The liquor acquires a yellow colour, which disappears on heating for a few minutes. After cooling, another addition of chlorine-water is made, and again the liquid is heated, and this is repeated until the strength is ascertained.

The liquid to be examined is then submitted to the same treatment, and the number of degrees of the burette read off, when the last addition of the chlorine water has ceased to cause any coloration.

William's Method is based on the decoloration of bromine by oil of turpentine, the bromine replacing the hydrogen of the oil ; and 34 parts of the oil decolorise 79 97 parts of bromine. Of this oil 20 grammes are dissolved in alcohol, so as to make 200 cc. of liquid. The bromine being liberated, the test-liquid is added drop by drop, the bottle being shaken after each addition, until the mixture is rendered colourless, when every 34 cc. of the test-liquid used represents 8 grammes of bromine.

ESTIMATION OF CHLORINE, IODINE, AND BROMINE IN MIXED SOLUTIONS.

Chlorine and Iodine.

Pisani's Method.—A few drops of the soluble iodide of starch are added to give a distinct blue tint to the solution of the mixed salts. A normal solution of silver is added until decoloration is complete, when the degrees of the burette being read off, the quantity of iodine can be ascertained. The silver test-liquid is then added in excess and the precipitate filtered. The excess of silver in the filtrate is determined, and this quantity being deducted from the second reading of the burette, gives the quantity of silver which corresponds with the chlorine.

This process requires care as to the exact point of decoloration of the starch-blue, which loses its colour immediately before the precipitation of the chlorine.

Lassaigne's Method.—The iodine is precipitated by proto-chloride of palladium, and the excess of the precipitant removed by sulphuretted hydrogen, which being in its turn expelled, the chlorine can be precipitated by silver in the usual way.

Moride's Method.—A few drops of fuming nitric acid, and two or three grammes of benzole are shaken up with the liquid under examination in a flask, when the iodine which is liberated dissolves in the benzole with a fine red colour. The benzole is removed, washed with water, and sulphurous acid added until the colour is destroyed. The iodine may be now determined by any of the plans already described.

Chlorine and Bromine.

These two bodies are precipitated together by silver, and after the usual precautions, the precipitate is weighed. The silver present in the precipitate may be ascertained, and with these data the proportion of each constituent can be calculated. Or, chlorine is passed through and over the precipitate until it ceases to lose weight, when the *loss of weight*, multiplied by 4·233 gives the quantity of bromide of silver originally present in the precipitate, and the proportions of the chlorine and bromine in the substance under examination are easily ascertained by calculation.

Iodine and Bromine.

The above process is applicable in the case of the mixture of these two substances.

Chlorine, Iodine, and Bromine.

Pisani's Method.—One portion of the liquid is treated with a silver solution, and the excess of silver used, is ascertained in the filtrate by the iodide of starch, the difference gives the quantity of silver in the precipitate, which has been weighed after taking the usual precautions.

In another portion of the original liquid, the iodine is determined by means of nitrate of palladium in the usual way.

The chloride and bromide of silver are treated as already described; and by this indirect method, the proportions of these three substances are ascertained.

Field's Method.—Chloride of silver is completely decomposed by digestion with solution of bromide of potassium, the chlorine and bromine changing places; and both chloride and bromide of silver are decomposed in like manner by iodide of potassium. Hence, if a solution containing chlorine, iodine, and bromine be divided into three equal parts, each portion precipitated by nitrate of silver; the first precipitate washed, dried, and weighed, the second digested with bromide of potassium, then washed, dried, and weighed, and the third with iodide of potassium, then washed, dried, and weighed,—the relative quantities of the three elements may be determined by the following calculation : Let the weights of the three precipitates be w, w', w'' ; also let the atomic weights of chloride, bromide, and iodide of silver be c, b, and i, respectively, and the unknown quantities of chloride, bromide, and iodide of silver x, y, and z; then we have the three equations :—

$$x + y + z = w$$

$$\frac{b}{c}x + y + z = w'$$

$$\frac{i}{c}x + \frac{i}{b}y + z = w''$$

The first and second give :— $x = \dfrac{c\,(w - w')}{b - c}$

Substituting this value in the second and third, they become—

$$y + z = w' - \frac{b\,(w - w')}{b - c} \qquad [= p]$$

$$\frac{i}{b}y + z = w'' - \frac{i\,(w - w')}{b - c} \qquad [= q]$$

$$\text{whence } y = \frac{b\,(q - p)}{i - b}$$

$$\text{and } z = w - (x + y).$$

Examples of this method are given in Mr. Field's paper, " On the separation of Iodine, Bromine, and Chlorine."—(*Chem. Soc. Qu. J.*, x. 234).

ALKALIMETRY.

I. ESTIMATION OF THE VALUE OF SODA AND POTASHES.

THESE two substances are employed in manufacturing operations, principally in the form of caustic or carbonated alkalies; and as a complete chemical analysis is, in most cases, of no importance, various methods have been proposed to determine, accurately and rapidly, the amount of alkali present in either of these forms.

These plans have received the general term of *Akalimetry*, and depend either on the known quantity of a strong acid required to neutralise a given weight, or on the weight of carbonic acid which is disengaged on the addition of an unknown weight of such an acid.

Sulphuric acid has been selected for this purpose, as it is always to be obtained, is cheap, and not volatile at common temperatures.

The equivalent numbers of the various substances furnish a ready means of ascertaining the relation between the re-agent and the substance under examination. Thus :—

100 parts of potash (KO) neutralise 104·2 parts hydrated sulphuric acid
100   „   soda (NaO)   „   158·1   „   „
100   „   carbonate of potash „   71·0   „   „
100   „   carbonate of soda „   92·45   „   „

And 1 part of carbonic acid disengaged corresponds with—

2·14 parts of potash, or 3·14 parts of carbonate of potash,
1·41 parts of soda, or 2·41 parts of carbonate of soda.

Or, stating these results in another form, 1 part of hydrated sulphuric acid corresponds with—

0·98 parts of potash, or 1·41 parts of carbonate of potash.
0·63 parts of soda, or 1·08 parts of carbonate of soda.

The earlier attempts to reduce these principles to practice, were made by Home, Richter, Berthollet, and Vauquelin ; but Dr. Lewis in this country, in 1767, and Decroizilles in France, in 1806, were the first chemists who successfully applied them to the valuation of potashes.

Gay-Lussac improved the process of Decroizilles, which has been rendered even more simple by Mohr.

Will and Fresenius, in the meantime, perfected the process depending on the weight of carbonic acid evolved, and their apparatus has been variously modified by different chemists.

Gay-Lussac's Method.

This plan is volumetric, dispensing with all weighing operations, and the efficiency of the test, depends chiefly upon the accuracy with which the amount of hydrated sulphuric acid is ascertained in the dilute acid (test-acid) used for saturation. Gay-Lussac attains this by carefully weighing 100 grammes of pure sulphuric acid of specific gravity 1·8427 at 15° C. (59° F.) and 962·09 grammes of water, which, on being mixed and allowed to cool (to 15° C. = 59° F.), exactly occupy the space of 1,000 grammes of distilled water, or 1,000 cc. = 1 litre. Two glass flasks are always sold with Gay-Lussac's apparatus, the one of which, when filled up to a certain mark in the narrow neck, contains at 15° C. exactly 54·268 cc. (= 100 grammes) of such sulphuric acid ; the other, filled in like manner to a certain mark, contains 1000 cc. If the contents of the first are poured into the second, and it is then filled up to 1000 cc. with distilled water, the necessary

quantity of acid of the proper strength is obtained. It is better
to add the acid gradually to a portion of the water, and not the
water to the acid : the whole is then filled up to 1000 cc. In
making the test-acid, the measure, which is divided from below
upwards, into half cubic centimetres, is filled to the uppermost
line, to the 0 therefore, with test-acid. As this acid contains 100
grammes of hydrated sulphuric acid in 1000 divisions of the
measure, the $\frac{100}{2}$ cc. of the glass will contain exactly 5 grammes
of acid, which require an equivalent, namely 4·8 of pure anhydrous potash for saturation. This quantity 4·8 is therefore
weighed from the mass to be examined ; if it were pure potash,
the whole of the 100 divisions of the test-acid would, of course,
be just sufficient ; if it only contained one half its weight of pure
potash, then 50 parts or divisions would be required ; if only
one-fourth, then 25 divisions. In short, there would be just as
much per cent. of pure potash contained in the specimen, as is
indicated by the number of volumes of test-acid required for
saturation.

In testing for soda, an equivalent to 5 grammes of acid must
also be weighed, which is 3·2 grammes.

Small quantities, like 3 or 4 grammes, cannot be weighed
with great accuracy upon a common balance ; Gay-Lussac,
therefore, recommends that ten times as much should be weighed,
therefore 47 potash, or 32 soda, and dissolved in so much water
that the whole shall occupy 500 cc. When, by means of a
pipette, one-fifth of this is taken as the test quantity, the quantity,
thus measured, will be a close approximation to 4·8 grammes. A
glass, graduated into 500 cc., and a pipette with a mark showing
50 cc. are required.*

The volumes of test-acid employed will also indicate the percentage, without further calculation, when the acid is so appor-
tioned that the measure shall contain 104·2 hydrated sulphuric
acid for potash, or 158·1 for soda, and when each experiment is
made with 100 parts of potash or soda.

Pure sulphuric acid, such as is requisite for Gay-Lussac's
method of testing, is not always to be procured, but Otto
has shown, that common sulphuric acid may be substituted,

---

* It need scarcely be noticed that any other measure or system of
weights may be used instead of the French, in precisely the same manner.

when its power of saturation has been previously determined.
This is then ascertained by a previous experiment, and for a
large quantity of acid, which can afterwards be used for a great
number of tests. A preliminary experiment of this kind is made
by testing the acid, diluted with about 12 parts of water, by
means of an alkali, or, what is the same thing, by means of a
carbonated alkali of known purity ; dry carbonate of soda, which
has been prepared by heating the bicarbonate to redness, is best
adapted for this purpose. If a quantity is taken equivalent to
100 parts of pure potash, 112·8 parts, therefore, the same result is
obtained as if carbonate of potash had been used, which latter
substance is much more difficult to obtain pure and in a dry
state. All that is now necessary, is to bring the ordinary sulphuric acid, diluted with about 12 parts of water, into a test-
glass, and to ascertain (observing the precautions to be mentioned
below) how many volumes (degrees) are required to complete the
decomposition of 112·8 parts of carbonate of soda. The number
of volumes employed will contain exactly 104·2 parts of hydrated
sulphuric acid, and if mixed with water so as to occupy 100
volumes, we shall have a test-acid similar to that of Gay-Lussac,
i.e , every volume employed to saturate a substance containing
100 parts of potashes, will indicate the presence of $\frac{1}{100}$ of potash.
The same quantity of acid (104·2 parts of hydrate) is just sufficient to decompose 65·9 parts of soda, and each volume will,
therefore, indicate $\frac{1}{100}$ of soda, when the test is made with 65·9
parts of the substance containing the soda.

The little instrument in which the test-acid is measured is
called *burette*, of which various forms are employed.

The Burette.

One form is shown in Fig. 289, which is somewhat similar to
that introduced by Decroizilles, and graduated into 100 equal
parts, commencing a little below the top, for the convenience of
pouring, and fitted with a perforated glass stopper, to insure
accuracy in filling with the test-acid.

It is not very easy to regulate very small additions of the acid
by this burette, and the next form (Fig. 290), while it obviates
this objection, is open to another, that the burette must be held
almost in a horizontal direction before all the acid can be run off.

The burette of Gay Lussac (Fig. 291) is superior, in both these
respects, to those above named. It consists of a glass tube,

FIG. 288.

FIG. 289.

FIG. 290.

FIG. 291.

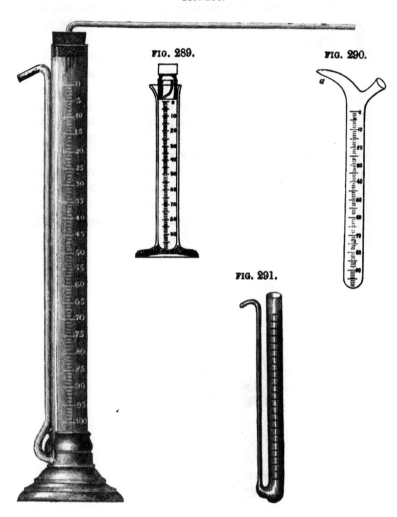

graduated from above, into which the test-acid is poured, and a long narrow tube attached below, from which the acid is poured into the liquid to be tested. The instrument is very delicate, and apt to be broken, and Mohr has greatly improved it, by attaching the escape tube to the burette itself, and fitting it into a wooden stand.

Mohr has also added a tightly-fitting cork through which a narrow tube passes, and bent at a right angle to the burette (Fig. 288). As the point of saturation approaches, the operator applies

38*

FIG. 292.

his mouth to the open end of this bent tube, and can then regulate the smallest addition of the test-acid, or suck back a drop hanging on the end of the pouring tube.

The attachment of this outside tube, even with Mohr's modification, is liable to accidents, and Geissler has finally got rid of this objection by placing the tube inside the burette, the upper end of which is slightly bent in the opposite direction, to avoid any danger of loss of acid when the burette must be inclined towards the horizon (Fig. 292).

The principle of these last two improvements has been introduced into the original burette, and in so simple a manner, that they can be added in any laboratory, and easily replaced when broken. This contrivance is seen in Fig. 293.

FIG. 293.

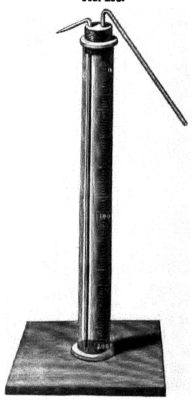

In using any of these instruments, however, an unsteady hand or a sudden surprise may lose a determination by a sudden or unexpected addition of the test-acid. This danger is avoided by using a burette-holder, the screw in which is only loosely turned, so as to allow a free movement to the burette, which is itself firmly held in the collar of the holder (Fig. 294).

FIG. 294.

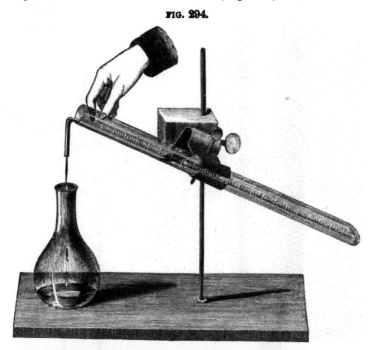

The most perfect instrument, however, is that invented by Mohr, and which may be called the *Pinch-cock burette*.

The wood-cuts on the next page convey so clear an idea of this beautiful instrument that the only description necessary, relates to the *pinch-cock*. To one end of the burette, which is slightly taper-ing, a caoutchouc tube is firmly bound, and in this a very small glass tube is inserted. The middle of the caoutchouc tube is fitted with the *pinch-cock*, which is made of brass wire in the form seen in section in Fig. 296, in which the small caoutchouc tube is represented as being open, in consequence of the operator pressing with a finger and thumb on the two plates. When not in use, the elasticity of the wire closes this tube so completely that not a trace of liquid can escape.

The bottom of the stand (Fig. 297) is made of white porcelain,

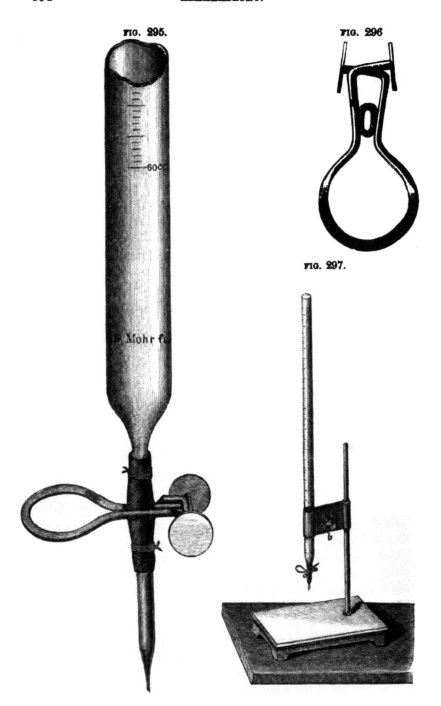

FIG. 295.

FIG. 296.

FIG. 297.

and the beaker, containing the solution of the substance for examination, is placed beneath the burette, on this white surface, so that the change and shade of colour are very easily recognised.

In large manufactories, where numbers of these determinations must be made every day, as quickly as possible, it is desirable to have several of these burettes ready for use, and Mohr has contrived a piece of apparatus for this purpose, as seen in Fig. 298.

FIG. 298.

Erdmaun has invented a species of aræometer for use with the burette, by means of which the reading off of the level of the liquids is rendered very accurate.

The float is a hollow glass instrument of the form shown in Fig. 299, and having a ring-shaped line drawn round its middle with the diamond. The ring at the top serves to remove it from the burette by means of a wire. It must be of such a diameter as to sink, without oscillation, in the burette when the liquid runs out, and to rise slowly after having been pressed down in the liquid with a glass rod or wire. Its weight is regulated by the introduction of mercury, so that, when introduced into the burette containing a liquid, the liquid may be level with its upper edge. The most essential condition of a good float is, that its axis shall coincide exactly with that of the tube, so that the divisions of the burette may be exactly parallel to the annular mark on the float.

FIG. 299.

The figure shows the float inside the burette. The burette being filled to above 0°, the float is introduced, and a quantity of liquid run out by the pinch-cock till the annular mark on the float coincides with the 0, or any other given division of the burette, the instrument being so placed that the half of the annulus turned towards the eye may exactly cover the other half. The same must be the case in reading off at the end of the experiment.

In this manner, the error of parallax is avoided, and a sharpness of reading obtained, which would be unattainable by any other method.

The Operation.

The method of conducting the operation requires some special precautions. The first point to be attended to iu alkilmetrical estimations, is the very unequal nature of the sample to be tested (particularly in potashes), tho determination of the mean quantity of potash in which, is the object of the test. It is best to take pieces from different parts of the mass, pound them together, and weigh out the test-quantity from this mixture. If this should contain so large a proportion of insoluble matter as to affect the volume of the solution, of which the tenth part has to be measured, then, the only plan is to filter and wash thoroughly in the first instance—a precaution which must always be adopted with wood-ashes. In this case, it is advisable, on account of the small quantity of alkali, to double the weight of the test-quantity and afterwards to halve the result.

On the addition of sulphuric acid, the carbonic acid is not immediately evolved, but forms, with the undecomposed portion of alkali, a bicarbonate, until the decomposition has extended to more than the half, and the fluid has become saturated with carbonic acid, which then escapes with violent effervescence; towards the end, however, this effervescence becomes so indistinct that it is impossible to know whether the operation is finished or not. The correctness of the experiment depending entirely upon the accurate determination of this point of saturation, it is necessary to employ a blue vegetable colouring matter,—tincture of litmus,—to indicate when the saturation is complete. The colour remains unchanged in the beginning; but when $\frac{1}{10}$ of the saturation has been effected (at that period, therefore, when the decomposition of the bicarbonate commences, and carbonic acid is set free), it becomes of a wine-red colour, and at length,—when the saturation is completed,—the red colour becomes very decided by a slight excess of sulphuric acid. Until the wine-red colour appears, the test-acid may be freely added, but it must then be done more cautiously; and towards the end, when the evolution of carbonic acid has nearly ceased, by two drops at a time, until no more gas escapes. At this period, the free carbonic acid in the fluid, renders it difficult to judge with accuracy of the colour. In order, therefore, to ascertain whether the reddening is attributable to it or to free sulphuric acid, after each addition, a streak upon blue litmus paper is made with a glass rod moistened in the test liquid. As soon as an excess of sulphuric acid has been employed, the paper remains red after having been dried, which is not the case if the reddening has been occasioned by carbonic acid. This excess of acid i. e. as many $\frac{1}{4}$th volumes (= 2 drops) as there are lasting red streaks, must now be deducted from the whole quantity used. As the action of the acid upon litmus paper is somewhat lessened by the presence of the sulphate of potash produced during the operation, an extra $\frac{1}{4}$ volume is deducted in addition.

A more correct plan consists in using a hot solution, in which the carbonic acid is less soluble, and when the wine-red colour appears, to boil the liquid, and frequently to shake the contents of the beaker, to facilitate the escape of the carbonic acid, when the blue colour will reappear, and then more acid must be added until the permanent red tint is obtained.

The action of sulphuric acid upon sulphide of sodium, sulphite and hyposulphite of soda, which are frequently present in artificial soda, as well as upon sulphide of potassium, is the same as upon the carbonates of the alkalies ; these compounds, however, do not add to the value of soda. A test of this kind, therefore, shows a higher value for soda than it actually possesses ; an error which (according to Gay-Lussac) may be avoided, by previously heating the test specimen to redness with chlorate of potash. By this means, these substances are converted into sulphates at the expense of the oxygen of the chlorate of potash, which remains behind as chloride, while the carbonated alkali is not acted upon. The hyposulphurous acid (S^2O^2), however, produces 2 equivalents of sulphuric acid, one of which decomposes an equivalent proportion of carbonate, which is consequently somewhat under-estimated. The opposite action which the silicates and phosphates exert upon the test-acid, and consequently upon the test itself, is unavoidable.

Method of Will and Fresenius.

The possibility of ascertaining the quantity of alkali by determining the amount of carbonic acid, depends upon the fact that only neutral carbonates are present in potashes and soda, so that each single equivalent of carbonic acid expelled, corresponds exactly to 2·14 of pure potash, and to 3·14 of carbonate, or to 1·41 equivalents of pure soda, and to 2·41 of carbonate of soda. For the sake of simplicity, the carbonic acid is estimated by the loss of weight which a previously weighed apparatus, containing the test-portion and the sulphuric acid, experiences when the carbonic acid is expelled. It is necessary in this method, that nothing except carbonic acid should escape during the operation; but this volatile acid cannot pass off from an aqueous solution without being charged with aqueous vapour. This evil is obviated in a very ingenious way by the apparatus of Will and Fresenius (already described under "Acidimetry," page 122), in which also the necessary temperature is generated in the fluid without the application of external heat. A (Fig. 300) is the larger flask, of about 2 oz. capacity, in which the decomposition is effected. B a somewhat smaller flask containing strong sulphuric acid. Both are

supplied with double-pierced corks for
the reception of the three tubes, *a, c,*
and *d.* The tube *a* is confined to the
flask *A,* being immersed below the
level of the fluid; in the same manner
d is connected with the flask *B,* and
only extends just below the cork.
Lastly, the tube *c* enters the neck of
A on the one side, but does not extend further, and by a double bend
is brought into connection with *B,*
which it enters, dipping into the sulphuric acid. The mouth of *a* is
closed by wax during the experiment
so that no orifice is left in the whole
apparatus except the mouth of the
tube *d.*

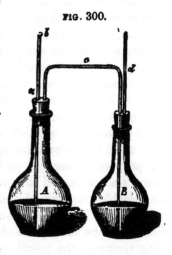

FIG. 300.

For weighing the test-sample and the whole apparatus, an
ordinary apothecary's scale is used. It is one of the great advantages of this method, that the experiment may be made upon a
much larger scale than is possible in ordinary analysis. By this
means, this cheap balance, when made to turn with the one-sixth
of a grain, affords quite as much accuracy as those which are
made for scientific purposes, and are at least twenty times as
costly,—particularly when such inaccuracies as may arise from
the unequal length of the arms are obviated, by placing the
weight upon the *same scale-pan* as the substance or apparatus to
be weighed, it having been previously exactly balanced by a
counterpoise—double weighing.

As a test-sample, several grammes of potash or soda, previously
thoroughly dried over a flame in a small metallic or porcelain
vessel, are weighed and introduced by means of a card into the
flask *A,* which is then filled with water to about one-third; the
apparatus is then closed by the wax stopper, and brought into
equilibrium, on the balance, by a counterpoise. The decomposition
is then induced by sucking out a small quantity of air, with the
mouth, from the tube *d.* The air is thus drawn not only from *B*
but also from *A,* both flasks being connected by the tube *c;*
bubbles of air are, therefore, seen passing from *A,* through the
sulphuric acid. The carbonic acid which is now evolved in *A,*
with effervescence and a rise in temperature, can only escape by

the tube c into the flask B, whence it must pass through the remainder of the sulphuric acid and the tube d, into the air. This sulphuric acid condenses with great energy all the aqueous vapour, and retains everything that the current of gas might possibly carry mechanically with it. When this operation has been repeated several times, the decomposition is completed. There is still, however, a portion of carbonic acid remaining in the apparatus, which was previously filled with air, and some still clings to the saline solution, which by this time has become cold. Both must be removed before the apparatus is re-weighed. For this purpose, by suction, as in the beginning at d, so much sulphuric acid is caused to pass over at once as will give rise to a considerable elevation of temperature in A, by which means the carbonic acid in solution is evolved, and with it that portion still clinging to the other parts of the apparatus. The wax stopper b being removed, the mouth of a is opened, and air may then be drawn through the apparatus from d, until all the carbonic acid is expelled. Here, too, all the moisture which is removed by the current of air in A, will remain in the sulphuric acid in B. When the whole apparatus has cooled, it is placed upon the scale, and the amount of carbonic acid is ascertained by the weights which must be added to re-establish the equilibrium.

It has been already stated that 3·14 parts of dry carbonate of potash contain exactly 1 part of carbonic acid ; and the calculation of the percentage of this salt from the result, is very much simplified, if 3·14 grammes of potashes are always taken as the test-quantity, as every centigramme of carbonic acid which has been evolved, will then indicate 1 per cent. of carbonate of potash ; if, however, $2 \times 3·14 = 6·28$ grammes are taken, as has been recommended, then 1 per cent. will be indicated by 2 centigrammes. In the same manner $2 \times 2·41 = 4·82$ grammes of soda must be used.

Precautions necessary in using this Method.

The presence of any salts which are decomposed by the sulphuric acid, but the acids of which are not volatile, has of course no effect whatever upon the result. This, however, is not the case with the sulphides of the metals, nor with the sulphites and hyposulphites ; the first of which evolve sulphuretted hydrogen, the second sulphurous acid, and the last hyposulphurous acid, which is immediately decomposed into sulphur and sulphurous acid. Sulphuretted hydrogen or sulphurous acid may, therefore, be evolved, which, calculated as carbonic acid, would erroneously augment the amount of potash or soda. This, however, is easily obviated by the addition of some neutral chromate of potash, which converts both the volatile acids into sulphuric acid and remains in combination with it, together with sulphur, as sulphate of chromium.

If the sample examined contains a large quantity of common salt, as is the case in *kelp-salts*, a great excess of sulphuric acid must be avoided, as an evolution of hydrochloric acid might ensue ; and when the chromate has been employed, the application of heat may cause chlorine to be disengaged.

The only precaution under such circumstances, is to employ the heat of the sand-bath instead of that generated by the sudden addition of the large excess of sulphuric acid, towards the end of the operation.

If there are carbonates of the earths in the insoluble portion of the potashes, the solution must be filtered and the residue well washed, and this must always be done in testing ashes and crude soda.

An error of an opposite kind, which rectifies itself in Gay-Lussac's process, may arise from the presence of caustic potash or soda. As these evolve no carbonic acid, they will not influence the result obtained by this process, although, both in potashes and soda, they are quite as valuable as the carbonated alkali. To ensure the estimation of these, the test-sample should be previously mixed, in a moist state, with carbonate of ammonia, and dried at a very high temperature, by which means all the caustic alkali becomes carbonated. If the sulphide of an alkali-metal is present, as is the case in crude soda, the mass is moistened with caustic ammonia instead of with water.

Two experiments are therefore necessary, when caustic and carbonate of soda or potash are present. One is made as already explained, and the other after the treatment with carbonate of ammonia, when the difference in the amount of the carbonic acid obtained, will correspond with the quantity of the caustic alkali present in the sample.

Or Gay-Lussac's process can be employed to determine the total alkali present, and Will and Fresenius's plan for the estimation of the carbonate, when the difference gives the percentage of the caustic alkali.

In all cases, especially when much caustic alkali is present, the samples must be previously dried, to expel the moisture; or the water must be estimated quantitively at the same time that the alkali is determined.

Modifications of the Method of Will and Fresenius.

Mohr's Apparatus consists of a wide-necked flask, closed with a cork bored with two perforations. One opening is fitted with a bulb-tube, containing strong nitric acid, and closed above with a pinch-cock, pressing a caoutchouc connecting piece; and the other opening is fitted with a bent chloride of calcium tube (Fig. 301). The weighed test-sample is placed in the flask with a little water, and the bulb-tube being filled with acid, the whole apparatus is adjusted and weighed. The acid is slowly admitted by pressing the pinch-cock, and when the decomposition is finished, the flask is gently heated, and atmospheric air is drawn through the apparatus by applying a suction bulb-pipette to the open end of the drying chloride of calcium tube, while the pinch-cock is kept open by the fingers. A second chloride of calcium tube can be attached to the nitric acid tube, to dry the air before entering the apparatus.

Geissler and Werther's Apparatus is shown in Fig. 302.

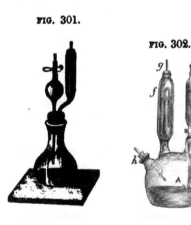

FIG. 301.

FIG. 302.

The test-sample is introduced through the neck h into the flask A. The acid for the decomposition fills the tube a, whence it escapes into the flask through the bent tube c, when the stopcock b is opened. The tube f is filled with sulphuric acid through the opening g, and is intended for drying the carbonic acid as it escapes through the bent tube e, among the acid, and finally through the small tube in the neck g. When the operation is finished, the air is drawn through the apparatus from h to g, by suction applied at g. The flask can be emptied of its contents without removing the acid from the tubes a and f.

Geissler's Apparatus, represented in Figs. 303 and 304, is one of the most convenient and handy modifications. The apparatus B and F contains the sulphuric acid for drying the escaping carbonic acid, and is so constructed that either part can hold all the acid, from whatever direction the pressure comes. The acid, sulphuric, nitric, or hydrochloric, for disengaging the carbonic acid, is placed in c, through which a tube, e, passes, which is ground into the lower part of the glass vessel c, and by raising which, the acid is admitted into the flask A. It is ground at a to fit the neck b of the flask. The working of this apparatus does not require any explanation after the previous descriptions.

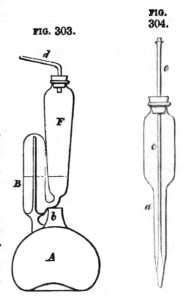

FIG. 303.

FIG. 304.

Reynold's apparatus consists of an ordinary flask, fitted with a cork bored for two tubes, as in Fig. 305. The tube-flask inside is also fitted in a similar manner, the larger having a tube which passes through the cork of the flask; the smaller bore has a curved quill tube, which reaches to the cone in the bottom of the tube-flask.

10 to 20 grammes of the test-sample is introduced into the larger flask, with a little water, and the tube flask is filled two-thirds with oil of vitriol. After all is weighed, a chloride of calcium tube is attached to the bent tube of the larger flask, and air blown through, which forces the acid through the quill tube. The carbonic acid evolved, then escapes through the sulphuric acid into the air. A fresh quantity of sulphuric acid is driven over into the large flask, the carbonic acid allowed to escape, and this is repeated until the decomposition is complete. The whole is gently warmed to expel the carbonic acid from the liquid, and

FIG. 305.

the wax plug being now removed from the quill-tube of the large
flask, air is drawn through in the usual way, and the whole again
weighed.[*]

All these modifications have this advantage over the original
one of Will and Fresenius, that the total weight of the apparatus
is much smaller, and it is not so much out of proportion to that of
the test-sample, as in their form.

Bauer has proposed to cover the surface of the drying acid
with small pieces of pumice-stone, in order to present a larger
absorbing surface to the carbonic acid gas, and to moderate the
agitation of the acid during any sudden evolution of this gas.

Mohr and Price's Method.—Mohr first suggested the substitution of oxalic acid for sulphuric acid in alkalimetrical operations,
in 1852, at the meeting of the German Association for Science at
Wiesbaden; and Dr. Price, without
knowing what Mohr had done, proposed the use of the same acid in
1854.

FIG. 306.

Mohr dissolves 63 grammes of
crystallised acid in water, so as
to form a litre; and prepares a
normal solution of caustic soda of
such strength that this liquid
exactly neutralises its own volume
of the test-acid. In order to preserve this caustic alkaline liquid
always in a state fit for use, Mohr
closes the bottle with a cork, soaked,
while heated, in wax, and fitted
with a tube containing a mixture
of dried sulphate of soda and quick
lime (Fig. 306). A known weight
of the alkali for examination, is dissolved in water, coloured blue with
litmus, and the test-solution of oxalic
acid added by a burette, with the
same precautions as in the case of
sulphuric acid, until the colour
changes permanently to red. The

[*] For the description of this apparatus, we are indebted to Mr. Crookes's
valuable "Chemical News."

blue colour is restored by adding the normal solution of caustic soda; and the degrees of the latter deducted from those of the test-acid, give the quantity of alkali present in the test-sample.

Price employs ammonia instead of soda, but in other respects the two proposals are very similar.

This plan has many advantages, but there is a serious drawback, in the difficulty of obtaining perfectly pure oxalic acid.

Wittstein has also proposed the use of crystallised tartaric acid, and Humbert that of bisulphate of potash, instead of sulphuric acid; but these suggestions have not met with much favour.

II. ESTIMATION OF THE ALKALINE SULPHATES.

In the manufacture of soda from common salt, it is of importance to ascertain the quality of the sulphate of soda produced or purchased in the form of chamber-sulphate or sulphate ash, and in some applications of potash, the sulphate is employed, so that a ready method of determining the quantity of these compounds is of some importance. Gay-Lussac first proposed the volumetric principle for these estimations.

His normal solution of the precipitant was prepared by dissolving 248 grammes of pure dry crystallised chloride of barium in a flask of 1000 cc. When this salt was dissolved, the volume of the solution was brought up to 1 litre. This solution at 68° F. has a specific gravity of 1·181, and is of such strength that each degree of the burette corresponds with 1 per cent. of potash in the form of sulphate.

The operation is conducted as follows: 48·07 grammes of the salt for examination, is dissolved in sufficient water to form half a litre, of which one-tenth is poured into a glass beaker. A little litmus is added, and coloured red with a few drops of nitric acid. The test-solution is poured cautiously into the solution from the graduated burette, until no more precipitate is formed, when the degrees of the burette are read off.

When operating on a salt, such as a bisulphate, the excess of acid must be previously saturated with a test-solution of carbonate of potash. This test-solution is prepared by dissolving 70 grammes of the alkaline carbonate in half a litre of water, which will exactly neutralise its own volume of the normal sulphuric acid. Every degree of this test-liquid used in neutralising the solution of the test-sample, must be deducted from the degrees used in estimating the sulphuric acid, and the balance then represents the potash present, combined with the acid found in the first determination by means of the test-solution of chloride of barium.

The same process is applicable to the sulphates of soda, substituting their equivalent numbers for those of potash.

When the two alkalies are present in the same sample as sulphates, a weighed quantity is taken, and the sulphuric acid estimated in the manner just described.

The sulphate of soda present, is then ascertained by the following short rule of Fresenius :

Multiply the sulphuric acid found by 2·1775, *deduct from the product the weight of the sulphates, and multiply the difference by* 4·4375.

III. ESTIMATION OF THE ALKALINE CHLORIDES.

Various plans have been proposed to ascertain the proportion in which potash and soda exist in a salt, when present as chlorides of their metals.

A weighed sample is taken, and the chlorine estimated by Mohr's process, described at page 425, and the proportion of the chloride of sodium is obtained by Fresenius's simple rule, viz :—

Multiply the quantity of chlorine found by 2·099, *deduct from the product the sum of the chlorides, and multiply the difference by* 3·656.

Gay-Lussac has proposed a very different process, which depends on the fall of temperature when these salts are dissolved in water. He found that 50 grammes of chloride of sodium, dissolved in 200 grammes of water, caused a fall of the temperature of the solution of 1·9° C., while a similar experiment with chloride of potassium was accompanied by a fall of 11·4° C.

He conducts the operation as follows : 100 grammes of the mixture are dissolved in water, filtered, washed, evaporated to dryness, and weighed. 50 grammes of this dry salt are rapidly dissolved in 200 grammes of water in a glass jar, taking the precaution that the salt and water have both the same temperature. A delicate thermometer is suspended in the solution, and the contents of the jar are kept in motion. The lowest point to which the temperature falls is noted, and on referring to the table in the appendix C, the figure corresponding with the fall of temperature gives the quantity of chloride of potassium which is present in the mixture.

Another expeditious and simple process, which has long been employed in some of our manufactories, to determine the quantity of potash present, whether it exists in the form of chloride or sulphate, in an acid or neutral solution, consists in adding sulphate of alumina to the test-solution of the salt, concentrating by evaporation, and crystallising out the alum. The mother-liquors are concentrated two or three times, to obtain all the alum, which is then dried and weighed, and the potash calculated in the usual way.

IV. ESTIMATION OF THE ALKALINE CARBONATES.

The relative proportions in which these two salts exist in any sample may be determined by a modification of some of the preceding processes.

They may be converted either into chlorides or sulphates, evaporated to dryness, and the chlorine or sulphuric acid determined in a weighed sample by the processes already described, and the soda-salt calculated by one of the rules above given: or, 50 grammes of dry mixed chlorides dissolved in water, and the chloride of potassium found from the fall in temperature by referring to the table in the appendix.

Mohr has recently proposed a very excellent process, which is founded on the conversion of the potash into cream of tartar, and is applicable to all these mixed salts of the two alkalies in neutral solutions.

In the case of carbonates, a weighed portion is saturated with tartaric acid, and when this point has been reached, as much more tartaric acid is added. The solution is evaporated to dryness, and the residue treated with a saturated alcoholic solution of cream of tartar. The insoluble potash-salt is filtered, and washed with the same alcoholic solution, dried and weighed.

When the alkalies are combined with the mineral acids, bitartrate of soda is employed instead of tartaric acid, when the soda-salt is converted into bitartrate of potash and a soluble salt of soda. The solution is evaporated to dryness, and the residue treated as already described.

Mohr found that a filter of 115 millimetres retained only

39*

one-sixth of the solution, and therefore that the quantity of salt retained by the paper was not of serious importance.

Mr. Dale, of Manchester, has devised a method likewise involving the use of tartaric acid, but more expeditious than the preceding, and dispensing with the necessity of weighing. It depends upon the fact that acid tartrate of potassium, though moderately soluble in water, is but very sparingly, if at all, soluble in a liquid containing acid tartrate of sodium. The method is as follows: Add to the mixture a standard solution of tartaric acid till an acid reaction becomes just perceptible: the alkalies are thereby converted into neutral tartrates; then add a second quantity of tartaric acid equal to the first, so as to convert them into acid tartrates: the whole, or nearly the whole, of the acid tartrate of potassium then separates. Next filter off the solution of acid tartrate of sodium, and add a standard solution of caustic soda till the liquid exhibits a faint alkaline reaction. The quantity of the soda-solution thus added is equal to the amount of soda present in the mixture. The quantity of tartaric acid required to form acid tartrate with the soda, subtracted from the total quantity added to the mixture of the two alkalies, gives the quantity required to form acid tartrate with the potash, and thus the amount of potash is determined. This method would scarcely be applicable where scientific accuracy is required; but, for rapid estimation in commercial practice, it is found to give good results.

A rough manufacturing process may also be employed for this purpose, when the operation can be performed on a tolerably large scale. This process has been applied by Mr. Bramwell for separating the carbonate of soda from the potashes used in making prussiate of potash.

The solution of a weighed quantity of the mixed salts is boiled up to 1·350 or 1·375 specific gravity, when the carbonate of soda begins to salt out, which is removed from time to time, and the drainings allowed to return to the pan. As long as the soda-salt continues to fall, the density of the liquid does not increase, but as soon as the specific gravity begins to rise, it indicates that all the soda-salt has separated, which, when drained, dried, and weighed, gives a rough approximation of the quantity of the carbonate of soda existing in the sample.

V. ESTIMATION OF THE SOLUBLE SULPHIDES.

M. Lestelle has proposed the following process, based on the insolubility of sulphide of silver, and the solubility of the other salts of silver, in the presence of ammonia.

He prepares a normal solution of ammoniacal nitrate of silver by dissolving 27·69 grammes of fine silver in pure nitric acid, to which 250 cubic centimetres of ammonia are added, when the whole is diluted with water, to bring up the volume to 1 litre. Each cc. of this solution corresponds to 0·010 grammes of monosulphide of sodium.

The substance to be examined is dissolved in water, and ammonia is added, after which the whole is boiled. The normal solution is added, drop by drop, from a burette divided into tenths of a cubic centimetre. When nearly all the sulphur has been precipitated in the form of a black sulphide of silver, it is filtered. A fresh quantity of the normal solution is poured into the filtrate, until, after repeated filtrations, a drop of the solution produces only a slight opacity.

The estimation is now complete, and the degrees read off from the burette can be compared with the weight. When small quantities of sulphide are present, the normal solution must be further diluted, until each cubic centimetre corresponds with 0·005 grammes of sulphide.

M. Lestelle states that the best sodas, tested by this method, are found to contain from 0·10 to 0·15 per cent. of sulphide, while those which have been furnaced too long, contain as much as 4, 5, and even 6 per cent.

The presence of common salt, caustic soda, sulphate or carbonate of soda, does not affect the accuracy of these determinations, as the precipitates formed by these compounds in nitrate of silver, are all soluble in ammonia.

Kynaston has shown that the presence of the sulphites and hyposulphites prevents the above process from giving accurate results. In a recent letter to the "Chemical News," he states that when a solution of sulphite of soda with ammonia is heated, and ammoniacal nitrate of silver is added, no permanent precipitate is formed at first; but in a few moments a grey precipitate of metallic silver appears. The change is thus represented :—

$$NaO.SO^2 + AgO.NO^3 = NaO.NO^3 + SO^3 + AgO.$$

Hyposulphite of soda, treated in the same way, gives no

permanent precipitate, but after some time, a brown coloration is visible, which rapidly increases, especially if more of the silver salt is added, until a deep black is produced. This change is shown by the following equation :—

$$NaO.S^2O^2 + AgO.NO^5 = NaO.NO^5 + SO^2 + AgS.$$

A solution containing sulphide of sodium, sulphite, and hyposulphite of soda, gives at once a black precipitate of sulphide of silver, and when more of the silver-salt is added, a dark brown precipitate is produced, which quickly becomes black, and continues until the other salts are decomposed.

The best method of determining the sulphide is by means of carbonate of cadmium, proposed by Werther.

The following is his method of analysing a mixture of sulphide of potassium with carbonate, hyposulphite, and sulphate of potash.*

a. The saline mass is treated in a closed vessel with water, in which a sufficient quantity of *carbonate of cadmium*, prepared by precipitation with carbonate of ammonia, is suspended ; sulphide of cadmium is thereby formed, which is to be collected on a filter, and freed from excess of carbonate of cadmium by treatment with dilute acetic (not hydrochloric) acid, and then digested with nitric acid and chlorate of potash ; the sulphuric acid resulting from the oxidation is precipitated from the solution by a baryta salt.

b. The solution, filtered from the sulphide of cadmium, is warmed, and mixed with nitrate of silver ; the precipitate, consisting of carbonate and sulphide of silver, is freed from the former salt by ammonia ; the silver is precipitated from the ammoniacal solution by chloride of sodium ; and the quantity of alkaline carbonate equivalent to the precipitated chloride of silver, is determined by calculation.

c. The sulphide of silver, containing half the sulphur of this alkaline hyposulphite, is dissolved in dilute nitric acid, then precipitated as *chloride*, and the quantity of hyposulphite determined therefrom by calculation :

$$AgCl : KO.S^2O^2 = 143\cdot5 : 95\cdot1.$$

d. The liquid filtered from the sulphide and carbonate of silver is next freed from excess of silver-salt by *hydrochloric acid*, and the sulphuric acid contained in it, is precipitated by a baryta

* J. pr. Chem. lv., 22.

salt. From the *amount* of sulphuric acid thence determined, a deduction is made for the quantity resulting from the oxidation of the hyposulphurous acid (0·28 part for every 1 part of chloride of silver formed from the sulphide) : the remainder is the quantity of sulphuric acid present in the original salt.

e. The amount of alkaline carbonate determined as in *b*, may be checked by *removing* the excess of baryta from the solution, evaporating to dryness, and igniting and weighing the residual alkaline sulphate ; deducting the quantity of sulphate originally present, and the quantities produced from the sulphide and hyposulphite ; and calculating the quantity of alkaline carbonate equivalent to the remainder.

If other acids of sulphur, especially sulphurous, trithionic, and tetrathionic acid, are present, this method is not applicable. It is given by Werther especially for the analysis of residues from the combustion of gunpowder. It is, of course, equally applicable to a mixture of soda-salts, and might also be applied to a mixture containing both potash and soda, if the proportion of those alkalies were determined by a separate experiment : *e. g.*, by converting a sample of the salt wholly into sulphate, and determining the quantity of sulphuric acid contained therein by precipitation with a baryta-salt.

Pesier's Natrometer.—M. Pesier employs, in analysing these salts, an instrument to which he has given the above name. His method is based on the following facts : A saturated solution of sulphate of potash at the same temperature, has a constant density ; but the sulphate of soda increases gradually with the density of this solution, and augments, at the same time, the solubility of the sulphate of potash.

The process is conducted in the following manner : 50 grammes of the salts are dissolved in as little water as possible, and the solution is filtered from any insoluble matter, neutralised with sulphuric acid, and evaporated to dryness. The mass is heated to redness, and dissolved in as small a quantity of water as possible. It is filtered into a graduated glass, and the saline deposit on the filter is washed with a reserve saturated solution of the same salt, until the jar is filled up to the mark of the 300 cc.

The solution is then mixed so as to render it uniform throughout, and the *natrometer* is immersed in the contents of the jar.

This instrument has two scales attached to it, one indicating

FIG. 307.   FIG. 308.

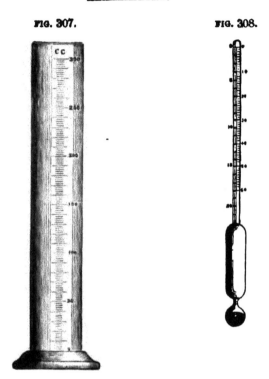

the temperature of the floating point of a saturated solution of
pure sulphate of potash; the other represents the percentage of
soda. The degrees of the temperature are deducted from the
degrees of the natrometer at which it floats, and the percentage
of soda is then read off. Thus, if the temperature is 20° and the
floating point 59, the former is deducted from the latter, and we
find that, as respects the balance, 39, the potash contains 13
per cent of soda.

Henry's Potassimeter.—Mr. Henry's process depends on the
property of perchloric acid forming with potash a salt which is
insoluble in alcohol. Thus, when acetate of potash is dissolved
in alcohol, and perchlorate of soda added, the potash is completely
separated.

The solution of perchlorate of soda is concentrated to the consistence of a syrup, diluted with its own weight of strong alcohol,
gently heated, and filtered. This alcoholic solution of perchlorate
of soda is then prepared, so that each degree of the instrument
represents one part of pure carbonate of potash.

The salts are converted into acetates and dissolved in alcohol, and any insoluble salts are separated by filtration.

The *potassimeter* is formed of a glass tube, a, d, 60 centimetres long and 4 millimetres in diameter. The upper extremity is made into a funnel and a copper cock is attached to the lower end. This tube has a scale, f, f, attached, divided into 100 parts, and the whole is supported by the holder g. The vessel e, below the tube, contains the salt for analysis. The graduated tube is filled between b and c, with the test-solution of perchlorate of soda, each degree corresponding with 1 per cent. of carbonate of potash.

50 grammes of the salt are dissolved in 1 decilitre of water, and, if necessary, filtered to separate any impurities. The solution is poured into a test glass divided into 50 equal parts. One part is then saturated with acetic acid in slight excess, evaporated to dryness, and dissolved in alcohol. The test-solution of perchlorate of soda is added by degrees to the solution of the test-sample, until no more precipitate is formed. The degrees used in the precipitation give the hundredths of pure carbonate of potash in 1 gramme of the salt.

FIG. 309.

VI. VOLUMETRIC EXAMINATION OF TANK-LIQUORS.

A weighed or measured quantity is neutralised with hydrochloric acid in slight excess, and the sulphuric acid is determined by means of the normal solution of chloride of barium, as explained at page 609.

A similar quantity of the liquid is slightly supersaturated with nitric acid, and the chlorine estimated by Mohr's process (p. 425).

Another portion is neutralised by acetic acid, and Bunsen's normal iodine-solution employed to determine the quantity of sulphide and hyposulphite which are present. (See p. 31.)

In a fourth portion, the sulphide is removed by sulphate of zinc, and in the filtrate the hyposulphite is estimated by the iodine and starch; when this determination is deducted from the former, the difference will represent the proportion of sulphide of sodium. The soda and carbonate of soda are determined respectively by the methods of Gay-Lussac and of Will and Fresenius; the former giving the total alkali present and the latter that of the carbonate.

Mohr analysed the tank-liquor from the manufactory of Rhenania, near Aix-la-Chapelle, by this method, and found—

Carbonate of soda	95·112
Chloride of sodium	2·154
Sulphate of soda	0·913
Hyposulphite of soda	1·207
Sulphide of sodium	0·547
	99·933

VII. VOLUMETRIC EXAMINATION OF POTASHES AND KELP.

The methods described for the analysis of crude soda are, of course, equally applicable in the examination of both potashes and kelp, with the addition of a process for the determination of the carbonate of soda in the former, and of the chloride of sodium in the latter.

These two mixtures of soda and potash may be analysed by any of the methods already explained at pages 610 and 612.

ANALYSES OF SODAS AND POTASHES.

The following determinations of the value of some of the sodas and potashes of commerce have been published.

Varieties of Soda.

Varieties.	Per cent. carbonate of soda.	Per cent. water.	Caustic soda.	Sulphide of sodium.	Sulphite of soda.	Hyposulphite of soda.
Yellow, calcined, Belgian......	83·5	24·0	little	none	much	much
White	42·8	4·0	none	—	—	—
Dieusé soda, very white	78·9	4·0	2·14	—	—	—
Cassel soda, white..............	84·5	undetermined	3·0 to 5·2	—	some	some
English soda	76·8		2·7 — 4·7	little	—	much
White calcined, from Büchner and Wilkins, in Darmstadt.	91·6		—	—	little	—
Soda from Debreczin............	89·2	15·6	—	—	—	—
White calcined, from Wesenfeld and Co., in Barmen ...	99·9	8·0	—	—	—	—

None of the varieties examined, contained carbonates of the earths in the insoluble residue.

Varieties of Potashes.

Varieties.	Per cent.		Varieties.	Per cent.	
	Dry carbonate of potash.	Water.		Dry carbonate of potash.	Water.
Bohemian potashes ...	94·6	10·0	Saxon potashes	·61·2	9·3
Illyrian I. ,, ...	95·9	7·6	Heidelberg, potash of Fries............... }	68·0	1·0
Illyrian II. ,, ...	93·8	14·0			

By means of Gay-Lussac's alkalimeter, the quantity of *pure potash* was estimated in the following varieties.

Varieties.	Deg ees.		Varieties.	Degrees.	
American potashes 1st variety	54 to 58		Kasan potashes	54 to 55	
,, ,, 2nd ,,	48	52	Polish ,, ...	55	60
,, ,, 3rd ,,	25	42	Riga ,, ...	30	50
New York ,, 1st ,,	55	60	Tuscany potashes 1st variety	50	55
,, ,, 2nd ,,	25	45	,, ,, 2nd ,,	55	60
,, ,, 3rd ,,	25	40	,, ,, 3rd ,,	50	55

SOAP.

Latin, *Sapo*; French, *Savon*; Spanish, *Sabon*; Portuguese, *Sabao*; Italian, *Sapone*; German, *Seife*; Dutch, *Zoep*; Danish, *Soebe*; Swedish, *Tvål*; Russian, *Mülo*; Polish, *Mylo.*

BY the term *soap*, in ordinary life, we understand the product resulting from the action of potash or soda upon the fats,—a substance in which these alkalies are combined with certain constituents of the latter class of bodies. In scientific language, this term has been extended to all similar combinations of the basic metallic oxides, so that in chemistry we meet with soaps of baryta, magnesia, iron, copper, &c. Besides *soap* in the limited sense of the word, or the soaps of the alkalies, there are only one or two others that have any technological interest; these are *lime-soap* in connection with the manufacture of stearin (Vol. I, Pt. II, p. 431) and *lead-soap* or *lead-plaster*, used in pharmacy.

History.—The use of soap is nearly as old as history ; but the assertion that soap is mentioned in the Old Testament is an error : for when we read in Jeremiah ii., 22, "For though thou wash thee with *nitre* and take thee much *soap,*" and in Malachi iii., 2, "For he is like a refiner's fire, and like fuller's *soap* ;" the words used in the Hebrew text are *borith* (vegetable lye salt, potashes), and *neter* (mineral lye salt, soda), which could never be intended to denote soap.   Pliny, on the contrary, makes particular and distinct mention of soap, which he declares to be a Gallic invention, and states that it is best prepared in Germany. He also distinguishes between hard and soft soap, and was acquainted with the mode of its preparation from the ashes of the beech-tree and goat's tallow.   Galen mentions soap with the like assurance, as do also Paulus Ægineta and Aëtius.   Homer describes fully and accurately what Nausicaa took with her for washing at the river, but makes no mention of soap, which was, doubtless, unknown to him.

MATERIALS FOR SOAP-BOILING.

The fats used by the soap-boiler are partly fluid and oily, as olive-oil, hemp-seed, rape-seed, linseed oil, train oil, and less frequently poppy-seed, beech-nut, nut and almond oil ; partly solid, as tallow, cocoa-nut and palm oil, and sometimes hogs'-lard, oil of Elipe, or Illipa, and Galam butter.

Tallow has already been fully described in treating of the materials for illumination (Pt. II, p. 422).   For the manufacture of soap, re-melted tallow is preferable to crude tallow, which how-ever can also be used, as the clarification is not absolutely neces-sary.   When the production of soap and candles is carried on in separate departments of the same manufactory, which is usually the case, the firmer kinds of tallow should be made into candles, and the softer varieties boiled for soap.   In northern countries, tallow is the chief fat used by the soap-boilers.

Lard.—*Hogs'-lard* produces a very soft soap, which is too readily dissolved by water.   For this reason, and on account of its high price, being much used as food, lard is seldom made into soap ; but when employed, it should always be previously melted.

Hemp-seed oil.—Hemp-seed oil is obtained chiefly in Russia from the seeds of *Cannabis sativa.*   It is of a greenish yellow colour, has a sharp smell, but mild taste. At $- 27°$ C. ($- 17°$ F.) it is solid. It is quite as soluble in alcohol as the following oil.

Linseed oil.—This oil, obtained from the seeds of *Linum usitatissimum*, is also a product of the North. When freshly pressed, it has a golden-yellow colour, which becomes brown on keeping; it has a peculiar smell, quite different from that of all the other oils, and becomes solid at a temperature of $- 16°$ to $- 20°$ C. ($+ 3°$ to $- 4°$ F.) From the experiments of Sacc, it appears to be a mixture of olein and margarin, and contains no stearin. It is frequently adulterated with colophony. Oil adulterated in this manner generally parts with the resin to alcohol, which then produces a white precipitate with acetate of lead.

Train oil.—*Fish oil* has already been mentioned in Part II, p. 419.

Olive oil.—Olive oil, which is used in large quantities for the production of soap, replaces tallow in the countries of the South. The manner in which it is obtained from the ripe fruit has already been described (Pt. II, p. 407).

The extensive trade carried on in olive oil, in the countries of southern Europe, renders it a matter of great importance to be able to detect the adulteration which is frequently practised in its preparation. The high price of this oil is a strong inducement to mix it with other vegetable oils; and poppy oil is generally selected for this purpose, on account of its low price, mild taste, and nearly imperceptible smell. Several methods have been suggested from time to time for ascertaining the amount and the nature of such adulterations, but there is only one which yields results of sufficient accuracy. The others, however, being frequently employed, require a passing notice. Poutet, a chemist and druggist in Marseilles, was the first to introduce a process which has been very generally adopted, and is founded upon the fact that pure olive oil, shaken with $\frac{1}{12}$th of a solution of 6 parts quicksilver and $7\frac{1}{2}$ parts of strong nitric acid, is converted into a solid white body. This conversion is dependent upon the presence of nitrous acid, and can be produced by that acid alone without the use of quicksilver. Nitrous acid converts the oleic acid, which is in combination with glycerin in the oil, into the isomeric compound, *Elaidic acid*, which remains, as *Elaidin*, in combination with the glycerin. The test of the purity of the olive oil is, therefore, by Poutet's method, the degree of firmness which it assumes after a short space of time. Whilst pure olive oil becomes solid and sonòrous, that which is mixed with $\frac{1}{50}$ of poppy oil only acquires the consistence of tallow, and with $\frac{1}{10}$ it is scarcely firmer than lard. These degrees of consistency, however, are so indefinite, and pass so easily the one into the other, that they afford no accurate measure of purity, when the amount of poppy oil is below $\frac{1}{10}$th. Boudet first discovered that the solidification was due to nitrous acid, and, he therefore advocated the use of a mixture of 3 parts nitric acid, and 1 part hyponitric acid or peroxide of nitrogen (NO^4), instead of the quicksilver, and made the time which a specimen took to become solid, or at the expiration of which it could no longer be poured from an inverted vessel at 10° C. (50° F.), the measure of its purity. He found that the time at which solidification occurred, with pure olive oil, was generally 55 to 60 minutes, and that this was retarded 40 minutes by the addition of $\frac{1}{100}$th poppy oil, 90 minutes by the $\frac{1}{50}$th, and very much more by the presence of $\frac{1}{10}$th. On examining this method of testing, Soubeiran and Blondeau found that pure oil required between 43 and 59 minutes to solidify ; oil containing $\frac{1}{50}$th of poppy oil, 45 to 59 minutes ; and that from 48 to 97 minutes elapsed before an oil containing $\frac{1}{10}$th of poppy oil became solid. The times of hardening are, therefore, so uncertain that many pure oils require more time to become solid, than others which are adulterated with poppy oil ; and all conclusions drawn from experiments of this kind, even when executed with the greatest care, can consequently never indicate the degree of purity of the oil; moreover, the method was intended only for the detection of poppy oil. Other adulterations are, however, seldom met with, on account of the smell and taste, or the costliness of the other oils. In tests of this kind, where two oils are compared, the same quantity of acid must, of course, be taken to each specimen : for the time required by one and the same oil to solidify, is different with a variable amount of acid. Thus, a mixture of 1 part hyponitric acid with

33 parts of olive oil becomes solid in   70 minutes.
50   „   „   „   „   „ 78   „
75   ;   „   „ 84   „
100   ;   „   „   „ 130   „
200   „   „   „   „   „ 435   „
400   „   „   will not solidify at all.

Fauré fancied that he had found a better test in ammonia,

$\frac{1}{10}$th of which, when shaken with pure olive oil, produces a milk-white, uniform thick soap (the volatile salve of the druggists in Germany), whilst poppy oil forms a granular mass Similar differences are observable in the combinations of ammonia with the pure and mixed oils ; but they are still less definite, as modes of testing, than those obtained by Boudet's process.

A very different method of examining oil has been introduced by Rousseau, and is carried out by means of his so-called *Diagomètre*. This test is founded upon the different powers of olive and poppy oil in conducting a galvanic current, the former conducting much less readily than the latter, without being a perfect non-conductor. The construction of the diagometer is such, that the power of conducting in the specimens—their purity, therefore, —may be seen by the deviation of a magnetic needle, under the influence of a galvanic current passing through the specimen of oil. This method is, however, of no practical utility, on account of a certain difficulty in making the observation, and still more because the power of different kinds of olive oil to conduct electricity is variable, and in some instances is even greater than in oil which has been mixed with poppy oil.

The old well-known plan of shaking, which depends upon the difference in the viscidity and consistence of olive and poppy oils, is still very much practised. When large bubbles of air are introduced by violent shaking below the olive oil, these are observed to vanish from the surface with a certain degree of rapidity. The duration of these bubbles gives an idea of the amount of poppy oil present. At least, Soubeiran and Blondeau found, in comparing 27 samples of olive oil which were in part pure, partly mixed with $\frac{1}{10}$th, and partly with $\frac{1}{20}$th of poppy oil, that $\frac{1}{10}$th could be pretty accurately detected in this manner, but not $\frac{1}{20}$th.[*]

---

[*] M. Heidenrich recommends three distinct tests for detecting the adulterations in commercial oils. The first test is the smell produced by the oil when gently heated, which is said to resemble that of the animal or plant from which the oil has been obtained. The second has reference to the change of colour experienced by the oil, when treated on a glass plate with sulphuric acid. The third experiment consists in ascertaining the density of the oil by Gay-Lussac's alcoholometer, the density of pure oils being changed by an admixture of the inferior kinds. For further particulars respecting the modes of testing, see *Chemical Gazette*, vol. i. 382. For other methods of detecting adulterations in expensive oils, see Pt. II, pp. 393—395.

Loutsodie uses bisulphide of carbon previously purified with acetate of lead, both for the extraction and for the purification of olive oil. The oil thus purified is said to possess a fine colour and its ordinary taste (*Compt. rend.* xlvi., 108).

Palm oil, Cocoa-nut oil, Oil of Illipa, Galam-butter.—All these oils are used in the manufacture of soap: the first two very extensively. Their properties have been already described (Vol. I., Pt. II, pp. 408—416).

As palm oil in its natural state has a strong yellow colour, the bleaching of it is a necessary preliminary to its use in soap-making. The methods of bleaching this oil by chloride of lime, by manganese and sulphuric acid, by bichromate of potash and sulphuric acid, by exposing it to the action of air alone, or of air and light together at a high temperature, have also been described. (See the reference just given.)

We have here only to add Pohl's method,* which consists in exposing the oil to a temperature of 240° C. (464° F.) *without* the action of air or light. From 10 to 20 cwt. of palm-oil are heated in a boiler, which should not be filled to more than two-thirds, as the oil undergoes great expansion in heating, and should be provided with a close-fitting cover and an escape-tube to carry off the pungent acid vapours which begin to escape at about 140° C. Ten minutes' exposure to the temperature of 240° C. is sufficient to bleach the oil completely. The expense of the process, labour included, is estimated at 7 to 9 kreuzers (farthings) per cwt., and the loss of oil does not exceed ¼ to 1 per cent.

Oleic Acid.—The mixture of oleic acid with portions of stearic and margaric acids, which is obtained in the manufacture of stearic acid by pressure, and generally goes by the name of "red oil" (Vol. I., Pt. II, p. 432), affords an excellent soap, the production of which is generally a secondary process in stearic acid manufactories: it is prepared so much the more readily, as the fat has not first to be decomposed, but merely neutralised with alkali.

There is also another variety of oleic acid occurring in commerce, called "distilled oleic acid," which is obtained by heating tallow or suet to 50° or 60° C., and treating it with half its weight of strong sulphuric acid. The fatty acids are thereby separated, and are subsequently washed with boiling water to remove the

* J. pr. Chem. lxiii, 248; Ding. J. cxxxv, 140.

sulphuric acid, and distilled with super-heated steam. They are then run into moulds and cooled, and afterwards submitted to pressure, to separate the oleic from the stearic acid. The oleic acid thus obtained, is much more strongly coloured than that obtained by saponification with lime : it has also a powerful empyreumatic odour. It is not so well adapted for soap-making as the acid obtained by the former process, the soap which it forms being deficient in coherence : the odour is also a great objection.

Rosin.—The resin, generally known as *rosin* or *colophony,* which is left after the distillation of turpentine is an equally important substance in the soap manufacture with the foreign vegetable fats; its origin and chemical properties have been described under the head of " Illuminating Materials " (Vol. I., Pt. II., p. 465).

A method of purifying and decolorising rosin by distillation with steam has lately been patented by Messrs. Hunt and Pochin of Manchester. The process is conducted in three different modes, according to the nature of the apparatus employed. " In each apparatus there is a still, which receives the substances to be purified, and one or more receivers communicating with it by a pipe or pipes. A steam-pipe in connection with a boiler is introduced into, and reaches to the bottom of, the still, where it radiates into various smaller pipes, perforated so as to allow the steam to escape. The size of the steam-pipe for a still capable of holding 6½ tons of rosin, should be 2 inches in diameter, when the steam is to be used at a pressure of 10 lbs. The exact proportions and arrangement of the apparatus are not material to the success of the process, but it is essential that the volume of steam should be about that which would be introduced under the conditions already mentioned.

· In the first mode of arrangement (apparatus No. 1), one receiver only is used, and the communication between the receiver and the still consists of a single pipe of ordinary construction. In the second mode (apparatus No. 2), several receivers are employed, the communication being direct from the still to only one of the receivers, and the communication being continued by pipes from one receiver to the other. The third mode (apparatus No. 3) is that in which two receivers are employed, the communication of both with the still being direct, and the pipes uniting and entering the still as one, at about 3 feet distance from the still.

These pipes are directly one above the other at their points of union.

The process, when conducted with apparatus No. 1, is as follows :—The substance to be operated upon is broken into convenient lumps, and thrown into the still. The still is then fired until the substance melts, when steam, at the pressure before mentioned, is turned into the steam-pipe, and, passing through the perforated pipes, thoroughly permeates the entire contents of the still. The still is meanwhile continuously fired until the temperature rises to about 600° F., at which point it is maintained until all such portions of the contents of the still as are capable of being volatilised, have passed from it into the receiver. The volatilisation commences at a temperature of about 390° F. The intromission of steam is continued during the entire period of the operation. The residue in the still consists principally of the foreign matters present in the substances before they were operated upon. The receiver is, in this arrangement, kept as cold as possible, by the application of water or other means, so as effectually to condense the whole of the volatile matter which has passed out of the still during the operation. The contents of the receiver will be found, on cooling, to consist of both fluids and solids ; the former (principally water condensed from the steam used in the process) is drawn off; the latter presents the characteristics of rosin, or other of the resinous substances operated upon, as found in its purest state. It is, however, at this stage still opaque, from containing a considerable amount of moisture. Various modes may be adopted to remove the moisture from the solid ; among others, that of placing it within a vacuum-pan, and heating it in the usual way, or placing it within a leaden vessel surrounded by a jacket, through which steam is passed, or melting it in an open leaden vessel by means of steam, and passing superheated steam through it when melted.

In the process as conducted with apparatus No. 2, the receiver nearest to the still, and in the case of more than two receivers being employed, every receiver except that furthest from the still, is maintained at a temperature a little above 212° F. The contents of the receivers nearest to the still are found to be transparent and almost free from moisture. The rosin and resinous matters condensed in the last receiver in the processes when conducted in this apparatus, alone contain any moisture which it is necessary to remove; this moisture may be removed in precisely the same manner as that indicated in the process before described with reference to apparatus No. 1.

In the process conducted with apparatus No. 3, the volatile matter arising from the substances placed in the still, arranges itself in two streams, the upper one composed principally of steam passing to receiver No. 1, and the lower one of rosin or resinous vapour passing to receiver No. 2. These receivers are kept at so low a temperature as to condense the whole of the products from the still. Various gases may be employed instead of the steam, with very similar effect; but the results are not equal to those obtained by steam.

The solid contents of the receivers during the distillation, as conducted by either of the above-mentioned processes, are found to vary in their characteristics at different stages of the process. At three stages of the process this variation is very marked, and the patentees distinguish the distillates obtained during the three periods, as follows:—The first fourth of the distilled product as "alpha-rosin," the second fourth as "beta-rosin," and the remainder as "gamma-rosin."

Yolk of Eggs.—Enormous quantities of this material are left as a hitherto useless residue in the preparation of albumen for the use of calico-printers. The print-works of Alsace consume annually 125,000 kilogrammes of dry albumen, and the preparation of 1 kilogramme of this substance requires the whites of 24 dozen eggs;—so that, as one hen lays, at most, 100 eggs in a year, the number of hens required to meet the demands of Alsace alone is 330,000. To find a profitable use for the yolk of this immense number of eggs, is, of course, a great desideratum, and it is now used for the preparation of soap. Both hard and soft soap may be obtained from it; the hard soap is very well adapted for washing woollen stuffs. M. Sichel, perfumer of Paris, uses yolk of eggs as an ingredient in several preparations for the toilet.*

Tallow, the olive, palm, and cocoa-nut oils, oleic acid, and rosin are used for the production of hard soaps; train oil, and the seed oils, on the contrary, for soft soaps.

* *Repertoire de Chimie appliquée,* 1861, p. 100.

40*

THE LYE.

The alkalies, as well as the fats, constitute one class of the crude materials employed in the soap manufacture ; and the former are of importance in a double point of view : first, as they themselves form constituents of the soap, and secondly, as they are the chemical agents concerned in the decomposition of the fats, which must always precede the actual production of soap. The soap-boiler, however, can only obtain wood-ashes, potashes, and soda, in which the alkalies are contained as carbonates. In this state of combination, neither potash nor soda is able to effect the change which the soap-boiler requires ;* his object cannot be attained by means of these salts, but only by the hydrates of potash and soda, and these are always employed in solutions of variable strength, known as *lye* or *caustic lye*. One of the most important operations of the soap-boiler, is that of rendering the alkalies caustic (*i. e.*, the conversion of the carbonates of the alkalies into hydrates), and this is effected by means of slaked lime or hydrate of lime. By a simple process of substitution, the carbonic acid combines with the lime, while the alkali assumes its water of hydration, and is converted into a compound possessing the most powerful chemical activity, which no substances of organic origin can withstand, and by which many others are decomposed. It is this property which is expressed by the term "*caustic.*" It has been observed, that very concentrated caustic lye will decompose carbonate of lime, abstracting the carbonic acid, so that hydrate of lime and carbonate of the alkali are produced, a result the very contrary of that which it is the object of the manufacturer to obtain. We therefore see why carbonate of potash is not at all affected by lime, when dissolved in 4 parts of water, and is but slightly acted upon when dissolved in 5 to 8 parts of water, and is rendered completely caustic only when dissolved in more than 10 parts of water. The best and most speedy method of obtaining caustic lye, is to slake 2 parts of burnt lime in 6 parts of hot water (converting it into hydrate), and add it gradually to a boiling solution of 3 parts carbonate of potash in 12 parts of water. When the fluid, which has become clear by the subsidence of the carbonate of lime, no longer effervesces with

* There are some few exceptions to this general rule, which will be mentioned in the sequel.

acids, nor renders lime-water turbid, all the carbonic acid has then entered into combination with the lime, and the lye is rendered perfectly caustic. It is necessary to use only half as much water with carbonate of soda, which is much more easily made caustic. Carbonates of the alkalies can even be rendered caustic in the cold, by means of lime; but more lime must then be employed, with weaker solutions of the carbonates, and the action must be allowed to continue for a longer time. In this manner, lyes of a certain strength, only indicating from 10° to 12° B. can be obtained, and with less expenditure of fuel. The loss of time is of no moment, as the materials may be left to themselves, and require no attention. According to theory, 100 parts of carbonate of potash should require 53 parts, and the same quantity of carbonate of soda, 40·5 parts of burnt lime; but it is advisable, in practice to use 50 parts for soda, 60 to 80 parts for potash, and from 8 to 10 parts of lime for crude wood-ash. Soda lyes cannot be obtained caustic at a higher gravity than 18° Twaddell = 1·090 specific gravity.

English soap manufacturers prepare their lyes invariably by *boiling together* soda and lime, but on the Continent the preparation of caustic lyes by digestion in the cold, is generally adopted, partly because strong lyes cannot be used in the ordinary method of boiling there practised, partly because crude ash was originally the only source of alkali (at least, in northern countries), and this, if treated with boiling water, would impart too large a proportion of foreign salts to the lye; and lastly, because the Continental soap-makers are accustomed to the process of boiling with lye prepared in the cold. The manipulations are as follows:

In purchasing and collecting ashes, the great object is to obtain pure wood-ashes without any mixture of peat-ash, which not only contains no carbonated alkali, but is injurious from the large amount of gypsum which frequently accompanies it. This substance exchanges its acid for that of the carbonate of the alkali, so that carbonate of lime is produced, and sulphate of potash, which is useless in the preparation of lye. It has been found that many kinds of peat-ash will, in this manner, spoil three times their weight of wood-ashes. In all common ashes there are portions of unburnt charcoal and wood, which would impart colouring and empyreumatic substances to the lye, and must consequently be separated by a sieve. The sifted ash is moistened upon a smooth stone floor with water or weak lye; it is then well raked about, and allowed to remain until all parts of the ash are thoroughly soaked. The heap is then hollowed out in the middle, so as to form a pit. In this pit the lime is slaked, and when it has fallen to pieces, is covered from all sides with ash, and afterwards both ash and lime are well mixed together. For lixiviating the mixture, wooden, or, better, cast-iron casks, with false bottoms, pierced with holes, and covered with straw, are used, and they are furnished with a stop-cock, as is represented in Fig. 310. In these *ash-tubs*,

FIG. 310.

the ingredients are mixed as thoroughly as possible by a stirring rod ; the mixture is then covered with straw, brushwood, or basket-work, and as much water is added as is necessary to induce the action of the lime. The cock is, at first, left open, that the air may easily escape from the interstices, and water (or weak lye) is poured very gradually over the ash, that it may slowly sink down. As soon as the lye begins to flow out from the cock, the latter must be closed, and the rest of the lye poured in. The action is now allowed to go on for twenty-four hours, and the lye is then drawn off; the strongest lye flows out first, containing potash of the strength of 20° or 25° B. ; a second lixiviation affords lye with 8° to 10° B. of potash ; and a third yields *weak* lye, indicating 3° to 4° B. The following portions are very weak lyes, and are employed as the first addition to a fresh quantity of ash.

The treatment of potashes is more simple, and consists in adding water to the mixture of dry potash and slaked lime in the ash-tubs. These ingredients are more easily and better mixed in the dry state. The lye is frequently not so clear on being drawn off, as in the former case ; and the lime is apt to pass through the straw and the perforated bottom. For this reason, potashes are often worked together with wood-ashes, whereby this difficulty is avoided. Both may be mixed with the lime, or the ashes only, and then the potashes should occupy the highest place in the ash-tubs, or may be dissolved in the lixiviating water.

The *soda-salt* is treated in the same manner, but never with an addition of ashes. It is best to spread it above and upon the lime in the ash-tub.

It is not convenient to render solutions of potash or soda caustic by stirring them together with milk of lime, for the carbonate of lime, which is in a state of extreme division, occupies a long time in subsiding from the liquid.

The production of lye from ashes is becoming less frequent in the soap manufactories, as the scarcity of wood increases, and as soda is being more generally introduced.

The production of lye from soda by boiling, in which case very little foreign matter has to be removed, is certainly a gain to the soap manufacturer, inasmuch as he can obtain at once strong lye, which is now so frequently needed—for instance, in the saponification of cocoa-nut oil—without being obliged to concentrate weak lyes by evaporation.

In the preparation of *steam lyes*, as they are called, the manipulations are the same, whether potash or soda lye is required. In making the first, however, 50 lbs. of lime are required to 100 lbs. of soda, whereas, for the same quantity of potash, 80 lbs. of lime are used. The proportion of water is 12 parts to 1 part of potash, and something less for soda, which is made caustic more readily than potash. The alkaline material (soda-ash or potash), and the proper quantity of water and lime, previously slaked into powder with hot water, are shovelled into the vat, and boiled by a current of steam for several hours, until a portion, taken out and left to repose, gives a supernatant liquor which does not effervesce with hydrochloric acid, nor form a cloud with clear lime-water. The intimate mixture of the lime and alkali greatly promotes the decomposition. Moreover, the lye thus prepared contains less lime than lye made in the cold, because lime is less soluble in hot than in cold water. After the boiling is completed, the mixture is allowed to repose till the carbonate of lime subsides, when the supernatant liquor is drawn off through a siphon into the reservoirs. The residue is then stirred up with fresh water, and the weak liquor thus obtained may be used for diluting stronger lyes, or instead of water in another boil.

It is not advisable to employ crystallised or even crude soda for the production of lye, as needless expense is incurred in carriage, and these contain but a small amount of soda; besides which, the latter is always more or less contaminated with sulphide of sodium, and this has to be removed by artificial means. A serious objection to the preparation of lye in the cold, is the uncertainty attending the action of the lime, which cannot be ascertained previously ; nor can the quantity added be subsequently increased or diminished at pleasure. Thus, an excess of lime has sometimes to be remedied, and at other times too little lime is employed. The first evil is corrected without any great difficulty ; but when the lye passes through in the state of carbonate, there is no other alternative than to pour it again through the lime.

Mr. Simpson, of Maidstone, proposes to use, for the saponification of fats and oils, the alkaline solutions which have been employed in manufacturing pulp from straw. Heretofore it has been usual to evaporate the water of this waste alkaline liquor, and burn the residue in order to use it with fresh quantities of straw ; but the process is objectionable, partly on account of the great cost of evaporation and burning, and partly from the impurity of the alkali thus obtained. This alkali may, however, according to the author, be beneficially used for the making of soaps, the impurities being got rid of in the ordinary cleansing of the soap. All the solutions of alkali, both weak and strong, drawn off after they have been used in acting on the straw, and those also obtained by washing the straw, are mixed together, and used as the first or weak alkaline lyes to act on the fats or oils used for soap-making. The rougher impurities are removed by filtering the liquid through a layer of unslaked lime and a layer of carbonate of soda, which process has also the advantage of rendering the lye more caustic.

The caustic lye of the soap-boiler, whether obtained from ashes or potashes, contains some sulphate of potash, a good deal of chloride of potassium, and some silicic acid in solution, which latter precipitates in combination with the lime on boiling, but not in the cold. Soda, particularly that obtained from marine plants, imparts common salt and sulphide of sodium to the lye. In preparing lye in the cold, hydrate of lime is found in solution, particularly in weak lyes ; and this, notwithstanding its slight degree of solubility, $\frac{1}{480}$ to $\frac{1}{700}$ of the water, is nevertheless of some moment, in consequence of the large quantity of water used, as it destroys, on boiling, a portion of the soap, converting it into insoluble lime-soap. Lastly, it must not be forgotten that caustic lye, at the boiling temperature, particularly when concentrated, attacks and dissolves iron.

C. N. Kottula, of Liverpool, purifies soda-lyes by the addition of alum. The lyes are first heated, and then the alum is added, in the proportion of

12 oz. alum to 1 cwt. of lye, marking 10° Baumé.

18	„	„	15	„
24	„	„	20	„
30	„	„	25	„

that is to say, $1\frac{1}{5}$ oz. alum for every degree of strength in each cwt. of lye. The liquid is agitated for half an hour, till all the alum is dissolved, and then left to settle and become clear. The lyes thus purified are said to be capable of saponifying all fatty matters and resins used for soap-making, at one operation.

Testing the Lye.—The soap-boiler is only in a position to judge correctly of his process, and proceed with safety, when, during the lixiviation in the ash-tubs, and whilst boiling the fat, he is enabled at any moment to ascertain accurately the strength of his lye. For this purpose various hydrometers are used, Twaddell's mostly in England, Baumé's on the Continent; Steppani's and Gay-Lussac's are also occasionally employed. In Baumé's hydrometer (that which is used for liquids heavier than water) the space between the point to which the instrument sinks in pure water,—the zero point,—and the point to which it sinks in a solution of 1 part common salt in 9 parts of water, is divided into 10 equal parts or degrees. This graduation is adopted throughout the whole length of the spindle. The indications of such a hydrometer have no relation to the chemical nature of the liquid ; and as the graduation is perfectly arbitrary, nothing more can be established by the instrument than a comparison of the densities of different liquids. When, therefore, the hydrometer sinks to the same depth in syrup of sugar and in a caustic lye, no other conclusion can be deduced from the observation, than that both fluids are of like density. But as the correspondence has already been established between the degrees of Baumé's hydrometer and the specific gravities, and it is also known what proportions of potash and soda correspond to the specific gravities of different solutions, the hydrometer may be used, with the aid of tables (see Appendix), as a means of ascertaining the amount of alkali present. A lye of 18° B., for instance, has a specific gravity equal to 1·138, which is equivalent to 15 per cent. of potash, or 12·8 per cent of soda, always supposing the lye to contain no other matters, or salts, which can influence the hydrometer. This, however, is never the case with soap-boilers' lye, so that the indications of Baumé's or Twaddell's hydrometer are mere approximations with reference to the amount of alkali, or are altogether fallacious. Nevertheless, they are useful to the operator in a variety of ways, and are, indeed, indispensable. In the first place, he can easily follow, with this instrument, the diminishing strength of the liquid flowing from the ash-tubs, and form a just estimate of its value, when, at the same time, a test with acid proves the absence of carbonic acid. But even in the process of soap-boiling itself, the test of the lye afforded by the hydrometer, combined with the personal experience of the workman, is a clue by which he is enabled to judge whether the proper strength has been attained, knowing, as he does, what degree the hydrometer ought to indicate from any particular soda or potash lye, for any particular purpose. The buoyant power of the lye was the old-fashioned and erroneous test: in applying this, it was ascertained whether an egg, or a piece of hard soap, would swim on the surface of the lye, and not sink, for which a certain density was of course necessary. By this method the respective densities of different lyes could never be ascertained, but only whether a certain lye had attained that particular density. In other words, weak lyes could never be examined upon this principle.

THEORY OF THE FORMATION OF SOAP.

A clear idea of the formation of soap, and a correct explanation of this remarkable process, can only be obtained from a thorough knowledge of the constituents of the fats, as well as of their mode of combination. The question respecting the theory of soap-formation, and the question concerning the constitution of the fats, have reference, therefore, to one and the same thing. The greater part of our knowledge concerning these bodies, is founded upon the extensive and trustworthy researches of Chevreul, dating from the year 1813 to 1823, the results of which were published by the author in his "*Recherches chimiques sur les Corps Gras,*" etc., *Paris*, 1823. More recent researches upon this subject have seldom found anything to correct, but have added new facts, and given rise to explanations of those already known, which are more in accordance with the advanced state of science. Very important additions to our knowledge of fatty bodies have been made by the recent investigations of Berthelot. These researches, conjointly, have proved that the fats, such as tallow, lard, olive oil, &c., are mixtures of two kinds of matters, which, taken singly, possess all the properties of the fats themselves. They are chiefly distinguished from each other by their state of aggregation at ordinary temperatures. Those which appear solid and hard were called *stearin* by Chevreul, the fluid kinds he termed *olein*. The consistence of a fat depends upon the predominance of one or other of these constituents; so that the fluidity of the oils is due to a predominance of olein, the solidity of the varieties of tallow to that of stearin. Pelouze and Boudet have observed that the olein from olive oil, hazel-nut oil, human and swine fat, is very different in solubility, and in its action with nitrous acid, to the olein from linseed oil, nut, poppy, hemp-seed oil, and cocoa-butter. There are, consequently, several distinct substances included under the collective name of olein. There are also several distinct solid fats commonly included under the name stearin, viz., the true stearin, which is the principal solid constituent of beef and mutton suet; margarin,* existing abundantly in human fat and olive oil; palmitin in palm-oil, &c.

These proximate principles,—stearin, palmitin, and olein,—which, either merely mixed or in chemical combination, constitute the fats,—are considered as true salts, or as combinations of a base with an acid, and this view has been arrived at by the appearances observed during their decomposition, and the similarity of their behaviour to other well-known combinations.

The base of the ordinary fats is *glycerin*, $C^6H^9O^6$. This substance, in the free state, is a colourless syrupy liquid, soluble in water, having a sweet taste, and not exhibiting any of the peculiar properties of the fats themselves, which properties are, on the contrary, retained by the fatty acids. These acids are fluid at common temperatures when derived from a fluid constituent —e. g., olein; but solid when obtained from a solid constituent—stearin.

Glycerin is a tri-acid base in the same sense as ordinary phosphoric acid is a tri-basic acid: that is to say, it is capable of forming three kinds of salts or fats, two acid, and one neutral, by taking up one, two, or three molecules of an acid. All these compounds of glycerin with the fatty acids may indeed be

* Margarin is perhaps not a distinct fat, but only a mixture of stearin and palmitin.

formed by simply heating the glycerin with the acid in a sealed
tube; the particular combination produced depending on the
temperature, the duration of the action, and the proportions in
which the ingredients are mixed. The natural fats,—stearin,
palmitin, &c.,—are all neutral salts of glycerin; that is to
say, they are formed from 3 molecules of acid and 1 molecule
of glycerin. They are exactly similar to the compounds obtained
by combining the glycerin and the acids artificially in the same
proportions.

The fats resemble ordinary salts in this respect, that they
do not contain all the elements of the acid together with all the
elements of the glycerin; but their formation is always attended
with the elimination of a certain number of molecules of water,
depending upon the number of molecules of base and acid which
enter into combination. The analogy thus existing between the
fats and the salts of inorganic bases will be seen from the following equations :—

$$C^4H^4O^4 + KHO^2 - H^2O^2 = C^4H^2KO^3.$$
Acetic acid.   Hydrate   Water.   Acetate of potash.
of potash.

$$C^8H^6O^{12} + 2KHO^2 - 2H^2O^2 = C^8H^4K^2O^{12}.$$
Tartaric acid.   Neutral
tartrate of potash.

$$PO^5H^3 + 3KHO^2 - 3H^2O^2 = PO^5K^3.$$
Phosphoric acid.   Phosphate of potash.
(Tri-basic.)   (Tri-basic.)

$$C^{36}H^{36}O^4 + C^6H^4O^6 - H^2O^2 = C^{42}H^{42}O^8.$$
Stearic acid.   Glycerin.   Monostearin.

$$2C^{36}H^{36}O^4 + C^6H^4O^6 - 2H^2O^2 = C^{78}H^{76}O^{10}.$$
Distearin.

$$3C^{36}H^{36}O^4 + C^6H^4O^6 - 3H^2O^2 = C^{114}H^{110}O^{12}.$$
Tristearin.

Now when these fats are boiled with caustic alkalies, or with
the earths and certain other metallic oxides, in presence of water,
the opposite decomposition takes place, a metallic salt of the
fatty acid being formed, and glycerin reproduced, thus :

$$C^{42}H^{42}O^8 + KHO^2 = C^{36}H^{35}KO^4 + C^6H^4O^6.$$
Monostearin.   Hydrate of   Stearate of   Glycerin.
potash.   potash.

$$C^{114}H^{110}O^{12} + 3KHO^2 = 3C^{36}H^{35}KO^4 + C^6H^4O^6.$$
Tristearin.

This is the process of *saponification*. A soap, in the widest
sense of the word, is a metallic salt of a fatty acid; but in its
more limited and practical sense, it is restricted to the potash
and soda salts of these acids.

When olive oil is boiled for some time with water and oxide of lead, a lead-soap is obtained, called *lead-plaster*, which is insoluble in water, and swims on the surface of an aqueous solution of glycerin. Indeed, the existence of glycerin was first observed in preparing this plaster, no other substance soluble in water, occurring in the process, to mask its presence.

The neutral fats may also be decomposed by heating them with water under pressure to about the melting point of lead. The products of the decomposition are a fatty acid and glycerin, the water acting at the higher temperature exactly in the same manner as the caustic alkali at the ordinary boiling heat. Thus, with stearin:

$$C'''H'''O'' + 3H'O' = 3C''H''O' + C'H'O'.$$

The apparatus by which this decomposition is effected will be described further on.

Rosin (colophony) is very differently affected, inasmuch as its constitution is very different from that of the fats. Commercial colophony is a mixture of a large quantity of pinic acid with a little silvic and colophonic acids (compare Pt. II. p. 466), a mixture which, from the nature of its ingredients, possesses the properties of a weak acid. In this case, therefore, no real saponification ensues, but the alkali is simply saturated with the resinous acids, and a substance obtained which, in a commercial point of view, is equivalent to soap. The production of this substance is still simpler and easier than the saponification of the acid vegetable fats; in fact resin decomposes carbonate of soda.

Thus far, in speaking of alkalies, reference has only been made to caustic soda and potash. It is, however, well known that soap can be prepared with the carbonates, and even with the bicarbonates of the alkalies; but the process, if carried on at the ordinary atmospheric pressure, is so tedious and imperfect, that it is never practised on a large scale. A solution of carbonate of potash, when boiled with fat, parts with one-half of its potash to form soap, whilst the other half becomes bicarbonate. The decomposition of this latter continues, with evolution of carbonic acid, so long as the boiling lasts, provided sufficient fat is present; but the process is so tardy, that a perfect soap is hardly procurable in this manner. Under high pressure, however, as in Tilghman's apparatus, the fatty acid is eliminated by the action

of water at the high temperature produced:—it is therefore
capable of decomposing carbonate of soda. For saponifying
(saturating) resins upon a large scale, the carbonate of the alkali
is quite as applicable as caustic lye.

An excess of alkali is requisite for saponificatiqn, and is so much
the more willingly employed, as it can be removed again without
difficulty. It must not be supposed that the production of soap
is a momentary process, or that it can be done with the same
exactness and rapidity as the decomposition of an ordinary salt.
On the contrary, the production of soap passes through a number
of stages, and these occupy a considerable length of time, from
the first mixing of the fat with the alkali—when a milky turbid
mixture (emulsion) is produced—to the formation of soap ready
for use, or to that point when the whole of the alkali is saturated
with fatty acid. Acid salts are first produced with the fatty acids,
and these hold the remainder of the fat in a state of solution
and division, until this portion also is enabled to combine with
alkali, and transform the acid into neutral salts, or into soap
ready for use. This reaction may be easily observed if the fat is
boiled with one-half the requisite quantity of alkali ; the whole
of the oil is at length dissolved, but the solution becomes turbid
on cooling, and when diluted with water and boiled, unsaponified
fat separates, which had been retained in the fluid only by the
stearate, margarate, &c., of the alkali that had been formed.

From what has been stated, it appears that ordinary soap is a
mixture of the compounds of the fatty and resinous acids with
potash or soda. The choice of the base is, however, by no means
a matter of indifference. The potash-soaps are of that nature
which in ordinary salts is termed deliquescent, i.e., they do not
dry up when exposed in solution to the air, but retain so much
water as to form a soft slimy jelly. On the other hand, arti-
ficially-dried potash-soaps absorb a large quantity of moisture,
and become converted into a soft jelly. This kind of soap is
called *soft soap* by the soap-boiler, in contradistinction to the
soda-soap, or *hard soap*. The latter neither retains so much
water, nor does it absorb so much as to render it soft, but hardens
when exposed to the air, and with a certain amount of water
forms a perfectly solid mass, in which it is difficult to make
impressions with the finger. The deliquescence of the former
kind of soap is derived from the stearate, margarate, palmitate,
and oleate of potash, whilst the properties of the latter are due

to the corresponding salts of soda.* Resin, in combination with either soda or potash, forms by itself a soft soap.

Soft soap is made from train oil and the drying vegetable oils; hard soap from the vegetable fats and oils which do not dry, or from tallow, and from mixtures of these fats and oils with resin. Every kind of soap found in commerce contains a variable quantity of water; part of this is in chemical combination. Hard soap becomes harder by drying, so that at last it can be pulverised. Potash-soap decomposes the salts of soda, e. g., common salt, or sulphate of soda; the potash, or the stronger base, unites with the more powerful (mineral) acid, and the fatty acid combines with the soda. There result, therefore, chloride of potassium, or sulphate of potash, and a soda-soap. It is, indeed, in this indirect manner that hard (soda) soap is sometimes manufactured in Germany.

The action of solvents upon soap is particularly interesting, and of the greatest importance in the different purposes for which it is employed. In alcohol and hot water, soap is perfectly soluble. The aqueous solution is more thickly fluid and slimy than the alcoholic solution, but both solidify to a jelly at a certain stage of concentration; opodeldoc is soap mixed with alcohol in this state of concentration. It has always been found that potash-soap is more readily soluble in water than soda-soap. This can be better seen with the salts of the pure fatty acids than with soap. Stearate of soda undergoes hardly any change when brought together with 10 parts of water, whilst stearate of potash is converted by it into a thick jelly. Oleate of soda is soluble in 10 parts of water, oleate of potash in 4 parts, and forms a jelly even with 2 parts of water; margarate of potash is converted by 10 parts of water into a transparent stiff jelly. From this, it will also be seen, that the salts of oleic acid are more soluble than those of stearic (or margaric) acid with the same base, so that the softness or hardness of the soap is not dependant solely upon the base that it used, but also upon the relative quantities of oleic and stearic acid which it contains. The fats mentioned as serving for the production of soft soap, are remarkable for the large proportion of oleic acid (olein) which they contain.

* 100 parts of dry oleate of potash absorb from the air 162 parts of water.

100	„	margarate of potash	„	55	„
100	„	stearate of potash		10	
100	„	stearate of soda		$7\frac{1}{2}$	

Cold water never dissolves the oleate, margarate, or stearate
of an alkali,—the soap of commerce, therefore,—without *decomposition.* The neutral salts are resolved into alkali, which dissolves, and into an acid salt, which is precipitated. The same
decomposition occurs when hot solutions of soap—particularly
weak solutions—are cooled.

This reaction is likewise exhibited by the neutral margarates
palmitates, and oleates of potash and soda, and it explains why
in using soap, even with the purest water, a whitish turbidity—
soap-suds—is always obtained ; the alkaline property of soap-
suds is due solely to the liberation of a portion of caustic potash
or soda, which removes the fatty impurities in water, which is the
sole object of washing with soap.

Every kind of soap, when it leaves the pan, and is afterwards
sold, is a more or less concentrated solution of soap in water,
which, when it has cooled, and become firm, should be subject to
the same phenomena of decomposition. In fact, common soap
presents a number of extremely slender crystalline fibres, but
slightly transparent, and having a silky lustre, which are surrounded by a more translucent matrix.

The physical reaction of soaps with different saline solutions,
as that of common salt, carbonate of soda, the corresponding
potash-compounds, sal-ammoniac, &c., is of the utmost importance
to the soap-boiler, because, although it may not be instrumental
in the formation, it is almost essential in separating the foreign
matters that render hard soap impure, and is also influential in
imparting to it the proper amount of water. In practice a *solution of common salt* is always employed for this purpose.

When soap, cut up into small pieces, is placed in a solution of
common salt saturated at the ordinary temperature, no action
whatever takes place. The pieces of soap, far from being dissolved or softened, swim on the surface of the solution without
even being wetted by it ; the solution of salt flows from their
surface as oil from ice. Even after long immersion, no other
result ensues than would occur if the soap were plunged into
mercury ; instead of softening, its hardness is rather increased.
If the solution of salt is boiled, the soap is softened by the heat,
and assumes the form of a gelatinous, or, rather, a thick and
doughy mass, which is equally insoluble in the saline solution,
keeping perfectly distinct from it, or, at most, separating into

flocks that swim upon the surface. These flocks harden when taken out, and cool down to hard soap. If the solution of salt is not saturated, but diluted to a certain extent, the soap and salt compete for the water after such a fashion that neither positively gets possession of it. The water is partly absorbed by the soap, but a part remains with the salt, so that a solution of soap is seen swimming upon the saline solution, which is now saturated, without mixing with it or dissolving, but still forming a distinct layer. It is only when the quantity of salt in solution is below the $\frac{1}{100}$ of the liquid, that the soap is not prevented by it from dissolving. If a solution of this kind is boiled for a length of time, the following appearances will be observed, as the water gradually evaporates.

The fluid, when steadily boiled, assumes, in the beginning, a thin, frothy character. The mass of soap and the froth become gradually thicker, until, on allowing a sample to run down the stirring rod, it is observed, by the manner of its descent, that the solution of soap, although still very soft and liquid, is nevertheless separated from the saline solution. At this period of the process, the solution of salt is so far concentrated, that the soap cannot remain any longer dissolved. It can easily be observed in what manner the stirring rod is wetted by a liquid (solution of salt) above or on the surface of which the solution of soap slides down in flat lumps or flocks, without attaching itself, or partially sticking to the rod. From this time, the solution of soap becomes constantly thicker; for the solution of salt takes water from it, in proportion as its own water is diminished by evaporation. The solution of salt collects more and more in the lower part of the vessel; the soap floating on the top boils, and throws up larger and larger froth-bubbles, until it becomes, at length, so tough and thick as to obstruct the passage of the vapours rising from below. The surface now splits up into several fields, separated from each other by deep furrows; these have not the fresh and soft appearance of the froth in the furrows, but present the appearance of dry slabs, which, being forced from side to side by the escaping vapour, slowly arrange themselves one above the other. The escape of the vapour soon becomes so retarded by the thick mass that it forces its way, as it were, through craters, and gives rise (particularly in covered boilers) to a peculiar sound (*Pfeifen*). At length, the period arrives when the attraction of the soap to the remainder of the water is so great, that it completely resists every effort of the salt to remove it, the soap and salt balancing each other with reference to their affinity for water. This state is attained when the soap, which, previous to this, was always covered with froth and bubbles, suddenly sinks, and the froth breaks up into roundish massive grains, distinctly separated from each other and from the saline solution. In this state, the mass no longer rises, even when a greater amount of heat is applied; but the saline solution is thrown up from time to time with much force from below, breaking through the granular mass (*curd*) on the surface, and then sinking down again. If soap is taken out during the boiling process, and allowed to cool, it solidifies to a more or less firm mass, depending upon the quantity of water it has retained; when it is removed in the granular state, it has the consistence and hardness of commercial soap.

Soap which contains a larger amount of water than curd soap, is called "*watered*" when water or weak lye is added and mixed with the curd in the boiler itself, or when the curd is treated subsequently with water whilst still in contact with the brine. It is called, on the contrary, "*filled*" when the water is added and stirred into the curd after its removal from the boiler, and immediately before it solidifies.

All varieties of soap are not separated with the same ease from their solutions, by means of salt. Thus, soap made from cocoanut oil requires a much larger quantity of salt to separate it from solution than soap made from tallow, the former being soluble in saline solutions in which the latter is perfectly insoluble. Rosin soap is affected by common salt in the same manner as the soaps from fat.

The same results as those obtained by the use of common salt, are also produced, although in a less energetic manner, by chloride of potassium (which acts but slightly), carbonates of the alkalies, sulphate of soda (also very weak in its action), acetate of potash, and sal-ammoniac. In weak caustic lye, soap is perfectly soluble; in strong lye, on the contrary, or when the concentration of the lye is increased by boiling, the soap separates in the same manner as from a solution of common salt. For this reason, the soap-boilers are in the habit of using weak lye, particularly in the beginning of their operations, as stronger lye, in separating the soap, would prevent the necessary amount of contact amongst the ingredients, and very much retard the process of saponification.

As glycerin has no kind of resemblance to soap, as regards its reaction with saline solutions and caustic lye, but dissolves in all with perfect ease, the use of common salt affords a ready means of separating this and other foreign matters from the soap.

It is perfectly impossible, during the preparation of soap, entirely to avoid the presence of the earths and metallic oxides. These, consequently, always decompose a small portion of the soap, combining with the fatty acid, which they take from the alkali. Portions of lime and magnesia constantly accompany the caustic lye, and are brought with it into the boiling pan, and the sides of this vessel are always sufficiently acted upon to impart a visible trace of iron or copper to the soap. The soaps formed with lime, magnesia, iron, and copper, are not soluble, and they are much less rapidly softened by heat than the corresponding alkaline compounds. They are disseminated, however, in such a minute state of division throughout the mass of hot soap, as almost amounts to solution. As the soaps of iron and copper possess the colours peculiar to the salts of those metals, the whole mass of soap acquires by their presence a uniform greenish or blue colour. This colour is partly caused by the sulphur contained in the lye (particularly in soda lye) forming sulphide of iron or copper. When a soap of this kind is allowed to cool rapidly, the cut surface presents a uniformly coloured appearance, something like wet slate. If the mass, however, cools slowly, the soaps of the earths and metallic oxides separate from the great bulk, and collect into larger or smaller groups in different parts with a certain degree of regularity, giving a *marbled* or *mottled* appearance to the cut surface. Coloured veins are then seen disseminated over a white ground; these are numerous and small when the soap is quickly cooled, and larger and farther apart when the cooling is less rapid. The soap, when it has been very rapidly cooled, assumes the appearance of granite. Sulphide of iron is the chief colouring matter of the veins; and when this is present, the speedy disappearance of the blue colour on exposure to the air, is explained by the oxidation of the sulphide of iron, nothing being left but the reddish yellow colour of the real iron soap. It is obvious, that the substances which impart the mottled appearance to soap, are only held in suspension in consequence of its thick state of fluidity. If the mottled appearance is to be entirely removed, and white soap produced, it is necessary to add a quantity of water, that those substances may subside, while the soap is still in a perfectly liquid state in the boiling-pan. This additional

41*

quantity of water is not again separated, but is sold with the soap. This is the reason why so much importance is attached in commerce to the mottled appearance of the soap; as it is a sure indication to the consumer, that the amount of water in the soap cannot exceed a certain limit, which is very much greater in white soap. It must, however, be borne in mind, that methods have lately been discovered of imparting any kind of mottled appearance to soap containing much more water than ordinary curd soap, by mixing mineral colours with it, when it has attained a certain stage of hardening.

It is obvious, that mottling is applicable only to hard and never to soft soap.

UTENSILS AND MANIPULATIONS EMPLOYED IN SOAP-MAKING.

Boiling Pans, or *Cauldrons.*—The mixture of fat and lye is boiled in iron cauldrons, which are heated either by the naked fire or by steam. The latter method is much to be preferred, as, besides effecting a great saving of labour and fuel, it obviates all danger of overheating, and gives the power of stopping the ebullition at any moment. It is, therefore, now adopted in all extensive and well-regulated soap factories. In smaller establishments, however, the old plan of boiling over the naked fire is still practised. The description and figures of the pans and fire-places will be given when describing the manufacture of olive-oil soap at Marseilles. We shall here describe the *steam-series.**

Fig. 311 represents an arrangement of three cauldrons; G designates the main-pipe or feeder, which is attached to the steam-boiler W of the establishment. It is stationary, and generally fitted against the wall, immediately above the cauldrons. These vessels are made either wholly of iron, or partly of iron and partly of wood; copper vessels are also sometimes used. In the figure, the upper portion or curb, A, is of wood, well hooped round by iron rings, and the lower portion, D, of cast iron, and so shaped that the worm may hug closely to the sides without loss of room, and the "blowpipe" fit snugly to the bottom. For the convenience of drawing off the spent lyes, a pipe and cock, I, are attached. Each of these kettles, resting on a hollow pile of brick or stone-work, M, is furnished with a welded iron worm which is connected with the main feeder at N. The steam is let

* This description is taken from Morfit's "Treatise on Soap and Candles," 1856, page 169.

FIG. 311.

on or off by opening or shutting the cock H, and the waste steam is conducted through the other end of the worm X, which passes upward by the side of its inlet, and thence out in any convenient way through the wall. Affixed to the main-feeder is another pipe, with a stopcock attached, and leading immediately downwards to the bottom of the kettle, where it is fixed to a circular iron tube pierced around its circumference with holes. It sets immediately below the worm, and is called the "blowpipe," serving to give additional heat occasionally to the contents of the kettle, as well as to stir it up when necessary—an operation more easily effected in this way than by a crutch in the hands of a workman. The whole interior arrangement of the boiling pan is seen in the figure at AD, the worm detached at K, and the blowpipe at L. These kettles are worked much in the same manner as the ordinary fire cauldrons, except that they require less attention. The charge of material is put in and melted by a rush of steam through both the blowpipe and the worm, the cock of the latter being shut off when it is necessary. The cock P serves to regulate the current from the generator.

The preceding figure represents an arrangement of three cauldrons. In large factories it is convenient to have this number; one, however, will answer in a small laboratory, though there will always be a loss of time in cleansing it, when the charge is to be changed from yellow to white soap. The curbs of conical form are preferable, though either shape may be used. Some manufacturers dispense with the iron bottoms entirely, and boil in water-tight vats or tubs made wholly of wooden-staves, hooped together with strong iron clamps.

In the steam-series above described, the steam is introduced directly into the material; but it is desirable for some soaps to apply the steam upon the outer surface of the kettle.

FIG. 312.

Fig. 312 represents an arrangement for this purpose. A is the interior of the cast-iron kettle, surrounded by brick-work; B is the outer cast-iron caldron, which should fit the inner kettle tightly, so as to prevent any escape of steam; D, D is the tube leading from the steam boiler and conveying the steam to the kettles. It is fitted with a cock, which is opened or shut according as the steam is to be let on or off, for accelerating or retarding the boiling of the soap. C, C is the tube by which the condensed vapour is discharged. The cock in this tube can be left slightly open, so as to operate as a safety-valve. E is the discharge-pipe of the cauldron.

St. John's Steam Jacket.—This apparatus, patented by J. R. St. John, of New York, accomplishes the mixing and boiling of the soap ingredients simultaneously. As the steam circulates around the kettle, and through tubes, instead of being admitted directly into the paste, a uniform temperature may readily be established. The whole arrangement is shown in longitudinal vertical section, by Fig. 313. The boiling pan, *a, a,* is enveloped by a steam casing or jacket, *b, b,* adjusted to which is the tube *k,* communicating with the steam generator, and leading the steam into the space *c, c,* between the pan and outer casing. The exit pipe

FIG. 313.

n, with its stopcock C, is for drawing off the condensed steam, as
may be necessary ; and the safety-valve *v* is a protection against
excessive pressure.

The stirring is accomplished by means of the revolving horizontal arm *d, d,* carrying teeth, *f, f, f,* and mounted upon a perpendicular shaft, *e.*

The stirring apparatus is put in motion by suitable gearing,
consisting of the bevel wheel *g,* mounted horizontally on the
vertical shaft *e,* and working into a similar wheel, *h,* on the
horizontal shaft *i,* which has a pulley, *j,* on its other end, driven
by a band or strap, E.

When the boiling is completed, the contents of the kettle or
pan are drawn off through the pipe *l,* and its branches, *m, m.*

The tubes *p, p, p,* closed at their upper ends, communicate
with the space between the pan and jacket, and by conveying
the steam throughout the contents of the pan, extend the
heating surface of the latter. They also serve the purpose of
stops, for breaking the mass as it is carried around by the stirrers
f, f, f.

The swivel or T joint *u* is so constructed that the arms *m, m* may be turned horizontally in a circle, so as to bring the cocks *x, x* over a range of receivers.

A is a cock for letting the charge into the branch pipes *m, m.* Another cock, B, is for regulating the admission of steam to the chamber *c* and the tubes *p, p.* The clutch lever D is for adjusting the cog-wheel *h* with the cog-wheel *g,* when the stirrers are to be put in motion. E is a driving band, connected with the pulley *j* on the shaft *i.* F, F are stay bolts for coupling the kettle and jacket.

Morfit's Steam-twirl.—This apparatus is used for mixing, stirring, and heating the contents of vessels of any size or form by "dry" steam at one and the same operation, and is well adapted both for the manufacture of soap and for the rendering of tallow, and similar operations.

V, V (Figs. 314, 315, 316) is the containing vessel, which may be a pan with a round bottom, as represented in the figure, or a wooden tub, the form and material being dependent upon the nature of the substance under process, and the kind of treatment to which it is to be subjected. Firmly fixed to the centre of the bottom of the containing vessel is a slightly conical metallic socket, *m,* communicating with a waste-pipe, *n;* and adjusted to and turning in this socket is the upright metallic shaft C, carrying the tubular arms or coils R, R, *r, r, r, r,* and *s, s, s, s,* arranged in horizontal and vertical series alternately, and which, together with the shaft C, constitute the heating twirl as shown in Figure 316. There may be two, four, or more of these series or branches of tubes according to the capacity of the containing vessel; and each is tightly joined with the shaft at W, the centre of which is hollow from that point upwards to the stuffing box D, as a passage way for the steam from the generator to the twirl. The lower end of the shaft is also hollow in the centre, *x,* to a height of several inches above the top of the socket, for the purpose of communication between the exit ends, *i,* of the series of branches or arms of the twirl with the waste pipe *n,* the connecting joints being made steam-tight.

The gearing by which the twirl is made to revolve, is shown at A and B, and it derives its motive power from an ordinary steam boiler or generator. The same boiler also supplies the steam for heating the twirl as it revolves; and the arrangement for this purpose consists of a piece of metal, D, adjusted to the shaft and

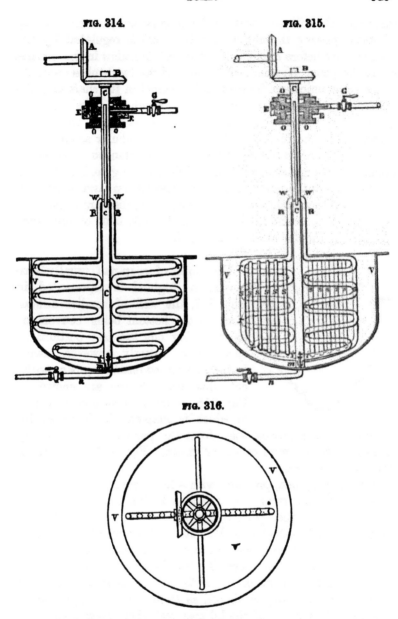

FIG. 314.

FIG. 315.

FIG. 316.

sunk in a wooden support E. This piece of metal, D, has on each
side stuffing boxes, *o, o,* and in its centre is a groove so cut as to
allow a free circulation of the steam, which is admitted through a
pipe, F, leading into it, and fixed by a screw joint. This pipe

communicates with the steam boiler or generator, and the current of steam passing through it into the twirl is regulated by the cock G. As before explained, the shaft C is hollow in the centre, above the point, W, where the mouths of the twirl-tubes enter it; and a communication between this channel in the shaft and the steam, is made by a hole penetrating through the metal into the groove. Following the course of this passage-way, the steam passes from the feed-pipe down into the tubular arms of the twirl, which twirl being made to revolve simultaneously by power applied to the gearing A, B, the mixing, stirring, and heating of the contents of the pan or containing vessel go on *pari passu* and uninterruptedly, the condensed steam being discharged through the waste-pipe *x* and *n*, into which the exit ends of the twirl-branches or tubular arms enter at *i, i.*

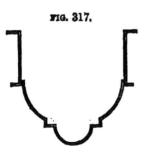

FIG. 317.

The form of soap-pan usually employed in England is represented in Fig. 317, the well of smaller dimensions at the bottom being convenient for pumping out the last portions of spent lyes. Latterly, however, manufacturers have adopted soap-pans without this well. These are shown in Fig. 318, which is a section of a pan to which is adapted Mr. Gossage's invention for transferring soap from the pans to the frames by means of the pressure of air or steam. Fig. 319 is a plan of the same apparatus.

A is the soap-pan, provided with a lid or cover, B, B', one part of which, B', is permanently attached to the upper part of the pan, while the other part, B, can be raised or lowered by the hinge or joint C, attached to B'. Round the edge of the movable part, B, of the lid is a groove, D, to which is adapted a continuous ring or rope of india-rubber, E. To the upper part of the pan and to the fixed portion, B', of the lid, is likewise attached an angular piece of iron, so shaped as to form a bearing, on which the movable portion of the lid may rest when lowered down on the soap-pan, and between which and the movable portion, the india-rubber ring E may be compressed by means of the binding screws F, F. This part of the arrangement is shown on a larger scale in Fig. 320.

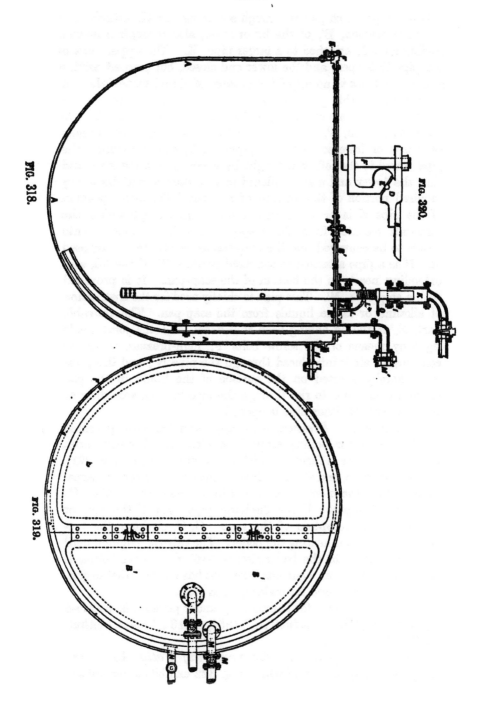

FIG. 318.

FIG. 320.

FIG. 319.

G is a pipe which passes through a stuffing-box H, attached to the fixed portion, B', of the lid or cover, also through a second stuffing-box, I, attached to a larger pipe, K. The upper end of the pipe G is open, and the lower end closed, but pierced with a number of holes to admit of the passage of liquid materials. A portion, G', of this pipe has on its outer surface a male screw fitting into a female screw, L', on the bridge-piece L, attached to the fixed portion, B', of the lid. By these means the pipe G may be raised or lowered in the soap-pan A, by simply turning the pipe round to the left or the right by means of the lever O, and thus the lower end may be placed in a position to withdraw any required portion of the contents of the pan. The upper portion of the pipe K is provided with a stop-cock, through which the materials elevated from the soap-pan may be conducted into spouts to be conveyed into the soap-frames or otherwise disposed of. M is a pipe attached to the fixed portion, B', of the lid, and extending nearly to the bottom of the soap-pan. It is provided with a stop-cock, M'. This pipe is chiefly used for the abstraction of alkaline or saline liquids from the soap-pan. The movable part, B, of the lid having been shut down, and the joint made tight by means of the screw bolts F, high-pressure steam or compressed air is introduced through the pipe N ; and the pressure thereby exerted on the surface of the liquid in the pan causes the contents to rise through the pipe M or G, according as the stop-cock of either pipe is opened.

Boiling under Pressure.—An apparatus has been patented by Mr. Dunn for boiling soaps in a close vessel under pressure. The boiler is furnished with a man-hole door, A (Fig. 321), a safety-valve, B, and all the ordinary appendages of such an apparatus, with a thermometer plunging into a mercury chamber, C ; D is the feed-pipe ; E the discharge-pipe, and F the pan or frame for the reception of the soap. G, G, G shows the fire-place and flues for heating the boiler. The valve is weighted until the temperature in the boiler rises to 310° F., and the boiling is complete in about an hour after the heat has attained that degree. This apparatus is used for making silica soap.

Another apparatus for boiling soap under pressure is described by Morveau (" *Génie industriel,*" 1854, p. 216 ; "*Polyt. Central-blatt,*" 1855, p. 60).

Tilghman's Apparatus for preparing Fatty Acids and Soap.—This apparatus, of which Fig. 322 exhibits a vertical and

FIG. 321.

FIG. 322.

Fig. 323 a horizontal section, consists of a continuous tube or channel, in which fatty bodies are subjected to the action of water or of alkalies at high temperatures and under pressure. It is used especially for two purposes :—1. To obtain fatty acids and glycerin by heating neutral fats with water to about the melting point of lead (p. 637). 2. To produce soap by heating neutral fats with *carbonated* alkalies.

The first of these processes belongs more especially to the candle-manufacture, but as it is not described in Pt. II. of this volume (having been published at a later date), we shall here give the description in the words of the specification. The mode of using the apparatus is indeed nearly the same, whether a fatty acid or a soap is to be prepared.

FIG. 323.

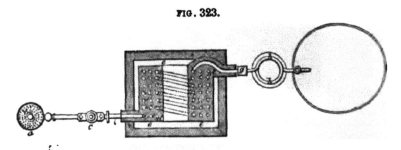

"I place the fat or oil in a fluid state in the vessel *a*, with from one third to half its bulk of warm water. The disc or piston *b*, perforated with numerous holes, being kept in rapid motion up and down in the vessel *a*, causes the fat or oil and water to form an emulsion or intimate mechanical mixture ; and a forcing pump, like those in common use for hydraulic presses, then drives the mixture through a long coil of very strong iron tube, *d, d, d, d*, which being placed in a furnace, *e, e*, is heated by a fire, *f*, to the temperature above mentioned. From the exit tube, *g*, of the heating tubes *d, d, d, d*, the mixture which has become converted into fatty acids and solution of glycerin, passes on through another coiled tube, *h, h, h*, immersed in water, by which it is cooled down from its high temperature to below 212° F., after which it makes its escape through the valve *i* into the receiving vessel. The iron tubes are about an inch in external diameter, and about half an inch in internal diameter, being such as are in common use for Perkins's hot-water apparatus. The ends of the tubes are joined together by welding, to make the requisite length ; but where welding is not practicable, I employ the kinds of joints used for Perkins's hot water apparatus, which are now well known. The heating tube, d, d, d, is coiled several times backwards and forwards, so as to arrange a considerable length of tube in a moderate space. The different coils of the tube are kept about one quarter of an inch apart from each other, and the interval between them is filled up solid with cast-iron, which also covers the outer coils or rows of tubes to the thickness of half to three-quarters of an inch, as shown in Fig. 323. This casing of metal ensures a considerable uniformity of temperature in the different parts of the coil, adding also to its strength, and protecting it from injury by the fire.

The exit valve, i, is so loaded, that when the heating tubes d, d, d, are at the desired working temperature, and the pump e is not in action, it will not be opened by the internal pressure produced by the application of heat to the mixture ; and, therefore, when the pump e is not in action, nothing escapes from the valve i, if the temperature be not too high. But when the pump forces fresh mixture into one end, j, of the heating tubes $d, d, d,$ the exit valve, i, is forced open, to allow an equal amount of the mixture which has been operated upon to escape out of the cooling tubes, h, h, at the valve i, placed at the other end of the apparatus. No steam or air should be allowed to accumulate in the tubes, which should be kept entirely full of the mixture. For this purpose, whenever it may be required, the speed of the pump should be increased, so that the current through the tubes may be made sufficiently rapid to carry out with it any air remaining in them. Although the decomposition of the neutral fats by water takes place with great quickness at the proper heat, yet I prefer that the pump e should be worked at such a rate, in proportion to the length or capacity of the heating tubes d, d, d, that the mixture, while flowing through them, should be maintained at the desired temperature for about ten minutes before it passes into the refrigerator or cooling parts, h, h, h, of the apparatus.

" The melting point of lead has been mentioned as the proper heat to be used in this operation, because it has been found to give good results ; but the change of fatty matters into fat acids and glycerin takes place with some materials (such as palm oil) at the melting point of bismuth ; yet the heat has been carried considerably above the melting point of lead without any apparent injury, and the decomposing power of the water becomes more powerful as the heat is increased. By starting the apparatus at a low heat, and gradually increasing it, the temperature giving products most suitable to the intended application of the fatty body employed, can easily be determined.

"To indicate the temperature of the tubes d, d, d, I have found the successive melting of metals and other substances of different and well-known degrees of fusibility to be convenient in practice. Several holes, half an inch in diameter and 2 or 3 inches deep, are bored into the solid parts of the casting surrounding the tubes, each hole being charged with a different substance. The series I have used consists of tin, melting at 440°; bismuth, at about 510°; lead, at 612°; and nitrate of potash, at about 660° F. A straight piece of iron wire passing through the sides of the furnace to the bottom of each of the holes, enables the workmen to feel which of the substances are melted, and to regulate the fire accordingly. It is important for the quickness and perfection of the decomposition, that the oil and water, during their entire passage through the heating tubes, should remain in the same state of intimate mixture in which they enter them. I, therefore, prefer to place the series of heating tubes in a vertical position, so that any partial separation which may take place while the liquids pass up one tube may be counteracted as they pass down the next. I believe that it will be found useful to fix, at intervals, in the heating tubes of such apparatus as may admit of such an addition, diaphragms pierced with numerous small holes, so that the liquids, being forced through these obstructions, may be thoroughly mixed together. I deem it prudent to test the strength of the apparatus to a pressure of 10,000 pounds to the square inch, before taking it into use; but I believe that the working pressure necessary for producing the heat I have mentioned will not be found to exceed 2,000 pounds to the square inch. When it is desired to diminish the contact of the liquid with iron, the tube or channels of the apparatus may be lined with copper.

"The hot fat-acids and solution of glycerin which escape from the exit-valve of the apparatus are separated from each other by subsidence. The fat acids may then be washed with water, and the solution of glycerin concentrated and purified by the usual means.

"The fat acids thus produced are applicable to their various uses, according to quality; but they may be further bleached and purified by distillation.

"To prevent injury to the pipes by the small portion of acetic or other acid generated during the process from the fatty material, it is necessary to add a little carbonated alkali to the water and oil before pumping into the tubes.

"This process is equally applicable for the conversion of the fatty matter into soap; and then it is necessary to dissolve the requisite amount of carbonated alkali in the water, and to keep the temperature from 350° to 400° F. The carbonic acid eliminated from the carbonated alkali will pass off with the soap through the exit valve. When the quantity of water used does not exceed that which it is intended the soap shall retain, the warm paste may be received into the frames, as it issues from the valve—it being ready, when cool, for the market."

A process similar in principle has been patented by Boullenger and Martin, of Paris. The fats are resolved into fatty acids and glycerin by placing them for some hours at a high temperature in contact with ordinary water, or with water containing 1 or 2 per cent. of alcohol or 5 to 6 per cent. of wood-spirit, with a mixture of bisulphide of carbon, oil of turpentine, and oil of lavender in equal parts.

Soap-frames.—After the fat has been saponified and the soap has been separated and clarified by processes to be here-after described, it is ladled or run out into moulds or frames of wood or iron, and there left to cool.

The moulds are so constructed as to afford a closed space, from which no soap can escape, and yet to be easily taken to pieces. The ordinary frames, represented in Figs. 324—327, are of

FIG. 324.

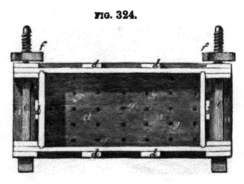

FIG. 325.

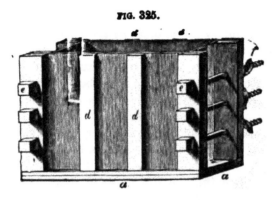

FIG. 326.   FIG. 327.

wood. The bottom, *a, a*, seen at full breadth in Fig. 326, and in longitudinal section in Fig. 327, is constructed of two layers of planks. In the upper one there are four grooves, *b, b*, into which the projections of the sides fit. The two narrow sides are also supported on the outside by the cross piece *c, c.* All the sides are strengthened by the supports, *d, d.* When all the parts are put together, the bolts *e, e*, having screws at the other end, have only to be inserted through the projecting parts of the longer sides, and made fast by the nuts *f, f*, at the ends, so as to form the whole into a solid box. A cloth spread over the bottom prevents any soap from passing the holes *g, g*, through which the lye drains off.

Another form of wooden frame is shown in Figs. 328, 329.

FIG. 328.   FIG. 329.

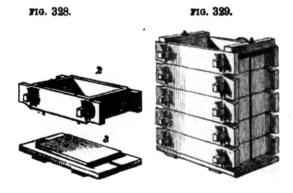

A number of these are piled together, so as to form a well, for receiving the soap. The bottom of the well and a single frame are shown in Fig. 328. This kind of frame is adopted almost universally in England where wooden frames are used.

When soap was subject to excise duty, the law required that the dimensions should be exactly 15 in. by 45 in. inside, and not less than 45 in. deep. The duty was charged by measurement. The same dimensions are still retained by manufacturers of soap.

When the soap has become solidified, the frames are lifted off, one by one, and the soap remains on the movable bottom, ready to be divided into slabs and subsequently into bars.

For common yellow soap, frames constructed of cast iron are extensively used. Fig. 330 is a perspective view of one of these frames when ready for use, the plates are held together by screwed pins, *d d*; the internal dimensions are 45 inches long by 15 inches wide, and 52 inches deep. Fig. 331 is a view of the bottom plate, and Fig. 332 of one of the end plates, showing the notches in which the pins are placed. These can be placed

FIG. 330.

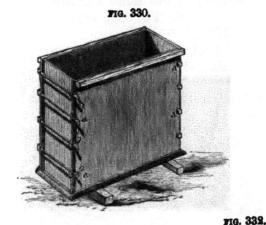

FIG. 332.

FIG. 331.

42*

on wheels, so as to be readily moved about from place to place. The good conducting power of the metal promotes the cooling and solidifying of the soap-paste.

Figures 333, 334, and 335 represent the iron soap-frames used in America.

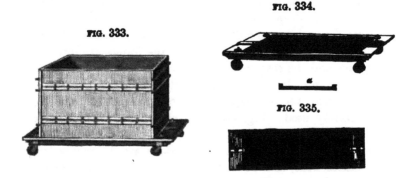

FIG. 334.

FIG. 333.

FIG. 335.

For stirring the soap-paste in the pans and frames, as in the operation of mottling, a tool called a *crutch* is used, composed of a long wooden handle (Fig. 336), adjusted at the end to a

FIG. 336.

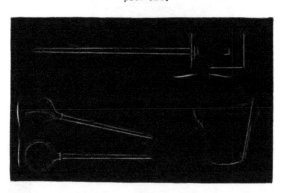

board. Large colandered iron ladles with long wooden handles, are employed for dipping the paste out of the kettles ; and copper buckets for conveying it to the frames.

Cutting Operation.—When the soap sets firmly, the frames are lifted off or unbound, according to their construction, and the solid mass of soap (Fig. 337) is divisioned off on the sides by

FIG. 337.

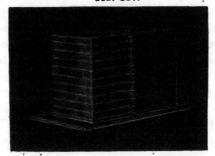

means of a scribe (Fig. 338), which is a wooden slab, carrying on its smooth side a number of slender iron teeth.
The workman then taking a brass wire (Fig. 339) directs
it in the track of the teeth, and thus cuts off a slab of the
required thickness, as shown at Fig. 340. When the
whole block is thus divided into slabs, the latter are in
their turn reduced to bars and lumps of smaller dimensions, the usual size of the bars being 15 inches long by

FIG.
338.

FIG. 339.

FIG. 340.

about 2½ inches every other way. The pound lumps are about 5 or 6 inches long, 3 inches wide, and 3 inches deep. It is convenient to have a scribe with several sets of teeth arranged at different distances. Such an instrument may be made of a piece of hard wood about two inches square, and bearing teeth on each of its four sides. On one side they may be set 1 inch from each other, and on the others 2, 2½, and 3½ inches respectively.

A much more rapid method of dividing the blocks into lumps is that invented by Van Haagen, of Cincinnati, U. S., which requires the use of two pieces of machinery (Figs. 341, 342).

FIG. 341.

The first is called the *slabbing and barring machine*, and consists of a carriage, A, which is so grooved at the top as to allow the wires to pass entirely through the block of soap. This carriage is then moved back to the driver, B, on which is placed a whole block of soap as it comes from the frame. This is done by a peculiar truck, constructed for the purpose, as shown in Fig. 334. The block of soap having been first cut from the bottom of the frame, this truck is run to the side of it, and by means of a rack and pinions, worked with levers, the block of soap is slipped on the truck, brought to the machine, and placed thereon. All this is done with great ease and dispatch, and by the same power.

The range of wires, C, is regulated by corresponding gauges in the upright posts, which allow it to be set to cut slabs of any desired thickness. The block of soap is forced up to these wires by the driver, B, propelled by means of racks and pinions and a winch. It will be seen that in this way the block will be converted into slabs. There is a similar arrangement of horizontal cutting wires, D, confined to a vertical motion by the posts of the frame. These wires are also arranged as above, so that any desired bars may be cut. It is caused to descend by the action of the rack and pinions and winch as above, and by this part of the machine the slabs are converted into bars without handling. They are consequently much smoother and neater than they could be when cut in any other manner.

The wires, being fastened at one end to a spring, E, E, easily yield and form the required loops at the beginning of the operation, and then both ends become fixed, so that the loops cannot get any larger if the soap is very hard; in which case the long loop is more apt to warp and cut uneven. The steady motion of this machine permits the use of a much smaller wire than will do for hand-cutting, and consequently the work is much smoother. This apparatus cuts the blocks of soap into bars as long as its width. To make pound lumps, or small cakes and tablets, the slabs must be transferred to the second, or *caking machine* (Fig. 342).

FIG. 342.

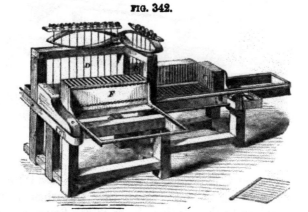

The slabs are arranged in as great a number as can be placed upon a range of rollers, A, and forced through the range of wires, B, by the driver, C, which is propelled by racks and pinions and a crank. The soap having been forced through lengthwise, and the crank being shifted, it is then forced through the range of

wires, D, by the driver, E. Both the drivers are connected with
the same crank, and by displacing it from the one, it gears itself
into the other. The wires are arranged in the same manner as
in the slabbing machine. They may be readily shifted so as to
cut any desired shape or size.

This mode of cutting gives great smoothness and uniformity of
weight and size to the bars and lumps, saves handling, scratching,
and bending, and effects a large gain over the usual method, in
time, labour, and expense.

Fig. 343 is a perspective view of a soap-cutting machine much
used in this country. It consists of a fixed frame of woodwork,
A, A, and a movable lever-frame, B, B, attached to A, A, by the
centre-pin C. The frames are wide enough to receive a slab of
soap 45 inches long by 15 inches wide. This is placed in an inclined position, as shown by the dotted lines, resting on the bar D
of the fixed frame, and against a number of wires forming part of
the movable frame.

FIG. 343.

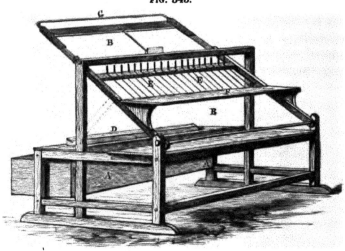

When the lever G is pressed down, the wires pass through the
slab of soap, dividing this into regular bars, and when the handle
is again raised up to the position shown in the figure, the bars of
soap are found on the table F, ready to be carried away.

Another machine for cutting soap, invented and patented by
Messrs. Davis and Hooper (1858, No. 668), is represented in
Figs. 344, 345, and 346; Fig. 344 being a side elevation, Fig.
345 an end elevation, and Fig. 346 a separate section of the

FIG. 344.

FIG. 345.

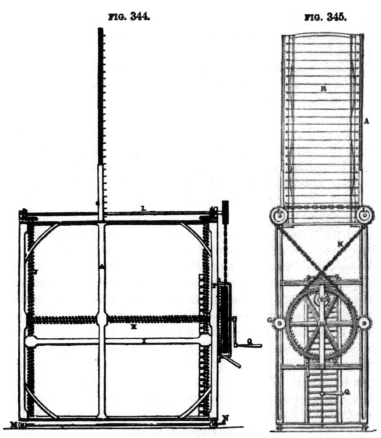

grooved head-piece and apparatus for moving it. A is a frame having transverse cutting wires, B, kept stretched apart by springs, C. By this frame and wires a block of soap is cut into slabs, the wires being fixed to the desired distance apart. The frame A is moved from end to end of the machine in a vertical position by means of two screws, H, H. The frame A, which carries the wires, slides or fits between the uprights A, one on each side of the machine, which at their upper and lower ends, fit and slide on V-guides on the upper and lower rails at the sides of the machine. Through these uprights, A, there are female screws, in which the screws H work. When the frame A arrives at the end of the machine, the wires, in passing out of the block of soap, enter grooves, F, F, in the

FIG. 346.

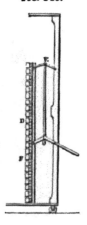

head-piece, D, which is pressed up to and back from the soap by means of the lever and parallel movements, E, shown in Fig. 346.

The soap is cut into bars by means of another frame of wires, similar to that above described, and this frame of wires slides and is fixed between two bars, I, which are simultaneously moved by means of four upright screws, J, working in female screws at the ends of the bars, I. Motion is communicated to the four screws by the endless chain, K, which gives motion to the two shafts, L, which drive the four screws, J, by bevel-toothed wheels. Motion is communicated to the two screws H, H, by means of a cog-wheel, P, which is driven by the crank-handle, Q. The endless chain receives motion by the chain-wheel, R, which is formed with internal cogs, and receives motion by the pinion S, put in motion by a crank-handle or key placed on its axis.

Lesage,* of Belleville, near Paris, has patented in France a soap-cutting machine, in which a cylinder, lying in a horizontal position, but capable of being raised into an inclined or vertical position for the convenience of filling it with soap, is closed in front by a plate perforated with a number of holes. The other end of the cylinder is closed by a piston, which, when urged forwards by the piston-rod, presses upon the mass of soap, and forces it through the holes in the front plate, thereby moulding it into sticks, whose transverse sections have the size and form of the holes. They are cut off from the cylinder by a rotating knife, or by a wire moving vertically up and down.

Drying.—The bars, lumps, tablets, and cakes, cut as directed by either of the preceding methods, are then carried to the

FIG. 347.

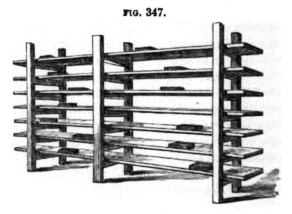

* "*Génie industriel,*" Jan. 1855, p. 26; *Polyt. Centralbl.* 1855, p. 895.

drying room, and placed on slat-work frames, made of wooden uprights and cross pieces (Fig. 347). The openings between the slats permit the free access of air, and thus promote the drying. Sometimes the bars of commoner soaps are placed on the floor crosswise, in piles several feet high, with intervening spaces for free circulation of air; but some injury is apt to ensue from the slipping of the bars and the tumbling down of the piles.

Stamping.[*]—When the soap is to be impressed with any device or lettering, it is subjected to this operation after it has acquired sufficient firmness to sustain pressure. For this purpose the press Fig. 348, with suitable boxes and dies, is used. It has two spiral springs, A and B, by which the cake of soap is immediately expelled from the box, C, as soon as it is pressed. The workman knocks it off the tablet that is to take its place, and so the pressing goes on without any delay in removing the tablets of soap as fast as they are finished. D is a rope, suspending a wooden rod, E, which serves as a support to the bottom of the die during the pressure. The box, C, is movable, being merely fastened by screws, and, when necessary, may be replaced by others of different sizes. The die from which the tablet is to receive a

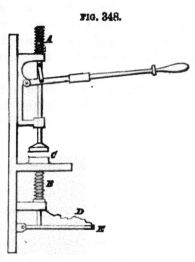

FIG. 348.

FIG. 349.

* Morfit, p. 185.

device or the impress of the manufacturer's name, is screwed to
the top of the box C, and may also be changed when required.

As the coarse soap in pound lumps is generally stamped only
with the maker's name, or words expressive of its kind and
quality, this is more expeditiously put on by means of a stereo-
type plate, worked by hand. The implement, and the mode of
using it, are shown in Fig. 349.

HARD SOAPS.

Hard soaps, as already observed, are those made with non-
drying oils or solid fats, and soda. Their hardness is in proportion to the amount of stearic and margaric or palmitic acid which
they contain. Soda soaps made with drying oils, such as linseed
oil, are pasty, and easily liquefied by a small quantity of water;
in fact, they approach to the character of soft soaps made with
potash.

The most important kinds of hard soap are those made with
tallow, and with olive oil, the former material being used in
England and other northern countries, the latter in the south of
Europe.

Marseilles, Venetian, or Olive-oil Soap.

The saponification of olive oil by means of caustic soda, forms
an extensive and flourishing branch of industry in the southern
countries of Europe (in the south of France and north of Italy),
and on the northern coasts of Africa; in other words, where the
olive-tree is a native of the soil. With reference to the oil itself,
the relative value of the different kinds depends upon their
amount of colour, and of the fatty substance recognised by
Pelouze and Boudet as a combination of olein and stearin. The
value of the oil for soap-making is greater in proportion as it is
less coloured, and contains a larger amount of stearin—in proportion, therefore, to the facility with which it solidifies in the cold.
The oils of Provence rank first for excellence. The finest salad
oil is too costly, and likewise too fluid for the manufacture of
soap; but the oil obtained by the second pressing is peculiarly
adapted for this purpose, on account of its cheapness and larger
amount of stearin. Sometimes a small quantity of seed-oil is
added to it, that the soap may acquire a softer consistence.

The lyes prepared by lime from soda are of three differen

strengths, the strongest is indicated by 20° to 25°, the lye of middle strength by 10° to 15°, and the weakest by 4° to 5° of Baumé's hydrometer. All three are preserved separately in walled cisterns. Formerly, when saponification was carried on only with natural soda-ash, the chloride of sodium contained in the ash was also dissolved in the lye ; but as the presence of salt is very objectionable in the first stage of the boiling process, and is required only in the subsequent additions of lye, an evil is incurred by employing it, which is not the case when artificial soda is used. Pure soda is therefore taken for the first lye, and afterwards soda containing chloride of sodium, which is especially prepared for this purpose.

The quantity of soap which is made at once, is technically called "a boil," and the operation is commenced in two boilers, and finished in one. The arrangement of the apparatus is shown in Fig. 350. The bottom and lower part of the boiling pan, A, are of cast-iron or copper. The sides are composed of brick-work, erected and lined with cement-mortar to resist the action of water. The upper part, b, b, b, b, which never comes into contact with the fire, and is intended to afford space for the soap to rise, expands in the form of a cone. The fire-place, f, is separated from the ash-pit, h, by the grate, a. The fire, after having heated the bottom of the pan, passes by the flue, x, x, x, half round the side of the pan into the chimney, E. This is accessible, for the purpose of cleaning, by the door P; the soot is thrown into the pit, K. A tube with a cock leads from the lowest part of the pan for the removal of the (under) lye. The

FIG. 350.

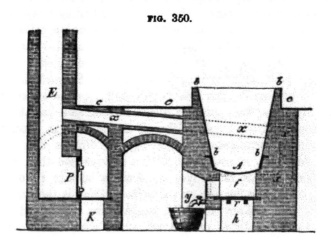

whole of the pan is sunk into the floor of the boiling-house, which is made of planks, stone, or iron plate, in such a manner that the brickwork of the upper part projects to about three feet above the floor.

The pans or boilers are calculated to contain 240 cwts. of soap. The upper part is frequently constructed of mere staves, without any bottom, and these are fixed, water-tight, into the rim of the boiler (Fig. 351). This arrangement is quite as convenient, and much cheaper. Sometimes a siphon is introduced between the bottom and the upper part, instead of the cock and tube, *y*. The plan of surrounding the pan with an external casing, and using the intermediate space for the purpose of applying steam heat, has been abandoned, as it is preferable to run some risk of burning, rather than forego the advantage of violent ebullition, which is valuable from its mechanical agency, and answers the purpose of a stirring apparatus: it can only be effected by an open fire, or by throwing a jet of steam into the mixture.

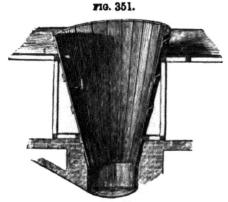

FIG. 351.

A boil, which comprises generally 120, 140, to 160 cwts. of oil, passes through three stages: the preliminary boiling (*empâtage*), the cutting up of the pan (*i. e.*, the addition of saline lye, *relargage*), and the clear boiling (*coction*), to which, for mottled soap, the marbling or mottling process (*madrage*) must be added.

For the preliminary boiling, a lye of from 8° to 11° is used, and this is prepared in the boiler by mixing together weaker and stronger lye, free from common salt. As soon as it has attained the boiling point, the whole of the oil is admitted at once, and combination is accelerated by stirring. A dirty whitish mass, like an emulsion, is produced; if this appears too thin, which results from an excess of lye, more oil is added; if, on the contrary, oil is seen floating on the surface, there is then need of more lye. In a short time, the temperature, which had been lowered by the addition of cold oil, again rises, and the mass begins to boil steadily, becoming thicker and thicker as the water gradually evaporates. This first action of the oil and alkali can only take place rapidly and completely, when the soap remains dissolved in the lye. An excess of lye of 8° to 11° may, however, produce a contrary effect; it then becomes necessary to added an extra quantity of weak lye, and stir until no further separation is visible. During this preliminary boiling, the mass of soap, inasmuch as it is dissolved in the lye, comes into immediate contact with the sides and bottom of the metallic vessel. Hence, when the whole has acquired a thicker consistence, a portion frequently attaches itself to the bottom of the vessel, becomes overheated, and burns. This burning is indicated by a black smoke (called by the Provençals, *tabaco*), which passes off here and there with the vapour, and is an indication to reduce the fire as quickly as possible, and arrest the mischief by adding some gallons of the strongest lye. By this means, a slight separation of the soap from the lye is occasioned, and the contact between the former and the metallic surface destroyed. In the ordinary process, when such interruptions do not occur, the soap boils steadily, throwing up froth as high as the breast of the boiling pan. Towards the end, the proper quantity (6 to 8 lbs.) of green vitriol is added, for the purpose of mottling; this is immediately precipitated by the lye, partly as sulphide of iron, and partly as iron soap. As soon as the mass of soap has become perfectly uniform, and has acquired the proper consistence, *i. e.*, is no longer like an emulsion, but has acquired more completely, though not entirely, the nature of soap, the fire is extinguished, and the salting process begins.

The lye which has been added up to this period, parts with the whole of its alkali to the oil, and remains merely as water, of which a small portion only is evaporated in the pan. If the stronger lye were added at once, it would become instantly diluted by the excess of water, and the object which it is intended to serve would be defeated. This water must therefore, be got rid of, not, however, by boiling, but by means of common salt, or rather by the lye above named, containing common salt. While one workman pours this gradually into the boiler, another is occupied in stirring the two thoroughly together, until the soap—having become insoluble in the solution of salt—begins to congeal, and separate in flocks. In the course of three hours, the soap has so far separated that the lye below it (*under-lye*) may be drawn off. The quantity drawn off must amount to double that of the saline lye employed. The soap is now all scooped into *one* pan to undergo the clarifying process, and soap and lye are allowed to boil with only just sufficient common salt to keep the soap perfectly distinct, as, if this were not done, the thick mass would certainly attach itself to the vessel. When the fire has again burnt up, about 220 gallons of a mixture of pure saline lye of from 18° to 20° are added, and the boiling is continued. The workmen keep the boil constantly stirred. particularly the surface, partly to hasten the combination of the caustic soda, and partly to prevent the liquid from boiling over. During the preliminary boiling, the caustic soda combines rapidly with the fat (the soap is then said to "eat well"), and this is also the case in the first period of the clarifying process, but the nearer the process approaches to a close, the more difficult it becomes to incorporate the last portions of soda with the soap. The further assimilation of soda is now chiefly dependent upon the state of concentration of the lye (solution of salt) in the pan ; if this is too concentrated, the soap becomes too dense and dry. To establish the proper relationship between the water in the soap and that of the lye, it is necessary that the amount of salt in the boiler should not exceed a certain limit. When, therefore, the caustic soda of the lye has been incorporated with the soap, or when the lye appears no longer caustic, and produces no burning sensation on the tongue, that which is in the lower part of the pan is drawn off, and is replaced by a fresh portion, about 275 gallons, of the same mixture of lye. The future stages of the process consist in four, five, or six repetitions of the same operation,* during which the soap acquires greater consistence, and the alkalinity of the lye is each time less easily destroyed. At length the lye retains its entire alkalinity, and the soap, now near its completion and highest consistence, begins to boil in jerks, and not smoothly and steadily, as before. The frothy mass begins to form itself into granules, the soap takes the form of curd, and the clarifying process is finished. The curd, when pressed against the flat part of the hand, should form a

---

* If the lye could be procured quite free from common salt, which is never the case, there would be no need of drawing it off, and the boiling might be completed with one and the same quantity of common salt. The lye as it is actually employed, however, would soon raise the amount of common salt to an injurious extent.

firm, granular, crumbly mass. The smell of oil is now no longer perceptible. According to the nature of the oil, and the state of the weather, the fire is kept up ten, twelve, or eighteen hours after the addition of the last portion of lye, and is then extinguished.

In consequence of the iron and sulphur in the soda, and the green vitriol which was added, the curd has a uniform slate-colour, produced by the sulphide of iron and iron soap. These substances must not, however, remain uniformly disseminated through the mass, but must be diffused as veins, in order to give the mottled appearance to the soap. But the proper consistence, which has been somewhat exceeded, must first be imparted to the curd, and this depends upon the strength of the under lye. With this object in view, the contents of the pan are left at rest for half an hour, until the under-lye has collected, and this is then drawn off. A tedious operation now begins : some workmen mount upon boards laid across the pans, and work the mass about with rakes, while others, at certain intervals, sprinkle the surface with weak lye. The motion of the stirrers must be carried on in such a manner as to bring the lower parts of the curd to the top, partly that all portions may be equally impregnated with lye, and partly to mingle the coloured parts, which have chiefly collected at the bottom in the interval, with the rest of the mass. During this operation, the roundish grains open, and unite to produce a uniform, connected, tough mass. As soon as the workman perceives that the proper consistence has been obtained, which very much depends upon the strength of the lye worked in, the mass is scooped into the *moulds*, where, in eight or ten days (with the above quantity), it cools down to mottled soap. Exposed to the air, the colour of the mottling changes and becomes lighter, so that the reddish tint, called *manteau Isabelle*, is diffused over the bluish mottled mass.

In the preparation of *white soap*, the same process is followed in the beginning, with the exception of the addition of green vitriol, which is of course omitted. Towards the end, however, certain deviations are necessary, with the view of separating by subsidence the iron-compounds which may have been formed, and the impurities. This is done by softening and liquefying the finished soap, by the addition of lye and heat to such an extent that the heavier iron-compounds sink below the lighter mass of soap. After standing for a sufficient length of time, the soap is pure, and of the ordinary yellowish-white colour; another pre-caution is, however, taken to pour it through sieves into the mould, to retain any casual impurities. After the removal of the white soap, there is always a residue of dark-coloured soap called "nigre" left behind. This amounts to ⅓th of the whole mass, and it may be used for the purpose of mottling.

It is evident, that the French oil-soap belongs to that class of soap which is always brought into the mould with a greater portion of water than is necessarily present in the curd. In mottled oil-soap, this excess of water is reduced within certain limits, which cannot be exceeded without causing the separation of the mottling. In white oil-soap, on the contrary, the quantity of water is left entirely to the option of the boiler; he can incorporate as much water with the soap as it will bear without impairing its firmness.

Curd Soap from Tallow.

1. *By the old German Method.*—The original German process of preparing hard (soda) soap from tallow is interesting and remarkable, because, by the use of certain ingenious contrivances—the result of very long experience—which have been confirmed, approved, and explained by modern theory—an exceedingly pure and perfectly hard curd soap is obtained from very impure materials, viz.: from crude tallow, and the lye of ashes, or from potash-lye.

An imperfect potash-soap is first prepared, which, in the subsequent boiling, is converted into soda-soap; this is supplied at the same time with an excess of alkali, and the neutral soda-soap produced is then boiled down to curd.

The indications of progressing saponification, through all the grades of the process, are so distinct, that it was never thought necessary in the first instance, to work with weighed quantities. All that was requisite was to add more lye or more fat, according to the appearance of a specimen, until the proper state had been attained.

The ashes are treated with lime exactly in the manner stated above, and as much tallow is added as the saponifying tub is thought capable of converting into soap; the boiling is begun with lye of 20°. The melted tallow immediately forms a kind of milk with the lye, in which the constituents, although not by any means in complete chemical union, are, nevertheless, in such a state of combination, that they cannot be distinguished from each other. On continued boiling, this mass becomes more and more translucent, clearer, and more thickly fluid, and the fire must consequently be regulated with care, to avoid burning. When finished, the result is a very uniform, slightly-coloured, clear transparent solution of potash-soap in the water of the lye (or in weak lye, when the whole of the potash is not in combination), and has altogether the appearance of a thick syrup. It is obvious, that this transparency cannot be brought about, as long as there is unsaponified fat present; it is this which, swimming about in small globules, gives the mixture the appearance of milk. This turbid appearance is a sure indication that more lye is requisite, and that the boiling must be continued; but it is always better to dissolve a few drops of the mixture in pure rain-water, that there may be no mistake. It is just possible, that the turbid appearance might arise from an excess of lye, or, as it is technically termed, from the soap being *overdone*. The water dilutes the lye, and the real nature of the soap then becomes apparent. If the lye contains free lime, the turbid appearance may arise from that cause, but it should then disappear on the addition of a little carbonated alkali. When, therefore, the milky nature of the solution continues after this addition, there is certainly a want of alkali, and the boiling must be continued. The lye employed for this purpose is much weaker than that used at first.

When the solution has become sufficiently clear, and flows from the spatula in a thick continuous stream (not in drops), which can be wound round it like treacle, and solidifies to a thick jelly when thrown upon a cold stone, the *salting process* begins, i. e., salt is thrown into the pan (previous solution being unnecessary), and is allowed to boil with the solution of soap until it is dissolved, and begins to act upon it.

The soap having become insoluble in the solution of salt, coagulates into a whitish mass consisting of small flocks, and on being removed from the pan, allows the mother-liquor to run off from the interstices. After remaining at rest for some time, the under-lye subsides, and can be drawn off from the pan. If the pan has no exit tube and stop-cock, as is the case in the German soap works, the soap is scooped in the meantime into the *cooling cistern*, and the under-lye can then be removed.

The process of salting answers several purposes. In the first

43*

place, it effects a chemical decomposition of the soap; for potash-soap and chloride of sodium mutually decompose each other, and produce soda-soap and chloride of potassium. This, together with the excess of common salt, causes the separation of the soda-soap, and the formation of the under-lye. In this separation, the excess of lye, the foreign salts in the ash, and the impurities—chiefly those produced by the action of the lye upon the mem-branous parts of the crude tallow—are all taken up by the under-lye. The salting process is, as it were, a means of washing the soap, by which it is obtained quite clean, although the crude materials used in its production were very impure.

Attention has already been called to the fact, that ashes—in consequence of the foreign salts which they contain—do not afford strong lye; indeed, the lye obtained from them is so dilute, that hardly any pan would be large enough to contain the whole quantity of fluid necessary to supply the amount of potash required to saponify an ordinary charge of fat. Another import-ant action, therefore, accompanies the salting process, that of separating from the soap the excess of water, which would other-wise render the later additions of lye inefficient (as was the case in the preparation of oil-soap). It is also worthy of notice, that many soap-boilers add the salt long before the potash-soap is completely formed. Although the soap is not spoilt by this mode of proceeding, inasmuch as it must always be subsequently boiled with alkali, and therefore acquires the deficient quantity; yet, it is not advantageous, for this reason, that an imperfect soap is not completely decomposed by the addition of salt, but remains viscid, and retains the under-lye in great part, which is quite opposed to the object in view.

When the under-lye has subsided sufficiently, the coagulated soap is again brought into the clean pan, and boiled with an addition of weak lye. After some time, it re-dissolves completely in this weak lye, forming a solution as at first, with this difference only, that it now consists chiefly of soda-soap. At the same time, more potash enters into combination, assisted by the viscid nature of the mixture, and this is promoted by an addition of lye from time to time. Lye is constantly added in this manner, until the soap is gradually saturated with alkali. Before this point is attained, the process must be often repeated; and, indeed, the number of repetitions must increase with the impurity of the ingredients and the dilute nature of the lye. Formerly, when crude tallow was commonly used, five repetitions of the process were hardly sufficient; under other circumstances, it may be finished in two. The boilings subsequent to the first, are performed precisely in the same manner, only less salt is required at each repetition, as there is naturally less potash-soap to be decomposed at the end than at the beginning.

When the last portion of water has been added, which may be the second or sixth portion, according to circumstances, or when the soap has become sufficiently firm and pure, it is then boiled clear, i. e. boiled until it has acquired the proper proportion of water. The same phenomena then occur as those already described, the solution of salt becomes more and more concentrated by boiling, and extracts the water from the soap. This first boils softly, throws up froth to a considerable height in small bubbles, until, at length, the frothy surface forms large brilliant bubbles, and the peculiar sound is produced under the boarded covering of the pan. Lastly, when the froth sinks down, and the curd itself boils in jerks,—when, on applying the thumb-test, a connected, flexible, shining thread is produced, which is neither brittle nor slimy,—when the lye has entirely lost its burning, alkaline taste, or nearly so: the fire is extinguished, and the soap is skimmed off from the under-lye into the coolers.* From thence the liquid mass is transferred to the mould, over the bottom of which a linen cloth is spread, that the lye may more easily drain away. It is necessary to stir the soap with an iron rod before it completely solidifies, in order that the curd may form a connected mass of soap, and separate the lye. By performing this operation with a certain degree of regularity, the mottling, which ensues spontaneously in the German soap, without any additional help, assumes the desired appearance.

The unmottled *clear-white soap*, which is more generally used in some parts than mottled soap, is prepared upon the same general principles. Instead of removing the coloured portion (the flux) by subsidence, as in France, white soap is fabricated in Germany by preventing the separation of the coloured portions

* In the German soap-works the pan has seldom an exit pipe for drawing off the lye; when this has to be removed, the soap is scooped from the pan into a vessel by the side of it—the cooler—until the pan is emptied. This cooler is placed in such a manner near the pan that any soap which may boil over shall fall into it, and not upon the ground.

in veins and groups by a quick process of cooling; this then remains disseminated throughout the mass of the soap, and gives it a grey tint. All additional matter, which is used to increase artificially the natural mottled appearance, is here, of course, omitted.

According to direct experiment, 100 lbs. of tallow will produce 150 to 155 lbs. of curd soap weighed as soon as it is cut, and the reason why some soap-boilers obtain more (sometimes as much as 200 lbs.) is that they are forced to add water, in order to compete with soap of an inferior quality.

The use of crude tallow is now abandoned as a useless hinderance in the boiling process, and *rendered* or purified tallow is used. When crude tallow is used, the membranous portions are converted into glue during the boiling, at the expense of a portion of the potash, and this portion is saved by the use of purified tallow. In the best manufactories, the nauseous odour which formerly infested the boiling house and its vicinity, is unknown.

It is obvious that it is a much safer process, and more rapidly completed, when the lye is prepared from *potash*, because there is then no necessity for making so weak a lye, and the workman is able to ascertain accurately the amount of alkali present, which, in the case of ashes, it is difficult to do, on account of the large amount of foreign salts. Although the price of pure potash is too high to admit of its being used alone, yet the addition of potash to a lye prepared from ash, enables the soap-boiler to work larger quantities of tallow at a time. The boiling is conducted precisely in the same manner as with the lye from ashes, only that it is unnecessary to add so many consecutive portions of salt.

2. *By the modern German Method.*—The boiling is conducted with much greater ease and rapidity when *caustic lye from soda-ash* is used, from 8 to 10 cwts. of tallow being then easily converted into soap with the same water. Less common salt is likewise necessary in this case, partly because soda-ash generally contains some of this salt, and partly because the portion previously required for the decomposition of the potash-soap is now no longer needful.

Another saving, not altogether inconsiderable, is effected in consequence of the equivalent of soda being smaller than that of potash,—2 parts of hydrate of soda combining with the same quantity of fatty acid as 3 parts of potash. The mode of conducting the boiling is as follows :—

The kettle is charged with 190 gallons of soda-lye of 13° B. and 2,000 lbs. of the best melted suet. When the mixture becomes milky, it is kept at a gentle ebullition for two hours, and then the fire is drawn, and, after two hours' repose, the lye is drawn off. The operations are continued as above. When the soap forms, by pressure, clean solid scales between the fingers, the magma is cooled by throwing in some buckets of lye, and soon after it has settled, drawn off clean. After the fire has been withdrawn, 9 or 10 gallons of lye are poured into the kettle, and when fusion is complete, a trial of the paste is made with the spatula. If it runs from the lye, water is added ; if it does not run, it must be boiled a little longer, adding a bucket of water containing a third of its weight of common salt in solution, in order to effect the separation of the soap from the water. When this separation appears to be perfect, after an hour or more of repose, the water which contains the greater part of the alkaline lye remaining from the first boiling, generally of a very deep bottle-green colour, is drawn off. The fire is replaced under the kettle, and about eight buckets of water turned in, and when, after a continued boiling, the incorporation is complete, an examination with the spatula must be made to ascertain whether the soap runs from the water. If it does, water is added in small portions until it ceases, and the pasty liquid moves tremulously like melted gelatin. To finish the operation, the contents of the kettle are well boiled, and unless the soap has a bluish tint (in which case it should have another washing) the fire must be extinguished entirely,—or the steam shut off,—the kettle carefully covered, and the whole left at rest for a day or more, and then ladled off into the frames.

Curd soap is generally very hard, and slow to dissolve and froth in water. It has, therefore, been proposed to add to the finished curd about 10 per cent. of tallow or oil soap, made with potash-lye of 6° to 8° Beaumé.

English Method of making Hard Soap.

The fatty and oily materials which are used for the production of hard soaps in this country are tallow, palm oil, and cocoa-nut oil, also resin ; the whole of which are saponified by soda.

The manufacturer provides himself with solutions of caustic soda of various strengths, called "lyes," these being obtained by boiling together a solution of carbonate of soda and slaked lime, running off the first solutions and washing the residual carbonate of lime with several affusions of water—the last liquors thus obtained being used for dissolving a fresh batch of carbonate of soda.

It is a well-known fact that lime, in whatever proportion it may be used, does not effect the perfect decomposition of carbonate of soda, unless the latter is present as a weak solution. And as any alkali in the state of carbonate which may be introduced into the soap copper is incapable of decomposing the neutral oils or fats, it becomes wasted; therefore the manufacturer uses solutions of carbonate of soda of such strength as will yield lyes (solutions of caustic soda) having a specific gravity not exceeding 1·090.

The construction of an ordinary soap-pan as used by English manufacturers of soap, is shown in Fig. 317, p. 650. These are made of various sizes, some being as large as 15 feet diameter and 15 feet deep, capable of yielding 25 to 30 tons of finished soap in one operation. They are at the present time constructed of wrought iron plates joined together by rivets. The contents of these pans are usually caused to boil by the injection of free steam through a number of small holes in a circular bent pipe which is in connection with a steam boiler. The pans are also sometimes heated by means of fires placed underneath the lower part, or by steam chambers formed round the lower part of the pans.

The manufacturer charges his soap-pan at the commencement with a quantity of neutral oil or fat, and adds to this weak lyes having a specific gravity of about 1·050. He causes steam to be injected to produce ebullition and mixing. If the process goes on properly, the oil or fat which was previously floating on the surface of the lye becomes speedily combined therewith, producing a uniform milky emulsion, from which no watery particles separate on cooling. If this combination does not take place, the operator adds either water or weaker lye, and continues the boiling until the perfect emulsion is produced: at this period all taste of alkali in the compound has passed away—the tongue being used by the operator in place of turmeric paper, to ascertain the presence of free alkali. The combination of the oil or fat with the mineral alkali, or the displacement of the glycerin, having been thus fairly put in progress, the operator makes repeated additions of stronger lyes, continuing the boiling, until he finds the presence of free alkali in the compound ; he then adds more oil or fat, or some rosin, if he is making soap containing rosin ; he also continues to make repeated additions of stronger lyes. In this part of the operation, he takes care that there shall be no excess of alkali present in the compound at the period when the soap-pan has become nearly filled by the repeated addition of oils or fats and lyes. He then adds common salt to the mixture, which, by its superior affinity for water, as compared with that of the saponaceous compound, effects the decomposition of the emulsion, and causes this to be separated into soap combined with a definite quantity of water, but not having its full proportion of alkali, and a solution of common salt—which latter contains the glycerin of the fats or oils employed. The soap floats in a granulated state on the surface of the solution, which is then called " spent lyes," and should contain no free alkali.

After a few hours for subsidence, the exhausted solution or spent lye is withdrawn from the soap-pan, from under the imperfectly-made soap, and is rejected as worthless.

The workman commences his second operation by the addition of some weak lye to the imperfect soap, and by boiling he brings the contents of his pan again into a state of homogeneous mixture,—called the " close state " as contradistinguished from the condition in which the soap is granulated and separated from the liquid contents of the pan,—and if he has not already added his full compliment of oils or fats, he completes the addition of these, adding also strong lyes until he finds the mixture has acquired a strongly alkaline taste. He then adds sufficient common salt to cause the separation of the soap from the alkaline solution, and boils the soap for some hours in the presence of the alkaline solution, to ensure the whole of the fatty matter being combined with alkali. If it is intended that the product should be framed as a " curd" soap, it is allowed to remain quiescent for a few hours, so that the lyes may subside, and the soap is then skimmed off and transferred to the " frames," in which it solidifies by cooling, and is then divided, by cutting with wires, into slabs and bars ready to be delivered to consumers. If the soap contains rosin as part of its acid constituents, it requires a further operation before framing. This consists in melting the curd soap (after abstraction of the alkaline lyes from the lower part of the pan) with the addition of water, and boiling the mixture by steam or fire, or both steam and fire, so as again to produce a homogeneous compound, containing an indefinite proportion of water, and allowing this compound to remain quiescent for two days, when a separation takes place, resulting in the elimination of a definite compound, containing in 100 parts about 65 of fatty acids, 6·5 of soda, and 28·5 of water. This forms the upper stratum of the contents of the soap-pan, underneath which is an indefinite compound, technically called "Nigre," containing a much larger proportion of water. When it is judged that the separation of these compounds has been well effected by subsidence, the upper portion, or the definite compound, is usually taken out of the soap-pan by ladling and transferred to frames, to solidify by cooling. The residual portion, which contains an excess of alkali, as well as of water, is boiled up with salt and the addition of fatty materials, to form part of the succeeding boil of soap.

Cocoa-nut-oil Soap; Marine Soap.

The consumption of cocoa-nut oil has increased enormously of late years. Differing in origin and constitution from the other oils and from tallow, its re-action with saponifying agents is also quite peculiar. It has been found by experience, that the soap prepared from this oil can only be separated from solution by very strong solutions of common salt. Weaker brine acts very imperfectly upon it; in fact, the soap is soluble in dilute brine, and is, therefore, serviceable for washing in salt water: hence the name marine soap. As this kind of soap can only be obtained by the use of a very large quantity of common salt, and then contains so very little water, or is so exceedingly hard that it cannot be cut with a knife, this mode of proceeding cannot be employed upon a large scale. For the same reasons, a clarifying process or boiling down to curd would be equally objectionable, and very difficult. Whilst tallow, for instance, when treated with very strong lye, comes to the surface and is but slightly acted upon, exactly the opposite occurs with cocoa-nut oil. It does not form that kind of milky mixture which generally indicates the commencement of saponification, but swims as a clear fat upon the surface; when, by continued boiling, the lye has attained a certain degree of concentration, saponification suddenly begins, and proceeds with extraordinary rapidity. For this reason the soap-boilers only employ the strongest soda lye, of at least 20° Baumé, and are able to dispense with the use of salt in purifying the soap, by employing pure and perfectly caustic lye, and avoiding any excess as much as possible. The saponification is also facilitated by the addition of potash-lye to the soda. Pure cocoa-nut-oil soap hardens much too quickly to exhibit any distinct formation of curd, and is consequently incapable of marbling by itself; it is very white, translucent like alabaster, exceedingly light, and forms a good lather, but always possesses a disagreeable savour. No means have as yet been made known to remove this smell, although it appears that several manufacturers are possessed of the secret. An important property of cocoa-nut-oil soap, is its power of combining with more water than can ever be communicated to tallow soap, and this property of the soap frequently gives rise to, or encourages, dishonest traffic. Cocoa-nut oil actually produces no greater quantity of soap than an equal weight of tallow; but the soap from the former can easily be made to absorb one-third more water or lye. Ordinary soap treated in the same manner, or containing the same quantity of water, would be so soft that it would yield easily to the pressure of the thumb; whilst cocoa-nut-oil soap, on the contrary, neither exhibits any want of consistence or softness, nor does its appearance in any way indicate the fraudulent practice which has been adopted in its manufacture.

It has been observed by soap-boilers, and the remarkable fact is by no means satisfactorily explained, that cocoa-nut oil is saponified with so much the more difficulty the more rancid it has become.

In general, cocoa-nut oil is not saponified alone, but is employed, as an addition to tallow, &c., for the purpose of producing quickly-solidifying soaps containing a large proportion of water, which could not be obtained from tallow alone. It is even possible to prepare soap on a large scale in a few hours, without salt, and almost without fire, by the use of cocoa-nut oil and tallow, which are merely warmed together with strong lye sufficiently to melt the fat, and kept in a constant state of agitation. Soap prepared in this manner has a fair appearance, and sets in the mould so

that it can be cut; it contains, however, nearly all the water of
the lye (there being hardly any evaporation in the pan) with the
entire amount of foreign salts, and in the fresh state has less
resemblance to soap than to stiff dough, which takes deep impressions of the thumb, and has a slimy consistence when squeezed
between the fingers. When dried for a length of time, during
which there is a copious efflorescence of foreign salts, it acquires
at last the consistence of ordinary soap.

The less cocoa-nut oil is used in the manufacture of soap, the
more imperceptible its smell becomes in the lather, but the
boiling of the cheap soaps, which harden quickly, and which
combine an excessive amount of water with the appearance of
ordinary tallow soap, is exceedingly difficult. When equal parts
of cocoa-nut oil and tallow are used, the soap has the smell of
common tallow soap. The boiling is continued until a specimen
exhibits the proper consistence under the thumb. It would be
impossible to saponify tallow under the same conditions, but the
saponification begins with the cocoa-nut oil, which induces the
action, and the saponification of the tallow is then effected by
means of the presence of the cocoa-nut soap.

The different kinds of cocoa-nut soap, which all belong to that
class of soaps which contain a considerable quantity of water, are
marbled artificially. Marbling or mottling of this kind is not
dependent upon the production of *Curd* and *Flux*, but is simply
a mechanical effect carried out in the following manner: The
blue or red colour (bolus, &c.) is rubbed up with a residue of the
soap, or better, with a separate portion of good cocoa-nut soap
until the whole acquires a uniform red or blue colour. This is
now scooped into the form in alternate layers with the colourless
soap, and by stirring the mass together, streaks and veins are
produced in all directions. It is evident, that this kind of mottling has nothing whatever to do with the natural mottling that
occurs in the other kinds of soap.

Castor-oil Soap.

Castor-oil is readily saponified by strong soda-lye. The product is white, amorphous, and translucent, and possesses considerable hardness, even when it contains as much as 70 per
cent. of water. (Soeber, *Dingl. polyt. J.* cxxxviii, 306.)

Palm-oil Soap.

This soap is boiled almost precisely in the same manner with caustic soda as tallow soap. It has an agreeable but powerful smell, and a yellow colour when the oil was used in an unbleached state, but is white, with a very slight odour when the oil has been bleached. Palm oil is seldom used as soap-stock without the addition of some other fat. 3 lbs. of palm oil to 1 lb. of tallow forms a good soap. An inferior article, called "demi-palme," is obtained from 1 lb. of palm oil with 4 lbs. of tallow.

Palm soap, when carefully prepared from pure materials, possesses emollient properties which, combined with an agreeable odour, render it well adapted for toilet purposes. In France, a good toilet soap is prepared from a mixture of 9 parts of palm oil, and 1 part of cocoa-nut oil; and a demi-palme soap, also for the toilet, with 11 parts of white tallow, 3 parts of palm oil, 1 part of cocoa-nut oil, and 1 part of purified yellow rosin.

Rosin, or Yellow Soap.

Colophony at the boiling temperature, when it is perfectly fluid, combines with the alkalies much more rapidly and with greater ease than the fats themselves. This kind of combination, which can scarcely be called saponification, ensues equally well with carbonated as with caustic alkali, and no particular precautions are necessary in conducting the process. The soap separates on the surface, when an excess of carbonate is used, or in the presence of common salt, as a thick slimy brown mass, smelling strongly of rosin, and containing 15·8 per cent. of dry soda. The amount of water in this soap, although not exceeding 27 to 30 per cent., is, nevertheless, sufficient to communicate to the soap a smeary viscid consistence, which is not altered by long exposure to the air. The attraction of the soap for water is so great, that it not only does not lose its water, but becomes liquid on exposure, after having previously been dried artificially. A portion of the rosin, however, appears to remain dissolved in a modified state in the under lye, judging from the deep brown colour of the latter.

Although rosin soap by itself is thus unfitted for use, an excellent and perfectly firm product is obtained by its combination in certain proportions with tallow and palm-oil soap. The amount

of rosin in this mixture should not exceed one-third of the fat ;
if equal parts are used, the properties of the rosin soap become
too prominent, and a soft bad soap results. 15 per cent. of rosin
makes a good soap, but beyond that limit, the soap is depreciated
in colour and firmness. The grease-stock of which this kind of
soap is generally made, consists of kitchen-fat, soap-grease, bone-
fat, and red oil.

The best plan of preparing this soap, is to saponify the rosin
and tallow separately, and to mix the two soaps in the boiler,
where they are retained in a state of ebullition for some time,
until a uniform mixture has been effected ; salt is then added, and
the soap brought into the moulds. It is not advisable to mix the
separate soaps in the moulds, which is a plan sometimes adopted,
In this country and America, where rosin soap was first manufactured, it is usual to add the rosin, in the form of coarse powder,
with the last quantity of lye, to the fatty (palm-oil or tallow) soap,
consequently, before the boiling is finished ; and to boil the
mixture with the necessary addition of lye for completing the
formation of the resinous compound. This point is attained
when a cooled specimen presents the proper consistence, and
leaves no film of rosin when used for washing the hands. The
boiling finishes with the addition of some weak lye to the soap
which has already separated from the under lye ; this addition is
made in order to facilitate the deposition of the impurities with
which commercial colophony is always mixed ; the soap is then
carefully filled into the moulds (pp. 659, 660).

Dunn's Process.—This plan, described by the author with
reference to steam as the heating agent, is as follows : Into each
of the ordinary boiling kettles, a circular ring of 1¼ inch pipe,
perforated with holes, is fixed just far enough above the bottom
to allow the free movement of a stirrer beneath it. This circular
ring of pipe is supplied with atmospheric air from a cylinder
blast, or other suitable forcing apparatus, being connected therewith by means of a pipe, which passes to the top of the kettle,
where it is finished with a stop-cock and union-joint, for the purpose of connecting or disconnecting the parts of the pipe within and
without the copper. For a clean yellow soap, put into the kettle 90
gallons of lyes of specific gravity 1·14, made from strong soda-ash.
The fire being rekindled, the kettle is charged in the usual way
with, say, 2050 lbs. of grease, and as soon as the lye is hot and on
the boil, or nearly so, the blast is set in action, keeping up a

good brisk fire, so as to keep the materials in the kettle as near ebullition as possible. When the lyes are exhausted, more lye is gradually added until the fatty matter is killed. 550 lbs. of fresh rosin are then added, a bucketful at a time, with more lye occasionally, until 300 gallons of the strength above mentioned have been used, keeping the blast in action the whole time, if the fire draws well; if not, it is advisable to stop the blast for a while before adding the rosin, to allow the contents of the kettle to approach to ebullition. When the whole of the rosin is melted and completely mixed with the soapy mass, and the strength of the lyes taken up, the blast must be stopped, and a brisk boiling given to the contents of the kettle; it is then left to rest, that the spent lyes may separate and settle, which being now drawn off, the soap is then brought to strength on fresh lyes, as in the ordinary process of soap boiling.

During the operation of the blast, the soap must be kept in what is technically called "an open or grained state," and for this purpose salt or brine is to be added when necessary. Experience proves that it is better not to make a change in the lye during the operation of the blast, where lye of the strength before mentioned is used; but if weaker lye is employed, one or more changes may be made, as is well understood. It is found desirable that the soap should be kept in what is called a weak state during the action of the streams of air through the materials, otherwise the soap is apt to swell up, from the air hanging in the grain, and this is found troublesome to get rid of, requiring long boiling. If dark-coloured materials are used, it is well to keep the blast in operation three or four hours after the rosin is melted, provided the soapy mass is kept weak and open or grained. When a charge is to be worked upon a nigre, such nigre should be grained, and the spent lye pumped or drawn off as usual, and the fresh charge added in the manner before mentioned, using less lye in proportion to the quantity and strength of the nigre, taking care not to turn on the blast until there is sufficient grease present to make the nigre weak.

Meinecke's Process.—This method requires that the soap-pan should be constructed with a still head and cooling worm, as the rosin is added in the form of white turpentine, which, during the boiling, gives off its volatile oil as a distillate, to be condensed and saved as an incidental product, thereby decreasing the expense of the soap. 1000 lbs. of white turpentine are melted in the kettle, by steam, with 800 lbs. of tallow or inferior fat, and when the mixture reaches 108° F., it must gradually receive, during constant stirring, 800 lbs. of caustic soda lye, containing 30 per cent. of dry soda. The union of the materials is very prompt at the above temperature, the acids of the resin and the grease being completely neutralised, and converted into liquid melted soap. The essential oil of turpentine is set free at the same time, and in order to promote its vaporisation, salt or brine is added. The head being then luted upon the pan and adjusted to the worm, and the mixture brought to a boil, the steam and spirits pass over into the worm, and are condensed. When all the essential oil is distilled over, the soap is finished in the usual way.

H. C. Jennings prepares rosin soap by adding to curd soap prepared with tallow or oil and caustic alkali in the usual manner, a large quantity, about 25 per cent., of colophone-rosin, then adding from 2 to 4 per cent. of bicarbonate of soda or soda of commerce, and 1 or. more per cent. of sulphate of alumina, common alum, or other double salt of alumina, the whole being boiled with water till perfect combination is effected. To prevent the rosin from precipitating, a small quantity of dilute sulphuric acid (1 part of acid to 9 parts of water) is stirred into the mixture, the quantity of this dilute acid being about 2 per cent. of the tallow or oils and rosin.

Rosin soap has a brownish colour, which passes into that of yellow wax when palm oil is used, and this, of course, requires no previous bleaching. It is very firm, somewhat rough to the touch, and very translucent. The best quality exhibits on the cut surface a fibrous structure like that of wood. It produces an excellent lather, but always retains the smell of rosin. The low price of rosin renders this soap cheap; and for common washing purposes, where the smell is not an objection, it is a highly useful product.

Red Oil, or Oleic Soap.

This oily acid is much used for the manufacture of soap, the saponification being easily effected, inasmuch as the acid is already separated from the glycerin, and has merely to enter into combination with the alkali.

It may be prepared by boiling 1300 lbs. of soda-lye of 18° B. in the kettle, and treating it in separate portions during stirring with 1000 lbs. of red oil or oleic acid. The oil is rapidly taken up by the lye, and the reaction is so lively that it is necessary to keep down the oil by uninterrupted raking. As long as the paste continues strongly caustic, it must have new additions of oil, to be continued until only slight alkalinity is left : for the paste must retain only a slight excess of soda. In case it should be too feeble in this respect, after two or three hours' repose in the kettle without fire, it must have 50 or 60 lbs. more lye. The fire is then extinguished, and the paste, after remaining in the pan twenty-four hours, is transvased into the cooling frames, which should be very shallow, as this soap sets slowly.

Oleic acid is also frequently used, together with tallow or beef marrow,—for instance, 3 parts of oleic acid to 2 parts of the neutral fat,—also with rosin.

Campbell Morfit has patented a process* for preparing soap with red oil and carbonate of soda.

He uses powdered carbonate of soda in the form of *barilla, kelp, trona, sal-soda, soda-ash, bicarbonate of soda*, &c., for saponifying other fatty acids generally, and more particularly the red oil, or red (oleic) acid oil, obtained in the manufacture of "adamantine, palmitic, palmatin, Belmont, sperm, and star candle stock," whether alone or mixed with rosin, and converting them, by direct combination, into soap in open pans or kettles, at temperatures between 32° and 500° F.

The red oil or other fatty acid, is poured into an open pan with a fire beneath, or preferably, into a kettle or tub, which it fills to one-third of the depth, and agitated and heated by the steam twirl, described at page 648. If it is desired to make a grade of soap lower than toilet soap, rosin is to be added in small lumps as soon as the oil has become hot. The proportion of rosin may range from 5 per cent. of the fat acid used and upwards. When, after continued heating and stirring, the rosin becomes entirely dissolved, carbonated alkali, finely powdered, is to be added portionwise to the homogeneous mixture of fat and rosin, while the twirl is kept slowly revolving. When all the alkali is in, and the intumescence has subsided, the paste will begin to thicken, and will promptly assume the condition of soap. At this stage, it is shovelled out into the frames, and left to settle.

* Patent granted to John Talbot Pitman (1858, No. 588), for improvements in the manufacture of soap, and for an apparatus connected therewith. A communication from Campbell Morfit, of Baltimore, Maryland, U. S.

For neutral soaps, the quantity of carbonated alkali should only slightly exceed the chemical equivalent proportion, and must be determined by calculation from the combining number of the fat-acid which constitutes the "stock." For strong soaps, the quantity may be increased several per cent. beyond that proportion.

The requisite amount of constitutional water of the soap being supplied in this process exclusively by the water of crystallization, or the accidental water of the alkali, it is to be remarked that, in the use of sal-soda alone, the quantity necessary to accomplish perfect saponification would introduce an excess of water, and thus render the paste soft, and slow to set and dry. To obviate this, a certain portion of the "sal-soda" is used in a lower state of hydration, to which it is reduced by previously subjecting it to air currents, or by the action of a "centrifugal mill." The same end may be attained by the use of a corresponding proportion of dry soda-ash. In either case, the requisite proportion of the dry sal-soda, or soda-ash, must be determined by calculation from the amount of normal "sal-soda employed." When two kinds of alkali are used, they must be added separately for the higher grades of soap, and that which contains most water, must go in first. In the lower grades of soap, they may be added separately or together, according to the judgment and experience of the operator. For toilet soap, the rosin is omitted. The relative proportions of fat, rosin, and alkali and water being adjusted at the beginning, there is no waste lye, or any other residue; and the soap is said to come out promptly, and in greater perfection than can be readily obtained by the usual method of boiling soap upon caustic lye. The soap, says the author, is always uniform in appearance and composition, and does not shrink or deteriorate by time and atmospheric influence. It wastes slowly in water, gives a rich lather, and whitens the clothes.

SOFT SOAP.

The substance commonly called "soft soap" is a more or less impure solution of potash-oil-soap in caustic lye, and not an actual soap; it forms, at ordinary temperatures, a transparent smeary jelly. If an attempt were made to separate this soap from the lye by means of salt, the potash-soap would be converted into hard soda-soap by the chloride of sodium, and the object in view would be frustrated. Recourse cannot therefore be had to a method of purification of this kind, and to render all purification less necessary, pure potash-lye is employed instead of the crude lye from ashes; this is kept ready, of different strengths, in separate cisterns. The weaker lye must be perfectly caustic, but this is neither necessary nor desirable in the stronger lye, which is prepared in the cold. Soap-boilers have found by experience, that the boiling is easier when these lyes are slightly carbonated. The oils employed in this country for making soft soap are whale oil, seal oil, linseed oil, and tallow; on the Continent, hemp, linseed, camelina oil, and poppy oil, which belong to the class of drying oils, and do not become solid at 0° C. (32° F.) are used; also the different varieties of rape-seed and train oils, which do not dry up, and become partially solid at 0° C. (32° F.). The former produce a softer kind of soap, the latter soap of firmer consistence; it is hence usual to mix the oils in different proportions, according to the season of the year, choosing the drying oils in the winter, and the others for the summer, when the market price admits of this selection being made.

In boiling soft soap, the weaker lyes from 9° to 11° B. are first used, and a moderate heat is applied, and kept up until complete combination is effected, i. e., until a thick sticky fluid falls in streaks from the stirrers; this ought to possess a shining appearance, and although it may be somewhat turbid, should not resemble soap separated by means of salt.

As soon as these characters clearly appear, the clarification commences with the gradual addition of the stronger lye. This is kept up at equal intervals, until the soap passes from the state of a turbid imperfect mixture to that of a clear transparent slime. When this change is too tardy in making its appearance, in consequence of too great dilution, the operation is assisted by an addition of very strong lye. The clarification completes the chemical combination of the constituents, which is indicated by a peculiar test; but the soap still requires the proper proportion of water to render it saleable. The tests employed for ascertaining these points are very empirical in their nature, but, nevertheless, they are distinctly marked, and of a peculiar character. The soap-boiler is enabled to ascertain, by the presence and extent of turbidity in a cooled drop, whether chemical combination has been effected, or what is the cause of failure. When a specimen

44*

free from scum is taken with a spoon from the middle of the soap-pan, and a drop of the size of a sixpenny-piece is allowed to fall and cool upon a piece of clean colourless glass, several appearances may be noticed : the drop either remains perfectly clear, in which case the soap has attained the proper state of mixture, or it exhibits on the margin only a grey rim, which indicates a want of lye, and this in proportion to the breadth of the rim. If this deficiency is very great, the rim is not only exceedingly broad, but the specimen itself is fluid and slimy. The opposite must, of course, occur when the proper quantity of lye has been exceeded, and the soap is overdone. This is easily recognised by the consistence, and by a grey skin, which spreads itself over the whole drop instead of forming a ring or rim only. The soap is then granular without lustre, and is easily detached in the wet state from the glass, and is technically said to be *vitreous*. It is advisable to follow the process closely, by constantly trying specimens from the period when the soap begins to boil clearly, until the end of the process. The lamination is the removal of the excess of water by evaporation, for in this process, as has been stated already, the use of salt for that purpose is inadmissible. The soap is kept in a frothy state by increasing the fire, and the evaporation of the water is accelerated by constantly beating the froth with the stirrers, thus causing a rapid exchange of the humid air over the surface of the pan : for dry air from without is seldom found in use in the boiling houses, although some plan for that purpose might easily be devised, and would effect a considerable saving of fuel.

As the evaporation proceeds, the soap becomes thicker and darker in colour and less froth is produced. At last the froth is so much diminished, that the soap sinks, and the bubbles are so far larger that they resemble films or lamellæ, which overlap and cover each other upon the surface. This is what the soap boilers term the *lamination*, and the noise occasioned by the process gives rise to the saying " the soap talks." The soap is now really finished, but another specimen or test is taken before it is scooped out into the moulds. When this no longer shows any opaque zone, after having cooled for some time, or only in a very slight degree, it may then be safely concluded that the proportions in the pan have been properly attained. A certain amount of experience is necessary, in order to form a correct judgment from these indications ; for when the specimen is left too long exposed to the air, the action of the moisture upon the small quantity of soap in the test, brings about changes in its appearance which interfere with the discriminating characters stated above. Thus, for instance, the opaque zone or rim spreads gradually over the whole surface of the specimen, until at length it entirely disappears. The mass, when completely cold, must never be fluid, but always of a thick consistence. As soon as the tests prove satisfactory, the fire is extinguished, the soap is left for some time longer in the pan to cool, and is then packed in small casks for sale.

Soft soap, intended for winter use, which is then liable to become thick or lumpy, should be clarified with stronger lye, or with such as contains a portion of carbonate of potash.

There are two kinds of soft soap known in commerce. The best quality contains—

Fatty acids	50·0
Potash	11·5
Water, &c.	38·5
	100·0

The second quality contains—

Fatty acids	40·0
Potash	9·5
Water, &c.	50·5
	100·0

In Russia, soft soap is prepared, from beginning to end, according to Kurrer, with one and the same species of lye, containing three-fourths caustic and one-fourth carbonate of potash. The half of this lye, which is brought up to 10° B., is added to the linseed, rape, or hemp-seed oil in the boiler. The other half is placed in a cistern by the side of the boiler, and is allowed to flow uninterruptedly from a cock in such a very thin stream throughout the process into the boiler, that the soap is kept constantly in a state of ebullition. When the soap flows from the stirrer as a clear slime, and can be drawn out in threads between the fingers, having thus attained a certain consistence, the process is considered finished. This mode of procedure, which is very uncertain, and never affords a very uniform product, can only be recommended in countries like Russia, where there is a great want of skilled workmen.

Some kinds of oil,—for instance, hemp-seed oil,—naturally communicate a green colour to the soap, which is much prized, and has, consequently, become a necessary property of soft soap. Other varieties of soap, therefore, which have a yellow colour, are rendered green by the addition of a little indigo. Indigo, however, is not soluble either in soap or in caustic lye; it is therefore necessary to pulverize it very finely, and even then it is difficult to diffuse it uniformly through the mass of soap, nor is it easy to moisten the powdered indigo, and it easily subsides to the bottom of the vessels. The usual mode of applying it consists in boiling the finely powdered indigo with a little water for some time, until it is uniformly moistened, and then adding it to the soap; or in precipitating a solution of indigo in oil of vitriol with lime, and using the precipitate as the colouring matter. Soft soap is often coloured black by the addition of inky fluids, which are prepared by making an infusion of gall-nuts, and mixing with it a solution of green vitriol.

A so-called corn or grain is sometimes produced in soft soap by the addition of tallow. The soap then retains its ordinary character, but fine granular particles of a crystalline structure are observed in it, consisting probably of salts of stearic and palmitic acids. The formation of this grain requires a certain degree of heat, and can only be effected, in the colder seasons of the year, at temperatures between 9° and 15° C. (48° and 59° F.). This process is called "figging;" it is practised merely from habit, and no useful object is gained by it. Attempts have been made to imitate this useless appearance in a manner calculated to injure the quality of the soap. Thus, for instance, slaked lime has been used, with the production of a lime-soap, and even starch has sometimes been mixed up with the soap.

All kinds of soft soap exhibit a strong alkaline reaction, and are characterised by a penetrating disagreeable odour, which, however, does not necessarily resemble that of the fats employed. The smell is most perceptible in soaps prepared from train oil, and is due to the presence of valerate of potash, produced by saponification.

Soft soap is used to some extent for washing coarse linen, but it is of far greater importance, as an indispensable and powerful detergent, in the linen-bleaching works.

The important distinction between soft soap and the ordinary soaps is very marked, the latter being easily and completely

separated by means of salt from all the impurities and the excess
of lye, which necessarily remain attached to the former. Soft
soap is consequently contaminated with all these substances, when
they are not removed in the beginning by the care of the workman; nor is there any guarantee, as is the case in curd soap,
that a certain amount of water is not exceeded, and, indeed, this
kind of soap is more hygroscopic, that is to say, more subject to
change with the amount of moisture in the atmosphere. Soft
soap never dries up completely to a solid mass, even after long
exposure to the air.

Soda Soft Soaps.—According to Gentele's recent experiments
on a large scale, the potash-base of soft soaps may in part be replaced by soda, without disadvantage to the resulting soap. The
product has the characteristics of soft soap, but contains a little
more water. The lyes must be free from salt and other saline
impurities, as they prevent the clarifying of the soap. The best
proportion is 1 part of soda to 4 parts of potash-lye, and the
former should be made from crystallised sal-soda. A mixture of
100 lbs. of red oil, 50 lbs. of tallow, and 3,750 lbs. of hemp-seed
oil makes a good stock for this soap.

All the soft soap (*savon vert*) made in Belgium and Germany
contains about equal portions of soda and potash soaps.

A soft soap of firm consistence (*white soft soap*) may also be
made by melting together 75 parts of tallow or tallow oil, and
25 parts of cocoa-oil, and boiling the mixture with lye till the
paste sharply bites the tongue. At this stage, salt solution of
20° B. is added, to give consistence to the paste, and the soap on
cooling is then ready to be barrelled.

Soap for Silks and Prints.—According to Calvert, the soft
soaps usually made for dyers' use are not indiscriminately applicable to all colours. To produce the maximum effect in brighten-
ing the shade, the soap should be composed of :—

	For madder purples.	For madder pinks.
Fatty matter.	60·4	59·23
Soda	5·6	6·77
Water	34·0	34·00

For removing the glutinous coating from silk, soap may be
partially replaced, according to Bolley, by borax. The silk, first
boiled with half its weight of this salt for an hour to an hour and

a half, and subsequently with soap, is white, soft, and unimpaired
in strength.   The borax may be recovered.

Gall Soap.—Ox-gall is converted, by treatment with caustic
alkalies, into a kind of soap which may be used for the removal
of grease-spots.   The gall itself in its natural state is also well
adapted, and indeed is much used for this purpose ; but it has a
very unpleasant odour, and quickly putrefies, in which state it
becomes filled with maggots.   These inconveniences are obviated
by mixing the gall with acetic ether, which destroys the odour
and kills the maggots.   This process, invented by Gagnage, is
carried out on a large scale in the works of Messrs. Pissand and
Meyer, in Paris.   They mix 32 litres of the gall with 225 grms.
of acetic ether (or 7 parts of the ether to 1,000 parts gall), and
the mixture, after being left to itself for some days, is decanted
from a black precipitate which separates.   This "disinfected
gall" removes grease stains just as well as the natural gall, and
may likewise be saponified with caustic alkalies.   A better product is, however, obtained by melting 2 parts of rosin or tallow
soap in 1 part of the disinfected gall.   The product thus obtained,
is very useful for removing grease spots, inasmuch as it acts as
well as the gall itself, and is much more convenient to work
with, since it does not run, and it is therefore easy to limit the
quantity placed on the stuff to what is required ; whereas, when
liquid gall is used, it is difficult to avoid taking too much.   The
gall-soap cannot, however, be applied to colours which cannot
stand the action of soap (Rep. Chim. App. I., 268).

SOAPS BY THE COLD PROCESS.

Saponification may in certain cases be effected by mechanical
agitation, without boiling.   The soap thus made is technically
called "little-pan soap," because the vessels used in its preparation are comparatively small.   The process is much used in preparing the fine qualities of toilet soap.   The fatty substances
selected are those which saponify most easily, viz.: lard, beef-
marrow, and sweet almond oil ; and the lyes used are of considerable strength.

Lard Soap.—To prepare this soap, weigh out 112 lbs. of lard
and 56 lbs. of caustic soda of 36° B.   Melt the lard by a gentle
heat, and add half the lye, stirring it well in, and continuing the
agitation without allowing the mixture to boil.   When all appears to be well incorporated, pour in by degrees, and during con-

stant stirring, the other portion of the lye, and be careful not to raise the temperature above 149° F. The paste should be well united, homogeneous, and of some consistence. Pressed between the fingers, it should be mild and unctuous, without being greasy. When it has reached this state, it is run into the frames, and in two days ought to have acquired sufficient solidity to be cut into tablets and pressed. The perfume must be added while the soap is yet in paste. This soap is of a very brilliant whiteness and flinty consistence.

Beef-Marrow Soap.—To 500 lbs. of beef-marrow add 250 lbs. of caustic soda-lye of 36° B., stir constantly and gently, and heat the mass till it becomes soluble in water. In this state, dilute with 2,000 parts of boiling water, and then pour in 1,000 parts of brine, holding in solution 180 parts of common salt, during uninterrupted raking, and then leave to repose.* After some time pour into frames and leave it for a day or two to set thoroughly.

Sweet Almond Oil Soap (Medicinal Soap).—This soap being manufactured only for the toilet or for medicinal purposes, should be made of the very best materials—select oil of almonds, free from rancidity, and pure carbonate of soda. Dissolve the latter in water, and add the third of its weight of hydrate of lime, frequently stirring the mixture during an interval of some hours, after which filter the clear lye running through, and concentrate it by evaporation, till it marks 36° B. Then take 12 to 25 parts of the oil, placing the lye first in the vessel, adding the oil by small quantities at a time, and stirring continually till the mixture has the appearance of a soft grease. In two or three days, it becomes thick enough to be run into porcelain moulds. When exposed to a temperature varying between 68° and 72° F., it acquires in a month sufficient solidity to enable it to be taken from the moulds. The temperature of the lye should be from 50° to 60° F. : but to prepare the soap quickly, it is necessary to place the mixture over hot coals, care being taken to prevent the undue concentration of the lye, by replacing the water as it evaporates.

This *amygdalin soap*, well prepared, is beautifully white and of a very mild odour. It becomes so hard that it may be easily pulverized when dry.

Hawes's Process.—The invention of Mr. Hawes has for its object the saponification of fat by mechanical means, without boiling. He uses 20 gallons of lye of specific gravity 1·125 for

* Morfit, p. 244.

every 100 lbs. of tallow. The apparatus (Fig. 352), consists of an
upright shaft A, from which the arms, a, a, a, a, radiate to the
sides of the kettle B; or a cylinder may be used with a shaft, C,

FIG. 352.

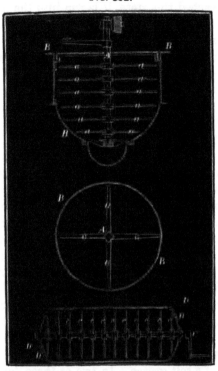

passing through it horizontally, with arms, c, c, c, c. An oscil-
lating or rotatory motion is communicated to these shafts in the
ordinary manner. A cylinder 6 ft. diameter and 12 feet long is
adapted for working 2½ tons of tallow. Convenient doors, D, D,
are attached for charging and emptying the cylinder. The con-
tainer is charged with tallow first, and the lye is afterwards
gradually added, while the agitation is continued for about three
hours. The whole is then allowed to stand for some time, varying according to the quantity of materials employed.

As the benefit of the process mainly arises from the saponification of the materials in a comparatively cold state, the contents
of the kettle or cylinder, as soon as they have thickened, are removed into an ordinary boiler, preparatory to being finished and
converted into yellow or white soaps, as the case may be.

The above process is described as for tallow, but it is equally applicable to the oils or other fats usually employed as ingredients of soaps.

A. P. Rochette prepares hard soap by combining caustic soda, without boiling, with the fatty or oily matters obtained from suds and soapy waters, which are a waste product in many branches of manufacture. These greasy and oily matters are simply melted and thoroughly mixed with soda-ley of sp. gr. 1·26. The proportions vary with the quality of the fat, but a good proportion, generally speaking, is about 60 lbs. of melted grease to 40 lbs. of soda-lye of the above strength.

The separation of the fatty matter from suds and soapy waters may be effected in various ways. Leslie, of Glasgow, boils the liquid, with addition of sulphuric acid, and purifies the oil which rises to the surface by the action of heat, caustic potash, &c. Schaeffer treats the wash-waters with a liquor containing chloride of calcium, viz. : the waste-liquor obtained in the manufacture of magnesia and its salts, by acting with hydrochloric acid on magnesian lime-stone, the lime then dissolving first and leaving the magnesia behind. When this liquor is added to water containing soap, an insoluble lime-soap is formed, which is easily separated and decomposed by acids : and by pressing this mixture in strong bags of woven fabric, and at the same time heating the bags by steam or otherwise, the fatty and oily matters are recovered. S. C. Lister effects the separation of the fats by means of revolving discs, which, being partly immersed in the soap-suds, carry up on their surfaces as they revolve, thin films of liquid which are acted upon by streams of air to facilitate the evaporation. W. Score purifies the fatty acids separated from soap-water with a solution of carbonate of soda and steam, and then reconverts them into soap in the usual way. (See List of Patents at the end of the volume.)

Kurten's Table,

Showing the composition and yield of Soap by the Cold Process, from Concentrated Lye and mixtures of Cocoa-oil with Palm-oil, Lard, and Tallow.

Kinds of Soap.	Tallow.	Cocoa-nut oil.	Palm-oil.	Lard.	Lye.	Degrees.	Salt-water.	Degrees.	Potash.	Degrees.	Product.
Cocoa-nut oil, No. 1	—	100	—	—	—	—	—	—	—	—	153
Paris toilette (round)	20	30	—	8	—	—	—	—	5	36	87
Windsor (square)	—	25	—	75	50—52	36	—	—	—	—	150
or	66	34	—	—	77	30	—	—	13	30	185
Shaving, No. 1	66	34	—	—	120	27	—	—	—	—	214
or	33	34	33	—	—	—	—	—	—	—	
Shaving, No. 2	33	34	33	—	120	27	12	12	—	—	226
or	60	40	—	—	—	—	—	—	—	—	
Washing, No. 1	30	40	30	—	125	27	25	12	—	—	244
or	40	60	—	—	—	—	—	—	—	—	
Washing, No. 2	—	60	40	—	135	27	50	15	—	—	278
or	—	100	—	—	—	—	—	—	—	—	
Ordinary cocoa-nut oil.	10	90	10	—	225	21	75	12	—	—	400
or	—	90	—	—	—	—	—	—	—	—	

TOILET SOAPS.

Toilet soap is essentially ordinary soap mixed with different aromatic oils, and diversified as to form to suit the fashion of the day. The fundamental material in this branch of industry, in which the French so much excel all other nations, is either one of the varieties of soap already mentioned, or a soap prepared specially for the purpose, of the best materials, and in a manner far too costly for any other. The kinds used are: olive-oil soap, tallow soap, palm-oil soap, and soap prepared from lard and almond oil (p. 697).

Purification of Soaps.

To refine an ordinary soap, which should, of course, be as free as possible from colour and impurity, for toilet purposes, it is reduced to shavings and melted over a water-bath with rose and orange-flower water and salt. To 24 lbs of soap take 4 pints of rose-water, 4 pints of orange-flower water, and two large handfuls of salt. The next day, if it is entirely cooled, cut it up into small bars and dry it in a shady place. Then melt it anew as before in the same quantities of rose and orange-flower water, and strain it; then cool and dry it again. This done, the soap will be free from bad odour. It must be powdered, and exposed for several days to the air, but not in the way of dust. It is then ready to receive the intended perfume, and to be moulded and pressed into the desired forms.

Another method is to melt 6 lbs. of best white soap in 3 pints of water, and when liquid to strain it through a linen cloth. It is then placed in a kettle with a pint of water and a tablespoonful of salt; a brisk fire is kindled under it, and the contents are whipped or stirred to make them foam and froth. The fire is then put out; the balling continued till the mass is sufficiently inflated; the fire again kindled; and the kettle kept on till its contents swell and foam. It is then emptied into the cooling frames, and, after solidification, taken out, cut into cakes, and pressed.

The following is the method of preparing toilet soaps at present adopted in the best regulated manufactories at Paris.[*] The soap is first reduced to very thin shavings by means of the

---

[*] "*Traité complet de la fabrication des savons;*" par G. E. Loviné, Paris, 1859, p. 319.

FIG. 353.　　　　　　FIG. 354.　　　　　　FIG. 355.

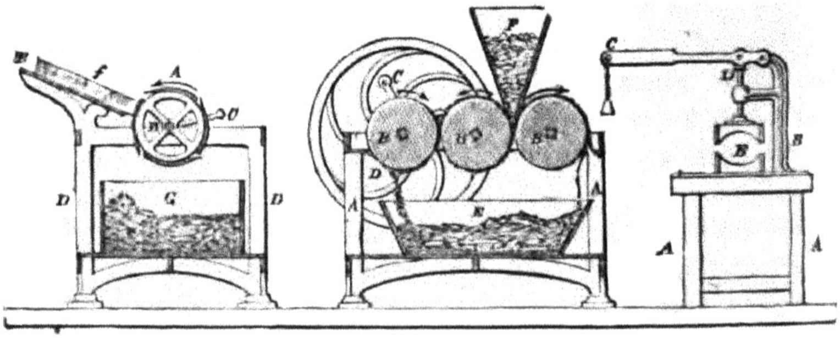

cutting machine represented in Figure 353. A is a wheel having on its circumference a number of cutting blades. The axis B, rests on the frame-work D, D, of wood or iron, and carries a handle C, by which the cylinder is turned. The block of soap to be cut is placed on the inclined plane E, and pressed against the cutting blades, and the shavings fall into a large box G. By means of this machine 100 kilogrammes (about 200 lbs.) of soap may be easily reduced to thin shavings in an hour.

The next process is to mix the soap with the requisite perfumes and colouring matters. The essences and colours are diluted with a little alcohol, and the whole well mixed by stirring with a clean wooden spatula, then added by small portions to the soap-shavings and incorporated as completely as possible by stirring. The mixture is then introduced into the hopper, F, of the mill, represented in Fig. 354, whence it passes between the three porphyry cylinders, B, B, B, by which the soap-shavings are reduced to thin laminæ, and at the same time are well incorporated with the perfume and colouring matter : lastly, they drop into the box E. Soaps coloured with ochre, vermilion, or metallic oxides require to be passed five or six times between the cylinders, to render the colour uniform. For white soaps, two or three passages are sufficient to produce a homogeneous mixture.

The soap having been thus reduced to thin laminæ is now to be well beaten in a marble mortar with a wooden pestle, in order to form it into a homogeneous mass. The mortar should be large enough to pound 10 or 12 lbs. of soap at once.

The next step in the preparation is to divide the mass of moist soap into lumps weighing about one-third more than the cakes of soap are to weigh after drying : thus, supposing the cakes are to weigh 100 grammes (4 oz.) each, the portions of wet soap weighed out must each contain about 135 grammes (or 5½ oz.) each. Each lump of soap is to be formed into a ball by pressure between the hands, then into a cylinder, by rolling it backwards and forwards on a marble slab, afterwards flattened and reduced to an oblong shape by beating it upon the slab in various directions, and lastly rounded at the corners. All these manipulations must be performed as quickly as possible, to prevent the soap from cracking.

The soap having been thus roughly formed into cakes, is to be dried on the rack (p. 666) either at ordinary temperatures or in a hot-air chamber, according to the state of the weather, care being taken to leave spaces between the cakes, to ensure a free circulation of air. They are left in the drying room for about a week, by which time a hard shining crust forms on the surface. They are then pressed in a mould E (Fig. 355), which gives them a sort of rounded rectangular form ; the ridge of soap which is pressed out at the junction of the two halves of the mould, is removed with a knife ; and the cakes are again placed in the drying chamber for several days till the soap has become quite dry and hard. The crust on the surface of each cake is carefully scraped off with a sharp knife, and the cakes are rubbed between the hands, slightly moistened with spirit of wine, to give them a smooth and uniform surface. After this they are placed in the hot-air chamber for 25 or 30 hours, and lastly pressed in a mould of the same shape as the former, but of somewhat smaller dimensions.

Some manufacturers adopt a less circuitous process of finishing toilet soaps, giving to the wet lumps of soap an excess of only 8 or 10 per cent. above the dried cakes, and drying them in the stove for only eight or ten days, without scraping or polishing with alcohol. This mode of proceeding is cheaper and more expeditious, but is not so well adapted as the former for the finishing of soaps of first-rate quality.

W. Clayton recommends the addition of bees-wax or vegetable wax to toilet-soap, in order to give it the property of softening the skin and prevent it from chapping in cold weather. 16 parts of soap are mixed with 1 to 2 parts of wax. The wax is added to the soap while hot, ultimately mixed by stirring. Soap thus prepared is also well adapted for getting up lace and muslin, imparting to them great softness and dispensing with the use of starch.*

* " Repertory of Patent Inventions," 1857, p. 9.

Perfuming and Colouring of Toilet Soaps.

The following formulæ are expressed in grammes and kilogrammes, as these serve better than the ordinary English weights to indicate the proportions of the several ingredients. The quantities are calculated for 10 kilogrammes of soap.

Bitter Almond Soap.—

          Pure white soap . . . .   10 kils.
          Oil of bitter almonds . . . . .   120 grms.
This soap is not coloured.

Savon à la Rose.—

     White tallow or lard soap . . . . . .   10 kils.
          Coloured with 60 or 80 grms. of vermilion.
          Perfumed with—Oil of roses . . . . . 40 grms.
                         „   cloves 15   „
                         „   cinnamon . . . 10   „
                         „   bergamot . . . 30   „
                         „   neroly 10   „
          Or with—Oil of roses. 25   „
                    „   geranium . . . 60   „
                    „   cloves 15   „
                    „   Chinese cinnamon 10   „

Toilet-palm Soap.—

     Pure palm soap . . . .   5 kils.
     Demi-palme soap   5   „
          Perfumed with—Oil of bergamot . . .   60 grms.
                         „   Chinese cinnamon 25   „
                         „   lavender (fine). . 30   „
                         „   cloves.   15   „

Mallow Soap.   Savon de Guimauve fin.—

     White tallow soap . . . .   5 kils.
     Pure palm soap   5   „
          Coloured with—Yellow ochre . . . .   30 grms.
                         Mine-orange   30   „
                         Gamboge   20   „

Perfumed with—Oil of lavender (fine) . .	75	grms.
„ lemons (pressed) . .	16	„
„ neroli (small grained)	16	„
„ verbena	10	„
„ English mint . . .	8	„
Or—Oil of Portugal	40	„
„ thyme	80	„
„ lavender (fine) . .	10	„
„ cinnamon	15	„
„ cloves	25	„
Or—Oil of lavender (fine) . .	50	„
„ mint (ordinary) . .	10	„
„ caraway	15	„
„ rosemary	10	„
„ lemons (pressed) . .	20	„
„ thyme	10	„

Savon au Bouquet.

White soap of tallow or lard	10	kils.
Perfume—Oil of bergamot	15	grms.
„ neroli	15	„
„ sassafras	10	„
„ thyme	10	„

Coloured with 100 grammes of brown ochre.

Another Perfume. Parfum au Bouquet des Alpes.

Oil of lavender (fine)	20	grms.
„ English mint	20	„
„ sage	20	„
„ wild thyme	10	„
„ rosemary	15	„
„ lemons (pressed)	15	„

Savon aux Fleurs d'Italie.

Oil of citronel	50	grms.
„ geranium	10	„
„ verbena	30	„
„ English mint	10	„

Coloured with 80 grammes of brown ochre.

Savon au Benzoin.

White soap of tallow or lard. . . . . . . 10 kils.
Tincture of benzoin. . . . . 800 grms.

As the introduction of this tincture into the 10 kilogrammes of soap would render it rather moist, the soap-shavings should be previously well dried by exposure in the drying chamber for several days. This soap is sometimes perfumed with powdered benzoin ; but the alcoholic tincture imparts a more penetrating odour. It may be coloured with brown ochre or with a small quantity of burnt sienna.

Savon à la Vanille.

White tallow soap . . . . 10 kils.
Tincture of vanilla . . . . 500 grms.
Oil of roses . . . . 5 „

It may be coloured with 100 grammes of burnt sienna. Vanilla soap is one of the most fragrant that can be prepared : its perfume is still further improved by a small quantity of oil of roses.

Savon à l'Huile de Cannelle.

Pure palm soap . . . . 5 kils.
Tallow soap . . . . 5 „
Perfumed with—Oil of Chinese cinnamon . 80 grms.
　　　　　　　 „　　 sassafras . . . . 20 „
　　　　　　　 „　　 bergamot . . . . 30 „

Coloured with 80 grammes of yellow ochre and 20 grammes of burnt sienna.

Musk Soap. Savon au Musc.

Pure palm soap . . . . 5 kils.
White tallow soap . . . . 5 „
Perfume—Oil of bergamot . . . . . . 50 grms.
　　　　 „　 roses . . . . 5 „
　　　　 „　 cloves . . . . 5 „
　　　　 „　 musk . . . . 10 „

The musk is usually employed in the form of an alcoholic solution. Pound 10 grammes of musk in a mortar with an equal weight of sugar and 5 grammes of pure potash, then add gradually 160 grammes of alcohol, triturate for a quarter of an hour, pour the mixture into a flask, and leave it to digest for two to four weeks, shaking it from time to time. Then filter; add the whole of the filtered liquid to the 10 kilogrammes of soap, and afterwards the other essences.

The soap may be coloured with 80 grammes of brown ochre.

Savon au Fleur d'Oranger.

White tallow soap	6 kils.
Pure palm soap	4 „
Perfume—Oil of Portugal	140 grms.
„ amber	10 „
Or—Oil of geranium	40 „
„ neroli	50 „

Savon de Crimée.

White tallow soap	8 kils.
Pure palm soap	2 „
Perfumed with—Oil of thyme	30 grms.
„ lavender (fine) . .	10 „
„ mint (English) . .	30 „
„ rosemary	30 „
„ cloves	5 „
„ tincture of benzoin .	40 „

Coloured with 10 grammes of vermilion, 30 grammes of brown ochre, and 2 grammes of ivory black.

Windsor Soap.—This soap was formerly made from mutton-suet; but experience has shown that an addition of lard or olive-oil in the proportion of 20 to 35 per cent. promotes saponification and improves the quality of the product. Windsor soap originated in this country, but is now made largely in France and, of inferior quality, in America. The several kinds are as follows:

Plain Windsor is made with lard in the manner of saponifying olive-oil, and perfumed, just before being poured into the frames, with oil of caraway, in the proportion of 9 parts of oil to 1,000 parts of soap-paste. Two parts of the oil of caraway may be replaced by a mixture of 1 part of oil of rosemary, and 1 part of oil of lavender.

Violet Windsor is composed of 50 parts of lard, 33 parts of palm-oil, and 17 parts of spermaceti, and perfumed with essence of Portugal, containing a little oil of cloves.

45*

Benzoin Windsor is the plain Windsor, to which, while in paste, flowers of benzoin are added in the proportion of 6 oz. to 100 lbs. of soap.

Palm Windsor is made with palm-oil containing a small proportion of lard, a little essence of Portugal and oil of cloves being added to improve the odour.

Rose Windsor is the plain Windsor, coloured with vermilion or oxide of iron, and aromatized with essence of roses. The perfume, however, must not be added until the soap has been poured into the frames, lest any loss should accrue from volatilization.

French Windsor is made with mutton-suet, mixed with olive oil or lard; if made with mutton-suet alone, it becomes rancid and yellow by exposure. It is perfumed with a mixture of 4 lbs. of oil of caraway, 1 lb. of oil of lavender (fine), and 1 lb. of oil of rosemary; 9 lbs. of this mixture being commonly used to every 1000 lbs. of soap-paste.

The following method of preparing Windsor soap is given by F. W. Weise :* 40 lbs. tallow and 15 to 20 lbs. olive-oil are saponified with soda-lye of 19° B., and the soap is finished with lye of 15°, and finally with lye of 20°, the process being conducted as for a curd-soap, excepting that no excess of alkali is to be used. The soap is then boiled clear, left to itself in the boiler for six or eight hours, then completely separated from the lye, placed in a flat mould, and pressed till it no longer exhibits any flux, to prevent it from mottling. The quantity above mentioned is perfumed with 10 oz. oil of cumin, 6 oz. of bergamot, 3 oz. oil of lavender, 1 oz. Spanish oil of hops (*oleum origani cretici*), and 3 oz. oil of thyme.

Toilet Soft Soap, or Shaving Cream.

This soap, which is admirably adapted for the beard, is made by mixing 50 lbs. of lard with 75 lbs. of caustic potash lye, marking 17° B. This mixture is heated slowly, and as soon as the pasting is complete, and the whole of the lye thoroughly united to the fatty matter, the mass is raised to the boiling heat, the ebullition however being arrested as soon as watery vapours cease to be given off. This indicates the completion of the operation. The soap, which has then become so thick that it

---

* Dingl. polyt. J. cxxxv., 237.

can hardly be stirred, is run into frames and cooled. These soaps are generally perfumed, but not coloured.

Glycerin Soap.

This soap, which is used as a toilet soap for softening the skin, is made by mixing glycerin with ordinary soap when transferred to the frames.

Transparent Soap.

Soap is more perfectly and readily soluble in alcohol than in water. A concentrated solution of soap in water becomes partially opaque on cooling, by the formation of an acid soap (p. 640), but this is not the case with a concentrated alcoholic solution. This fact is applied to the manufacture of transparent soaps. In preparing soap of this description, ordinary soap is thoroughly dried in a stove, and dissolved in hot alcohol. All foreign matters not consisting of soap will then remain undissolved, and must be carefully removed, since they cannot remain concentrated in an opaque mass, as in ordinary soap. They are removed by deposition, or by a filter, supported by a funnel, surrounded on the outside with hot water. The alcohol is then separated from the soap by distillation, until the residue is capable of forming a solid mass, when cooled in the metallic moulds. When colours are required for transparent soaps, an alcoholic tincture of alkanet is used for red; a tincture of turmeric for yellow ; a mixture of those two tinctures for orange; alcoholic tincture of chlorophyll for green ; and an alcoholic tincture of indigo-carmine for blue.

Savonettes or *Wash-balls.*—These are toilet soaps, which, instead of being shaped into cakes, are moulded in spherical forms. The necessary implement is shown in Fig. 356. It is merely a brass handle with a hollow bowl-shaped trimmer, the edges of which are smooth and thin. The soap-ball being roughly reduced to the proper size and form, is then, with a rotatory motion, pressed against these edges, until it assumes a true spherical shape and is perfectly smooth on the surface.

FIG. 356.

Light or *Flotant Soaps.*—Soap bearing this name is merely ordinary toilet soap in a different form ; and any of the kinds mentioned above, with the exception of those made from hog's lard and suet, may be used in its preparation. It is obtained by threshing or agitating a solution of soap, to which one-fifth or one-sixth part of water has been added, with a rouser or paddle-wheel (Fig 357), until the lather has risen to twice the height of

FIG. 357.

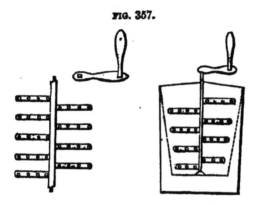

the soap-solution, and then transferring it to the moulds. A soap is thus obtained inflated with air, which gives it such a buoyancy that it floats on the water. It is a favourite soap, both on account of its lightness and mildness, and of the facility with which it gives a lather.

A very clear and full description of the preparation of perfumed soaps is given in Hirzel's " *Toilettenchemie*," Leipzig, 1857.

SILICATED SOAPS.

Silicate of soda (soluble glass) is a compound possessing considerable detergent power; in fact, the alkali contained in it is held in a state of feeble combination, similar to that in which it exists in soap made with fat. When mixed with ordinary soap, it forms a mixed soap of greatly reduced price, and very useful as a cleansing agent for domestic use, and in many manufacturing processes. Some of the so-called silicated soaps, however, are mere mechanical mixtures of siliceous substances, such as fine sand, alumina, fuller's-earth, &c., with ordinary soap: these are comparatively worthless.

The value of the true silicated soaps was recognised at the Great International Exhibition of 1862, by the award of a prize medal, by the jurors of Class IV., to W. Gossage & Sons, for "Excellent Samples of Silicated Soaps." The manufacture of these soaps is now carried out on a most extensive scale.

The mode of incorporating silica with soap, practised by Sheridan, is as follows: A solution of silica is obtained by boiling ordinary flints,—which are first reduced to powder by being heated to redness and then thrown into water, and afterwards ground,—with a lye of caustic potash or soda. This mass is then disseminated, by mechanical stirring, throughout soap previously prepared in the ordinary manner, and brought into the proper state for solidifying; the whole mixture is then placed in the moulds. It must not be forgotten, in practising this method, that powdered quartz is only dissolved with the greatest difficulty, and very slowly, in a boiling solution of potash, and that a red-heat is the proper temperature at which the combination of silica with potash is effected. An exception to this rule has been pointed out by Fuchs, who finds that the amorphous varieties of quartz,—for example, powdered opal,—dissolve easily in solutions of caustic alkali. The neglect of these facts involves the production of soap containing a large quantity of water instead of siliceous soap; for the fluid thus obtained holds but little silica in solution, and consists for the greater part of caustic lye. Girardin examined a soap of this sort, which is called silica-soap. It contained resin and palm oil, saponified with soda, and was easily dissolved by water with production of a copious lather; a deposit of siliceous matter subsided, however, from the solution, the ash of which, when treated with hydrochloric acid, left 19 per cent. of residue. That the silica may be thoroughly incorporated with the soap in a state of solution, it is necessary to prepare the siliceous fluid, or soluble glass, in the ordinary manner. In Sheridan's soap the insoluble and soluble portions are mixed with the soap in equal quantities, so that the greater part is evidently only in a state of mechanical admixture.

Gossage's Process.—This method consists of a mechanical mixing of soluble glass with the soap paste. For hard soaps, the soluble glass is prepared by melting together in a reverberatory furnace, 9 parts of soda-ash (containing 50 per cent. of caustic soda) with 11 parts of clean sand; and for soft soaps, equal weights of carbonate of potash and sand. To effect solution, the glass, after being drawn off into moulds and quenched with water, is either ground in a mill, and then boiled with alkaline water, or it is placed in lumps in an iron vessel having a false bottom, water is poured into the vessel in sufficient quantity

to cover the lumps, and steam is blown into the liquid.
The solution, when complete, is evaporated down to a syrup
having a density corresponding to 49° B. It is then ready to
mix with the soap-paste in the pan just as it has reached the
condition in which it is transferred to the frames. The temperature of the glass and of the soap-paste should be about 160° F.
at the moment of mixing, and to promote perfect homogeneity,
the mixture is stirred up by machinery, which will be described
immediately. When the mixture has cooled to 150° F., it is put
into the frames and again stirred with the crutch till it begins to
stiffen.

Rosin soap which is to be treated by this process, may contain
as much as 1 part of rosin to 2 parts of fat. The solution of
glass used for this soap must mark 51° B., and must be added to
the paste when it is fitted and about to be cleansed.

The stirring apparatus consists of a large tub or vessel, A (Fig.
358), having the shape of an inverted cone, of about 2 ft. 2 in.
internal diameter at its lowest part, 3 feet 6 inches at the upper
part, and 6 feet deep. To this vessel is adapted a central upright shaft, B, supported by a foot-step, C, fixed to the bottom of
the tub, and by a journal, D, adapted to a metallic bridge-piece,
E, which is fixed over the tub, and secured by screw-bolts to its
sides. To the upper part of this upright shaft is adapted a
bevelled cog-wheel, working in gear with another bevelled cog-

FIG. 358.

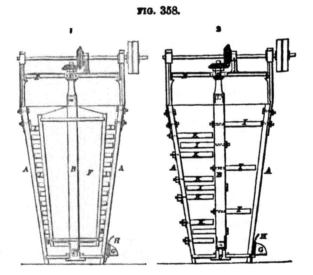

wheel, fixed on a horizontal shaft, S, which is made to revolve by a band passing round the driving-pulley, P, and also round another driving-pully, worked by some mechanical power. The upright shaft revolves at the rate of 60 to 80 revolutions per minute.

To the upright shaft B, is fixed a closed tub or vessel, F (Fig 1), of such a diameter as to admit of its being placed within the larger vessel A, leaving a space of about 2 inches at the lower and 6 inches at the upper part; and to the outside of this inner vessel are attached, by means of screws or otherwise, a number of projecting blades, I I, made by preference of sheet iron, of such a length as to approach within about half an inch of the inside of the vessel A. A spout, G, having a movable stopper, H, is adapted to the lower part of the vessel A, for the purpose of running off its contents. The projecting blades I, I, instead of being attached to an inner vessel, may also be fixed to the inside of the upright shaft, and in that case, it is best to attach other projecting blades, K, K, to the inside of the vessel A, in such a manner as to allow the blades, I, I, to revolve between them.

When this apparatus is to be used for the production of compound soap, by mixing genuine soap with a viscous solution of soluble glass, it is necessary to ascertain previously the highest temperature at which the mixture will become too thick to run from the mixing apparatus. For this purpose, a preparatory mixing of the genuine soap with the viscous solution is made by means of paddles or crutches in a vessel capable of containing about half a ton of soap, the soap and viscous solution being added at such temperatures as will yield a mixture having a mean temperature at least 10° higher than the temperature above referred to. The contents of the preparatory vessel are then transferred to the mixing apparatus, and a rapid revolving motion is communicated to the projecting blades. The stopper of the spout G is then withdrawn, so as to allow the compound soap in the state of perfect mixture to flow from the mixing apparatus, and further quantities of genuine soap and viscous solution of soluble glass, which have undergone a preparatory mixing, are then supplied. The mixed compound soap is transferred to the ordinary frames in which it becomes solid on cooling.

In mixing viscous solutions of soluble glass with genuine soap, whether such mixing be subsequently completed by crutching in frames, or by means of the mixing apparatus just described, it is best to commence the mixing by adding a portion of the viscous solution, at a specific gravity of about 1·300, and to add the remaining portions required for the mixing, at increasing specific gravities, so that the average specific gravity of the whole solution used may be equal to that which has been found by previous trials to yield a compound soap of proper hardness when using a genuine soap of the composition employed.

When it is desired to produce a compound soap having less detergent power than the compound soaps obtained by mixing genuine soaps of ordinary quality with solutions of soluble glass, a portion of the alkali contained in the solution may be neutralised by combining it with rosin, or with fatty or oily acids. The combination is effected by boiling the resin or the fatty or oily acids with solution of soluble glass, in the same manner as rosin and other soap-making materials are combined with alkali in the ordinary process of soap making.

Gossage* also reduces the quantity of free alkali in his silicated soap by neutralising it with sulphuric, hydrochloric, sulphurous, or carbonic acid, added either before or after the incorporation of the glass with the soap. The carbonic and sulphurous acids are passed into the liquid in the gaseous form.

Dunn prepares soluble glass by boiling 1 cwt. of crushed flint or quartz with 100 gallons of soda or potash lye of 21° B. in the steam boiler represented in Fig. 321, page 653. The whole is heated to about 310° F., and kept under steam pressure of 50 to 70 pounds to the square inch for three or four hours, when it is discharged and cooled down. The alkaline silicate thus obtained is added to the proper per-centage of the soap-paste in the pan, and before it has cooled down.

Other siliceous matters, such as powdered soap-stone, porcelain earth, pipe-clay, and fuller's earth, are also used for mixing with soap instead of soluble glass.

Davis's Alkalumino-Silicic Soap is a mixture of ordinary soap with fuller's-earth, pipe-clay, and pearl-ash, or soda, by which the cost of the soap is said to be much diminished, while its detergent and normal properties are improved. It is prepared by adding to every 126 lbs. of soap already made and in paste, 56 lbs. of fuller's-earth, slaked or dried, 56 lbs. of dried pipe-clay, and 112 lbs. of calcined soda or pearl-ash, all reduced to powder and sieved as finely as possible, and thoroughly incorporating the whole by stirring or crutching. The mixing must be very perfect, and

* Date of patent, Sept. 9th, 1856.

performed as quickly as possible, before the pasty soap cools. To obviate any objection against the use of this soap for washing white linens, a modification of the above process is proposed by which the use of fuller's-earth is entirely omitted, leaving the proportions for every 120 lbs. of soap, 112 lbs. of dried pipe-clay, and 96 lbs. of calcined alkali. A soap thus prepared is said by the patentee to be useful for general purposes at sea, and for washing white linen in salt water. For washing white linens in fresh water, the process is still further modified by using 112 lbs. of soap, 28 lbs. of dried pipe-clay, and 36 lbs. of calcined soda; and as a toilet soap, either for fresh or salt water, by employing 28 lbs. of fuller's-earth, slaked or dried, and 20 lbs of calcined soda, to 112 lbs. of perfumed curd-soap.

Way's Silica Soap.—J. P. Way prepares soluble glass for mixing with soap, from a rock or clay found in Surrey, which contains a considerable amount of soluble silica, sometimes as much as 70 per cent. The following is the process as described in the specification :—

"I put into a suitable pan, heated by steam, or in any convenient manner, a quantity of caustic alkaline lye (potash or soda, or both, as the case may be), of about 18° Twaddell, so that the silica solution, when made, shall have a specific gravity of as nearly 36° as possible, and having raised this lye to the boiling point, I add by degrees the rock or clay before mentioned, either in small pieces or ground to powder, until the alkali has taken up as much silica as it will dissolve. The heat is now withdrawn, and the undissolved earthy matter is allowed to settle. The clear liquor is run off, and a fresh quantity of water added to the sediment to wash out further portions of soluble matter.

" The liquors so obtained are solutions of alkaline silicates. The quantity of rock or clay required will vary with the percentage of soluble silica which it contains. . . . . I find it necessary, for every 31 parts of actual soda, or 53 parts of carbonate of soda, rendered caustic, to employ as much of the rock or clay as contains 78 parts of soluble silica.

"I produce similar alkaline silicates from the rock or clay before mentioned, by gently heating it in a furnace with alkalies or alkaline carbonates. In this case, I find that combination of the materials and production of the alkaline silicates takes place at a temperature much below that which is necessary when other forms of siliceous matter are used, and though preferring the method formerly mentioned in the treatment of the rock or clay, the one last described may be employed. The alkaline silicate is dissolved out from the furnaced materials by water or alkaline lye.

"I prefer, in either case, to saturate the alkali as fully as possible with silica, but this is not absolutely necessary. The silicates so produced are more suitable for the soap-maker than those ordinarily employed, from the following amongst other reasons:—They are more economically produced. The caustic property of the alkali contained in them is more perfectly neutralised, and they contain no iron, alumina, or other matter injurious to soap. The soap produced by them is therefore of superior quality, as well as cheaper. The alkaline silicate produced by either of these processes may be employed in any of the modes now used by soap-makers when using the silicates of the alkalies in soap-making.

"As a practical illustration of my invention, I will now state what I have found to give satisfactory results for the manufacture of 100 lbs. of soap. I put into the soap-pan 11·5 per cent. of bleached palm oil, 11·5 per cent. of cocoa-nut oil, 30·6 per cent. of soda lye of 36° Twaddell. I boil till the soap becomes stiff, then add 44 per cent. of the solution of silicate, prepared as before described, and of 36° Twaddell, continue the boiling till the soap becomes thin and limpid, at which stage I add 2·4 per cent. of common salt, and boil for three or four hours, when the soap will be ready to cleanse, which may be done at once, or after it has been allowed to stand for a few hours. If open steam is used, that is, blowing steam into the soap-pan, I consider it best to have the solution of silicate and the lye of greater strength than that mentioned, in proportion to the quantity of water which is condensed from such steam into the soap-pan."

PATENT SOAPS OF VARIOUS KINDS.

Bone Soap.—This is nothing more than ordinary soap from tallow, rosin, or palm oil, mixed with gelatin or bones, disintegrated, dissolved, or partially decomposed by potash. It was at one time extensively sold under the name of *Liverpool poor man's soap*, but it is a worthless article, and appears to have fallen into disuse.

Patents have likewise been obtained for the so-called saponifi-

cation of the intestines of animals, and refuse of a like character,
for instance, skins, sinews, hoofs, &c. Cheap fish of various kinds
have also been recommended as material for the manufacture of
similar kinds of soap. These varieties of soap, however, have
been known for a much longer period than the patents which
have recently brought them again into notice: for Hermbstädt
mentions a soap made with fish, and a similar one from refuse
wool.

Later processes for employing naphtha, hair, fish, dextrin
(starch), seeds, glue, &c., have been patented in this country
by Caldecott, Chauvier, Poole, Snell, Partridge, Gedge, and
others, but none of them are of any practical use.

Snell's Dextrin Soap.—Snell's so called Dextrin Soap is
merely ordinary soap mixed with potato-starch containing a
certain quantity of the fibrous matter of the potato. This
admixture is said to increase the detergent power of the soap.
To obtain the starch, the washed potatoes are reduced to as fine
a pulp as possible, by a revolving cylinder furnished with parallel
plates of steel or iron, having fine teeth like saws, the teeth rising
just above the surface of the cylinder, which revolves rapidly
under a hopper placed to receive the potatoes. The pulp falls on
sieves or screens below, and passes over the upper sieve, which is
a plane 5 feet broad and 8 feet long, slightly inclined, with four
bars the whole length, and one or two cross-bars, covered with a
wire-cloth of thirty holes to the inch. At the top and about
midway down the sieve, are pipes containing water, so placed that
the water shall run from the pipes through small holes, the whole
breadth of the sieve, the finer parts of the potato passing through
the meshes of the sieve with the water, and the coarser parts
running over the sieve into a vat. The finer parts,—starch, called
dextrin by the patentee,—which fall on a wooden plane placed
under the second sieve, but inclined in a contrary direction, are
received in vats, where they are repeatedly washed over sieves of
fine wire cloth until cleansed from all impurities. The grosser
parts (fibre) are washed over coarse sieves.

The usual proportions of material are from 300 to 350 lbs. of
dry starch, mixed with 150 lbs. of water, to every ton of saponified
oil or tallow. The white soap should be made with tallow and
by steam heat; and when as much alkali has been added as can
be taken up in the usual way, and the soap is in a fit state for
cleansing in a frame, each ton of saponified matter in the caldron

is mixed with from 450 to 500 lbs. of dextrin (starch) in its wet
state, and after it has stood for an hour, 600 lbs. of boiling
water, or enough to make it into a thick paste, are poured by
pailfuls into the soap, stirring all the time. When the mixture
is complete, and has been heated, it is fit to be transferred to the
frames. Before the starch is added, the spent lye must first be
pumped or drawn off. When yellow soap is made with rosin in
the usual way, and is what is called "fitted," it should be heated
with steam and dosed with starch in the same manner and proportion as already described. If no rosin is employed, a larger
quantity of dextrin may be used with or without fitting the soap.

In making common soap, either yellow or brown, the product,
especially if it contains excess of alkali, will be much improved in
quality and colour, by using the fibre mixed up with as much
water as will form a thick mush ; the soap will then waste much
less quickly in hot water.

Partridge's Soap.—J. W. Partridge, of Birmingham, mixes
recently made soap with any quantity of linseed, rape-seed,
hemp-seed, cotton-seed, or any similar seed possessing oleaginous
and mucilaginous properties to the extent of 100 per cent. and
upwards. For common soaps, the seeds are bruised to a pulp,
either before or after the oil has been extracted from them, and
this ingredient having been thoroughly mixed with the soap, the
whole is transferred to the frames and left to cool. For better
qualities, an extract of the seeds is prepared by boiling, infusion,
&c., which is filtered through a linen or flannel bag, and boiled
down to a jelly. It is then mixed with the soap in the proportion of 50 per cent.

Gedge's Glue-soap.—*For Yellow Soap.*—Place in a copper
heated by fire or steam, 375 lbs. of cocoa-nut oil and about 650
lbs. of rosin ; when these are melted, add about 1,200 lbs. of
alkali of the strength of 1·260 or thereabouts ; boil ; then pour
in by three operations 1,000 lbs. of water, boiling the mass ; at
each operation add 200 lbs. of common salt to separate the
lye. Then place in a copper, about 750 lbs. of cocoa-nut
oil : when well melted, add 300 lbs. of alkali of the strength
above mentioned, and then add, as in the former operation, 700
lbs. of water. Add the product to that of the first operation ;
mix ; boil ; add 25 lbs. of glue, and when it is thoroughly dissolved, run the soap into the moulds.

For White Soap.—Place in the copper about 700 lbs. of cocoa-nut oil and about 100 lbs. of rosin, upon which, when melted, throw about 900 lbs. of alkali: after boiling, add about 900 lbs. of water, and separate the lyes with 300 lbs. of salt. Then place in a copper 700 lbs. of cocoa-nut oil, and when melted add 700 lbs. of alkali, and 700 lbs. of water; after the last boiling, add 200 lbs. of cocoa-nut oil and boil again: then mix the soap obtained by the first operation with the above, adding about 800 lbs. of water to the whole: when all is thoroughly melted, add 85 to 90 lbs. of glue, and proceed as already explained.

Bousfield adds to soap made in the ordinary way, before it assumes the solid form, shreds of animal and vegetable fibre, or fibrous substances, such as wool, silk, cotton, linen, sponge, hair, &c., with the view of increasing the detergent power of the soap.

Löwissohn proposes to mix soap with Irish moss, adding borax to facilitate the mixture.

Normandy's (Salinated) Soap.—This soap contains the sulphates and carbonates of soda and potash, added while the soap is yet in paste. The advantages gained, are said to be hardness and economy. It is prepared by adding to 80 lbs. of soap 28 lbs. of sulphate of soda and 4 lbs. of carbonate of potash, or 2 lbs. of carbonate of potash and 2 lbs. of carbonate of soda. Or, if the substances are used singly, then to 80 lbs. of soap there are added only 32 lbs. of sulphate of soda, 15 lbs. of carbonate of potash, or 10 lbs. of carbonate of soda. As this soap contains an excess of alkali, it may perhaps answer well for marine use.

Another process for the preparation of salinated soaps has been patented by Normandy and Simpson. Soap is prepared by boiling tallow, bone-fat, lard, palm oil, &c., with soda-lye in the usual way, and after the soap has been "curded" by means of salt, or strong lye, the lye is allowed to settle down, and after it has been drawn off, a certain quantity of fresh lye and cocoa-nut oil is added, and the whole well boiled together, till it has become a homogeneous well saponified mass, having the appearance of well finished and fitted yellow soap, except as to colour. This being done, a certain quantity of sulphate, sulphite, or hyposulphite of soda is added, the whole boiled together, and the soap transferred to the frames. This process is said to yield a mottled soap of better consistence than could be obtained from the same materials in the ordinary way, and without separation of lyes in the frames.

Hyposulphite of soda crutched into the soap in the frames, increases its hardness, like the sulphate, and is not so liable to effloresce. It has likewise the property of removing the chlorine which bleached fabrics have a tendency to retain, and by which they are exposed to deterioration.

Normandy has also patented a process for the use of sulphate, sulphite, or hyposulphite of soda, separately or jointly, in combination with the rock or clay found in Surrey, mentioned in Way's specification (p. 715).

Chlorine Soap (Savon chloruré).—This substance is intended to realize the idea of the union of the cleansing properties of soap with the bleaching effects of certain compounds of chlorine. The method of preparing it proposed by the inventor, is, however, so much at variance with the laws of chemical combination, that the product must necessarily be very different from that which is desired. It is proposed to saponify the oils or fats with chloride (hypochlorite) of lime, or with the corresponding potash or soda compound, or to soften the ready prepared soap by immersion in a solution of this nature. With reference to chloride of lime, it is obvious that the greater part of the valuable alkaline soap will be converted by its agency into a useless lime-soap. With regard to the other bleaching compounds, soap may possibly be obtained which is mixed with a greater or smaller proportion of the bleaching salt, but this cannot exert any appreciable amount of action when the soap is used for washing. An attempt has also been made to saturate the fats themselves with chlorine before saponification; but it is well known that these bodies are decomposed by the continued action of chlorine, with formation of hydrochloric acid; and if the action of the chlorine is modified, the oil will be bleached; but soap can never absorb such an amount of chlorine as will communicate bleaching properties to it.

Rowland's Soap.—This is soap mixed with some liquid hydrocarbon, as turpentine, coal tar, naphtha, camphine, &c., whereby the detergence of the soap is said to be greatly increased. It is prepared by dissolving 3 lbs. of soap in 1 lb. of water with the aid of heat; adding to another pound of water 2 oz. of farina, flour, dextrin, starch, oatmeal, or other analogous substance; mixing and boiling to the consistence of a paste; adding this paste to the dissolved soap; and stirring till a perfect incorporation of the materials is effected. Or the whole of the water may be mixed with the flour and converted into a pasty soap by boiling, and the soap may be dissolved therein instead of being melted separately. The mixture, when well incorporated, is to be removed from the fire and stirred till it has cooled to about 140° F., when 8 oz. of turpentine, mineral naphtha, camphine, benzole, or some other equivalent substance must be added, together with a saturated solution of carbonate of ammonia. The whole must then be again well stirred and the mixture run off into suitable vessels, closed air-tight. About half an ounce of the liquor ammoniæ of the London Pharmacopœia may be added to every 4 oz. of the saturated solution of the carbonate.

In place of the dextrin, flour, or other farinaceous substances above mentioned, an equivalent quantity of gelatin, glue, or other mucilaginous or gelatinous substance may sometimes be used. For making a detergent composition for common use, common mineral or coal-tar naphtha, or turpentine may be the hydrocarbon used, and a larger quantity of caustic ammonia may then be advantageously added; but for a detergent preparation for toilet purposes, or for washing and cleansing body-linen, it is preferable to use the best rectified camphine or some analogous material of more agreeable odour.

The carbonate of ammonia may also be added to the pasty saponaceous mixture in the dry pulverized state, in the proportion of a quarter of an ounce to every pound of soap. The mass after cooling will have the consistence of soft soap. When it is to be used for washing linen, about 4 oz. of the mixture should be added to about 12 gallons of water, and stirred till quite dissolved. The clothes may be put in and stirred with a stick, after which they may be boiled for a few minutes, and then rinsed in fresh warm water.

Kottula's Compact Neutral Soap.—This soap is prepared by combining any of the usual fats or oils with concentrated soda-lyes and lime-liquor. The soda-lye is concentrated to about 28° Baumé, and purified by boiling for half an hour with alum, in the proportion of 4 to 4½ lbs. to every hundredweight of lye. The vessel is then removed from the fire, alum again added, in the proportion of about 2 to 2½ lbs. to each cwt. of lyes, and the liquid is stirred till the alum dissolves, after which the vessel is covered, and the whole is left to settle and become clear. The lime-liquor is prepared by combining water and lime, and then adding to each cwt. of lime-liquor about 1½ to 1¾ lbs. of sal-ammoniac. The liquid is boiled for about half an hour, and then allowed to settle and become clear, or the sal-ammoniac is added to the lime-liquor while hot, and stirred for about half an hour.

Ten tons of fatty matter, with or without resin, 9 tons of lye prepared as above, and 13 tons of lime-liquor, will produce a superior compact neutral soap, which may be coloured, mottled, or perfumed by the usual processes.

Kottula's Hand or Skin Soap.—This invention consists in mixing fatty matters with certain quantities of highly concentrated soda-lyes purified with a certain quantity of alum and sal-ammoniac, whereby a neutral hand or skin soap is said to be obtained cheaper and better than by any means heretofore adopted. The process is thus described : " I prepare the highly concentrated lyes by boiling until they reach, say about 30° to 33° B., add about 5 lbs. of alum to each cwt. of lyes, and boil together for about half an hour. I remove the lyes and alum from the heat, and add to each cwt. of lyes 1 lb. of sal-ammoniac, stir for half an hour, cover, and allow the mass to settle and become perfectly clear. To obtain the lyes stronger than 33°, I make a second addition of alum, but in smaller proportion. To obtain lyes of 42°, I make a third addition of alum, and then add the sal-ammoniac. I melt a quantity of any fatty matter used in soap-making, and while still hot, stir and add the highly concentrated lyes prepared as before described—say, to every 100 lbs. of fatty matter, about 100 lbs. of lyes of 30° B., or 90 lbs. of 33° B, or 80 lbs. of 36° B., or 70 lbs. of 39° B., or 60 lbs. of 42° B. ; continue to agitate the mass till it becomes thick, and when thick it can be transferred to the frames. After the soap is finished, it may be coloured, mottled, or perfumed in the manner well known to soap-makers."

Blake and Maxwell's Process of Soap-making.—This invention consists :—Firstly, in forming a soap by combining soap or saponified matter in the state in which it is commonly called soft curd, with a neutral soap which has not been deprived of its water. The curd soap may be formed of any of the oils or fatty matters usually employed in soap-making, and such oils or fatty matters may be saponified by means of alkaline lyes in the usual way ; but the patentees prefer to use curd made by means of the soda-lyes of the strength and in the quantity hereafter mentioned, for the purpose of obtaining a soft curd better adapted for combining with a neutral soap ; the saponified matter thus formed may be separated from the water or excess of lyes with which it is mixed, by means of salt, salt water, or concentrated soda-lyes in the usual way.

The neutral soap which has not been deprived of its water, is a soap which retains in its constitution the soda-lyes or soda and water employed in its manufacture; it may be called a *hydrated* soap. In calling this soap neutral, it is to be understood that it is made of fatty or oily matters and soda-lyes in such proportion that they may be effectually saponified or combined by means of boiling or heat, so as to make a soap or combination in which neither the fatty or oily matter nor the alkali shall be in excess. The proportions of fatty or oily matter and soda-lye used in making both the soft curd and the hydrated soap will be found from the tables given below (p. 727).

The invention consists :—

Secondly, of a mode of combining rosin or resinous substances with alkali or alkaline lyes, so as to obtain saponification.

To effect this combination, first mix a portion, say about one-third of the rosin, intended to be used with a small quantity of fatty matter, equal to from 6 to 10 per cent. of the rosin. A similar proportion, say one-third of the alkaline lyes, is also mixed with the rosin, and the mixture is slowly melted. The remainder of the resinous substance is then added gradually by small portions at a time, as the added portions melt, and when the entire quantity is melted, the residue of the alkaline lye is added. Increased heat is then applied till the material boils, and this is continued till the saponification of the whole mass has been effected, when it will have assumed an appearance or consistence somewhat similar to thick glue or paste; and this effect may be produced, under the application of heat in the manner above described and careful management, in about three hours. The saponified rosin is then separated from the excess of lyes and water by means of salt or concentrated lyes, and is thus obtained in the form of soft curd. In saponifying rosin in this way, it is deprived of much of its peculiar smell, and is considerably bleached, and the soft resinous curd soap thus obtained is fit for mixing with other soaps. To regulate the temperature, an easily manageable fire must be used, or still better, a steam-heat.

The following are examples of the application of these products in the manufacture of various kinds of soap (similar to soaps in ordinary use).

46*

"To make mottled soap, we take a known quantity of tallow, fatty matter, or bleached palm-oil, and boil it with sufficient soda-lyes to produce saponification, and then obtain the saponified fatty matters in the form of soft curd, as before mentioned; at the same time, we prepare in another soap-pan the hydrated soap, which may be formed of any of the oils or fatty matters mentioned in the second table of equivalents or proportions, either singly or in combination.

"For a combination of oils or fatty matters to be used in making this soap, we prefer to use cocoa-nut oil, tallow, and lard in the proportion of about 80 per cent. of cocoa-nut oil, 14 per cent. of tallow, and 6 per cent. of lard. The quantity of lye to be used for the saponification of these oils and fats, either singly or in combination, should be carefully calculated by means of the second table, in order that a suitable hydrated soap may be formed. We then transfer the lard-soap to the pan containing the hydrated soap made as last described, and boil both soaps together by means of a slow fire until they are perfectly united, and an appearance is presented of a soap finished in the ordinary manner. A mottle may be produced in this soap by adding to it, when in a finished state, colouring matter, to impart to the mottle such colour as may be desired in the usual way; in about half an hour after the soaps and colouring matter have been thoroughly incorporated, the soap may be transferred to the frames. To manufacture the best descriptions of mottled soaps, we use a quantity or weight of oil or fatty matter converted into hydrated soap, equal to from one-fourth to one-half of the oil or fatty matters converted, used in making the curd soap intended to be combined with it; and for cheaper descriptions of soap, the proportion of hydrated soap may be increased to proportions in which the oil or fatty matters converted into hydrated soap shall be equal to from two-thirds to one and a half the weight of oil and fatty matter used to make the curd soap.

"To make white soap, we use a combination of tallow, oils, and fats to form both the curd soap and the hydrated soap, and mix them together, as indicated in the process for making mottled soap.

"To prepare a soap similar to that made from tallow or rosin, and known in the trade as "pale soap," we put into the soap-pan two-thirds of the entire quantity of tallow intended to be used in the operation, and then add about one-third of the quantity of lye calculated to be required for its saponification; a slow heat must be applied, and when the tallow and lyes have united, a quantity of rosin, saponified as before described, must be added to the soap under the process of manufacture. We do not confine ourselves to any proportion, but prefer to use, for the better qualities of pale soap, saponified rosin in the proportion of one-third to the weight of the tallow used in making the soap. When the tallow, lyes, and saponified rosin are thoroughly united, we add the remainder of the lyes, gradually increasing the heat at the same time, until the whole of the calculated quantity is used. After saponification, the excess of lyes and water is separated in the usual manner, and the mixture of tallow and rosin soap obtained in the state of soft curd. In another soap-pan we make the hydrated soap by saponifying the remaining quantity of tallow, that is the third part left from the first operation, by using soda-lyes of the strength and in the quantity which may be ascertained by reference to the second table of equivalents, and into the pan containing such hydrated tallow soap we transfer the soft curd of rosin and tallow soap made as already mentioned. And after boiling the mixture together for about two hours, the soaps will become thoroughly united, and the compound soap will have assumed the appearance similar to the ordinary soap in the process of finishing. The soap should be removed to the frames within two or three hours after it is finished, and the frames should be covered, so as to retain the heat as long as practicable.

" To make a palm-oil and rosin soap, similar to that known in the trade as 'crown soap,' we prepare a curd soap from palm-oil and rosin in a similar manner to that described in making 'pale soap' from tallow and rosin; the weight of the lyes required for palm-oil being ascertained by reference to the first table of equivalents. In another soap-pan we form a hydrated soap by saponifying a quantity of palm-oil or tallow, equal to one-third the weight of palm-oil firstly used for the curd soap, by using lyes of the strength and in the quantity which may be ascertained by means of the second table of equivalents; and to such hydrated soap we add the palm-oil and rosin curd soap, and proceed in the completion of this soap in like manner as in the production of tallow pale soap above described.

" To manufacture soaps at a reduced cost, we use a greater proportion of saponified rosin curd for admixture with the curd soap obtained from fatty matters, or we use a greater proportion of hydrated soap in proportion to the curd soap used. When the curd soap to be combined with a hydrated soap contains any excess of fatty or oily matters, so that the combination of the two soaps would not produce a neutral soap, we add such a quantity of lyes to the hydrated soap or to the mixture as will make the combination neutral.

" In the preparation of soda-lyes, we prefer to use soda-ash of the strength containing from 40 to 48 per cent. of pure soda, and to every 100 lbs. of such soda-ash we use 60 to 70 lbs. of freshly calcined lime slacked in its own weight of water, and 1,000 lbs. of water to dissolve the soda-ash.

" We slack the lime firstly, and then add the required weight of water, and when the lime and water have been thoroughly mixed, we add the soda-ash. We employ free steam, blown through the mixture of water, lime, and soda-ash, in order to boil it, and to assist in the operation of rendering the lyes caustic. The lyes produced from the above proportions of materials will be of the specific gravity of 11° of Baumé's hydrometer. Stronger lyes may be produced by using less proportions of water or larger proportions of soda-ash in the quantities necessary to produce the lyes desired.

" In the following table of equivalents or proportions, we state the oils and fatty matters we prefer to use for making the soft curd soap, and the strength of soda-lyes (as indicated by Baumé's hydrometer), and the quantities of the same which we deem to be best adapted for speedily effecting the saponification of such matter, such quantities being stated in degrees of Baumé's hydrometer, with reference to a combination with 100 lbs. of each fatty matter, so that the number of pounds of lyes required for the saponification of any particular fatty matter may be ascertained by dividing such number of degrees by the strength of the lyes stated to be applicable to such fatty matters, and the quotient will be the number of pounds of lyes required to be used to saponify 100 lbs. of fatty matter."

First Table of Equivalents or Proportions.

Fatty matters to be used.	Quantity of lyes in degrees of Baumé's hydrometer.	Strength of lyes in degrees of Baumé's hydrometer.
100 lbs. tallow require . . .	3,800°	14° to 15°
100 lbs. palm-oil require . .	3,200	16 to 18
100 lbs. tallow olein require .	2,800	16 to 18
100 lbs. rosin require . . .	2,700	16 to 22

The second table of equivalents or proportions states the oils and fatty matters used for making the hydrated soap, and exhibits in like manner the quantities and strength of the lyes required for that purpose.

Second Table of Equivalents or Proportions.

Fatty matters to be used.	Quantity of lyes in degrees of Baumé's hydrometer.	Strength of lyes in degrees of Baumé's hydrometer.
100 lbs. of tallow require . .	3,800°	11°
cocoa-nut oil . .	4,100	16 to 20°
palm-oil	3,200	18 to 22
lard	3,400	13
tallow olein . . .	2,800	18 to 22
olive oil	3,000	16
rape-seed oil . .	2,400	24 to 28
linseed oil . . .	2,400	24 to 28

To give an example of the mode in which to calculate the quantity of lyes required for an operation, we take that first mentioned for the manufacture of mottled soap, assuming palm-oil to be used to form the curd soap. On referring to the first table of proportions, it appears that 100 lbs. of palm-oil require 3,200° of lyes which may be of a strength from 16° to 18° B. This number of degrees, 3,200, is divided by the number of degrees of strength of the lyes intended to be used; and if the strength be 16° B. we pour $\frac{3200}{16} = 200$ lbs. for the quantity of lye required to saponify 100 lbs. of palm-oil. If lyes of 18° B. be used, the quantity required will be $\frac{3200}{18} = 107$ lbs. Reference is then to be made to the second table of proportions, in order to ascertain the quantities and strength of the lyes to be used in making the hydrated soap. For cocoa-nut oil, either of the two strengths 16° or 20° mentioned in the second table may be used. The quantities may be stated as follows :

For 80 lbs. of cocoa-nut oil {if lyes of 16° be used, we take the proportion} as $100 : \dfrac{4100}{16} :: 80 : 205 =$ lbs. weight of lyes of 16°.

For 80 lbs. of cocoa-nut oil {if lyes of 20° be used, we say} as $100 : \dfrac{4100}{20} :: 80 : 164 =$ lbs. weight of lyes of 20°.

For 14 lbs. of tallow {we take the proportion} as $100 : \dfrac{3800}{11} :: 14 : 48 =$ lbs. weight of lyes of 11°.

For 6 lbs. of tallow {we take the proportion} as $100 : \dfrac{3400}{18} :: 6 : 15\frac{1}{2} =$ lbs. weight of lyes of 13°.

Kottula has also patented a process for combining "fitted" soaps with "hydrated" soaps. The fitted soap is either put into a pan and a quantity of cocoa-nut oil, tallow or other oil added, together with the quantity of soda-lye necessary for saponification, and the whole is boiled together till complete combination takes place ; or the neutral hydrated soap is made in a separate pan, and then boiled with the fitted soap.

A patent* has been taken out for the decomposition of certain oils and fats which contain constituents of different degrees of hardness, by the use of alkalies, alkaline carbonates, or alkaline earths, in such proportion as to effect only a partial saponification of such oils or fats, the softer constituents then uniting with the alkali and forming soap, while the harder and less fusible fat swims on the surface of the saponaceous compound, and being separated, may be used for the manufacture of candles, &c.

Thus when 1,000 parts by weight of palm-oil are heated in an ordinary soap-pan by steam or otherwise to 180° Fah., and well mixed by stirring with paddles or crutches with a caustic soda solution of specific gravity about 1·200, the mixture, after remaining quiet for some hours, separates into two portions, the upper

* Patent granted to W. E. Newton of the Patent Office, Aug. 2, 1858, No. 290.

being fatty matter which may be removed into casks ready for sale or use, and the lower a soap which may be finished in the usual way. The process is conducted in the same manner with potash, the saponaceous product being of course a soft soap.

Solutions of carbonate of soda or potash may also be used, such solutions having a specific gravity of 1·250.

The saponification may also be effected by caustic lime. A fine cream of lime is prepared, containing 60 parts by weight of caustic·lime mixed with 940 parts by weight of water; and this quantity of cream of lime is added to 1,000 parts by weight of palm-oil or other fat, heated to 180° F. The mixing is effected by crutching, also by injecting steam into the mass, the process being continued for six or eight hours, after which the mixture is left at rest. It then separates into two portions, the upper being a hard fat, and the lower a soap of lime which is transferred to a wooden vessel, and decomposed by hydrochloric acid or other suitable acids. The softer portion of the oil or fat is thus obtained in a state fit for use as liquid oil or soft fat.

COMPOSITION AND VALUE OF SOAP.

There are few things which are so ill understood, in practical life, as the *real value*, or what is the same thing, the proper price of soap. The value of soap is mainly dependent upon the amount of dry soap (the dry combination of alkali with the fatty acid) which it contains, and this is very easily ascertained. To establish this point, a weighed specimen, in the form of thin shavings, is exposed to the heat of a drying stove or over sulphuric acid, until its weight is no longer diminished. The loss of weight is hygroscopic water, and what remains is dry soap. The latter invariably retains the water which is in chemical combination, or water of hydration; but this quantity is so small in proportion to the equivalent of the soap-compound, that it is practically of no moment.

Greater sources of deception are naturally traceable to an excess of alkali, salt, &c., which are peculiar to soap not in the form of curd, and particularly to such as contains much water, or has been prepared without boiling. These admixtures are indicated by the general appearance of the soap, but can only be accurately estimated by a careful chemical analysis.

The following method serves for the complete analysis of soaps.

A portion of the soap (say 100 grains) in very thin scrapings is
first dried over sulphuric acid or by exposing it to a current of
air heated to 100° C., in order to determine the water. The
dried soap is then introduced into a glass flask, having a long
neck and wide mouth, about 3 ounces of 90 per cent. alcohol are
added, and the whole is heated to boiling over a water-bath.
All the true soap then dissolves in the alcohol, together with any
free fat or free resin that may be present, while all mineral salts,
silica, pipe-clay, glue, starch, dextrin, or in short anything added
for the purpose of adulteration, will remain undissolved. To
make the separation complete, the liquid is filtered through a
weighed filter, surrounded by a hot water bath, and washed with
alcohol.

a.—The hot alcoholic filtrate, containing the true soap,
together with free fat and resin, and perhaps a little ammonia, is
freed from alcohol by careful evaporation over a water-bath,
and then diluted with water, which throws to the surface any
free resin or grease that may be present. These latter are collected on a weighed filter, washed with hot water, then dried in
vacuo and weighed, together with the filter. The weight, less
that of the filter, gives the quantity of free fat and resin in the
soap. The filtrate now contains only the fat soap and resin soap,
if any, and must be treated with a graduated solution of sulphuric acid (see ALKALIMETRY) to determine the amount of
potash or soda in combination with the fat acids. At the same
time, the soap-solution is decomposed; the fat and resin acids,
being displaced by the sulphuric acid, rise to the surface, and
may be collected on a weighed filter, washed with hot water,
dried in vacuo, and weighed. The weight expresses the joint
amount of fat and resin-acids in the soap. Cold alcohol will dissolve out all the fat-acid, together with a small portion of the
resin from the filter ; and the filter, dried in vacuo and weighed
as before, gives approximately the amount of resin in the soap.

To determine whether the base of the soap is soda or potash,
the solution of the sulphates, filtered from the fat-acids, is concentrated by evaporation and treated in the usual manner with
alcohol and bichloride of platinum or tartaric acid, either of which
will give a precipitate with potash, but not with soda.

The nature of the fat-acids may be determined by their consistency and their melting point; the odour developed during
fusion frequently serves also to detect the kind of fat which is

present in the largest quantity. When the oleic acid predominates, as in some soaps made with vegetable fats, and consequently the fat-acids show but little tendency to solidify, the quantitative estimation may be facilitated by the addition of a weighed quantity of white wax. If heat be then applied, the wax and the fat melt and solidify together on cooling, and the separation from the watery liquid is thereby rendered easier.

It must not be forgotten that the fat-acids separated as above, are in the hydrated state, and if estimated in that form will make the result of the analysis exceed 100 ; for in combination they are anhydrous, and the excess represents the water which they have taken up when separated from the alkali (p. 634). This water of hydration must therefore be deducted.

b.—The filter containing the matter insoluble in alcohol is dried at the heat of the water-bath—after having been thoroughly washed with alcohol—and then weighed. The weight thus found, diminished by the weight of the filter itself, gives the weight of the matter insoluble in alcohol. In genuine soap this insoluble matter is of very small amount, not exceeding 1 per cent. for mottled, and even less for white soap. Any greater quantity indicates the presence of alkaline salts, silica, or other foreign matters. Generally speaking it is sufficient to determine the total amount of these foreign substances, but in some cases it may be desirable to ascertain their nature and the amount of each.

The foreign substances likely to be present in soap may be classified as follows :—

Soluble Salts and Organic Matters.	*Insoluble Inorganic Matters.*
Sugar	Silica
Borax	Sulphate of baryta
Carbonate of soda	Sulphate of lime
Sulphate of soda	Pipe clay
Chloride of sodium	Chalk
Carbonate of ammonia	Lime
Alum	Magnesia.
Starch	
Glue	
Dextrin	
Bran.	

To separate these substances, the filter with its contents is carefully replaced on the funnel and washed with cold water, to dissolve sugar, borax, alum, carbonate of soda, &c. The washed filter is then dried and re-weighed, the loss of weight indicating the joint amount of these soluble substances present in the soap. In the filtrate, sugar may be detected by its sweet taste, or by its reaction with copper salts ; the carbonates of the alkalies by their alkaline reaction ; sulphates by their behaviour with baryta salts, &c. It is seldom necessary to make separate determinations of the amounts of these salts; if required, it may be done by the ordinary methods of mineral analysis. The quantity of carbonate of soda or potash may be determined by the alkalimetric method.

The filter, freed from the salts soluble in cold water, is next treated with boiling water to dissolve starch, dextrin, and glue, then dried and weighed as before. The loss indicates the joint amount of these substances. Glue may be detected in the filtrate by its peculiar odour ; starch and dextrin by the blue or purple colour which they respectively produce with iodine.

The filter is now to be burned to ash and the residue weighed: and thence by allowing for the weight of the filter and that of its ash, previously determined by burning a filter of the same weight, the amount of bran or other insoluble organic matter burned off may be determined.

The weight of the ignited residue, diminished by that of the filter, gives the total quantity of mineral matter insoluble in water. Lastly, the chalk, lime, and magnesia may be separated from the silica and sulphate of baryta, if present, by solution in hydrochloric acid.

It is seldom, however, that so complete an analysis of soap is required as that afforded by the method just described. Generally speaking all that is necessary is to determine the amount of water, fat-acids, and alkali, as the proportions of these ingredients serve at once to indicate whether the soap is genuine. For this purpose more expeditious methods may be adopted, avoiding the numerous weighings of the preceding process, and more especially the use of weighed filters.

A weighed portion of the soap is first desiccated in the manner already described, to determine the amount of water. The dried soap is then decomposed by a measured volume of sulphuric acid of known strength, and the fatty acids hereby separated are weighed, with or without addition of wax, according to their consistence. The quantity of sulphuric acid neutralized by the alkali in the soap is determined by means of the quantity of a standard solution of caustic soda required to neutralize the excess of acid remaining after the decomposition (see ALKALIMETRY). This determines the quantity of alkali in the soap; and the total amount of water, alkali, and fatty acids deducted from the weight of soap analyzed, gives the quantity of foreign matter present. This, as already observed, should not exceed 1 per cent. if the soap is unadulterated.

The amount of alkali in soap may also be determined by incinerating the soap in a muffle. The residue then consists of carbonate of soda or potash, which may be weighed and the quantity of real alkali thence calculated if the soap is genuine, or determined by the alkalimetric method if other substances are likewise present.

M. Cailletet* has propsed a method of analysis by which only one weighing is required, viz.: that of the quantity of soap analyzed. The soap (10 grammes), in thin shavings, is decomposed by a measured volume of sulphuric acid of known strength in presence of a measured volume of oil of turpentine. The fatty acids thus liberated dissolve in the oil of turpentine and increase its volume, and this increase, multiplied by the specific gravity of the fat-acid, determines its weight. The amount of alkali is then determined in the manner just described, and the water is estimated by difference; this, of course, implies that the soap is genuine, otherwise the difference will include the foreign matter as well as the water.

The standard acid is prepared by mixing 189·84 grammes of the strongest sulphuric acid (SO^4H) with sufficient distilled water to make up the volume to a litre at 15° C. Of this acid 10 cubic centimetres neutralise 1·2 grammes of soda, and are therefore sufficient to decompose 10 grammes of soap, the amount of alkali in which never exceeds 12 per cent.

Ten cubic centimetres of the standard acid and 20 cubic centimetres of oil of turpentine are poured into a tube containing 50 cubic centimetres, and divided into 100 equal parts; 10 grammes of the soap in thin shavings are then added, the tube is closed with a good cork, well shaken for a few minutes till the soap is dissolved, and then left at rest for a quarter of an hour, till the

* "*Bulletin de la Société industrielle de Mulhouse,*" No. 144, Tome. XXIX, p. 8. This method received the prize offered by the Society for a ready mode of analyzing soap without weight-analysis.

oily solution of the fatty acids has completely separated from the watery liquid.

The volume of this oily solution is now to be read off, a deduction of half a division, or $\frac{1}{4}$ cubic centimetre, being made to allow for the diminution of the capacity of the tube, and consequent rise of the level of the oil, occasioned by the thin film of watery liquid which adheres to the inner surface of the tube.

In an experiment made with olive oil soap, the total volume of liquid in the tube, after agitation and standing, was 79·5 divisions, and that of the watery liquid 26 divisions : hence the volume of the oily solution was 79·5 − 26·5 = 53 divisions, or 26·5 cubic centimetres. Deducting from this the volume of turpentine oil added, the remainder, = 6·5 cubic centimetres, is the volume of the fatty acids in the soap.

To determine the specific gravity of these fatty acids, 10 grammes of the same soap were dissolved and decomposed by sulphuric acid, 10 grammes of white wax were then added; the union of this wax with the separated fatty acids was promoted by heating, the whole then left to cool, the cake of wax separated from the liquid, dried between filtering paper, and weighed. Its weight was 15·97 grammes, which, diminished by 10 grammes, the weight of the wax added, gives 5·97 grammes as that of the fatty acids in the soap, and this divided by 6·5 cubic centimetres, the volume of the fatty acids obtained from the same quantity of soap, gives 0·918 for the specific gravity of these fatty acids. By similar experiments with different kinds of soap, the specific gravities of the fatty acids contained in them were found to be, on the average,—

	Specific gravity of Fatty Acids.
Olive oil (Marseilles) soap . .	0·9188
Cocoa-nut oil soap.	0·940
Palm-oil soap	0·922
Tallow soap	0·9714
Soap from oleic acid	0·9003

When rosin-soap is shaken up with dilute sulphuric acid and oil of turpentine, scarcely any of the rosin is dissolved. If fatty acids are likewise present, these dissolve in the oil of turpentine, but the quantity of rosin dissolved is only sufficient to increase the volume of the turpentine oil (20 cubic centimetres) by $\frac{1}{100}$ cubic centimetres. The undissolved rosin collects below the turpentine as a bulky layer, so that in this way the presence of rosin in a soap can easily be detected.

Bolley's Method.—1 gramme of the soap is decomposed in a small beaker-glass with ether and. acetic acid. Two layers of liquid are then quickly formed, the upper being an ethereal solution of the fatty acid (or rosin), and the lower an aqueous solution of alkaline acetate and of the salts contained in the soap, whilst insoluble admixtures are left in various forms, according to their particular nature. The two liquids are separated by a pipette; the ethereal solution is evaporated in a tared glass over the water-bath, and the residual fatty acid (or rosin) is weighed. The watery liquid is evaporated to dryness in a platinum dish, the residue is ignited, and the amount of alkali (remaining as chloride and sulphate) determined by the usual method.

Gräger's Method.—The soap, cut in thin shavings, is dissolved in alcohol of 30 per cent., the quantity of solvent used being such that 100 grammes of soap shall yield a litre of solution. All impurities are then left behind, together with a certain quantity of carbonate of potash or carbonate of soda, 10 cubic centimetres of the solution, clarified by standing, are then diluted with water, and precipitated by chloride of calcium. The precipitate, consisting of the lime-salts of the fatty acids, is washed, dried at 100° C., and weighed. 100 parts of this precipitate correspond, according to Gräger, with 100·5 parts of anhydrous soap, that is, of stearate of soda, the result not being perceptibly affected by the fact that the acid of soap is not pure stearic acid, but likewise contains palmitic and oleic acids.

The extreme limit of water in genuine hard soap is 20 per cent. for mottled, 25 for white, and 30 for yellow soap. The proportion of alkaline bases is mostly from 8 to 9 per cent., and that of the fat-acids from 60 to 70 per cent. In yellow soap part of the fat is replaced by 10 to 20 per cent. of rosin, and soaps made from cocoa-nut oil contain a much larger amount of normal water than those made from tallow or olive oil. But the above proportions may be regarded as standards of comparison; and any deviation from them indicates a deterioration in the quality of the soap, either from excess of water, or from the substitution of some foreign substances for the normal constituents of the soap.

The consumer should not be satisfied with a soap merely because the amount of the dry fatty alkaline compound accords with the price, unless the soap fulfils another condition of equal importance. The problem to be solved by the soap-boiler is the production of an article from which no more is dissolved or washed away, when it is employed, for washing linen or the hands for instance, than is absolutely necessary for cleansing purposes. When this is not the case, much soap is uselessly wasted. This property of soap is carefully watched by the laundress; and for purposes of domestic economy, the quantity of soap employed is much more materially affected by it, than by the state of the materials to be cleansed. If the soap contains too much water, or its consistence is rendered too loose by reason of an excess of lye or salt disseminated through it, as is the case with many of the varieties of soap described, the waste from this cause will be proportionally great. On the contrary, if the soap is over dry, much laborious exertion will be required to detach a sufficient quantity for the purposes required. Curd soap is the only kind which maintains the proper mean between these two opposing characters, and this is due to the quantity of water which it contains, and its state of solidity. Soap which has become too dry is improved by being kept in a moist place, and an opposite treatment improves soap that contains too much water. But soaps of the latter kind are not altered or improved by keeping, as regards the foreign salts which they contain. Curd soap is therefore the proper form in which soap should be sold; no deception need be practised, however, when soap containing a larger amount of water, as that from cocoa-nut oil, is manufactured, provided a corresponding reduction be made in the price. The same applies to soap prepared with bones, fuller's earth, &c., which cannot be positively called adulterations, but are very much calculated to promote dishonest traffic.

The statements made above with reference to the too rapid solution of soap, are, of course, not applicable in cases where the substances to be cleansed are boiled with the soap, as in bleaching and dyeing; they must not therefore be extended to soft soap.

The following table exhibits the composition of various kinds of soap :—

Hard Soaps.

Kind of Soap.	Fatty acids*	Dry potash	Dry soda	Water.	Common salt, lime, and insoluble residue.	Name of the Analyst.
Castile soap, sp. gr. = 1·0705	76·5	—	9·0	14·5	—	Ure.
Ditto sp. gr. = 0·9669	75·2	—	10·5	14·3	—	,,
Fine white toilet soap	75·0	—	9·0	16·0	—	,,
Ordinary white soap from Glasgow	60·0	—	6·4	33·6	—	,,
Mottled tallow soap of good quality, prepared by an able soap-boiler from potashes, after having been kept for several years	81·25	1·77	8·55	8·43	—	Heeren.
Brown rosin soap from Glasgow	70·0	—	6·5	23·5	3·8	Ure.
Colgate's yellow soap	56·2	—	9·0	31·0	—	Morfit.
Ditto do.	64·0	—	7·29	28·71	—	Kent.
London cocoa-nut soap†	22·0	—	4·5	73·5	—	Ure.
Hard poppy-oil soap	76·0	—	7·0	17·0	—	,,
French soap (savon en tables blanc)	50·2	—	4·6	45·2	—	Thénard.
Marseilles soap (savon marbré)	64·0	—	6·0	30·0	—	,,
Ditto do. do.	60·0	—	6·0	34·0	—	D'Arcet.
White Marseilles soap	68·4	—	10·24	21·36	—	Braconnot.
White tallow soap, Leipzig, prepared by a company ...	76·3	8·8		14·7	—	Abendroth.
Ditto, Leipzig, privileged manufactory	50·0	9·4		29·8	—	,,
Marbled soap from the same manufactory	45·0	9·8		38·0	—	,,
Soap from hazelnut oil	64·0	7·0	—	28·0	1·0	—
Beatty's soap (patent)	37·0	—	5·5	52·0	0·5	Morfit.
Butterfield's soap	32·5	—	9·2	55·0	3·7	,,
Nonpareil soap	36·2	—	5·2	51·9	5·25	—
Holt's yellow soap...............	74·0	—	7·08	18·92	—	Kent.
American Soap Co.'s soap ...	35·0	—	5·0	20·0	40·0	,,

* Including rosin in the case of yellow soap.

† This soap, consisting of three-fourths water, was tolerably hard, but dissolved very easily in boiling water. It is called *marine soap*, and is said to be applicable to washing with sea water.

Soft Soaps.

Kind of Soap.	Fatty acids.	Dry potash.	Water.	Name of the Analyst.
Soft soap (*savon vert*)	44·0	9·5	46·5	Thénard.
London soft soap	45·0	8·5	46·5	Ure.
Belgian soft, or green soap	36·0	7·0	57·0	,,
Scotch soft soap.........................	47·0	8·0	45·0	,,
Another kind of good green soap	34·0	9·0	57·0	,,
Scotch, soft rape-oil soap...	51·66	10·0	38·33	,,
Scotch, soft olive-oil soap.....................	48·0	10·0	42·0	,,
Semi-hard soap for fulling	62·0	11·5	26·5	Verviers.
Ordinary soft soap, 1st sample.................	44·0	9·5	46·5	Chevreul.
Ditto 2nd ,, ...	42·8	9·1	48·0	,,
Ditto 3rd ,, ...	39·2	8·8	52·0	,,
Soft soap, 1st quality	50·0	11·5	38·5	Gossage.
,, 2nd ,, ...	40·0	9·5	50·5	,,

Analyses of Three Samples of Marseilles Soap, by P. Bolley.

	1	2	3
Fatty acid	66·99	67·16	68·01
Soda in combination with fatty acids. .	7·80	7·82	7·25
Sulphate of soda and Chloride of sodium	4·00	1·08	1·33
Unsaponified fat . .	—	trace	—
Water	21·21	23·94	23·41
	100·00	100·00	100·00

Analyses of Hard and Soft Soaps used in the trade of Manchester (O'Neill).

	Fatty matter.	Alkali.	Water.
Soft soap, made from cod oil . . .	42·00	8·25	47·00
,, origin unknown	37·50	8·81	44·00
,, good commercial, figged .	36·75	9·57	44·75
Linseed-oil soap, broken by salt . .	54·50	6·50	39·00
Hydrated palm-oil soap, not broken .	28·50	3·50	63·50
Hard yellow palm-oil soap	56·60	7·20	36·00
Hard olive-oil soap.	68·80	7·50	23·70
Do. do.	64·27	7·73	28·00

*Analyses of Soapers' Waste from Heilbronn, used for Manure;
by Stein.*

PORTION SOLUBLE IN WATER:—

Potash	0·056
Soda	0·648
Chloride of sodium	0·151
Carbonate of lime, &c.	0·724
Silica and organic matters	0·359—1·938

PORTION SOLUBLE IN HYDROCHLORIC ACID:—

Potash	0·188
Soda	0·279
Sulphuric acid	0·456
Phosphoric acid	1·144
Oxide of iron	2·186
Alumina	3·075
Lime	29·511
Magnesia	5·656
Carbonic acid	22·666
Organic matters	4·570—69·731

PORTION INSOLUBLE IN ACID . . . . 28·331

100·000

STATISTICS OF THE SOAP TRADE.
Export of Soap from France.

1850	5,487,456 kils.	1854	6,576,156 kils.
1851	6,286,387 „	1855	7,524,012 „
1852	6,463,224 „	1856	7,851,523 „
1853	6,064,942 „	1857	6,713,369 „

*Details of the Exports from France in 1858 and 1859.
Soap made with Palm oil.*

	1858.	1859.
Sardinian States	3,118 kils.	3,754 kils.
Switzerland	8,496 „	8,629 „
Algeria	32,733 „	— „
Other parts of Africa	— „	2,808 „
Senegal, St. Louis	1,703 „	1,994 „
„ Gorée	1,602 „	2,025 „
Other countries	1,572 „	5,423 „
	49,224 „	24,663 „

47*

Other descriptions of Soap.

	1858.	1859.
Sweden	—	34,483 kils.
Zollverein	66,887 kils.	73,383 „
Holland	72,128 „	69,407 „
Belgium	139,565 „	166,373 „
Hanseatic Towns .	40,341 „	44,154 „
England	324,421 „	305,403 „
Portugal	99,591 „	—
Austria	59,393 „	20,574 „
Spain	80,311 „	198,093 „
Sardinian States .	677,862 „	480,824 .,
Tuscany	67,493 „	76,415 „
Switzerland . .	1,365,493 „	1,507,011 „
Roman States . .	31,738 „	21,629 „
West Coast of Africa	60,712 „	—
Mauritius	48,767 „	147,373 „
United States. .	1,492,130 „	1,869,276 „
Peru	124,753 „	293,738 „
Venezuela . . .	—	25,083 „
Brazil	—	134,337 „
Chili	—	32,994 „
Cuba and Porto Rico	14,576 „	24,827 „
St. Thomas . . .	35,032 „	47,974 „
Algeria. . . .	2,362,470 „	2,549,205 „
Gaudaloupe . . .	323,179 „	157,723 „
Martinique . .	424,537 „	242,777 „
Island of Réunion .	370,273 „	266,729 „
Senegal, St. Louis .	34,393 „	92,816 „
„ Gorée . .	34,026 „	30,495 „
Cayenne	79,275 „	86,325 „
St. Pierre and Pêche	14,627 „	22,561 „
Other countries . .	90,753 „	104,761 „
	8,534,726 „	9,126,743 „

Soap is manufactured in France principally at Marseilles, where the annual production is said to amount to 60½ millions of kilogrammes. Manufactories of soap also exist at Rouen, Nantes, Paris, Elbeuf, Rheims, Draguignau, Honfleur, Lyons, &c.

Exports from England.

The total exports of Soap and Candles, in 1836, were 15,813,406 lbs., which had increased in 1860 to—

Soap . . . . . . 21,838,656 lbs.
Candles . . . . . 4,943,769 „

The following table shows, however, that this trade has been falling off within the last few years :—

1856 . . . . 207,925 cwts.
1857 . . . . 180,172 „
1858 . . . . 163,162 „
1859 . . . . 174,410 „
1860 . . . . 194,988 „

presenting a striking contrast to the French export trade, which has nearly doubled since 1850.

Mr. Gossage has kindly given us the following information :— The manufacture of soap was subjected to a heavy fiscal impost during the long-continued war with France. Up to the year 1833, the duty charged on hard soap was at the rate of £28 per ton, or 3d. per lb., and on soft soap £18 13s. 4d., or 2d. per lb. At this date, the duty was decreased one-half, but was again increased in 1840 by the addition of 5 per cent. to all excise duties. In 1853, the whole duty was finally repealed.

The following table exhibits the quantity of soap on which excise duty was paid, the rate of duty, and the amount collected in 1832, before the first reduction of duty, and at subsequent quinqennial periods, up to 1853, when the duty was repealed :—

Year.	Hard soap, in tons.	Duty per lb.	Soft soap, in tons.	Duty per lb.	Duty charged.
1832	53,346	3d.	4,621	2d.	£1,569,170
1837	62,642	1½d.	5,265	1d.	926,141
1842	69,382	1½d., with 5 per additional.	5,564	1d., with 5 per cent. additional.	1,070,863
1847	72,330	ditto	6,375	ditto	1,125,733
1852	90,641	ditto	9,300	ditto	1,423,566

It will be seen from this table that the effect of reducing the duty one-half was to cause an increase in the production, and subsequent consumption, of soap, of 80 per cent. in 1852 as compared with 1832 ; and it is doubtless a fact that a similar increase has taken place since this manufacture has been entirely relieved from fiscal impost, although no reliable evidence of this fact can now be obtained.

During the existence of this obnoxious impost, the manufacturer had to carry on his trade under the strictest surveillance of excise officers, who were in constant attendance on his premises, and dictated the mode in which his business should be conducted. The regulations to be observed were such as entirely to preclude any application of science to the conduct of the operation, and, as a consequence, the manufacture was conducted entirely by the "rule of thumb." Since the soap manufacture has been released from this thraldom, many parties who are possessed of chemical knowledge have directed their attention to the subject, and the advance which has been consequent upon this change may be indicated by the fact that soap can now be obtained by the consumer at 2d. per lb., when the same article would have been chargeable with an excise duty of 3d. per lb. under the former regulations.

These changes have been effected in a great measure by the introduction of other materials possessing detergent properties into the soap made from fatty acids; also by improved modes of effecting the combination of fatty acids and alkalies. Amongst the new compounds which have obtained the most extensive application, is the silicate of soda, which is analogous as a detergent with the combinations of the fatty acids with soda, as we have before explained, at page 710.

RAILWAY AND WAGGON GREASE.

THE grease or soap used for diminishing friction on the axles of carriages may be divided into two classes : one, ordinarily termed "locomotive grease," has tallow or fixed oils, or a mixture of both, for its basis, diffused through a weak solution of carbonate of soda; while the other, "anti-friction" or "waggon" grease, is a soap of lime and rosin-oil, or certain cheaper substitutes mentioned below. This subject is of more importance than might at first sight appear; for while the grease-account of railway companies forms no inconsiderable item of their expenditure, the waste of motive power, and wear and tear of plant entailed by the use of an unsuitable grease, is enormous. The difficulties

experienced from this cause have compelled the leading companies to manufacture their own grease, cost being a secondary
consideration to uniformity of quality.

Locomotive Grease.

This is the only article suited for high velocities; it is invariably used for passenger carriages, and latterly for goods and
mineral waggons when these are provided with axle-boxes. As
ordinarily prepared, it is of a yellow colour, and of various degrees
of consistence. In its preparation some makers use palm-oil
alone, while others employ tallow also, the extra cost of the latter
being more than compensated by a reduction of the *total* quantity
of fatty matter employed, since a grease containing 35 per cent. of
the mixed fats will go as far as another with 42 per cent. or 45
per cent. of oil alone. The grease richest in material is not
necessarily the best or most enduring lubricant, the characteristics
of which ought to be :—

1. A suitable consistency :

2. Good lasting power, attended with but little increase of
temperature, even at the highest speeds ; and,

3. The smallest possible residue left in the axle-boxes.

A grease which is too firm will remain comparatively solid in
the box, and allow the axle to become hot; while, on the other
hand, one too thin becomes exhausted by a few miles' running.

A few years ago, a series of trials was made upon two lines of
railway, the grease under experiment being placed in locked
boxes, and the number of miles travelled before replenishing,
carefully noted ; the minimum was 46, and the maximum within
a few miles of 1200. In the former case, the grease contained
30 per cent. of palm oil alone, and in the latter 35 per cent. of
tallow and oil ; while a grease containing 46 per cent. of material
(almost entirely tallow) ran about 800 miles, being too firm, and
deficient in cooling power. That which gave the best results
became, after a few revolutions, gradually softened, from the axle
upwards, to the consistence of very thick cream, and by continued
upward and downward currents, maintained a comparatively low
temperature until exhausted.

It is desirable that as little residue as possible should be left in
the boxes, for which purpose the smallest available quantity of
alkaline carbonate should be used ; 1·10 per cent. to 1·20 per cent.
actual soda, gives the best practical results ; a smaller proportion

fails to ensure consistency and durability. The soda is used in the form of crystals.

The Process of Manufacture is exceedingly simple; the fats are melted in a boiler and brought up to 180° to 190° F.; the water and soda-crystals are heated in another vessel to 200° F., and both run into a wooden tub, with constant stirring, which is continued at intervals until the mixture is cold. Slow cooling ensures a firm article, and large batches, which retain their heat for a considerable time, are therefore preferable. Care is taken, for obvious reasons, to avoid the introduction of sand or gritty matter. The proportions of materials vary in different establishments and in the same establishment at different seasons of the year, the grease for July use being too firm for December, and *vice versâ*; 25 per cent. is the lowest quantity of fat that can be safely used in the coldest weather, while 35 per cent. is amply sufficient for our warmest summer use; the following proportions have been used with excellent results:—

| | Winter. | | | Summer. | | |
	cwt.	qr.	lbs.	cwt.	qr.	lbs.
Tallow	3	3	0	4	2	0
Palm oil . . .	2	2	0	2	2	0
Sperm oil . . .	0	1	7	0	0	27
Soda crystals . .	1	0	14	1	0	8
Water	12	3	12	12	0	26

These proportions produced one ton of grease (allowing 2½ per cent. for loss), that marked "summer" ran 1200 miles; the sperm oil, although small in quantity, has an excellent effect; but some makers, for economy, use rosin-oil instead, although the result is questionable.

From the foregoing remarks, it will be seen that a simple analysis of a fatty grease, although indicating cost of production, affords little or no test of its practical value, running power being the true and only criterion.

Antifriction, or Waggon Grease.

The greatly increased price of rosin, occasioned by the American blockade, has led to a diminished use of this article. It is well suited for low speeds; it was extensively employed for goods and mineral traffic when its cost was less than one-half that of locomotive grease, but now that the value is considerably greater than the latter, its use is confined to waggons having no axle-boxes.

This grease is of two kinds, with or without water; both should be free from lumps or gritty particles, smooth, and of such consistency as to be easily spread on the bare axle; the former should not part with its water when moderately agitated. The materials used are rosin-oil (obtained by distillation) and caustic lime in a very fine state of division: this latter is obtained by the well-known process of running milk of lime through a series of overflow tubs, in the last of which, receiving the finest particles, it is allowed to subside, and the precipitate drained on canvas to a pulpy condition. If grease free from water is wanted, this pulp of lime is agitated with rosin spirit, which, seizing the lime, expels the water; the latter, floating on the surface, is run off, and a further quantity of spirit is added until the mass is reduced to a cream, when it is ready for use. If a watery grease is wanted, the finely-divided lime, without spirit, is simply used in a milky condition.

The Process of Manufacture consists in adding to any convenient quantity of rosin-oil, without the application of heat, either of the above preparations of lime, until the mixture begins to thicken, avoiding an excess, lest the grease should be too hard. The quantity of lime varies with the quantity of the oil used; in the case of the finer grease, there is no gain in adding an excess, as the spirit is the most costly ingredient; a cream is therefore used; while for the lower qualities, the watery milk of lime is made very thin, so that, in some cases, as much as one-fourth of the bulk of the oil is added. The materials are well stirred together for about fifteen minutes in a tight box or a barrel in which a shaft furnished with blades or stirrers is made to revolve. The grease thus formed is run into barrels or receptacles before it solidifies or "sets."

Any written directions for the preparation of this grease must necessarily be general rather than specific, as so much depends upon daily practical experience; a manufacturer, when in doubt about a new oil or lime mixture, generally experiments upon a small quantity first, and is guided thereby in his operations on the large scale.

The present high price of rosin has compelled manufacturers to employ several cheaper substitutes for it, such as paraffin residues mixed with coal-tar, residues from candle-making, from eotton-seed oil, fish oils or footes, pitch oil, the heavier parts of the American petroleum, &c. ; all these produce grease of a certain quality, but none with the lubricating power of rosin grease.

· For abstracts of the patents which have been taken out for the manufacture of railway or waggon grease, see the Appendix to this volume.

Statistics of the Trade.

Any attempt to ascertain the annual consumption of grease for railways, waggons, and carts is attended with great difficulties, and the results at which we have arrived, must be regarded as merely an approximation. As regards railway consumption, we are indebted for the following statements to the kindness of Mr. Fletcher, of the North-Eastern line, which, from the varied character of its traffic, may be fairly taken as an average of the railways in this country ; from the data he has furnished, the quantity used would appear to be 11·7 cwt. per mile per annum.

The consumption for waggons in connection with the production of the ores and minerals in the kingdom, has been estimated from the data kindly furnished by Messrs. Pease, of Darlington, Mr. I. L. Bell, of Washington, Mr. Hayes, of Runcorn, Mr. Cowen, of Blaydon, Mr. Liddell, of Newcastle, Mr. Sopwith, of Allenheads, and other gentlemen. The quantities of the different ores, &c., are taken from the valuable mineral statistics of Mr. Robert Hunt.

· As regards the use of grease for carts, we have found, from information obtained from various agriculturists, that twenty carts are necessary in cultivating 1000 acres of land, and that each cart requires 4 lbs. of grease per annum ; whence it follows that the consumption may be taken at 0·71 cwt. per 1000 acres.

It has not been so easy to ascertain the consumption of grease for carts used in the ordinary traffic of the country ; but from the information which has been very courteously given us by Major Graham of the Census Office, the officers of the Board of Inland Revenue, and Mr. McCrie, of Newcastle, we are led to believe that there are 112 carts for every 10,000 of the town population, and that each cart requires 18 lbs. of grease per annum.

Tabulating the information thus obtained, we have the following result as the annual consumption of grease in this country :

Railways.

Tons.

10·869 miles at 11·7 cwt. per mile . . . . = 6,358

Ores and Minerals.

Cwt. per 1000 tons.

Coals . . . . 86,039,214 tons at 1·42 . = 6,108

Iron ore . . .	8,024,205
Copper ore . .	236,696
Sulphur ore. .	135,669
Lead ore . . .	90,696
Zinc ore . . .	15,770
Tin ore . . .	10,462
Gossan . . .	3,016
Arsenic ore . .	1,600
Manganese ore.	932
Silver ore . .	125
Tungsten ore .	19
Nickel ore . .	15
Antimony ore .	15

8,519,220 tons, at 0·20 . = 85

Clays, fire and china . . . } 407,960 tons, at 1·75 . = 354

Ditto, brick and tile . . . . } 6,257,511 „ 0·30 . = 94

Building and other stones . } 15,764,200 „ 0·10 . = 178

Agriculture.

60,000,000 acres of cultivated land, at 0·71 cwt.
per 1,000 acres per annum . . . . . = 2,130

Trade.

257,600 carts at 18 lbs. per annum . . = 2,070

Total tons . 17,277

This estimate is under the actual consumption; and we have no means of ascertaining how much grease is used for engines and other purposes.

GLYCERIN.

THIS substance, which, as already observed, is the basis of most of the natural fats, was discovered in 1776, by Scheele, who gave it the name of *Sweet Principle of Oils* (*Principe doux des Huiles—Oelsüss*). Its chemical relations were afterwards investigated by Chevreul, Pelouze, and Redtenbacher, and of late years, our knowledge of its compounds and reactions has been largely increased by the admirable researches of Berthelot.

Glycerin, when concentrated as much as possible by evaporation, is a colourless, syrupy, uncrystallisable liquid, inodorous, and having a pure sweet taste. It deliquesces in damp air. Its specific gravity at 60° Fah. is 1·260. It dissolves in all proportions in water and alcohol, but is insoluble in ether. It dissolves all deliquescent salts, as well as several nitrates, chlorides, and sulphates, and even oxide of lead. It likewise dissolves vegetable acids, and many of the alkaloids and their salts (see page 766).

Glycerin is not altered by exposure to the air. When submitted to simple distillation, a small portion of it passes over without alteration, but by far the greater part is completely decomposed, giving off gaseous hydrocarbons, and an intensely pungent volatile compound called *acrolein*, which violently attacks the eyes, and leaves a residue of carbonaceous matter. In an atmosphere of aqueous vapour, however, glycerin may be distilled entirely without alteration, and this property is now made available for its preparation and purification on the manufacturing scale.

Glycerin, when heated with acids, organic or inorganic, in sealed tubes, forms oily compounds exactly analogous in constitution to the compound ethers produced by the action of acids upon alcohols: for, just as acetic ether or acetate of ethyl is formed by the union of acetic acid and alcohol (hydrate of ethyl) with elimination of the elements of water, so acetic acid unites with glycerin, also with elimination of the elements of water, forming an oily compound, or ether, called *acetin*, thus :—

$$C^4H^4O^4 + C^4H^6O^2 - H^2O^2 = C^8H^8O^4,$$

Acetic acid.　　Alcohol.　　　　　　Acetic ether.

$$C^4H^4O^4 + C^6H^8O^6 - H^2O^2 = C^{10}H^{10}O^8.$$

Acetic acid.　　Glycerin.　　　　　　Acetin.

Glycerin is, in fact, an alcohol; but it differs from common (ethylic) alcohol in this respect, that it is capable of uniting with monobasic acids in three different proportions, the combination being attended with the elimination of 1, 2, or 3 molecules of water (H^2O^2), and resulting in the formation of three different fats or compound ethers, whereas common alcohol forms but one ether with monobasic acids, such as acetic acid. This difference is expressed by saying that common alcohol is *monatomic*, and glycerin *triatomic*. The formation of the three compounds of acetic acid and glycerin is represented by the following equations :—

$$C^4H^4O^4 + C^6H^8O^6 - H^2O^2 = C^{10}H^{10}O^8,$$
Acetic acid.   Glycerin.   Acetin.

$$2C^4H^4O^4 + C^6H^8O^6 - 2H^2O^2 = C^{14}H^{12}O^{10},$$
                               Diacetin.

$$3C^4H^4O^4 + C^6H^8O^6 - 3H^2O^2 = C^{18}H^{14}O^{12}.$$
                               Triacetin.

Glycerin unites also in a similar manner with bibasic and tribasic acids, thus :—

$$C^8H^6O^8 + C^6H^8O^6 - 2H^2O^2 = C^{14}H^{10}O^{10},$$
Succinic acid.   Succinin.
(bibasic.)

$$C^{12}H^8O^{14} + C^6H^8O^6 - 3H^2O^2 = C^{18}H^{10}O^{14}.$$
Citric acid.   Citrin.
(tribasic.)

All these compounds, and many others of similar constitution, may be obtained by the simple process of heating glycerin and acids, in various proportions, in sealed tubes. The compounds thus produced by heating 1 equivalent of glycerin with 3 equivalents of a monobasic organic acid, are identical in every respect with the natural fats, stearin, palmitin, olein, &c. These are the only compounds of the class which at present possess any interest in a technical point of view, and it is from them that glycerin is prepared, as we now proceed to explain.

Manufacture of Glycerin.

Glycerin is obtained from the neutral fats by the process of *saponification*, which, as explained in the chapter on "Soap," consists in restoring to the radicals of the acid and glycerin the elements of water which have been abstracted from them in the formation of the fat. In the ordinary process of soap-making, and in the preparation of lead-plaster, the change is brought about by the action of a metallic base, which unites with the acid of the fat, while the elements of the water present are taken up by the remaining elements of the fat, to form glycerin; but, as we have already had occasion to observe, and shall presently explain more fully, the conversion of the fat into acid and glycerin may be effected by the action of water alone.

Glycerin was originally prepared by saponifying oils with oxide of lead, having been, in fact, first introduced into pharmacy as a bye-product obtained in the preparation of lead-plaster. Equal parts of olive oil and finely-pounded massicot are placed in a basin together with water, and the whole is boiled, hot water being added as fast as it evaporates, and the mixture being continually agitated to prevent it from charring at the bottom of the basin. As soon the formation of the lead-plaster (which is a mixture of oleate and margarate of lead) is complete, more hot water is added, and the aqueous liquid is decanted, filtered, treated with sulphuretted hydrogen to remove dissolved lead, again filtered, and evaporated to a syrup over the water-bath.

For many years, all the glycerin known in commerce was obtained by this method: for, although immense quantities of glycerin are constantly separated in the process of soap-making, and remain in the waste liquor, no economical method was known of separating it from the salt with which it is there so largely mixed, and obtaining it in the pure state. Processes have, however, been recently devised for effecting this separation, which will be presently noticed. The glycerin obtained in the preparation of lead-plaster was too small in quantity to be employed in any manufacturing operation, and it was very apt to retain small quantities of lead, the presence of which is highly objectionable in any therapeutic application of glycerin. Hence the use of glycerin was for a long time very limited.

The first important step that was made in obtaining glycerin in a state of purity, and on the large scale, was its formation as a secondary product in the manufacture of stearin candles. In this manufacture, tallow is saponified by lime, the stearin being thereby converted into stearate of lime, which is afterwards decomposed by sulphuric acid, yielding stearic acid to be made into candles; while the glycerin remains dissolved in the aqueous mother-liquor, together with a certain quantity of lime, from which it is freed by the following process, devised by M. Cap.

The liquid containing the glycerin from lime-saponification is concentrated, and afterwards treated with sulphuric acid, to remove the remaining lime. It is then boiled and agitated in a close vessel, to expel volatile fatty acids; and on being cooled at a certain density (14° Twaddell), it is rendered neutral, if necessary, by the addition of carbonate of lime. After this, the boiling is renewed, and continued till the liquid marks 37·5° Tw., when it is again cooled, and any further deposit of sulphate of lime removed by filtration. Finally, the liquid is concentrated by evaporation to 49° Tw., and passed through washed animal charcoal to remove the colour. This reduces the density to 45° Tw., but, by careful concentration, a fourth part of the water may be removed, and the density raised to 51·2° Tw.

This method, when properly carried out, yields a very pure product; but it is complicated, and, unless great attention is paid to every part of it, small quantities of lime are apt to remain in the glycerin, rendering it unfit for many purposes, especially for use in medicine and pharmacy.

The only perfectly unobjectionable mode of obtaining glycerin, inasmuch as it alone ensures the entire absence of mineral impurities, is the decomposition of the fats by vapour of water at a high temperature. This mode of decomposition was first adopted as a means of obtaining fatty acids and glycerin by Mr. T. Tilghman, in 1854. His process, which consists in pumping a mixture of fat and water through a coil of pipe heated to about 612° Fah., kept under a pressure of about 2000 lbs. to the square inch, has already been fully described in our chapter on Soap (p. 651).

This process, simple and beautiful as it is, and capable of yielding perfectly pure products, has not been carried out on a very large scale; but it appears to have led Mr. G. F. Wilson, of the Belmont Works, Vauxhall, to the invention of the process by which nearly all the pure glycerin of commerce is now obtained, viz. the decomposition of the fats by distillation in an atmosphere of steam. In a paper read before the Society of Arts, Mr. Wilson says: "I went with my chemist-assistant, Mr. Payne, to see Mr. Tilghman's little apparatus at work; and in the course of some experiments which it led us to try, or rather to try over again, it struck me that steam passed into the fat at a high temperature should effect by a gentle process what Mr. Tilghman aimed at effecting by a violent process, the resolving of the neutral fat into glycerin and fat acids. We proved that this was so, and that the glycerin distilled over in company with the fat acids, but no longer combined with them. In July, 1854, we took out a patent for this process, by which many hundred tons of palm oil and other fats have now been worked, and which has given to the arts and medicine a body never known before, either in France or here, even in the chemist's laboratory,—glycerin, which had passed over in the form of *vapour*, without a trace of decomposition."

The distillation of fatty bodies is not altogether a new process. Chevreul and Gay-Lussac, indeed, as early as 1825, sketched out the idea of it, and even noticed the introduction of steam into the apparatus; and the process was afterwards carried out, with various improvements and modifications, by several chemists and manufacturers, both in this country and in France. Among these we may especially notice the process patented by E. Price & Co., in 1842, in the name of Jones and Wilson, for the distillation of fats, after they have been acted upon by sulphuric acid and nitrous gases. But all these processes had for their special object, the preparation and purification of the fat acids, not of the glycerin ; indeed, the temperature employed was generally so high that the glycerin was completely decomposed, giving off intensely acrid fumes of acrolein, and leaving a charred residue. The distillation of fats in an atmosphere of steam in such a manner as to obtain the glycerin as well as the fatty acids, was first effected by the process above mentioned, patented by George Ferguson Wilson and George Payne, on the 24th of July, 1854 (No. 1624). The process is described in the specification as follows :—

" An ordinary still and condensing or refrigerating apparatus is employed, preference being given to one with an ample refrigerating surface, such as are now resorted to for distilling saponified fatty matters when using heated steam, with a view to exclude the atmospheric air, and also in order that the steam may act as a carrier to remove the fatty vapours. The bottom of the still is heated by a fire provided with a damper to the flue or chimney, that the heat of the matters under process may not exceed the desired temperature. In charging a still, a quantity of neutral or partially neutral fats is introduced into the heated still, and heated steam is introduced below the fats or oils, so as to rise up through them in numerous streams, care being taken that the temperature of the matters in the still shall not rise to the temperature which will decompose the glycerin. A thermometer is used in the still to indicate the temperature of the contents, and it is desirable in all cases, with neutral fats and oils, to keep the heat below 600° and above 550° Fah., when the glycerin is not decomposed, but comes over very pure. When the fats or oils are only partially neutral, which is very commonly the case with palm oil, the draught of the fire may be quickened and the process hastened so long as there are fat-acids in the still; but as soon as the fat acids have passed over, if the temperature is much above 600° Fah., acrolein will probably be formed, particularly if steam be not freely supplied, its production being quickly indicated by its pungent smell and its action on the eyes of persons near the condenser, from which the distilled products constantly flow.

Although it is recommended, as a general rule, that the temperature should be kept below 600° Fah., charges have nevertheless been worked off at higher temperatures, keeping up a very plentiful supply of steam, and the glycerin has not been decomposed; still, there is no superior result obtained, whilst there is greater hazard of decomposing the glycerin.

It is, however, most convenient to retain the contents of the still rather under than above 600° Fah. (keeping up a free supply of steam) during the whole process, whether the fats or oils be neutral or partially neutral. With respect to the use of fire or other external means of heating a still, it is desirable that the external heat immediately under the still should be very moderate, and that the heated steam admitted into the still should, by preference, be depended on for maintaining the higher temperature. Different neutral fatty and oily substances appear to vary in some degree, but not to a great extent, in regard to the temperature at which they may be distilled most quickly in an atmosphere of steam or vapour of water, without decomposing the glycerin; a workman, however, with attention, when acting on the first charge of a neutral oil or fat with which he has not before operated, will, by raising the temperature gradually, and noticing at what temperature the matters come over most freely, and yet avoid the production of acrolein, ascertain at what temperature the particular fat or oil may be distilled most favourably; and then, in distilling subsequent charges of the same fat or oil, he will retain the contents of the still as near as may be at the most favourable temperature, that is, at a temperature which will bring over the products most quickly, without allowing the heat to rise so high as to produce acrolein.

When using a still with a refrigerator or condenser in compartments, each being more and more distant from the still, and each compartment provided with a cock to draw off the distilled and condensed products, which is the most convenient arrangement, it is found that the products which flow from the hottest of the condensers are for the most part free from water and glycerin, most of the glycerin passing off with the products which condense in the compartment of the condenser more distant from the still, and where the condenser is kept lower in temperature; and in all the receivers, the fat-acids quickly separate from the glycerin and water, when allowed to stand and cool for a short time. It may be proper to state that the compartment of the condenser most distant from the still, is open to the atmosphere, no pressure being necessary within the still and condenser.

When the glycerin is required in a more concentrated state than when it comes over and is condensed, the water contained in it may be more or less separated by evaporation, and in this manner very pure glycerin will be obtained, as well as fat-acids.

The process of distillation just described may also be employed to purify glycerin prepared by either of the old processes, such as saponification by lime or oxide of lead.

The following apparatus has been patented by Messrs. Wright & Fouché (2 September, 1857, No. 894) for decomposing fats into fat-acids and glycerin by means of highly or superheated water and steam. It consists of two boilers, a and h (Fig. 359); the first, a, containing water, is heated over a naked fire (the fire bars are shown at b, the ash-pit at c), and the second, h, containing the water and the fat to be decomposed, is heated by steam passed into it from a. A continuous circulation is thus kept up without mechanical agitation; d is a dipping pipe, furnished with a cock to empty the apparatus by pressure; e, a man-hole, serving for cleaning the cylindrical vessel a, and for the introduction of substances if required; f, a metal tube (of iron or copper) connecting the bottom of the boiler a with the bottom of cylinder h; g, a metal tube of ascension, conducting the superheated or highly-heated water from the boiler a to the upper part of cylinder h

FIG. 359.

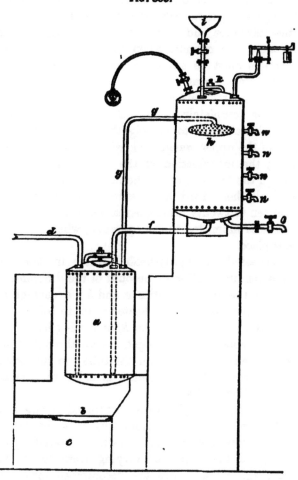

This tube terminates in the interior of the cylinder h by a rose jet, or, more simply, holes are made in the extremity, so as to distribute the water uniformly over the area of the cylinder h, and to ensure a molecular or finely subdivided contact between the superheated or highly-heated water and the substance submitted to the operation; i, a funnel, furnished with a tube and cock, serving for the introduction of the substance to be treated into the cylinder h, that is, when this substance is of such a nature as may be introduced through a small aperture; k, a man-hole, serving for cleaning the cylinder h, and for the introduction of substances to be treated which cannot pass through the funnel i;

48*

l, a safety valve; *m*, a pressure-gauge, indicating the pressure in the whole apparatus; *n*, *n*, cocks, serving to indicate the height or level of the substance and of the water in cylinder *h*; *o*, a cock serving to empty the cylinder when the operation is completed. Both the boilers must be strong enough to resist a pressure of about twenty atmospheres.

To decompose fatty substances into fatty acids and glycerin, the boiler *a* is completely filled with water, and the cylinder *h* is filled with water up to one-eighth of its height, and then to the level of the upper gauge-cock with the fatty bodies to be decomposed; the introduction of the fatty bodies takes place, either through the funnel *i* or by the man-hole *k*. The boiler *a* is then gradually heated till the pressure-gauge indicates a pressure of from ten to twenty atmospheres, according to the nature of the substances submitted to the operation, when the following action takes place.

The superheated or highly-heated water in the boiler *a* acquires an ascending motion on account of the difference of the temperature of the two capacities *a* and *h*; a current is thus created, whence it results that the heated water in boiler *a* ascends through the tube *g* into the cylinder *h*, and being forcibly driven out through the holes in the rose-jet, passes through the fatty bodies and descends again through the tube *f* to the bottom of the boiler *a*, where it is again heated in order to recommence its ascending motion, and so on.

When this operation has been thus continued during a length of time, which may vary, say from five to eight hours, according to the nature of the fatty bodies operated upon, and also according to the pressure (varying from ten to twenty atmospheres), the fatty bodies are decomposed into glycerin, which remains dissolved in the water, and into fatty acids which float in the cylinder *h*; the contents are now emptied out and separated from each other.

Figs. 360, 361, 362, represent modifications of the heating boiler *a*. In Fig. 360 the boiler is heated by a continuous current of steam. For this purpose, *p* is a double casing receiving the steam; *r* and *q* are the entering and exit pipes. This same apparatus, having the double casing, may be placed on a furnace with a naked fire, so as to form a water-bath at a high temperature; it is furnished with a man-hole and safety valve, not represented in the figure.

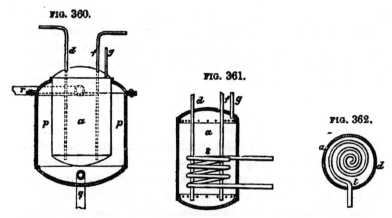

FIG. 360.

FIG. 361.

FIG. 362.

In Fig. 361 the water in the boiler a is heated by a worm s, having one branch leading from any steam generator, and returning to the same generator by the other branch.

In Fig 362 the water in boiler $\dot{a}$ is heated by a horizontal worm t, having one branch coming from any suitable steam-generator, and returning to this same generator by the other branch.

Fig. 363 represents a second apparatus, based on the same

FIG. 363.

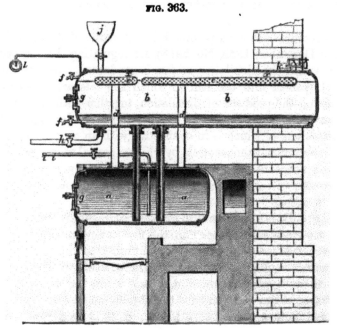

principles and producing the same results; *a* is the first boiler, heated by a naked fire, or by a water or sand-bath, by a worm-pipe, or by a current of high-pressure steam ; *b*, the second and upper boiler, containing the water and fatty matters, and having the vacant space *x, x*, above the fatty matters; *c, c*, tubes leading from the bottom of the second boiler *b* to the bottom of the first boiler *a* ; *d, d*, tubes communicating from the upper part of the first boiler to the upper part of the second boiler ; *e, e*, tubes or chambers pierced with holes terminating the upper part of the tubes *d, d*, and extending over the area of the second boiler, *b* ; *f, f*, gauge-cocks to indicate the levels of the contents ; *g, g*, man-holes ; *h*, pipe and emptying cock ; *i, i*, pipe for the supply of water ; *j*, funnel for the introduction of the fatty matters; *k*, safety-valve ; *l*, pressure-gauge.

The apparatus being heated, and the pressure within raised to from ten to twenty atmospheres, continuous and automatic currents will be generated, which cause the water to ascend through the tubes *d, d*, whence it flows through the tube pierced like a rose-jet on to the liquefied fatty bodies, and then returns to boiler *a* through the tubes *c, c*, somewhat reduced in temperature ; when it arrives in boiler *a*, it will again be heated, and will resume its action, make another ascent, and so on.

An apparatus has been patented in the name of W. A. Gilbee (27th of December, 1858, No. 2958), for regulating the temperature at which the distillation of fats is conducted. Superheated steam is passed into the still in which the fats are previously melted, and the products of distillation, together with the excess of watery vapour, pass from the still through a condensing worm, the farther end of which is in connection with an air-pump working with an intermittent motion, by which a partial vacuum is kept up in the apparatus, the temperature consequently lowered, and the tendency to the formation of acrolein and gaseous hydrocarbons diminished.

A (Fig. 364) is a furnace with two compartments, each provided with a fire-place and ash-box; in the first compartment, there is a worm, composed of a double row of iron pipes, *a, a*, arranged between two vaults; in the second compartment the still B is fixed with its appliances ; *a, a, a, a*, is a worm for superheating the water or steam necessary for the distillation, the water being supplied from a steam-boiler placed as close as possible to the

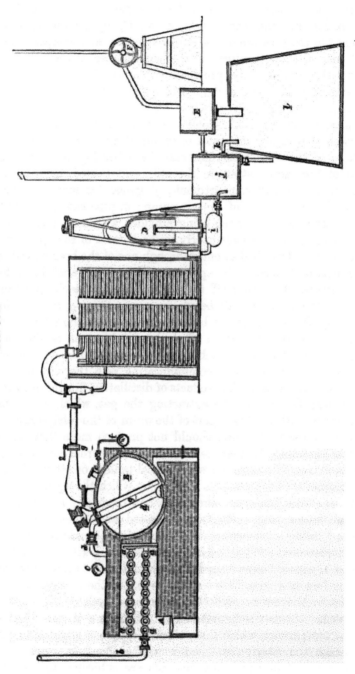

FIG. 364.

furnace A ; b, pipe with a cock for introducing the steam from
the boiler into the worm a, a, a, a ; C, pyrometer for indicating the temperature of the furnace in the first compartment,
and also the temperature of the superheated steam when it
enters the still ; d, pipe provided with a tap for conveying the
superheated steam from the worm a, a, into the still B ; this
pipe descends obliquely into the still, and has, at its lower extremity, a rose or head, which is serrated at the edges for the
purpose of dividing the water or steam, this rose resting on the
bottom of the still ; e thermometer for indicating the degree of
heat of the fatty matters in the still, and heated by the furnace in the second compartment ; f, gauge for indicating and
regulating the action of the air-pump ; g, pipe having a double
branch at top, each branch being provided with a tap ; the lower
end of the pipe g extends nearly to the bottom of the apparatus.
   The liquid fatty bodies enter through one of the branches when
it is desired to charge the apparatus, the other branch being for
the purpose of drawing off the tar produced at each operation,
which it is necessary to do before recommencing ; g, g, valve for
shutting off communication between the still and the refrigerator
when the tar is being drawn off ; c, large tub or vat, serving as a
refrigerator.   This tub, which should always be filled with water,
contains a metallic worm communicating with the still, for the
purpose of condensing the products of distillation drawn up by the
air-pump ; D, air-pump for extracting the gas, which is in communication with the lower part of the worm of the refrigerator, its
action is intermittent, and should not produce more than about
half a vacuum.   It may receive its motion from any suitable
prime mover, in connection with conical pulleys or other suitable
mechanism for regulating its motion, by which the egress of the
gases of distillation may be retarded or expedited ; i, receiver of
the air-pump, into which the extracted and condensed gases
enter, holding in suspension the water of the condensed steam ;
j, a double box, lined with lead, communicating with the receiver
i, and in which the condensed products collect.   At the lower part
of the box is a cock to let off the water at the same time that
the fat acids are running off through the pipe k, k, into the vat
or tub E ; this box is furnished with a wooden chimney or funnel
at the top, through which the acrolein escapes ; e, large vat or tub,
in which the fatty acids, freed from their condensed water, are
collected, and in which the fat-acids are subjected to the first

acid washing; this vat is provided with a cover made in two parts, one of which is fixed, and the other movable; the movable part serves for emptying and cleaning the vat, and the fixed part communicates by means of a wooden conducting pipe with a ventilator F, which expels the acrolein remaining in the fat acids disengaged by ebullition during the washing process.

The still B is two-thirds filled with neutral fats or fat-acids, first reduced to a fluid state, and freed from foreign bodies by washing or filtration. The two fires of the furnace A are lighted, and as soon as the pyrometer marks 660° to 678° Fah. in the compartment containing the worm a, a, a, a, and the thermometer placed in the still marks 460° to 487° Fah.,—at which temperature the neutral fats are near the point of distillation,—superheated steam is introduced into the still, and at the same time the airpump is set in action. The distilling operation then commences, and the pump will draw off the gases as soon as they are formed, it being merely necessary to take care that the two temperatures should equilibrate, so that the vapour should not exceed 660° to 678° Fah.

It will be observed that the temperature at which this process is directed to be conducted is 70° or 80° Fah. higher than that at which the distillation is effected in Wilson & Payne's apparatus. In this respect, then, Gilbee's apparatus does not appear to possess any advantage to compensate for its increased complexity; but it might, perhaps, be worked at a lower temperature.

Another distillatory apparatus, invented by C. Leroy and J. J. Marie Durand of Paris, and patented in the name of E. T. Hughes (July 17th, 1859, No. 1459) is represented in Fig. 365, and some of its details in Figs. 366, 367. There are three methods of working with it, which will be understood from the following description :—

The tallow or other fatty substance is first introduced at a temperature of 60° C. into a vessel or receiver A, made of wood or metal lined with sheet lead, and provided with a cover B, which is removed at the commencement of the process, or, if desirable, the fatty substance can be introduced cold, and heated to the above temperature in the vessel itself by means of steam circulating in the serpentine H, the temperature being ascertained by the thermometer O. The cock C is then partially opened, ad-

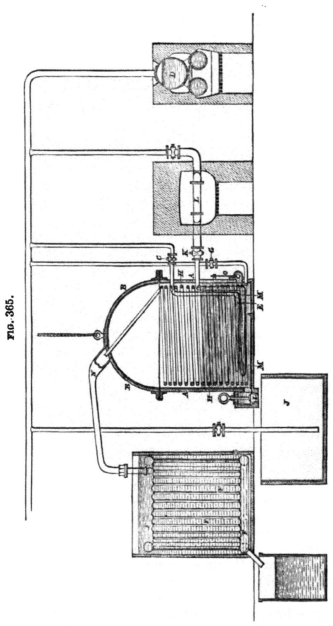

FIG. 365.

mitting a certain quantity of steam from the boiler D, through the horizontal serpentine E, into the fatty substances, which prevents carbonization. Concentrated sulphuric acid at a den-

FIG. 366.

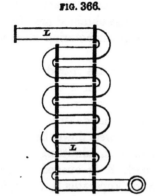

FIG. 367.

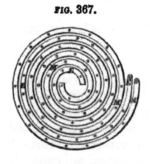

sity of 60° B. is then introduced. In most cases, for good tallow, 10 per cent. will suffice; but for some kinds of fatty substances from 20 to 25 per cent. will be required. This acid is added very gradually, so as to occupy two hours before the whole quantity is introduced, and during that time the tallow must be well agitated; the cover B is then lowered and secured hermetically to the vessel A by means of a joint made in any convenient manner, and then the cock C is still further opened, the steam thereby admitted acting in two different ways, viz., mechanically as an agitator, and chemically as a dissolving agent, on the glycerin, which is carried off in solution with the steam through the pipe N to the refrigerator F, where they are condensed. This part of the operation is carried on slowly for the first four hours, and then the cock C is thrown full open, and the process continued uninterruptedly for from 24 to 36 hours, according as a partial or complete decomposition is to be effected; and during the last six hours, as water will have been added to the tallow by the condensation of the steam, and the sulphuric acid will have diminished in density to 12° or 15° B., the steam-cock G is turned on, admitting steam from the boiler D through the vertical serpentine H, which steam acts merely as a heating agent, and does not mix with the fatty substances; at the expiration of 24 to 36 hours, both the cocks C and G are turned off, and a small quantity of the fatty acid is drawn off through the tap I, in order to ascertain whether the decomposition is thoroughly effected; and in case it is not perfect, this operation is continued until it is complete, and then the whole of the fatty acids are drawn off into the vat J, where

they are washed in acidulated water in the usual manner, run
into moulds, left till cold, and the blocks thus obtained are afterwards compressed in the usual manner for manufacturing stearin
candles.

Or, the tallow or other fatty substances are introduced in
the vessel A as above stated, but instead of being acted upon
by concentrated sulphuric acid, they are treated with an equal
weight of water acidulated with that acid to the density of 2° or
3° B. The cover B is then lowered and screwed down hermetically on the receiver, and the cock G opened, admitting steam
into the serpentine H, and then the cock K is opened, which
admits superheated steam from the heating apparatus L, through
the horizontal serpentine M in direct contact with the fatty substances contained in the receiver A. This part of the operation
is continued for eight hours, the cock K admitting superheated
steam at 200° C. and the cock G, ordinary steam to augment the
heating surface and maintain the required temperature. During
this part of the operation, the tallow which is in contact with the
slightly acidulated water, parts with its glycerin, which is dissolved
in the water and carried off with the rush of superheated steam
through the pipe N to the refrigerator F, where they are condensed.
This decomposing operation is accelerated by adding to the
tallow from 20 to 25 per cent. of fatty acid which has already
undergone distillation, and contains the empyreumatic oils
thereby produced.

Thirdly, by the use of superheated and ordinary steam, acting
in the manner last described, the process of *calcareous saponification* may be effected with quantities of lime smaller than those
hitherto employed.

A modification of the distillation process has been patented by
George Gwynne (2nd of April, 1856, No. 798), in which the
operation is assisted by mechanical agitation. A still similar to
those generally used for the distillation of fatty and oily bodies,
has a mechanical agitator ; a quantity of palm or other oil or
fatty body is to be introduced into this still, and heat and steam
are applied in the usual manner. As soon as the temperature
has attained 250° Fah., the throttle valve of the still is shut, the
agitator set to work, and the heat raised to the point most
favourable for effecting the separation of the fatty acids and
glycerin from each other (about 530° Fah.), and there kept for

some hours, the steam and agitation being continued all the time, when the contents of the still are gradually discharged into a vessel containing boiling water holding in solution a little oxalic acid, in a state of violent agitation, and there thoroughly washed.

The acidulated water in which the fatty matters have been boiled, contains glycerin dissolved in it, but the quantity of this latter body is too trifling to repay the expense of a separate evaporation. By the following arrangement, however, an economical result can be obtained. The fatty matters, after they are discharged from the still, are to be boiled in the acidulated water, in the manner before described. Instead of throwing away this water, it is to be retained, and used again and again for a similar purpose, as in the first instance, a little oxalic acid being added from time to time, as may be required. After a certain period, the length of which will depend upon the number of times it may have been used, the acidulated water is to be brought to the boil, and thin "cream of lime" added to it until it is slightly alkaline to test-paper. After boiling a few minutes, the mixture is to be allowed to repose until the supernatant fluid has become transparent, when it is to be drawn off from the dregs, and evaporated in a still or other suitable vessel to the consistence of a syrup. The glycerin thus obtained may be purified by processes already described.

Separation of Glycerin from "Spent Soap-lees."—Spent soap-lees, as we have several times had occasion to observe, consist of a weak aqueous solution of glycerin, mixed with free or carbonated alkali, common salt, and other impurities. The glycerin may be separated in a tolerable state of purity by saturating the alkali with sulphuric acid, evaporting almost to dryness, dissolving the glycerin from the residue with alcohol, and distilling off the alcohol, the glycerin then remaining. This process is, however, too troublesome and expensive for use on the large scale.

The separation of the glycerin may be more economically effected by means of superheated steam, a method of doing which has been patented by H. Reynolds (10th of June, 1858, No. 1322). The liquid may be first reduced by evaporation, and raised nearly to the boiling point by being passed through a heated still-like vessel, or by being heated in any other convenient vessel. It is then passed into another vessel, where it is met by a jet or jets of superheated steam, at a temperature of 380° to 400° Fah., which raises it to the boiling point and evaporates the glycerin, carrying the vapour upwards with it, and leaving the salts to deposit in the vessel. If greater purity is required, it may be obtained by repeating the process; and the little colour that remains can easily be removed in the ordinary manner, by animal charcoal or chlorine.

Applications of Glycerin.

The solvent power of glycerin, its unctuous character, and its property of remaining liquid, even at low temperatures, together with its perfect harmlessness, render it a valuable agent in pharmacy as an excipient for various substances. The following remarks on its pharmaceutical applications are taken from a paper by M. Cap (*Journal de Pharmacie et de Chimie*, 3rd series, vol. xxv, p. 81)

Glycerin blends with wonderful ease with every kind of medicament. It may be used either alone or with other therapeutic agents. It mixes in all proportions with water for baths, injections, fomentations, and lotions of all descriptions. Applied to burns or other wounds, it keeps them from the air, and maintains the suppleness of their edges. When added to cataplasms, it preserves their softness and—which is very important—it prevents them from adhering by their edges to the surfaces on which they are applied. This combination of properties renders glycerin a most valuable excipient to be added to the scanty list of bodies of this nature now at the disposal of art, an excipient which appears to hold a place between water and oil, for it participates in most of the qualities of both.

Glycerin unites as well with aqueous and alcoholic liquids as with lard, ointments, pomades, and soaps. It will serve as a base for liniments, ointments, and embrocations; it will mix with extracts, tinctures, alcoholates, and medicinal wines; a few drops of glycerin added to a pillular mass, will prevent it from drying, &c.; it will consequently assist in most uses of medicine, surgery, and veterinary science, imparting to any preparation with which it is mixed, the advantage of its lenitive, sedative properties, softening the tissues, and preparing them for the absorption of the medicinal substances with which it is united.

But this is not the limit of its pharmaceutical uses. Glycerin dissolves the vegetable acids, the deliquescent salts, the sulphates of potash, soda, and copper, the nitrates of potash and silver, the alkaline chlorides, potash, soda, baryta, strontia, bromine, iodine, and even oxide of lead. It dissolves or suspends the vegetable alkaloids in the same manner as the aqueous liquids, and at the same time the resulting products may be used for the same purposes as though mixed with oil. Thus the salts of morphia dissolve in it completely, even cold, in all proportions. Sulphate of quinine, in the proportion of one-tenth, dissolves in it when hot, but when cold, separates in clots, which, triturated with the supernatant liquid, give it the consistence of a cerate, very useful for frictions and embrocations. It is the same with the salts of bromine, strychnine, veratrine, and most preparations of the same order, which enables us to consider that we have now, if not medicinal oils with a vegetable alkaloid base, at least a series of preparations fulfilling a perfectly analogous use in therapeutics.

Glycerin has been found very useful as an outward application in skin diseases and inflammations of the mucous membranes of the nose, mouth, &c., and has been used, with partial success in diseases of the eye and ear, and as a substitute for cod-liver oil in the treatment of phthisis; but its efficacy in these latter cases is not yet satisfactorily proved.

The same qualities which render glycerin so useful in pharmacy, likewise make it available for many other purposes, as in perfumery and the manufacture of cosmetics. Thus it may be mixed with toilette vinegar, eau de Cologne, hydrolates, alcoholates, &c., to prevent the too rapid drying of these liquids on the skin; also with cold cream, pomades, oils, &c., to which it imparts the power of holding in suspension certain alcoholic dyes, or aqueous extracts, which otherwise separate from them after admixture.* Its use as an addition to soap has been already mentioned (p. 709).

Glycerin may also be used for preserving articles of food, especially those which require to be kept moist, as sugar, fruits, chocolate, &c., enclosed in tin cases. Meat keeps well in it, retaining its flavour and softness. A very useful application of it is to mix with tobacco. Many sweet substances are used for this purpose, such as treacle and sugar; but they are apt to spoil the tobacco by fermenting. Extracts of roots, such as

---

* Patent granted to J. H. Johnson, 21st of January, 1854.

liquorice, are also used, but they do not keep the tobacco moist, so that it requires to be pressed and wrapped in tin-foil; these inconveniences may be obviated by the employment of glycerin.

Barreswil recommends the use of glycerin for keeping modelling clay moist. It is also useful as a solvent for gum-arabic and white of egg, the solutions remaining unaltered for a long time, and for *aniline-violet* (*anileine* or *indisine*); the solubility of this colouring matter in glycerin has not been exactly ascertained, but it appears to dissolve more freely in that liquid than in alcohol or acetic acid (C. Gros-Renaud, *Rép. Chim. app.*, 1859, p. 427).

Amadon (*Technologiste*, January 1858, p. 191), describes a process of madder-dyeing with the aid of glycerin. Alizarin and alcoholic madder-extract dissolve in glycerin, even in the cold; at higher temperatures more quickly and in larger quantity, so that the liquid acquires a deep scarlet colour. The alcoholic extract of madder does not yield any deposit on cooling or on addition of water, but the solution of alizarin in glycerin deposits red flocks on being mixed with water.

Messrs. Vasseurs and Houbigant have patented[*] the application of glycerin for the preparation of ink, paper, and other materials.

Glycerin, diluted with four or five times its weight of water, imparts to paper the peculiarity of retaining a permanently damp condition, so that, in taking copies of letters, &c., written on paper so prepared, pressure and damping of the copying paper will not be required. The writing paper may be prepared either by introducing glycerin into the pulp of which it is to be manufactured, or by damping it therewith after it has been made up into books or otherwise. Or, ink may be prepared or combined with glycerin; and writings effected with the ink so treated will remain for a long period in a sufficiently damp state to allow of copies or impressions being taken without pressure or damping of the copying paper.

A good ink may be produced by mixing 6 parts of glycerin with 2 parts of crystallized sugar, adding sufficient pure water to melt the sugar (or sugar-of-milk or honey may be used instead of crystallized sugar). The liquid should be mixed with an equal quantity of common ink.

[*] Patent granted to Michael Henry, 21st of May, 1858, No. 1132.

An alkaline copying ink, preserving the pen from oxidation, may be manufactured of—

  Decoction of logwood, at 8° . . . 5 parts
  Sugar. . . . . 3 „
  Gum arabic. . . . . 2 „
  Glycerin. . . . . 5 „

adding, as a colouring solution, till a violet hue (*couleur pensée*) is obtained—

  Water . . . . 100 parts
  Common potassa . . . . . . 20 „
  Flowers of sulphur . . . . . 3 „

The materials are melted in a cast-iron vessel, and 10 parts of tanned skin-parings or cuttings are added ; the whole is then boiled to dryness, stirring well to keep it from burning, and after taking it off the fire, 200 parts of water are added, and the liquid is expressed and filtered.

The action of the glycerin on the paper, combined with that of a chemical agent, such as prussiate of potassa, turnsol dye, prussian blue, &c., will, owing to the damp condition of the paper, produce certain colouring effects. Thus, in the first instance, if the ink contains iron in excess, dark blue writing characters will be produced on the paper; in the second instance, if the deliquescent ink has acetate of iron in excess, the writing will be red, while an alkaline ink will produce green; in the third instance, with an alkaline deliquescent ink, the characters will be white. Such effects of chemical reaction may be extensively varied.

It is necessary to observe, that all the inks of commerce are not equally adapted for the production of copying ink; but copying inks, properly so called, are the best suited for the purpose, especially the black inks of two English chemists, viz., Mr. Almonds and Mr. Robertson (3, Maiden Lane); next come violet-black inks and those of other colours. The following recipes are suggested as examples of permanent inks :—

Recipe No. 1.

 3 parts, by weight, of white glycerin,
 3   „   purified white honey, best quality,
 10   „   violet-black, or other coloured ink.

Mix up well, and leave the mixture to settle two or three days before using.

Recipe No. 2.

4 parts, by weight, of white glycerin,
4   ,,   purified white honey, best quality,
10   ,,   Robertson's ink,
0¼   ,,   powdered gum arabic.

Add one or two drops of strong solution of bichloride of mercury to prevent deterioration of the ink; stir up well, and leave to settle for two or three days before using.

If it should be found, in taking copies from inks according to recipe No. 1, that thicker characters are produced than those of the original, the proportions of glycerin and honey may be respectively reduced to 2 parts each, or another quarter by weight of 1 part powdered gum arabic may be added. The following mode is recommended for purifying honey before employing it in manufacturing these improved inks :— Dissolve 2 parts of powdered magnesia in 16 parts by weight of pure water, and add it to 100 parts by weight of good honey; heat the whole in a water-bath, but do not allow it to boil; skim carefully. The honey so purified should be kept in closely-stoppered jars. The copying paper should not be too thin or too highly glazed, as in the former case it would not absorb all the ink, and in the latter would hardly absorb any, and would cause the characters to appear jagged or spluttered. The writing paper should not be so porous as not to absorb the ink. Two copies of one letter or writing may be readily taken; but when one copy only is required, it is of importance, after taking it, and before folding the original, &c., to apply a blotter consisting of about three sheets of unsized paper firmly cemented by pressure in the hot-pressing or glazing process; this is a very useful precaution; the blotter is a good absorbent, and may serve for hundreds of pages. If the ink produce pale or indistinct copies, all that need be done is to increase its colouring ingredient by a fourth. Honey is preferred to any other saccharine matter in manufacturing these improved inks.

Glycerin may also be employed, instead of common salt, for preserving untanned skins and hides, especially when intended for exportation, and therefore requiring rapid means of applying a preserving agent; it may also be applied for the purpose of preserving food. Paste, cement, mortar, mastic, and other matters, especially when intended for daily use, may be treated with glycerin, in order to keep them in a suitably damp condition ; this treatment will also have the effect of preserving them from frost. Vesicatory or blister paper, lint, and textile fabrics, particularly cloths, rags, and bandages, intended for medical, surgical, or therapeutical purposes, may also be treated with glycerin to render them absorbent and consequently better adapted for the desired object.

Unsized paper, as generally used for the purpose of taking copies of writings, will be rendered more effective by previously glazing, pressing, or satining it. This pressing, glazing, &c., may be applied on any sort of copying paper, whether containing salts or not.

As glycerin never freezes, and is not altered by exposure to the air, it may be advantageously applied as a lubricator for delicate machinery, such as clockwork. For this purpose it is superior to the purest olein or oleic acid, the lubricants commonly used, as the former thickens by oxidation, and the latter solidifies at a few degrees below the freezing point of water.

Glycerin may also be added to the water of gas-meters, to prevent freezing and evaporation.

LONDON : *October* 1865.
219 REGENT STREET.

PUBLICATIONS

OF

H. BAILLIÈRE

IN

MEDICAL AND THE COLLATERAL SCIENCES.

TO WHICH IS ADDED

A LIST OF SCIENTIFIC AMERICAN AND FRENCH WORKS.

NEW YORK.	MELBOURNE (AUSTRALIA).
BAILLIÈRE BROTHERS,	F. F. BAILLIÈRE,
440 BROADWAY.	COLLINS STREET EAST.
PARIS.	MADRID.
J. B. BAILLIÈRE ET FILS,	BAILLY BAILLIÈRE,
RUE HAUTEFEUILLE.	CALLE DEL PRINCIPE.

Anatomy.—Muscular Anatomy of the Horse. By James I. Lupton, M.R.C.V.S. 8vo. with 12 Woodcuts. 1862. 8s. 6d.

Anatomy of the External Form of the Horse, with Explanations. By James I. Lupton, M.R.C.V.S. Part I., with 9 Plates, large folio. Price, plain, £1 11s. 6d.; India paper, £2 5s. Part II., with 4 Plates, price, plain, 15s.; India paper, 20s. *The Plates of the Mouth and Leg are coloured.*

** This Work will be completed in Three Parts, consisting of 18 to 20 Plates, with Explanations, and a volume 8vo. of Text, giving the Study of the External Form of the Horse and the Physiology of Loco-motion.

Ashley (W. H.)—A Practical Treatise on Vesicular Hydatids of the Uterus. 1856. 2s. 6d.

Atkinson.—Physics. (*See* Ganot.)

Austin (Thos.)—The Millstone Grift, its Fossils in the Bristol District. 8vo. with 4 Plates. London, 1865. 4s.

Baudens.—On Military and Camp Hospitals, and the Health of Troops in the Field. Translated by F. B. Hough, M.D. 12mo. New York, 1862. 6s.

Berkeley (Rev. J. M.)—Introduction to Cryptogamic Botany. 8vo. illustrated with 127 Engravings. London, 1857. £1.

Bernard and Huette.—Illustrated Manual of Operative Surgery and Surgical Anatomy. Edited, with Notes and Additions, by W. H. Van Buren, M.D., Professor of Anatomy, University Medical College, and C. E. Isaacs, M.D. Complete in 1 vol. 8vo. with 113 coloured Plates, half-bound morocco, gilt tops. New York, 1864. £3 4s.

Birch.—On the Therapeutic Action of Oxygen; with Cases proving its singular Efficacy in various Intractable Diseases. 8vo. 1857. 2s. 6d.

Boenninghausen. — Manual of Homœopathic Therapeutics, Guide to the Study of Materia Medica Pura. Translated, with Additions, by J. Laurie, M.D. 8vo. 1848. 12s.

Boenninghausen.—Essay on the Homœopathic Treatment of Intermittent Fevers. 8vo. New York, 1845. 2s. 6d.

Boussingault.—Rural Economy; in its Relation with Chemistry, Physics, and Meteorology. By J. B. Boussingault, Member of the Institute of France. 2nd Edition. 8vo. boards. London, 1845. 18s.

Campbell.—A Practical Text-Book of Inorganic Chemistry, including the Preparation of Substances, and their Qualitative and Quantitative Analyses, with Organic Analyses. By D. Campbell, Demonstrator of Practical Chemistry to the University College. 12mo. 5s. 6d.

Canton (A.)—The Teeth and their Preservation, in Infancy and Manhood to Old Age. 12mo. with Woodcuts. 4s.

Chapman.—A Brief Description of the Characters of Minerals. By Edward J. Chapman, Professor in the University of Toronto (Canada). 12mo. with 8 Plates. London, 1844. 4s.

Chapman.—Practical Mineralogy; or, a Compendium of the distinguishing Characters of Minerals; by which the Name of any Species may be speedily ascertained. 8vo. with 13 Engravings, showing 270 specimens. London, 1843. 7s.

Chemical Society (Quarterly Journal of the). 14 vols. 8vo. London, 1848-61. Vols. I. to IX. have been reduced to 6s. each; vols. X. to XIV., 18s. each. Or in Quarterly Parts, price 8s. each.

* This Journal, beginning with Vol. XV., is now issued Monthly. Price (the Year), 12s.

Cruveilhier and Bonamy.—Atlas of the Descriptive Anatomy of the Human Body. By J. Cruveilhier. With Explanations by C. Bonamy. Containing 82 Plates of Osteology, Syndemology, and Myology. 4to. London, 1844. Plain, £3. Coloured Plates, £5 15s.

Vol. II. On the Circulation of the Blood, French Text, 4to. coloured Plates, £3 4s.

Vol. III. On the Digestive Organs and Viscera, French Text, coloured Plates, £3 6s.

Curie (P. F., M.D.)—Principles of Homœopathy. 1 vol. 8vo. London, 1837. 5s.

Day (G.)—Chemistry in its Relation to Physiology and Medicine. 8vo. with 5 coloured Plates. London, 1860. £1. Reduced to 10s.

Deleuze.—Practical Instruction in Animal Magnetism. 4th Edition, 1850. 4s. 6d.

Dick (H.)—Gleet: its Pathology and Treatment. By H. Dick, M.D. 8vo. London, 1858. 3s. 6d.

Drysdale (C. R.)—On the Treatment of Syphilis and other Diseases without Mercury. 8vo. London, 1868. 2s. 6d.

Drysdale (C. R.)—Remarks on the An cedents and Treatment of Consumpti 8vo. 1s.

Dumas and Boussingault.—The Chemi and Physiological Balance of Organic N ture: an Essay. 1 vol. 12mo. Lond 1844. 4s.

Edwards (A. M.)—Life beneath the Waters or, the Aquarium in America. 12mo. beau tifully illustrated. New York, 1858. 7s. 6d

Ethnographical Library. Vol. I. Th Native Races of the Indian Archipelago Papuans. By G. W. Earl. Post 8v Illustrated with 5 coloured Plates, 2 Ma and Woodcuts. London, 1853. 10s. Reduced to 5s.

Ditto. Vol. II. The Russian Races. B R. Latham, M.D. With a Map a coloured Plates. 1854. 8s. Reduced to

Fau.—The Anatomy of the External Fo of Man, for Artists, Painters, and Sculp tors. Edited by R. Knox, M.D. 8vo Text, and 28 4to. Plates. London, 1849 Plain, £1 4s. Coloured £2 2s.

Fielding and Gardner.—Sertum Plan rum; or Drawings and Descriptions of Rar and Undescribed Plants from the Author Herbarium. 8vo. London, 1843. £1 1 Reduced to 10s. 6d.

Flourens (P.)—On Human Longevity, an the Amount of Life upon the Globe. Edit by C. Martel. 12mo. London, 1855. 3s.

Gamgee (J. S.)—Researches in Pathologi cal Anatomy and Clinical Surgery. Witl 6 Plates. London, 1856. 9s.

Ganot.—Elementary Treatise on Physi Experimental and Applied. Translated or the 9th Original Edition by Dr. Atkinson of the Royal Military College, Sandhurs With 600 Illustrations. Post 8vo. 1862 12s. 6d.

(*Adapted for the use of Colleges and Schools.*)

Gerber and Gulliver.—Elements of th General and Microscopical Anatomy of M and the Mammalia; chiefly after Origina Researches. To which is added an Appen dix, comprising Researches on the Ana tomy of the Blood, Chyle, Lymph, Thy mous Fluid, Tubercle, by C. Gulliver, F.R. 8vo. and an Atlas of 34 Plates. 2 vols. 8v 1842. £1 4s. Reduced to 15s.

Gesner (A.)—A Practical Treatise on Coa Petroleum, and other Distilled Oils. 2n Edition, enlarged. Illustrated with 4 Figures, and a View on Oil Creek, in Penn sylvania. 8vo. New York, 1865. 10s.

Gill (J. B.)—An Epitome of Surgery. 18mo. London, 1860. 1s.

Gordon (S.)—A Synopsis of Lectures on Civil Engineering and Mechanics. 4to. London, 1849. 7s. 6d.

Grant.—General View of the Distribution of Extinct Animals. In the 'British Annual,' 1839. 18mo. London, 1839. 8s. 6d.

Grant.—Outlines of Comparative Anatomy. 8vo. 148 Woodcuts. London. £1 8s. Reduced to £1.

Grant.—On the Principles of Classification, as applied to the Primary Divisions of the Animal Kingdom. In the 'British Annual,' 1838. 18mo. illustrated with 28 Woodcuts. London, 1838. 8s. 6d.

Graham. — Elements of Chemistry; including the application of the Science in the Arts. By T. Graham, F.R.S. L. & E., Master of the Mint. 2nd Edition, revised and enlarged, illustrated with Woodcuts, 2 vols. 8vo. £2.
Ditto. Vol. II. Edited by H. Watts, M.C.S. Separately. 1857. £1.

Greatest (The) of our Social Evils: Prostitution as it now exists in London, Liverpool, Manchester, Glasgow, Edinburgh, and Dublin ; an Inquiry into the Cause and Means of Reformation, based on Statistical Documents. By a Physician. 12mo. London, 1857. 5s. Reduced to 3s. 6d.

Hahnemann.—Lesser Writings. Collected and Translated by R. E. Dudgeon. 8vo. London, 1857. £1 1s.

Hall (Marshall).—On the Diseases and Derangements of the Nervous System, in their Primary Forms, and in their Modifications by Age, Sex, Constitution, Hereditary Predisposition, Excesses, General Disorder, and Organic Disease. 8vo. with 8 Engraved Plates. London, 1841. 15s.

Hall.—On the Mutual Relations between Anatomy, Physiology, Pathology, Therapeutics, and the Practice of Medicine: being the Gulstonian Lectures for 1842. 8vo. with 2 coloured Plates and 1 plain. London, 1842. 5s.

Hall.—New Memoir of the Nervous System, True Spinal Marrow, and its Anatomy, Physiology, Pathology, and Therapeutics. 4to. with 5 Plates. London. £1.

Hamilton.—A Guide to the Practice of Homœopathy. Translated and Compiled in Alphabetical Order from the German of Ruoff, Haas, and Ruckert, with Additions. 12mo. 1844. 5s.

Hamilton.—Flora Homœopathica; or Illustrations and Descriptions of the Plants used as Homœopathic Remedies. 2 vols. 8vo. with 66 coloured Plates. 1851. £3 10s.

Hamilton.—Practical Treatise on Military Surgery and Hygiene. 8vo. with 127 Engravings. New York, 1865. £1.

Hooker.—Icones Plantarum. By Sir W. J. Hooker. New Series. Vols. I.–IV. containing 100 Plates each, with Explanations. 8vo. cloth. London, 1842–1844. Each vol. £1 8s. Reduced to 10s.

Hooker.—The London Journal of Botany. Vols. I.–VI., with 24 Plates each, board 1842–1847. Now reduced to £4.

Hooker.—Niger Flora; or, an Enumeration of the Plants of Western Tropica Africa. Collected by the late Dr. T. Vogel Botanist to the Voyage of the Expeditio sent by Her Britannic Majesty to the Rive Niger in 1841, including Spicilegia Gor gonea, by P. B. Webb, and Flora Nigritian by Dr. J. D. Hooker and George Bentham With 2 Views, a Map, and 60 Plates. 8vo London, 1849. £1 1s. Reduced to 10s. 6d

Hooker. — Notes on the Botany of th Antarctic Voyage conducted by Captai James Clark Ross, R.N., in H.M.S. Ere and Terror; with Observations on th Tussac Grass of the Falkland Islands. 8vo With 2 Plates. London, 1843. 4s.

Horse (See Anatomy of the).

Hufeland.—Manual of the Practice o Medicine; the Result of Fifty Years' Ex perience. Translated by C. Bruchhause and R. Nelson. 2nd Edition. 8vo. Londo 1858. 7s. 6d.

Hulme (E. T.)—Contributions to Dent Pathology. Calcification of the Dent Pulp. Three Lectures on the Diseases the Dental Periosteum. 3 Plates. 8v London, 1862. 2s. 6d.

Humboldt.—Kosmos: a General Surve of the Physical Phenomena of the Uni verse. The original English Edition. 2 vol post 8vo. Reduced to 10s. 6d.

Isherwood (B. E.)—Engineering Prece dents for Steam Machinery: embracing th Performances of Steamships, Experiment with Propelling Instruments, Condense Boilers, &c., accompanied by Analyses the same; the whole being original ma ter, and arranged in the most practical an useful manner for Engineers. 2 vols. 8v With Plates and Tables. New York. 15s

Kæmtz.—A Complete Course of Meteorology. With Notes by Ch. Martins, and an Appendix by L. Lalanne. Translated, with Additions, by C. V. Walker. Post 8vo. With 15 Plates. 12s. 6d.

Knox (R.)—Man; his Structure and Physiology, popularly explained and demonstrated, by the aid of 8 movable dissected coloured Plates and 5 Woodcuts. 2nd Edition, revised. Post 8vo. London, 1858. 10s. 6d.

⁎ This work is written especially for the use of Students who intend graduating in Arts at the English Universities.

Latham (R. G.)—The Native Races of the Russian Empire. 12mo. With a Map and coloured Plates. London, 1854. 5s.

Lebaudy.—The Anatomy of the Regions interested in the Surgical Operations performed upon the Human Body; with Occasional Views of the Pathological Condition which render the interference of the Surgeon necessary. 24 Plates. Folio. London, 1845. £1 4s.

Lee.—The Anatomy of the Nerves of the Uterus. Folio, with 2 Plates. London, 1845. 8s.

Liebig.—Chemistry and Physics, in relation to Physiology and Pathology. 2nd Edition, 8vo. London, 1847. 3s.

Lobb.—Curative Treatment of Paralysis and Neuralgia with the aid of Galvanism. 12mo. 2nd Edition. 1859. 5s.

Lupton.—(*See* Anatomy of the Horse.)

Massy.—Analytical Ethnology; the Mixed Tribes in Great Britain and Ireland Examined, and the Political, Physical, and Metaphysical Blunderings on the Celt and the Saxon exposed. 12mo. Plates. 5s.

Memoirs of the Literary and Philosophical Society of Manchester. 3rd Series. Vol. I. 8vo. or Vol. XX. of the Collection. With Woodcuts and Plates. London, 1862. £1.

Miers (J.)—Illustrations of South American Plants. 2 vols. 4to. 84 Plates. £3 14s.

Ditto. Vol. II. 42 Plates. 1857. £1 17s.

Mitchell (J.)—Manual of Practical Assaying, intended for the use of Metallurgists, Captains of Mines, and Assayers in General. With copious Tables, for the purpose of ascertaining in Assays of Gold and Silver the precise amount, in Ounces, Pennyweights, and Grains, of noble metal contained in one ton of Ore from a given quantity. 2nd Edition, 8vo. much enlarged, with 360 Illustrations. London, 1854. £1 1s.

Mitchell (J.)—Treatise on the Adulterations of Food, and the Chemical Means employed to detect them. Containing Water, Flour, Bread, Milk, Cream, Beer, Cider, Wines, Spirituous Liquors, Coffee, Tea, Chocolate, Sugar, Honey, Lozenges, Cheese, Vinegar, Pickles, Anchovy Sauce and Paste, Catsup, Olive Oil, Pepper, Mustard. 12mo. 1848. 6s. Reduced to 3s.

Moquin-Tandon.—Elements of Medical Zoology; a Description of the Animals used in Medicine, as well as of those Species which are Injurious to or are Parasitic upon Man. Edited by R. T. Hulme. With 124 Illustrations. 1861. 12s. 6d.

Moreau (Professor).—Icones Obstetricæ; a Series of 60 Plates and Text, illustrative of the Art and Science of Midwifery in all its Branches. By M. Moreau, Professor of Midwifery to the Faculty of Medicine, Paris. Edited by J. S. Streeter, M.R.C.S. Folio. 1841. Plain, £3 3s.

Morel (C.)—Compendium of Human Histology. Edited by W. H. Van Buren, M.D. 8vo. with 28 Plates. New York, 1861. 14s.

Mueller (F.)—Fragmenta Phytographiæ Australiæ. 3 vols. 8vo. with Plates. Melbourne, 1860-64. £2 14s.

Mueller (F.)—The Plants Indigenous to the Colony of Victoria. Vol. I. 4to. *Thalamiflora*. With 23 Plates. Melbourne, 1860-62. £3 3s.

Muller.—Principles of Physics and Meteorology. By J. Muller, M.D. Illustrated with 530 Woodcuts and 2 coloured Plates. 8vo. London, 1847. 18s.

Newman (George).—Homœopathic Family Assistant. 3rd Edition. 18mo. 1859. 3s. 6d.

Nichol.—The Architecture of the Heavens. By J. P. Nichol, Professor of Astronomy in the University of Glasgow. 9th Edition, entirely revised and greatly enlarged. Illustrated with 23 Steel Engravings and numerous Woodcuts. 8vo. London, 1851. 16s.

Noeggerath and Jacobi.—Contributions to Midwifery, and Diseases of Women and Children; with a Report on the Progress of Obstetrics, and Uterine and Infantile Pathology. 8vo. New York, 1858. 15s.

Norris (E.)—(*See* Ethnographical Library, and Prichard.)

Otto (J.)—Manual of the Detection of Poisons by Medico-Chemical Analysis. By J. Otto, Professor of Chemistry in Brunswick, Germany. With Illustrations. 12mo. New York, 1857. 6s.

Owen (R.)—Odontography; or, a Treatise on the Comparative Anatomy of the Teeth, their Physiological Relations, Mode of Development, and Microscopical Structure in the Vertebrate Animals. This splendid work is now completed. 2 vols. royal 8vo. containing 168 Plates, half-bound russia. London, 1840–45.

Phillips (B.)—Scrofula; its Nature, Prevalence, Causes, and the Principles of Treatment. 8vo. with an Engraved Plate. London, 1846. 12s. Reduced to 5s.

Pietra Santa.—The Climate of Algiers in reference to the Chronic affections of the Chest. Being a Report of a Medical Mission to Algeria. London, 1862. 1s. 6d.

Prescriber's (The) Complete Handbook. (See Trousseau.)

Prichard.—Six Ethnographical Maps. Supplement to the Natural History of Man, and to the Researches into the Physical History of Mankind. Folio, coloured, and 1 sheet of letterpress. 2nd Edition. London, 1860. £1 4s.

Prichard.—On the Different Forms of Insanity, in relation to Jurisprudence. 12mo. London, 1842. 5s.

Prichard (J. C.)—The Natural History of Man; comprising Inquiries into the Modifying Influences of Physical and Moral Agencies on the different Tribes of the Human Family. 4th Edition, revised and enlarged, by Edwin Norris, of the Royal Asiatic Society. With 62 Plates, coloured, engraved on Steel, and 100 Engravings on Wood. 2 vols. royal 8vo. 1855. £1 18s.

Quekett (J.)—Practical Treatise on the use of the Microscope. Illustrated with 11 Steel Plates and 300 Wood Engravings. 3rd Edition, 8vo. £1 1s. Reduced to 12s. 6d.

Quekett (J.)—Lectures on Histology, delivered at the Royal College of Surgeons of England—Elementary Tissues of Plants and Animals. On the Structure of the Skeletons of Plants and Invertebrate Animals. 2 vols. 8vo. Illustrated by 340 Woodcuts. London, 1852–54. £1 3s. 6d. Reduced to 15s.

Rayer (P.)—A Theoretical and Practical Treatise on the Diseases of the Skin. Translated by R. Willis, M.D. 2nd Edition, remodelled and much enlarged, in 1 thick vol. 8vo. with Atlas, royal 4to. of 26 coloured Plates, exhibiting 400 varieties of Cutaneous Affections. £4 8s.

Richardson and Henry Watts.—Chemistry in its Application to the Arts and Manufactures. 8vo. Fuel and its Applications.

Vol. I. Parts 1 and 2, contain Fuel and its Applications. Profusely Illustrated with 433 small Engravings and 4 Plates. £1 16s.

Vol. I. Part 3, contains Acids, Alkalies, Salts, Soap, Soda, Railway Grease, &c., their Manufacture and Applications. With numerous Illustrations. 1864. £1 13s.

Vol. I. Part 4, contains Phosphorus, Mineral Waters, Gunpowder, Gun Cotton, Fire Works, Aluminium, Lucifer Matches. With Illustrations. 1865. £1 1s.

Vol. II. contains Glass, Alum, Potteries, Cements, Gypsum, &c. With numerous Illustrations. £1 1s.

Vol. III. contains Food generally, Bread, Cheese, Tea, Coffee, Tobacco, Milk, Sugar. With numerous Illustrations and Coloured Plates. £1 2s.

‚ The Authors of this Edition in their Preface say: ' So rapid has been the growth and so great the development of the branches of Manufacture more intimately connected with Fuel, that, in preparing a Second Edition, we have found it necessary, not only to re-write much of the original, but to extend so considerably the limits of the first group as to occupy the entire of this volume, which may, therefore, with far greater propriety be called a New Work than a Second Edition.'

Reichenbach.— Physico-Physiological Researches on the Dynamics of Magnetism, Electricity, Heat, Light, Crystallization, and Chemism, in their Relations to Vital Force. 2nd Edition, with Additions, Preface, and Critical Notes, by John Ashburner, M.D. 8vo. with Woodcuts, and 1 Plate. London, 1850. 15s.

Reid. — Rudiments of Chemistry, with Illustrations of the Chemistry of Daily Life. 4th Edition, with 130 Woodcuts. 12mo. 1850. 2s. 6d.

Regnault.—An Elementary Treatise on Crystallography, illustrated with 108 Wood Engravings. 8vo. 1848. 8s.

Richardson.—Geology for Beginners; comprising a Familiar Exposition of the Elements of Geology and its Associate Sciences, Mineralogy, Fossil Conchology, Fossil Botany, and Palæontology. 2nd Edition, 8vo. with 251 Woodcuts. 10s. 6d.

Richardson (B.)—Medical History and Treatment of Diseases of the Teeth, and the adjacent structures. 8vo. 8s.

Say (Thos.)—The Complete Writings of, on the Conchology of the United States. 8vo. with 75 Plates, half-bound. New York. Coloured, £2 10s.; Plain, £1 5s.

Say (Thos.)—The Complete Writings of, on the Entomology of North America. Edited by J. Le Conte, with a Memoir by G. Ord. 8vo. with 54 coloured Plates. 2 vols. New York, 1859. £4.

Schleiden.—The Plant, a Biography, in a Series of Fourteen Popular Lectures on Botany. Edited by A. Henfrey. 2nd Edition, 8vo. with 7 coloured Plates and 16 Woodcuts. London, 1853. 15s.

Shuckard.—Elements of British Entomology. Part 1. 1839. 8vo. with Figures. 8s.

Shuckard.—Essay on the Indigenous Fossorial Hymenoptera; comprising a Description of the British Species of Burrowing Sand Wasps contained in all the Metropolitan Collections; with their Habits as far as observed. 8vo. with 4 Plates. (Plate I. is wanting.) 10s.

Simpson (M. D.) — Practical View of Homœopathy. 8vo. London. 1836. 10s. 6d.

Smith (S. E.)—Diagnostics of Aural Disease. 8vo. Illustrated with a Plate and 22 Woodcuts. London, 1861. 8s. 6d.

Smith (Stephen)—Handbook of Surgical Operations. 12mo. with 257 Woodcuts. New York, 1865. 8s.

Stars and the Earth.—The Stars and the Earth; or, Thoughts upon Space, Time, and Eternity. 10th Thousand. 18mo. London, 1862. Cloth, 1s. 4d., or paper cover, 1s.

Stevens (W.)—Observations on the Nature and Treatment of the Asiatic Cholera. 8vo. London, 1853. 10s.

Stockhardt.—Chemical Field Lectures for Agriculturists. Translated from the German, with Notes. 12mo. 1853. 4s.

Teste.—A Practical Manual of Animal Magnetism: containing an Exposition of the Methods employed in producing the Magnetic Phenomena, with its Application to the Treatment and Cure of Diseases. Translated by C. Spillan. 18mo. 6s.

Thomson.—Chemistry of Organic Bodies —Vegetables. By Thomas Thomson,M.D., F.R.S. L. & E. 8vo. pp. 1,092. 1838. £1 4s. Reduced to 12s.

Thomson. — Heat and Electricity. 2nd Edition. 8vo. Illustrated. 15s. Reduced to 7s. 6d.

Thomson (R. D.)—British Annual and Epitome of the Progress of Science. 3 vols. 1837-38-89. 18mo. each 8s. 6d.

Townshend.—Facts in Mesmerism; with Reasons for a Dispassionate Inquiry into it. 8vo. 9s.

Trousseau and Reveil.—The Prescriber's Complete Handbook to the Principles of the Art of Prescribing, with a List of Diseases and their Remedies, a Materia Medica of the Medicines employed, classified according to their Natural Families, with their Properties, Preparations, and Uses, and a Sketch of Toxicology. Edited by J. B. Nevins, M.D. 1858. 2nd Edition. Roan, limp, 6s. 6d.

Turnbull (L.) — Electro-Magnetic Telegraph; with an Historical Account. 2nd Edition. 8vo. with Plates. 1853. 12s.

Unity of Medicine—(The), its Corruptions and Divisions as by LAW Established in England and Wales: with their Causes, Effects, and Remedies. By a Fellow of the Royal College of Surgeons. 8vo. with a coloured Chart. London, 1858. 7s.

Vogel and Day.—The Pathological Anatomy of the Human Body. Translated, with Additions, by G. E. Day. Illustrated with 100 plain and coloured Engravings. 8vo. 1847. 18s. Reduced to 10s.

Waterhouse.—A Natural History of the Mammalia: Marsupiata or Pouched, and of the Rodentia or Gnawing Animals. 2 vols. 8vo. Coloured Plates, £3 9s.

Plain, £2 18s.

Weisbach (J.)—Principles of the Mechanics of Machinery and Engineering. 2 vols. 8vo. illustrated with 1,000 Wood Engravings. London, 1848. £1 19s.

Williams (R.)—Elements of Medicine: Morbid Poisons. 2 vols. 8vo. 1836-41. £1 8s. 6d. Reduced to 15s.

Vol II. 1841. 18s. Reduced to 8s.

Willis.—Illustrations of Cutaneous Diseases. A Series of Delineations of Affections of the Skin in their more frequent form. 94 col. Plates. Folio. 1841. £8 8s.

Wilmott.—Glycerin and Cod Liver Oil; their History, Introduction, and Therapeutic Value. 12mo. London, 1859. 3s. 6d.

Wilson (Dr., of Malvern.)—How to Regain Health and Keep it. Popular Lectures on the Preservation of Health and Cure of Chronic Disease. No. 1, 8vo. with Engravings. 1859. 1s. 6d.

Yeldam (S.)—Homœopathy in Acute Diseases. 8vo. 2nd Edition. 1850. 6s. 6d.

AMERICAN
SCIENTIFIC PUBLICATIONS.

Mr. C. E. BAILLIÈRE offers his services to the Profession and the Public for the filling of orders for American Books.

Having just returned after a residence of fifteen years in New York, during which time he was engaged exclusively in the Scientific Book business, he is enabled to give full information relative to American publications; and having his own establishment (not an agency) in New York, he can promise a prompt and faithful execution of all orders entrusted to him.

Gentlemen about to Publish in any of the departments of Science, and wishing to find an American outlet, are requested to communicate with him.

REDUCTION IN THE PRICE OF AMERICAN BOOKS.

Mr. BAILLIÈRE will be happy to execute orders for American Books at the rate of 4s. 2d. to the Dollar nett Cash.

Adams. — Contributions to Conchology. Vol. I. (all published). 8vo. 12s.

Agassiz, L.—Contributions to the Natural History of the United States. Vols. I. to IV. 4to. Boston. Each £2 2s.

American (The) Journal of Insanity. Edited by the Officers of the New York State Lunatic Asylum, Utica. 12 vols. 8vo. cloth, boards. Utica, 1844–56. £7 4s.

American Historical and Literary Curiosities. Consisting of Fac-similes of Documents relating to the Revolution, &c. First and Second Series. 4to. New York. Each £1 12s.

American Association for the Advancement of Science—Proceedings of the. Vols. 1 to 18. Philadelphia, 1849 to 1859. £6 10s.

American Journal of Dental Science. New Series. Vols. II.-IX. Neat half calf. 8vo. Very scarce. Philadelphia, 1851–59. £9.

American Journal of Conchology. Published in Quarterly Numbers of about 100 pages each. With Notes. Conducted by G. W. Tryon. New York, 1865.

American Medical Times (The). Edited by S. Smith. 1860 to 1864. 8 vols. 4to. in cloth, boards. New York. £2 2s.

Australia.—A Complete Description of the whole of the Colony of Victoria. With new Map. The most recent and accurate information as to every Township, Village, and Hamlet; every River, Creek, Mountain, Lake, Goldfield, Road, and Railway in Victoria; with its Botanical, Geological, and Physical features; the division of the Colony into Counties, Ridings, and Electoral, Police, Squatting, Mining, and Municipal Districts; the Names, Areas, Occupiers' Position nearest Post Town, and grazing capabilities of all the Squatting Stations, and the best means of reaching them. 8vo. pp. 442, cloth. £1 1s.

Baird.—Mammals of North America, Descriptions of Species, based chiefly on collections of the Museum of Smithsonian Institution. 4to. 764 pages, and 87 Plates. Philadelphia, 1859. £4 4s.

Barton.—The Cause and Prevention of Yellow Fever at New Orleans and other Cities in America. 3rd Edition. 8vo. with Maps. New York, 1857. 7s. 6d.

Baudens.—On Military and Camp Hospitals, and the Health of Troops in the Field. Translated by F. B. Hough, M.D. 12mo. New York, 1862. 5s.

Beck, T. R.—Elements of Medical Jurisprudence. 12th Edition. By C. R. Gilman. 2 vols. 8vo. Philadelphia. £2 10s.

Bedford.—Principles and Practice of Obstetrics. 3rd Edition. 8vo. 1861. £1.

Bedford.—The Diseases of Women and Children. 7th Edition. New York. 16s.

Bell (A. M.)—A Knowledge of Living Things, with the Laws of their Existence. 12mo. with 2 coloured Plates and 60 Woodcuts. New York, 1860. 5s.

Bernard and Huette.—Illustrated Manual of Operative Surgery. With 113 coloured Plates, royal 8vo. half-bound. 1861. £3.

Blake (W. S.)—Observations on the Physical Geography and Geology of the Coast of California, from Bodega Bay to San Diego. 1 vol. 4to. coloured Plates. New York. £2.

Bland and Cooper. — Papers on North American Helicide and on the Geographical Distribution of West India Land Shells. Also Notice of Land and Freshwater Shells from the Rocky Mountains. 8vo. with Plates. New York, 1862. 10s.

Blowpipe.—A System of Instruction in the Practical Use of the: being a Gradual Course of Analyses for the use of Students. 12mo. New York, 1852. 6s.

Brown-Sequard.—Lectures on the Physiology and Pathology of the Central Nervous System, delivered at the Royal College of Surgeons of England, in May 1858. 8vo. with 8 Plates. Philadelphia, 1860. 14s.

Bumstead.—Pathology and Treatment of Venereal Diseases. 8vo. with Illustrations. Philadelphia, 1861. £1.

Canada, Geological Survey of; Figures and Descriptions of Canadian Organic Remains. Sir W. E. Logan, Director. Parts 1-4, with Plates. Montreal, 1858-64. £1 5s.

Canada, Geological Survey of, Report of Progress from its Commencement to 1863. Illustrated by 498 Woodcuts. An Atlas of Maps and Sections will shortly be published. 4to. Montreal, 1863. £1 4s.

Canada, Geological Survey of, for the Years 1853-54-55-56. 8vo. And Atlas of Plates and Maps. 4to. £1 10s.

Canada and the Western States of America. Illustrated with 50 Engravings, forming a complete Emigrant's Handbook. 4s.

Canadian (The) Naturalist and Geologist. 8vo. with Illustrations. Vol. 1 to 8. Montreal, 1857-1862. £3 15s. 2nd Series. Vol. 1 to 8. £2 5s.

California, Geological Survey of. Palæontology. Vol. 1, with 82 Plates. 4to. By J. D Whitney, State Geologist. 1864. £1 7s. 6d.

Cooper and Suckley.—The Natural History of Washington Territory; with much relating to Minnesota, Nebrasca, Kansas, Oregon, and California. 55 Plates of Botany and Zoology, and Isothermal Chart. 4to. New York, 1859. £2 2s.

Da Costa.—Medical Diagnosis; with special reference to Practical Medicine. Illustrated. 8vo. Philadelphia. £1 1s.

Dalton's Treatise on Human Physiology. 2nd Edition. New York, 1861. £1 1s.

Dana.—A Manual of Geology. 8vo. Philadelphia. 16s. 6d.

Dawson.—Air-Breathers of the Coal Period: a Descriptive Account of the Remains of Land Animals found in the Coal Formations of Nova Scotia. 8vo. with 6 Plates. Montreal, 1863. 6s.

Draper.—Human Physiology, Statical and Dynamical, &c. 8vo. 1856. £1 1s.

Durkee (S.)—A Treatise on Gonorrhœa and Syphilis. 2nd Edition. 8vo. with 8 coloured Plates. Philadelphia, 1864. £1 1s.

Dunglison.—Dictionary of Medical Science. 2nd Edition. 8vo. Philadelphia, 1865. £1 5s.

Emory.—Report of the United States Mexican Boundary. 2 vols. 4to. with 849 coloured and plain Geological Plates. New York, 1858. £5 5s.

Gesner.—A Practical Treatise on Coal, Petroleum, and other Distilled Oils. 2nd Edition. 8vo. with 40 Woodcuts. New York, 1865. 10s. 6d.

Gibbes (L. R.)—Rules for the Accentuation of Names in Natural History; with Examples, Zoological and Botanical. 4to. Charleston, 1860. 2s.

Grace (W.)—The Army Surgeon's Manual, for the Use of Medical Officers, Cadets, Chaplains, and Hospital Stewards. 12mo. New York, 1864. 6s.

Gray (Asa). — Chloris Boreali-Americana. Illustrations of New and Rare North American Plants. 4to. 7s. 6d.

Gross.—Practical Treatise on the Urinary Organs. 8vo. 1851. £1 1s.

Gross.—Elements of Pathological Anatomy. 8vo. Illustrated. New Edition. 1857. £1 1s.

Gross.—System of Surgery; Pathological, Diagnostic, Therapeutic, and Operative. With 936 Illustrations. 2nd Edition. 2 vols. 8vo. Philadelphia. 1862. £2 16s.

Hall (J.) and Whitney.—Report on the Geological Survey of the State of Iowa. Vol. I. 4to. In two Parts, illustrated with 27 steel Engravings, and numerous Woodcuts. Iowa, 1858. £2 10s.

Hall and Whitney.—Report on the Geological Survey of the State of Wisconsin. Vol. I. 4to. 1862. £1 10s.

Hamilton.—Practical Treatise on Fractures and Dislocations. 8vo. with 289 Woodcuts. Philadelphia, 1860. £1 1s.

Harris.—The Principles and Practice of Dental Surgery. 5th Edition, 8vo. with 200 Woodcuts. £1 5s.

Hitchcock.—Ichnology of New England; a Report on the Sandstone of the Connecticut Valley, especially its Fossil Footmarks. 4to. with 60 Plates. £1 15s.

Hitchcock.—An Attempt to Discriminate and Describe the Animals that made the Fossil Footmarks of the United States, and especially of New England. 4to. 15s.

Hitchcock and Hager.—Report on the Geology of Vermont. 2 vols. 4to. 1861. £2 5s.

Hodge.—On Diseases Peculiar to Women. 8vo. with Illustrations. Philadelphia. 14s.

Hodge (H. L.)—The Principles and Practice of Obstetrics. With 149 Plates and Woodcuts. 4to. Philadelphia, 1864. £3.

Isherwood, —, Engineer-in-Chief, United States Navy.—Experimental Researches in Steam Engineering. Vol. I. 4to. 1863. £2 2s.

Jay (J. C.)—A Catalogue of the Shells arranged according to the Lamarkian System, with their Authorities, Synonyma, and references to Works where figured or described. 4th Edition. 4to. with Supplement. New York, 1852. 14s.

Johnson (W.)—On the Correlation, Conversion, or Allotropism of the Physical and Vital Forces. 8vo. Melbourne, 1864. 2s. 6d.

La Roche.—Pneumonia : its Supposed Connection, Pathological and Etiological, with Autumnal Fevers. Philadelphia. 12s. 6d.

La Roche.—Yellow Fever, considered in its Historical, Pathological, and Therapeutical Relations. 2 vols. 8vo. £1 10s.

Lawson.—Practical Treatise on Phthisis Pulmonalis; Pathology, Causes, Symptoms, and Treatment. 8vo. Cincinnati, 1861. £1 1s.

Lehmann.—Physiological Chemistry. Complete in 2 vols. Plates. Phila. £1 5s.

Malgaigne.—Treatise on Fractures. With 106 Illustrations. Philadelphia, 1859. 17s. 6d.

Maury, Nott, and Gliddon.—Indigenous Races of the Earth; or, New Chapters of Ethnological Inquiry. Coloured Plates and Woodcuts. Royal 8vo. Phila. £1 1s.

McCrady (J.)—Gymnopthalmata of Charleston Harbour. 8vo. 5 Plates. 1857. 5s.

Meigs (C.)—Treatise on Acute and Chronic Diseases of the Neck of the Uterus. 8vo. with 22 coloured Plates. Phila. £1 2s.

Meigs (C.)—Obstetrics; the Science and Art. 8vo. 121 Woodcuts. Phila. 17s.

Meigs (C.)—Woman; her Diseases and Remedies. 4th Edition, 8vo. Phila. 17s.

Meigs (C.)—A Practical Treatise on the Diseases of Children. 8vo. Phila. 18s.

Mitchell, Morehouse, and Keen.—Gunshot Wounds and other Injuries of the Nerves. 12mo. Philadelphia, 1864. 6s. 6d.

Morel.—Compendium of Human Histology. Illustrated with 28 Plates. Edited by W. H. Van Buren, M.D. 8vo. New York, 1861. 12s.

Morland (W.)—Diseases of the Urinary Organs; a Compendium of their Diagnosis, Pathology, and Treatment. 8vo. with Illustrations. 1858. 15s.

Morris.—An Essay on the Pathology and Therapeutics of Scarlet Fever. Philadelphia, 1858. Royal 8vo. 6s.

Natural History of the State of New York. 21 vols. 4to. Albany, 1849 to 1861.

Part 1. Zoology. By J. De Kay. 5 vols.

Part 2. Botany. By J. Torrey. 2 vols.

Part 3. Mineralogy. By Beck. 1 vol.

Part 4. Geology. By Mather, Emmons, Vanuxem, and Hall. 4 vols.

Part 5. Agriculture. 5 vols.

Part 6. Palæontology. By Hall. 4 vols.

Neumayer (G.)—Results of the Meteorological Observations taken in the Colony of Victoria in 1859-62, and of the Nautical Observations. 4to. With Charts. Melbourne, 1864. £3 3s.

Noeggerath and Jacobi.—(*See* page 4.)

Nott and Gliddon.—Types of Mankind. Royal 8vo. Plates. £1 1s.

O'Reilly (J.)—The Nervous and Vascular Connection between the Mother and Fœtus in Utero. 8vo. New York, 1864. 3s. 6d.

Otto.—A Manual of the Detection of Poisons by Medico-Chemical Analysis. 12mo. New York, 1857. 6s.

Overman.—A Treatise on Metallurgy, comprising Mining, and General and Particular Metallurgy. With 377 Illustrations. 8vo. New York. £1 10s.

Owen. — Geological Reconnaissance of Arkansas, made during 1857-60. 2 vols. 8vo. Illustrated. £1 10s.

Owen.—Geological Survey in Kentucky, made during 1854-59. 4 vols. 4to. and Plates. Very scarce. £6 6s.

Owen.—Geological Survey of Wisconsin and Minnesota. 2 vols. 4to. Plates. £2 2s.

Parrish.—A Treatise on Pharmacy, designed as a Text-Book for the Student, and as a Guide for the Physician and Pharmaceutist. 3rd Edition, 8vo. Philadelphia. £1 1s.

Regnault.—Elements of Chemistry. 2 vols. 1853. £1 10s.

Reid (David Boswell).—Quarterly Clinical Reports on Surgery. No. 1. Geelong, 1860. 2s. 6d.

Reports of Explorations and Surveys for a Railroad from the Mississippi River to the Pacific Ocean. 12 vols. in 13. 4to. with numerous Plates of Birds, Mammals, Reptiles, Plants, Shells, and Fossils. 1855-1860.

Report upon the Colorado River, expl(in 1857-58. By J. C. Ives and A. H phreys. 4to. with numerous Maps, Vi(and Geological Plates. 1860. £1.

Report exhibiting the Experience of Mutual Life Insurance Company of N York. 4to. half morocco. 1859. 12s.

Richardson.—Practical Treatise on M nical Dentistry. 8vo. with 110 Illustra Philadelphia, 1860. 14s.

Rokitansky.—Manual of Pathological tomy. 4 vols. in 2. 8vo. Philadel 1855. £1 4s.

Say.—The Complete Writings of Th(Say, on the Entomology of North Am

Say.—Conchology of the United Sta Edited by W. G. Binney. (*See* page 6.)

Scanzoni.—Diseases of Females. 8 New York, 1861. 17s. 6d.

Schoolcraft.—The History, Condition, a Prospects of the Indian Tribes of the Unit States. Illustrated by Eastman. 6 v(4to. Plates. 1853-56.

Smith (Stephen).—Handbook of Surgi Operations, with 257 Figures. 12mo. N York, 1865. 8s.

Smithsonian Contributions to Knowl Vols. I.—V. 4to. with Maps and Pl Washington, 1848-55. £5.

Stille.—Therapeutics and Materia Medi 2 vols. 8vo. Philadelphia, 1860. £2 2s.

Storer.—A Dictionary of Solubilities. 8v Cambridge, Massachusetts. £1 5s.

Sullivant.—Contributions to the Bryol and Hepaticology of North America. 4s. 6d.

The New World in 1859; being the Uni States and Canada. Illustrated and scribed in Five Parts.—Part I. The Unit States. II. Scenes and Scenery. III. U per and Lower Canada. IV. Things as th(Are in 1859. V. Emigration, Land, a Agriculture, together with Routes of Trav Fares, Distances, &c. With upwards of 1 Engravings, showing the Rivers, Lak and Mountains. 8vo. cloth. 8s.

Thomas.—A Comprehensive Medical D tionary. Post 8vo. Philadelphia, 1 15s.

Transactions of the Medical Society of the State of New York. 8vo. 1832 to 1860; and 1 vol. of Addresses, 1817 to 1831, in all 20 vols. half-bound in 11. Albany, 1833–61. £8 8s.

Transactions of the American Medical Association. Vols. I.—XIII. 8vo. with Plates. Philadelphia.

Tryon.—Synonymy of the species of Strepomatidæ (Melanians) of the United States; with descriptions of Land, Fresh Water, and Marine *Mollusca*. 8vo. New York. 8s.

. *See* American Journal of Conchology.

Tryon.—List of American Writers on Recent Conchology, with the Titles of their Memoirs, and Dates of Publication. 8vo. New York, 1861. 5s.

Tuomey and Holmes.—The Post-Pleiocene Fossils of South Carolina. 4to. Text and Plates. Charleston, 1858. £4 4s.

Tuomey and Holmes.—Fossils of South Carolina. 4to. Text and Plates. Charleston, 1856. £6 6s.

United States Pharmacopœia; Fourth Decennial Revision. 12mo. 1864. 6s.

Victorian Gazetteer and Road Guide, containing the most Recent and Accurate Information as to every Place in the Colony; with a Map. Compiled by Robt. Whitworth. 8vo. Melbourne, 1865. £1 1s.

Wood and Bache.—The Dispensatory of the United States of America. 12th Edition. 8vo. Philadelphia, 1865. £1 16s.

Wood.—Therapeutics and Pharmocology, or Materia Medica. 2 vols. 8vo. Philadelphia, 1857. £2.

Wood.—Treatise on the Practice of Medicine. 4th Edition. 2 vols. 8vo. Philadelphia, 1855. £2.

Woodward.—Outlines of the Camp Diseases of the U. S. A. 8vo. 1863. 10s.

Wynne (J.)—Report on the Vital Statistics of the United States. 1 vol. 4to. (not printed for sale). New York, 1857. £2 10s.

Wynne.—Private Libraries of New York. 8vo. New York, 1860. £1.

BIBLIOTHÈQUE DE PHILOSOPHIE CONTEMPORAINE

Volumes in-18 à 2s.

—◦◦◦—

Ouvrages parus.

H. TAINE . . .	Le Positivisme anglais, étude sur Stuart Mill.
H. TAINE . . .	L'Idéalisme anglais, étude sur Carlyle.
PAUL JANET . . .	Le Matérialisme contemporain. Examen du système du docteur Büchner.
ODYSSE-BAROT . .	Philosophie de l'histoire.
ALAUX	La philosophie de M. Cousin.
AD. FRANCK . .	Philosophie du droit pénal.
AD. FRANCK . .	Philosophie du droit ecclésiastique : des rapports de la religion et de l'État.
ÉMILE SAISSET .	L'Ame et la vie, suivi d'une étude sur l'Esthétique française.
ÉMILE SAISSET .	Critique et histoire de la philosophie (fragments et discours).
CHARLES LEVÊQUE .	Le Spiritualisme dans l'art.
AUGUSTE LAUGEL .	Les Problèms de la nature.
CHALLEMEL LACOUR .	La Philosophie individualiste, étude sur Guillaume de Humboldt.
CHARLES DE RÉMUSAT	Philosophie religieuse.
ALBERT LEMOINE .	Le Vitalisme et l'animisme de Stahl.
MILSAND	L'Esthétique anglaise, étude sur John Ruskin.
A. VERA	Essais de philosophie hégelienne.
BEAUSSIRE . . .	Origines françaises du panthéisme allemand.
BOST	Le Protestantisme libéral.

FRENCH
SCIENTIFIC PUBLICATION

—◦◦⦂✕⦂◦◦—

REDUCTION IN THE PRICE OF FRENCH BOOKS.

Mr. BAILLIÈRE will be happy to execute orders for French Books at 10d. to the nett Cash.

Abeille.—Traité des Maladies à Urines Albumineuses et Sucrées. 1863. 8vo., avec une planche et figures. 6s. 6d.

Anger (B.)—Traité Iconographique des Maladies Chirurgicales. Précédé d'une Introduction par M. Velpeau. Première Monographie. *Luxations et Fractures.* Avec 8 planches coloriées. 4to. Livr. 1, 2, 3. Paris 1865. Each 10s.

Annales d'Hygiène Publique et de Médicine. Première série, collection complète (1829 à 1853), dont il ne reste que peu d'exemplaires, 50 vols. in-8, avec figures et planches et tables alphabétiques des 50 vols. £15. La 2e série commence avec 1854 à 1865. Prix de chaque année £1 4s.

Axenfeld—Des Névroses. Fol. 1864. 6s.

Barth et Roger.—Traité Pratique d'Auscultation. 7e édit. 18mo. 1865. 5s. 6d.

Bazin.—Leçons sur la Scrofule. 8vo. 2e édit. 1861. 6s. 6d.

Bazin.—Leçons Théoriques et Cliniques sur les Affections Cutanées Parasitaires. 2e édit. 8vo., orné de 5 pl. 1862. 4s.

Bazin.—Leçons Théoriques et Cliniques sur les Syphilides. 1859. 8vo. 8s. 6d.

Bazin.—Leçons Théoriques et Cliniques sur les Affections Cutanées de Nature Arthritique et Dartreuse. 1860. 8vo. 4s.

Bazin.—Leçons Théoriques et Cliniques sur les Affections Cutanées Artificielles et sur la Lèpre, les Diathèses, le Purpura, les Difformités de la Peau etc. 1862. 8vo. 5s.

Bazin.—Leçons sur les Affections Génériques de la Peau. 1862. 8vo. 4s.

Beclard.—Éléments d'Anatomie Générale. 4e édit. 8vo. 1865. 8s. 6d.

Beclard.—Traité Élémentaire de Physiologie. 5e édit. 8vo. 1865. 12s. 6d.

Belot (C.)—La Fièvre Jaune à la Havane sa Nature et Traitement. 8vo. 1865.

Beraud (B. J.)—Atlas Complet d'Anatomi Chirurgicale Topographique, pouvant de complément à tous les ouvrages d'anatomie chirurgicale, composé d'environ 1 planches représentant plus de 200 figu dessinées d'après nature, par M. Bi Figures noires, £2. 10s. Col. £5.

Bernard (C.)—Introduction a l'Étude de Médecine Expérimentale. 8vo. Paris, 1 6s.

Bernard (Cl.)—Leçons sur les Proprié des Tissus Vivants. 8vo., illustré. 18 6s. 6d.

Bossu.—Nouveau Compendium Médical l'usage des Médecins-praticiens. 1862. édit. 18mo. 6s.

Bouchardat.—Manuel de Matière Médical de Thérapeutique Comparée et de P macie. 1864. 2 vol. 18mo. 4e édit. 1

Bouchardat.—Le Travail, son Influence su la Santé. 1863. 1 vol. 18mo. 2s.

Bourguignon et Sandras.—Traité Pratiq des Maladies Nerveuses. 2e édit. 1 1863. 2 vol. 8vo. 10s.

Briand et Chaudé.—Manuel Complet d Médecine Légale, et contenant un Traité Chimie Légale, par H. GAULTIER D CLAUBRY. 7e édit. 1864. 8vo., avec 8 gravées et 64 fig. 10s.

Brierre de Boismont.—Des Hallucination 8vo. 3e édit. Paris, 1862. 6s.

Brierre de Boismont.—Du Suicide et de Folie Suicide. 2e édit. Paris, 1864.

Casper.—Traité Pratique de Médicine Légale, traduit de l'allemand sous les yeux de l'auteur, par M. Gustave Germer Baillière. 1862. 2 vol. 8vo. 12s.
Atlas colorié se vendant séparément 10s.

Chauffard.—Principes de Pathologie Générale. 8vo. Paris, 1862. 7s. 6d.

Charpentier.—Des Accidents Fébriles qui surviennent chez les Nouvelles Accouchées. 8vo. Paris, 1863. 3s.

Chenu.—Manuel de Conchyliologie et de Paléontologie Conchylio'ogique. 1862. 2 vol. 4to., avec 4943 figures dans le texte, dont les principales coloriées. £1 7s.

Churchill (J. F.)—De la Cause Immédiate de la Phthisie Pulmonaire et des Maladies Tuberculeuses et de leur Traitement spécifique par les Hypophosphites. 2e édit. Paris, 1864. 8vo. 15s.

Clemenceau.—De la Génération des Éléments Anatomiques. 8vo. Paris, 1865. 3s. 6d.

Comte (A.)—Cours de Philosophie Positive. 2e édit. Revue par E. Littré. 6 vols. 8vo. Paris, 1864. £1 17s. 6d.

Cruveilhier.—Traité d'Anatomie Descriptive. Vol. 1, 8vo. 1865. 12s. 6d.

Cruveilhier.—Traité d'Anatomie Pathologique Générale. Vol. v. 8vo. *Completing the Work.* Paris, 1864. 6s.

Cullerier.—Précis Iconographique des Maladies Vénériennes. Liv. 1 à 8. 18mo. Paris, 1861-5. Chaque livraison 4s. 6d.

Davaine.—Traité des Entozoaires et des Maladies Vermineuses de l'Homme et des Animaux Domestiques. 1860. 8vo., avec 88 fig. 10s.

Deshayes.—Description des Animaux sans Vertèbres découverts dans le Bassin de Paris. 4to., avec planches. 48 livr. parus. Paris, 1863-5. Chaque 5s.

Dictionnaire Encyclopédique des Sciences Médicales. Publié sous la Direction par MM. Raige-Delorme et A. Dechambre. *L'ouvrage sera publié en 25 vols. 8vo. avec fig. 8 vols. sont en vente.* Chaque vol. 10s.

Dictionnaire de Médicine, de Chirurgie, de Pharmacie, des Sciences Accessoires et de l'Art Vétérinaire. Par Nysten. 12e édit. par Littré. Grand 8vo. demi-rel. Paris, 1865. £1.

Dictionnaire (Nouveau) de Médicine, et de Chirurgie Pratique, illustré de figures intercalées dans le texte, par une *Société de Médecins et Chirurgiens de Paris.* Publié en 15 vols. Paris, 1865. L'ouvrage paraît par vol. tous les trois mois. 8 vols. sont en vente. Chaque 8s. 6d.

Donders.—L'Astigmatisme et les Verres Cylindriques. 1862. 8vo. 4s.

Edwards (Milne.)—Leçons de Physiologie et d'Anatomie comparée de l'Homme et des Animaux. 8 vols. 8vo. Paris. £3. L'ouvrage aura 10 vols.

Eliphas (Levi).—Dogme et Rituel de la haute Magie. 1861. 2e édit. 2 vol. 8vo., avec 24 fig. 15s.

Eliphas (Levi).—Histoire de la Magie. 1860. 1 vol. 8vo., avec 90 fig. 10s.

Eliphas (Levi).—La Clef des Grands Mystères. 1861. 1 vol. 8vo., avec 22 pl. 10s.

Eliphas (Levi). — Philosophie Occulte. Fables et Symboles. 8vo. 1863. 6s.

Eliphas (Levi).—Science des Esprits. 8vo. 1865. 6s.

Falret.—Des Maladies Mentales et des Asiles d'Aliénés. Leçons Cliniques et Considérations Générales. 8vo. Paris, 1864. 12s.

Follin.—Leçons sur les Principales Méthodes d'Exploration de l'Œil Malade et en principal sur l'Application de l'Ophthalmoscope. 8vo., illustrated. 1863. 6s.

Frerichs.—Traité Pratique des Maladies du Foie. 1862. 8vo., avec 80 fig. 9s.

Garnier.—Dictionnaire Annuel des Progrès des Sciences et Institutions Médicales. 12mo. Paris, 1865. 4s.

Goldfuss (A.)—Petrifacta Germaniæ Iconibus et Descriptionibus Illustrata. 2e édit. 4to. Texte et Atlas de 199 planches. fol. Leipz., 1862. £7 10s.

Gosselin.—Leçons sur les Hernies Abdominales faites à la Faculté. 8vo., fig. Paris, 1865. 6s.

Griesinger (W.)—Traité des Maladies Mentales Pathologique et Thérapeutique, traduit par Doumic. 8vo. Paris, 1865. 7s. 6d.

Grisolle.—Traité de Pathologie Interne. 9e édit. 1865. 2 vol. grand 8vo. 15s.

Grisolle.—Traité de la Pneumonie. 2e édit. 8vo. Paris, 1864. 7s. 6d.

Guardia.—La Médecine à travers les Siècles. *Histoire et Philosophie.* 8vo. Paris, 1865. 8s. 6d.

Guérin.—Leçons Cliniques sur les Maladies des Organes Génitaux Externes de la Femme 8vo. 1864. 11s.

Guersent.—Notices sur la Chirurgie des Enfants. 4 parts. 8vo. Paris, 1865. 4s.

Hardy.—Leçons sur les Maladies de la Peau. 4 vols. 8vo. Paris, 1863 à 1865. 16s.

Houel.—Manuel d'Anatomie Pathologique Générale et Appliquée. 2e edition. 1862. 18mo. 6s.

Hugueir (P. C.)—De l'Hystérométrie et du Cathétérisme Utérin. 8vo., avec 4 planches. Paris, 1865. 5s.

Husson.—Étude sur les Hôpitaux considérés sous le rapport de la construction, de la distribution de leurs bâtiments, de l'ameublement, de l'hygiène et du service des malades. Paris, 1863. 4to., avec 24 pl., tableaux et figures. £1.

Jaccoud.—Les Paraplégies et l'Ataxie du Mouvement. 8vo. Paris, 1865. 7s. 6d.

Jamain. — Nouveau Traité Élémentaire d'Anatomie Descriptive et de Préparations Anatomiques, suivi d'un Précis d'Embryologie, par M. Verneuil. 2e édition. 1861. 18mo., avec 200 fig. 10s.

Jamain.—Manuel de Petite Chirurgie. 1860. 3e édition. 1 vol. 18mo., avec 307 fig. 6s.

Jobert (de Lamballe.)—De la Réunion en Chirurgie. 8vo., avec 7 planches coloriées. Paris, 1864. 10s.

Lebert.—Traité d'Anatomie Pathologique Générale et Spéciale, ou Description et Iconographie Pathologique des Affections Morbides, tant liquides que solides, observées dans le corps humain. *Ouvrage complet.* Paris, 1855—1861. 2 vol. in-fol. de texte, et 2 vol. in-fol. comprenant 200 planches dessinées d'après nature, gravées et coloriées. £25.

Legouest.—Traité de Chirurgie d'Armée. 1863. 8vo., avec 128 figures. 10s.

Les Arts Somptuaires.—Histoire du Costume et de l'Ameublement, et des Arts et Industries qui s'y rattachent. 2 vol. de planches coloriées et 2 de texte, 4to. Demi-maroquin. Paris, 1858. £18 18s.

Lévy. — Traité d'Hygiène Publique et Privée. 4e édition. 1862. 2 vol. 8vo. 15s.

Liebreich (R.)—Atlas d'Ophthalmoscopie, représentant l'état normal et les modifications pathologiques du fond de l'œil visibles à l'ophthalmoscope; composé de 12 planches représentant 57 figures tirées en chromolithographie accompagnées d'un texte explicatif. In-folio. £2.

Luys.—Recherches sur le Système Nerveux Cérébro-spinal, sa Structure, ses Fonctions et ses Maladies. 8vo., avec atlas de 40 planches. 4to. Fig. noires £1 10s. Coloriées £3.

Maire.—Nouveou Guide des de Mères Famille. 8vo. 2nd édition. Paris, 1865. 4s.6d.

Malgaigne.—Manuel de Médecine Opératoire. 7e édition. 1861. 18mo. 6s.

Mandl.—Anatomie Microscopique. Paris, 1838-1857, *ouvrage complet.* 2 vols. in-folio, avec 92 planches. £13.

Martin et Collineau.—De la Coxalgie, de sa Nature, de son Traitement. 8vo., avec 30 fig. Paris, 1865. 6s.

Maunoury et Salmon.—Manuel de l'Art des Accouchements. 1861. 2e édition. 8vo. avec 32 fig. 6s.

Menière.—Les Consultations de Madame de Sévigné. 8vo. Paris, 1864. 2s. 6d.

Meunier.—La Science et les Savants en 1864-65. 2 vols. 12mo. Paris, 1865. 6s.

Niemayer.—Traitéde Pathologie Interne et de Thérapeutique. Vol. 1. 8vo. 1865. 7s. 6d.

Nysten.—Dictionnaire de Médecine, de Chirurgie, de Pharmacie, des Sciences accessoires et de l'Art Vétérinaire. 12e édition. Ouvrage contenant la synonymie grecque, latine, anglaise, allemande, italienne et espagnole, et le Glossaire de ces diverses langues. 1864. Grand 8vo., avec plus de 500 figures intercalées dans le texte. Demi-relié £1.

Peclet (E.)—Traité de la Chaleur considérée dans ses Applications. 3e édit'on. Paris, 1860-61. 3 vol. grand 8vo. £1 16s.

Racle.—Traité de Diagnostic Médical, ou Guide Clinique pour l'Étude des Signes caractéristiques des Maladies. 3e édition. 1864. 1 vol. 18mo., avec 17 fig. 5s.

Requin.—Éléments de Pathologie Médicale. Vol. 5, 8vo. ——

Revue des Cours Littéraires.—Littérature. — Physiologie. — Théologie. — Eloquence.—Histoire.—Législation.—Esthétique.—Archéologie.

Revue des Cours Scientifiques.—Physique. —Chimie — Botanique.—Zoologie.—Anatomie.— Physiologie. — Géologie.—Paléontologie.—Médecine.

Directeur, M. Eug. Yung; chef de la rédaction, M. Em. Alglave.

Ces deux journaux reproduisent les cours des Facultés de Paris, des départements et de l'étranger, et paraissent tous les samedis depuis le 5 décembre 1863. On peut s'abonner séparément à la partie littéraire ou à la partie scientifique.

Prix de chaque journal isolément, 15s. par an.; prix des deux journax réunis, 26s. par an.

Subscriptions commence on 1st December and 1st June. ——

Ricord.—Lettres sur la Syphilis. 3e édition. 18mo. 3s. 6d.

Robert.—Nouveau Traité sur les Maladies Vénériennes. 1861. 8vo. 7s, 6d.

Robin (Ch.)—Programme du Cours d'Histologie professe à la Faculté de Médecine. Paris, 1864. 4s.

Rollet.—Recherches Cliniques et Expérimentales sur la Syphilis. 1861. 1 vol. 8vo., avec atlas de 20 figures, dont 10 coloriées. 12s.

Rollet (F.)—Traité des Maladies Vénériennes. Vol. 1. 8vo. Paris, 1865. 6s. 6d.

Rose (H.)—Traité Complet de Chimie Analytique; édition française originale. Paris, 1859-1862. 2 vol. grand 8vo. £1.

Saurel.—Traité de Chirurgie Navale, suivi d'un Résumé de leçons sur le Service Chirurgical de la Flotte, par le docteur J. Rochard. Paris, 1861. 8vo., avec 106 fig. 6s. 6d.

Schweigger.—Leçons d'Ophthalmoscopie. 8vo., avec 8 planches, 8vo. Paris, 1865. 3s.

Sichel.—Iconographic Opthalmologique, ou Description avec Figures coloriées des Maladies de l'Organe de la Vue. 1852-1859. 2 vol. grand in-4, dont 1 volume de texte, et 1 volume de 80 planches coloriées. Rel. £9 10s.

Sous (G.)—Manuel d'Ophthalmoscopie. 8vo., avec figures. Paris, 1865. 3s. 6d.

Tardieu.—Manuel de Pathologie et de Clinique Médicale. 3e édition. 1864. 18mo. 6s.

Tardieu.—Dictionnaire d'Hygiène Publique et de Salubrité. 2e édition. 1862. 4 vol. gr. 8vo. £1 7s.

Thomas.—Éléments d'Ostéologie. Descriptive et Comparée de l'Homme et des Animaux Domestiques. 8vo., et avec Atlas de 12 planches. Paris, 1865. 10s.

Topinard (P.)—De l'Ataxie Locomotrice. 8vo. Paris, 1864. 6s. 6d.

Trousseau.—Clinique Médicale de l'Hôtel-Dieu de Paris. *Deuxieme édition*, corrigée et augmentée. Paris, 1864. 3 vol. 8vo. £1 5s.

Vernois.—Traité Pratique d'Hygiène Industrielle et Administrative, comprenant l'étude des établissements insalubres, dangereux et incommodes. Paris, 1860. 2 vol. 8vo. 18s. 6d.

Virchow.—Traité des Tumeurs. 2 vol. 8vo., avec fig. Sous presse.

Voisin (A.)—De l'État Mental dans l'Æcoolisme Aigu et Chronique et dans l'Absinthisme. 8vo. Paris, 1864. 2s. 6d.

Wecker.—Traité Pratique des Maladies des Yeux. Vol. 1. 8vo., avec 6 planches. Paris, 1865. 10s.

Watelot (A.)—Description des Plantes Fossiles du Bassin de Paris. 4to., avec planches. Livraisons 1 et 2. Prix de chaque 5s.

Wurtz.—Traité Élémentaire de Chimie Médicale *Organique et Inorganique*. 2 vols. 8vo. 18s. 6d.

Just published, price 10s. 6d.

A PRACTICAL TREATISE ON COAL, PETROLEUM, AND OTHER DISTILLED OILS.

By ABRAHAM GESNER, M.D., F.G.S.

Second Edition, Revised and Enlarged. By GEORGE WELTDEN GESNER, Consulting Chemist and Engineer.

THE ANATOMY OF THE EXTERNAL FORM OF THE HORSE.

With Explanations by JAMES I. LUPTON, M.R.C.V.S.

The Plates by Bagg and Stanton. Part I. containing 9 Plates and Explanations. Large Folio. Price, plain, £1 11s. 6d.; on India Paper, £2 5s. Part II., with 4 Plates (2 of which, the Leg and Mouth, are coloured). Plates Plain, 14s; coloured, £1.

. The last Part of this important work will be published soon with the 8vo. vol. of Text, giving the Study of the External Form of the Horse, and the Physiology of Locomotion.

LIBRARY OF ILLUSTRATED
STANDARD SCIENTIFIC WORKS.

The following Volumes are now published:

MULLER'S PRINCIPLES of PHYSICS and METEOROLOGY.
With 530 Woodcuts and Two Coloured Engravings. 8vo. 18s.

WEISBACH'S MECHANICS of MACHINERY and ENGINEERING. 2 vols. 8vo. With 900 Woodcuts. £1 19s.

KNAPP, RONALDS, RICHARDSON, and HENRY WATTS'
CHEMICAL TECHNOLOGY; or, Chemistry in its Application to the Arts and Manufactures.

Vol. I., Parts 1 and 2, 8vo. contains—FUEL and its APPLICATIONS. Profusely Illustrated with 45 Engravings and 4 Plates. £1 16s.

Vol. I., Parts 3 and 4, in 2 vols. 8vo., contain—The ALKALIES and ACIDS. With numerous Illustrations on Wood. 1863. £2 14s.
These 2 vols. are entirely by Messrs. Richardson and Watts.

PART V. IN THE PRESS.

Vol. II., 8vo. contains—GLASS, ALUM, POTTERIES, CEMENTS, GYPSUM, &c. With numerous Illustrations. £1 1s.

Vol. III., 8vo. contains—FOOD GENERALLY: BREAD, CHEESE, TEA, COFFEE, TOBACCO, MILK, SUGAR. With numerous Illustrations and Coloured Plates. £1 2s.

QUEKETT'S (JOHN) PRACTICAL TREATISE on the USE of
the MICROSCOPE. Third Edition, with 11 Steel and numerous Wood Engravings. 8vo. £1 1s. Reduced to 12s. 6d.

FAU'S ANATOMY of the EXTERNAL FORMS of MAN. For
Artists. Edited by R. KNOX, M.D. 8vo. and an Atlas of 28 Plates, 4to. Plain, £1 4s.; Coloured, £2 2s.

GRAHAM'S ELEMENTS of CHEMISTRY, including the APPLICATIONS
of the SCIENCE in the ARTS. Second Edition, with numerous Woodcuts. 2 vols. 8vo. £2. Vol. II. separately. Edited by H. WATTS, Esq. £1.

NICHOL'S ARCHITECTURE of the HEAVENS. Ninth Edition.
8vo. with 23 Steel Plates and many Woodcuts. 1851. 16s.

MITCHELL'S (J.) MANUAL of PRACTICAL ASSAYING. For
the Use of Metallurgists, Captains of Mines, and Assayers in General. Second Edition, much enlarged, with Illustrations, &c. 8vo. £1 1s.

BERKELEY'S (Rev. J.) INTRODUCTION to CRYPTOGAMIC
BOTANY. Illustrated with numerous Wood Engravings. 8vo. 1857. £1.

ANATOMY of the EXTERNAL FORM of the HORSE. With
Explanations by J. I. LUPTON, M.R.C.V.S., and Plates by BAGG and STANTON. This Work will be completed in Three Parts, consisting of 18 to 20 Plates, with Explanations, and One Volume of Octavo Text, giving the Study of the External Form of the Horse and the Physiology of Locomotion. Part I., with Nine Plates and Explanations. Large folio. Price, plain, £1 11s. 6d.; on India Paper, £2 5s. Part II. with Four Plates (Two of which, the Leg and Mouth, are coloured). Plain, 15s.; Coloured, £1.

*** The last Part of this important Work will be published in the course of 1863, with the volume (8vo.) of Text.

GANOT.—ELEMENTARY TREATISE ON PHYSICS,
EXPERIMENTAL AND APPLIED.

Translated on the Ninth Original Edition by Dr. Atkinson, of the Royal Military College, Sandhurst. With 600 Illustrations. Post 8vo. 1862. 12s. 6d.

(Adapted for the use of Colleges and Schools.)

Lightning Source UK Ltd.
Milton Keynes UK
UKHW020656211218
334381UK00011B/614/P